Log on to gpscience.com

ONLINE STUDY TOOLS

- Section Self-Check Quizzes
- Interactive Tutor
- Chapter Review Tests
- Standardized Test Practice
- Vocabulary PuzzleMaker

ONLINE RESEARCH

- WebQuest Projects
- Prescreened Web Links
- Career Links
- Internet Labs

INTERACTIVE ONLINE STUDENT EDITION

- Complete Interactive Student Edition available at mhln.com

FOR TEACHERS

- Teacher Bulletin Board
- Teaching Today—Professional Development

SAFETY SYMBOLS

SAFETY SYMBOLS	HAZARD	EXAMPLES	PRECAUTION	REMEDY
DISPOSAL	Special disposal procedures need to be followed.	certain chemicals, living organisms	Do not dispose of these materials in the sink or trash can.	Dispose of wastes as directed by your teacher.
BIOLOGICAL	Organisms or other biological materials that might be harmful to humans	bacteria, fungi, blood, unpreserved tissues, plant materials	Avoid skin contact with these materials. Wear mask or gloves.	Notify your teacher if you suspect contact with material. Wash hands thoroughly.
EXTREME TEMPERATURE	Objects that can burn skin by being too cold or too hot	boiling liquids, hot plates, dry ice, liquid nitrogen	Use proper protection when handling.	Go to your teacher for first aid.
SHARP OBJECT	Use of tools or glassware that can easily puncture or slice skin	razor blades, pins, scalpels, pointed tools, dissecting probes, broken glass	Practice common-sense behavior and follow guidelines for use of the tool.	Go to your teacher for first aid.
FUME	Possible danger to respiratory tract from fumes	ammonia, acetone, nail polish remover, heated sulfur, moth balls	Make sure there is good ventilation. Never smell fumes directly. Wear a mask.	Leave foul area and notify your teacher immediately.
ELECTRICAL	Possible danger from electrical shock or burn	improper grounding, liquid spills, short circuits, exposed wires	Double-check setup with teacher. Check condition of wires and apparatus.	Do not attempt to fix electrical problems. Notify your teacher immediately.
IRRITANT	Substances that can irritate the skin or mucous membranes of the respiratory tract	pollen, moth balls, steel wool, fiberglass, potassium permanganate	Wear dust mask and gloves. Practice extra care when handling these materials.	Go to your teacher for first aid.
CHEMICAL	Chemicals can react with and destroy tissue and other materials	bleaches such as hydrogen peroxide; acids such as sulfuric acid, hydrochloric acid; bases such as ammonia, sodium hydroxide	Wear goggles, gloves, and an apron.	Immediately flush the affected area with water and notify your teacher.
TOXIC	Substance may be poisonous if touched, inhaled, or swallowed.	mercury, many metal compounds, iodine, poinsettia plant parts	Follow your teacher's instructions.	Always wash hands thoroughly after use. Go to your teacher for first aid.
FLAMMABLE	Flammable chemicals may be ignited by open flame, spark, or exposed heat.	alcohol, kerosene, potassium permanganate	Avoid open flames and heat when using flammable chemicals.	Notify your teacher immediately. Use fire safety equipment if applicable.
OPEN FLAME	Open flame in use, may cause fire.	hair, clothing, paper, synthetic materials	Tie back hair and loose clothing. Follow teacher's instruction on lighting and extinguishing flames.	Notify your teacher immediately. Use fire safety equipment if applicable.

 Eye Safety Proper eye protection should be worn at all times by anyone performing or observing science activities.

 Clothing Protection This symbol appears when substances could stain or burn clothing.

 Animal Safety This symbol appears when safety of animals and students must be ensured.

 Handwashing After the lab, wash hands with soap and water before removing goggles.

Glencoe Science

Physical Science

NATIONAL GEOGRAPHIC

gpscience.com

 Glencoe

New York, New York   Co___s, Ohio   Chicago, Illinois   Peoria, Illinois   Woodland Hills, California

Physical Science

The VLA (Very Large Array) is Y-shaped arrangement of 27 radio antennas on the Plains of San Agustin in New Mexico. Each 25-m antenna can move on tracks. The electronic data collected from the dishes can give a resolution of a 36-km antenna and sensitivity of a 130-m dish, depending on the distance between antennas.

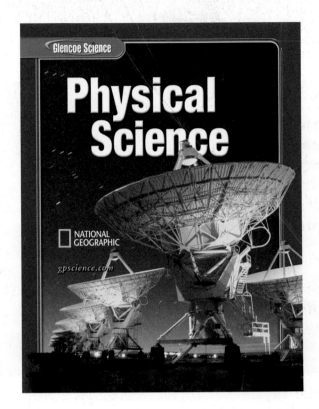

 **Glencoe**

Copyright © 2005 by The McGraw-Hill Companies, Inc. All rights reserved. Except as permitted under the United States Copyright Act, no part of this publication may be reproduced or distributed in any form or by any means, or stored in a database or retrieval system, without prior written permission of the publisher.

The National Geographic features were designed and developed by the National Geographic Society's Education Division. Copyright © National Geographic Society. The name "National Geographic Society" and the Yellow Border Rectangle are trademarks of the Society, and their use, without prior written permission, is strictly prohibited.

The "Science and Society" and the "Science and History" features that appear in this book were designed and developed by TIME School Publishing, a division of TIME Magazine. TIME and the red border are trademarks of Time Inc. All rights reserved.

Send all inquiries to:
Glencoe/McGraw-Hill
8787 Orion Place
Columbus, OH 43240-4027

ISBN: 0-07-860051-0

Printed in the United States of America.

4 5 6 7 8 9 10 071/055 09 08 07 06

Contents In Brief

Authors

Education Division
Washington, D.C.

Charles William McLaughlin, PhD
Senior Lecturer
University of Nebraska
Lincoln, NE

Marilyn Thompson, PhD
Assistant Professor, College of Education
Arizona State University
Tempe, AZ

Dinah Zike
Educational Consultant
Dinah-Might Activities, Inc.
San Antonio, TX

Contributing Authors

Nancy Ross-Flanigan
Science Writer
Detroit, MI

Margaret K. Zorn
Science Writer
Yorktown, VA

Science Consultants

Jack Cooper
Ennis High School
Ennis, TX

David G. Haase, PhD
North Carolina State University
Raleigh, NC

Michael A. Hoggarth, PhD
Department of Life and
Earth Sciences
Otterbein College
Westerville, OH

Madelaine Meek
Physics Consultant Editor
Lebanon, OH

Cheryl Wistrom
St. Joseph's College
Rensselaer, IN

Carl Zorn, PhD
Staff Scientist
Jefferson Laboratory
Newport News, VA

Series Consultants

MATH

Michael Hopper, DEng.
Manager of Aircraft Certification
L-3 Communications
Greenville, TX

Teri Willard, EdD
Mathematics Curriculum Writer
Belgrade, MT

READING

Elizabeth Babich
Special Education Teacher
Mashpee Public Schools
Mashpee, MA

Barry Barto
Special Education Teacher
John F. Kennedy Elementary
Manistee, MI

Carol A. Senf, PhD
School of Literature, Communication, and Culture
Georgia Institute of Technology
Atlanta, GA

Rachel Swaters-Kissinger
Science Teacher
John Boise Middle School
Warsaw, MO

SAFETY

Aileen Duc, PhD
Science 8 Teacher
Hendrick Middle School, Plano ISD
Plano, TX

Sandra West, PhD
Department of Biology
Texas State University-San Marcos
San Marcos, TX

ACTIVITY TESTERS

Nerma Coats Henderson
Pickerington Lakeview Jr. High School
Pickerington, OH

Mary Helen Mariscal-Cholka
William D. Slider Middle School
El Paso, TX

Science Kit and Boreal Laboratories
Tonawanda, NY

Reviewers

Sharla Adams
IPC Teacher
Allen High School
Allen, TX

Desiree Bishop
Environmental Studies Center
Mobile County Public Schools
Mobile, AL

Tom Bright
Concord High School
Charlotte, NC

Nora M. Prestinari Burchett
Saint Luke School
McLean, VA

Mary Helen Mariscal-Cholka
William D. Slider Middle School
El Paso, TX

Obioma Chukwu
J.H. Rose High School
Greenville, NC

Inga Dainton
Merrilvills High School
Merrilville, IN

Robin Dillon
Hanover Central High School
Cedar Lake, IN

Anthony J. DiSipio, Jr.
8th Grade Science
Octorana Middle School
Atglen, PA

Dwight Dutton
East Chapel Hill High School
Chapel Hill, NC

Carolyn Elliott
South Iredell High School
Statesville, NC

Sueanne Esposito
Tipton High School
Tipton, IN

George Gabb
Great Bridge Middle School
Chesapeake Public Schools
Chesapeake, VA

Nerma Coats Henderson
Pickerington Lakeview Jr.
High School
Pickerington, OH

Maria E. Kelly
Principal
Nativity School
Catholic Diocese of Arlington
Burke, VA

Annette Parrott
Lakeside High School
Atlanta, GA

Darcy Vetro-Ravndal
Hillsborough High School
Tampa, FL

Clabe Webb
Permian High School
Ector County ISD
Odessa, TX

Alison Welch
William D. Slider Middle School
El Paso, TX

Kim Wimpey
North Gwinnett High School
Suwanee, GA

Teacher Advisor Board

The Teacher Advisory Board gave the authors, editorial staff, and the design team feedback on the content and design of the Student Edition. They were instrumental in providing valuable input toward the development of the 2005 edition of *Glencoe Physical Science.* We thank these teachers for their hard work and creative suggestions.

The Teacher Advisory Board gathers for a photo at the Glencoe/McGraw-Hill headquarters in Columbus, Ohio.

Field Test Schools

Glencoe/McGraw-Hill wishes to thank the following schools that field-tested prepublication manuscript. They were instrumental in providing feedback and verifying the effectiveness of this program.

Contents

In each chapter, look for these opportunities for review and assessment:

- Reading Checks
- Caption Questions
- Section Review
- Chapter Study Guide
- Chapter Review
- Standardized Test Practice
- Online practice at **gpscience.com**

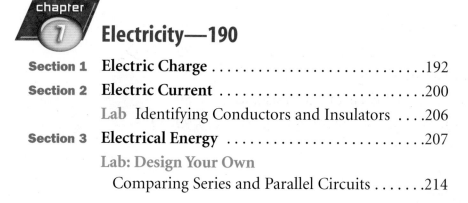

Contents

In each chapter, look for these opportunities for review and assessment:
- Reading Checks
- Caption Questions
- Section Review
- Chapter Study Guide
- Chapter Review
- Standardized Test Practice
- Online practice at gpscience.com

x

Contents

In each chapter, look for these opportunities for review and assessment:
- Reading Checks
- Caption Questions
- Section Review
- Chapter Study Guide
- Chapter Review
- Standardized Test Practice
- Online practice at **gpscience.com**

Contents

unit 5 Diversity of Matter—566

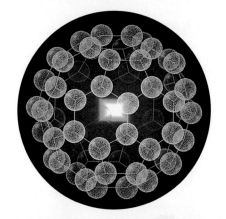

Contents

Student Resources—786

Science Skill Handbook—788

Extra Try at Home Labs—800

Technology Skill Handbook—813

Math Skill Handbook—817

Extra Math Problems—834

Reference Handbook—846

English/Spanish Glossary—850

Index—871

Credits—891

In each chapter, look for these opportunities for review and assessment:
- **Reading Checks**
- **Caption Questions**
- **Section Review**
- **Chapter Study Guide**
- **Chapter Review**
- **Standardized Test Practice**
- **Online practice at gpscience.com**

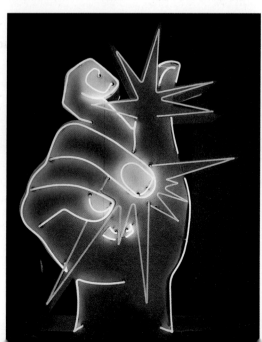

Cross-Curricular Readings

TIME SCIENCE AND Society

TIME SCIENCE AND HISTORY

Oops! Accidents in SCIENCE

Science and Language Arts

SCIENCE Stats

Content Details

Mini LAB Try at Home

Content Details

available as a video lab

One-Page Labs

Two-Page Labs

Content Details

Design Your Own Labs

Model and Invent Labs

Use the Internet Labs

Content Details

Activities

Applying Math

Applying Science

Content Details

INTEGRATE

Astronomy: 39, 76, 331, 480, 524, 733
Career: 30, 208, 240, 325, 370, 520, 576, 743
Chemistry: 136, 202, 554
Earth Science: 11, 17, 45, 79, 162, 176, 208, 227, 267, 275, 295, 456, 463, 489, 549
Environment: 111, 364, 459, 637, 650, 772
Health: 84, 115, 205, 309, 343, 363, 429, 573, 608, 685
History: 9, 48, 299, 403, 429, 462, 482, 540, 617, 650, 713, 760
Language Arts: 104
Life Science: 179, 228, 324, 391, 392, 417, 427, 514, 609, 678, 700
Physics: 408
Social Studies: 266, 299, 549

Science Online

7, 12, 18, 43, 53, 69, 76, 102, 113, 160, 174, 196, 212, 237, 242, 268, 275, 295, 307, 329, 334, 371, 372, 396, 401, 428, 434, 464, 479, 507, 519, 521, 523, 540, 553, 591, 605, 610, 639, 674, 684, 701, 710, 733, 740, 744, 762, 773, 775

Standardized Test Practice

34–35, 64–65, 96–97, 122–123, 154–155, 186–187, 220–221, 252–253, 284–285, 318–319, 350–351, 380–381, 412–413, 444–445, 472–473, 502–503, 532–533, 564–565, 598–599, 628–629, 658–659, 692–593, 722–723, 754–755, 784–785

How Are
Waffles &
Running Shoes
Connected?

For centuries, shoes were made mostly of leather, cloth, or wood. These shoes helped protect feet, but they didn't provide much traction on slippery surfaces. In the early twentieth century, manufacturers began putting rubber on the bottom of canvas shoes, creating the first "sneakers." Sneakers provided good traction, but the rubber soles could be heavy—especially for athletes. One morning in the 1970s, an athletic coach stared at the waffles on the breakfast table and had an idea for a rubber sole that would be lighter in weight but would still provide traction. That's how the first waffle soles were born. Waffle soles soon became a world standard for running shoes.

unit ⚡ projects

Visit **gpscience.com/unit_project** to find project ideas and resources. Projects include:

- **Career** Explore the field of mechanical engineering through sports. Create an advertisement for your aerodynamic sports equipment.
- **Technology** Create a brochure showing five different shoe treads with reference to particular sports and environmental conditions.
- **Model** Design and build a Rube Goldberg machine with at least 20 steps to complete a simple task using physics principles.

WebQuest Using virtual programming, *Roller Coaster Physics* provides an opportunity to engineer, test, and evaluate roller coaster design, and then build your own 3-dimensional coaster.

The Nature of Science

Out of This World

The space program was developed in response to many unanswered questions. Scientists have worked together to develop ways in which to answer those questions. In this chapter, you will learn how scientists learn about the natural world.

Science Journal Look at the picture above. Write in your Science Journal why scientists study space.

Start-Up Activities

Understanding Measurements

During a track meet, one athlete ran 1 mile in 5 min and another athlete ran 5,000 m in 280 s. The two runners used different units to describe their races, so how can you compare them? Do the following lab to explore how using different units can make it difficult to compare measurements.

1. Measure the distance across your classroom using your foot as a measuring device.

2. Record your measurement and name your measuring unit.

3. Now, have your partner measure the same distance using his or her foot as the measuring device. Record this measurement and make up a different name for the unit.

4. **Think Critically** Explain why you think it might be important to have standard, well-defined units to make measurements.

 Preview this chapter's content and activities at
gpscience.com

 **Scientific Processes** Make the following Foldable to help identify what you already know, what you want to know, and what you learned about science.

STEP 1 **Fold** a vertical sheet of paper from side to side. Make the front edge about 1.25 cm shorter than the back edge.

STEP 2 **Turn** lengthwise and **fold** into thirds.

STEP 3 **Unfold and cut** only the top layer along both folds to make three tabs. **Label** each tab.

Identify Questions Before you read the chapter, write what you already know about science under the left tab of your Foldable, and write questions about what you'd like to know under the center tab. After you read the chapter, list what you learned under the right tab.

The Methods of Science

Reading Guide

What You'll Learn

- **Identify** the steps scientists often use to solve problems.
- **Describe** why scientists use variables.
- **Compare and contrast** science and technology.

Why It's Important

Using scientific methods will help you solve problems.

⊙ Review Vocabulary

investigation: to observe or study by close examination

New Vocabulary

- scientific method
- hypothesis
- experiment
- variable
- dependent variable
- independent variable
- constant
- control
- bias
- model
- theory
- scientific law
- technology

What is science?

Science is not just a subject in school. It is a method for studying the natural world. After all, science comes from the Latin word *scientia,* which means "knowledge." Science is a process that uses observation and investigation to gain knowledge about events in nature.

Nature follows a set of rules. Many rules, such as those concerning how the human body works, are complex. Other rules, such as the fact that Earth rotates about once every 24 h, are much simpler. Scientists ask questions to learn about the natural world.

Major Categories of Science Science covers many different topics that can be classified according to three main categories. (1) Life science deals with living things. (2) Earth science investigates Earth and space. (3) Physical science deals with matter and energy. In this textbook, you will study mainly physical science. Sometimes, though, a scientific study will overlap the categories. One scientist, for example, might study the motions of the human body to understand how to build better artificial limbs. Is this scientist studying energy and matter or how muscles operate? She is studying both life science and physical science. It is not always clear what kind of science you are using, as shown in **Figure 1.**

Figure 1 Astronaut Michael Lopez-Alegria uses a pistol grip tool on the *International Space Station.*
Observe *What evidence do you see of the three main branches of science in the photograph?*

Science Explains Nature Scientific explanations help you understand the natural world. Sometimes these explanations must be modified. As more is learned about the natural world, some of the earlier explanations might be found to be incomplete or new technology might provide more accurate answers.

For example, look at **Figure 2.** In the late eighteenth century, most scientists thought that heat was an invisible fluid with no mass. Scientists observed that heat seemed to flow like a fluid. It also moves away from a warm body in all directions, just as a fluid moves outward when you spill it on the floor.

However, the heat fluid idea did not explain everything. If heat were an actual fluid, an iron bar that had a temperature of 1,000°C should have more mass than it did at 100°C because it would have more of the heat fluid in it. The eighteenth-century scientists thought they just were not able to measure the small mass of the heat fluid on the balances they had. When additional investigations showed no difference in mass, scientists had to change the explanation.

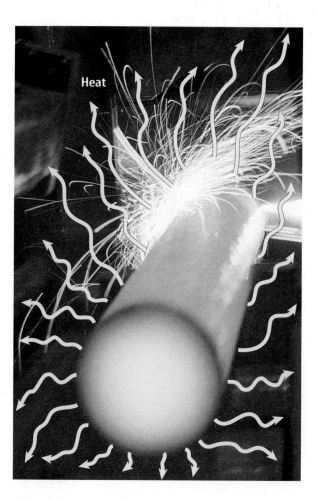
Heat

Investigations Scientists learn new information about the natural world by performing investigations, which can be done many different ways. Some investigations involve simply observing something that occurs and recording the observations, perhaps in a journal. Other investigations involve setting up experiments that test the effect of one thing on another. Some investigations involve building a model that resembles something in the natural world and then testing the model to see how it acts. Often, a scientist will use something from all three types of investigation when attempting to learn about the natural world.

Figure 2 Many years ago, scientists thought that heat, such as in this metal rod, was a fluid. **Infer** *how heat acts like a fluid.*

✔ **Reading Check** *Why do scientific explanations change?*

Scientific Methods

Although scientists do not always follow a rigid set of steps, investigations often follow a general pattern. An organized set of investigation procedures is called a **scientific method.** Six common steps found in scientific methods are shown in **Figure 3.** A scientist might add new steps, repeat some steps many times, or skip steps altogether when doing an investigation.

Science online

Topic: Prediction
Visit gpscience.com for Web links to information about why leaves change color in the autumn.

Activity Fill a glass with cold water and add a few drops of blue or red food coloring. Cut a piece of celery and place it in the glass. Over the next few days observe what happens to the celery. Make a prediction about why this occurs. Support your answer with evidence.

Figure 3 The series of procedures shown here is one way to use scientific methods to solve a problem.

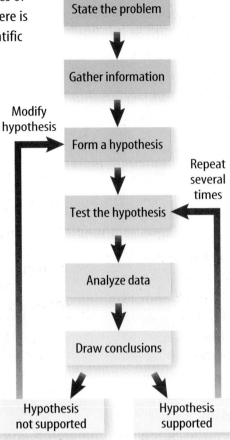

State the problem

Gather information

Modify hypothesis

Form a hypothesis

Test the hypothesis

Repeat several times

Analyze data

Draw conclusions

Hypothesis not supported

Hypothesis supported

Stating a Problem Many scientific investigations begin when someone observes an event in nature and wonders why or how it occurs. Then the question of "why" or "how" is the problem. Sometimes a statement of a problem arises from an activity that is not working. Some early work on guided missiles showed that the instruments in the nose of the missiles did not always work. The problem statement involved finding a material to protect the instruments from the harsh conditions of flight.

Later, National Aeronautics and Space Administration (NASA) scientists made a similar problem statement. They wanted to build a new vehicle—the space shuttle—that could carry people to outer space and back again. Guided missiles did not have this capability. NASA needed to find a material for the outer skin of the space shuttle that could withstand the heat and forces of reentry into Earth's atmosphere.

Researching and Gathering Information Before testing a hypothesis, it is useful to learn as much as possible about the background of the problem. Have others found information that will help determine what tests to do and what tests will not be helpful? The NASA scientists gathered information about melting points and other properties of the various materials that might be used. In many cases, tests had to be performed to learn the properties of new, recently created materials.

Forming a Hypothesis A **hypothesis** is a possible explanation for a problem using what you know and what you observe. NASA scientists knew that a ceramic coating had been found to solve the guided missile problem. They hypothesized that a ceramic material also might work on the space shuttle.

Testing a Hypothesis Some hypotheses can be tested by making observations. Others can be tested by building a model and relating it to real-life situations. One common way to test a hypothesis is to perform an experiment. An **experiment** tests the effect of one thing on another using controlled conditions.

Variables An experiment usually contains at least two variables. A **variable** is a quantity that can have more than a single value. You might set up an experiment to determine which of three fertilizers helps plants to grow the biggest. Before you begin your tests, you would need to think of all the factors that might cause the plants to grow bigger. Possible factors include plant type, amount of sunlight, amount of water, room temperature, type of soil, and type of fertilizer.

In this experiment, the amount of growth is the **dependent variable** because its value changes according to the changes in the other variables. The variable you change to see how it will affect the dependent variable is called the **independent variable.**

Constants and Controls To be sure you are testing to see how fertilizer affects growth, you must keep the other possible factors the same. A factor that does not change when other variables change is called a **constant.** You might set up one trial, using the same soil and type of plant. Each plant is given the same amount of sunlight and water and is kept at the same temperature. These are constants. Three of the plants receive a different amount of fertilizer, which is the independent variable.

The fourth plant is not fertilized. This plant is a control. A **control** is the standard by which the test results can be compared. Suppose that after several days, the three fertilized plants grow between 2 and 3 cm. If the unfertilized plant grows 1.5 cm, you might infer that the growth of the fertilized plants was due to the fertilizers.

How might the NASA scientists set up an experiment to solve the problem of the damaged tiles shown in **Figure 4?** What are possible variables, constants, and controls?

✔ Reading Check *Why is a control used in an experiment?*

Classification Systems
Through observations of living organisms, Aristotle designed a classification system. Systems used today group organisms according to variables such as habits and physical and chemical features. Research to learn recent reclassifications of organisms. Share your findings with your class.

Figure 4 NASA has had an ongoing mission to improve the space shuttle. A technician is replacing tiles damaged upon reentry into Earth's atmosphere.

Figure 5 An exciting and important part of investigating something is sharing your ideas with others, as this student is doing at a science fair.

Analyzing the Data An important part of every experiment includes recording observations and organizing the test data into easy-to-read tables and graphs. Later in this chapter you will study ways to display data. When you are making and recording observations, you should include all results, even unexpected ones. Many important discoveries have been made from unexpected occurrences.

Interpreting the data and analyzing the observations is an important step. If the data are not organized in a logical manner, wrong conclusions can be drawn. No matter how well a scientist communicates and shares that data, someone else might not agree with the data. Scientists share their data through reports and conferences. In **Figure 5** a student is displaying her data.

Drawing Conclusions Based on the analysis of your data, you decide whether or not your hypothesis is supported. When lives are at stake, such as with the space shuttle, you must be very sure of your results. For the hypothesis to be considered valid and widely accepted, the experiment must result in the exact same data every time it is repeated. If your experiment does not support your hypothesis, you must reconsider the hypothesis. Perhaps it needs to be revised or your experiment needs to be conducted differently.

Being Objective Scientists also should be careful to reduce bias in their experiments. A **bias** occurs when what the scientist expects changes how the results are viewed. This expectation might cause a scientist to select a result from one trial over those from other trials. Bias also might be found if the advantages of a product being tested are used in a promotion and the drawbacks are not presented.

Scientists can lessen bias by running as many trials as possible and by keeping accurate notes of each observation made. Valid experiments also must have data that are measurable. For example, a scientist performing a global warming study must base his or her data on accurate measures of global temperature. This allows others to compare the results to data they obtain from a similar experiment. Most importantly, the experiment must be repeatable. Findings are supportable when other scientists perform the same experiment and get the same results.

✔ **Reading Check** *What is bias in science?*

Visualizing with Models

Sometimes, scientists cannot see everything that they are testing. They might be observing something that is too large, too small, or takes too much time to see completely. In these cases, scientists use models. A **model** represents an idea, event, or object to help people better understand it.

Models in History Models have been used throughout history. One scientist, Lord Kelvin, who lived in England in the 1800s, was famous for making models. To model his idea of how light moves through space, he put balls into a bowl of jelly and encouraged people to move the balls around with their hands. Kelvin's work to explain the nature of temperature and heat still is used today.

High-Tech Models Scientific models don't always have to be something you can touch. Today, many scientists use computers to build models. NASA experiments involving space flight would not be practical without computers. The complex equations would take far too long to calculate by hand, and errors could be introduced much too easily.

Another type of model is a simulator, like the one shown in **Figure 6.** An airplane simulator enables pilots to practice problem solving with various situations and conditions they might encounter when in the air. This model will react the way a plane does when it flies. It gives pilots a safe way to test different reactions and to practice certain procedures before they fly a real plane.

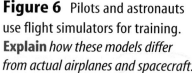

Figure 6 Pilots and astronauts use flight simulators for training. **Explain** *how these models differ from actual airplanes and spacecraft.*

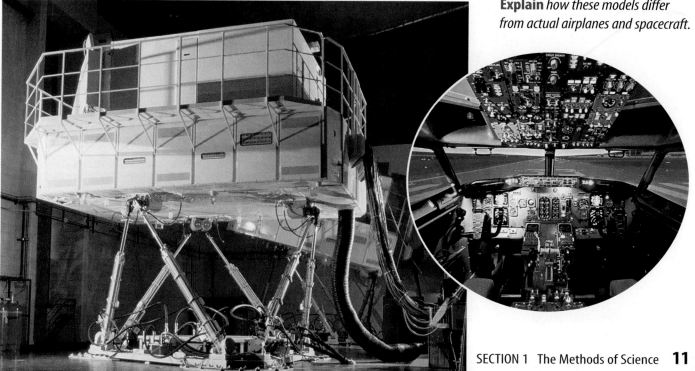

Figure 7 Science can't answer all questions.
Analyze *Can anyone prove that you like artwork? Explain.*

Scientific Theories and Laws

A scientific **theory** is an explanation of things or events based on knowledge gained from many observations and investigations. It is not a guess. If scientists repeat an investigation and the results always support the hypothesis, the hypothesis can be called a theory. Just because a scientific theory has data supporting it does not mean it will never change. Recall that the theory about heat being a fluid was discarded after further experiments. As new information becomes available, theories can be modified. A theory accepted today might at some time in the future also be discarded.

A **scientific law** is a statement about what happens in nature and that seems to be true all the time. Laws tell you what will happen under certain conditions, but they don't explain why or how something happens. Gravity is an example of a scientific law. The law of gravity says that any one mass will attract another mass. To date, no experiments have been performed that disprove the law of gravity.

A theory can be used to explain a law. For example, many theories have been proposed to explain how the law of gravity works. Even so, there are few theories in science and even fewer laws.

 **Reading Check** *What is the difference between a scientific theory and a scientific law?*

The Limitations of Science

Science can help you explain many things about the world, but science cannot explain or solve everything. Although it's the scientist's job to make guesses, the scientist also has to make sure his or her guesses can be tested and verified. But how do you prove that people will like a play or a piece of music? You cannot and science cannot.

Most questions about emotions and values are not scientific questions. They cannot be tested. You might take a survey to get people's opinions about such questions, but that would not prove that the opinions are true for everyone. A survey might predict that you will like the art in **Figure 7,** but science cannot prove that you or others will.

Using Science—Technology

Many people use the terms *science* and *technology* interchangeably, but they are not the same. **Technology** is the application of science to help people. For example, when a chemist develops a new, lightweight material that can withstand great amounts of heat, science is used. When that material is used on the space shuttle, technology is applied. **Figure 8** shows other examples of technology.

Technology doesn't always follow science, however. Sometimes the process of discovery can be reversed. One important historic example of science following technology is the development of the steam engine. The inventors of the steam engine had little idea of how it worked. They just knew that steam from boiling water could move the engine. Because the steam engine became so important to industry, scientists began analyzing how it worked. Lord Kelvin, James Prescott Joule and Sadi Carnot, who lived in the 1800s, learned so much from the steam engine that they developed revolutionary ideas about the nature of heat.

Science and technology do not always produce positive results. The benefits of some technological advances, such as nuclear technology and genetic engineering, are subjects of debate. Being more knowledgeable about science can help society address these issues as they arise.

Figure 8 Technology is the application of science.
State *the type of science (life, Earth, or physical) that is applied in these examples of technology.*

section 1 review

Summary

What is science?

- Scientists ask questions and perform investigations to learn more about the natural world.

Scientific Methods

- Scientists perform the six-step scientific method to test their hypotheses.

Visualizing with Models

- Models help scientists visualize concepts.

Scientific Theories and Laws

- A theory is a possible explanation for observations while a scientific law describes a pattern but does not explain why things happen.

Using Science—Technology

- Technology is the application of science into our everyday lives.

Self-Check

1. **Define** the first step a scientist usually takes to solve a problem.

2. **Explain** why a control is needed in a valid experiment.

3. **Think Critically** What is the dependent variable in an experiment that shows how the volume of gas changes with changes in temperature?

Applying Math

4. **Find the Average** You perform an experiment to determine how many breaths a fish takes per minute. Your experiment yields the following data: minute 1: 65 breaths; minute 2: 73 breaths; minute 3: 67 breaths; minute 4: 71 breaths; minute 5: 62 breaths. Calculate the average number of breaths that a fish takes per minute.

Standards of Measurement

What You'll Learn

- **Name** the prefixes used in SI and indicate what multiple of ten each one represents.
- **Identify** SI units and symbols for length, volume, mass, density, time, and temperature.
- **Convert** related SI units.

Why It's Important

By using uniform standards, nations can exchange goods and compare information easily.

🔍 Review Vocabulary

measurement: the dimensions, capacity, or amount of something

New Vocabulary

- standard
- SI
- volume
- mass
- density

Units and Standards

A **standard** is an exact quantity that people agree to use to compare measurements. Look at **Figure 9.** Suppose you and a friend want to make some measurements to find out whether a desk will fit through a doorway. You have no ruler, so you decide to use your hands as measuring tools. Using the width of his hands, your friend measures the doorway and says it is 8 hands wide. Using the width of your hands, you measure the desk and find it is $7\frac{3}{4}$ hands wide. Will the desk fit through the doorway? You can't be sure. What went wrong? Even though you both used hands to measure, you didn't check to see whether your hands were the same width as your friend's. In other words, you didn't use a measurement standard, so you can't compare the measurements.

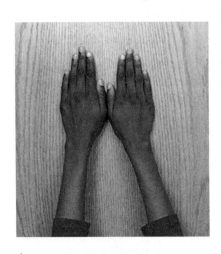

Figure 9 Hands are a convenient measuring tool, but using them can lead to misunderstanding.

Measurement Systems

Suppose the label on a ball of string indicates that the length of the string is 150. Is the length 150 feet, 150 m, or 150 cm? For a measurement to make sense, it must include a number and a unit.

Your family might buy lumber by the foot, milk by the gallon, and potatoes by the pound. These measurement units are part of the English system of measurement, which is commonly used in the United States. Most other nations use the metric system—a system of measurement based on multiples of ten.

International System of Units In 1960, an improved version of the metric system was devised. Known as the International System of Units, this system is often abbreviated SI, from the French *Le Systeme Internationale d'Unites.* All **SI** standards are universally accepted and understood by scientists throughout the world. The standard kilogram, which is kept in Sèvres, France, is shown in **Figure 10.** All kilograms used throughout the world must be exactly the same as the kilogram kept in France.

Each type of SI measurement has a base unit. The meter is the base unit of length. Every type of quantity measured in SI has a symbol for that unit. These names and symbols for the seven base units are shown in **Table 1.** All other SI units are obtained from these seven units.

Figure 10 The standard for mass, the kilogram, and other standards are kept at the International Bureau of Weights and Measures in Sèvres, France. **Explain** *the purpose of a standard.*

SI Prefixes The SI system is easy to use because it is based on multiples of ten. Prefixes are used with the names of the units to indicate what multiple of ten should be used with the units. For example, the prefix *kilo-* means "1,000." That means that one kilometer equals 1,000 meters. Likewise, one kilogram equals 1,000 grams. Because *deci-* means "one-tenth," one decimeter equals one tenth of a meter. A decigram equals one tenth of a gram. The most frequently used prefixes are shown in **Table 2.**

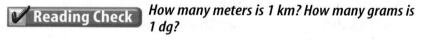 **Reading Check** *How many meters is 1 km? How many grams is 1 dg?*

Table 1 SI Base Units

Quantity Measured	Unit	Symbol
Length	meter	m
Mass	kilogram	kg
Time	second	s
Electric current	ampere	A
Temperature	kelvin	K
Amount of substance	mole	mol
Intensity of light	candela	cd

Table 2 Common SI Prefixes

Prefix	Symbol	Multiplying Factor
Kilo-	k	1,000
Deci-	d	0.1
Centi-	c	0.01
Milli-	m	0.001
Micro-	μ	0.000 001
Nano-	n	0.000 000 001

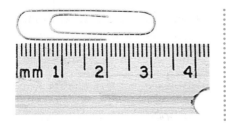

Figure 11 One centimeter contains 10 mm.
Determine *the length of the paper clip in centimeters and millimeters.*

Converting Between SI Units Sometimes quantities are measured using different units as shown in **Figure 11.** A conversion factor is a ratio that is equal to one and is used to change one unit to another. For example, there are 1,000 mL in 1 L, so 1,000 mL = 1 L. If both sides in this equation are divided by l L, the equation becomes:

$$\frac{1{,}000 \text{ mL}}{1 \text{ L}} = 1$$

To convert units, you multiply by the appropriate conversion factor. For example, to convert 1.255 L to mL, multiply 1.255 L by a conversion factor. Use the conversion factor with new units (mL) in the numerator and the old units (L) in the denominator.

$$1.255 \text{ L} \times \frac{1{,}000 \text{ mL}}{1 \text{ L}} = 1{,}255 \text{ mL}$$

Applying Math   Convert Units

CALCULATING CENTIMETERS How long in centimeters is a 3,075 mm rope?

IDENTIFY known values and the unknown value

Identify the known values:
The rope measures 3,075 mm; 1 m = 100 cm = 1,000 mm

Identify the unknown value:
How long is the rope in cm?

SOLVE the problem

This is the equation you need to use:

$$? \text{ cm} = 3{,}075 \text{ mm} \times \frac{100 \text{ cm}}{1{,}000 \text{ mm}}$$

Cancel units and multiply:

$$3{,}075 \text{ \cancel{mm}} \times \frac{100 \text{ cm}}{1{,}000 \text{ \cancel{mm}}} = 307.5 \text{ cm}$$

CHECK your answer

Does your answer seem reasonable? Check your answer by multiplying the answer by $\frac{1{,}000 \text{ mm}}{100 \text{ cm}}$. Did you calculate the original length in millimeters?

Practice Problems

1. Your pencil is 11 cm long. How long is it in millimeters?

2. The Bering Land Bridge National Preserve is a summer home to birdlife. Some birds migrate 20,000 miles. Assume 1 mile equals 1.6 kilometers. Calculate the distance birds fly in kilometers.

For more practice problems go to page 834, and visit gpscience.com/extra_problems.

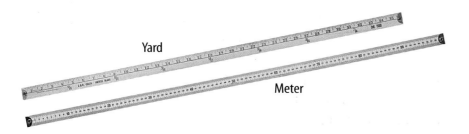

Yard

Meter

Figure 12 One meter is slightly longer than 1 yard and 100 m is slightly longer than a football field. **Predict** *whether your time for a 100-m dash would be slightly more or less than your time for a 100-yard dash.*

Measuring Distance

The word *length* is used in many different ways. For example, the length of a novel is the number of pages or words it contains. In scientific measurement, however, length is the distance between two points. That distance might be the diameter of a hair or the distance from Earth to the Moon. The SI base unit of length is the meter, m. A baseball bat is about 1 m long. Metric rulers and metersticks are used to measure length. **Figure 12** compares a meter and a yard.

Choosing a Unit of Length As shown in **Figure 13,** the size of the unit you measure with will depend on the size of the object being measured. For example, the diameter of a shirt button is about 1 cm. You probably also would use the centimeter to measure the length of your pencil and the meter to measure the length of your classroom. What unit would you use to measure the distance from your home to school? You probably would want to use a unit larger than a meter. The kilometer, km, which is 1,000 m, is used to measure these kinds of distances.

By choosing an appropriate unit, you avoid large-digit numbers and numbers with many decimal places. Twenty-one kilometers is easier to deal with than 21,000 m. And 13 mm is easier to use than 0.013 m.

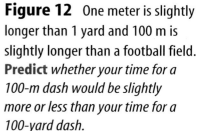

INTEGRATE Earth Science

Astronomical Units The standard measurement for the distance from Earth to the Sun is called the astronomical unit, AU. The distance is about 150 billion (1.5×10^{11}) m. In your Science Journal, calculate what 1 AU would equal in km.

Figure 13 The size of the object being measured determines which unit you will measure in. A tape measure measures in meters. The micrometer, shown at the left, measures in small lengths. **State** *what unit you think it measures in.*

Measuring Volume

The amount of space occupied by an object is called its **volume.** If you want to know the volume of a solid rectangle, such as a brick, you measure its length, width, and height and multiply the three numbers and their units together ($V = l \times w \times h$). For a brick, your measurements probably would be in centimeters. The volume would then be expressed in cubic centimeters, cm^3. To find out how much a moving van can carry, your measurements probably would be in meters, and the volume would be expressed in cubic meters, m^3, because when you multiply you add exponents.

Measuring Liquid Volume How do you measure the volume of a liquid? A liquid has no sides to measure. In measuring a liquid's volume, you are indicating the capacity of the container that holds that amount of liquid. The most common units for expressing liquid volumes are liters and milliliters. These are measurements used in canned and bottled foods. A liter occupies the same volume as a cubic decimeter, dm^3. A cubic decimeter is a cube that is 1 dm, or 10 cm, on each side, as in **Figure 14.**

Look at **Figure 14.** One liter is equal to 1,000 mL. A cubic decimeter, dm^3, is equal to 1,000 cm^3. Because 1 L = 1 dm^3, it follows that:

$$1 \text{ mL} = 1 \text{ cm}^3$$

Sometimes, liquid volumes such as doses of medicine are expressed in cubic centimeters.

Suppose you wanted to convert a measurement in liters to cubic centimeters. You use conversion factors to convert L to mL and then mL to cm^3.

$$1.5 \text{ L} \times \frac{1,000 \text{ mL}}{1 \text{ L}} \times \frac{1 \text{ cm}^3}{1 \text{ mL}} = 1,500 \text{ cm}^3$$

Science Online

Topic: International System of Units
Visit gpscience.com for Web links to information about the International System of Units.

Activity Measure four different things in your classroom using a different unit—distance, volume, time, or temperature—for each. Use each unit only once. Write the measurements down and using the common prefixes in **Table 2,** convert them to different units.

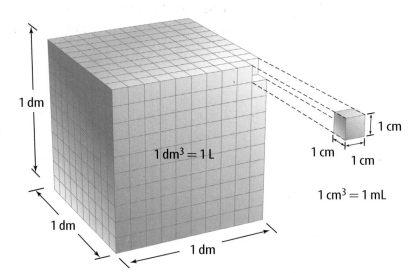

Figure 14 The large cube has a volume of 1 dm^3, which is equivalent to 1 L. **Calculate** *the cubic centimeters (cm^3) in the large cube.*

Table 3 Densities of Some Materials at 20°C			
Material	Density (g/cm³)	Material	Density (g/cm³)
hydrogen	0.000 09	aluminum	2.7
oxygen	0.001 4	iron	7.9
water	1.0	gold	19.3

Measuring Matter

A table-tennis ball and a golf ball have about the same volume. But if you pick them up, you notice a difference. The golf ball has more mass. **Mass** is a measurement of the quantity of matter in an object. The mass of the golf ball, which is about 45 g, is almost 18 times the mass of the table-tennis ball, which is about 2.5 g. A bowling ball has a mass of about 5,000 g. This makes its mass roughly 100 times greater than the mass of the golf ball and 2,000 times greater than the table-tennis ball's mass. To visualize SI units, see **Figure 15** on the following page.

Density A cube of polished aluminum and a cube of silver that are the same size not only look similar but also have the same volume. The mass and volume of an object can be used to find the density of the material the object is made of. **Density** is the mass per unit volume of a material. You find density by dividing an object's mass by the object's volume. For example, the density of an object having a mass of 10 g and a volume of 2 cm³ is 5 g/cm³. **Table 3** lists the densities of some familiar materials.

Derived Units The measurement unit for density, g/cm³, is a combination of SI units. A unit obtained by combining different SI units is called a derived unit. An SI unit multiplied by itself also is a derived unit. Thus the liter, which is based on the cubic decimeter, is a derived unit. A meter cubed, expressed with an exponent—m³—is a derived unit.

Measuring Time and Temperature

It is often necessary to keep track of how long it takes for something to happen, or whether something heats up or cools down. These measurements involve time and temperature.

Time is the interval between two events. The SI unit for time is the second. In the laboratory, you will use a stopwatch or a clock with a second hand to measure time.

Mini LAB

Determining the Density of a Pencil

Procedure
1. Find a **pencil** that will fit in a 100-mL graduated cylinder below the 90-mL mark.
2. Measure the mass of the pencil in grams.
3. Put 90 mL of **water** (initial volume) into a 100-mL **graduated cylinder.** Lower the pencil, eraser first, into the cylinder. Push the pencil down until it is just submerged. Hold it there and record the final volume to the nearest tenth of a milliliter.

Analysis
1. Determine the water displaced by the pencil by subtracting the initial volume from the final volume.
2. Calculate the pencil's density by dividing its mass by the volume of water displaced.
3. Is the density of the pencil greater than or less than the density of water? How do you know?

Figure 15

The characteristics of most of these everyday objects are measured using an international system known as SI dimensions. These dimensions measure length, volume, mass, density, and time. Celsius is not an SI unit but is widely used in scientific work.

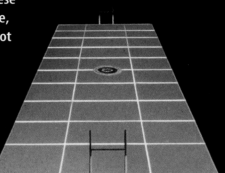

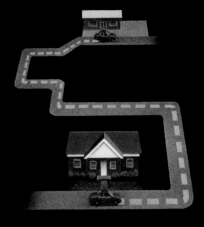

MILLIMETERS A dime is about 1 mm thick.

METERS A football field is about 91 m long.

KILOMETERS The distance from your house to a store can be measured in kilometers.

LITERS This carton holds 1.98 L of frozen yogurt.

MILLILITERS A teaspoonful of medicine is about 5 mL.

GRAMS/METER This stone sinks because it is denser—has more grams per cubic meter—than water.

GRAMS The mass of a thumbtack and the mass of a textbook can be expressed in grams.

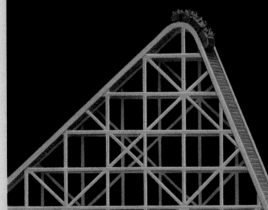

METERS/SECOND The speed of a roller-coaster car can be measured in meters per second.

CELSIUS Water boils at 100°C and freezes at 0°C.

What's Hot and What's Not You will learn the scientific meaning of the word *temperature* in a later chapter. For now, think of temperature as a measure of how hot or how cold something is.

Look at **Figure 16.** For most scientific work, temperature is measured on the Celsius (C) scale. On this scale, the freezing point of water is 0°C, and the boiling point of water is 100°C. Between these points, the scale is divided into 100 equal divisions. Each one represents 1°C. On the Celsius scale, average human body temperature is 37°C, and a typical room temperature is between 20°C and 25°C.

Kelvin and Fahrenheit The SI unit of temperature is the kelvin (K). Zero on the Kelvin scale (0 K) is the coldest possible temperature, also known as absolute zero. Absolute zero is equal to −273°C, which is 273° below the freezing point of water.

Most laboratory thermometers are marked only with the Celsius scale. Because the divisions on the two scales are the same size, the Kelvin temperature can be found by adding 273 to the Celsius reading. So, on the Kelvin scale, water freezes at 273 K and boils at 373 K. Notice that degree symbols are not used with the Kelvin scale.

The temperature measurement you are probably most familiar with is the Fahrenheit scale, which was based roughly on the temperature of the human body, 98.6°.

 **Reading Check** *What is the relationship between the Celsius scale and the Kelvin scale?*

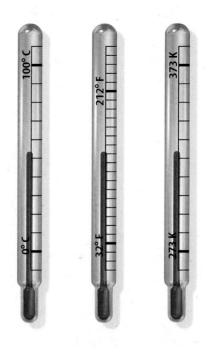

Figure 16 These three thermometers illustrate the scales of temperature between the freezing and boiling points of water. **Compare** *the boiling points of the three scales.*

---

section 2 review

Summary

Units and Standards

- When making measurements, it is important to be accurate.

Measurement Systems

- The International System of Units, or SI, was established to provide a standard of measurement and reduce confusion.

- Conversion factors are used to change one unit to another and involve using a ratio equal to 1.

Measuring

- The size of an object being measured determines which unit you will measure in.

Self Check

1. **Explain** why it is important to have exact standards of measurement.

2. **Explain** why density is a derived unit.

3. **Think Critically** Using a metric ruler, measure a shoe box and a pad of paper. Find the volume of each in cubic centimeters. Then convert the units to mL.

Applying Math

4. **Convert Units** Make the following conversions: 27°C to Kelvin, 20 dg to milligrams, and 3 m to decimeters.

5. **Calculate Density** What is the density of an unknown metal that has a mass of 158 g and a volume of 20 mL? Use **Table 3** to identify this metal.

Communicating with Graphs

A Visual Display

Scientists often graph the results of their experiments because they can detect patterns in the data easier in a graph than in a table. A **graph** is a visual display of information or data. **Figure 17** is a graph that shows a girl walking her dog. The horizontal axis, or the *x*-axis, measures time. Time is the independent variable because as it changes, it affects the measure of another variable. The distance from home that the girl and the dog walk is the other variable. It is the dependent variable and is measured on the vertical axis, or *y*-axis.

Graphs are useful for displaying numerical information in business, science, sports, advertising, and many everyday situations. Different kinds of graphs—line, bar, and circle—are appropriate for displaying different types of information.

✔ **Reading Check** *What are three common types of graphs?*

Graphs are useful for displaying numerical information in business, science, sports, advertising, and many other everyday situations. Graphs make it easier to understand complex patterns by displaying data in a visual manner. Scientists often graph their data to detect patterns that would not have been evident in a table. Different graphs use a different methods for displaying information. The conclusions drawn from graphs must be based on accurate information and reasonable scales.

Figure 17 This graph tells the story of the motion that takes place when a girl takes her dog for an 8-min walk.

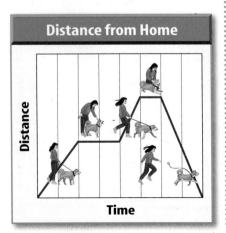

Distance from Home

Distance

Time

Line Graphs

A line graph can show any relationship where the dependent variable changes due to a change in the independent variable. Line graphs often show how a relationship between variables changes over time. You can use a line graph to track many things, such as how certain stocks perform or how the population changes over any period of time—a month, a week, or a year.

You can show more than one event on the same graph as long as the relationship between the variables is identical. Suppose a builder had three choices of thermostats for a new school. He wanted to test them to know which was the best brand to install throughout the building. He installed a different thermostat in classrooms A, B, and C. He set each thermostat at 20°C. He turned the furnace on and checked the temperatures in the three rooms every 5 min for 25 min. He recorded his data in **Table 4.**

The builder then plotted the data on a graph. He could see from the table that the data did not vary much for the three classrooms. So he chose small intervals for the *y*-axis and left part of the scale out (the part between 0° and 15°). See **Figure 18.** This allowed him to spread out the area on the graph where the data points lie. You can see easily the contrast in the colors of the three lines and their relationship to the black horizontal line. The black line represents the thermostat setting and is the control. The control is what the resulting room temperature of the classrooms should be if the thermostats are working efficiently.

Table 4 Room Temperature			
Time*	**Classroom Temperature (C°)**		
	A	**B**	**C**
0	16	16	16
5	17	17	16.5
10	19	19	17
15	20	21	17.5
20	20	23	18
25	20	25	18.5

*minutes after turning on heat

Figure 18 The room temperatures of classrooms A, B, and C are shown in contrast to the thermostat setting of 20°C.
Identify *the thermostat that achieved its temperature setting the quickest.*

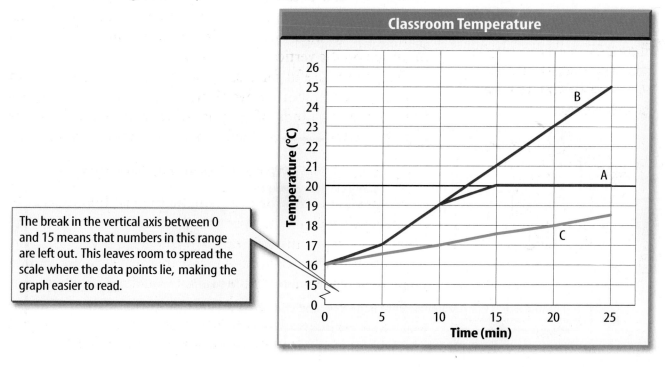

The break in the vertical axis between 0 and 15 means that numbers in this range are left out. This leaves room to spread the scale where the data points lie, making the graph easier to read.

Figure 19 Graphing calculators are valuable tools for making graphs.

Constructing Line Graphs Besides choosing a scale that makes a graph readable, as illustrated in **Figure 18,** other factors are involved in constructing useful graphs. The most important factor in making a line graph is always using the *x*-axis for the independent variable. The *y*-axis always is used for the dependent variable. Because the points in a line graph are related, you connect the points.

Another factor in constructing a graph involves units of measurement. For example, you might use a Celsius thermometer for one part of your experiment and a Fahrenheit thermometer for another. But you must first convert your temperature readings to the same unit of measurement before you make your graph.

In the past, graphs had to be made by hand, with each point plotted individually. Today, scientists use a variety of tools, such as computers and graphing calculators like the one shown in **Figure 19,** to help them draw graphs.

Applying Math Make and Use Graphs

GRAPHING TEMPERATURE In an experiment, you checked the air temperature at certain hours of the day. At 8 A.M., the temperature is 27°C; at noon, the temperature is 32°C; and at 4 P.M., the temperature is 30°C. Graph the results of your experiment.

IDENTIFY known values

        time = independent variable which is the *x*-axis

    temperature = dependent variable which is the *y*-axis

GRAPH the problem

Graph time on the *x*-axis and temperature on the *y*-axis. Mark the equal increments on the graph to include all measurements. Plot each point on the graph by finding the time on the *x*-axis and moving up until you find the recorded temperature on the *y*-axis. Place a point there. Continue placing points on the graph. Then connect the points from left to right.

Practice Problems

As you train for a marathon, you compare your previous times. In year one, you ran it in 5.2 h; in year two, you ran it in 5 h; in year three, you ran it in 4.8 h; in year four, you ran it in 4.3 h; and in year five, you ran it in 4 h.

1. Make a table of your data.

2. Graph the results of your marathon races.

3. Calculate your percentage of improvement from year 1 to year 5.

For more practice problems, go to page 834, and visit gpscience.com/extra_problems.

Bar Graphs

A bar graph is useful for comparing information collected by counting. For example, suppose you counted the number of students in every classroom in your school on a particular day and organized your data as in **Table 5.** You could show these data in a bar graph like the one shown in **Figure 20.** Uses for bar graphs include comparisons of oil, or crop productions, costs, or as data in promotional materials. Each bar represents a quantity counted at a particular time, which should be stated on the graph. As on a line graph, the independent variable is plotted on the *x*-axis and the dependent variable is plotted on the *y*-axis.

Recall that you might need to place a break in the scale of the graph to better illustrate your results. For example, if your data were 1,002, 1,010, 1,030, and 1,040 and the intervals on the scale were every 100 units, you might not be able to see the difference from one bar to another. If you had a break in the scale and started your data range at 1,000 with intervals of ten units, you could make a more accurate comparison.

Reading Check *Describe possible data where using a bar graph would be better than using a line graph.*

Table 5 Classroom Size	
Number of Students	**Number of Classrooms**
20	1
21	3
22	3
23	2
24	3
25	5
26	5
27	3

Figure 20 The height of each bar corresponds to the number of classrooms having a particular number of students.

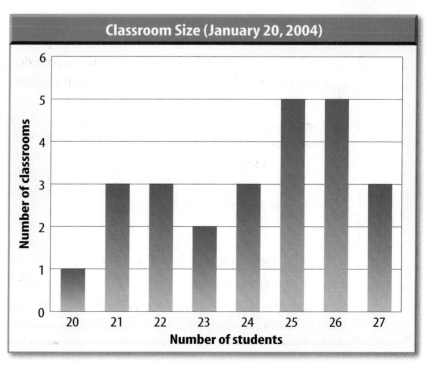

Mini LAB

Observing Change Through Graphing

Procedure
1. Place a **thermometer** in a **plastic foam cup** of hot, but not boiling, **water.**
2. Measure and record the temperature every 30 s for 5 min.
3. Repeat the experiment with freshly heated water. This time, cover the cup with a **plastic lid** in between measurements.

Analysis
1. Make a line graph of the changing temperature from step 2, showing time on the *x*-axis and temperature on the *y*-axis. Using a different color pen, plot the changing temperature from step 3 on the same graph.
2. Use the graph to describe the cooling process in each of the trials.

Try at Home

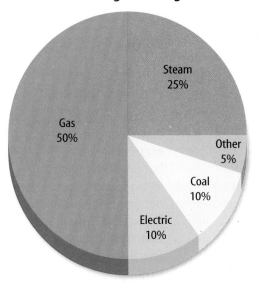

Heating Fuel Usage

Gas 50%
Steam 25%
Other 5%
Coal 10%
Electric 10%

Figure 21 A circle graph shows the different parts of a whole quantity.

Circle Graphs

A circle graph, or pie graph, is used to show how some fixed quantity is broken down into parts. The circular pie represents the total. The slices represent the parts and usually are represented as percentages of the total.

Figure 21 illustrates how a circle graph could be used to show the percentage of buildings in a neighborhood using each of a variety of heating fuels. You easily can see that more buildings use gas heat than any other kind of system. What else does the graph tell you?

To create a circle graph, you start with the total of what you are analyzing. **Figure 21** starts with 72 buildings in the neighborhood. For each type of heating fuel, you divide the number of buildings using each type of fuel by the total (72). You then multiply that decimal by 360° to determine the angle that the decimal makes in the circle. Eighteen buildings use steam. Therefore, $18 \div 72 \times 360° = 90°$ on the circle graph. You then would measure 90° on the circle with your protractor to show 25 percent.

When you use graphs, think carefully about the conclusions you can draw from them. You want to make sure your conclusions are based on accurate information and that you use scales that help make your graph easy to read.

section 3 review

Summary

A Visual Display

- Graphs are a visual representation of data.
- Scientists often graph their data to detect patterns.
- The type of graph used is based on the conclusions you want to identify.

Line Graphs

- A line graph shows how a relationship between two variables changes over time.

Bar Graphs

- Bar graphs are best used to compare information collected by counting.

Circle Graphs

- A circle graph shows how a fixed quantity is broken down into parts.

Self Check

1. **Identify** the kind of graph that would best show the results of a survey of 144 people where 75 ride a bus, 45 drive cars, 15 carpool, and 9 walk to work.

2. **State** which type of variable is plotted on the *x*-axis and which type is plotted on the *y*-axis.

3. **Explain** why the points in a line graph are connected.

4. **Think Critically** How are line, bar, and circle graphs similar? How are they different?

Applying Math

5. **Percentage** In a survey, it was reported that 56 out of 245 people would rather drink orange juice in the morning than coffee. Calculate what percentage of a circle graph this data would occupy.

Science online gpscience.com/self_check_quiz

Converting Kitchen Measurements

Look through a recipe book. Are any of the ingredient amounts stated in SI? How can you convert English measurements to SI measurements?

▶ Real-World Question

How do kitchen measurements compare with SI measurements?

Goals

- Determine a relationship between two systems of measurements.
- Calculate the conversion factors for converting English units to SI units.

Materials

balance	dried beans
100-mL graduated cylinder	dried rice
measuring cup	potato flakes
measuring teaspoon	water
measuring tablespoon	vinegar
cornmeal	salad oil

Safety Precautions

▶ Procedure

1. Copy the data table into your Science Journal and record each SI measurement.

2. Use the appropriate English measuring cup or spoon to measure the amounts of each ingredient shown in the table.

3. Use a balance to measure each dry ingredient. Use a graduated cylinder to measure each liquid ingredient.

English to SI Conversions		
Ingredient	English Measure	SI Measure
Water	1/2 cup	
Cornmeal	2 cups	
Salad oil	4 tablespoons	Do not write in this book.
Dried rice	1/2 cup	
Potato flakes	3 cups	
Vinegar	1 teaspoon	
Dried beans	3 cups	

▶ Conclude and Apply

1. **Calculate** the number of grams in one cup of each dry ingredient. Calculate the number of milliliters in one cup, one teaspoon, and one tablespoon of each liquid ingredient.

2. **Write** conversion factors that will convert each English unit to an SI unit for each ingredient.

3. **Calculate** how many milliliters you would measure if a recipe called for three tablespoons of salad oil.

4. **Compare and contrast** your conversion factors for the dry ingredients and your conversion factors for the liquid ingredients.

Communicating Your Data

Write a recipe used in your home converting all the English units to SI units.

LAB Design Your Own

Setting High Standards for Measurements

▶ Real-World Question

To develop the International System of Units, people had to agree on set standards and basic definitions of scale. If you had to develop a new measurement system, people would have to agree with your new standards and definitions. In this activity, your team will use string to devise and test its own SI (String International) system for measuring length. What are the requirements for designing a new measurement system using string?

Goals

- **Design** an experiment that involves devising and testing your own measurement system for length.
- **Measure** various objects with the string measurement system.

Possible Materials

string
scissors
marking pen
masking tape
miscellaneous objects for standards

Safety Precautions

▶ Form a Hypothesis

Based on your knowledge of measurement standards and systems, form a hypothesis that explains how exact units help keep measuring consistent.

▶ Test Your Hypothesis

Make a Plan

1. As a group, agree upon and write out the hypothesis statement.
2. As a group, list the steps that you need to take to test your hypothesis. Be specific, describing exactly what you will do at each step.
3. Make a list of the materials that you will need.

4. **Design** a data table in your Science Journal so it is ready to use as your group collects data.

5. As you read over your plan, be sure you have chosen an object in your classroom to serve as a standard. It should be in the same size range as what you will measure.

6. Consider how you will mark scale divisions on your string. Plan to use different pieces of string to try different-sized scale divisions.

7. What is your new unit of measurement called? Come up with an abbreviation for your unit. Will you name the smaller scale divisions?

8. What objects will you measure with your new unit? Be sure to include objects longer and shorter than your string. Will you measure each object more than once to test consistency? Will you measure the same object as another group and compare your findings?

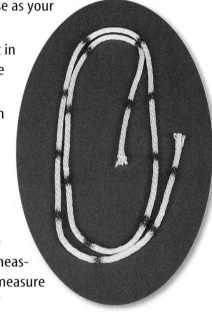

Follow Your Plan

1. Make sure your teacher approves your plan before you start.

2. Carry out the experiment as it has been planned.

3. **Record** observations that you make and complete the data table in your Science Journal.

▶ Analyze Your Data

1. Which of your string scale systems will provide the most accurate measurement of small objects? Explain.

2. How did you record measurements that were between two whole numbers of your units?

▶ Conclude and Apply

1. When sharing your results with other groups, why is it important for them to know what you used as a standard?

2. **Infer** how it is possible for different numbers to represent the same length of an object.

Communicating Your Data

Compare your conclusions with other students' conclusions. Are there differences? Explain how these may have occurred.

Thinking in Pictures: and other reports from my life with autism[1]

By Temple Grandin

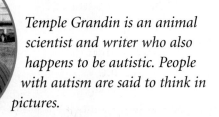

Temple Grandin is an animal scientist and writer who also happens to be autistic. People with autism are said to think in pictures.

I think in pictures. Words are like a second language to me. I translate both spoken and written words into full-color movies, complete with sound, which run like a VCR tape in my head. When somebody speaks to me, his words are instantly translated into pictures. Language-based thinkers often find this phenomenon difficult to understand, but in my job as equipment designer for the livestock industry, visual thinking is a tremendous advantage.

. . . I credit my visualization abilities with helping me understand the animals I work with. Early in my career I used a camera to help give me the animals' perspective as they walked through a chute for their veterinary treatment. I would kneel down and take pictures through the chute from the cow's eye level. Using the photos, I was able to figure out which things scared the cattle.

Every design problem I've ever solved started with my ability to visualize and see the world in pictures. I started designing things as a child, when I was always experimenting with new kinds of kites and model airplanes.

1 Autism is a complex developmental disability that usually appears during the first three years of life. Children and adults with autism typically have difficulties in communicating with others and relating to the outside world.

Understanding Literature

Identifying the Main Idea The most important idea expressed in a paragraph or essay is the main idea. The main idea in a reading might be clearly stated, but sometimes the reader has to summarize the contents of a reading in order to determine its main idea. What do you think is the main idea of the passage?

Respond to the Reading

1. How do people with autism think differently than other people?
2. What did the author use to see from a cow's point of view?
3. What did the author use for models to design things when she was a child?
4. **Linking Science and Writing** Research the use of a scientific model. Write a paragraph stating the main ideas and listing supporting details.

Models enable scientists to see things that are too big, too small, or are too complex. Scientists might build models of DNA, airplanes, or other equipment. Temple Grandin's visual thinking and ability to make models enables her to predict how things will work when they are put together.

Reviewing Main Ideas

The Methods of Science

1. Science is a way of learning about the natural world, such as the hurricane shown below, through investigation.

2. Scientific investigations can involve making observations, testing models, or conducting experiments.

3. Scientific experiments investigate the effect of one variable on another. All other variables are kept constant.

4. Scientific laws are repeated patterns in nature. Theories attempt to explain how and why these patterns develop.

Section 2 **Standards of Measurement**

1. A standard of measurement is an exact quantity that people agree to use as a basis of comparison. The International System of Units, or SI, was established to provide a standard and reduce confusion.

2. When a standard of measurement is established, all measurements are compared to the same exact quantity—the standard. Therefore, all measurements can be compared with one another.

3. The most commonly used SI units include: length—meter, volume—liter, mass—kilogram, and time—second.

4. In SI, prefixes are used to make the base units larger or smaller by multiples of ten.

5. Any SI unit can be converted to any other related SI unit by multiplying by the appropriate conversion factor. These towers are 45,190 cm in height, which is equal to 451.9 m.

Section 3 **Communicating With Graphs**

1. Graphs are a visual representation of data that make it easier for scientists to detect patterns.

2. Line graphs show continuous changes among related variables. Bar graphs are used to show data collected by counting. Circle graphs show how a fixed quantity can be broken into parts.

3. To create a circle graph, you have to determine the angles for your data.

4. In a line graph, the independent variable is always plotted on the horizontal *x*-axis. The dependent variable is always plotted on the vertical *y*-axis.

FOLDABLES Use the Foldable that you made at the beginning of this chapter to help you review scientific processes.

Using Vocabulary

bias p.10
constant p.9
control p.9
density p.19
dependent variable p.9
experiment p.8
graph p.22
hypothesis p.8
independent variable p.9
mass p.19

model p.11
scientific law p.12
scientific method p.7
SI p.15
standard p.14
technology p.13
theory p.12
variable p.9
volume p.18

Match each phrase with the correct term from the list of vocabulary words.

1. the modern version of the metric system

2. the amount of space occupied by an object

3. an agreed-upon quantity used for comparison

4. the amount of matter in an object

5. a variable that changes as another variable changes

6. a visual display of data

7. a test set up under controlled conditions

8. a variable that does NOT change as another variable changes

9. mass per unit volume

10. an educated guess using what you know and observe

Checking Concepts

Choose the word or phrase that best answers the question.

11. Which of the following questions CANNOT be answered by science?
    A) How do birds fly?
    B) Is this a good song?
    C) What is an atom?
    D) How does a clock work?

12. Which of the following is an example of an SI unit?
    A) foot
    B) second
    C) pound
    D) gallon

13. One one-thousandth is expressed by which prefix?
    A) kilo-
    B) nano-
    C) centi-
    D) milli-

14. Which of the following is SI based on?
    A) inches
    B) powers of five
    C) English units
    D) powers of ten

15. What is the symbol for deciliter?
    A) dL
    B) dcL
    C) dkL
    D) Ld

16. Which of the following is NOT a derived unit?
    A) dm^3
    B) m
    C) cm^3
    D) g/ml

17. Which of the following is NOT equal to 1,000 mL?
    A) 1 L
    B) 100 cL
    C) $1\ dm^3$
    D) $1\ cm^3$

Interpreting Graphics

Use the photo below to answer question 18.

18. The illustrations above show the items needed for an investigation. Which item is the independent variable? Which items are the constants? What might a dependent variable be?

Science Online gpscience.com/vocabulary_puzzlemaker

19. Copy and complete this concept map on scientific methods.

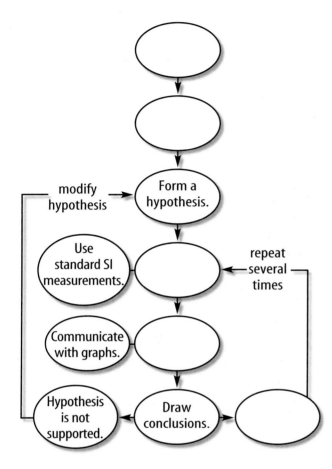

Thinking Critically

20. Communicate Standards of measurement used during the Middle Ages often were based on such things as the length of the king's arm. How would you go about convincing people to use a different system of standard units?

21. Analyze What are some advantages and disadvantages of adopting SI in the United States?

22. Identify when bias occurs in scientific experimentation. Describe steps scientists can take to reduce bias and validate experimental data.

23. Demonstrate Not all objects have a volume that is measured easily. If you were to determine the mass, volume, and density of your textbook, a container of milk, and an air-filled balloon, how would you do it?

24. Apply Suppose you set a glass of water in direct sunlight for 2 h and measure its temperature every 10 min. What type of graph would you use to display your data? What would the dependent variable be? What would the independent variable be?

25. Form a Hypothesis A metal sphere is found to have a density of 5.2 g/cm^3 at 25°C and a density of 5.1 g/cm^3 at 50°C. Form a hypothesis to explain this observation. How could you test your hypothesis?

26. List the SI units of length you would use to express the following.
   a. diameter of a hair
   b. width of your classroom
   c. width of a pencil lead
   d. length of a sheet of paper

27. Compare and contrast the ease with which conversions can be made among SI units versus conversions among units in the English system.

Applying Math

28. Convert Units Make the following conversions.
   A) 1,500 mL to L   **C)** 5.8 dg to mg
   B) 2 km to cm   **D)** 22°C to K

29. Calculate the density of an object having a mass of 17 g and a volume of 3 cm^3.

30. Solve A block of wood is 0.4 m by 0.2 m by 0.7 m. Find its dimensions in centimeters. Then find its volume in cubic centimeters.

Part 1 Multiple Choice

Record your answers on the answer sheet provided by your teacher or on a sheet of paper.

Use the graph below to answer questions 1 and 2.

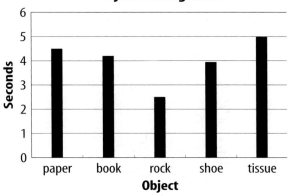

Object Falling Time

1. Students drop objects from a height and measure the time it takes each to reach the ground. What is the dependent variable in this experiment?
   A. falling time   C. drop height
   B. shoe   D. paper

2. What is a constant in this experiment?
   A. throwing some objects and dropping others
   B. measuring different falling times for each object
   C. dropping each object from the same height
   D. dropping a variety of objects

3. Which of the following is a statement about something that happens in nature which seems to be true all the time?
   A. theory   C. hypothesis
   B. scientific law   D. conclusion

Test-Taking Tip

Recheck Your Answers Double check your answers before turning in the test.

4. What does the symbol *ns* represent?
   A. millisecond   C. microsecond
   B. nanosecond   D. kelvin

5. Which of these best defines mass?
   A. the amount of space occupied by an object
   B. the distance between two points
   C. the quantity of matter in an object
   D. the interval between two events

Use the graph below to answer questions 6 and 7.

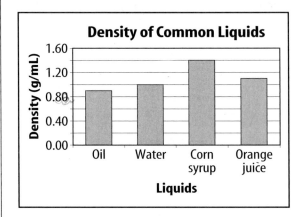

Density of Common Liquids

6. Which two liquids have the highest and the lowest densities?
   A. oil and water
   B. oil and corn syrup
   C. orange juice and water
   D. corn syrup and orange juice

7. What is the density of oil in units of mg/cm^3?
   A. 850 mg/cm^3 C. 0.085 mg/cm^3
   B. 85 mg/cm^3 D. 8500 mg/cm^3

8. Which type of graph is most useful for showing how the relationship between independent and dependent variables changes over time?
   A. circle graph   C. pictograph
   B. bar graph   D. line graph

Part 2 | Short Response/Grid In

Record your answers on the answer sheet provided by your teacher or on a sheet of paper.

9. Describe several ways scientists use investigations to learn about the natural world.

10. Define the term *technology*. Identify three ways that technology makes your life easier, safer, or more enjoyable.

11. Describe the three major categories into which science is classified. Which branches of science would be most important to an environmental engineer? Why?

12. Make the following conversions:
   a. 615 mg to g
   b. 75 dL to mL
   c. 0.95 km to cm

Use the illustration below to answer questions 13 and 14.

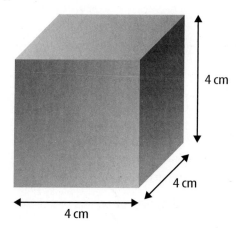

4 cm

4 cm

4 cm

13. Define the term *volume*. Calculate the volume of the cube shown above. Give your answer in cm^3 and mL.

14. Define the term *density*. If the mass of the cube is 96 g, what is the density of the cube material?

15. Why do scientists use graphs when analyzing data?

Part 3 | Open Ended

Record your answers on a sheet of paper.

16. A friend frequently misses the morning school bus. Use the scientific method to address this problem.

Use the illustration below to answer questions 17 and 18.

17. What is the standard unit shown in this picture? Why is it kept under cover, in a vacuum-sealed container?

18. Why is this standard important to anyone who makes measurements? Explain why valid experimental results must be based on standards.

19. You must decide what items to pack for a hiking and camping trip. Space is limited, and you must carry all items during hikes. What measurements are important in your preparation?

Use the table below to answer question 20.

Animal Life Span			
	Cow	**Dog**	**Horse**
Resting Heart Rate	52 beats per min	95 beats	48 beats per min
Average Life Span	18 years	16 years	27 years

20. Create a graph to display the data shown above.

Motion

Taking the Plunge

You probably don't think of an amusement park as a physics laboratory. The fast speeds, quick turns, and plunging falls are a great place to study the laws of gravity and motion. In fact, engineers use these laws when they are designing and building amusement park rides.

Science Journal Write a paragraph describing three rides in an amusement park and how the rides cause you to move.

Start-Up Activities

Compare Speeds

A cheetah can run at a speed of almost 120 km/h and is the fastest runner in the world. A horse can reach a speed of 64 km/h; an elephant's top speed is about 40 km/h; and the fastest snake slithers at a speed of about 3 km/h. The speed of an object is calculated by dividing the distance the object travels by the time it takes it to move that distance. How does your speed compare to the speeds of these animals?

1. Use a meterstick to mark off 10 m.
2. Have your partner use a stopwatch to determine how fast you run 10 m.
3. Divide 10 m by your time in seconds to calculate your speed in m/s.
4. Multiply your answer by 3.6 to determine your speed in km/h.
5. **Think Critically** Write a paragraph in your Science Journal comparing your speed with the maximum speed of a cheetah, horse, elephant, and snake. Could you win a race with any of them?

 **Preview this chapter's content and activities at** gpscience.com

Motion Many things are in motion in your everyday life. Make the following Foldable to help you better understand motion as you read the chapter.

STEP 1 Fold a sheet of paper in half lengthwise. Make the back edge about 1.25 cm longer than the front edge.

STEP 2 Fold in half, then fold in half again to make three folds.

STEP 3 Unfold and cut only the top layer along the three folds to make four tabs.

STEP 4 Label the tabs.

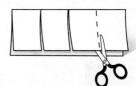

Identify Questions Before you read the chapter, select a motion you can observe and write it under the left tab. As you read the chapter, write answers to the other questions under the appropriate tabs.

Describing Motion

Reading Guide

What You'll Learn

■ **Distinguish** between distance and displacement.
■ **Explain** the difference between speed and velocity.
■ **Interpret** motion graphs.

Why It's Important

Understanding the nature of motion and how to describe it helps you understand why motion occurs.

Review Vocabulary

instantaneous: occurring at a particular instant of time

New Vocabulary
● distance
● displacement
● speed
● average speed
● instantaneous speed
● velocity

Figure 1 This mail truck is in motion.

Infer *How do you know the mail truck has moved?*

Motion

Are distance and time important in describing running events at the track-and-field meets in the Olympics? Would the winners of the 5-km race and the 10-km race complete the run in the same length of time?

Distance and time are important. In order to win a race, you must cover the distance in the shortest amount of time. The time required to run the 10-km race should be longer than the time needed to complete the 5-km race because the first distance is longer. How would you describe the motion of the runners in the two races?

Motion and Position You don't always need to see something move to know that motion has taken place. For example, suppose you look out a window and see a mail truck stopped next to a mailbox. One minute later, you look out again and see the same truck stopped farther down the street. Although you didn't see the truck move, you know it moved because its position relative to the mailbox changed.

A reference point is needed to determine the position of an object. In **Figure 1,** the reference point might be a tree or a mailbox. Motion occurs when an object changes its position relative to a reference point. The motion of an object depends on the reference point that is chosen. For example, the motion of the mail truck in **Figure 1** would be different if the reference point were a car moving along the street, instead of a mailbox.

Frame of Reference After a reference point is chosen, a frame of reference can be created. A frame of reference is a coordinate system in which the position of the objects is measured. The *x-axis* and *y-axis* of the reference frame are drawn so that they intersect the reference point.

Distance In track-and-field events, have you ever run a 50-m dash? A distance of 50 m was marked on the track or athletic field to show you how far to run. An important part of describing the motion of an object is to describe how far it has moved, which is **distance.** The SI unit of length or distance is the meter (m). Longer distances are measured in kilometers (km). One kilometer is equal to 1,000 m. Shorter distances are measured in centimeters (cm). One meter is equal to 100 centimeters.

Displacement Suppose a runner jogs to the 50-m mark and then turns around and runs back to the 20-m mark, as shown in **Figure 2.** The runner travels 50 m in the original direction (north) plus 30 m in the opposite direction (south), so the total distance she ran is 80 m. How far is she from the starting line? The answer is 20 m. Sometimes you may want to know not only your distance but also your direction from a reference point, such as from the starting point. **Displacement** is the distance and direction of an object's change in position from the starting point. The runner's displacement in **Figure 2** is 20 m north.

The length of the runner's displacement and the distance traveled would be the same if the runner's motion was in a single direction. If the runner ran from the starting point to the finish line in a straight line, then the distance traveled would be 50 m and the displacement would be 50 m north.

Reading Check *How do distance and displacement differ?*

Speed

Think back to the example of the mail truck's motion in **Figure 1.** You could describe the movement by the distance traveled and by the displacement from the starting point. You also might want to describe how fast it is moving. To do this, you need to know how far it travels in a given amount of time. **Speed** is the distance an object travels per unit of time.

Figure 2 Distance and displacement are not the same. The runner's displacement is 20 m north of the starting line. However, the total distance traveled is 80 m.

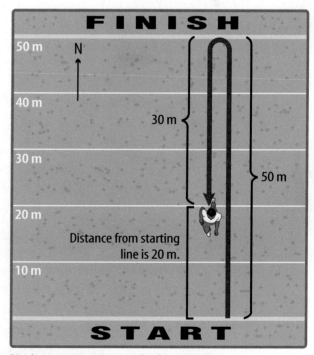

Displacement = 20 m north of starting line
Distance traveled = 50 m + 30 m = 80 m

Calculating Speed Any change over time is called a rate. If you think of distance as the change in position, then speed is the rate at which distance is traveled or the rate of change in position.

Speed Equation

speed (in meters/second) = $\dfrac{\text{distance (in meters)}}{\text{time (in seconds)}}$

$$s = \frac{d}{t}$$

The SI unit for distance is the meter and the SI unit of time is the second (s), so in SI, units of speed are measured in meters per second (m/s). Sometimes it is more convenient to express speed in other units, such as kilometers per hour (km/h). **Table 1** shows some convenient units for certain types of motion.

Applying Math — Solve a One-Step Equation

CALCULATING SPEED A car traveling at a constant speed covers a distance of 750 m in 25 s. What is the car's speed?

IDENTIFY known values and the unknown value

Identify the known values:

| covers a distance of 750 m | means | $d = 750$ m |
| in 25 s | means | $t = 25$ s |

Identify the unknown value:

| What is the car's speed? | means | $s = ?$ m/s |

SOLVE the problem

Substitute the given values of distance and time into the speed equation:

$$s = \frac{d}{t} = \frac{750 \text{ m}}{25 \text{ s}} = 30 \text{ m/s}$$

CHECK the answer

Does your answer seem reasonable? Check your answer by multiplying the time by the speed. The result should be the distance given in the problem.

Practice Problems

1. A passenger elevator operates at an average speed of 8 m/s. If the 60th floor is 219 m above the first floor, how long does it take the elevator to go from the first floor to the 60th floor?

2. A motorcyclist travels an average speed of 20 km/h. If the cyclist is going to a friend's house 5 km away, how long does it take the cyclist to make the trip?

For more practice problems go to page 834, and visit gpscience.com/extra_problems.

Table 1 Examples of Units of Speed		
Unit of Speed	**Examples of Uses**	**Approximate Speed**
km/s	rocket escaping Earth's atmosphere	11.2 km/s
km/h	car traveling at highway speed	100 km/h
cm/yr	geological plate movements	2cm/yr–17 cm/yr

Motion with Constant Speed Suppose you are in a car traveling on a nearly empty freeway. You look at the speedometer and see that the car's speed hardly changes. If the car neither slows down nor speeds up, the car is traveling at a constant speed. If you are traveling at a constant speed, you can measure your speed over any distance interval.

Changing Speed Usually speed is not constant. Think about riding a bicycle for a distance of 5 km, as in **Figure 3.** As you start out, your speed increases from 0 km/h to 20 km/h. You slow down to 10 km/h as you pedal up a steep hill and speed up to 30 km/h going down the other side of the hill. You stop for a red light, speed up again, and move at a constant speed for a while. Finally, you slow down and then stop. Checking your watch, you find that the trip took 15 min. How would you express your speed on such a trip? Would you use your fastest speed, your slowest speed, or some speed between the two?

Science Online

Topic: Running Speeds
Visit gpscience.com for Web links to information about the running speeds of various animals.

Activity In your Science Journal, describe how running fast benefits the survival of animals in the wild.

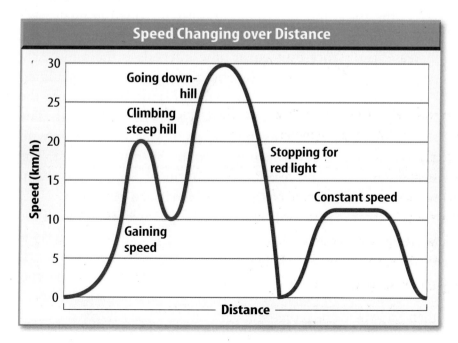

Figure 3 The graph shows how the speed of a cyclist changes during a trip.
Explain *how you describe the speed of an object when the speed is changing.*

Figure 4 The speed shown on the speedometer gives the instantaneous speed—the speed at one instant in time.

Describing the Motion of a Car

Procedure

1. Mark your starting point on the floor with **tape.**
2. At the starting line, give your **toy car** a gentle push forward. At the same time, start your **stopwatch.**
3. Stop timing when the car comes to a complete stop. Mark the spot on the floor at the front of the car with a **pencil.** Record the time for the entire trip.
4. Use a **meterstick** to measure the distance to the nearest tenth of a centimeter and convert it to meters.

Analysis
Calculate the speed. How would the speed differ if you repeated your experiment in exactly the same way but the car traveled in the opposite direction?

Average Speed Average speed describes speed of motion when speed is changing. **Average speed** is the total distance traveled divided by the total time of travel. It can be calculated using the relationships among speed, distance, and time. For the bicycle trip just described, the total distance traveled was 5 km and the total time was 1/4 h, or 0.25 h. The average speed was:

$$s = \frac{d}{t} = \frac{5\ \text{km}}{0.25\ \text{h}} = 20\ \text{km/h}$$

Instantaneous Speed Suppose you watch a car's speedometer, like the one in **Figure 4,** go from 0 km/h to 60 km/h. A speedometer shows how fast a car is going at one point in time or at one instant. The speed shown on a speedometer is the instantaneous speed. **Instantaneous speed** is the speed at a given point in time.

Changing Instantaneous Speed When something is speeding up or slowing down, its instantaneous speed is changing. The speed is different at every point in time. If an object is moving with constant speed, the instantaneous speed doesn't change. The speed is the same at every point in time.

 **Reading Check** *What are two examples of motion in which the instantaneous speed changes?*

Graphing Motion

The motion of an object over a period of time can be shown on a distance-time graph. Time is plotted along the horizontal axis of the graph and the distance traveled is plotted along the vertical axis of the graph. If the object moves with constant speed, the increase in distance over equal time intervals is the same. As a result, the line representing the object's motion is a straight line.

For example, the graph shown in **Figure 5** represents the motion of three swimmers during a 30-min workout. The straight red line represents the motion of Mary, who swam with a constant speed of 80 m/min over the 30-min workout. The straight blue line represents the motion of Kathy, who swam with a constant speed of 60 m/min during the workout.

The graph shows that the line representing the motion of the faster swimmer is steeper. The steepness of a line on a graph is the slope of the line. The slope of a line on a distance-time graph equals the speed. A horizontal line on a distance-time graph has zero slope, and represents an object at rest. Because Mary has a larger speed than Kathy, the line representing her motion has a larger slope.

Changing Speed The green line represents the motion of Julie, who did not swim at a constant speed. She covered 400 m at a constant speed during the first 10 min, rested for the next 10 min, and then covered 800 m during the final 10 min. During the first 10 min, her speed was less than Mary's or Kathy's, so her line has a smaller slope. During the middle period her speed is zero, so her line over this interval is horizontal and has zero slope. During the last time interval she swam as fast as Mary, so that part of her line has the same slope.

Plotting a Distance-Time Graph On a distance-time graph, the distance is plotted on the vertical axis and the time on the horizontal axis. Each axis must have a scale that covers the range of numbers to be plotted. In **Figure 5** the distance scale must range from 0 to 2,400 m and the time scale must range from 0 to 30 min. Then, each axis can be divided into equal time intervals to represent the data. Once the scales for each axis are in place, the data points can be plotted. After plotting the data points, draw a line connecting the points.

Science nline

Topic: Olympic Swimming Speeds
Visit gpscience.com for Web links to information about the speeds of Olympic swimmers over the past 60 years.

Activity Make a speed-year graph showing the swimming speeds over time. Are there any trends in the speed data?

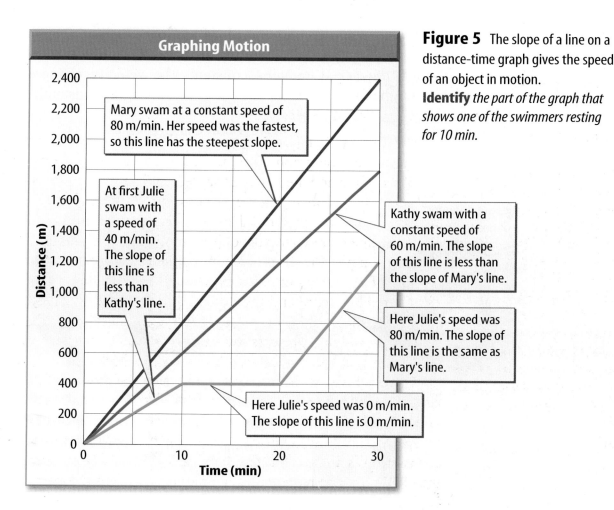

Figure 5 The slope of a line on a distance-time graph gives the speed of an object in motion.
Identify *the part of the graph that shows one of the swimmers resting for 10 min.*

Figure 6 The speed of a storm is not enough information to plot the path. The direction the storm is moving must be known, too.

Figure 7 For an object to have constant velocity, speed and direction must not be changing.

Velocity

You turn on the radio and hear the tail end of a news story about a hurricane, like the one in **Figure 6,** that is approaching land. The storm, traveling at a speed of 20 km/h, is located 100 km east of your location. Should you be worried?

Unfortunately, you don't have enough information to answer that question. Knowing only the speed of the storm isn't much help. Speed describes only how fast something is moving. To decide whether you need to move to a safer area, you also need to know the direction that the storm is moving. In other words, you need to know the velocity of the storm. **Velocity** includes the speed of an object and the direction of its motion.

Escalators like the one shown in **Figure 7** are found in shopping malls and airports. The two sets of passengers pictured are moving at constant speed, but in opposite directions. The speeds of the passengers are the same, but their velocities are different because the passengers are moving in different directions.

Because velocity depends on direction as well as speed, the velocity of an object can change even if the speed of the object remains constant. For example, look at **Figure 7.** The race car has a constant speed and is going around an oval track. Even though the speed remains constant, the velocity changes because the direction of the car's motion is changing constantly.

✔ Reading Check *How are velocity and speed different?*

The people on these two escalators have the same speed. However, their velocities are different because they are traveling in opposite directions.

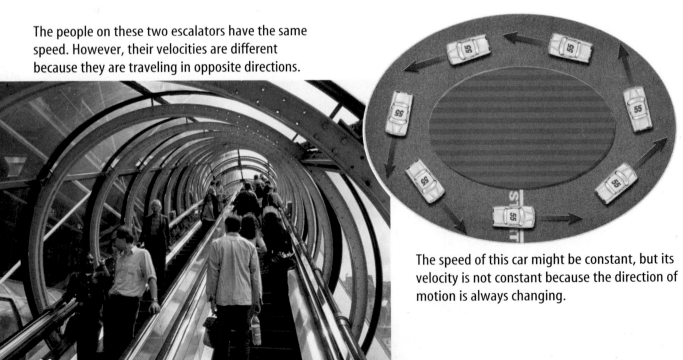

The speed of this car might be constant, but its velocity is not constant because the direction of motion is always changing.

Motion of Earth's Crust

INTEGRATE Earth Science Can you think of something that is moving so slowly you cannot detect its motion, yet you can see evidence of its motion over long periods of time? As you look around the surface of Earth from year to year, the basic structure of the planet seems the same. Mountains, plains, lakes, and oceans seem to remain unchanged over hundreds of years. Yet if you examined geological evidence of what Earth's surface looked like over the past 250 million years, you would see that large changes have occurred. **Figure 8** shows how, according to the theory of plate tectonics, the positions of landmasses have changed during this time. Changes in the landscape occur constantly as continents drift slowly over Earth's surface. However, these changes are so gradual that you do not notice them.

About 250 million years ago, the continents formed a supercontinent called Pangaea.

Figure 8 Geological evidence suggests that Earth's continents have moved slowly over time.

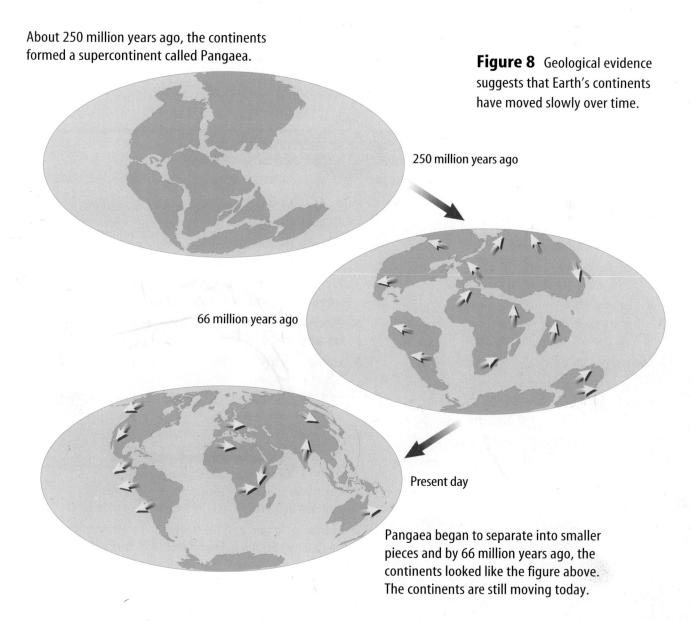

250 million years ago

66 million years ago

Present day

Pangaea began to separate into smaller pieces and by 66 million years ago, the continents looked like the figure above. The continents are still moving today.

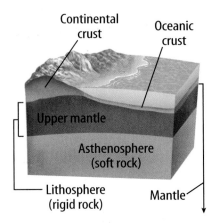

Continental
crust
Oceanic
crust

Upper mantle

Asthenosphere
(soft rock)

Lithosphere
(rigid rock)

Mantle

Figure 9 Earth's crust floats over a puttylike interior.

Moving Continents How can continents move around on the surface of Earth? Earth is made of layers, as shown in **Figure 9**. The outer layer is the crust, and the layer just below the crust is called the upper mantle. Together the crust and the top part of the upper mantle are called the lithosphere. The lithosphere is broken into huge sections called plates that slide slowly on the puttylike layers just below. If you compare Earth to an egg, these plates are about as thick as the eggshell. These moving plates cause geological changes such as the formation of mountain ranges, earthquakes, and volcanic eruptions.

The movement of the plates also is changing the size of the oceans and the shapes of the continents. The Pacific Ocean is getting smaller while the Atlantic Ocean is getting larger. The movement of the plates also changes the shape of the continents as they collide and spread apart.

Plates move so slowly that their speeds are given in units of centimeters per year. In California, two plates slide past each other along the San Andreas Fault with an average relative speed of about 1 cm per year. The Australian Plate's movement is one of the fastest, pushing Australia north at an average speed of about 17 cm per year.

section 1 review

Summary

Position and Motion

- The position of an object is determined relative to a reference point.
- Motion occurs when an object changes its position relative to a reference point.
- Distance is the length of the path an object has traveled. Displacement is the distance and direction of a change in position.

Speed and Velocity

- Speed is the distance an object travels per unit time and is given by this equation:

$$s = \frac{d}{t}$$

- The velocity of an object includes the object's speed and its direction of motion relative to a reference point.

Graphing Motion

- On a distance-time graph, time is the horizontal axis and distance is the vertical axis.
- The slope of a line plotted on a distance-time graph is the speed.

Self Check

1. **Infer** whether the size of an object's displacement could be greater than the distance the object travels.

2. **Describe** the motion represented by a horizontal line on a distance-time graph.

3. **Explain** whether, during a trip, a car's instantaneous speed can ever be greater than its average speed.

4. **Describe** the difference between average speed and constant speed.

5. **Think Critically** You are walking toward the back of a bus that is moving forward with a constant velocity. Describe your motion relative to the bus and relative to a point on the ground.

Applying Math

6. **Calculate Speed** Michiko walked a distance of 1.60 km in 30 min. Find her average speed in m/s.

7. **Calculate Distance** A car travels at a constant speed of 30.0 m/s for 0.8 h. Find the total distance traveled in km.

Science online gpscience.com/self_check_quiz

Acceleration

Reading Guide

What You'll Learn
- **Identify** how acceleration, time, and velocity are related.
- **Explain** how positive and negative acceleration affect motion.
- **Describe** how to calculate the acceleration of an object.

Why It's Important
Acceleration occurs all around you as objects speed up, slow down, or change direction.

◑ Review Vocabulary
speed: rate of change of position; can be calculated by dividing the distance traveled by the time taken to travel the distance

New Vocabulary
- acceleration

Acceleration, Speed, and Velocity

You're sitting in a car at a stoplight when the light turns green. The driver steps on the gas pedal and the car starts moving faster and faster. Just as speed is the rate of change of position, **acceleration** is the rate of change of velocity. When the velocity of an object changes, the object is accelerating.

Remember that velocity includes the speed and direction of an object. Therefore, a change in velocity can be either a change in how fast something is moving or a change in the direction it is moving. Acceleration occurs when an object changes its speed, its direction, or both.

Speeding Up and Slowing Down When you think of acceleration, you probably think of something speeding up. However, an object that is slowing down also is accelerating.

Imagine a car traveling through a city. If the speed is increasing, the car has positive acceleration. When the car slows down its speed is decreasing and the car has negative acceleration. In both cases the car is accelerating because its speed is changing.

Acceleration also has direction, just as velocity does. If the acceleration is in the same direction as the velocity, as in **Figure 10,** the speed increases and the acceleration is positive. If the speed decreases, the acceleration is in the opposite direction from the velocity, and the acceleration is negative for the car shown in **Figure 10.**

Figure 10 These cars are both accelerating because their speed is changing.

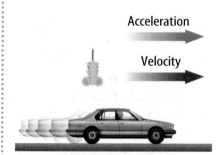

The speed of this car is increasing. The car has positive acceleration.

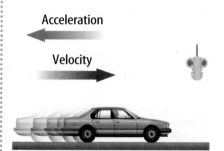

The speed of this car is decreasing. The car has negative acceleration.

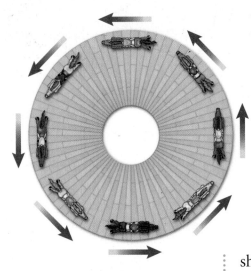

Figure 11 The speed of the horses in this carousel is constant, but the horses are accelerating because their direction is changing constantly.

Changing Direction A change in velocity can be either a change in how fast something is moving or a change in the direction of movement. Any time a moving object changes direction, its velocity changes and it is accelerating. Think about a horse on a carousel. Although the horse's speed remains constant, the horse is accelerating because it is changing direction constantly as it travels in a circular path, as shown in **Figure 11.** In the same way, Earth is accelerating constantly as it orbits the Sun in a nearly circular path.

Graphs of speed versus time can provide information about accelerated motion. The shape of the plotted curve shows when an object is speeding up or slowing down. **Figure 12** shows how motion graphs are constructed.

Calculating Acceleration

Acceleration is the rate of change in velocity. To calculate the acceleration of an object, the change in velocity is divided by the length of the time interval over which the change occurred.

To calculate the change in velocity, subtract the initial velocity—the velocity at the beginning of the time interval—from the final velocity—the velocity at the end of the time interval. Let v_i stand for the initial velocity and v_f stand for the final velocity. Then the change in velocity is:

$$\text{change in velocity} = \text{final velocity} - \text{initial velocity}$$
$$= v_f - v_i$$

Using this expression for the change in velocity, the acceleration can be calculated from the following equation:

Acceleration Equation

$$\text{acceleration (in meters/second}^2) = \frac{\text{change in velocity (in meters/second)}}{\text{time (in seconds)}}$$

$$a = \frac{v_f - v_i}{t}$$

Recall that velocity includes both speed and direction. However, if the direction of motion doesn't change and the object moves in a straight line, the change in velocity is the same as the change in speed. The change in velocity then is the final speed minus the initial speed.

The unit for acceleration is a unit for velocity divided by a unit for time. In SI units, velocity has units of m/s, and time has units of s, so acceleration has units of m/s^2.

Figure 12

Acceleration can be positive, negative, or zero depending on whether an object is speeding up, slowing down, or moving at a constant speed. If the speed of an object is plotted on a graph, with time along the horizontal axis, the slope of the line is related to the acceleration.

A The car in the photograph on the right is maintaining a constant speed of about 90 km/h. Because the speed is constant, the car's acceleration is zero. A graph of the car's speed with time is a horizontal line.

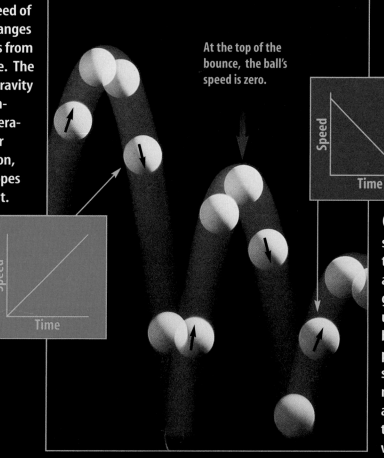

B The green graph shows how the speed of a bouncing ball changes with time as it falls from the top of a bounce. The ball speeds up as gravity pulls the ball downward, so the acceleration is positive. For positive acceleration, the plotted line slopes upward to the right.

At the top of the bounce, the ball's speed is zero.

C The blue graph shows the change with time in the speed of a ball after it hits the ground and bounces upward. The climbing ball slows as gravity pulls it downward, so the acceleration is negative. For negative acceleration, the plotted line slopes downward to the right.

Calculating Positive Acceleration

How is the acceleration for an object that is speeding up different from that of an object that is slowing down? Suppose the jet airliner in **Figure 13** starts at rest at the end of a runway and reaches a speed of 80 m/s in 20 s. The airliner is traveling in a straight line down the runway, so its change in speed and velocity are the same. Because it started from rest, its initial speed was zero. Its acceleration can be calculated as follows:

$$a = \frac{(v_f - v_i)}{t} = \frac{(80 \text{ m/s} - 0 \text{ m/s})}{20 \text{ s}} = 4 \text{ m/s}^2$$

The airliner is speeding up, so the final speed is greater than the initial speed and the acceleration is positive.

Calculating Negative Acceleration

Now imagine that the skateboarder in **Figure 13** is moving in a straight line at a constant speed of 3 m/s and comes to a stop in 2 s. The final speed is zero and the initial speed was 3 m/s. The skateboarder's acceleration is calculated as follows:

$$a = \frac{(v_f - v_i)}{t} = \frac{(0 \text{ m/s} - 3 \text{ m/s})}{2 \text{ s}} = -1.5 \text{ m/s}^2$$

The skateboarder is slowing down, so the final speed is less than the initial speed and the acceleration is negative. The acceleration always will be positive if an object is speeding up and negative if the object is slowing down.

Figure 13 A speed-time graph tells you if acceleration is a positive or negative number.

If acceleration is a positive number, the line slopes upward to the right.

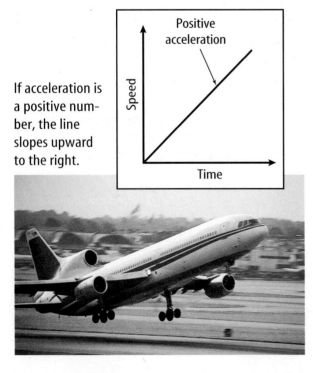

If acceleration is a negative number, the line slopes downward to the right.

Amusement Park Acceleration

Riding roller coasters in amusement parks can give you the feeling of danger, but these rides are designed to be safe. Engineers use the laws of physics to design amusement park rides that are thrilling, but harmless. Roller coasters are constructed of steel or wood. Because wood is not as rigid as steel, wooden roller coasters do not have hills that are as high and steep as some steel roller coasters have. As a result, the highest speeds and accelerations usually are produced on steel roller coasters.

Steel roller coasters can offer multiple steep drops and inversion loops, which give the rider large accelerations. As the rider moves down a steep hill or an inversion loop, he or she will accelerate toward the ground due to gravity. When riders go around a sharp turn, they also are accelerated. This acceleration makes them feel as if a force is pushing them toward the side of the car. **Figure 14** shows the fastest roller coaster in the United States.

 **Reading Check** *What happens when riders on a roller coaster go around a sharp turn?*

Figure 14 This roller coaster can reach a speed of about 150 km/h in 4 s.

section 2 review

Summary

Acceleration, Speed, and Velocity

- Acceleration is the rate of change of velocity.
- A change in velocity occurs when the speed of an object changes, or its direction of motion changes, or both occur.
- The speed of an object increases if the acceleration is in the same direction as the velocity.
- The speed of an object decreases if the acceleration and the velocity of the object are in opposite directions.

Calculating Acceleration

- Acceleration can be calculated by dividing the change in velocity by the time according to the following equation:
  $$a = \frac{v_f - v_i}{t}$$
- The SI unit for acceleration is m/s^2.
- If an object is moving in a straight line, the change in velocity equals the final speed minus the initial speed.

Self Check

1. **Describe** three ways to change the velocity of a moving car.
2. **Determine** the change in velocity of a car that starts at rest and has a final velocity of 20 m/s north.
3. **Explain** why streets and highways have speed limits rather than velocity limits.
4. **Describe** the motion of an object that has an acceleration of 0 m/s^2.
5. **Think Critically** Suppose a car is accelerating so that its speed is increasing. Describe the plotted line on a distance-time graph of the motion of the car.

Applying Math

6. **Calculate Time** A ball is dropped from a cliff and has an acceleration of 9.8 m/s^2. How long will it take the ball to reach a speed of 24.5 m/s?
7. **Calculate Speed** A sprinter leaves the starting blocks with an acceleration of 4.5 m/s^2. What is the sprinter's speed 2 s later?

Motion and Forces

Reading Guide

***What* You'll Learn**

- **Explain** how force and motion are related.
- **Describe** what inertia is and how it is related to Newton's first law of motion.
- **Identify** the forces and motion that are present during a car crash.

***Why* It's Important**

Force and motion are directly linked—without force, you cannot have motion.

Review Vocabulary

scientific law: statement about something that happens in nature that seems to be true all the time

New Vocabulary

- force
- net force
- balanced force
- inertia

What is force?

Passing a basketball to a team member or kicking a soccer ball into the goal are examples of applying force to an object. A **force** is a push or pull. In both examples, the applied force changes the movement of the ball. Sometimes it is obvious that a force has been applied. But other forces aren't as noticeable. For instance, are you conscious of the force the floor exerts on your feet? Can you feel the force of the atmosphere pushing against your body or gravity pulling on your body? Think about all the forces you exert in a day. Every push, pull, stretch, or bend results in a force being applied to an object.

Changing Motion What happens to the motion of an object when you exert a force on it? A force can cause the motion of an object to change. Think of hitting a ball with a racket, as in **Figure 15.** The racket strikes the ball with a force that causes the ball to stop and then move in the opposite direction. If you have played billiards, you know that you can force a ball at rest to roll into a pocket by striking it with another ball. The force of the moving ball causes the ball at rest to move in the direction of the force. In these cases, the velocities of the ball and the billiard ball were changed by a force.

Figure 15 This ball is hit with a force. The racket strikes the ball with a force in the opposite direction of its motion. As a result, the ball changes the direction it is moving.

Figure 16 Forces can be balanced and unbalanced.

Net Force = 0

A These students are pushing on the box with an equal force but in opposite directions. Because the forces are balanced, the box does not move.

Net Force = ⟹

B These students are pushing on the box with unequal forces in opposite directions. The box will be moved in the direction of the larger force.

Net Force = ⟹

C These students are pushing on the box in the same direction. The combined forces will cause the box to move.

Balanced Forces Force does not always change velocity. In **Figure 16A,** two students are pushing on opposite sides of a box. Both students are pushing with an equal force but in opposite directions. When two or more forces act on an object at the same time, the forces combine to form the **net force.** The net force on the box in **Figure 16A** is zero because the two forces cancel each other. Forces on an object that are equal in size and opposite in direction are called **balanced forces.**

Unbalanced Forces Another example of how forces combine is shown in **Figure 16B.** When two students are pushing with unequal forces in opposite directions, a net force occurs in the direction of the larger force. In other words, the student who pushes with a greater force will cause the box to move in the direction of the force. The net force that moves the box will be the difference between the two forces because they are in opposite directions. They are considered to be unbalanced forces.

In **Figure 16C,** the students are pushing on the box in the same direction. These forces are combined, or added together, because they are exerted on the box in the same direction. The net force that acts on this box is found by adding the two forces together.

 *Give another example of an unbalanced force.*

Topic: Forces and Fault Lines
Visit gpscience.com for Web links to information about the unbalanced forces that occur along Earth's fault lines.

Activity Use inexpensive materials such as bars of soap to model the forces and movements along the fault lines. Share your models and demonstrations with your class.

Inertia and Mass

The dirt bike in **Figure 17** is sliding on the track. This sliding bike demonstrates the property of inertia. **Inertia** (ih NUR shuh) is the tendency of an object to resist any change in its motion. If an object is moving, it will have uniform motion. It will keep moving at the same speed and in the same direction unless an unbalanced force acts on it. The velocity of the object remains constant unless a force changes it. If an object is at rest, it tends to remain at rest. Its velocity is zero unless a force makes it move.

Does a bowling ball have the same inertia as a table-tennis ball? Why is there a difference? You couldn't change the motion of a bowling ball much by swatting it with a table-tennis paddle. However, you easily could change the motion of the table-tennis ball. A greater force would be needed to change the motion of the bowling ball because it has greater inertia. Why is this? Recall that mass is the amount of matter in an object, and a bowling ball has more mass than a table-tennis ball does. The inertia of an object is related to its mass. The greater the mass of an object is, the greater its inertia.

Newton's Laws of Motion Forces change the motion of an object in specific ways. The British scientist Sir Isaac Newton (1642–1727) was able to state rules that describe the effects of forces on the motion of objects. These rules are known as Newton's laws of motion. They apply to the motion of all objects you encounter every day such as cars and bicycles, as well as the motion of planets around the Sun.

Figure 17 This racer is skidding because of inertia. The bike tends to move in a straight line with constant speed despite the efforts of the rider to steer the bike around the curve.

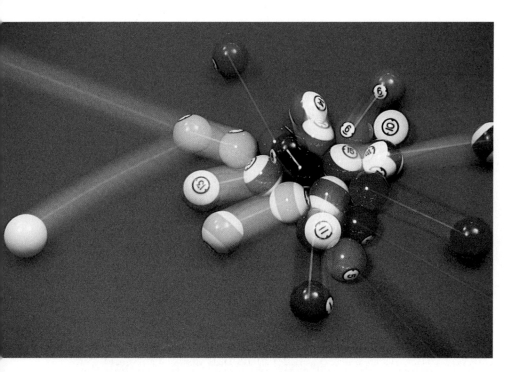

Figure 18 The inertia of the billiard balls causes them to remain at rest until a force is exerted on them by the cue ball.

Newton's First Law of Motion Newton's first law of motion states that an object moving at a constant velocity keeps moving at that velocity unless an unbalanced net force acts on it. If an object is at rest, it stays at rest unless an unbalanced net force acts on it. Does this sound familiar? It is the same as the earlier discussion of inertia. This law is sometimes called the law of inertia. You probably have seen and felt this law at work without even knowing it. **Figure 18** shows a billiard ball striking the other balls in the opening shot. What are the forces involved when the cue ball strikes the other balls? Are the forces balanced or unbalanced? How does this demonstrate the law of inertia?

✔ **Reading Check** *What is Newton's first law of motion?*

What happens in a crash?

The law of inertia can explain what happens in a car crash. When a car traveling about 50 km/h collides head-on with something solid, the car crumples, slows down, and stops within approximately 0.1 s. Any passenger not wearing a safety belt continues to move forward at the same speed the car was traveling. Within about 0.02 s (1/50 of a second) after the car stops, unbelted passengers slam into the dashboard, steering wheel, windshield, or the backs of the front seats, as in **Figure 19.** They are traveling at the car's original speed of 50 km/h—about the same speed they would reach falling from a three-story building.

Figure 19 The crash dummy is not restrained in this low-speed crash. Inertia causes the dummy to slam into the steering wheel. **Explain** *how safety belts can help keep passengers from being seriously injured.*

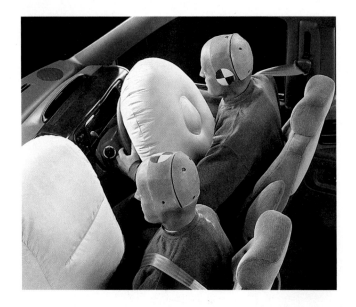

Safety Belts The crash dummy wearing a safety belt in **Figure 20** is attached to the car and slows down as the car slows down. The force needed to slow a person from 50 km/h to zero in 0.1 s is equal to 14 times the force that gravity exerts on the person. The belt loosens a little as it restrains the person, increasing the time it takes to slow the person down. This reduces the force exerted on the person. The safety belt also prevents the person from being thrown out of the car. Car-safety experts say that about half the people who die in car crashes would survive if they wore safety belts. Thousands of others would suffer fewer serious injuries.

Figure 20 These crash dummies were restrained safely with safety belts in this low-speed crash. Usually humans would have fewer injuries if they were restrained safely during an accident.

Air bags also reduce injuries in car crashes by providing a cushion that reduces the force on the car's occupants. When impact occurs, a chemical reaction occurs in the air bag that produces nitrogen gas. The air bag expands rapidly and then deflates just as quickly as the nitrogen gas escapes out of tiny holes in the bag. The entire process is completed in about 0.04 s.

section ③ review

Summary

What is Force?

- A force is a push or a pull on an object.
- The net force on an object is the combination of all the forces acting on the object.
- When the forces on an object are balanced, the net force on the object is zero.
- Unbalanced forces cause the motion of objects to change.

Inertia and Newton's First Law of Motion

- The inertia of an object is the tendency of an object to resist a change in motion.
- The larger the mass of an object, the greater its inertia.
- Newton's first law of motion states that the motion of an object at rest or moving with constant velocity will not change unless an unbalanced net force acts on the object.
- In a car crash, inertia causes an unrestrained passenger to continue moving at the speed of the car before the crash.

Self Check

1. **Infer** whether the inertia of an object changes as the object's velocity changes.
2. **Explain** whether or not there must be an unbalanced net force acting on any moving object.
3. **Explain** Can there be forces acting on an object if the object is at rest?
4. **Infer** the net force on a refrigerator if you push on the refrigerator and it doesn't move.
5. **Think Critically** Describe three situations in which a force changes the velocity of an object.

Applying Math

6. **Calculate Net Force** Two students push on a box in the same direction, and one pushes in the opposite direction. What is the net force on the box if each pushes with a force of 50 N?
7. **Calculate Acceleration** The downward force of gravity and the upward force of air resistance on a ball are both 5 N. What is the ball's acceleration?

Science Online gpscience.com/self_check_quiz

Force AND ACCELERATION

If you stand at a stoplight, you will see cars stopping for red lights and then taking off when the light turns green. What makes the cars slow down? What makes them speed up? Can a study of unbalanced forces lead to a better understanding of these everyday activities?

◉ Real-World Question

How does an unbalanced force on a book affect its motion?

Goals

- **Observe** the effect of force on the acceleration of an object.
- **Interpret** the data collected for each trial.

Materials

tape	this science book
paper clip	triple-beam balance
10-N spring scale	*electronic balance
large book	*Alternate materials

Safety Precautions

Proper eye protection should be worn at all times while performing this lab.

◉ Procedure

1. With a piece of tape, attach the paper clip to your textbook so that the paper clip is just over the edge of the book.
2. Prepare a data table with the following headings: *Force, Mass.*
3. If available, use a large balance to find the mass of this science book.
4. Place the book on the floor or on the surface of a long table. Use the paper clip to hook the spring scale to the book.

5. Pull the book across the floor or table at a slow but constant velocity. While pulling, read the force you are pulling with on the spring scale and record it in your table.
6. Repeat step 5 two more times, once accelerating slowly and once accelerating quickly. Be careful not to pull too hard. Your spring scale will read only up to 10 N.
7. Place a second book on top of the first book and repeat steps 3 through 6.

◉ Conclude and Apply

1. **Organize** the pulling forces from greatest to least for each set of trials. Do you see a relationship between force and acceleration? Explain your answer.
2. **Explain** how adding the second book changed the results.

Compare your conclusions with those of other students in your class. **For more help, refer to the Science Skill Handbook.**

LAB Design Your Own

Comparing Motion from Different Forces

Goals

- **Identify** several forces that you can use to propel a small toy car across the floor.
- **Demonstrate** the motion of the toy car using each of the forces.
- **Graph** the position versus time for each force.
- **Compare** the motion of the toy car resulting from each force.

Possible Materials

small toy car
ramps or boards
  of different lengths
springs or rubber bands
string
stopwatch
meterstick or tape measure
graph paper

Safety Precautions

▶ Real-World Question

Think about a small ball. How many ways could you exert a force on the ball to make it move? You could throw it, kick it, roll it down a ramp, blow it with a large fan, etc. Do you think the distance and speed of the ball's motion will be the same for all of these forces? Do you think the acceleration of the ball would be the same for all of these types of forces?

▶ Form a Hypothesis

Based on your reading and observations, state a hypothesis about how a force can be applied that will cause the toy car to go fastest.

▶ Test Your Hypothesis

Make a Plan

1. As a group, agree upon the hypothesis and decide how you will test it. Identify which results will confirm the hypothesis that you have written.
2. **List** the steps you will need to test your hypothesis. Be sure to include a control run. Be specific. Describe exactly what you will do in each step. List your materials.
3. **Prepare** a data table in your Science Journal to record your observations.

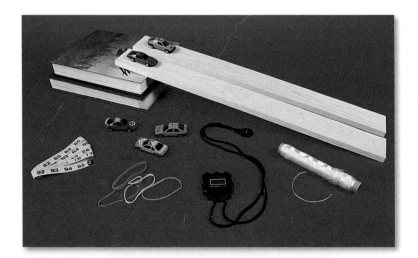

4. **Read** the entire experiment to make sure all steps are in logical order and will lead to a useful conclusion.

5. **Identify** all constants, variables, and controls of the experiment. Keep in mind that you will need to have measurements at multiple points. These points are needed to graph your results. You should make sure to have several data points taken after you stop applying the force and before the car starts to slow down. It might be useful to have several students taking measurements, making each responsible for one or two points.

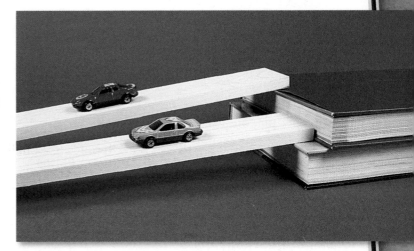

Follow Your Plan

1. Make sure your teacher approves your plan before you start.

2. Carry out the experiment as planned.

3. While doing the experiment, record your observations and complete the data tables in your Science Journal.

▶ *Analyze Your Data*

1. **Graph** the position of the car versus time for each of the forces you applied. How can you use the graphs to compare the speeds of the toy car?

2. **Calculate** the speed of the toy car over the same time interval for each of the forces that you applied. How do the speeds compare?

▶ *Conclude and Apply*

1. **Evaluate** Did the speed of the toy car vary depending upon the force applied to it?

2. **Determine** For any particular force, did the speed of the toy car change over time? If so, how did the speed change? Describe how you can use your graphs to answer these questions.

3. **Draw Conclusions** Did your results support your hypothesis? Why or why not?

Communicating
Your Data

Compare your data with those of other students. **Discuss** how the forces you applied might be different from those others applied and how that affected your results.

Science and Language Arts

A Brave and Startling Truth
by Maya Angelou

We, this people, on a small and lonely planet
Traveling through casual space
Past aloof stars, across the way of indifferent suns
To a destination where all signs tell us
It is possible and imperative that we learn
A brave and startling truth …

When we come to it
Then we will confess that not the Pyramids
With their stones set in mysterious perfection …
Not the Grand Canyon
Kindled into delicious color
By Western sunsets
These are not the only wonders of the world …

When we come to it
We, this people, on this minuscule and kithless[1]
globe …
We this people on this mote[2] of matter

When we come to it
We, this people, on this wayward[3], floating body
Created on this earth, of this earth
Have the power to fashion for this earth
A climate where every man and every woman
Can live freely without sanctimonious piety[4]
Without crippling fear

When we come to it
We must confess that we
are the possible
We are the miraculous, the
true wonder of the world
That is when, and only
when
We come to it.

Understanding Literature

Descriptive Writing The poet names some special places on Earth. These places, although marvelous, fall short of being really wonderful. How does Angelou contrast Earth's position within the universe to emphasize the importance of people?

Respond to the Reading

1. What adjectives does the poet use to describe Earth?
2. What does the poet believe are the true wonders of the world?
3. **Linking Science and Writing** Write a six-line poem that describes Earth's movement from the point of view of the Moon.

INTEGRATE Physics Sometimes a person doesn't need to see movement to know that something has moved. Even though we don't necessarily see Earth's movement, we know Earth moves relative to a reference point such as the Sun. If the Sun is the reference point, Earth moves because the Sun appears to change its position in the sky. The poem describes Earth's movement from a reference point outside of Earth, somewhere in space.

1 to be without friends or neighbors

2 small particle

3 wanting one's own way in spite of the advice or wishes of another

4 a self-important show of being religious

Reviewing Main Ideas

Describing Motion

1. Motion is a change of position of a body. Distance is the measure of how far an object moved. Displacement is the distance and direction of an object's change in position from the starting point.

2. A reference point must be specified in order to determine an object's position.

3. The speed of an object can be calculated from this equation:

$$s = \frac{d}{t}$$

4. The slope of a line on a distance-time graph is equal to the speed.

5. Velocity describes the speed and direction of a moving object.

Acceleration

1. Acceleration occurs when an object changes speed or changes direction.

2. An object speeds up if its acceleration is in the direction of its motion.

3. An object slows down if its acceleration is opposite to the direction of its motion.

4. Acceleration is the rate of change of velocity, and is calculated from this equation:

$$a = \frac{v_f - v_i}{t}$$

Motion and Forces

1. A force is a push or a pull.

2. The net force acting on an object is the combination of all the forces acting on the object.

3. The forces on an object are balanced if the net force is zero.

4. Inertia is the resistance of an object to a change in motion.

5. According to Newton's first law of motion, the motion of an object does not change unless an unbalanced net force acts on the object.

FOLDABLES Use the Foldable that you made at the beginning of this chapter to help you review motion.

Using Vocabulary

acceleration p. 47
average speed p. 42
balanced force p. 53
displacement p. 39
distance p. 39
force p. 52

inertia p. 54
instantaneous speed p. 42
net force p. 53
speed p. 39
velocity p. 44

Compare and contrast the following pairs of vocabulary words.

1. speed—velocity

2. distance—displacement

3. average speed—instantaneous speed

4. balanced force—net force

5. force—inertia

6. acceleration—velocity

7. velocity—instantaneous speed

8. force—net force

9. force—acceleration

Checking Concepts

Choose the word or phrase that best answers the question.

10. Which of the following do you calculate when you divide the total distance traveled by the total travel time?
    A) average speed
    B) constant speed
    C) variable speed
    D) instantaneous speed

11. Which term below best describes the forces on an object with a net force of zero?
    A) inertia
    B) balanced forces
    C) acceleration
    D) unbalanced forces

12. Which of the following is a proper unit of acceleration?
    A) s/km^2
    B) km/h
    C) m/s^2
    D) cm/s

13. Which of the following is not used in calculating acceleration?
    A) initial velocity
    B) average speed
    C) time interval
    D) final velocity

14. In which of the following conditions does the car NOT accelerate?
    A) A car moves at 80 km/h on a flat, straight highway.
    B) The car slows from 80 km/h to 35 km/h.
    C) The car turns a corner.
    D) The car speeds up from 35 km/h to 80 km/h.

15. What is the tendency for an object to resist any change in its motion called?
    A) net force
    B) acceleration
    C) balanced force
    D) inertia

16. How can speed be defined?
    A) acceleration/time
    B) change in velocity/time
    C) distance/time
    D) displacement/time

Interpreting Graphics

Use the table below to answer question 17.

Distance-Time for Runners				
Time (s)	1	2	3	4
Sally's Distance (m)	2	4	6	8
Alonzo's Distance (m)	1	2	2	4

17. Make a distance-time graph that shows the motion of both runners. What is the average speed of each runner? Which runner stops briefly? Over what time interval do they both have the same speed?

Science online gpscience.com/vocabulary_puzzlemaker

18. Copy and complete this concept map on motion.

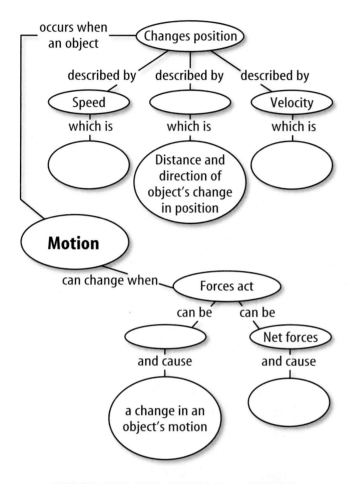

Thinking Critically

19. Evaluate Which of the following represents the greatest speed: 20 m/s, 200 cm/s, or 0.2 km/s?

20. Recognize Cause and Effect Acceleration can occur when a car is moving at constant speed. What must cause this acceleration?

21. Explain why a passenger who is not wearing a safety belt will likely hit the windshield in a head-on collision.

22. Determine If you walked 20 m, took a book from a library table, turned around and walked back to your seat, what are the distance traveled and displacement?

23. Explain When you are describing the rate that a race car goes around a track, should you use the term *speed* or *velocity* to describe the motion?

Applying Math

24. Calculate Speed A cyclist must travel 800 km. How many days will the trip take if the cyclist travels 8 h/day at an average speed of 16 km/h?

25. Calculate Acceleration A satellite's speed is 10,000 m/s. After 1 min, it is 5,000 m/s. What is the satellite's acceleration?

26. Calculate Displacement A cyclist leaves home and rides due east for a distance of 45 km. She returns home on the same bike path. If the entire trip takes 4 h, what is her average speed? What is her displacement?

27. Calculate Velocity The return trip of the cyclist in question 13 took 30 min longer than her trip east, although her total time was still 4 h. What was her velocity in each direction?

Use the graph below to answer question 28.

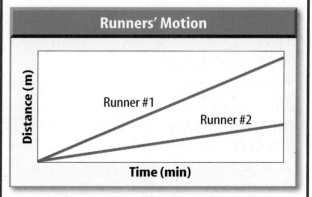

28. Interpret a Graph Use the graph to determine which runner had the greatest speed.

Part 1 Multiple Choice

Record your answers on the answer sheet provided by your teacher or on a sheet of paper.

1. Sound travels at a speed of 330 m/s. How long does it take for the sound of thunder to travel 1485 m?
   A. 45 s   **C.** 4,900 s
   B. 4.5 s   **D.** 0.22 s

Use the graph below to answer questions 2–4.

Speed Changing Over Distance

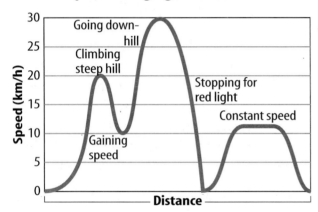

2. The graph shows how a cyclist's speed changed over distance of 5 km. What is the cyclist's average speed if the trip took 0.25 h?
   A. 2 km/h   **C.** 20 km/h
   B. 30 km/h   **D.** 8 km/h

3. Once the trip was started, how many times did the cyclist stop?
   A. 0   **C.** 2
   B. 4   **D.** 5

4. What was the fastest speed the cyclist traveled?
   A. 20 km/h   **C.** 12 km/h
   B. 30 km/h   **D.** 10 km/h

Test-Taking Tip

Read Carefully Read each question carefully for full understanding.

5. A skier is going down a hill at a speed of 9 km/s. The hill gets steeper and her speed increases to 18 m/s in 3 s. What is her acceleration?
   A. 9 m/s^2 **C.** 27 m/s^2
   B. 3 m/s^2 **D.** 6 m/s^2

6. Which of the following best describes an object with constant velocity?
   A. It is changing direction.
   B. Its acceleration is increasing.
   C. Its acceleration is zero.
   D. Its acceleration is negative.

7. Which of the following is a force?
   A. friction   **C.** inertia
   B. acceleration   **D.** velocity

Use the table below to answer questions 8 and 9.

Runner	Distance covered (km)	Time (min)
Daisy	12.5	42
Jane	7.8	38
Bill	10.5	32
Joe	8.9	30

8. What is Daisy's average speed?
   A. 0.29 km/min   **C.** 2.9 km/min
   B. 530 km/min   **D.** 3.4 km/min

9. Which runner has the fastest average speed?
   A. Daisy   **C.** Bill
   B. Jane   **D.** Joe

10. The movement of the Australian plate pushes Australia north at an average speed of about 17 cm per year. What will Australia's displacement be in meters in 1,000 years?
    A. 170 m north   **C.** 1,700 m north
    B. 170 m south   **D.** 1,700 m south

Record your answers on the answer sheet provided by your teacher or on a sheet of paper.

Use the graph below to answer questions 11 and 12.

Graphing Motion

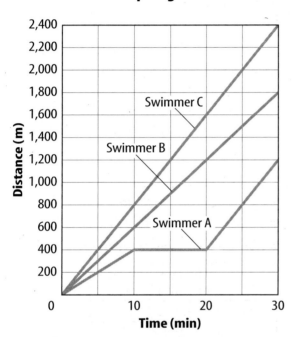

11. The graph shows the motion of three swimmers during a 30-min workout. Which swimmer had the highest average speed over the 30-min time interval?

12. Did all the swimmers swim at a constant speed? Explain how you know.

13. Why is knowing just the speed at which a hurricane is traveling toward land not enough information to be able to warn people to evacuate?

14. If the speedometer on a car indicates a constant speed, can you be sure the car is not accelerating? Explain.

15. If a car is traveling at a speed of 40 km/h and then comes to a stop in 5 s, what is its acceleration in m/s^2?

Record your answers on a sheet of paper.

16. Describe three ways that your acceleration could change as you jog along a path through a park.

17. An object in motion slows down and comes to a stop. Use Newton's first law of motion to explain why this happens.

18. Give an example of a force applied to an object that does not change the object's velocity.

19. In an airplane flying at a constant speed, the force exerted by the engine pushing the airplane forward is equal to the opposite force of air resistance. Describe how these forces compare when the plane speeds up and slows down. In which direction is the net force on the airplane in each case?

20. Where would you place the location of a reference point in order to describe the motion of a space probe traveling from Earth to Jupiter? Explain your choice.

Use the table below to answer question 21.

Car	Mass (kg)	Stopping distance(m)
A	1000	80
B	1250	100
C	1500	120
D	2000	160

21. What is the relationship between a car's mass and its stopping distance? How can you explain this relationship?

22. Two cars approach each other. How does the speed of one car relative to the other compare with speed of the car relative to the ground?

Forces

Who's a dummy?

Crunch! This test dummy would have some explaining to do if this were a traffic accident. But in a test crash, the dummy plays an important role. The forces acting on it during a crash are measured and analyzed in order to learn how to make cars safer.

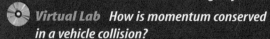 Explain which would be a safer car—a car with a front that crumples in a crash, or one with a front that doesn't crumple.

Start-Up Activities

The Force of Gravity

The force of Earth's gravity pulls all objects downward. However, objects such as rocks seem to fall faster than feathers or leaves. Do objects with more mass fall faster?

1. Measure the mass of a softball, a tennis ball, and a flat sheet of paper. Copy the data table below and record the masses.

2. Drop the softball from a height of 2.5 m and use a stopwatch to measure the time it takes for the softball to hit the floor. Record the time in your data table.

3. Repeat step 2 using the tennis ball and the flat sheet of paper. Record the times in your data table.

4. Crumple the flat sheet of paper into a ball, and measure the time for the crumpled paper to fall 2.5 m. Record the time in your data table.

5. **Think Critically** Write a paragraph comparing the times it took each item to fall 2.5 m. From your data, infer if the speed of a falling object depends on the object's mass.

Object	Mass	Time
Softball		
Tennis ball		
Flat paper		
Crumpled paper		

The Force of Friction One of the forces you encounter every day is friction. Make the following Foldable to help you compare the three types of friction—static friction, sliding friction, and rolling friction.

STEP 1 Fold the top of a vertical piece of paper down and the bottom up to divide the paper into thirds.

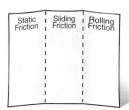

STEP 2 Unfold and label the rows *Static Friction, Sliding Friction,* and *Rolling Friction.*

Read and Write As you read, write the definition and give examples of each type of friction.

Preview this chapter's content and activities at gpscience.com

Newton's Second Law

Reading Guide

What You'll Learn

- **Define** Newton's second law of motion.
- **Apply** Newton's second law of motion.
- **Describe** the three different types of friction.
- **Observe** the effects of air resistance on falling objects.

Why It's Important

Newton's second law explains how forces cause the motion of objects to change.

⊙ **Review Vocabulary**

net force: the combination of all forces acting on an object

New Vocabulary

- Newton's second law of motion
- friction
- static friction
- sliding friction
- air resistance

Force, Mass, and Acceleration

The previous chapter discussed Newton's first law of motion which states that the motion of an object changes only if an unbalanced force acts on the object. Newton's second law of motion describes how the forces exerted on an object, like the volleyball in **Figure 1,** its mass, and its acceleration are related.

Force and Acceleration What's different about throwing a ball horizontally as hard as you can and tossing it gently? When you throw hard, you exert a much greater force on the ball. The ball has a greater velocity when it leaves your hand than it does when you throw gently. Thus, the hard-thrown ball has a greater change in velocity, and the change occurs over a shorter period of time. Recall that acceleration is the change in velocity divided by the time it takes for the change to occur. So, a hard-thrown ball has a greater acceleration than a gently thrown ball.

Mass and Acceleration If you throw a softball and a baseball as hard as you can, why don't they have the same speed? The difference is due to their masses. A softball has a mass of about 0.20 kg, but a baseball's mass is about 0.14 kg. The softball has less velocity after it leaves your hand than the baseball does, even though you exerted the same force. If it takes the same amount of time to throw both balls, the softball would have less acceleration. The acceleration of an object depends on its mass as well as the force exerted on it. Force, mass, and acceleration are related.

Figure 1 A volleyball's motion changes when an unbalanced force acts on it.

Newton's Second Law

Newton's second law of motion states that the acceleration of an object is in the same direction as the net force on the object, and that the acceleration can be calculated from the following equation:

Science Online

Topic: Motion in Sports
Visit gpscience.com for Web links to information about methods used to analyze the motions of athletes.

Activity Choose a sport and write a report on how analyzing the motions involved in the sport can improve performance and reduce injuries.

Newton's Second Law of Motion

$$\text{acceleration (in meters/second}^2) = \frac{\text{net force (in newtons)}}{\text{mass (in kilograms)}}$$

$$a = \frac{F_{net}}{m}$$

Applying Math — Solve a Simple Equation

THE ACCELERATION OF A SLED You push a friend on a sled. Your friend and the sled together have a mass of 70 kg. If the net force on the sled is 35 N, what is the sled's acceleration?

IDENTIFY known values and the unknown value

Identify the known values:

The net force on the sled is 35 N ⟹ means ⟹ $F_{net} = 35$ N

Your friend and the sled together have a mass of 70 kg ⟹ means ⟹ $m = 70$ kg

Identify the unknown value:

What is the sled's acceleration? ⟹ means ⟹ $a = ?$ m/s^2

SOLVE the problem

Substitute the known values $F_{net} = 35$ N and $m = 70$ kg into the equation for Newton's second law of motion:

$$a = \frac{F_{net}}{m} = \frac{35\ \text{N}}{70\ \text{kg}} = 0.5\ \frac{\text{N}}{\text{kg}} = 0.5\ \frac{\text{kg m}}{\text{s}^2} \times \frac{1}{\text{kg}} = 0.5\ \text{m/s}^2$$

CHECK the answer

Does your answer seem reasonable? Check your answer by multiplying the acceleration you calculated by the mass given in the problem. The result should be the net force given in the problem.

Practice Problems

1. If the mass of a helicopter is 4,500 kg, and the net force on it is 18,000 N, what is the helicopter's acceleration?

2. What is the net force on a dragster with a mass of 900 kg if its acceleration is 32.0 m/s^2?

3. A car is being pulled by a tow truck. What is the car's mass if the net force on the car is 3,000 N and it has an acceleration of 2.0 m/s^2?

For more practice problems go to page 834, and visit gpscience.com/extra_problems.

Calculating Net Force with the Second Law

Newton's second law also can be used to calculate the net force if mass and acceleration are known. To do this, the equation for Newton's second law must be solved for the net force, F. To solve for the net force, multiply both sides of the above equation by the mass:

$$\cancel{m} \times \frac{F_{net}}{\cancel{m}} = ma$$

The mass, m, on the left side cancels, giving the equation:

$$F_{net} = ma$$

For example, when the tennis player in **Figure 2** hits a ball, the ball might be in contact with the racket for only a few thousandths of a second. Because the ball's velocity changes over such a short period of time, the ball's acceleration could be as high as 5,000 m/s^2. The ball's mass is 0.06 kg, so the net force exerted on the ball would be:

$$F_{net} = ma = (0.06 \text{ kg}) (5,000 \text{ m/s}^2) = 300 \text{ kg m/s}^2 = 300 \text{ N}$$

Figure 2 The tennis racket exerts a force on the ball that causes it to accelerate.

Friction

Suppose you give a skateboard a push with your hand. According to Newton's first law of motion, if the net force acting on a moving object is zero, it will continue to move in a straight line with constant speed. Does the skateboard keep moving with constant speed after it leaves your hand?

You know the answer. The skate board slows down and finally stops. Recall that when an object slows down it is accelerating. By Newton's second law, if the skateboard is accelerating, there must be a net force acting on it.

The force that slows the skateboard and brings it to a stop is friction. **Friction** is the force that opposes the sliding motion of two surfaces that are touching each other. The amount of friction between two surfaces depends on two factors—the kinds of surfaces and the force pressing the surfaces together.

Reading Check *What does the force of friction between two objects in contact depend on?*

What causes friction? Would you believe the surface of a highly polished piece of metal is rough? **Figure 3** shows a microscopic view of the dips and bumps on the surface of a polished silver teapot. If two surfaces are in contact, welding or sticking occurs where the bumps touch each other. These microwelds are the source of friction.

Figure 3 While surfaces might look and even feel smooth, they can be rough at the microscopic level.

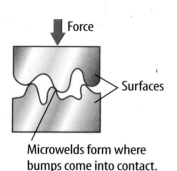

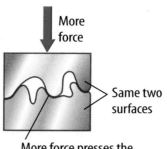

Force

Surfaces

Microwelds form where
bumps come into contact.

More
force

Same two
surfaces

More force presses the
bumps closer together.

Figure 4 Friction is due to
microwelds formed between
two surfaces. The larger the force
pushing the two surfaces together
is, the stronger the microwelds
will be.

Explain *how the area of contact
between the surfaces changes when
they are pushed together.*

Sticking Together The larger the force pushing the two surfaces together is, the stronger these microwelds will be, because more of the surface bumps will come into contact, as shown in **Figure 4.** To move one surface over the other, a force must be applied to break the microwelds.

Static Friction Suppose you have filled a cardboard box, like the one in **Figure 5,** with books and want to move it. It's too heavy to lift, so you start pushing on it, but it doesn't budge. Is that because the mass of the box is too large? If the box doesn't move, then it has zero acceleration. According to Newton's second law, if the acceleration is zero, then the net force on the box is zero. Another force that cancels your push must be acting on the box. That force is friction due to the microwelds that have formed between the bottom of the box and the floor. This type of friction is called static friction. **Static friction** is the frictional force that prevents two surfaces from sliding past each other. In this case, your push is not large enough to break the microwelds, and the box does not move.

Applied force

Static friction

Figure 5 The box doesn't move because static friction is equal to the applied force.

Infer *the net force on the box.*

Figure 6 Sliding friction acts in the direction opposite the motion of the sliding box.

Applied force

Sliding friction

Sliding Friction You ask a friend to help you move the box, as in **Figure 6.** Pushing together, the box moves. Together you and your friend have exerted enough force to break the microwelds between the floor and the bottom of the box. But if you stop pushing, the box quickly comes to a stop. This is because as the box slides across the floor, another force—sliding friction—opposes the motion of the box. **Sliding friction** is the force that opposes the motion of two surfaces sliding past each other. Sliding friction is caused by microwelds constantly breaking and then forming again as the box slides along the floor. To keep the box moving, you must continually apply a force to overcome sliding friction.

✔ **Reading Check** *What causes sliding friction?*

Rolling Friction You may have watched a car stuck in snow, ice, or mud spin its wheels. The driver steps on the gas, but the wheels just spin without the car moving. To make the car move, sand or gravel may be spread under the wheels. When a wheel is spinning there is sliding friction between the wheels and surface. Spreading sand or gravel on the surface increases the sliding friction until the wheel stops slipping and begins rolling.

As a wheel rolls over a surface, the wheel digs into the surface, causing both the wheel and the surface to be deformed. Static friction acts over the deformed area where the wheel and surface are in contact, producing a frictional force called rolling friction. Rolling friction is the frictional force between a rolling object and the surface it rolls on. Rolling friction would cause the train in **Figure 7** to slow down and come to a stop, just as sliding friction causes a sliding object to slow down and come to a stop.

Figure 7 Rolling friction between the train's wheels and the track is reduced by making both from steel. This reduces the deformation that occurs as the wheel rolls on the track.

Air Resistance

When an object falls toward Earth, it is pulled downward by the force of gravity. However, a friction-like force called **air resistance** opposes the motion of objects that move through the air. Air resistance causes objects to fall with different accelerations and different speeds. If there were no air resistance, then all objects, like the apple and the feather shown in **Figure 8,** would fall with the same acceleration.

Air resistance acts in the opposite direction to the motion of an object through air. If the object is falling downward, air resistance acts upward on the object. The size of the air resistance force also depends on the size and shape of an object. Imaging dropping two identical plastic bags. One is crumpled into a ball and the other is spread out. When the bags are dropped, the crumbled bag falls faster than the spread out-bag. The downward force of gravity on both bags is the same, but the upward force of air resistance on the crumpled bag is less. As a result, the net downward force on the crumpled bag is greater, as shown in **Figure 9.**

The amount of air resistance on an object depends on the speed, size, and shape of the object. Air resistance, not the object's mass, is why feathers, leaves, and pieces of paper fall more slowly than pennies, acorns, and apples.

Figure 8 This photograph shows an apple and feather falling in a vacuum. The photograph was taken using a strobe light that flashes on and off at a steady rate. Because there is no air resistance in a vacuum, the feather and the apple fall with the same acceleration.

Figure 9 Because of its greater surface area, the bag on the left has more air resistance acting on it as it falls.

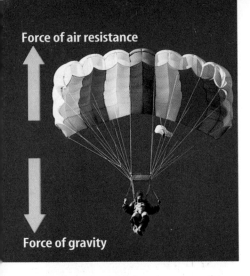

Force of air resistance

Force of gravity

Figure 10 The force of air resistance on an open parachute balances the force of gravity on the sky diver when the parachute is falling slowly.

Terminal Velocity As an object falls, the downward force of gravity causes the object to accelerate. For example, after falling 2,000 m, without the effects of air resistance the sky diver's speed would be almost 200 m/s, or over 700 km/h.

However, as an object falls faster, the upward force of air resistance increases. This causes the net force on a sky diver to decrease as the sky diver falls. Finally, the upward air resistance force becomes large enough to balance the downward force of gravity. This means the net force on the object is zero. Then the acceleration of the object is also zero, and the object falls with a constant speed called the terminal velocity. The terminal velocity is the highest speed a falling object will reach.

The terminal velocity depends on the size, shape, and mass of a falling object. The air resistance force on an open parachute, like the one in **Figure 10,** is much larger than the air resistance on the sky diver with a closed parachute. With the parachute open, the terminal velocity of the sky diver becomes small enough that the sky diver can land safely.

section 1 review

Summary

Force, Mass, and Acceleration

- The greater the force on an object, the greater the object's acceleration.
- The acceleration of an object depends on its mass as well as the force exerted on it.

Newton's Second Law

- Newton's second law of motion states that the acceleration of an object is in the direction of the net force on the object, and can be calculated from this equation:

$$a = \frac{F_{net}}{m}$$

Friction

- Friction is the force that opposes motion between two surfaces that are touching each other.
- Friction depends on the types of surfaces and the force pressing the surfaces together.
- Friction results from the microwelds formed between surfaces that are in contact.

Air Resistance

- Air resistance is a force that acts on objects that move through the air.

Self Check

1. **State** Newton's second law of motion.
2. **Infer** why an object with a smaller mass has a larger acceleration than a larger mass if the same force acts on each.
3. **Explain** why the frictional force between two surfaces increases if the force pushing the surfaces together increases.
4. **Compare** the force of air resistance and the force of gravity on an object falling at its terminal velocity.
5. **Think Critically** Why does coating surfaces with oil reduce friction between the surfaces?

Applying Math

6. **Convert Units** show that the units N/kg can be written using only units of meters (m) and seconds (s). Is this a unit of mass, acceleration or force?
7. **Calculate Mass** You push yourself on a skateboard with a force of 30 N and accelerate at 0.5 m/s². Find the mass of the skateboard if your mass is 58 kg.
8. **Calculate Sliding Friction** You push a 2-kg book with a force of 5 N. Find the force of sliding friction on the book if it has an acceleration of 1.0 m/s².

Science online gpscience.com/self_check_quiz

Gravity

Reading Guide

What You'll Learn

- **Describe** the gravitational force.
- **Distinguish** between mass and weight.
- **Explain** why objects that are thrown will follow a curved path.
- **Compare** circular motion with motion in a straight line.

Why It's Important

There is a gravitational force between you and every other object in the universe.

❓ Review Vocabulary

acceleration: the rate of change of velocity which occurs when an object changes speed or direction

New Vocabulary

- gravity
- weight
- centripetal acceleration
- centripetal force

What is gravity?

There's a lot about you that's attractive. At this moment, you are exerting an attractive force on everything around you—your desk, your classmates, even the planet Jupiter millions of kilometers away. It's the attractive force of gravity.

Anything that has mass is attracted by the force of gravity. **Gravity** is an attractive force between any two objects that depends on the masses of the objects and the distance between them. This force increases as the mass of either object increases, or as the objects move closer, as shown in **Figure 11.**

You can't feel any gravitational attraction between you and this book because the force is weak. Only Earth is both close enough and has a large enough mass that you can feel its gravitational attraction. While the Sun has much more mass than Earth, the Sun is too far away to exert a noticeable gravitational attraction on you. And while this book is close, it doesn't have enough mass to exert an attraction you can feel.

Gravity—A Basic Force Gravity is one of the four basic forces. The other basic forces are the electromagnetic force, the strong nuclear force, and the weak nuclear force. The two nuclear forces only act on particles in the nuclei of atoms. Electricity and magnetism are caused by the electromagnetic force. Chemical interactions between atoms and molecules also are due to the electromagnetic force.

Figure 11 The gravitational force between two objects depends on their masses and the distance between them.

If the mass of either of the objects increases, the gravitational force between them increases.

If the objects are closer together, the gravitational force between them increases.

The Law of Universal Gravitation

For thousands of years people everywhere have observed the stars and the planets in the night sky. Gradually, data were collected on the motions of the planets by a number of observers. Isaac Newton used some of these data to formulate the law of universal gravitation, which he published in 1687. This law can be written as the following equation.

> **The Law of Universal Gravitation**
>
> $$\text{gravitational force} = (\text{constant}) \times \frac{(\text{mass 1}) \times (\text{mass 2})}{(\text{distance})^2}$$
>
> $$F = G\frac{m_1 m_2}{d^2}$$

In this equation G is a constant called the universal gravitational constant, and d is the distance between the two masses, m_1 and m_2. The law of universal gravitation enables the force of gravity to be calculated between any two objects if their masses and the distance between them are known.

The Range of Gravity According to the law of universal gravitation, the gravitational force between two masses decreases rapidly as the distance between the masses increases. For example, if the distance between two objects increases from 1 m to 2 m, the gravitational force between them becomes one fourth as large. If the distance increases from 1 m to 10 m, the gravitational force between the objects is one hundredth as large.

However, no matter how far apart two objects are, the gravitational force between them never completely goes to zero. Because the gravitational force between two objects never disappears, gravity is called a long-range force.

 **Finding Other Planets** Earth's motion around the Sun is affected by the gravitational pulls of the other planets in the solar system. In the same way, the motion of every planet in the solar system is affected by the gravitational pulls of all the other planets.

In the 1840s the most distant planet known was Uranus. The motion of Uranus calculated from the law of universal gravitation disagreed slightly with its observed motion. Some astronomers suggested that there must be an undiscovered planet affecting the motion of Uranus. Using the law of universal gravitation and Newton's laws of motion, two astronomers independently calculated the orbit of this planet. As a result of these calculations, the planet Neptune, shown in **Figure 12,** was found in 1846.

Figure 12 The location of the planet Neptune in the night sky was correctly predicted using Newton's laws of motion and the law of universal gravitation.

Earth's Gravitational Acceleration

If you dropped a bowling ball and a marble at the same time, which would hit the ground first? Suppose the effects of air resistance are small enough to be ignored. When all forces except gravity acting on an a falling object can be ignored, the object is said to be in free fall. Then all objects near Earth's surface would fall with the same acceleration, just like the two balls in **Figure 13.**

Close to Earth's surface, the acceleration of a falling object in free fall is about 9.8 m/s^2. This acceleration is given the symbol g and is sometimes called the acceleration of gravity. By Newton's second law of motion, the force of Earth's gravity on a falling object is the object's mass times the acceleration of gravity. This can be expressed by the equation:

> **Force of Earth's Gravity**
>
> **force of gravity** (N) = **mass** (kg) $\times$ acceleration of gravity (m/s^2)
> $$F = mg$$

For example, the gravitational force on a sky diver with a mass of 60 kg would be

$$F = mg = (60 \text{ kg}) (9.8 \text{ m/s}^2) = 588 \text{ N}$$

Weight Even if you are not falling, the force of Earth's gravity still is pulling you downward. If you are standing on a floor, the net force on you is zero. The force of Earth's gravity pulls you downward, but the floor exerts an upward force on you that balances gravity's downward pull.

Whether you are standing, jumping, or falling, Earth exerts a gravitational force on you. The gravitational force exerted on an object is called the object's **weight.** Because the weight of an object on Earth is equal to the force of Earth's gravity on the object, weight can be calculated from this equation:

> **Weight Equation**
>
> **weight** (N) = **mass** (kg) $\times$ acceleration of gravity (m/s^2)
> $$W = mg$$

On Earth where g equals 9.8 m/s^2, a cassette tape weighs about 0.5 N, a backpack full of books could weigh 100 N, and a jumbo jet weighs about 3.4 million N. A sky diver with a mass of 60 kg has a weight of 588 N. Under what circumstances would the net force on the sky diver equal the sky diver's weight?

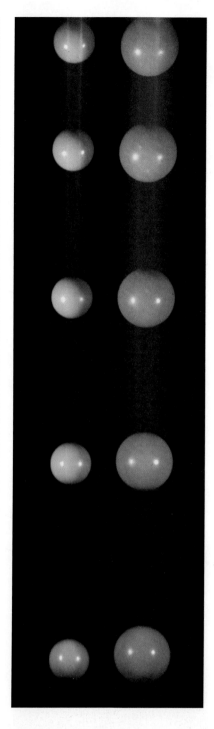

Figure 13 Time-lapse photography shows that two balls of different masses fall at the same rate.

Figure 14 On the Moon, the gravitational force on the astronaut is less than it is on Earth. As a result, the astronaut can take longer steps and jump higher than on Earth.

Weight and Mass Weight and mass are not the same. Weight is a force and mass is a measure of the amount of matter an object contains. However, according to the weight equation on the previous page, weight and mass are related. Weight increases as mass increases.

The weight of an object usually is the gravitational force between the object and Earth. But the weight of an object can change, depending on the gravitational force on the object. For example, the acceleration of gravity on the Moon is 1.6 m/s², about one sixth as large as Earth's gravitational acceleration. As a result, a person, like the astronaut in **Figure 14,** would weigh only about one sixth as much on the Moon as on Earth. **Table 1** shows how various weights on Earth would be different on the Moon and some of the planets.

✔ **Reading Check** *How are weight and mass related?*

Weightlessness and Free Fall

You've probably seen pictures of astronauts and equipment floating inside the space shuttle. Any item that is not fastened down seems to float throughout the cabin. They are said to be experiencing the sensation of weightlessness.

However, for a typical mission, the shuttle orbits Earth at an altitude of about 400 km. According the law of universal gravitation, at 400-km altitude the force of Earth's gravity is about 90 percent as strong as it is at Earth's surface. So an astronaut with a mass of 80 kg still would weigh about 700 N in orbit, compared with a weight of about 780 N at Earth's surface.

Table 1 Weight Comparison Table					
Weight on Earth (N)	**Weight on Other Bodies in the Solar System (N)**				
	Moon	**Venus**	**Mars**	**Jupiter**	**Saturn**
75	12	68	28	190	87
100	17	90	38	254	116
150	25	135	57	381	174
500	84	450	190	1,270	580
700	119	630	266	1,778	812
2,000	333	1,800	760	5,080	2,320

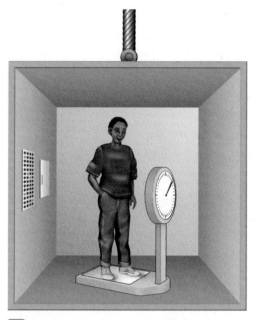

A When the elevator is stationary, the scale shows the boy's weight.

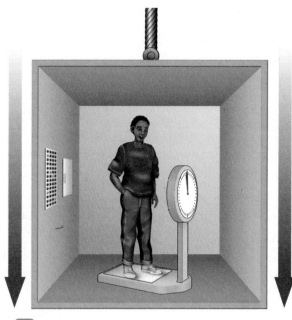

B If the elevator were in free fall, the scale would show a zero weight.

Floating in Space So what does it mean to say that something is weightless in orbit? Think about how you measure your weight. When you stand on a scale, as in **Figure 15A,** you are at rest and the net force on you is zero. The scale supports you and balances your weight by exerting an upward force. The dial on a scale shows the upward force exerted by the scale, which is your weight. Now suppose you stand on a scale in an elevator that is falling, as in **Figure 15B.** If you and the scale were in free fall, then you no longer would push down on the scale at all. The scale dial would say you have zero weight, even though the force of gravity on you hasn't changed.

A space shuttle in orbit is in free fall, but it is falling around Earth, rather than straight downward. Everything in the orbiting space shuttle is falling around Earth at the same rate, in the same way you and the scale were falling in the elevator. Objects in the shuttle seem to be floating because they are all falling with the same acceleration.

Projectile Motion

If you've tossed a ball to someone, you've probably noticed that thrown objects don't always travel in straight lines. They curve downward. That's why quarterbacks, dart players, and archers aim above their targets. Anything that's thrown or shot through the air is called a projectile. Earth's gravity causes projectiles to follow a curved path.

Figure 15 The boy pushes down on the scale with less force when he and the scale are falling at the same rate.

INTEGRATE Earth Science

Gravity and Earth's Atmosphere Apart from simply keeping your feet on the ground, gravity is important for life on Earth for other reasons, too. Because Earth has a sufficient gravitational pull, it can hold around it the oxygen/nitrogen atmosphere necessary for sustaining life. Research other ways in which gravity has played a role in the formation of Earth.

Figure 16 The pitcher gives the ball a horizontal motion. Gravity, however, is pulling the ball down. The combination of these two motions causes the ball to move in a curved path.

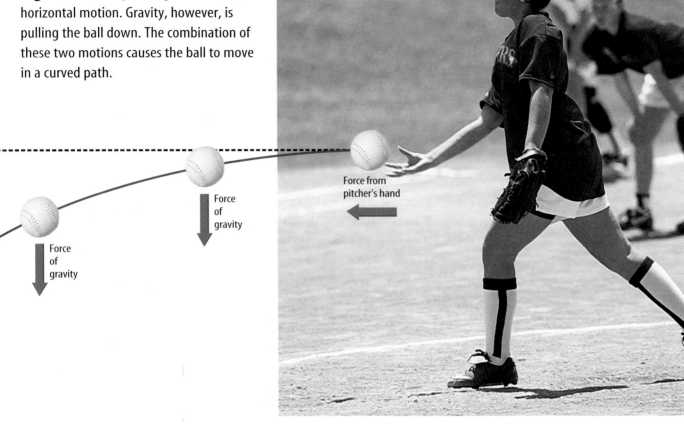

Force from pitcher's hand

Force of gravity

Force of gravity

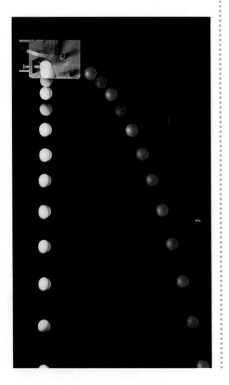

Figure 17 Multiflash photography shows that each ball has the same acceleration downward, whether it's thrown or dropped.

Horizontal and Vertical Motions When you throw a ball, like the pitcher in **Figure 16,** the force exerted by your hand pushes the ball forward. This force gives the ball horizontal motion. After you let go of the ball, no force accelerates it forward, so its horizontal velocity is constant, if you ignore air resistance.

However, when you let go of the ball, gravity can pull it downward, giving it vertical motion. Now the ball has constant horizontal velocity but increasing vertical velocity. Gravity exerts an unbalanced force on the ball, changing the direction of its path from only forward to forward and downward. The result of these two motions is that the ball appears to travel in a curve, even though its horizontal and vertical motions are completely independent of each other.

Horizontal and Vertical Distance If you were to throw a ball as hard as you could from shoulder height in a perfectly horizontal direction, would it take longer to reach the ground than if you dropped a ball from the same height? Surprisingly, it won't. A thrown ball and one dropped will hit the ground at the same time. Both balls in **Figure 17** travel the same vertical distance in the same amount of time. However, the ball thrown horizontally travels a greater horizontal distance than the ball that is dropped.

Centripetal Force

Look at the path the ball follows as it travels through the curved tube in **Figure 18.** The ball may accelerate in the straight sections of the pipe maze if it speeds up or slows down. However, when the ball enters a curve, even if its speed does not change, it is accelerating because its direction is changing. When the ball goes around a curve, the change in the direction of the velocity is toward the center of the curve. Acceleration toward the center of a curved or circular path is called **centripetal acceleration.**

According to the second law of motion, when the ball has centripetal acceleration, the direction of the net force on the ball also must be toward the center of the curved path. The net force exerted toward the center of a curved path is called a **centripetal force.** For the ball moving through the tube, the centripetal force is the force exerted by the walls of the tube on the ball.

Centripetal Force and Traction When a car rounds a curve on a highway, a centripetal force must be acting on the car to keep it moving in a curved path. This centripetal force is the frictional force, or the traction, between the tires and the road surface. If the road is slippery and the frictional force is small, the centripetal force might not be large enough to keep the car moving around the curve. Then the car will slide in a straight line. Anything that moves in a circle, such as the people on the amusement park ride in **Figure 19,** is doing so because a centripetal force is accelerating it toward the center.

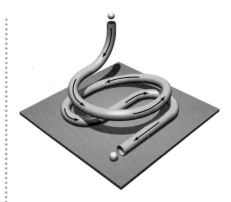

Figure 18 When the ball moves through the curved portions of the tube, it is accelerating because its velocity is changing. **Identify** *the forces acting on the ball as it falls through the tube.*

Mini LAB

Observing Centripetal Force

Procedure
1. Thread a **string** about 1 m long through the holes of a **plastic, slotted golf ball.**
2. Swing the ball in a vertical circle.
3. Swing the ball at different speeds and observe the motion of the ball and the tension in the string.

Analysis
1. What force does the string exert on the ball when the ball is at the top, sides, and bottom of the swing?
2. How does the tension in the string depend on the speed of the ball?

Try at Home

Figure 19 Centripetal force keeps these riders moving in a circle.

Figure 20 The Moon would move in a straight line except that Earth's gravity keeps pulling it toward Earth. This gives the Moon a nearly circular orbit.

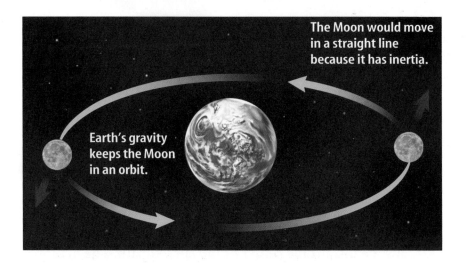

The Moon would move in a straight line because it has inertia.

Earth's gravity keeps the Moon in an orbit.

Gravity Can Be a Centripetal Force Imagine whirling an object tied to a string above your head. The string exerts a centripetal force on the object that keeps it moving in a circular path. In the same way, Earth's gravity exerts a centripetal force on the Moon that keeps it moving in a nearly circular orbit, as shown in **Figure 20.**

section 2 review

Summary

Gravity

- According to the law of universal gravitation, the gravitational force between two objects depends on the masses of the objects and the distance between them.
- The acceleration due to gravity near Earth's surface has the value 9.8 m/s^2.
- Near Earth's surface, the gravitational force on an object with mass, m, is given by:
$$F = mg$$

Weight

- The weight of an object is related to its mass according to the equation:
$$W = mg$$
- An object in orbit seems to be weightless because it is falling around Earth.

Projectile Motion and Centripetal Force

- Projectiles follow a curved path because their horizontal motion is constant, but gravity causes the vertical motion to be changing.
- The net force on an object moving in a circular path is called the centripetal force.

Self Check

1. **Describe** how the gravitational force between two objects depends on the mass of the objects and the distance between them.
2. **Distinguish** between the mass of an object and the object's weight.
3. **Explain** what causes the path of a projectile to be curved.
4. **Describe** the force that causes the planets to stay in orbit around the Sun.
5. **Think Critically** Suppose Earth's mass increased, but Earth's diameter didn't change. How would the acceleration of gravity near Earth's surface change?

Applying Math

6. **Calculate Weight** On Earth, what is the weight of a large-screen TV that has a mass of 75 kg?
7. **Calculate Gravity on Mars** Find the acceleration of gravity on Mars if a person with a mass of 60.0 kg weighs 222 N on Mars.
8. **Calculate Force** Find the force exerted by a rope on a 10-kg mass that is hanging from the rope.

Science Online gpscience.com/self_check_quiz

The Third Law of Motion

Newton's Third Law

Push against a wall and what happens? If the wall is sturdy enough, usually nothing happens. If you pushed against a wall while wearing roller skates, you would go rolling backwards. Your action on the wall produced a reaction—movement backwards. This is a demonstration of Newton's third law of motion.

Newton's third law of motion describes action-reaction pairs this way: When one object exerts a force on a second object, the second one exerts a force on the first that is equal in strength and opposite in direction. Another way to say this is "to every action force there is an equal and opposite reaction force."

Action and Reaction When a force is applied in nature, a reaction force occurs at the same time. When you jump on a trampoline, for example, you exert a downward force on the trampoline. Simultaneously, the trampoline exerts an equal force upward, sending you high into the air.

Action and reaction forces are acting on the two skaters in **Figure 21.** The male skater is pulling upward on the female skater, while the female skater is pulling downward on the male skater. The two forces are equal, but in opposite directions.

Figure 21 According to Newton's third law of motion, the two skaters exert forces on each other. The two forces are equal, but in opposite directions.

Space Astronauts who stay in outer space for extended periods of time may develop health problems. Their muscles, for example, may begin to weaken because they don't have to exert as much force to get the same reaction as they do on Earth. A branch of medicine called space medicine deals with the possible health problems that astronauts may experience. Research some other health risks that are involved in going into outer space. Do trips into outer space have any positive health benefits?

Action and Reaction Forces Don't Cancel If action and reaction forces are equal, you might wonder how some things ever happen. For example, how does a swimmer move through the water in a pool if each time she pushes on the water, the water pushes back on her? According to the third law of motion, action and reaction forces act on different objects. Thus, even though the forces are equal, they are not balanced because they act on different objects. In the case of the swimmer, as she "acts" on the water, the "reaction" of the water pushes her forward. Thus, a net force, or unbalanced force, acts on her so a change in her motion occurs. As the swimmer moves forward in the water, how does she make the water move?

Reading Check *Why don't action and reaction forces cancel?*

Rocket Propulsion Suppose you are standing on skates holding a softball. You exert a force on the softball when you throw the softball. According to Newton's third law, the softball exerts a force on you. This force pushes you in the direction opposite the softball's motion. Rockets use the same principle to move even in the vacuum of outer space. In the rocket engine, burning fuel produces hot gases. The rocket engine exerts a force on these gases and causes them to escape out the back of the rocket. By Newton's third law, the gases exert a force on the rocket and push it forward. The car in **Figure 22** uses a rocket engine to propel it forward. **Figure 23** shows how rockets were used to travel to the Moon.

Figure 22 If more gas is ejected from the rocket engine, or expelled at a greater velocity, the rocket engine will exert a larger force on the car.

Figure 23

On the afternoon of July 16, 1969, *Apollo 11* lifted off from Cape Kennedy, Florida, bound for the Moon. Eight days later, the spacecraft returned to Earth, splashing down safely in the Pacific Ocean. The motion of the spacecraft to the Moon and back is governed by Newton's laws of motion.

◀ *Apollo 11* roars toward the Moon. At launch, a rocket's engines must produce enough force and acceleration to overcome the pull of Earth's gravity. A rocket's liftoff is an illustration of Newton's third law: For every action there is an equal and opposite reaction.

▲ As *Apollo* rises, it burns fuel and ejects its rocket booster engines. This decreases its mass, and helps *Apollo* move faster. This is Newton's second law in action: As mass decreases, acceleration can increase.

▶ The lunar module uses other engines to slow down and ease into a soft touchdown on the Moon. A day later, the same engines lift the lunar module again into outer space.

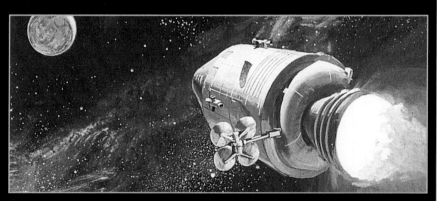

▲ After the lunar module returns to *Apollo,* the rocket fires its engines to set it into motion toward Earth. The rocket then shuts off its engines, moving according to Newton's first law. As it nears Earth, the rocket accelerates at an increasing rate because of Earth's gravity.

Momentum

A moving object has a property called momentum that is related to how much force is needed to change its motion. The **momentum** of an object is the product of its mass and velocity. Momentum is given the symbol p and can be calculated with the following equation:

> **Momentum Equation**
>
> **momentum (kg m/s) = mass (kg) × velocity (m/s)**
>
> $$p = mv$$

The unit for momentum is kg·m/s. Notice that momentum has a direction because velocity has a direction.

Applying Math   Solve a Simple Equation

THE MOMENTUM OF A SPRINTER At the end of a race, a sprinter with a mass of 80 kg has a speed of 10 m/s. What is the sprinter's momentum?

IDENTIFY known values and the unknown value

Identify the known values:

a sprinter with a mass of 80 kg means→ $m = 80$ kg

has a speed of 10 m/s means→ $v = 10$ m/s

Identify the unknown value:

What is the sprinter's momentum? means→ $p = ?$ kg·m/s

SOLVE the problem

Substitute the known values $m = 80$ kg and $v = 10$ m/s into the momentum equation:

$$p = mv = (80 \text{ kg}) (10 \text{ m/s}) = 800 \text{ kg·m/s}$$

CHECK the answer

Does your answer seem reasonable? Check your answer by dividing the momentum you calculated by the mass given in the problem. The result should be the speed given in the problem.

Practice Problems

1. What is the momentum of a car with a mass of 1,300 kg traveling at a speed of 28 m/s?

2. A baseball thrown by a pitcher has a momentum of 6.0 kg·m/s. If the baseball's mass is 0.15 kg, what is the baseball's speed?

3. What is the mass of a person walking at a speed of 0.8 m/s if their momentum is 52.0 kg·m/s?

For more practice problems go to page 834, and visit gpscience.com/extra_problems.

Force and Changing Momentum If you catch a baseball, your hand might sting, even if you use a baseball glove. Your hand stings because the baseball exerted a force on your hand when it came to a stop, and its momentum changed.

Recall that acceleration is the difference between the initial and final velocity, divided by the time. Also, from Newton's second law, the net force on an object equals its mass times its acceleration. By combining these two relationships, Newton's second law can be written in this way:

$$F = (mv_f - mv_i)/t$$

In this equation mv_f is the final momentum and mv_i is the initial momentum. The equation says that the net force exerted on an object can be calculated by dividing its change in momentum by the time over which the change occurs. When you catch a ball, your hand exerts a force on the ball that stops it. The force you exert on the ball equals the force the ball exerts on your hand. This force depends on the mass and initial velocity of the ball, and how long it takes the ball to stop.

Law of Conservation of Momentum The momentum of an object doesn't change unless its mass, velocity, or both change. Momentum, however, can be transferred from one object to another. Consider the game of pool shown in **Figure 24.**

When the cue ball hits the group of balls that are motionless, the cue ball slows down and the rest of the balls begin to move. The momentum the group of balls gained is equal to the momentum that the cue ball lost. The total momentum of all the balls just before and after the collision would be the same. If no other forces act on the balls, their total momentum is conserved—it isn't lost or created. This is the law of conservation of momentum—if a group of objects exerts forces only on each other, their total momentum doesn't change.

Figure 24 Momentum is transferred in collisions. **A** Before the collision, only the cue ball has momentum. **B** When the cue ball strikes the other balls, it transfers some of its momentum to them.

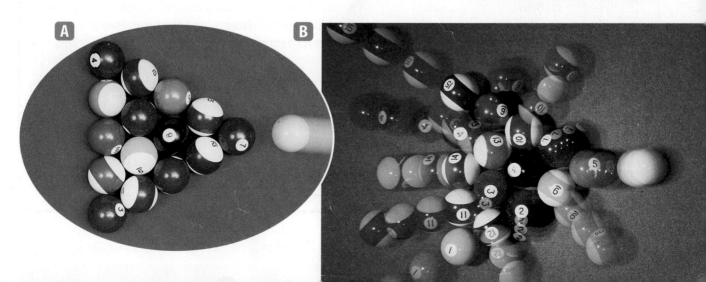

Figure 25 The results of a collision depend on the momentum of each object. **A** When the first puck hits the second puck from behind, it gives the second puck momentum in the same direction. **B** If the pucks are speeding toward each other with the same speed, the total momentum is zero.

Predict *how the pucks will move after they collide.*

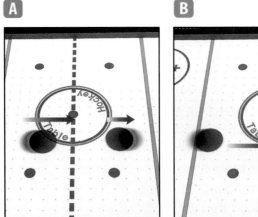 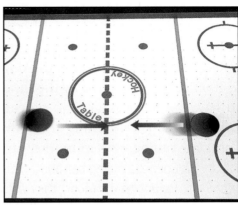

When Objects Collide Collisions of two air hockey pucks are shown in **Figure 25.** Suppose one of the pucks was moving in one direction and another struck it from behind. The first puck would continue to move in the same direction but more quickly. The second puck has given it more momentum in the same direction. What if two pucks of equal mass were moving toward each other with the same speed? They would have the same momentum, but in opposite directions. So the total momentum would be zero. After the pucks collide, each would reverse direction, and move with the same speed. The total momentum would be zero again.

section 3 review

Summary

Newton's Third Law

- According to Newton's third law of motion, for every action force, there is an equal and opposite reaction force.
- Action and reaction forces act on different objects.

Momentum

- The momentum of an object is the product of its mass and velocity:
$$p = mv$$
- The net force on an object can be calculated by dividing its change in momentum by the time over which the change occurs.

The Law of Conservation of Momentum

- According to the law of conservation of momentum, if objects exert forces only on each other, their total momentum is conserved.
- In a collision, momentum is transferred from one object to another.

Self Check

1. **Determine** You push against a wall with a force of 50 N. If the wall doesn't move, what is the net force on you?
2. **Explain** how a rocket can move through outer space where there is no matter for it to push on.
3. **Compare** the momentum of a 6,300-kg elephant walking 0.11 m/s and a 50-kg dolphin swimming 10.4 m/s.
4. **Describe** what happens to the momentum of two billiard balls that collide.
5. **Think Critically** A ballet director assigns slow, graceful steps to larger dancers, and quick movements to smaller dancers. Why is this plan successful?

Applying Math

6. **Calculate Momentum** What is the momentum of a 100-kg football player running at a speed of 4 m/s?
7. **Calculate Force** What is the force exerted by a catcher's glove on a 0.15-kg baseball moving at 35 m/s that is stopped in 0.02 s?

 Science online gpscience.com/self_check_quiz

MEASURING THE EFFECTS OF AIR RESISTANCE

If you dropped a bowling ball and a feather from the same height on the Moon, they would both hit the surface at the same time. All objects dropped on Earth are attracted to the ground with the same acceleration. But on Earth, a bowling ball and a feather will not hit the ground at the same time. Air resistance slows the feather down.

Real-World Question

How does air resistance affect the acceleration of falling objects?

Goals

- **Measure** the effect of air resistance on sheets of paper with different shapes.
- **Design** and create a shape from a piece of paper that maximizes air resistance.

Materials

paper (4 sheets of equal size)   stopwatch
scissors   masking tape
meterstick

Safety Precautions

Procedure

1. Copy the data table above in your Science Journal, or create it on a computer.
2. Measure a height of 2.5 m on the wall and mark the height with a piece of masking tape.
3. Have one group member drop the flat sheet of paper from the 2.5-m mark. Use the stopwatch to time how long it takes for the paper to reach the ground. Record your time in your data table.

Effects of Air Resistance	
Paper Type	Time
Flat paper	Do not write in this book.
Loosely crumpled paper	
Tightly crumpled paper	
Your paper design	

4. Crumple a sheet of paper into a loose ball and repeat step 3.
5. Crumple a sheet of paper into a tight ball and repeat step 3.
6. Use scissors to shape a piece of paper so that it will fall slowly. You may cut, tear, or fold your paper into any design you choose.

Conclude and Apply

1. **Compare** the falling times of the different sheets of paper.
2. **Explain** why the different-shaped papers fell at different speeds.
3. **Explain** how your design caused the force of air resistance on the paper to be greater than the air resistance on the other paper shapes.

*C*ommunicating Your Data

Compare your paper design with the designs created by your classmates. As a class, compile a list of characteristics that increase air resistance.

The Momentum of Colliding Objects

Goals

- **Observe** and calculate the momentum of different balls.
- **Compare** the results of collisions involving different amounts of momentum.

Materials

meterstick
softball
racquetball
tennis ball
baseball
stopwatch
masking tape
balance

Safety Precautions

Real-World Question

Many scientists hypothesize that dinosaurs became extinct 65 million years ago when an asteroid collided with Earth. The asteroid's diameter was probably no more than 10 km. Earth's diameter is more than 12,700 km. How could an object that size change Earth's climate enough to cause the extinction of animals that had dominated life on Earth for 140 million years? The asteroid could have caused such damage because it may have been traveling at a velocity of 50 m/s, and had a huge amount of momentum. The combination of an object's velocity and mass will determine how much force it can exert. How do the mass and velocity of a moving object affect its momentum?

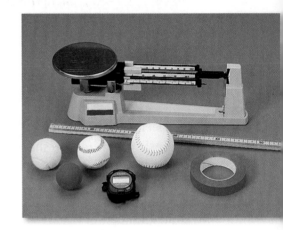

Momentum of Colliding Balls

Action	Time	Velocity	Mass	Momentum	Distance softball moved
Racquetball rolled slowly				Do not write in this book.	
Racquetball rolled quickly					
Tennis ball rolled slowly					
Tennis ball rolled quickly					
Baseball rolled slowly					
Baseball rolled quickly					

Procedure

1. Copy the data table on the previous page in your Science Journal.

2. Use the balance to measure the mass of the racquetball, tennis ball, and baseball. Record these masses in your data table.

3. Measure a 2-m distance on the floor and mark it with two pieces of masking tape.

4. Place the softball on one piece of tape. Starting from the other piece of tape, slowly roll the racquetball toward the center of the softball.

5. Use a stopwatch to time how long it takes the racquetball to roll the 2-m distance and hit the softball. Record this time in your data table.

6. Measure and record the distance the racquetball moved the softball.

7. Repeat steps 4–6, rolling the racquetball quickly.

8. Repeat steps 4–6, rolling the tennis ball slowly, then quickly.

9. Repeat steps 4–6, rolling the baseball slowly, then quickly.

Analyze Your Data

1. **Calculate** the momentum for each type of ball and action using the formula $p = mv$. Record your calculations in the data table.

2. **Graph** the relationship between the momentum of each ball and the distance the softball was moved using a graph like the one shown.

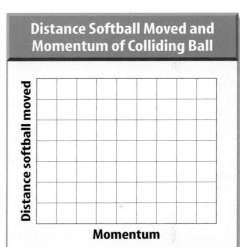

Distance Softball Moved and Momentum of Colliding Ball

Distance softball moved

Momentum

Conclude and Apply

1. **Infer** from your graph how the distance the softball moves after each collision depends on the momentum of the ball that hits it.

2. **Explain** how the motion of the balls after they collide can be determined by Newton's laws of motion.

Communicating
Your Data

Compare your graph with the graphs made by other students in your class. **Discuss** why the graphs might look different.

Newton and the Plague

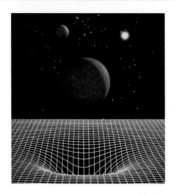

In Einstein's theory of general relativity, gravity is due to a distortion in space-time.

In 1665, the bubonic plague swept through England and other parts of Europe. Isaac Newton, then a 23-year-old university student, returned to his family's farm until Cambridge university reopened. To occupy his time, Newton made a list of 22 questions. During the next 18 months, Newton buried himself in the search for answers. And in that brief time, Newton developed calculus, the three laws of motion, and the universal law of gravitation!

The Laws of Motion

Earlier philosophers thought that force was necessary to keep an object moving. By analyzing the data collected by Galileo and others, Newton realized that forces did not cause motion. Instead, forces cause a change in motion. Newton came to understand that force and acceleration were related and that objects exert equal and opposite forces on each other. Newton's three laws of

motion were able to explain how all things moved, from an apple falling from a tree to the motions of the moon and the planets, in terms of force, mass, and acceleration.

Isaac Newton was a university student when he developed the laws of motion.

What is gravity?

Using the calculus and data on the motion of the planets, Newton deduced the law of universal gravitation. This law enabled the force of gravity between any two objects to be calculated, if their masses and the distance between them were known.

Newton was able to show mathematically that the law of universal gravitation predicted that the planets' orbits should be ellipses, just as Johannes Kepler had discovered two generations earlier. The application of Newton's laws of motion and the law of universal gravitation also were able to explain phenomena such as tides, the motion of the moon and the planets, and the bulge at the Earth's equator.

A Different View of Gravity

In 1916, Albert Einstein proposed a different model for gravity called the general theory of relativity. In Einstein's model, objects create distortions in space-time, like a bowling ball dropped on a sheet. What we see as the force of gravity is the motion of an object on distorted space-time. Today, Einstein's theory is used to help explain the nature of the big bang and the structure of the universe.

Investigate Research how both Newton's law of gravitation and Einstein's general theory of relativity have been used to develop the current model of the universe.

Science online

For more information, visit gpscience.com/time

Reviewing Main Ideas

Section 1 **Newton's Second Law**

1. Newton's second law of motion states that a net force causes an object to accelerate in the direction of the net force and that the acceleration is given by

$$a = \frac{F_{net}}{m}$$

2. Friction is a force opposing the sliding motion of two surfaces in contact. Friction is caused by microwelds that form where the surfaces are in contact.

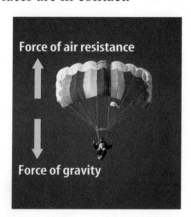

Force of air resistance

Force of gravity

3. Air resistance opposes the motion of objects moving through the air.

Section 2 **Gravity**

1. Gravity is an attractive force between any two objects with mass. The gravitational force depends on the masses of the objects and the distance between them.

2. The gravitational acceleration, g, near Earth's surface equals 9.8 m/s². The force of gravity on an object with mass, m, is:

$$F = mg$$

3. The weight of an object near Earth's surface is:

$$W = mg$$

4. Projectiles travel in a curved path because of their horizontal motion and vertical acceleration due to gravity.

5. The centripetal force is the net force on an object in circular motion and is directed toward the center of the circular path.

Section 3 **The Third Law of Motion**

1. Newton's third law of motion states that for every action there is an equal and opposite reaction.

2. The momentum of an object can be calculated by the equation $p = mv$.

3. When two objects collide, momentum can be conserved. Some of the momentum from one object is transferred to the other.

FOLDABLES Use the Foldable that you made at the beginning of this chapter to help you review the different types of friction.

Using Vocabulary

air resistance p.73
centripetal acceleration
   p.81
centripetal force p.81
friction p.70 ✓
gravity p.75 ✓
momentum p.86 ✓

Newton's 2nd law of
   motion p.69 ✓
Newton's 3rd law of
   motion p.83 ✓
sliding friction p.72
static friction p.71
weight p.77

Complete each statement using a word(s) from the vocabulary list above.

1. The way in which objects exert forces on each other is described by _____.

2. _____ prevents surfaces in contact from sliding past each other.

3. The _____ of an object is different on other planets in the solar system.

4. When an object moves in a circular path, the net force is called a(n) _____.

5. The attractive force between two objects that depends on their masses and the distance between them is _____.

6. _____ relates the net force exerted on an object to its mass and acceleration.

Checking Concepts

Choose the best answer for each question.

7. What is the gravitational force exerted on an object called?
   A) centripetal force **C)** momentum
   B) friction **D)** weight

8. Which of the following best describes why projectiles move in a curved path?
   A) They have horizontal velocity and vertical acceleration.
   B) They have momentum.
   C) They have mass.
   D) They have weight.

9. Which of the following explains why astronauts seem weightless in orbit?
   A) Earth's gravity is much less in orbit.
   B) The space shuttle is in free fall.
   C) The gravity of Earth and the Sun cancel.
   D) The centripetal force on the shuttle balances Earth's gravity.

10. Which of the following exerts the strongest gravitational force on you?
    A) the Moon **C)** the Sun
    B) Earth **D)** this book

Use the graph below to answer question 11.

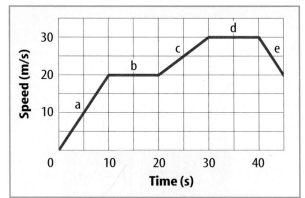

11. The graph shows the speed of a car moving in a straight line. Over which segments are the forces on the car balanced?
    A) A and C **C)** C and E
    B) B and D **D)** D only

12. Which of the following is true about an object in free fall?
    A) Its acceleration depends on its mass.
    B) It has no inertia.
    C) It pulls on Earth, and Earth pulls on it.
    D) Its momentum is constant.

13. The acceleration of an object is in the same direction as which of the following?
    A) net force **C)** static friction
    B) air resistance **D)** gravity

14. Which of the following is NOT a force?
    A) weight **C)** momentum
    B) friction **D)** air resistance

Science*online* gpscience.com/vocabulary_puzzlemaker

Interpreting Graphics

15. Copy and complete the following concept map on forces.

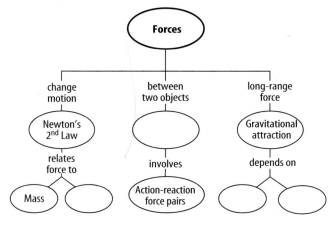

Use the table below to answer questions 16–18.

Time of Fall for Dropped Objects

Object	Mass (g)	Time of Fall (s)
A	5.0	2.0
B	5.0	1.0
C	30.0	0.5
D	35.0	1.5

16. If the objects in the data table above all fell the same distance, which object fell with the greatest average speed?

17. Explain why all four objects don't fall with the same speed.

18. On which object was the force of air resistance the greatest?

Thinking Critically

19. **Determine** the direction of the net force on a book sliding on a table if the book is slowing down.

20. **Explain** whether there can be any forces acting on a car moving in a straight line with constant speed.

21. **Explain** You pull a door open. If the force the door exerts on you is equal to the force you exert on the door, why don't you move?

22. **Predict** Suppose you are standing on a bathroom scale next to a sink. How does the reading on the scale change if you push down on the sink?

23. **Describe** the action and reaction force pairs involved when an object falls toward Earth. Ignore the effects of air resistance.

Applying Math

24. **Calculate Mass** Find your mass if a scale on Earth reads 650 N when you stand on it.

25. **Calculate Acceleration of Gravity** You weigh yourself at the top of a high mountain and the scale reads 720 N. If your mass is 75 kg, what is the acceleration of gravity at your location?

Use the figure below to answer question 26.

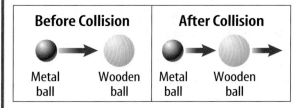

26. **Calculate Speed** The 2-kg metal ball moving at a speed of 3 m/s strikes a 1-kg wooden ball that is at rest. After the collision, the speed of the metal ball is 1 m/s. Assuming momentum is conserved, what is the speed of the wooden ball?

27. **Calculate Mass** Find the mass of a car that has a speed of 30 m/s and a momentum of 45,000 kg,m/s.

28. **Calculate Sliding Friction** A box being pushed with a force of 85 N slides along the floor with a constant speed. What is the force of sliding friction on the box?

Part 1 | Multiple Choice

Record your answers on the answer sheet provided by your teacher or on a sheet of paper.

1. The net force on an object moving with constant speed in circular motion is in which direction?
- **A.** downward
- **B.** opposite to the object's motion
- **C.** toward the center of the circle
- **D.** in the direction of the object's velocity

Use the table below to answer questions 2–4.

Object in Free Fall with Air Resistance	
Time (s)	**Speed (m/s)**
0	0
1	9.1
2	15.1
3	18.1
4	19.3
5	19.9

2. According to the trend in these data, which of the following values is most likely the speed of the object after falling for 6 s?
- **A.** 26.7 m/s
- **B.** 15.1 m/s
- **C.** 20.1 m/s
- **D.** 0 m/s

3. Over which of the following time intervals is the acceleration of the object the greatest?
- **A.** 0 s to 1 s
- **B.** 1 s to 2 s
- **C.** 4 s to 5 s
- **D.** The acceleration is constant.

4. Over which of the following time intervals is the force on the object the smallest?
- **A.** 0 s to 1 s
- **C.** 4 s to 5 s
- **B.** 1 s to 2 s
- **D.** The force is constant.

5. Which of the following would cause the gravitational force between object A and object B to increase?
- **A.** Decrease the distance between them.
- **B.** Increase the distance between them.
- **C.** Decrease the mass of object A.
- **D.** Decrease the mass of both objects.

Use the graphs below to answer questions 6–7.

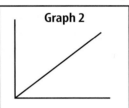
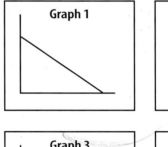
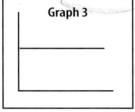
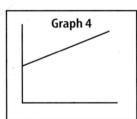

6. Which of the graphs above shows how the force on an object changes if the mass increases and the acceleration stays constant?
- **A.** Graph 1
- **C.** Graph 3
- **B.** Graph 2
- **D.** Graph 4

7. Which of the graphs above shows how the force on an object changes if the acceleration increases and the mass stays constant?
- **A.** Graph 1
- **C.** Graph 3
- **B.** Graph 2
- **D.** Graph 4

Test-Taking Tip

Keep Track of Time If you are taking a timed test, keep track of time during the test. If you find you are spending too much time on a multiple-choice question, mark your best guess and move on.

Part 2 | Short Response/Grid In

Record your answers on the answer sheet provided by your teacher or on a sheet of paper.

8. You are pushing a 30-kg wooden crate across the floor. The force of sliding friction on the crate is 90 N. How much force must you exert on the crate to keep it moving with a constant velocity?

9. A sky diver with a mass of 60 kg jumps from an airplane. Five seconds after jumping the force of air resistance on the sky diver is 300 N. What is the sky diver's acceleration five seconds after jumping?

10. A pickup truck is carrying a load of gravel. The driver hits a bump and gravel falls out, so that the mass of the truck is one half as large after hitting the bump. If the net force on the truck doesn't change, how does the truck's acceleration change?

Use the figure below to answer question 11.

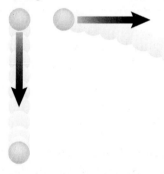

11. Two balls are at the same height. One ball is dropped and the other initially moves horizontally, as shown in the figure above. After one second, which ball has fallen the greatest vertical distance?

12. How does the acceleration of gravity 5,000 km above Earth's surface compare with the acceleration of gravity at Earth's surface?

Part 3 | Open Ended

Record your answers on a sheet of paper.

13. You push on a rolling ball so that the direction of your push is different from the direction of the ball's motion. After you push, will the ball be moving in the same direction that you pushed? Explain.

14. Two balls are dropped from an airplane. Both balls are the same size, but one has a mass ten-times greater than the other. The force of air resistance on each ball depends on the ball's speed. Explain whether both balls will reach the same terminal velocity.

Use the graph below to answer question 15.

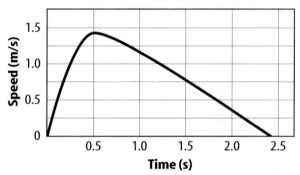

15. The graph above shows the how the speed of a book changes as it slides across a table. Over what time interval is the net force on the book in the opposite direction of the book's motion?

16. When the space shuttle is launched from Earth, the rocket engines burn fuel and apply a constant force on the shuttle until they burn out. Explain why the shuttles acceleration increases while the rocket engines are burning fuel.

17. Catching a hard-thrown baseball with your bare hand can cause your hand to sting. Explain why using a baseball glove reduces the sting when you catch a ball.

Energy

A Big Lift

How does this pole vaulter go from standing at the end of a runway to climbing through the air? The answer is energy. During her vault, energy originally stored in her muscles is converted into other forms of energy that help her achieve her goal.

Science Journal Which takes more energy: walking up stairs, or taking an escalator? Explain your reasoning.

Start-Up Activities

Energy Conversions

One of the most useful inventions of the nineteenth century was the electric lightbulb. Being able to light up the dark has enabled people to work and play longer. A lightbulb converts electrical energy to heat energy and light, another form of energy. The following lab shows how electrical energy is converted into other forms of energy.

WARNING: *Steel wool can become hot—connect to battery only for a brief time.*

1. Obtain two D-cell batteries, two non-coated paper clips, tape, metal tongs and some steel wool. Separate the steel wool into thin strands and straighten the paper clips.

2. Tape the batteries together and then tape one end of each paper clip to the battery terminals.

3. While holding the steel wool with the tongs, briefly complete the circuit by placing the steel wool in contact with both the paper clip ends.

4. **Think Critically** In your Science Journal, describe what happened to the steel wool. What changes did you observe?

 **Preview this chapter's content and activities at** gpscience.com

 Energy Make the following Foldable to help you identify what you already know, what you want to know, and what you learned about energy.

STEP 1 **Fold** a vertical sheet of paper from side to side. Make the front edge about 1 cm shorter than the back edge.

STEP 2 **Turn** lengthwise and **fold** into thirds.

STEP 3 **Unfold and cut** only the top layer along both folds to make three tabs.

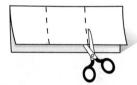

STEP 4 **Label** each tab as shown.

Question Before you read this chapter, write what you already know about energy under the left tab of your Foldable, and write questions about what you'd like to know under the center tab. After you read the chapter, list what you learned under the right tab.

The Nature of Energy

Reading Guide

What You'll Learn

- **Distinguish** between kinetic and potential energy.
- **Calculate** kinetic energy.
- **Describe** different forms of potential energy.
- **Calculate** gravitational potential energy.

Why It's Important

All of the changes that occur around you every day involve the conversion of energy from one form to another.

⟳ Review Vocabulary

gravity: the attractive force between any two objects that have mass

New Vocabulary

- kinetic energy
- joule
- potential energy
- elastic potential energy
- chemical potential energy
- gravitational potential energy

What is energy?

Wherever you are sitting as you read this, changes are taking place—lightbulbs are heating the air around them, the wind might be rustling leaves, or sunlight might be glaring off a nearby window. Even you are changing as you breathe, blink, or shift position in your seat.

Every change that occurs—large or small—involves energy. Imagine a baseball flying through the air. It hits a window, causing the glass to break as shown in **Figure 1.** The window changed from a solid sheet of glass to a number of broken pieces. The moving baseball caused this change—a moving baseball has energy. Even when you comb your hair or walk from one class to another, energy is involved.

Change Requires Energy When something is able to change its environment or itself, it has energy. Energy is the ability to cause change. The moving baseball had energy. It certainly caused the window to change. Anything that causes change must have energy. You use energy to arrange your hair to look the way you want it to. You also use energy when you walk down the halls of your school between classes or eat your lunch. You even need energy to yawn, open a book, and write with a pen.

Figure 1 The baseball caused changes to occur when it hit the window.
Describe *the changes that are occurring.*

Different Forms of Energy

Turn on an electric light, and a dark room becomes bright. Turn on your CD player, and sound comes through your headphones. In both situations, energy moves from one place to another. These changes are different from each other, and differ from the baseball shattering the window in **Figure 1.** This is because energy has several different forms—electrical, chemical, radiant, and thermal.

Figure 2 shows some examples of everyday situations in which you might notice energy. Is the chemical energy stored in food the same as the energy that comes from the Sun or the energy stored in gasoline? Radiant energy from the Sun travels a vast distance through space to Earth, warming the planet and providing energy that enables green plants to grow. When you make toast in the morning, you are using electrical energy. In short, energy plays a role in every activity that you do.

Figure 2 Energy can be stored and it can move from place to place.
Infer *which materials are storing chemical energy.*

Reading Check *What are some different forms of energy?*

An Energy Analogy Money can be used in an analogy to help you understand energy. If you have $100, you could store it in a variety of forms—cash in your wallet, a bank account, travelers' checks, or gold or silver coins. You could transfer that money to different forms. You could deposit your cash into a bank account or trade the cash for gold. Regardless of its form, money is money. The same is true for energy. Energy from the Sun that warms you and energy from the food that you eat are only different forms of the same thing.

Topic: Glacier Flow
Visit gpscience.com for Web links to information about the speeds at which glaciers flow.

Activity For the five fastest glaciers, calculate the kinetic energy a 1 kg block of ice in each glacier would have.

Kinetic Energy

When you think of energy, you might think of action—or objects in motion, like the baseball that shatters a window. An object in motion does have energy. **Kinetic energy** is the energy a moving object has because of its motion. The kinetic energy of a moving object depends on the object's mass and its speed.

Kinetic Energy Equation

kinetic energy (in joules) = $\frac{1}{2}$ mass (in kg) × [speed (in m/s)]^2

$$KE = \frac{1}{2}mv^2$$

The SI unit of energy is the **joule,** abbreviated J. If you dropped a softball from a height of about 0.5 m, it would have a kinetic energy of about one joule before it hit the floor.

Applying Math Solve a Simple Equation

CALCULATE KINETIC ENERGY A jogger whose mass is 60 kg is moving at a speed of 3 m/s. What is the jogger's kinetic energy?

IDENTIFY known values and the unknown value

Identify the known values:

a jogger whose mass is 60 kg means⟩ $m = 60$ kg

is moving at a speed of 3 m/s means⟩ $v = 3$ m/s

Identify the unknown value:

what is the jogger's kinetic energy means⟩ $KE = ?$ J

SOLVE the problem

Substitute the known values $m = 60$ kg and $v = 3$ m/s into the kinetic energy equation:

$$KE = \frac{1}{2}mv^2 = \frac{1}{2}(60 \text{ kg})(3 \text{ m/s})^2 = \frac{1}{2}(60)(9) \text{ kg m}^2/\text{s}^2 = 270 \text{ J}$$

CHECK your answer

Does your answer seem reasonable? Check your answer by dividing the kinetic energy you calculate by the square of the given velocity, and then multiplying by 2. The result should be the mass given in the problem.

Practice Problems

1. What is the kinetic energy of a baseball moving at a speed of 40 m/s if the baseball has a mass of 0.15 kg?

2. A car moving at a speed of 20 m/s has a kinteic energy of 300,000 J. What is the car's mass?

3. A sprinter has a mass of 80 kg and a kinetic energy of 4,000 J. What is the sprinter's speed?

For more practice problems go to page 834, and visit gpscience.com/extra_problems.

Figure 3 As natural gas burns, it combines with oxygen to form carbon dioxide and water. In this chemical reaction, chemical potential energy is released.

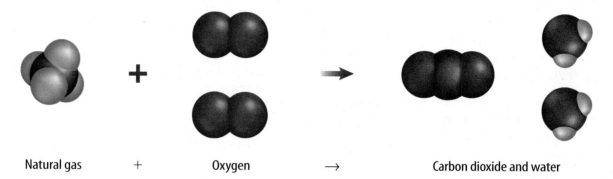

Natural gas   +   Oxygen   →   Carbon dioxide and water

Potential Energy

Energy doesn't have to involve motion. Even motionless objects can have energy. This energy is stored in the object. Therefore, the object has potential to cause change. A hanging apple in a tree has stored energy. When the apple falls to the ground, a change occurs. Because the apple has the ability to cause change, it has energy. The hanging apple has energy because of its position above Earth's surface. Stored energy due to position is called **potential energy.** If the apple stays in the tree, it will keep the stored energy due to its height above the ground. If it falls, that stored energy of position is converted to energy of motion.

Elastic Potential Energy Energy can be stored in other ways, too. If you stretch a rubber band and let it go, it sails across the room. As it flies through the air, it has kinetic energy due to its motion. Where did this kinetic energy come from? Just as the apple hanging in the tree had potential energy, the stretched rubber band had energy stored as elastic potential energy. **Elastic potential energy** is energy stored by something that can stretch or compress, such as a rubber band or spring.

Chemical Potential Energy The cereal you eat for breakfast and the sandwich you eat at lunch also contain stored energy. Gasoline stores energy in the same way as food stores energy—in the chemical bonds between atoms. Energy stored in chemical bonds is **chemical potential energy. Figure 3** shows a molecule of natural gas. Energy is stored in the bonds that hold the carbon and hydrogen atoms together and is released when the gas is burned.

 **Reading Check** *How is elastic potential energy different from chemical potential energy?*

Interpreting Data from a Slingshot

Procedure

1. Using two fingers, carefully stretch a **rubber band** on a table until it has no slack.
2. Place a **nickel** on the table, slightly touching the mid-point of the rubber band.
3. Push the nickel back 0.5 cm into the rubber band and release. Measure the distance the nickel travels.
4. Repeat step 3, each time pushing the nickel back an additional 0.5 cm.

Analysis

1. How did the takeoff speed of the nickel seem to change relative to the distance that you stretched the rubber band?
2. What does this imply about the kinetic energy of the nickel?

The Myth of Sisyphus In Greek mythology, a king named Sisyphus angered the gods by attempting to delay death. As punishment, he was doomed for eternity to endlessly roll a huge stone up a hill, only to have it roll back to the bottom again. Explain what caused the potential energy of the stone to change as it moved up and down the hill.

Gravitational Potential Energy Anything that can fall has stored energy called gravitational potential energy. **Gravitational potential energy** (GPE) is energy stored by objects due to their position above Earth's surface. The GPE of an object depends on the object's mass and height above the ground. Gravitational potential energy can be calculated from the following equation.

> **Gravitational Potential Energy Equation**
>
> gravitational potential energy (J) =
> **mass** (kg) × **acceleration of gravity** (m/s²) × **height** (m)
>
> $$GPE = mgh$$

On Earth, the acceleration of gravity is 9.8 m/s² and has the symbol g. Like all forms of energy, gravitational potential energy is measured in joules.

Applying Math   Solve a Simple Equation

CALCULATE GRAVITATIONAL POTENTIAL ENERGY What is the gravitational potential energy of a ceiling fan that has a mass of 7 kg and is 4 m above the ground?

IDENTIFY known values and the unknown value

Identify the known values:

has a mass of 7 kg means⟩ $m = 7$ kg

is 4 m above the ground means⟩ $h = 4$ m

Identify the unknown value:

what is the gravitational potential energy means⟩ $GPE = ?$ J

SOLVE the problem

Substitute the known values $m = 7$ kg, $h = 4$ m, and $g = 9.8$ m/s² into the gravitational potential energy equation:

$$GPE = mgh = (7 \text{ kg})(9.8 \text{ m/s}^2)(4 \text{ m}) = (274) \text{ kg m}^2/\text{s}^2 = 274 \text{ J}$$

CHECK your answer

Does your answer seem reasonable? Check your answer by dividing the gravitational potential energy you calculate by the given mass, and then divide by 9.8 m/s². The result should be the height given in the problem.

Practice Problems

1. Find the height of a baseball with a mass of 0.15 kg that has a GPE of 73.5 J.

2. Find the GPE of a coffee mug with a mass of 0.3 kg on a 1-m high counter top.

3. What is the mass of a hiker 200 m above the ground if her GPE is 117,600 J?

For more practice problems go to page 834, and visit gpscience.com/extra_problems.

Changing GPE Look at the objects in the bookcase in **Figure 4.** Which of these objects has the most gravitational potential energy? According to the equation for gravitational potential energy, the GPE of an object can be increased by increasing its height above the ground. If two objects are at the same height, then the object with the larger mass has more gravitational potential energy.

In **Figure 4,** suppose the green vase on the lower shelf and the blue vase on the upper shelf have the same mass. Then the blue vase on the upper shelf has more gravitational potential energy because it is higher above the ground.

Imagine what would happen if the two vases were to fall. As they fall and begin moving, they have kinetic energy as well as gravitational potential energy. As the vases get closer to the ground, their gravitational potential energy decreases. At the same time, they are moving faster, so their kinetic energy increases. The vase that was higher above the floor has fallen a greater distance. As a result, the vase that initially had more gravitational potential energy will be moving faster and have more kinetic energy when it hits the floor.

Figure 4 An object's gravitational potential energy increases as its height increases.

---

section 1 review

Summary

Energy
- Energy is the ability to cause change.
- Forms of energy include electrical, chemical, thermal, and radiant energy.

Kinetic Energy
- Kinetic energy is the energy a moving object has because of its motion.
- The kinetic energy of a moving object can be calculated from this equation:

$$KE = \frac{1}{2}mv^2$$

Potential Energy
- Potential energy is stored energy due to the position of an object.
- Different forms of potential energy include elastic potential energy, chemical potential energy, and gravitational potential energy.
- Gravitational potential energy can be calculated from this equation:

$$GPE = mgh$$

Self Check

1. **Explain** whether an object can have kinetic energy and potential energy at the same time.
2. **Describe** three situations in which the gravitational potential energy of an object changes.
3. **Explain** how the kinetic energy of a truck could be increased without increasing the truck's speed.
4. **Think Critically** The different molecules that make up the air in a room have on average the same kinetic energy. How does the speed of the different air molecules depend on their masses?

Applying Math

5. **Calculate Kinetic Energy** Find the kinetic energy of a ball with a mass of 0.06 kg moving at 50 m/s.
6. **Use Ratios** A boulder on top of a cliff has potential energy of 8,800 J, and has twice the mass of a boulder next to it. What is the GPE of the smaller boulder?
7. **Calculate GPE** An 80-kg diver jumps from a 10-m high platform. What is the gravitational potential energy of the diver halfway down?

Bouncing Balls

What happens when you drop a ball onto a hard, flat surface? It starts with potential energy. It bounces up and down until it finally comes to a rest. Where did the energy go?

⬤ Real-World Question

Why do bouncing balls stop bouncing?

Goals

■ **Identify** the forms of energy observed in a bouncing ball.

■ **Infer** why the ball stops bouncing.

Materials

tennis ball	masking tape
rubber ball	cardboard box
balance	*shoe box
meterstick	*Alternate materials

Safety Precautions

⬤ Procedure

1. **Measure** the mass of the two balls.

2. Have a partner drop one ball from 1 m. Measure how high the ball bounced. Repeat this two more times so you can calculate an average bounce height. Record your values on the data table.

3. Repeat step 2 for the other ball.

4. **Predict** whether the balls would bounce higher or lower if they were dropped onto the cardboard box. Design an experiment to measure how high the balls would bounce off the surface of a cardboard box.

Bounce Height			
Type of Ball	Surface	Trial	Height (cm)
Tennis	Floor	1	
Tennis	Floor	2	
Tennis	Floor	3	
Rubber	Floor	1	
Rubber	Floor	2	
Rubber	Floor	3	
Tennis	Box	1	

⬤ Conclude and Apply

1. **Calculate** the gravitational potential energy of each ball before dropping it.

2. **Calculate** the average bounce height for the three trials under each condition. Describe your observations.

3. **Compare** the bounce heights of the balls dropped on a cardboard box with the bounce heights of the ballls dropped on the floor. Hint: *Did you observe any movement of the box when the balls bounced?*

4. **Explain** why the balls bounced to different heights, using the concept of elastic potential energy.

*C*ommunicating Your Data

Meet with three other lab teams and compare average bounce heights for the tennis ball on the floor. Discuss why your results might differ. **For more help, refer to the** Science Skill Handbook.

Conservation of Energy

What You'll Learn

- ■ **Describe** how energy can be transformed from one form to another.
- ■ **Explain** how the mechanical energy of a system is the sum of the kinetic and potential energy.
- ■ **Discuss** the law of conservation of energy.

Why It's Important

All the energy transformations that occur inside you and around you obey the law of conservation of energy.

Review Vocabulary

friction: a force that opposes the sliding motion of two surfaces that are touching each other

New Vocabulary

- ● mechanical energy
- ● law of conservation of energy

Changing Forms of Energy

Unless you were talking about potential energy, you probably wouldn't think of the book on top of a bookshelf as having much to do with energy—until it fell. You'd be more likely to think of energy as race cars roar past or as your body uses energy from food to help it move, or as the Sun warms your skin on a summer day. These situations involve energy changing from one form to another form.

Transforming Electrical Energy You use many devices every day that convert one form of energy to other forms. For example, you might be reading this page in a room lit by lightbulbs. The lightbulbs transform electrical energy into light so you can see. The warmth you feel around the bulb is evidence that some of that electrical energy is transformed into thermal energy, as illustrated in **Figure 5.** What other devices have you used today that make use of electrical energy? You might have been awakened by an alarm clock, styled your hair, made toast, listened to music, or played a video game. What form or forms of energy is electrical energy converted to in these examples?

Figure 5 A lightbulb is a device that transforms electrical energy into light energy and thermal energy.
Identify *other devices that convert electrical energy to thermal energy.*

Light energy out

Thermal energy out

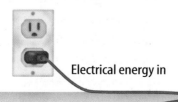

Electrical energy in

Spark plug fires

Gases expand

In a car, a spark plug fires, initiating the conversion of chemical potential energy into thermal energy.

As the hot gases expand, thermal energy is converted into kinetic energy.

Figure 6 In the engine of a car, several energy conversions occur.

Transforming Chemical Energy Fuel stores energy in the form of chemical potential energy. For example, the car or bus that might have brought you to school this morning probably runs on gasoline. The engine transforms the chemical potential energy stored in gasoline molecules into the kinetic energy of a moving car or bus. Several energy conversions occur in this process, as shown in **Figure 6.** An electric spark ignites a small amount of fuel. The burning fuel produces thermal energy. So chemical energy is changed to thermal energy. The thermal energy causes gases to expand and move parts of the car, producing kinetic energy.

Some energy transformations are less obvious because they do not result in visible motion, sound, heat, or light. Every green plant you see converts light energy from the Sun into energy stored in chemical bonds in the plant. If you eat an ear of corn, the chemical potential energy in the corn is transformed into other forms of energy by your body.

Conversions Between Kinetic and Potential Energy

You have experienced many situations that involve conversions between potential and kinetic energy. Systems such as bicycles, roller coasters, and swings can be described in terms of potential and kinetic energy. Even launching a rubber band or using a bow and arrow involves energy conversions. To understand the energy conversions that occur, it is helpful to identify the mechanical energy of a system. **Mechanical energy** is the total amount of potential and kinetic energy in a system and can be expressed by this equation.

mechanical energy = potential energy + kinetic energy

In other words, mechanical energy is energy due to the position and the motion of an object or the objects in a system. What happens to the mechanical energy of an object as potential and kinetic energy are converted into each other?

Falling Objects Standing under an apple tree can be hazardous. An apple on a tree, like the one in **Figure 7,** has gravitational potential energy due to Earth pulling down on it. The apple does not have kinetic energy while it hangs from the tree. However, the instant the apple comes loose from the tree, it accelerates due to gravity. As it falls, it loses height so its gravitational potential energy decreases. This potential energy is transformed into kinetic energy as the velocity of the apple increases.

Look back at the equation for mechanical energy. If the potential energy is being converted into kinetic energy, then the mechanical energy of the apple doesn't change as it falls. The potential energy that the apple loses is gained back as kinetic energy. The form of energy changes, but the total amount of energy remains the same.

 **Reading Check** *What happens to the mechanical energy of the apple as it falls from the tree?*

Earth pulls an apple down

Figure 7 Objects that can fall have gravitational potential energy.
Apply *What objects around you have gravitational potential energy?*

Energy Transformations in Projectile Motion Energy transformations also occur during projectile motion when an object moves in a curved path. Look at **Figure 8.** When the ball leaves the bat, it has mostly kinetic energy. As the ball rises, its velocity decreases, so its kinetic energy must decrease, too. However, the ball's gravitational potential energy increases as it goes higher. At its highest point, the baseball has the maximum amount of gravitational potential energy. The only kinetic energy it has at this point is due to its forward motion. Then, as the baseball falls, gravitational potential energy decreases while kinetic energy increases as the ball moves faster. However, the mechanical energy of the ball remains constant as it rises and falls.

Figure 8 Kinetic energy and gravitational potential energy are converted into each other as the ball rises and falls.

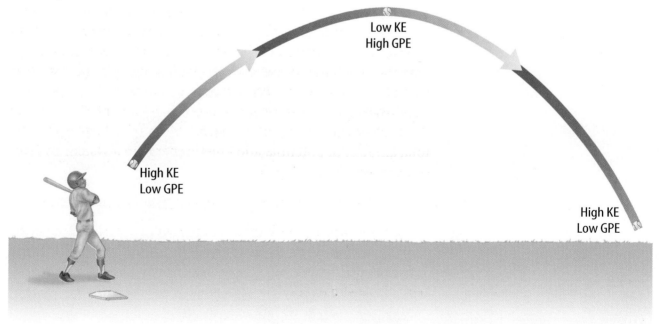

Low KE
High GPE

High KE
Low GPE

High KE
Low GPE

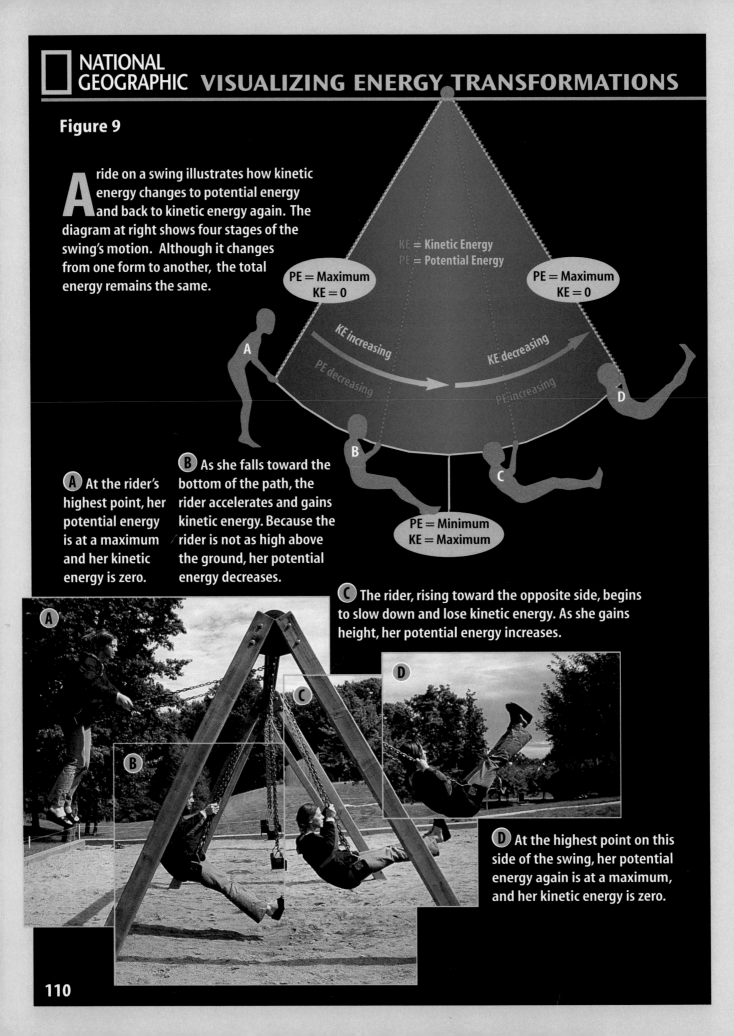

Figure 9

A ride on a swing illustrates how kinetic energy changes to potential energy and back to kinetic energy again. The diagram at right shows four stages of the swing's motion. Although it changes from one form to another, the total energy remains the same.

KE = Kinetic Energy
PE = Potential Energy

PE = Maximum
KE = 0

PE = Maximum
KE = 0

KE increasing

PE decreasing

KE decreasing

PE increasing

PE = Minimum
KE = Maximum

A At the rider's highest point, her potential energy is at a maximum and her kinetic energy is zero.

B As she falls toward the bottom of the path, the rider accelerates and gains kinetic energy. Because the rider is not as high above the ground, her potential energy decreases.

C The rider, rising toward the opposite side, begins to slow down and lose kinetic energy. As she gains height, her potential energy increases.

D At the highest point on this side of the swing, her potential energy again is at a maximum, and her kinetic energy is zero.

Energy Transformations in a Swing When you ride on a swing, like the one shown in **Figure 9,** part of the fun is the feeling of almost falling as you drop from the highest point to the lowest point of the swing's path. Think about energy conservation to analyze such a ride.

The ride starts with a push that gets you moving, giving you kinetic energy. As the swing rises, you lose speed but gain height. In energy terms, kinetic energy changes to gravitational potential energy. At the top of your path, potential energy is at its greatest. Then, as the swing accelerates downward, potential energy changes to kinetic energy. At the bottom of each swing, the kinetic energy is at its greatest and the potential energy is at its minimum. As you swing back and forth, energy continually converts from kinetic to potential and back to kinetic. What happens to your mechanical energy as you swing?

The Law of Conservation of Energy

When a ball is thrown into the air or a swing moves back and forth, kinetic and potential energy are constantly changing as the object speeds up and slows down. However, mechanical energy stays constant. Kinetic and potential energy simply change forms and no energy is destroyed.

This is always true. Energy can change from one form to another, but the total amount of energy never changes. Even when energy changes form from electrical to thermal and other energy forms as in the hair dryer shown in **Figure 10,** energy is never destroyed. Another way to say this is that energy is conserved. This principle is recognized as a law of nature. The **law of conservation of energy** states that energy cannot be created or destroyed. On a large scale, this law means that the total amount of energy in the universe does not change.

✔️ **Reading Check** *What law states that the total amount of energy never changes?*

Conserving Resources You might have heard about energy conservation or been asked to conserve energy. These ideas are related to reducing the demand for electricity and gasoline, which lowers the consumption of energy resources such as coal and fuel oil. The law of conservation of energy, on the other hand, is a universal principle that describes what happens to energy as it is transferred from one object to another or as it is transformed.

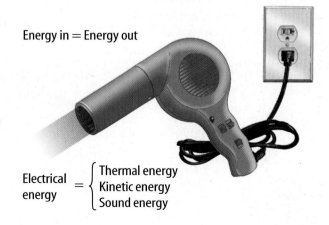

INTEGRATE Environment

Energy and the Food Chain One way energy enters ecosystems is when green plants transform radiant energy from the Sun into chemical potential energy in the form of food. Energy moves through the food chain as animals that eat plants are eaten by other animals. Some energy leaves the food chain, such as when living organisms release thermal energy to the environment. Diagram a simple biological food chain showing energy conservation.

Figure 10 The law of conservation of energy requires that the total amount of energy going into a hair dryer must equal the total amount of energy coming out of the hair dryer.

Energy in = Energy out

Electrical energy = { Thermal energy, Kinetic energy, Sound energy }

Figure 11 In a swing, mechanical energy is transformed into thermal energy because of friction and air resistance.
Infer *how the kinetic and potential energy of the swing change with time.*

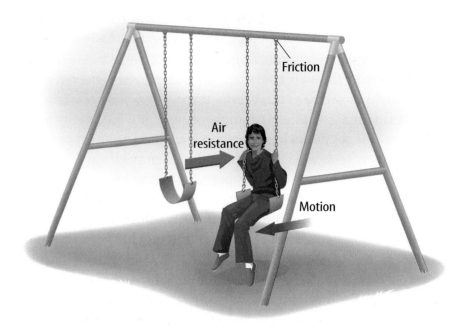

Mini LAB

Energy Transformations in a Paper Clip

Procedure

1. Straighten a **paper clip.** While holding the ends, touch the paper clip to the skin just below your lower lip. Note whether the paper clip feels warm, cool, or about room temperature.
2. Quickly bend the paper clip back and forth five times. Touch it below your lower lip again. Note whether the paper clip feels warmer or cooler than before.

Analysis

1. What happened to the temperature of the paper clip? Why?
2. Explain the energy conversions that take place as you bend the paper clip.

Try at Home

Is energy always conserved? You might be able to think of situations where it seems as though energy is not conserved. For example, while coasting along a flat road on a bicycle, you know that you will eventually stop if you don't pedal. If energy is conserved, why wouldn't your kinetic energy stay constant so that you would coast forever? In many situations, it might seem that energy is destroyed or created. Sometimes it is hard to see the law of conservation of energy at work.

The Effect of Friction You know from experience that if you don't continue to pump a swing or be pushed by somebody else, your arcs will become lower and you eventually will stop swinging. In other words, the mechanical (kinetic and potential) energy of the swing seems to decrease, as if the energy were being destroyed. Is this a violation of the law of conservation of energy?

It can't be—it's the law! If the energy of the swing decreases, then the energy of some other object must increase by an equal amount to keep the total amount of energy the same. What could this other object be that experiences an energy increase? To answer this, you need to think about friction. With every movement, the swing's ropes or chains rub on their hooks and air pushes on the rider, as illustrated in **Figure 11.** Friction and air resistance cause some of the mechanical energy of the swing to change to thermal energy. With every pass of the swing, the temperature of the hooks and the air increases a little, so the mechanical energy of the swing is not destroyed. Rather, it is transformed into thermal energy. The total amount of energy always stays the same.

Converting Mass into Energy You might have wondered how the Sun unleashes enough energy to light and warm Earth from so far away. A special kind of energy conversion—nuclear fusion—takes place in the Sun and other stars. During this process a small amount of mass is transformed into a tremendous amount of energy. An example of a nuclear fusion reaction is shown in **Figure 12.** In the reaction shown here, the nuclei of the hydrogen isotopes deuterium and tritium undergo fusion.

Nuclear Fission Another process involving the nuclei of atoms, called nuclear fission, converts a small amount of mass into enormous quantities of energy. In this process, nuclei do not fuse—they are broken apart, as shown in **Figure 12.** In either process, fusion or fission, mass is converted to energy. In processes involving nuclear fission and fusion, the total amount of energy is still conserved if the energy content of the masses involved are included. Then the total energy before the reaction is equal to the total energy after the reaction, as required by the law of conservation of energy. The process of nuclear fission is used by nuclear power plants to generate electrical energy.

Science Online

Topic: Nuclear Fusion
Visit gpscience.com for Web links to information about using nuclear fusion as a source of energy in electric power plants.

Activity Make a table listing the advantages and disadvantages of using nuclear fusion as an energy source.

Figure 12 Mass is converted to energy in the processes of fusion and fission.

Nuclear fusion

He

$^2_1$H $^3_1$H

+

Radiant energy

Nuclear fission

Xe

U

+

Radiant energy   Sr

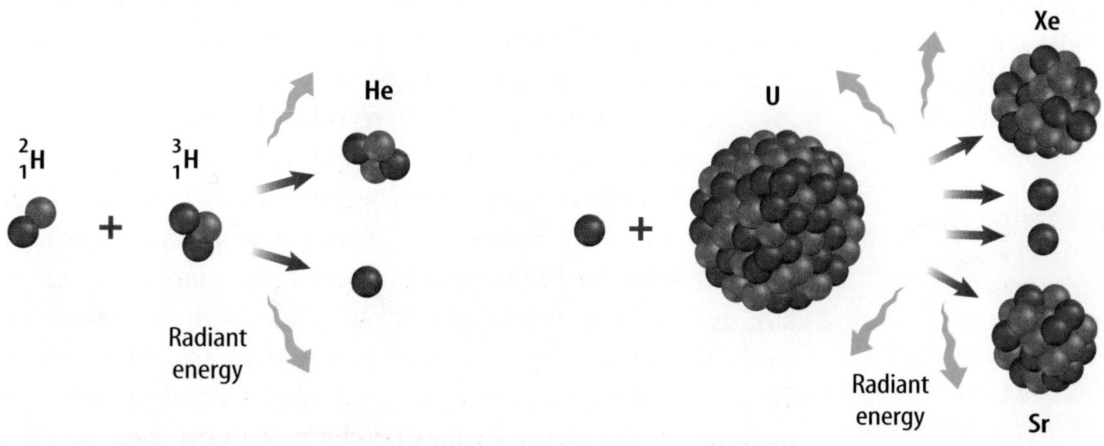

Mass $^2_1$H + Mass $^3_1$H > Mass **He** + Mass **neutron**

In this fusion reaction, the combined mass of the two hydrogen nuclei is greater than the mass of the helium nucleus, He, and the neutron.

Mass **U** + Mass **neutron** > Mass **Xe** + Mass **Sr** + Mass **neutrons**

In nuclear fission, the mass of the large nucleus on the left is greater than the combined mass of the other two nuclei and the neutrons.

The Human Body—Balancing the Energy Equation

What forms of energy discussed in this chapter can you find in the human body? With your right hand, reach up and feel your left shoulder. With that simple action, stored potential energy within your body was converted to the kinetic energy of your moving arm. Did your shoulder feel warm to your hand? Some of the chemical potential energy stored in your body is used to maintain a nearly constant internal temperature. A portion of this energy also is converted to the excess heat that your body gives off to its surroundings. Even the people shown standing in **Figure 13** require energy conversions to stand still.

Energy Conversions in Your Body The complex chemical and physical processes going on in your body also obey the law of conservation of energy. Your body stores energy in the form of fat and other chemical compounds. This chemical potential energy is used to fuel the processes that keep you alive, such as making your heart beat and digesting the food you eat. Your body also converts this energy to heat that is transferred to your surroundings, and you use this energy to make your body move. **Table 1** shows the amount of energy used in doing various activities. To maintain a healthy weight, you must have a proper balance between energy contained in the food you eat and the energy your body uses.

Figure 13 The runners convert the energy stored in their bodies more rapidly than the spectators do. **Calculate** *Use Table 1 to calculate how long a person would need to stand to burn as much energy as a runner burns in 1 h.*

Food Energy Your body has been busy breaking down your breakfast into molecules that can be used as fuel. The chemical potential energy in these molecules supplies the cells in your body with the energy they need to function. Your body also can use the chemical potential energy stored in fat for its energy needs. The food Calorie (C) is a unit used by nutritionists to measure how much energy you get from various foods—1 C is equivalent to about 4,184 J. Every gram of fat a person consumes can supply 9 C of energy. Carbohydrates and proteins each supply about 4 C of energy per gram.

Look at the labels on food packages. They provide information about the Calories contained in a serving, as well as the amounts of protein, fat, and carbohydrates.

Table 1 Calories Used in 1 h

Type of Activity	Body Frames		
	Small	Medium	Large
Sleeping	48	56	64
Sitting	72	84	96
Eating	84	98	112
Standing	96	112	123
Walking	180	210	240
Playing tennis	380	420	460
Bicycling (fast)	500	600	700
Running	700	850	1,000

section 2 review

Summary

Energy Transformations

- Energy can be transformed from one form to another.
- Devices such as lightbulbs, hair dryers, and automobile engines convert one form of energy into other forms.
- The mechanical energy of a system is the sum of the kinetic and potential energy in the system:

  mechanical energy = KE + PE

- In falling, projectile motion, and swings, kinetic and potential energy are transformed into each other and the mechanical energy doesn't change.

The Law of Conservation of Energy

- According to the law of conservation of energy, energy cannot be created or destroyed.
- Friction converts mechanical energy into thermal energy.
- Fission and fusion are nuclear reactions that convert a small amount of mass in a nucleus into an enormous amount of energy.

Self Check

1. **Explain** how friction affects the mechanical energy of a system.
2. **Describe** the energy transformations that occur as you coast down a long hill on a bicycle and apply the brakes, causing the brake pads and bicycle rims to feel warm.
3. **Explain** how energy is conserved when nuclear fission or fusion occurs.
4. **Think Critically** A roller coaster is at the top of a hill and rolls to the top of a lower hill. If mechanical energy is conserved, on the top of which hill is the kinetic energy of the roller coaster larger?

Applying Math

5. **Calculate Kinetic Energy** The potential energy of a swing is 200 J at its highest point and 50 J at its lowest point. If mechanical energy is conserved, what is the kinetic energy of the swing at its lowest point?
6. **Calculate Thermal Energy** The mechanical energy of a bicycle at the top of a hill is 6,000 J. The bicycle stops at the bottom of the hill by applying the brakes. If the potential energy of the bicycle is 2,000 J at the bottom of the hill, how much thermal energy was produced?

Swinging Energy

⊙ Real-World Question

Imagine yourself swinging on a swing. What would happen if a friend grabbed the swing's chains as you passed the lowest point? Would you come to a complete stop or continue rising to your previous maximum height? How does the motion and maximum height reached by a swing change if the swing is interrupted?

⊙ Form a Hypothesis

Examine the diagram on this page. How is it similar to the situation in the introductory paragraph? An object that is suspended so that it can swing back and forth is called a pendulum. Hypothesize what will happen to the pendulum's motion and final height if its swing is interrupted.

Goals

■ **Construct** a pendulum to compare the exchange of potential and kinetic energy when a swing is interrupted.

■ **Measure** the starting and ending heights of the pendulum.

Possible Materials

ring stand
test-tube clamp
support-rod clamp, right angle
30-cm support rod
2-hole, medium rubber stopper
string (1 m)
metersticks
graph paper

Safety Precautions

WARNING: *Be sure the base is heavy enough or well anchored so that the apparatus will not tip over.*

◉ *Test Your Hypothesis*

Make a Plan

1. As a group, write your hypothesis and list the steps that you will take to test it. Be specific. Also list the materials you will need.

2. **Design** a data table and place it in your Science Journal.

3. Set up an apparatus similar to the one shown in the diagram.

4. **Devise** a way to measure the starting and ending heights of the stopper. Record your starting and ending heights in a data table. This will be your control.

5. **Decide** how to release the stopper from the same height each time.

6. Be sure you test your swing, starting it above and below the height of the cross arm. How many times should you repeat each starting point?

Follow Your Plan

1. Make sure your teacher approves your plan before you start.

2. Carry out the approved experiment as planned.

3. While the experiment is going on, write any observations that you make and complete the data table in your Science Journal.

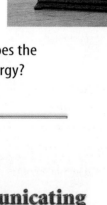

◉ *Analyze Your Data*

1. When the stopper is released from the same height as the cross arm, is the ending height of the stopper exactly the same as its starting height? Use your data to support your answer.

2. **Analyze** the energy transfers. At what point along a single swing does the stopper have the greatest kinetic energy? The greatest potential energy?

◉ *Conclude and Apply*

1. **Explain** Do the results support your hypothesis?

2. **Compare** the starting heights to the ending heights of the stopper. Is there a pattern? Can you account for the observed behavior?

3. **Discuss** Do your results support the law of conservation of energy? Why or why not?

4. **Infer** What happens if the mass of the stopper is increased? Test it.

Communicating
Your Data

Compare your conclusions with those of the other lab teams in your class. **For more help, refer to the** Science Skill Handbook.

The Impossible Dream

A machine that keeps on going? It has been tried for hundreds of years.

Many people have tried throughout history—and failed—to build perpetual-motion machines. In theory, a perpetual-motion machine would run forever and do work without a continual source of energy. You can think of it as a car that you could fill up once with gas, and the car would run forever. Sound impossible? It is!

Science Puts Its Foot Down

For hundreds of years, people have tried to create perpetual-motion machines. But these machines won't work because they violate two of nature's laws. The first law is the law of conservation of energy, which states that energy cannot be created or destroyed. It can change form—say,

from mechanical energy to electrical energy—but you always end up with the same amount of energy that you started with.

How does that apply to perpetual-motion machines? When a machine does work on an object, the machine transfers energy to the object. Unless that machine gets more energy from somewhere else, it can't keep doing work. If it did, it would be creating energy.

The second law states that heat by itself always flows from a warm object to a cold object. Heat will only flow from a cold object to a warm object if work is done. In the process, some heat always escapes.

To make up for these energy losses, energy constantly needs to be transferred to the machine. Otherwise, it stops. No perpetual motion. No free electricity. No devices that generate more energy than they use. No engine motors that run forever without refueling. Some laws just can't be broken.

Visitors look at the Keely Motor, the most famous perpetual-motion machine fraud of the late 1800s.

Analyze Using your school or public-library resources, locate a picture or diagram of a perpetual-motion machine. Figure out why it won't run forever. Explain to the class what the problem is.

Science Online

For more information, visit gpscience.com/time

Reviewing Main Ideas

The Nature of Energy

1. Energy is the ability to cause change.

2. Energy can have different forms, including kinetic, potential, and thermal energy.

3. Moving objects have kinetic energy that depends on the object's mass and velocity, and can be calculated from this equation:

$$KE = \frac{1}{2}\,mv^2$$

4. Potential energy is stored energy. An object can have gravitational potential energy that depends on its mass and its height, and is given by this equation:

$$GPE = mgh$$

Conservation of Energy

1. Energy can change from one form to another. Devices you use every day transform one form of energy into other forms that are more useful.

2. Falling, swinging, and projectile motion all involve transformations between kinetic energy and gravitational potential energy.

3. The total amount of kinetic energy and gravitational potential energy in a system is the mechanical energy of the system:

$$mechanical\ energy = KE + GPE$$

4. The law of conservation of energy states that energy never can be created or destroyed. The total amount of energy in the universe is constant.

5. Friction converts mechanical energy into thermal energy, causing the mechanical energy of a system to decrease.

6. Mass is converted into energy in nuclear fission and fusion reactions. Fusion and fission occur in the nuclei of certain atoms, and release tremendous amounts of energy.

FOLDABLES Use the Foldable you made at the beginning of this chapter to review what you learned about energy.

Using Vocabulary

chemical potential energy
p. 103
elastic potential energy
p. 103
gravitational potential
energy p. 104

joule p. 102
kinetic energy p. 102
law of conservation of
energy p. 111
mechanical energy p. 108
potential energy p. 103

Complete each statement using a word(s) from the vocabulary list above.

1. If friction can be ignored, the _____ of a system doesn't change.

2. The energy stored in a compressed spring is _____.

3. The _____ is the SI unit for energy.

4. When a book is moved from a higher shelf to a lower shelf, its _____ changes.

5. The muscles of a runner transform chemical potential energy into _____.

6. According to the _____ the amount of energy in the universe doesn't change.

Checking Concepts

Choose the word or phrase that best answers the question.

7. What occurs when energy is transferred from one object to another?
   A) an explosion
   B) a chemical reaction
   C) nuclear fusion
   D) a change

8. For which of the following is kinetic energy converted into potential energy?
   A) a boulder rolls down a hill
   B) a ball is thrown upward
   C) a swing comes to a stop
   D) a bowling ball rolls horizontally

9. The gravitational potential energy of an object changes when which of the following changes?
   A) the object's speed
   B) the object's mass
   C) the object's temperature
   D) the object's length

10. Friction causes mechanical energy to be transformed into which of these forms?
    A) thermal energy   C) kinetic
    B) nuclear energy   D) potential

11. The kinetic energy of an object changes when which of the following changes?
    A) the object's chemical potential energy
    B) the object's volume
    C) the object's direction of motion
    D) the object's speed

12. When an energy transformation occurs, which of the following is true?
    A) Mechanical energy doesn't change.
    B) Mechanical energy is lost.
    C) The total energy doesn't change.
    D) Mass is converted into energy.

Interpreting Graphics

13. Copy and complete the following concept map on energy.

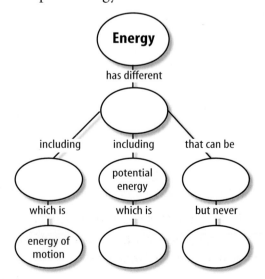

Science Online   gpscience.com/vocabulary_puzzlemaker

Use the table below to answer question 14.

Toy Cars Rolling Down Ramps			
Ramp Height (m)	Speed at Bottom (m/s)	GPE (J)	KE (J)
0.50	3.13		
0.75	3.83		
1.00	4.43		

14. **Make and Use Tables** Three toy cars, each with a mass of 0.05 kg, roll down ramps with different heights. The height of each ramp and the speed of each car at the bottom of each ramp is given in the table. Copy and complete the table by calculating the GPE of each car at the top of the ramp and the KE for each car at the bottom of the ramp to two decimal places. How do the values of GPE and KE you calculate compare?

Use the graph below to answer questions 15–17.

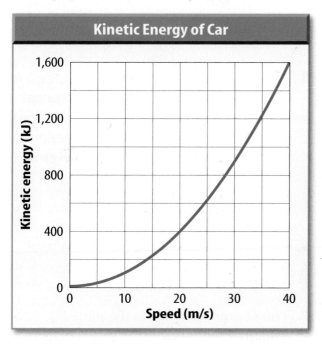

Kinetic Energy of Car

15. When the car's speed doubles from 20 m/s to 40 m/s, by how many times does the car's kinetic energy increase?

16. Using the graph, estimate the car's kinetic energy at a speed of 50 m/s.

17. If the car's kinetic energy at a speed of 20 m/s is 400 kJ, what is the car's kinetic energy at a speed of 10 m/s?

Thinking Critically

18. **Describe** the energy changes that occur in a swing. Explain how energy is conserved as the swing slows down and stops.

19. **Explain** why the law of conservation of energy must also include changes in mass.

20. **Infer** why the tires of a car get hot when the car is driven.

21. **Diagram** On a cold day you rub your hands together to make them warm. Diagram the energy transformations that occur, starting with the chemical potential energy stored in your muscles.

Applying Math

22. **Calculate Kinetic Energy** What is the kinetic energy of a 0.06-kg tennis ball traveling at a speed of 150 m/s?

23. **Calculate Potential Energy** A boulder with a mass of 2,500 kg rests on a ledge 200 m above the ground. What is the boulder's potential energy?

24. **Calculate Mechanical Energy** What is the mechanical energy of a 500-kg roller-coaster car moving with a speed of 3 m/s at the top of hill that is 30 m high?

25. **Calculate Speed** A boulder with a mass of 2,500 kg on a ledge 200 m above the ground falls. If the boulder's mechanical energy is conserved, what is the speed of the boulder just before it hits the ground?

Part 1 | Multiple Choice

Record your answers on the answer sheet provided by your teacher or on a sheet of paper.

1. What is the potential energy of a 5.0-kg object located 2.0 m above the ground?
A. 2.5 J
C. 98 J
B. 10 J
D. 196 J

Use the figure below to answer questions 2–4.

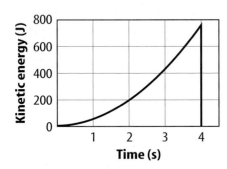

Kinetic Energy of Falling Rock

2. According to the graph, which of the following is the best estimate for the kinetic energy of the rock after it has fallen for 1 s?
A. 100 J
C. 200 J
B. 50 J
D. 0 J

3. According to the graph, which of the following is the best estimate for the potential energy of the rock before it fell?
A. 400 J
C. 200 J
B. 750 J
D. 0 J

4. If the rock has a mass of 1 kg, which of the following is the speed of the rock after it has fallen for 2 s?
A. 10 m/s
C. 20 m/s
B. 100 m/s
D. 200 m/s

5. Which of the following describes the energy conversions in a car's engine?
A. chemical to thermal to mechanical
B. chemical to electrical to mechanical
C. thermal to mechanical to chemical
D. kinetic to potential to mechanical

6. What is the difference in the gravitational potential energy of a 7.75 kg book that is 1.50 m above the ground and a 9.53 kg book that is 1.75 m above the ground?
A. 0.28 J
C. 11.7 J
B. 5.1 J
D. 49.5 J

7. A box with a mass of 14.8 kg sits on the floor. How high would you have to lift the box to for it to have a gravitational potential energy of 355 J?
A. 1.62 m
C. 2.45 m
B. 2.40 m
D. 4.90 m

Use the figure below to answer questions 8 and 9.

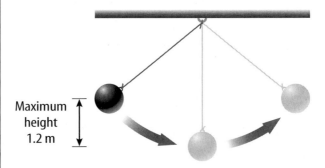

Maximum height 1.2 m

8. At its highest point, the pendulum is 1.2 m above the ground and has a gravitational potential energy of 62 J. If the gravitational potential energy is 10 J at its lowest point, what is the pendulum's kinetic energy at this point?
A. 0 J
C. 62 J
B. 31 J
D. 52 J

9. What is the mass of the pendulum bob?
A. 2.7 kg
C. 6.3 kg
B. 5.3 kg
D. 52 kg

10. The SI unit of energy is the joule (J). Which of the following is an equivalent way of expressing this unit?
A. $kg \cdot m$
C. $kg \cdot m^2/s^2$
B. $kg \cdot m/s$
D. $kg \cdot m/s^2$

Part 2 | Short Response/Grid In

Record your answers on the answer sheet provided by your teacher or on a sheet of paper.

11. Explain why the law of conservation of energy also includes mass when applied to nuclear reactions.

12. A student walks to school at a speed of 1.2 m/s. If the student's mass is 53 kg, what is the student's kinetic energy?

13. A book sliding across a horizontal table slows down and comes to a stop. The book's kinetic energy was converted into what form of energy?

14. Electrical energy was converted into which forms of energy by a hair dryer?

Use the figure below to answer questions 15 and 16.

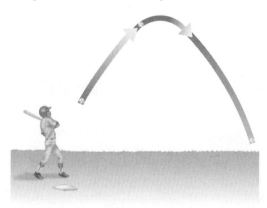

15. At what point on the ball's path is the ball's kinetic energy lowest but its gravitational potential energy highest?

16. How does the mechanical energy of the ball change from the moment just after the batter hits it to the moment just before it touches the ground?

17. Explain whether it is possible for an object at rest to have energy.

18. Find the speed of a 5.6-kg bowling ball that has a kinetic energy of 25.2 J.

Part 3 | Open Ended

Record your answers on a sheet of paper.

Use the figure below to answer questions 19 and 20.

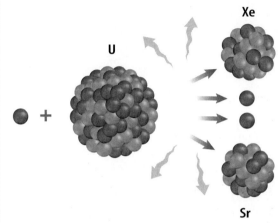

19. Describe the process shown in the figure above, and explain how it obeys the law of conservation of energy.

20. Describe how the total mass of the particles before the reaction occurs compares to the total mass of the particles produced by the reaction.

21. Is the mechanical energy of a liter of water at the top of a waterfall greater than, the same as, or less than the mechanical energy of a liter of water just before it reaches the bottom of the waterfall? Explain.

22. Name and describe some examples of how different forms of energy can be stored.

23. Describe a process in which energy travels through the environment and changes from one form to another.

Test-Taking Tip

Show All your Work For constructed response questions, show all your work and any calculations on your answer sheet.

Question 19 On your answer sheet, list the energy changes that occur during each for each step of the process that you can think of.

Work and Machines

It Seemed Longer Going Up

Have you ever thought of a mountain bike as a machine? A mountain bike actually is a combination of simple machines. Like all machines, a bicycle makes doing work easier. A mountain bike, for example, gets you uphill and downhill faster than you can travel by walking or running.

Science Journal Diagram a bicycle and identify the parts you think are simple machines.

Start-Up Activities

Doing Work with a Simple Machine

Did you know you can lift several times your own weight with the help of a pulley? Before the hydraulic lift was invented, a car mechanic used pulleys to raise a car off the ground. In this lab you'll see how a pulley can increase a force.

1. Tie a rope several meters in length to the center of a broom handle. Have one student hold both ends of the handle.
2. Have another student hold the ends of a second broom handle and face the first student as shown in the photo.
3. Have a third student loop the free end of the rope around the second handle, making six or seven loops.
4. The third student should stand to the side of one of the handles and pull on the free end of the rope. The two students holding the broom handles should prevent the handles from coming together.
5. **Think Critically** Describe how the applied force was changed. What would happen if the number of rope loops were increased?

Work and Machines Make the following Foldable to help you understand how machines make doing work easier.

STEP 1 Fold a vertical sheet of paper in half from top to bottom.

STEP 2 Fold in half from side to side with the fold at the top.

STEP 3 Unfold the paper once. Cut only the fold of the top flap to make two tabs.

STEP 4 Turn the paper vertically and label the front tabs as shown.

Work Without Machines

Work With Machines

List Before you read the chapter, list five examples of work you do without machines, and five examples of work you do with machines. Next to each example, rate the effort needed to do the work on a scale of 1 (little effort) to 3 (much effort).

Preview this chapter's content and activities at gpscience.com

Work

***What* You'll Learn**

- **Explain** the meaning of work.
- **Describe** how work and energy are related.
- **Calculate** work.
- **Calculate** power.

***Why* It's Important**

Doing work is another way of transferring energy from one place to another.

Review Vocabulary

energy: the ability to cause change

New Vocabulary

- work
- power

What is work?

Have you done any work today? To many people, the word *work* means something they do to earn money. In that sense, work can be anything from filling fast-food orders or loading trucks to teaching or doing word processing on a computer. The word *work* also means exerting a force with your muscles. Someone might say they have done work when they push as hard as they can against a wall that doesn't move. However, in science the word *work* is used in a different way.

Work Makes Something Move Press your hand against the surface of your desk as hard as you can. Have you done any work? The answer is no, no matter how tired your effort makes you feel. Remember that a force is a push or a pull. In order for work to be done, a force must make something move. **Work** is the transfer of energy that occurs when a force makes an object move. If you push against the desk and nothing moves, then you haven't done any work.

Doing Work There are two conditions that have to be satisfied for work to be done on an object. One is that the applied force must make the object move, and the other is that the movement must be in the same direction as the applied force.

For example, if you pick up a pile of books from the floor as in **Figure 1,** you do work on the books. The books move upward, in the direction of the force you are applying. If you hold the books in your arms without moving the books, you are not doing work on the books. You're still applying an upward force to keep the books from falling, but no movement is taking place.

Figure 1 When you lift a stack of books, your arms apply a force upward and the books move upward. Because the force and distance are in the same direction, your arms have done work on the books.

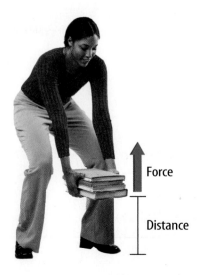

Force

Distance

Force and Direction of Motion When you carry books while walking, like the student in **Figure 2,** you might think that your arms are doing work. After all, you are exerting a force on the books with your arms, and the books are moving. Your arms might even feel tired. However, in this case the force exerted by your arms does no work on the books. The force exerted by your arms on the books is upward, but the books are moving horizontally. The force you exert is at right angles to the direction the books are moving. As a result, your arms exert no force in the direction the books are moving.

Reading Check *How are an applied force and an object's motion related when work is done?*

Work and Energy

How are work and energy related? When work is done, a transfer of energy always occurs. This is easy to understand when you think about how you feel after carrying a heavy box up a flight of stairs. Remember that when the height of an object above Earth's surface increases, the potential energy of the object increases. You transferred energy from your moving muscles to the box and increased its potential energy by increasing its height.

You may recall that energy is the ability to cause change. Another way to think of energy is that energy is the ability to do work. If something has energy, it can transfer energy to another object by doing work on that object. When you do work on an object, you increase its energy. The student carrying the box in **Figure 3** transfers chemical energy in his muscles to the box. Energy is always transferred from the object that is doing the work to the object on which the work is done.

Force

Distance

Figure 2 If you hold a stack of books and walk forward, your arms are exerting a force upward. However, the distance the books move is horizontal. Therefore your arms are not doing work on the books.

Figure 3 By carrying a box up the stairs, you are doing work. You transfer energy to the box. **Explain** *how the energy of the box changes as the student climbs the stairs.*

Calculating Work The amount of work done depends on the amount of force exerted and the distance over which the force is applied. When a force is exerted and an object moves in the direction of the force, the amount of work done can be calculated as follows.

Work Equation

work (in joules) = applied force (in newtons) $\times$ distance (in meters)

$$W = Fd$$

In this equation, force is measured in newtons and distance is measured in meters. Work, like energy, is measured in joules. One joule is about the amount of work required to lift a baseball a vertical distance of 0.7 m.

Applying Math   Solve a Simple Equation

CALCULATING WORK You push a refrigerator with a force of 100 N. If you move the refrigerator a distance of 5 m, how much work do you do?

IDENTIFY known values and the unknown value

Identify the known values:

you push a refrigerator with a force of 100 N   means⟩ $F = 100$ N

you move the refrigerator a distance of 5 m   means⟩ $d = 5$ m

Identify the unknown value:

how much work do you do?   means⟩ $W = ?$ J

SOLVE the problem

Substitute the known values $F = 100$ N and $d = 5$ m into the work equation.

$$W = Fd = (100 \text{ N}) (5 \text{ m}) = 500 \text{ Nm} = 500 \text{ J}$$

CHECK your answer

Does your answer seem reasonable? Check your answer by dividing the work you calculated by the force given in the problem. The result should be the distance given in the problem.

Practice Problems

1. A force of 75 N is exerted on a 45-kg couch and the couch is moved 5 m. How much work is done in moving the couch?

2. A lawn mower is pushed with a force of 80 N. If 12,000 J of work are done in mowing a lawn, what is the total distance the lawn mower was pushed?

3. The brakes on a car do 240,000 J of work in stopping the car. If the car travels a distance of 50 m while the brakes are being applied, what is the total force the brakes exert on the car?

For more practice problems go to page 834, and visit gpscience.com/extra_problems.

Figure 4 A pitcher exerts a force on the ball to throw it to the catcher. After the ball leaves her hand, she no longer is exerting any force on the ball. She does work on the ball only while it is in her hand.

When is work done? Suppose you give a book a push and it slides along a table for a distance of 1 m before it comes to a stop. The distance you use to calculate the work you did is how far the object moves while the force is being applied. Even though the book moved 1 m, you do work on the book only while your hand is in contact with it. The distance in the formula for work is the distance the book moved while your hand was pushing on the book. As **Figure 4** shows, work is done on an object only when a force is being applied to the object.

Power

Suppose you and another student are pushing boxes of books up a ramp to load them into a truck. To make the job more fun, you make a game of it, racing to see who can push a box up the ramp faster. The boxes weigh the same, but your friend is able to push a box a little faster than you can. She moves a box up the ramp in 30 s. It takes you 45 s. You both do the same amount of work on the books because the boxes weigh the same and are moved the same distance. The only difference is the time it takes to do the work.

In this game, your friend has more power than you do. **Power** is the amount of work done in one second. It is a rate—the rate at which work is done.

✔ Reading Check *How is power related to work?*

Mini LAB

Calculating Your Work and Power

Procedure

1. Find a set of **stairs** that you can safely walk and run up. Measure the vertical height of the stairs in meters.
2. Record how many seconds it takes you to walk and run up the stairs.
3. Calculate the work you did in walking and running up the stairs using $W = Fd$. For force, use your weight in newtons (your weight in pounds times 4.5).
4. Use the formula $P = W/t$ to calculate the power you needed to walk and run up the stairs.

Analysis

1. Is the work you did walking and running the steps the same?
2. Which required more power—walking or running up the steps? Why?

Calculating Power Power is the rate at which work is done. To calculate power, divide the work done by the time that is required to do the work.

Power Equation

$$\text{Power (in watts)} = \frac{\text{work (in joules)}}{\text{time (in seconds)}}$$

$$P = \frac{W}{t}$$

The SI unit for power is the watt (W). One watt equals one joule of work done in one second. Because the watt is a small unit, power often is expressed in kilowatts. One kilowatt (kW) equals 1,000 W.

Applying Math   Solve a Simple Equation

CALCULATING POWER You do 900 J of work in pushing a sofa. If it took 5 s to move the sofa, what was your power?

IDENTIFY known values and the unknown value

Identify the known values:

| You do 900 J of work | means | $W = 900$ J |
| it took 5 s to move the sofa | means | $t = 5$ s |

Identify the unknown value:

| what was your power? | means | $P = ?$ W |

SOLVE the problem

Substitute the known values $W = 900$ J and $t = 5$ s into the power equation.

$$P = \frac{W}{t} = \frac{900 \text{ J}}{5 \text{ s}} = 180 \text{ J/s} = 180 \text{ W}$$

The symbol for work, *W*, is usually italicized. However, the abbreviation for watt, W, is not italicized.

CHECK your answer

Does your answer seem reasonable? Check your answer by multiplying the power you calculated by the time given in the problem. The result should be the work given in the problem.

Practice Problems

1. To lift a baby from a crib 50 J of work are done. How much power is needed if the baby is lifted in 0.5 s?

2. If a runner's power is 130 W, how much work is done by the runner in 10 minutes?

3. The power produced by an electric motor is 500 W. How long will it take the motor to do 10,000 J of work?

For more practice problems go to page 834, and visit gpscience.com/extra_problems.

Power and Energy Doing work is a way of transferring energy from object to another. Just as power is the rate at which work is done, power is also the rate at which energy is transferred. When energy is transferred, the power involved can be calculated by dividing the energy transferred by the time needed for the transfer to occur.

Power Equation for Energy Transfer

$$\text{power (in watts)} = \frac{\text{energy transferred (in joules)}}{\text{time (in seconds)}}$$

$$P = \frac{E}{t}$$

For example, when the lightbulb in **Figure 5** is connected to an electric circuit, energy is transferred from the circuit to the lightbulb filament. The filament converts the electrical energy supplied to the lightbulb into heat and light. The power used by the lightbulb is the amount of electrical energy transferred to the lightbulb each second.

Figure 5 This 100 W lightbulb converts electrical energy into light and heat at a rate of 100 J/s.

---

section 1 review

Summary

Work and Energy

- Work is done on an object when a force is exerted on the object and it moves in the direction of the force.

- If a force, *F*, is exerted on object while the object moves a distance, *d*, in the direction of the force, the work done is

$$W = Fd$$

- When work is done on an object, energy is transferred to the object.

Power

- Power is the rate at which work is done or energy is transferred.

- When work is done, power can be calculated from the equation

$$P = \frac{W}{t}$$

- When energy is transferred, power can be calculated from the equation

$$P = \frac{E}{t}$$

Self Check

1. **Explain** how the scientific definition of work is different from the everyday meaning.

2. **Describe** a situation in which a force is applied, but no work is done.

3. **Explain** how work and energy are related.

4. **Think Critically** In which of the following situations is work being done?
   a. A person shovels snow off a sidewalk.
   b. A worker lifts bricks, one at a time, from the ground to the back of a truck.
   c. A roofer's assistant carries a bundle of shingles across a construction site.

Applying Math

5. **Calculate Force** Find the force a person exerts in pulling a wagon 20 m if 1,500 J of work are done.

6. **Calculate Work** A car's engine produces 100 kW of power. How much work does the engine do in 5 s?

7. **Calculate Energy** A color TV uses 120 W of power. How much energy does the TV use in 1 hour?

Using Machines

Reading Guide

What **You'll Learn**

- **Explain** how machines make doing work easier.
- **Calculate** the mechanical advantage of a machine.
- **Calculate** the efficiency of a machine.

Why **It's Important**

Cars, stairs, and teeth are all examples of machines that make your life easier.

🔎 **Review Vocabulary**

force: a push or a pull

New Vocabulary

- machine
- input force
- output force
- mechanical advantage
- efficiency

What is a machine?

A **machine** is a device that makes doing work easier. When you think of a machine you may picture a device with an engine and many moving parts. However, machines can be simple. Some, like knives, scissors, and doorknobs, are used every day to make doing work easier.

Making Work Easier

Machines can make work easier by increasing the force that can be applied to an object. A screwdriver increases the force you apply to turn a screw. A second way that machines can make work easier is by increasing the distance over which a force can be applied. A leaf rake is an example of this type of machine.

Machines also can make work easier by changing the direction of an applied force. A simple pulley changes a downward force to an upward force.

Increasing Force A car jack, like the one in **Figure 6,** is an example of a machine that increases an applied force. The upward force exerted by the jack is greater than the downward force you exert on the handle. However, the distance you push the handle downward is greater than the distance the car is pushed upward. Because work is the product of force and distance, the work done by the jack is not greater than the work you do on the jack. The jack increases the applied force, but it doesn't increase the work done.

Figure 6 A car jack is an example of a machine that increases an applied force.

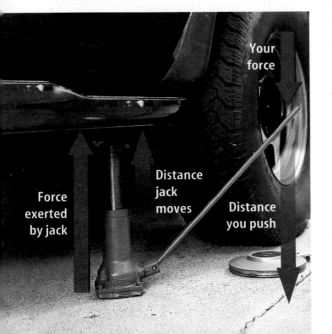

Your force

Force exerted by jack

Distance jack moves

Distance you push

Figure 7 Whether the mover slides the chair up the ramp or lifts it directly into the truck, she will do the same amount of work. Doing the work over a longer distance allows her to use less force.

Force and Distance Why does the mover in **Figure 7** push the heavy furniture up the ramp instead of lifting it directly into the truck? It is easier for her because less force is needed to move the furniture.

The work done in lifting an object depends on the change in height of the object. The same amount of work is done whether the mover pushes the furniture up the long ramp or lifts it straight up. If she uses a ramp to lift the furniture, she moves the furniture a longer distance than if she just raised it straight up. If work stays the same and the distance is increased, then less force will be needed to do the work.

Figure 8 An ax blade changes the direction of the force from vertical to horizontal.

✔ **Reading Check** *How does a ramp make lifting an object easier?*

Changing Direction Some machines change the direction of the force you apply. When you use the car jack, you are exerting a force downward on the jack handle. The force exerted by the jack on the car is upward. The direction of the force you applied is changed from downward to upward. Some machines change the direction of the force that is applied to them in another way. The wedge-shaped blade of an ax is one example. When you use an ax to split wood, you exert a downward force as you swing the ax toward the wood. As **Figure 8** shows, the blade changes the downward force into a horizontal force that splits the wood apart.

Resulting force

Applied force

Figure 9 A crowbar increases the force you apply and changes its direction.

Machines Multiplying Force

Procedure

1. Open a **can of food** using a **manual can opener.** *WARNING: Do not touch can opener's cutting blades or cut edges of the can's lid.*
2. Use a **metric ruler** to measure the diameter of the cutting blade of the can opener.
3. Measure the length of the handle you turn.

Analysis

1. Compare how difficult it is to open the can using the can opener with how difficult it would have been to open the can by turning the cutting blade with a smaller handle.
2. Compare the diameter of the cutting blade with the diameter of the circle formed by turning the can opener's handle.
3. Infer why a can opener makes it easier to open a metal can.

Try at Home

The Work Done by Machines

To pry the lid off a wooden crate with a crowbar, you'd slip the end of the crowbar under the edge of the crate lid and push down on the handle. By moving the handle downward, you do work on the crowbar. As the crowbar moves, it does work on the lid, lifting it up. **Figure 9** shows how the crowbar increases the amount of force being applied and changes the direction of the force.

When you use a machine such as a crowbar, you are trying to move something that resists being moved. For example, if you use a crowbar to pry the lid off a crate, you are working against the friction between the nails in the lid and the crate. You also could use a crowbar to move a large rock. In this case, you would be working against gravity—the weight of the rock.

Input and Output Forces Two forces are involved when a machine is used to do work. You exert a force on the machine, such as a bottle opener, and the machine then exerts a force on the object you are trying to move, such as the bottle cap. The force that is applied to the machine is called the **input force.** F_{in} stands for the input force. The force applied by the machine is called the **output force,** symbolized by F_{out}. When you try to pull a nail out with a hammer as in **Figure 10,** you apply the input force on the handle. The output force is the force the claw applies to the nail.

Two kinds of work need to be considered when you use a machine—the work done by you on the machine and the work done by the machine. When you use a crowbar, you do work when you apply force to the crowbar handle and make it move. The work done by you on a machine is called the input work and is symbolized by W_{in}. The work done by the machine is called the output work and is abbreviated W_{out}.

Conserving Energy Remember that energy is always conserved. When you do work on the machine, you transfer energy to the machine. When the machine does work on an object, energy is transferred from the machine to the object. Because energy cannot be created or destroyed, the amount of energy the machine transfers to the object cannot be greater than the amount of energy you transfer to the machine. A machine cannot create energy, so W_{out} is never greater than W_{in}.

However, the machine does not transfer all of the energy it receives to the object. In fact, when a machine is used, some of the energy transferred changes to heat due to friction. The energy that changes to heat cannot be used to do work, so W_{out} is always smaller than W_{in}.

Ideal Machines Remember that work is calculated by multiplying force by distance. The input work is the product of the input force and the distance over which the input force is exerted. The output work is the product of the output force and the distance over which that force is exerted.

Suppose a perfect machine could be built in which there was no friction. None of the input work or output work would be converted to heat. For such an ideal machine, the input work equals the output work. So for an ideal machine,

$$W_{in} = W_{out}$$

Suppose the ideal machine increases the force applied to it. This means that the output force, F_{out}, is greater than the input force, F_{in}. Recall that work is equal to force times distance. If F_{out} is greater than F_{in}, then W_{in} and W_{out} can be equal only if the input force is applied over a greater distance than the output force is exerted over.

For example, suppose the hammer claw in **Figure 10** moves a distance of 1 cm to remove a nail. If an output force of 1,500 N is exerted by the claw of the hammer, and you move the handle of the hammer 5 cm, you can find the input force as follows.

$$W_{in} = W_{out}$$
$$F_{in}\, d_{in} = F_{out}\, d_{out}$$
$$F_{in}\, (0.05\ \text{m}) = (1{,}500\ \text{N})\, (0.01\ \text{m})$$
$$F_{in}\, (0.05\ \text{m}) = 15\ \text{N·m}$$
$$F_{in} = 300\ \text{N}$$

Because the distance you move the hammer is longer than the distance the hammer moves the nail, the input force is less than the output force.

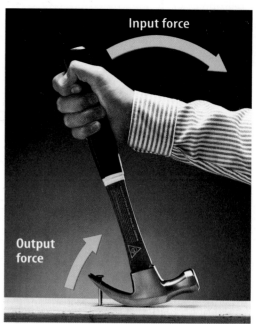

Figure 10 When prying a nail out of a piece of wood with a claw hammer, you exert the input force on the handle of the hammer, and the claw exerts the output force. **Describe** *how the hammer changes the input force.*

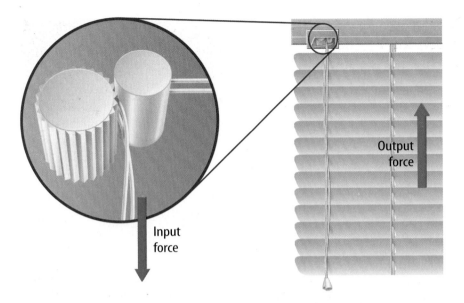

Mechanical Advantage

Machines like the car jack, the ramp, the crow bar, and the claw hammer make work easier by making the output force greater than the input force. The ratio of the output force to the input force is the **mechanical advantage** of a machine. The mechanical advantage of a machine can be calculated from the following equation.

Figure 11 Window blinds use a machine that changes the direction of an input force. A downward pull on the cord is changed to an upward force on the blinds. The input and output forces are equal, so the MA is 1.

Mechanical Advantage Equation

$$\text{mechanical advantage} = \frac{\text{output force (in newtons)}}{\text{input force (in newtons)}}$$

$$MA = \frac{F_{out}}{F_{in}}$$

Figure 11 shows that the mechanical advantage equals one when only the direction of the input force changes.

Ideal Mechanical Advantage The mechanical advantage of an a machine without friction is called the ideal mechanical advantage, or IMA. The IMA can be calculated by dividing the input distance by the output distance. For a real machine, the IMA would be the mechanical advantage of the machine if there were no friction.

Efficiency

For real machines, some of the energy put into a machine is always converted into heat by frictional forces. For that reason, the output work of a machine is always less than the work put into the machine.

Efficiency is a measure of how much of the work put into a machine is changed into useful output work by the machine. A machine with high efficiency produces less heat from friction so more of the input work is changed to useful output work.

Graphite Lubricant
Graphite is a solid that sometimes is used as a lubricant to increase the efficiency of machines. Find out what element graphite is made of and why graphite is a good lubricant.

 *Why is the output work always less than the input work for a real machine?*

Calculating Efficiency To calculate the efficiency of a machine, the output work is divided by the input work. Efficiency is usually expressed as a percentage by this equation:

Efficiency Equation

$$\text{efficiency (\%)} = \frac{\text{output work (in joules)}}{\text{input work (in joules)}} \times 100\%$$

$$efficiency = \frac{W_{out}}{W_{in}} \times 100\%$$

In an ideal machine there is no friction and the output work equals the input work. So the efficiency of an ideal machine is 100 percent. In a real machine, friction causes the output work to always be less than the input work. So the efficiency of a real machine is always less than 100 percent.

Increasing Efficiency Machines can be made more efficient by reducing friction. This usually is done by adding a lubricant, such as oil or grease, to surfaces that rub together, as shown in **Figure 12.** A lubricant fills in the gaps between the surfaces, enabling the surfaces to slide past each other more easily.

Figure 12 Oil reduces the friction between two surfaces. Oil fills the space between the surfaces so high spots don't rub against each other.

---

section 2 review

Summary

Work and Machines

- Machines make doing work easier by changing the applied force, changing the distance over which the force is applied, or changing the direction of the applied force.

- Because energy cannot be created or destroyed, the output work cannot be greater than the input work.

- In a real machine, some of the input work is converted into heat by friction.

Mechanical Advantage and Efficiency

- The mechanical advantage of a machine is the output force divided by the input force:
$$MA = \frac{F_{out}}{F_{in}}$$

- The efficiency of a machine is the output work divided by the input work times 100%:
$$efficiency = \frac{W_{out}}{W_{in}} \times 100\%$$

Self Check

1. **Describe** the circumstances for which the output work would equal the input work in a machine.

2. **Infer** how lubricating a machine affects the output force exerted by the machine.

3. **Explain** why in a real machine the output work is always less than the input work.

4. **Think Critically** The mechanical advantage of a machine is less than one. Compare the distances over which the input and output forces are applied.

Applying Math

5. **Calculate** the mechanical advantage of a hammer if the input force is 125 N and the output force is 2,000 N.

6. **Calculate Efficiency** Find the efficiency of a machine that does 800 J of work if the input work is 2,400 J.

7. **Calculate Force** Find the force needed to lift a 2,000-N weight using a machine with a mechanical advantage of 15.

Simple Machines

Reading Guide

What You'll Learn

- **Describe** the six types of simple machines.
- **Explain** how the different types of simple machines make doing work easier.
- **Calculate** the ideal mechanical advantage of the different types of simple machines.

Why It's Important

All complex machines, such as cars, are made of simple machines. Even your body contains simple machines.

Review Vocabulary

compound: composed of separate elements or parts

New Vocabulary

- simple machine
- lever
- pulley
- wheel and axle
- inclined plane
- screw
- wedge
- compound machine

Figure 13 There are three classes of levers.

First-class Lever
The fulcrum is between the input force and the output force.

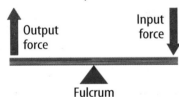

Second-class Lever
The output force is between the fulcrum and the input force.

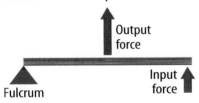

Third-class Lever
The input force is between the fulcrum and the output force.

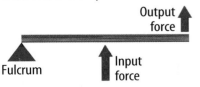

Types of Simple Machines

If you cut your food with a knife, or use a screwdriver, or even chew your food, you are using a simple machine. A **simple machine** is a machine that does work with only one movement of the machine. There are six types of simple machines: lever, pulley, wheel and axle, inclined plane, screw, and wedge. The pulley and the wheel and axle are modified levers, and the screw and the wedge are modified inclined planes.

Levers

You've used a lever if you've used a wheelbarrow, or a lawn rake, or swung a baseball bat. A **lever** is a bar that is free to pivot or turn around a fixed point. The fixed point the lever pivots on is called the fulcrum. The input arm of the lever is the distance from the fulcrum to the point where the input force is applied. The output arm is the distance from the fulcrum to the point where the output force is exerted by the lever.

The output force produced by a lever depends on the lengths of the input arm and the output arm. If the output arm is longer than the input arm, the law of conservation of energy requires that the output force be less than the input force. If the output arm is shorter than the input arm, then the output force is greater than the input force.

There are three classes of levers, as shown in **Figure 13.** The differences among the three classes of levers depend on the locations of the fulcrum, the input force, and the output force.

Figure 14 Levers are classified by the location of the input force, output force, and the fulcrum.

A The screwdriver is being used as a first-class lever. The fulcrum is the paint can rim.

B A wheelbarrow is a second-class lever. The fulcrum is the wheel.

First-Class Lever

The screwdriver used to open the paint can in **Figure 14A** is an example of a first-class lever. For a first-class lever, the fulcrum is located between the input and output forces. The output force is always in the opposite direction to the input force in a first-class lever.

Second-Class Lever

For a second-class lever, the output force is located between the input force and the fulcrum. Look at the wheelbarrow in **Figure 14B.** You apply an upward input force on the handles, and the wheel is the fulcrum. The output force is exerted between the input force and fulcrum. For a second-class lever, the output force is always greater than the input force.

Third-Class Lever

Many pieces of sports equipment, such as a baseball bat, are third-class levers. For a third-class lever, the input force is applied between the output force and the fulcrum. The right-handed batter in **Figure 14C,** applies the input force with the right hand and the left hand is the fulcrum. The output force is exerted by the bat above the right hand. The output force is always less than the input force in a third-class lever. Instead, the distance over which the output force is applied is increased.

Every lever can be placed into one of these classes. Each class can be found in your body, as shown in **Figure 15** on the next page.

C A baseball bat is a third-class lever. The fulcrum here is this batter's left hand.

Figure 15

▲ Fulcrum
▼ Input force
▲ Output force

All three types of levers—first-class, second-class, and third-class—are found in the human body. The forces exerted by muscles in your body can be increased by first-class and second-class levers, while third-class levers increase the range of movement of a body part. Examples of how the body uses levers to help it move are shown here.

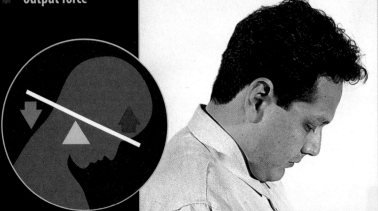

▲ **FIRST-CLASS LEVER** The fulcrum lies between the input force and the output force. Your head acts like a first-class lever. Your neck muscles provide the input force to support the weight of your head.

◀ **SECOND-CLASS LEVER** The output force is between the fulcrum and the input force. Your foot becomes a second-class lever when you stand on your toes.

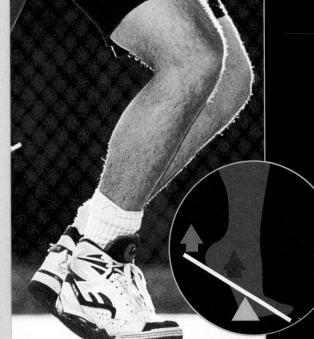

▶ **THIRD-CLASS LEVER** The input force is between the fulcrum and the output force. A third-class lever increases the range of motion of the output force. When you do a curl with a dumbbell, your forearm is a third-class lever.

Figure 16 A fixed pulley is another form of the lever. **Infer** *the lengths of the input arm and output arm in a pulley.*

Ideal Mechanical Advantage of a Lever The ideal mechanical advantage, or IMA, can be calculated for any machine by dividing the input distance by the output distance. For a lever, the input distance is the length of the input arm and the output distance is the length of the output arm. The IMA of a lever can be calculated from this equation:

> **Ideal Mechanical Advantage of a Lever**
>
> $$\text{ideal mechanical advantage} = \frac{\text{length of input arm (m)}}{\text{length of output arm (m)}}$$
>
> $$\text{IMA} = \frac{L_{\text{in}}}{L_{\text{out}}}$$

Pulleys

A **pulley** is a grooved wheel with a rope, chain, or cable running along the groove. A fixed pulley is a modified first-class lever, as shown in **Figure 16.** The axle of the pulley acts as the fulcrum. The two sides of the pulley are the input arm and output arm. A pulley can change the direction of the input force or increase input force, depending on whether the pulley is fixed or movable. A system of pulleys can change the direction of the input force and make it larger.

Fixed Pulleys The cable attached to an elevator passes over a fixed pulley at the top of the elevator shaft. A fixed pulley, such as the one in **Figure 17,** is attached to something that doesn't move, such as a ceiling or wall. Because a fixed pulley changes only the direction of force, the IMA is 1.

Figure 17 A fixed pulley changes only the direction of your force. You need to apply an input force of 4 N to lift the 4-N weight.

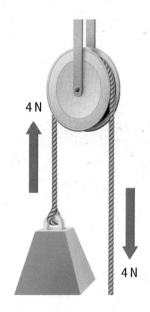

4 N

4 N

Figure 18 A movable pulley and a pulley system called a block and tackle reduce the force needed to lift a weight.

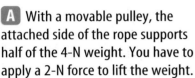

A With a movable pulley, the attached side of the rope supports half of the 4-N weight. You have to apply a 2-N force to lift the weight.

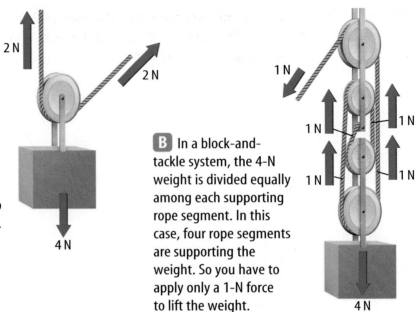

2 N

2 N

4 N

B In a block-and-tackle system, the 4-N weight is divided equally among each supporting rope segment. In this case, four rope segments are supporting the weight. So you have to apply only a 1-N force to lift the weight.

1 N

1 N   1 N

1 N   1 N

4 N

Movable Pulleys A pulley in which one end of the rope is fixed and the wheel is free to move is called a movable pulley. Unlike a fixed pulley, a movable pulley does multiply force. Suppose a 4-N weight is hung from the movable pulley in **Figure 18A.** The ceiling acts like someone helping you to lift the weight. The rope attached to the ceiling will support half of the weight—2 N. You need to exert only the other half of the weight—2 N—in order to support and lift the weight. The output force exerted on the weight is 4 N, and the applied input force is 2 N. Therefore the IMA of the movable pulley is 2.

For a fixed pulley, the distance you pull the rope downward equals the distance the weight moves upward. For a movable pulley, the distance you pull the rope upward is twice the distance the weight moves upward.

✔ Reading Check *How does a movable pulley reduce the input force needed to lift a weight?*

The Block and Tackle A system of pulleys consisting of fixed and movable pulleys is called a block and tackle. **Figure 18B** shows a block and tackle made up of two fixed pulleys and two movable pulleys. If a 4-N weight is suspended from the movable pulley, each rope segment supports one fourth of the weight, reducing the input force to 1 N. The IMA of a pulley system is equal to the number of rope segments that support the weight. The block and tackle shown in **Figure 18B** has a IMA of 4. The IMA of a block and tackle can be increased by increasing the number of pulleys in the pulley system.

Figure 19 The handle on a pencil sharpener is part of a wheel and axle. You apply a force to the handle. This force is made larger by the wheel and axle, making it easy to turn the sharpening mechanism.

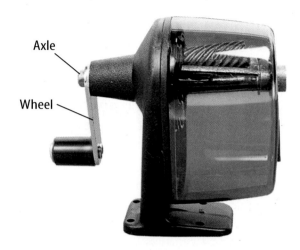

Axle

Wheel

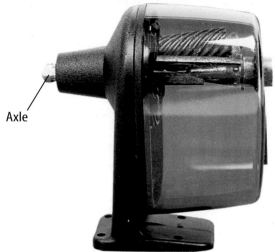

Axle

Wheel and Axle

Could you use the pencil sharpener in **Figure 19** if the handle weren't attached? The handle on the pencil sharpener is part of a wheel and axle. A **wheel and axle** is a simple machine consisting of a shaft or axle attached to the center of a larger wheel, so that the wheel and axle rotate together. Doorknobs, screwdrivers, and faucet handles are examples of wheel and axles. Usually the input force is applied to the wheel, and the output force is exerted by the axle.

Mechanical Advantage of the Wheel and Axle A wheel and axle is another modified lever. The center of the axle is the fulcrum. The input force is applied at the rim of the wheel. So the length of the input arm is the radius of the wheel. The output force is exerted at the rim of the axle. So the length of the output arm is the radius of the axle. The ideal mechanical advantage of a lever is the length of the input arm divided by the length of the output arm. So the IMA of a wheel and axle is given by this equation:

Ideal Mechanical Advantage of Wheel and Axle

$$\text{ideal mechanical advantage} = \frac{\text{radius of wheel (m)}}{\text{radius of axle (m)}}$$

$$\text{IMA} = \frac{r_w}{r_a}$$

According to this equation, the IMA of a wheel and axle can be increased by increasing the radius of the wheel.

Gears A gear is a wheel and axle with the wheel having teeth around its rim. When the teeth of two gears interlock, turning one gear causes the other gear to turn.

When two gears of different sizes are interlocked, they rotate at different rates. Each rotation of the larger gear causes the smaller gear to make more than one rotation. If the input force is applied to the larger gear, the output force exerted by smaller gear is less than the input force.

Gears also may change the direction of the force as shown in **Figure 20.** When the larger gear in is rotated clockwise, the smaller gear rotates counterclockwise.

Figure 20 If an input force is applied to the larger gear, and it rotates clockwise, the smaller gear rotates counterclockwise. The output force exerted by the smaller gear is less than the input force applied to the larger gear.

Inclined Planes

Why do the roads and paths on mountains zigzag? Would it be easier to climb directly up a steep incline or walk a longer path gently sloped around the mountain? A sloping surface, such as a ramp that reduces the amount of force required to do work, is an **inclined plane.**

Mechanical Advantage of an Inclined Plane You do the same work by lifting a box straight up or pushing it up an inclined plane. But by pushing the box up an inclined plane, the input force is exerted over a longer distance compared to lifting the box straight up. As a result the input force is less than the force needed to lift the box straight upward. The IMA of an inclined plane can be calculated from this equation.

Science_online_

Topic: Nanorobots
Visit gpscience.com for Web links to information about how tiny robots called nanorobots might be used as microsurgical instruments.

Activity Use the information you find to make a diagram of a nanorobot that might be used to perform surgery.

Ideal Mechanical Advantage of Inclined Plane

$$\text{ideal mechanical advantage} = \frac{\text{length of slope (m)}}{\text{height of slope (m)}}$$

$$\text{IMA} = \frac{l}{h}$$

The IMA of an inclined plane for a given height is increased by making the plane longer.

When you think of an inclined plane, you normally think of moving an object up a ramp—you move and the inclined plane remains stationary. The screw and the wedge, however, are variations of the inclined plane in which the inclined plane moves and the object remains stationary.

Figure 21 A screw has an inclined plane that wraps around the post of the screw.

The thread gets thinner farther from the post. This helps the screw force its way into materials.

Many lids, such as those on peanut butter jars, also contain threads.

The Screw

A **screw** is an inclined plane wrapped in a spiral around a cylindrical post. If you look closely at the screw in **Figure 21,** you'll see that the threads form a tiny ramp that runs upward from its tip. You apply the input force by turning the screw. The output force is exerted along the threads of the screw. The IMA of a screw is related to the spacing of the threads. The IMA is larger if the threads are closer together. However, if the IMA is larger, more turns of the screw are needed to drive it into some material.

How do you remove the lid off a jar of peanut butter, like in **Figure 21?** If you look closely, you see threads similar to the ones on the screw in **Figure 21.** Where else can you find examples of a screw?

The Wedge

Like the screw, the wedge is also a simple machine where the inclined plane moves through an object or material. A **wedge** is an inclined plane with one or two sloping sides. It changes the direction of the input force.

Look closely at the knife in **Figure 22.** One edge is sharp, and it slopes outward at both sides, forming an inclined plane. As it moves through the apple, the downward input force is changed to a horizontal force, forcing the apple apart.

Figure 22 A knife blade is a wedge. As you cut through the apple, it pushes the halves of the apple apart.

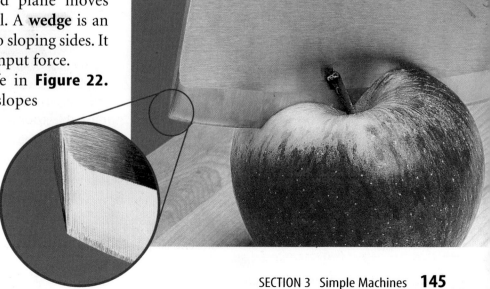

Figure 23 A compound machine, such as a can opener, is made up of simple machines.

Wheel and axle

Lever

Wedge

Compound Machines

Some of the machines you use every day are made up of several simple machines. Two or more simple machines that operate together form a **compound machine.**

Look at the can opener in **Figure 23.** To open the can you first squeeze the handles together. The handles act as a lever and increase the force applied on a wedge, which then pierces the can. You then turn the handle, a wheel and axle, to open the can.

A car is also a compound machine. Burning fuel in the cylinders of the engine causes the pistons to move up and down. This up-and-down motion makes the crankshaft rotate. The force exerted by the rotating crankshaft is transmitted to the wheels through other parts of the car, such as the transmission and the differential. Both of these parts contain gears, that can change the rate at which the wheels rotate, the force exerted by the wheels, and even reverse the direction of rotation.

section 3 review

Summary

The Lever Family

- A lever is a bar that is free to pivot about a fixed point called the fulcrum.
- There are three classes of levers based on the relative locations of the input force, output force, and the fulcrum.
- A pulley is a grooved wheel with a rope, chain, or cable placed in the groove and is a modified form of a lever.
- The IMA of a lever is the input arm divided by the output arm.
- A wheel and axle consists of a shaft or axle attached to the center of a larger wheel.

The Inclined Plane Family

- An inclined plane is a ramp or sloping surface that reduces the force needed to do work.
- The IMA of an inclined plane is the length of the plane divided by the height of the plane.
- A screw consists of an inclined plane wrapped around a shaft.
- A wedge is an inclined plane that moves and can have one or two sloping surfaces.

Self Check

1. **Classify** a screwdriver as one of the six types of simple machine. Explain how the IMA of screw driver could be increased.

2. **Determine** for which class of lever the output force is always greater than the input force. For which class is the output force always less than the input force?

3. **Make a diagram** of a bicycle and label the parts of a bicycle that are simple machines.

4. **Think Critically** Use the law of conservation of energy to explain why in a second-class lever the distance over which the input force is applied is always greater than the distance over which the output force is applied.

Applying Math

5. **Calculate IMA** What is the IMA of a car's steering wheel if the wheel has a diameter of 40 cm and the shaft it's attached to has a diameter of 4 cm?

6. **Calculate Output Arm Length** A lever has an IMA of 4. If the length of the input arm is 1.0 m, what is the length of the output arm?

7. **Calculate IMA** A 6.0 m ramp runs from a sidewalk to a porch that is 2.0 m above the sidewalk. What is the ideal mechanical advantage of this ramp?

Levers

Have you ever tried to balance a friend on a seesaw? If your friend was lighter, you had to move toward the fulcrum. In this lab, you will use the same method to measure the mass of a coin.

▶ Real-World Question

How can a lever be used to measure mass?

Goals

- ■ **Measure** the lengths of the input arm and output arm of a lever.
- ■ **Calculate** the ideal mechanical advantage of a lever.
- ■ **Determine** the mass of a coin.

Materials

stiff cardboard, 3 cm by 30 cm
coins (one quarter, one dime, one nickel)
balance
metric ruler

Safety Precautions 🥽 🧤

▶ Procedure

1. **Measure** the mass of each coin.
2. Mark a line 2 cm from one end of the cardboard strip. Label this line *Output*.
3. Slide the other end of the cardboard strip over the edge of a table until the strip begins to tip. Mark a line across the strip at the table edge and label this line *Input*.
4. **Measure** the mass of the strip to the nearest 0.1 g. Write this mass on the input line.
5. Center a dime on the output line. Slide the cardboard strip until it begins to tip. Mark the balance line. Label it *Fulcrum 1*.

6. **Measure** the lengths of the output and input arms to the nearest 0.1 cm.
7. Calculate the IMA of the lever. Multiply the IMA by the mass of the lever to find the approximate mass of the coin.
8. Repeat steps 5 through 7 with the nickel and the quarter. Mark the fulcrum line *Fulcrum 2* for the nickel and *Fulcrum 3* for the quarter.

▶ Conclude and Apply

1. **Explain** why there might be a difference between the mass of each coin measured by the balance and the mass measured using the lever.
2. **Explain** what provides the input and output force for the lever.
3. **Explain** why the IMA of the lever changes as the mass of the coin changes.

*C*ommunicating
Your Data

Compare your results with those of other students in your class. **For more help, refer to the** Science Skill Handbook.

Model and Invent

Using Simple Machines

Real-World Question

You are the contractor on a one-story building with a large air-conditioner. How can you get the air conditioner to the roof? How can you minimize the force needed to lift an object? What machines could you use? Consider a fixed pulley with ideal mechanical advantage (IMA) = 1, a movable pulley with IMA = 2, a block and tackle with one fixed double pulley and one movable double pulley with IMA = 4, and an inclined plane with IMA = slope / height = 4. How can you find the efficiency of machines?

Make the Model

1. Work in teams of at least two. **Collect** all the needed equipment.
2. Sketch a model for each lifting machine. **Model** the inclined plane with a board 40-cm long and raised 10 cm at one end. Include a control in which the weight is lifted while being suspended directly from the spring scale.
3. Make a table for data.
4. Is the pulley support high enough that the block and tackle can lift a weight 10 cm?
5. Obtain your teacher's approval of your sketches and data table before proceeding.

Goals

- **Model** lifting devices based on a block and tackle and on an inclined plane.
- **Calculate** the output work that will be accomplished.
- **Measure** the force needed by each machine to lift a weight.
- **Calculate** the input work and efficiency for each model machine.
- **Select** the best machine for your job based on the force required.

Possible Materials

spring scale, 0–10 N range
9.8-N weight (1 kg mass)
two double pulleys
string for pulleys
stand or support for the pulleys
wooden board, 40 cm long
support for board, 10 cm high

Safety Precautions

Problem Data			
	Control	Inclined Plane	Block and Tackle
Ideal Mechanical Advantage, IMA	1	4	4
Input force, F_{in}, N			
Input distance, d_{in}, m	0.10		
Output force, F_{out}, N	9.8	9.8	9.8
Output distance, d_{out}, m	0.10	0.10	0.10
$Work_{in} = F_{in} \, d_{in}$, Joules			
$Work_{out} = F_{out} \, d_{out}$, Joules	0.98	0.98	0.98
$Efficiency = (Work_{out}/Work_{in}) \times 100\%$			

⊙ Test the Model

1. Tie the weight to the spring scale and measure the force required to lift it. Record the input force in your data table under *Control,* along with the 10-cm input distance.

2. Assemble the inclined plane so that the weight can be pulled up the ramp at a constant rate. The 40-cm board should be supported so that one end is 10 cm higher.

3. Tie the string to the spring scale and measure the force required to move the weight up the ramp at a constant speed. Record this input force under *Inclined Plane* in your data table. Record 40 cm as the input distance for the inclined plane.

4. Assemble the block and tackle using one fixed double pulley and one movable single pulley.

5. Tie the weight to the single pulley and tie the spring scale to the string at the top of the upper double pulley.

6. **Measure** the force required to lift the weight with the block and tackle. Record this input force.

7. **Measure** the length of string that must be pulled to raise the weight 10 cm. Record this input distance.

⊙ Analyze Your Data

1. **Calculate** the output work for all three methods of lifting the 9.8-N weight a distance of 10 cm.

2. **Calculate** the input work and the efficiency for the control, the inclined plane, and the block and tackle.

3. **Compare** the efficiencies of each of the three methods of lifting.

⊙ Conclude and Apply

1. **Explain** how you might improve the efficiency of the machine in each case.

2. **Infer** what types of situations would require use of a ramp over a pulley to help lift something.

3. **Infer** which machines would be most likely to be affected by friction.

𝒞ommunicating Your Data

Make a poster showing how the best machine would be used to lift the air conditioner to the roof of your building.

The Science of very, very small

Imagine an army of tiny robots, each no bigger than a bacterium swimming through your bloodstream.

Welcome to the world of nanotechnology—the science of creating molecular-sized machines. These machines are called nanobots.

This is the smallest guitar in the world. It is about as big as a human white blood cell. Each of its six silicon strings is 100 atoms wide. You can see the guitar only with an electron microscope.

The smallest of these machines are only billionths of a meter in size. They are so tiny that they can do work on the molecular scale.

Small, Smaller, Smallest

Nanotechnologists are predicting that within a few decades they will be creating nanobots that can do just about anything, as long as it's small. Already, nanotechnologists have built gears 10,000 times thinner than a human hair. They've also built tiny molecular "motors" only 50 atoms long. At Cornell University, nanotechnologists created the world's smallest guitar. It is appoximately the size of a white blood cell and it even has six strings.

In the future, they might transmit your internal vital signs to a nanocomputer, which might be implanted under your skin. There the data could be analyzed for signs of disease. Other nanomachines then could be sent to scrub your arteries clean of dangerous blockages, or mop up cancer cells, or even vaporize blood clots with tiny lasers. These are just some of the possibilities in the imaginations of those studying the new science of nanotechnology.

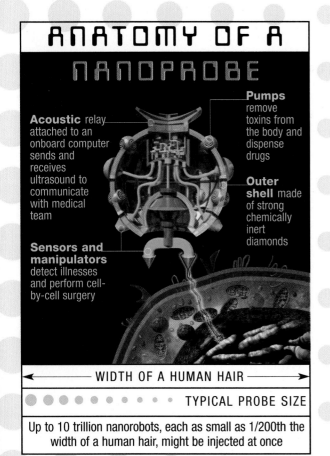

ANATOMY OF A NANOPROBE

Acoustic relay attached to an onboard computer sends and receives ultrasound to communicate with medical team

Pumps remove toxins from the body and dispense drugs

Outer shell made of strong chemically inert diamonds

Sensors and manipulators detect illnesses and perform cell-by-cell surgery

◄——— WIDTH OF A HUMAN HAIR ———►

• • • • • • • TYPICAL PROBE SIZE

Up to 10 trillion nanorobots, each as small as 1/200th the width of a human hair, might be injected at once

Design Think up a very small simple or complex machine that could go inside the body and do something. What would the machine do? Where would it go? Share your diagram or design with your classmates.

Science online

For more information, visit gpscience.com/time

Reviewing Main Ideas

Work

1. Work is the transfer of energy when a force makes an object move.

2. Work is done only when force produces motion in the direction of the force.

3. Power is the amount of work, or the amount of energy transferred, in a certain amount of time.

Using Machines

1. A machine makes work easier by changing the size of the force applied, by increasing the distance an object is moved, or by changing the direction of the applied force.

2. The number of times a machine multiplies the force applied to it is the mechanical advantage of the machine. The actual mechanical advantage is always less than the ideal mechanical advantage.

3. The efficiency of a machine equals the output work divided by the input work.

4. Friction always causes the output work to be less than the input work, so no real machine can be 100 percent efficient.

Simple Machines

1. A simple machine is a machine that can do work with a single movement.

2. A simple machine can increase an applied force, change its direction, or both.

3. A lever is a bar that is free to pivot about a fixed point called a fulcrum. A pulley is a grooved wheel with a rope running along the groove. A wheel and axle consists of two different-sized wheels that rotate together. An inclined plane is a sloping surface used to raise objects. The screw and wedge are special types of inclined planes.

4. A combination of two or more simple machines is called a compound machine.

FOLDABLES Use the Foldable that you made at the beginning of this chapter to help you review how machines make doing work easier.

Using Vocabulary

compound machine p.146	output force p.134
efficiency p.136	power p.129
inclined plane p.144	pulley p.141
input force p.134	screw p.145
lever p.138	simple machine p.138
machine p.132	wedge p.145
mechanical advantage p.136	wheel and axle p.143
	work p.126

Complete each statement using a word(s) from the vocabulary list above.

1. A combination of two or more simple machines is a(n) _____.

2. A wedge is another form of a(n) _____.

3. The ratio of the output force to the input force is the _____ of a machine.

4. A(n) _____ is a grooved wheel with a rope, chain, or cable in the groove.

5. The force exerted by a machine is the _____.

6. Energy is transferred when _____ is done.

7. _____ is the rate at which work is done or energy is transferred.

Checking Concepts

Choose the word or phrase that best answers the question.

8. Using the scientific definition, which of the following is true of work?
   A) It is difficult.
   B) It involves levers.
   C) It involves a transfer of energy.
   D) It is done with a machine.

9. How many types of simple machines exist?
   A) three   C) eight
   B) six   D) ten

10. In an ideal machine, which of the following is true?
    A) Work input is equal to work output.
    B) Work input is greater than work output.
    C) Work input is less than work output.
    D) The IMA is always equal to one.

11. Which of these is not done by a machine?
    A) multiply force
    B) multiply energy
    C) change direction of a force
    D) work

12. What term indicates the number of times a machine multiplies the input force?
    A) efficiency
    B) power
    C) mechanical advantage
    D) resistance

13. How could you increase the IMA of an inclined plane?
    A) increase its length
    B) increase its height
    C) decrease its length
    D) make its surface smoother

14. In a wheel and axle, which of the following usually exerts the output force?
    A) the axle   C) the fulcrum
    B) the wheel   D) the input arm

15. What is the IMA of screwdriver with a shaft radius of 3 mm and a handle radius of 10 mm?
    A) 0.3   C) 30
    B) 3.3   D) 13

16. What is the IMA of an inclined plane that is 2.1 m long and 0.7 m high?
    A) 0.3   C) 1.5
    B) 2.8   D) 3.0

17. Which of the following increases as the efficiency of a machine increases?
    A) work input   C) friction
    B) work output   D) IMA

Interpreting Graphics

18. Copy and complete the concept map of simple machines using the following terms: *compound machines, mechanical advantage, output force, work.*

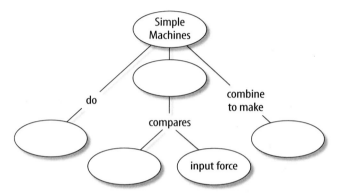

Use the table below to answer questions 19 and 20.

Lever Input and Output Arms		
Lever	Input arm (cm)	Output arm (cm)
A	25	75
B	53	42
C	36	36
D	32	99
E	10	30

19. Determine which of the levers listed in the table above has the largest IMA.

20. An input force of 50 N is applied to lever B. If the lever is 100 percent efficient, what is the output force?

Thinking Critically

21. Describe how the effort force, resistance force and fulcrum should be arranged so that a child can lift an adult using his own body weight on a see saw.

22. Determine what arrangement of movable and fixed pulleys would give a mechanical advantage of 3.

23. Explain which would give the best mechanical advantage for driving a screw down into a board: a screwdriver with a long, thin handle, or a screwdriver with a short, fat handle.

Applying Math

24. Calculate Work Find the work needed to lift a book weighing 20.0 N 2.0 m.

25. Calculate Axle Radius A doorknob has an IMA equal to 8.5. If the diameter of the doorknob is 8.0 cm, what is the radius of the shaft the doorknob is connected to?

26. Calculate Input Work A machine has an efficiency of 61 percent. Find the input work if the output work is 140 J.

27. Calculate Efficiency Using a ramp 6 m long, workers apply an input force of 1,250 N to move a 2,000-N crate onto a platform 2 m high. What is the efficiency of the ramp?

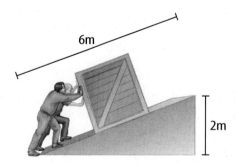

28. Calculate Power A person weighing 500 N climbs 3 m. How much power is needed to make the climb in 5 s?

Part 1 Multiple Choice

Record your answers on the answer sheet provided by your teacher or on a sheet of paper.

Use the figure below to answer questions 1 and 2

1. The figure above shows a doorknob with a radius of 4.8 cm and a mechanical advantage of 4.0. What is the radius of the inner rod that connects the knob to the door?
   A. 0.6 cm   C. 1.8 cm
   B. 1.2 cm   D. 2.4 cm

2. What would happen to the mechanical advantage if the radius of the doorknob were doubled?
   A. It would be multiplied by 4.
   B. It would be multiplied by 2.
   C. It would be divided by 4.
   D. It would be divided by 2.

3. A ramp is 2.8 m long and 1.2 m high. How much power is needed to push a box up the ramp in 4.6 s with a force of 96 N?
   A. 21 W   C. 58 W
   B. 25 W   D. 270 W

4. How much work is done in lifting a 9.10-kg box onto a shelf 1.80 m high?
   A. 5.06 J   C. 49.5 J
   B. 16.4 J   D. 161 J

5. What type of simple machine is your foot when you stand on your toes?
   A. first-class lever
   B. second-class lever
   C. third-class lever
   D. inclined plane

6. An input force of 80 N is used to lift an object weighing 240 N with a system of pulleys. How far down must the rope around the pulleys be pulled in order to lift the object a distance of 1.4 m?
   A. 0.47 m   C. 2.8 m
   B. 1.4 m   D. 4.2 m

Use the figure below to answer questions 7 and 8.

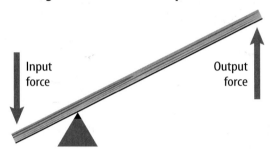

Input force

Output force

7. If the distance between the lever's input force and the fulcrum is 8 cm, and the distance between the fulcrum and the output force is 24 cm, what is the ideal mechanical advantage of the lever?
   A. 4   C. 0.33
   B. 3   D. 0.25

8. Which device uses the same class of lever as that shown in the figure?
   A. baseball bat   C. shovel
   B. scissors   D. wheelbarrow

9. How much more work is done to push a box 2.5 m with a force of 30 N than to push a box 2.0 m with a force of 26 N?
   A. 28 J   C. 4 J
   B. 23 J   D. 56 J

Test-Taking Tip

Don't Panic Stay calm during the test. If you feel yourself getting nervous, close your eyes and take five slow, deep, breaths.

Part 2 | Short Response/Grid In

Record your answers on the answer sheet provided by your teacher or on a sheet of paper.

10. As you throw a ball, you exert a force on ball of 4.2 N. You exert this force on the ball while the ball moves a distance of 0.45 m. The ball leaves your hand and travels a horizontal distance of 8.5 m to your friend. How much work have you done on the ball?

Use the illustration below to answer questions 11–13.

11. What is the ideal mechanical advantage of the pulley system shown in the figure to the right?

12. If the block supported by the pulley system shown above has a weight of 20 N, what is the input force on the rope?

13. If an additional pulley were included in the system shown in the illustration, what would the input force be?

14. A machine does 760 J of work in 32 s. What is the machine's power?

15. Write an equation for the efficiency of a lever if you know the lever's input force and distance as well as the output force and distance.

16. A centripetal force is exerted in a direction perpendicular to the motion of an object in circular motion. Is work done by a centripetal force? Why or why not?

Part 3 | Open Ended

Record your answers on a sheet of paper.

17. Is the following statement true? If so, give an example that supports it. If not, explain why.

    When work is done, a transfer of energy always occurs, but a transfer of energy does not always mean that work has been done.

18. What are three ways that simple machines can make work easier? For each one, give an example of a machine that makes work easier in that way.

19. Explain what causes friction in a machine and how a lubricant reduces a machine's friction. Describe the change a lubricant would make in the efficiency of a machine.

20. Use the law of conservation of energy to explain why it is impossible for the output work of a machine to be greater than the machine's input work.

Use the illustration below to answer questions 21 and 22.

21. The boy in the photograph to the right is carrying a box to the top of the stairs. Describe how the work that the boy does on the box is related to the energy transfer that occurs. How does the energy of the box change form as the boy carries the box up the stairs?

22. Explain how the work, power, and energy would change if the boy walked faster. How would the work, power, and energy change if the steps were the same height but steeper?

Thermal Energy

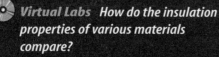
Hot Stuff

This hot, glowing liquid will become solid steel when it cools. The difference between liquid and solid steel is energy—thermal energy. Increasing the thermal energy of solid steel can cause it to melt and change into a fiery liquid.

Science Journal Describe things you do to make yourself feel warmer and cooler.

Start-Up Activities

Temperature and Kinetic Energy

Hot water can burn your skin, but warm water doesn't. How is hot water different from warm water? You know that the temperature of hot water is higher. The temperature depends on the energy of the water molecules. In hot water, molecules of water are moving faster than they are in warm water. As a result, the kinetic energy of water molecules in hot water is larger. The difference in kinetic energy also has other effects, as you'll see in this lab.

1. Pour 200 mL of room-temperature water into a beaker.
2. Pour 200 mL of water into a beaker and add some ice.
3. Put one drop of food coloring into each beaker.
4. Compare how quickly the food coloring causes the color of the water to change in each beaker.
5. **Think Critically** Write a paragraph describing the results of your experiment. Infer why the food coloring spread throughout the water in the two beakers at different rates.

Thermal Energy and Heat Make the following Foldable to help you understand thermal energy and heat.

STEP 1 Fold a vertical sheet of paper in half from top to bottom.

STEP 2 Fold in half from side to side with the fold at the top.

STEP 3 Unfold the paper once. Cut only the fold of the top flap to make two tabs.

STEP 4 Turn the paper vertically and label the front tabs as shown.

Find Main Ideas As you read the chapter, write the main ideas you find about thermal energy and heat under the appropriate tab.

Preview this chapter's content and activities at
gpscience.com

157

Temperature and Heat

Temperature

You use the words hot and cold to describe temperature. Something is hot when its temperature is high. When you heat water on a stove, its temperature increases. How are temperature and heat related?

Matter in Motion The matter around you is made of tiny particles—atoms and molecules. In all materials these particles are in constant, random motion; moving in all directions at different speeds. Because these particles are moving, they have kinetic energy. The faster they move, the more kinetic energy they have. **Figure 1** shows that particles move faster in hot objects than in cooler objects.

Figure 1 The particles in an object are in constant random motion.

When the horseshoe is hot, the particles in it move faster.

When the horseshoe has cooled, its particles are moving more slowly.

Temperature The temperature of an object and the kinetic energy of its atoms and molecules are related. The **temperature** of an object is a measure of the average kinetic energy of the particles in the object. As the temperature of an object increases, the average speed of the particles in random motion increases. The temperature of hot tea is higher than the temperature of iced tea because the particles in the hot tea are moving faster on average. In SI units, temperature is measured in kelvins (K). A more commonly used temperature scale is the Celsius scale. One kelvin degree is the same size as one Celsius degree.

Thermal Energy

If you let cold butter sit at room temperature for a while, it warms and becomes softer. Because the air in the room is at a higher temperature than the butter, particles in air have more kinetic energy than butter particles. Collisions between particles in butter and particles in air transfer energy from the faster-moving particles in air to the slower-moving butter particles. The butter particles then move faster and the temperature of the butter increases.

Particles in the butter can exert attractive forces on each other. Recall that Earth exerts an attractive gravitational force on a ball. When the ball is above the ground, the ball and Earth are separated, and the ball has potential energy. In the same way, atoms and particles that exert attractive forces on each other have potential energy when they are separated. The sum of the kinetic and potential energy of all the particles in an object is the **thermal energy** of the object. Because the kinetic energy of the butter particles increased as it warmed, the thermal energy of the butter increased.

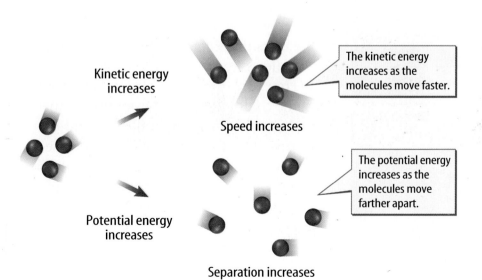

Kinetic energy increases

The kinetic energy increases as the molecules move faster.

Speed increases

The potential energy increases as the molecules move farther apart.

Potential energy increases

Separation increases

Figure 2 The thermal energy of a substance is the sum of the kinetic and potential energy of its molecules.
Infer *why increasing the temperature of an object increases its thermal energy.*

Thermal Energy and Temperature Thermal energy and temperature are related. When the temperature of an object increases, the average kinetic energy of the particles in the object increases. Because thermal energy is the total kinetic and potential energy of all the particles in an object, the thermal energy of the object increases when the average kinetic energy of its particles increases. Therefore, the thermal energy of an object increases as its temperature increases.

Thermal Energy and Mass Suppose you have a glass and a beaker of water that are at the same temperature. The beaker contains twice as much water as the glass. The water in both containers is at the same temperature, so the average kinetic energy of the water molecules is the same in both containers. But there are twice as many water molecules in the beaker as there are in the glass. So the total kinetic energy of all the molecules is twice as large for the water in the beaker. As a result, even though they are at the same temperature, the water in the beaker has twice as much thermal energy as the water in the glass does. If the temperature doesn't change, the thermal energy in an object increases if the mass of the object increases.

Heat

Can you tell if someone has been sitting in your chair? Perhaps you've noticed that your chair feels warm, and maybe you concluded that someone has been sitting in it recently. The chair feels warmer because thermal energy from the person's body flowed to the chair and increased its temperature.

Heat is thermal energy that flows from something at a higher temperature to something at a lower temperature. Heat is a form of energy, so it is measured in joules—the same units that energy is measured in. Heat always flows from warmer to cooler materials. How did the ice cream in **Figure 3** become cold? Heat flowed from the warmer liquid ingredients to the cooler ice-and-salt mixture. The liquid ingredients released enough thermal energy to become cold enough to form solid ice cream. Meanwhile, the ice-and-salt solution absorbed thermal energy, causing some of the ice to melt.

Figure 3 Heat flows from the warmer ingredients inside the container to the ice-and-salt mixture.

 **Reading Check** *How are heat and thermal energy related?*

Specific Heat

If you are at the beach in the summertime, you might notice that the ocean seems much cooler than the air or sand. Even though energy from the Sun is falling on the air, sand, and water at the same rate, the temperature of the water has changed less than the temperature of the air or sand has.

As a substance absorbs heat, its temperature change depends on the nature of the substance, as well as the amount of heat that is added. For example, compared to 1 kg of sand, the amount of heat that is needed to raise the temperature of 1 kg of water by 1°C is about six times greater. So the ocean water at the beach would have to absorb six times as much heat as the sand to be at the same temperature. The amount of heat that is needed to raise the temperature of 1 kg of some material by 1°C is called the **specific heat** of the material. Specific heat is measured in joules per kilogram degree Celsius [J/(kg °C)]. **Table 1** shows the specific heats of some familiar materials.

✔ Reading Check *What is the specific heat of a material?*

Water as a Coolant Compared with the other common materials in **Table 1,** water has the highest specific heat. **Figure 4** shows why this is. Because water can absorb heat without a large change in temperature, it is useful as a coolant. A coolant is a subtance that is used to absorb heat. For example, water is used as the coolant in the cooling systems of automobile engines. As long as the water temperature is lower than the engine temperature, heat will flow from the engine to the water. Compared to other materials, the temperature of water will increase less.

Table 1 Specific Heat of Some Common Materials

Substance	Specific Heat [J/(kg°C)]
Water	4,184
Wood	1,760
Carbon (graphite)	710
Glass	664
Iron	450

Figure 4 The specific heat of water is high because water molecules form strong bonds with each other.

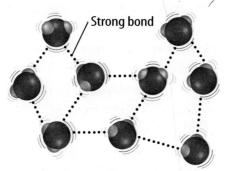

When heat is added, some of the added heat has to break some of these bonds before the molecules can start moving faster.

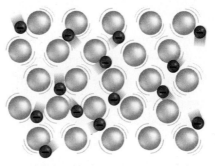

In metals, electrons can move freely. When heat is added, no strong bonds have to be broken before the electrons can start moving faster.

Coastal Climates The high specific heat of water causes large bodies of water to heat up and cool down more slowly than land masses. As a result, the temperature changes in coastal areas tend to be less extreme than they are farther inland.

Changes in Thermal Energy The thermal energy of an object changes when heat flows into or out of the object. If Q is the change in thermal energy and C is specific heat, the change in thermal energy can be calculated from the following equation:

Thermal Energy Equation

change in thermal energy (J) =
mass (kg) $\times$ change in temperature (°C) $\times$ specific heat $\left(\dfrac{J}{kg\,°C}\right)$

$$Q = m(T_f - T_i)C$$

Applying Math   Solve a Simple Equation

CHANGE IN THERMAL ENERGY Find the change in thermal energy of a 20-kg wooden chair that warms from 15°C to 25°C if the specific heat of wood is 1,700 J/(kg °C).

IDENTIFY known values and the unknown value

Identify the known values:

A wooden chair with a mass of 20 kg   [means]→ $m = 20$ kg

warms from 15°C to 25°C   [means]→ $T_i = 15°C$ and $T_f = 25°C$

the specific heat of wood is 1,700 J/(kg °C)   [means]→ $C = 1,700$ J/(kg °C)

Identify the unknown value:

What is the change in thermal energy   [means]→ $Q = ?$ J

SOLVE the problem

Substitute the known values into the thermal energy equation:

$$Q = m(T_f - T_i)C = (20\text{ kg})(25°C - 15°C)(1,700\frac{J}{kg\ °C})$$

$$= (20\text{ kg})(10°C)(1,700\frac{J}{kg\ °C}) = 340,000\text{ kg °C}\frac{J}{kg\ °C} = 340,000\text{ J}$$

CHECK your answer

Does your answer seem reasonable? Check your answer by dividing the change in thermal energy you calculated by the mass and the specific heat given in the problem. The result should be the difference in temperature given in the problem.

Practice Problems

The air in a living room has a mass of 72 kg and a specific heat of 1,010 J/(kg °C). What is the change in thermal energy of the air when it warms from 20°C to 25°C?

For more practice problems go to page 834, and visit gpscience.com/extra_problems.

Measuring Specific Heat

The specific heat of a material can be measured using a device called a calorimeter, shown in **Figure 5.** The specific heat of a material can be determined if the mass of the material, its change in temperature, and the amount of heat absorbed or released are known. In a calorimeter, a heated sample transfers heat to a known mass of water. The energy absorbed by the water can be calculated by measuring the water's temperature change. Then the thermal energy released by the sample equals the thermal energy absorbed by the water.

Using a Calorimeter To measure the specific heat of a material, the mass of a sample of the material is measured, as is the initial temperature of the water in the calorimeter. The material is then heated, its temperature measured, and the sample is placed in the water in the inner chamber of the calorimeter. The sample cools as heat is transferred to the water, and the temperature of the water increases. The transfer of heat continues until the sample and the water are at the same temperature. Then the initial and final temperatures of the water are known, and the amount of heat gained by the water can be calculated.

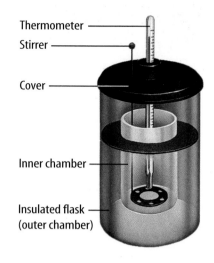

Figure 5 A calorimeter can be used to measure the specific heat of materials. The sample is placed in the inner chamber.

---

section 1 review

Summary

Temperature

- The temperature of an object is a measure of the average kinetic energy of the particles that make up the object.

Thermal Energy and Heat

- Thermal energy is the sum of the kinetic and potential energy of all the particles in an object.
- If temperature is constant, the thermal energy increases when the mass increases.
- Heat is thermal energy that is transferred from an object at a higher temperature to an object at a lower temperature.

Specific Heat

- The specific heat of a material is the amount of heat needed to raise the temperature of 1 kg of the material 1°C.
- The change in thermal energy of an object can be calculated from this equation

$$Q = m(T_f - T_i)C$$

Self Check

1. **Explain** how energy moves when you touch a block of ice with your hand.

2. **Describe** how the thermal energy of an object changes when the object's temperature changes.

3. **Infer** When heat flows between two objects, does the temperature increase of one object always equals the temperature decrease of the other object? Explain.

4. **Explain** why the specific heat of water is higher than the specific heat of most other substances.

5. **Think Critically** Explain whether or not the following statement is true: for any two objects, the one with the higher temperature always has more thermal energy.

Applying Math

6. **Calculate** the change in thermal energy of the water in a pond with a mass of 1,000 kg and a specific heat of 4,184 J/(kg °C) if the water cools by 1°C.

7. **Calculate** the specific heat of a metal if 0.5 kg of the metal absorb 9,000 J of heat as it warms by 10°C.

Transferring Thermal Energy

Reading Guide

What You'll Learn

- **Compare and contrast** the transfer of thermal energy by conduction, convection, and radiation.
- **Compare and contrast** thermal conductors and insulators.
- **Explain** how insulators are used to control the transfer of thermal energy.

Why It's Important

You must be able to control the flow of thermal energy to keep from being too hot or too cold.

🔎 Review Vocabulary

density: the mass per unit volume of a substance

New Vocabulary

- conduction
- convection
- radiation
- insulator

Conduction

Thermal energy is transferred from place to place by conduction, convection, and radiation. **Conduction** is the transfer of thermal energy by collisions between particles in matter. Conduction occurs because particles in matter are in constant motion.

Collisions Transfer Thermal Energy Thermal energy is transferred when one end of a metal spoon is heated by a Bunsen burner, as shown in **Figure 6.** The kinetic energy of the particles near the flame increases. Kinetic energy is transferred when these particles collide with neighboring particles. Thermal energy is transferred by collisions between particles with more kinetic energy and particles with less kinetic energy. As these collisions continue, thermal energy is transferred from one end of the spoon to the other end of the spoon. When heat is transferred by conduction, thermal energy is transferred from place to place without transferring matter. Thermal energy is transferred by the collisions between particles, not by movement of matter.

Figure 6 Conduction occurs within a material as faster-moving particles transfer thermal energy by colliding with slower-moving particles.

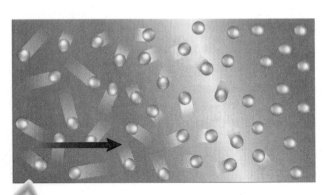

Heat Conductors Although heat can be transferred by conduction in all materials, the rate at which heat moves depends on the material. Heat moves faster by conduction in solids and liquids than in gases. In gases, particles are farther apart, so collisions with other particles occur less frequently than they do in solids or liquids.

The best conductors of heat are metals. In a piece of metal, there are electrons that are not bound to individual atoms, but can move easily through the metal. Collisions between these electrons and other particles in the metal enable thermal energy to be transferred more quickly than in other materials. Silver, copper and aluminum are among the best conductors of heat.

Convection

Unlike solids, liquids and gases can flow and are classified as fluids. In fluids, thermal energy can be transferred by convection. **Convection** is the transfer of thermal energy in a fluid by the movement of warmer and cooler fluid from place to place. When conduction occurs, more energetic particles collide with less energetic particles and transfer thermal energy. When convection occurs, more energetic particles move from one place to another.

As the particles move faster, they tend to be farther apart. As a result, a fluid expands as its temperature increases. Recall that density is the mass of a material divided by its volume. When a fluid expands, its volume increases, but its mass doesn't change. As a result, its density decreases. The same is true for parts of a fluid that have been heated. The density of the warmer fluid, therefore, is less than that of the surrounding cooler fluid.

Heat Transfer by Currents How does convection occur? Look at the lamp shown in **Figure 7.** Some of these lamps contain oil and alcohol. When the oil is cool, its density is greater than the alcohol, and it sits at the bottom of the lamp. When the two liquids are heated, the oil becomes less dense than the alcohol. Because it is less dense than the alcohol, it rises to the top of the lamp. As it rises, it loses heat by conduction to the cooler fluid around it. When the oil reaches the top of the lamp, it has become cool enough that it is denser than the alcohol, and it sinks. This rising-and-sinking action is a convection current. Convection currents transfer heat from warmer to cooler parts of the fluid. In a convection current, both conduction and convection transfer thermal energy.

✔ Reading Check *How are conduction and convection different?*

Figure 7 The heat from the light at the bottom of the lamp causes one fluid to expand more than the other. This creates convection currents in the lamp.
Explain *why the substances in the lamp rise and sink.*

Figure 8

When the Sun beats down on the equator, warm, moist air begins to rise. As it rises, the air cools and loses its moisture as rain that sustains rain forests near the equator. Convection currents carry the now dry air farther north and south. Some of this dry air descends at the tropics, where it creates a zone of deserts.

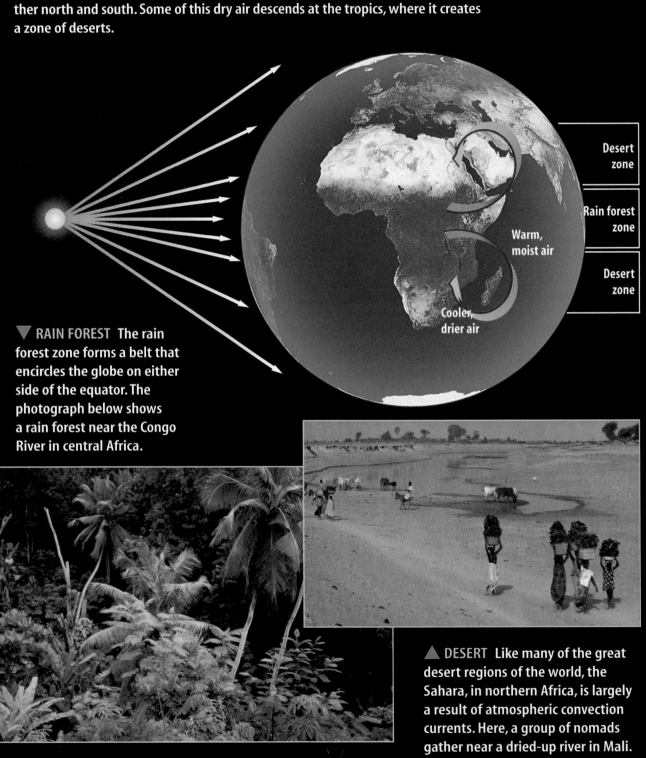

Desert zone

Rain forest zone

Desert zone

Warm, moist air

Cooler, drier air

▼ RAIN FOREST The rain forest zone forms a belt that encircles the globe on either side of the equator. The photograph below shows a rain forest near the Congo River in central Africa.

▲ DESERT Like many of the great desert regions of the world, the Sahara, in northern Africa, is largely a result of atmospheric convection currents. Here, a group of nomads gather near a dried-up river in Mali.

Desert and Rain Forests Earth's atmosphere is made of various gases and is a fluid. The atmosphere is warmer at the equator than it is at the north and south poles. Also, the atmosphere is warmer at Earth's surface than it is at higher altitudes. These temperature differences create convection currents that carry heat to cooler regions. **Figure 8** shows how these convection currents create rain forests and deserts over different regions of Earth's surface.

Radiation

Earth gets heat from the Sun, but how does that heat travel through space? Almost no matter exists in the space between Earth and the Sun, so heat cannot be transferred by conduction or convection. Instead, the Sun's heat reaches Earth by radiation.

Radiation is the transfer of energy by electromagnetic waves. These waves can travel through space even when no matter is present. Energy that is transferred by radiation often is called radiant energy. When you stand near a fire and warm your hands, much of the warmth you feel has been transferred from the fire to your hands by radiation.

Radiant Energy and Matter When radiation strikes a material, some of the energy is absorbed, some is reflected, and some may be transmitted through the material. **Figure 9** shows what happens to radiant energy from the Sun as it reaches Earth. The amount of energy absorbed, reflected, and transmitted depends on the type of material. Materials that are light-colored reflect more radiant energy, while dark-colored materials absorb more radiant energy. When radiant energy is absorbed by a material, the thermal energy of the material increases.

For example, when a car sits outside in the Sun, some of the radiation from the Sun passes through the transparent car windows. Materials inside the car absorb some of this radiation and become hot. Radiation can pass through solids, liquids, and gases.

Radiation in Solids, Liquids, and Gases The transfer of energy by radiation is most important in gases. In a solid, liquid or gas, radiant energy can travel through the space between molecules. Molecules can absorb this radiation and emit some of the energy they absorbed. This energy then travels through the space between molecules, and is absorbed and emitted by other molecules. Because molecules are much farther apart in gases than in solids or liquids, radiation usually passes more easily through gases than through solids or liquids.

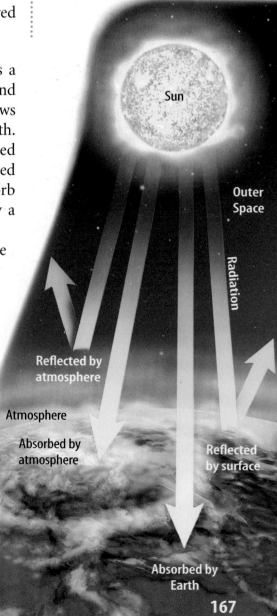

Figure 9 Not all of the Sun's radiation reaches Earth. Some of it is reflected by the atmosphere. Some of the radiation that does reach the surface is also reflected.

Sun

Outer Space

Radiation

Reflected by atmosphere

Atmosphere

Absorbed by atmosphere

Reflected by surface

Absorbed by Earth

167

Controlling Heat Flow

You might not realize it, but you probably do a number of things every day to control the flow of heat. For example, when it's cold outside, you put on a coat or a jacket before you leave your home. When you reach into an oven to pull out a hot dish, you might put a thick, cloth mitten over your hand to keep from being burned. In both cases, you used various materials to help control the flow of heat. Your jacket kept you from getting cold by reducing the flow of heat from your body to the surrounding air. And the oven mitten kept your hand from being burned by reducing the flow of heat from the hot dish.

As shown in **Figure 10,** almost all living things have special features that help them control the flow of heat. For example, the antarctic fur seal's thick coat and the emperor penguin's thick layer of blubber help keep them from losing heat. This helps them survive in a climate in which the temperature is often below freezing. In the desert, however, the scaly skin of the desert spiny lizard has just the opposite effect. It reflects the Sun's rays and keeps the animal from becoming too hot. An animal's color also can play a role in keeping it warm or cool. The black feathers on the penguin's back, for example, allow it to absorb radiant energy. Can you think of any other animals that have special adaptations for cold or hot climates?

✔ **Reading Check** *What are two animal adaptations that control the flow of heat?*

Figure 10 Animals have different features that help them control heat flow.

The antarctic fur seal grows a coat that can be as much as 10 cm thick.

The emperor penguin has a thick layer of blubber and thick, closely spaced feathers, which help reduce the loss of body heat.

The scaly skin of the desert spiny lizard not only reflects sunlight but it also prevents water loss. This is important in a dry environment.

Figure 11 The tiny pockets of air in fleece make it a good insulator. They help reduce the flow of the jogger's body heat to the colder outside air.

Insulators

A material in which heat flows slowly is an **insulator.** Examples of materials that are insulators are wood, some plastics, fiberglass, and air. Materials, such as metals, that are good conductors of heat are poor insulators. In these materials, heat flows more rapidly from one place to another.

Gases, such as air, are usually much better insulators than solids or liquids. Some types of insulators contain many pockets of trapped air. These air pockets conduct heat poorly and also keep convection currents from forming. Fleece jackets, like the one shown in **Figure 11,** work in the same way. When you put the jacket on, the fibers in the fleece trap air and hold this air next to you. This air slows down the flow of your body heat to the colder air outside the jacket. Gradually, the air trapped by the fleece is warmed by your body heat, and underneath the jacket you are wrapped in a blanket of warm air.

 *Why does trapped air make a material like fleece a good insulator?*

Insulating Buildings Insulation, or materials that are insulators, helps keep warm air from flowing out of buildings in cold weather and from flowing into buildings in warm weather. Building insulation is usually made of some fluffy material, such as fiberglass, that contains pockets of trapped air. The insulation is packed into a building's outer walls and attic, where it reduces the flow of heat between the building and the surrounding air.

Insulation helps furnaces and air conditioners work more effectively, saving energy. In the United States, about 55 percent of the energy used in homes is used for heating and cooling.

Mini LAB

Comparing Thermal Conductors

Procedure

1. Obtain a **plastic spoon,** a **metal spoon,** and a **wooden spoon** with similar lengths.
2. Stick a small **plastic bead** to the handle of each spoon with a dab of **butter or wax.** Each bead should be the same distance from the tip of the spoon.
3. Stand the spoons in a **beaker,** with the beads hanging over the edge of the beaker.
4. Carefully pour **boiling water** to a depth of about 5 cm in the beaker holding the spoons.

Analysis

1. In what order did the beads fall from the spoons?
2. Describe how heat was transferred from the water to the beads.
3. Rank the spoons in their

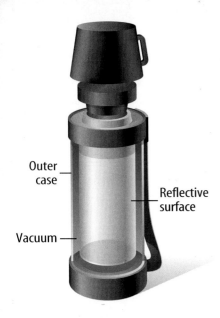

Figure 12 A thermos bottle uses a vacuum and reflective surfaces to reduce the flow of heat into and out of the bottle. The vacuum prevents heat flow by conduction and convection. The reflective surfaces reduce the heat transfer by radiation.

Outer case

Reflective surface

Vacuum

Reducing Heat Flow in a Thermos You might have used a thermos bottle, like the one in **Figure 12,** to carry hot soup or iced tea. A thermos bottle reduces the flow of heat into and out of the liquid in the bottle, so that the temperature of the liquid hardly changes over a number of hours. To do this, a thermos bottle has two glass walls. The air between the two walls is removed so there is a vacuum between the glass layers. Because the vacuum contains almost no matter, it prevents heat transfer by conduction or convection between the liquid and the air outside the thermos.

To further reduce the flow of heat into or out of the liquid, the inside and outside glass surface of a thermos bottle is coated with aluminum to make each surface highly reflective. This causes electromagnetic waves to be reflected at each surface. The inner relective surface prevents radiation from transferring heat out of the liquid. The outer reflective surface prevents radiation from transferring heat into the liquid.

Think about the things you do to stay warm or cool. Sitting in the shade reduces the heat transferred to you by radiation. Opening or closing windows reduces heat transfer by convection. Putting on a jacket reduces the heat transferred from your body by conduction. In what other ways do you control the flow of heat?

section ② review

Summary

Conduction

- Conduction is the transfer of thermal energy by collisions between more energetic and less energetic particles.
- Conduction occurs in solids, liquids, and gases. Metals are the best conductors of heat.

Convection

- Convection is the tranfer of thermal energy by the movement of warmer and cooler material.
- Convection occurs in fluids. Rising of warmer fluid and sinking of cooler fluid forms a convection current.

Radiation

- Radiation is the transfer of energy by electromagnetic waves.

Controlling Heat Flow

- Insulators are used to reduce the rate of heat transfer from one place to another.

Self Check

1. **Explain** why materials that are good conductors of heat are poor insulators.
2. **Explain** why the air temperature near the ceiling of a room tends to be warmer than near the floor.
3. **Predict** whether plastic foam, which contains pockets of air, would be a good conductor or a good insulator.
4. **Describe** how a convection current occurs.
5. **Think Critically** Several days after a snowfall, the roofs of some homes on a street have almost no snow on them, while the roofs of other houses are still snow-covered. Describe what would cause this difference.

Applying Math

6. **Calculate Solar Radiation** Averaged over a year in the central United States, radiation from the sun transfers about 200 W to each square meter of Earth's surface. If a house is 10 m long by 10 m wide, how much solar energy falls on the house each second?

Science online gpscience.com/self_check_quiz

Convection in Gases and Liquids

A hawk gliding through the sky will rarely flap its wings. Hawks and some other birds conserve energy by gliding on columns of warm air rising up from the ground. These convection currents form when gases or liquids are heated unevenly, and the warmer, less dense fluid is forced upward.

▶ Real-World Question

How can convection currents be modeled and observed?

Goals

- ■ **Model** the formation of convection currents in water.
- ■ **Observe** convection currents formed in water.
- ■ **Observe** convection currents formed in air.

Materials

burner or hot plate   500-mL beaker
water   black pepper
candle

Safety Precautions

WARNING: *Use care when working with hot materials. Remember that hot and cold glass appear the same.*

▶ Procedure

1. Pour 450 mL of water into the beaker.
2. Use a balance to measure 1 g of black pepper.
3. Sprinkle the pepper into the beaker of water and let it settle to the bottom of the beaker.

4. Heat the bottom of the beaker using the burner or by placing it on the hotplate.
5. **Observe** how the particles of pepper move as the water is heated, and make a drawing showing their motion in your Science Journal.
6. Turn off the hot plate or burner. Light the candle and let it burn for a few minutes.
7. Blow out the candle, and observe the motion of the smoke.
8. Make a drawing of the movement of the smoke in your Science Journal.

▶ Conclude and Apply

1. **Describe** how the particles of pepper moved as the water became hotter.
2. **Explain** how the motion of the pepper particles is related to the motion of the water.
3. **Explain** how a convection current formed in the beaker.
4. **Explain** why the motion of the pepper changed when the heat was turned off.
5. **Predict** how the pepper would move if the water were heated from the top.
6. **Describe** how the smoke particles moved when the candle was blown out.
7. **Explain** why the smoke moved as it did.

Communicating Your Data

Compare your conclusions with other students in your class. **For more help, refer to the** Science Skill Handbook.

Using Heat

Reading Guide

What You'll Learn
- **Describe** common types of heating systems.
- **Describe** the first and second laws of thermodynamics.
- **Explain** how an internal combustion engine works.
- **Explain** how a refrigerator transfers thermal energy from a cool to a warm temperature.

Why It's Important
Imagine your life without heating systems, cooling systems, and cars.

Review Vocabulary
work: the product of the force exerted on an object and the distance the object moves in the direction of the force

New Vocabulary
- solar collector
- thermodynamics
- first law of thermodynamics
- second law of thermodynamics
- heat engine
- internal combustion engine

Heating Systems

Almost everywhere in the United States air temperatures at some time become cold enough that a source of heat is needed. As a result, most homes and public buildings contain some type of heating system. The best heating system for any building depends on the local climate and how the building is constructed.

All heating systems require some source of energy. In the simplest and oldest heating system, wood or coal is burned in a stove. The heat that is produced by the burning fuel is transferred from the stove to the surrounding air by conduction, convection, and radiation. One disadvantage of this system is that heat transfer from the room in which the stove is located to other rooms in the building can be slow.

Figure 13 In forced-air systems, air heated by the furnace gets blown through ducts that usually lead to every room.

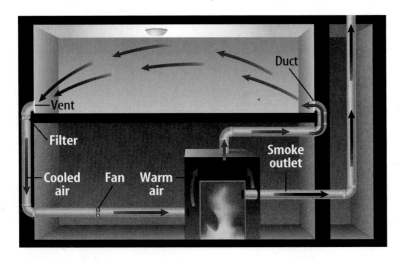

Forced-Air Systems The most common type of heating system in use today is the forced-air system, shown in **Figure 13.** In this system, fuel is burned in a furnace and heats a volume of air. A fan then blows the warm air through a series of large pipes called ducts. The ducts lead to openings called vents in each room. Cool air returns through additional vents to the furnace, where it is reheated.

Radiator Systems Before forced-air systems were widely used, many homes and buildings were heated by radiators. A radiator is a closed metal container that contains hot water or steam. The thermal energy contained in the hot water or steam is transferred to the air surrounding the radiator by conduction. This warm air then moves through the room by convection.

In radiator heating systems, fuel burned in a central furnace heats a tank of water. A system of pipes carries the hot water to radiators in the rooms of the building. After the water cools, it flows through the pipes back to the water tank and is reheated. In some radiator systems, the water is heated to produce steam that flows through the pipes to the radiators. As the steam cools, it condenses into water and flows back to the tank.

Electric Heating Systems An electric heating system has no central furnace. Instead, electrically heated coils placed in floors and in walls heat the surrounding air by conduction. Heat is then distributed through the room by convection. Electric heating systems are not as widely used as forced-air systems. However, in warmer climates the walls and floors of some buildings may not be thick enough to contain pipes and ducts. Then an electric heating system might be the only practical way to provide heat.

Solar Heating

The Sun emits an enormous amount of radiant energy that strikes Earth every day. The radiant energy from the Sun can be used to help heat homes and buildings. There are two types of systems that use the Sun's energy for heating—passive solar heating systems and active solar heating systems.

Passive Solar Heating In passive solar heating systems, materials inside a building absorb radiant energy from the Sun during the day and heat up. At night when the building begins to cool, thermal energy absorbed by these materials helps keep the room warm. **Figure 14** shows a room in a house that uses passive solar heating. Walls of windows receive the maximum amount of sunlight during the day. The other walls are heavily insulated and have few or no windows to reduce heat loss at night.

Figure 14 In a passive solar heating system, radiant energy from the Sun is transferred to the room through windows. Windows also prevent air inside from mixing with cooler air outside.
Infer *in which regions of the United States passive solar systems would be practical.*

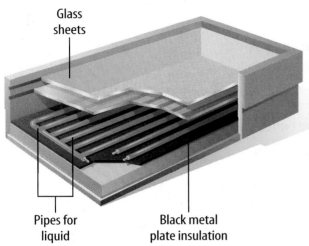

Glass sheets

Pipes for liquid

Black metal plate insulation

Figure 15 This active solar heating system uses solar collectors mounted on the roof to absorb solar energy. The absorbed energy heats a liquid that is circulated throughout the house.

Topic: Solar Heating
Visit gpscience.com for Web links to information about systems that use solar energy to heat buildings.

Activity Draw a diagram showing how an active solar heating system is used to heat a home.

Active Solar Heating Active solar heating systems use **solar collectors** that absorb radiant energy from the Sun. The collectors usually are installed on the roof or south side of a building. Radiant energy from the Sun heats air or water in the solar collectors. One type of active solar collector is shown in **Figure 15.** The black metal plate absorbs radiant energy from the Sun. The absorbed energy heats water in pipes just above the plate. A pump circulates the hot water to radiators in rooms of the house. The cooled water then is pumped back to the collector to be reheated.

Thermodynamics

There is another way to increase the thermal energy of an object besides adding heat. Have you ever rubbed your hands together to warm them on a cold day? Your hands get warmer and their thermal energy and temperature increase, even though there is no heat flowing to them. You did work on your hands by rubbing them together. The work you did caused the thermal energy of your hands to increase. Thermal energy, heat, and work are related, and the study of the relationship among them is **thermodynamics.**

Heat and Work Increase Thermal Energy You can warm your hands by placing them near a fire, so that heat is added to your hands by radiation. If you rub your hands and hold them near a fire, the increase in thermal energy of your hands is even greater. Both the work you do and the heat transferred from the fire increase the thermal energy of your hands.

In the example above your hands can be considered as a system. A system can be a group of objects such as a galaxy, or a car's engine, or something as simple as a ball. In fact, a system is anything you can draw a boundary around, as shown in **Figure 16.** The heat transferred to a system is the amount of heat flowing into the system that crosses the boundary. The work done on a system is the work done by something outside the system's boundary.

The First Law of Thermodynamics According to the **first law of thermodynamics,** the increase in thermal energy of a system equals the work done on the system plus the heat transferred to the system. Doing work on a system is a way of adding energy to a system. As a result, the temperature of a system can be increased by adding heat to the system, doing work on the system, or both. The first law of thermodynamics is another way of stating the law of conservation of energy. The increase in energy of a system equals the energy added to the system.

Closed and Open Systems A system is an open system if heat flows across the boundary or if work is done across the boundary. Then energy is added to the system. If no heat flows across the boundary and there is no outside work done, then the system is a closed system. According to the first law of thermodynamics, the thermal energy of a closed system doesn't change. There may be processes going on in the system that are converting one form of energy into another, but the total energy of the system doesn't change. Because energy cannot be created or destroyed, the total energy stays constant in a closed system.

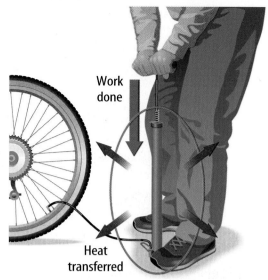

Figure 16 A bicycle air pump can be a system. Work is done on the system by pushing down on the handle. This causes the pump to become warm, and heat is transferred from the system to the outside air.

The Second Law of Thermodynamics When heat flows from a warm object to a cool object the thermal energy of the warm object decreases and the thermal energy of the cool object increases. According to the law of conservation of energy or the first law of thermodynamics, the increase in thermal energy of the cool object equals the decrease in thermal energy of the warm object.

Can heat flow spontaneously from a cold object to warm object? This process never happens, but it wouldn't violate the first law of thermodynamics. The first law would require only that the decrease in thermal energy of the cool object would be equal to the increase in thermal energy of the warm object.

Reading Check *How does heat flow from a warm to a cool object satisfy the first law of thermodynamics?*

However, the flow of heat spontaneously from a cool object to a warm object never happens because it violates another law—the second law of thermodynamics. One way to state the **second law of thermodynamics** is that it is impossible for heat to flow from a cool object to a warmer object unless work is done. For example, if you hold an ice cube in your hand, no work is done. As a result, heat flows only from your warmer hand to the colder ice.

**INTEGRATE
Earth Science**

Nature's Heat Engines
Hurricanes are storms that form over the ocean in regions of low pressure. Because hurricanes use heat from warm ocean water to produce strong winds, they are sometimes called nature's heat engines. Research hurricanes and draw a diagram showing how they are like a heat engine.

Figure 17 Burning fuel in the engine's cylinders produces thermal energy that is converted into work as the pistons move up and down. The crankshaft, transmission, and differential convert the up and down motion of the pistons into rotation of the wheels.

Converting Heat to Work

If you give a book sitting on a table a push, the book will slide and come to a stop. Friction between the book and the table converted the work you did on the book to heat. As a result, the book and the table became slightly warmer.

In the example above, work was converted completely into heat. Is it possible to do the reverse, and convert heat completely into work? Even though this process would not violate the first law of thermodynamics, it also is forbidden by the second law of thermodynamics. The second law of thermodynamics makes it impossible to build a device that converts heat completely into work.

✔ Reading Check *Why can't heat be convered completely into work?*

A device that converts heat into work is a **heat engine.** A car's engine is an example of a heat engine. A car's engine converts the chemical energy in gasoline into heat. The engine then transforms some of the thermal energy into work by rotating the car's wheels, as shown in **Figure 17.** However, only about 25 percent of the heat released by the burning gasoline is converted into work, and the rest is transferred to the engine's surroundings.

Internal Combustion Engines The heat engine in a car is an **internal combustion engine** in which fuel is burned inside the engine in chambers or cylinders. Automobile engines usually have four, six, or eight cylinders. Each cylinder contains a piston that moves up and down. Each up-and-down movement of the piston is called a stroke. Automobile and diesel engines have four different strokes. **Figure 18** shows the four-stroke cycle in an automobile engine.

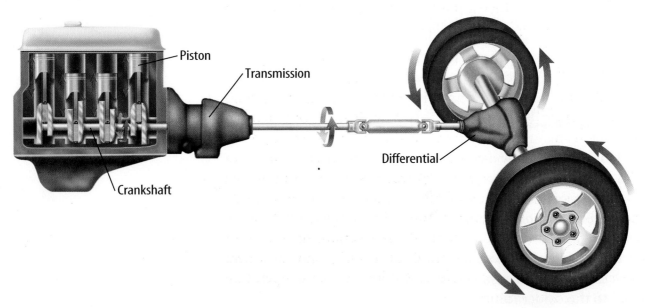

Piston

Transmission

Differential

Crankshaft

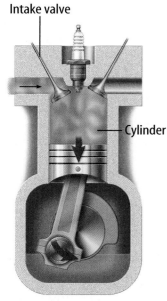
Intake valve — Cylinder — Piston

Fuel-air mixture

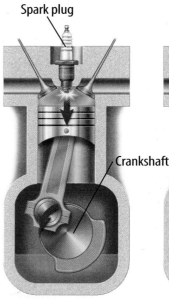
Spark plug — Crankshaft

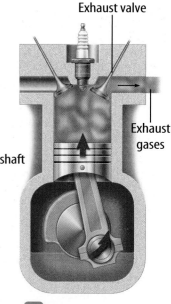
Exhaust valve — Exhaust gases

A Intake stroke
The intake valve opens as the piston moves downward, drawing a mixture of gasoline and air into the cylinder.

B Compression stroke
The intake valve closes as the piston moves upward, compressing the fuel-air mixture.

C Power stroke
A spark plug ignites the fuel-air mixture. As the mixture burns, hot gases expand, pushing the piston down.

D Exhaust stroke
As the piston moves up, the exhaust valve opens, and the hot gases are pushed out of the cylinder.

Figure 18 The up-and-down movement of a piston in an automobile engine consists of four separate strokes. These four strokes form a cycle that is repeated many times a second by each piston. **Determine** *whether eliminating friction would make the engine 100 percent efficient.*

Friction and the Efficiency of Heat Engines Almost three fourths of the heat produced in an internal combustion engine is not converted into useful work. Friction between moving parts causes some of the work done by the engine to be converted into heat. However, even if friction were totally eliminated, a heat engine still could not convert heat completely into work and be 100 percent efficient. Instead, the efficiency of an internal combustion engine depends on the difference in the temperature of the burning gases in the cylinder and the temperature of the air outside the engine. Increasing the temperature of the burning gases makes the engine more efficient.

Heat Movers

How can the inside of a refrigerator stay cold? The second law of thermodynamics prevents heat from spontaneously flowing from inside the refrigerator to the warmer room. However, the second law of thermodynamics allows heat to move from a cold to a warm object if work is done in the process. A refrigerator does work as it moves heat from inside the refrigerator to the warmer room. The energy to do the work comes from the electrical energy the refrigerator obtains from an electrical outlet. You can think of a refrigerator as a heat mover that does work to move heat from a cooler temperature to a warmer temperature.

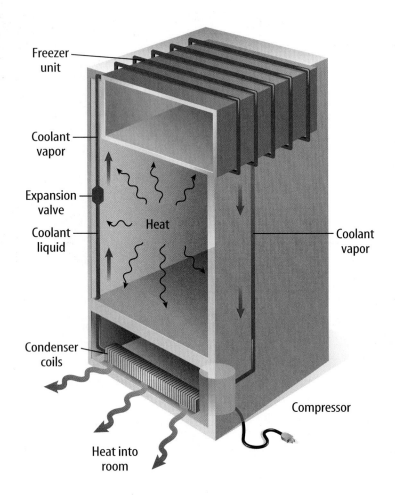

Freezer
unit

Coolant
vapor

Expansion
valve

Coolant
liquid

Heat

Coolant
vapor

Condenser
coils

Compressor

Heat into
room

Figure 19 A refrigerator must do work on the coolant in order to transfer heat from inside the refrigerator to the warmer air outside. Work is done when the compressor compresses the coolant vapor, causing its temperature to increase.

Refrigerators A refrigerator contains a coolant that is pumped through pipes on the inside and outside of the refrigerator. The coolant is a special substance that evaporates at a low temperature. **Figure 19** shows how a refrigerator operates. Liquid coolant is pumped through an expansion valve and changes into a gas. When the coolant changes to a gas, it cools. The cold gas is pumped through pipes inside the refrigerator, where it absorbs thermal energy. As a result, the inside of the refrigerator cools.

The gas then is pumped to a compressor that does work by compressing the gas. This makes the gas warmer than the temperature of the room. The warm gas is pumped through the condenser coils. Because the gas is warmer than the room, thermal energy flows from the gas to the room. Some of this heat is the thermal energy that the coolant gas absorbed from the inside of the refrigerator. As the gas gives off heat, it cools and changes to a liquid. The liquid coolant then is changed back to a gas, and the cycle is repeated.

Reading Check *How does a refrigerator do work on the coolant?*

Air Conditioners and Heat Pumps An air conditioner is another type of heat mover. It operates like a refrigerator, except that warm air from the room is forced to pass over tubes containing the coolant. The warm air is cooled and is forced back into the room. The thermal energy that is absorbed by the coolant is transferred to the air outdoors. Refrigerators and air conditioners are heat engines working in reverse—they use mechanical energy supplied by the compressor motor to move thermal energy from cooler to warmer areas.

A heat pump is a two-way heat mover. In warm weather, it operates as an air conditioner. In cold weather, a heat pump operates like an air conditioner in reverse. The coolant gas is cooled and is pumped through pipes outside the house. There, the coolant absorbs heat from the outside air. The coolant is then compressed and pumped back inside the house, where it releases heat.

The Human Coolant After exercising on a warm day, you might feel hot and be drenched with sweat. But your temperature would be close to your normal body temperature of 37°C. Your body uses evaporation to keep its internal temperature constant. When a liquid changes to a gas, energy is absorbed from the liquid's surroundings. As you exercise, your body generates sweat from tiny glands within your skin. As the sweat evaporates, it carries away heat, as shown in **Figure 20,** making you cooler.

Energy Transformations Produce Heat Every day many energy transformations occur around you that convert one form of energy into a more useful form. However, usually when these energy transformations occur, some heat is produced. For example, friction converts mechanical energy into thermal energy when the shaft of an electric motor or an electric generator rotates. The thermal energy produced in these energy transformations is no longer in a useful form and is transferred into the surroundings by conduction and convection.

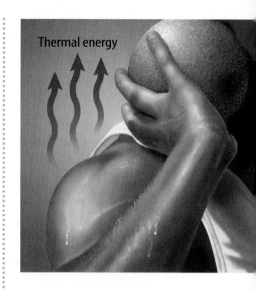
Thermal energy

Figure 20 As perspiration evaporates from your skin, it carries heat away, cooling your body.

---

section ③ review

Summary

Heating Systems
- A forced-air heating system uses a fan to force air heated by a furnace through a system of ducts.
- Radiator and electric heating systems transfer heat to rooms by conduction and convection.
- Solar heating systems convert radiant energy from the Sun to thermal energy.

Thermodynamics
- The first law of thermodynamics states that the increase in thermal energy of a system equals the work done on the system plus the heat added to the system.
- One way to state the second law of thermodynamics is that heat will not flow from a hot to a cold object unless work is done.

Converting Heat to Work
- The second law of thermodynamics states that heat cannot be converted completely into work.
- A heat engine converts heat into work.
- A refrigerator moves heat by doing work on the coolant.

Self Check

1. **Explain** how the thermal energy of a closed system changes with time.
2. **Compare and contrast** an active solar heating system with a radiator system.
3. **Explain** whether or not a heat engine could be made 100 percent efficient by eliminating friction.
4. **Diagram** how the thermal energy of the coolant changes as it flows in a refrigerator.
5. **Think Critically** Suppose you vigorously shake a bottle of fruit juice. Predict how the temperature of the juice will change. Explain your reasoning.

Applying Math

6. **Calculate Change in Thermal Energy** You push down on the handle of a bicycle pump with a force of 20 N. The handle moves 0.3 m, and the pump does not absorb or release any heat. What is the change in thermal energy of the bicycle pump?
7. **Calculate Work** The thermal energy released when a gallon of gasoline is burned in a car's engine is 140 million J. If the engine is 25 percent efficient, how much work does it do when one gallon of gasoline is burned?

C⦿nduction in Gases

Goals
- **Measure** temperature changes in air near a heat source.
- **Observe** conduction of heat in air.

Materials
thermometers (3)
foam cups (2)
400-mL beakers (2)
burner or hot plate
paring knife
thermal mitts (2)

Safety Precautions

WARNING: *Use care when handling hot water. Pour hot water using both hands.*

⦿ Real-World Question

Does smog occur where you live? If so, you may have experienced a temperature inversion. Usually the Sun warms the ground, and the air above it. When the air near the ground is warmer than the air above, convection occurs. This convection also carries smoke and other gases emitted by cars, chimneys, and smokestacks upward into the atmosphere. If the air near the ground is colder than the air above, convection does not occur. Then smoke and other pollutants can be trapped near the ground, sometimes forming smog. How does the insulating properties of air cause a temperature inversion to occur?

⦿ Procedure

1. Using the paring knife, carefully cut the bottom from one foam cup.
2. Use a pencil or pen to poke holes about 2 cm from the top and bottom of each foam cup.
3. Turn both cups upside down, and poke the ends of the thermometers through the upper holes and lower holes, so both thermometers are supported horizontally. The bulb end of both thermometers should extend into the middle of the bottomless cup.

4. Heat about 350 mL of water to about 80°C in one of the beakers.

5. Place an empty 400-mL beaker on top of the bottomless cup. Record the temperature of the two thermometers in your data table.

6. Add about 100 mL of hot water to the empty beaker. After one minute, record the temperatures of the thermometers in a data table like the one shown here.

7. Continue to record the temperatures every minute for 10 min. Add hot water as needed to keep the temperature of the water at about 80°C.

Air Temperatures in Foam Cup		
Time (min)	Upper Thermometer (°C)	Lower Thermometer (°C)
0		
1	Do not write in this book.	
2		
3		
4		
5		

Analyze Your Data

1. **Graph** the temperatures measured by the upper and lower thermometers on the same graph. Make the vertical *y*-axis the temperature and the horizontal *x*-axis the time.

2. **Calculate** the total temperature change for both thermometers by subtracting the initial temperature from the final temperature.

3. **Calculate** the average rate of temperature change for each thermometer by dividing the total temperature change by 10 min.

Conclude and Apply

1. **Explain** whether convection can occur in the foam cup if it's being heated from the top.

2. **Describe** how heat was transferred through the air in the foam cup.

3. **Explain** why the average rate of temperature change was different for each thermometer.

Communicating Your Data

Compare your results with other students in your class. **Identify** the factors caused the average rate of temperature change to be different for different groups.

SCIENCE Stats

Surprising Thermal Energy

Did you know...

...The average amount of solar energy that reaches the United States each year is about 600 times greater than the nation's annual energy demands.

...When a space shuttle reenters Earth's atmosphere at more than 28,000 km/h, its outer surface is heated by friction to nearly 1,650°C. This temperature is high enough to melt steel.

...A lightning bolt heats the air in its path to temperatures of about 25,000°C. That's about 4 times hotter than the average temperature on the surface of the Sun.

Applying Math

1. The highest recorded temperature on Earth is 58°C and the lowest is −89°C. What is the range between the highest and lowest recorded temperatures?
2. What is the average temperature of the surface of the Sun? Draw a bar graph comparing the temperature of a lightning bolt to the temperature of the surface of the Sun.
3. The Sun is almost 150 million km from Earth. How long does it take solar energy to reach Earth if it travels at 300,000 km/s?

Reviewing Main Ideas

Section 1 Temperature and Heat

1. The temperature of a material is a measure of the average kinetic energy of the molecules in the material.

2. Heat is thermal energy that flows from a higher to a lower temperature.

3. The thermal energy of an object is the total kinetic and potential energy of the molecules in the object.

4. The specific heat is the amount of heat needed to raise the temperature of 1 kg of a substance by 1°C.

Section 2 Transferring Thermal Energy

1. Conduction occurs when thermal energy is transferred by collisions between particles. Matter is not transferred when conduction occurs.

2. Convection occurs in a fluid as warmer and cooler fluid move from place to place.

3. Radiation is the transfer of energy by electromagnetic waves. Radiation can transfer energy through empty space.

4. Heat flows more easily in materials that are conductors than in insulators.

5. Some insulating materials contain pockets of trapped air that reduce the flow of heat.

Section 3 Using Heat

1. Conventional heating systems use air, hot water, and steam to transfer thermal energy through a building.

2. A solar heating system converts radiant energy from the Sun to thermal energy. Active solar systems use solar collectors to absorb the thermal radiant energy.

3. According to the first law of thermodynamics, the increase in the thermal energy of a system equals the work done on the system and the amount of heat added to the system.

4. The second law of thermodynamics states that heat cannot flow from a colder to a hotter temperature unless work is done, and that heat cannot be converted completely into work.

5. Heat engines convert heat into work. The efficiency of a heat engine can never be 100 percent. Refrigerators transfer heat from a cooler to a warmer temperature by doing work on the coolant.

FOLDABLES Use the Foldable that you made at the beginning of this chapter to help you review thermal energy.

Using Vocabulary

conduction p. 164
convection p. 165
first law of thermodynamics
  p. 175
heat p. 160
heat engine p. 176
insulator p. 169
internal combustion
  engine p. 176

radiation p. 167
second law of
  thermodynamics p. 175
solar collector p. 174
specific heat p. 161
temperature p. 159
thermal energy p. 159
thermodynamics p. 174

Complete each statement using a word(s) from the vocabulary list above.

1. A _____ is a device that converts thermal energy into mechanical energy.

2. _____ is energy that is transferred from warmer to cooler materials.

3. A _____ is a device that absorbs the Sun's radiant energy.

4. The energy required to raise the temperature of 1 kg of a material 1°C. is a material's _____.

5. _____ is a measure of the average kinetic energy of the particles in a material.

6. Heat flows easily in a(n) _____.

Checking Concepts

Choose the word or phrase that best answers the question.

7. Which is NOT a method of heat transfer?
   A) conduction
   B) specific heat
   C) radiation
   D) convection

8. In which of the following devices is fuel burned inside chambers called cylinders?
   A) internal combustion engine
   B) radiator
   C) heat pump
   D) air conditioner

9. During which phase of a four-stroke engine are waste gases removed?
   A) power stroke
   B) intake stroke
   C) compression stroke
   D) exhaust stroke

10. Which of the following materials is a poor insulator of heat?
    A) iron
    B) feathers
    C) air
    D) plastic

11. Which of the following devices is an example of a heat mover?
    A) solar panel
    B) refrigerator
    C) internal combustion engine
    D) diesel engine

12. Which term describes the measure of the average kinetic energy of the particles in an object?
    A) potential energy
    B) thermal energy
    C) temperature
    D) specific heat

13. Which of these is NOT used to calculate change in thermal energy?
    A) volume
    B) temperature change
    C) specific heat
    D) mass

14. Which of the following processes does NOT require the presence of particles of matter?
    A) radiation
    B) conduction
    C) convection
    D) combustion

15. Which of the following is the name for thermal energy that is transferred only from a higher temperature to a lower temperature?
    A) potential energy
    B) kinetic energy
    C) heat
    D) solar energy

Science online   gpscience.com/vocabulary_puzzlemaker

Interpreting Graphics

16. Complete the following events-chain concept map to show how an active solar heating system works.

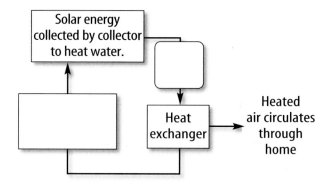

17. Copy and complete this concept map.

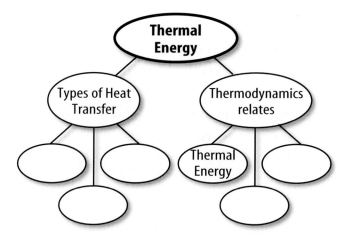

Thinking Critically

18. **Explain** On a hot day a friend suggests that you can make your kitchen cooler by leaving the refrigerator door open. Explain whether leaving the refrigerator door open would cause the air temperature in the kitchen to decrease.

19. **Explain** Which has the greater amount of thermal energy, one liter of water at 50°C or two liters of water at 50°C?

20. **Explain** whether or not the following statement is true: If the thermal energy of an object increases, the temperature of the object must also increase.

21. **Predict** Suppose a beaker of water is heated from the top. Predict which is more likely to occur in the water—heat transfer by conduction or convection. Explain.

22. **Classify** Order the events that occur in the removal of heat from an object by a refrigerator. Draw the complete cycle, from the placing of a warm object in the refrigerator to the changes in the coolant.

Applying Math

Use the table below to answer questions 23 to 25.

Specific Heat of Materials	
Material	**Specific Heat (J/kg°C)**
Water	4,184
Copper	385
Silver	235
Graphite	710
Iron	450

23. **Calculate Thermal Energy** How much thermal energy is needed to raise the temperature of 4.0 kg of water from 25°C to 75°C?

24. **Calculate Temperature Change** How does the temperature of 33.0 g of graphite change when it absorbs 350 J of thermal energy?

25. **Calculate Mass** A hot iron ball is dropped into 200.0 g of cooler water. The water temperature increases by 2.0°C and the temperature of the ball decreases by 18.6°C. What is the mass of the iron ball?

Record your answers on the answer sheet provided by your teacher or on a sheet of paper.

1. The difference between the boiling point and the freezing point of potassium is 695.72 K. What is the difference between the two points on the Celsius temperature scale?
   A. 100.00°C
   C. 422.57°C
   B. 275.15°C
   D. 695.72°C

Use the table below to answer questions 2 and 3.

Material	Specific Heat [J/(kg °C)]
Copper	385
Gold	449
Lead	129
Tin	228
Zinc	388

2. According to the table above, a 2-kg block of which of the following materials would require 898 joules of heat to increase its temperature by 1°C?
   A. gold
   C. tin
   B. lead
   D. zinc

3. Which of the following materials would require the most heat to raise a 5-kg sample of the material from 10°C to 50°C?
   A. gold
   C. tin
   B. lead
   D. zinc

4. Automobile engines usually are four-stroke engines. During which stroke does the spark from a spark plug ignite the fuel-air mixture?
   A. the intake stroke
   B. the compression stroke
   C. the power stroke
   D. the exhaust stroke

5. A refrigerator is an example of what type of device that removes thermal energy from one location and transfers it to another location at a different temperature?
   A. condenser
   C. heat mover
   B. conductor
   D. heat pump

6. The temperature of a 24.5-g block of aluminum decreases from 30.0°C to 21.5°C. If aluminum has a specific heat of 897 J/(kg°C), what is the change in thermal energy of the block of aluminum?
   A. 187 J
   C. 5,820 J
   B. 2,590 J
   D. 187,000 J

Use the figure below to answer question 7.

7. The photograph above shows a pot of boiling water. What type of heat transfer causes the water at the top of the pot to become hot?
   A. conduction
   B. convection
   C. convection and radiation
   D. conduction and convection

Test-Taking Tip

Read All the Information On a bar graph, line up each bar with its corresponding value by laying your pencil between the two points.

Part 2 | Short Response/Grid In

Record your answers on the answer sheet provided by your teacher or on a sheet of paper.

8. Define the term *heat* and tell what units are used to measure heat.

9. Give an example of how waves transfer energy by radiation. Give an example of how energy is transferred by conduction.

10. How is the color of a material related to its absorption and reflection of radiant energy?

11. What property of water makes it useful as a coolant?

Use the figure below to answer questions 12 and 13.

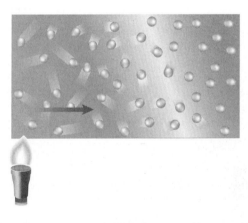

12. The illustration above shows a heat source below one end of a material. Heat is transferred from the warmer part of the material to the cooler part. Name and describe the type of heat transfer illustrated in the figure.

13. If the heat source were replaced by a block of ice, would cold be transferred through the material in the same way? Explain why or why not.

14. How is a heat pump different from an air conditioner?

Part 3 | Open Ended

Record your answers on a sheet of paper.

Use the figure below to answer question 15.

15. The calorimeter in the illustration above is composed of inner and outer chambers that surround a thick layer of air. Describe the process by which the calorimeter is used to measure the specific heat of materials.

16. Define the terms *temperature* and *thermal energy*. Explain how the temperature and thermal energy of an object are related.

17. Conduction can occur in solids, liquids, and gases. Explain why solids and liquids are better conductors of heat than gases.

18. Explain why radiation usually passes more easily through gases than through solids or liquids.

19. Suppose you have a glass half-filled with 250 mL of water at a temperature of 30°C. You then add 250 mL of water at the same temperature to the glass. Explain any changes in the water's temperature and thermal energy.

20. Explain how changes in a fluid's density enables convection to occur.

How Are Clouds & Toasters Connected?

NATIONAL GEOGRAPHIC

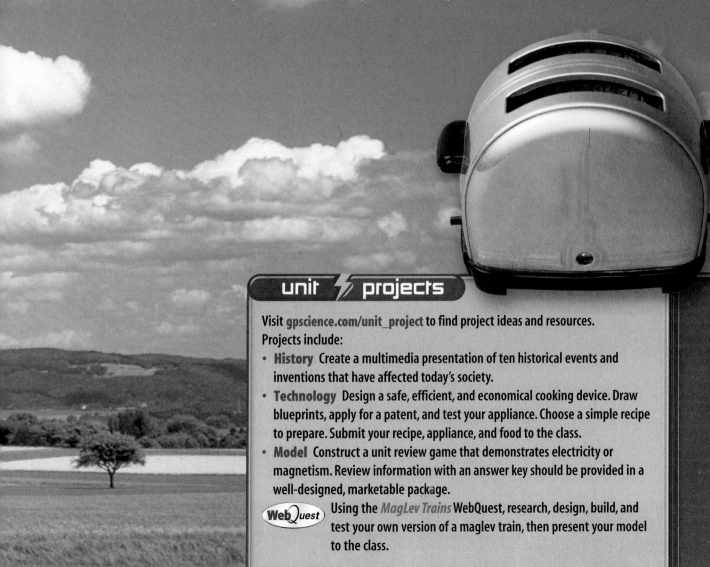

In the late 1800s, a mysterious form of radiation called X rays was discovered. One French physicist wondered whether uranium would give off X rays after being exposed to sunlight. He figured that if X rays were emitted, they would make a bright spot on a wrapped photographic plate. But the weather turned cloudy, so the physicist placed the uranium and the photographic plate together in a drawer. Later, on a hunch, he developed the plate and found that the uranium had made a bright spot anyway. The uranium was giving off some kind of radiation even without being exposed to sunlight! Scientists soon determined that the atoms of uranium are radioactive—that is, they give off particles and energy from their nuclei. In today's nuclear power plants, this energy is harnessed and converted into electricity. This electricity provides some of the power used in homes to operate everything from lamps to toasters.

unit ⚡ projects

Visit **gpscience.com/unit_project** to find project ideas and resources. Projects include:

- **History** Create a multimedia presentation of ten historical events and inventions that have affected today's society.
- **Technology** Design a safe, efficient, and economical cooking device. Draw blueprints, apply for a patent, and test your appliance. Choose a simple recipe to prepare. Submit your recipe, appliance, and food to the class.
- **Model** Construct a unit review game that demonstrates electricity or magnetism. Review information with an answer key should be provided in a well-designed, marketable package.

WebQuest Using the *MagLev Trains* WebQuest, research, design, build, and test your own version of a maglev train, then present your model to the class.

Electricity

chapter preview

Shine on Brightly

Electricity lights these city lights so that people can continue to work and have fun, day or night. Electric lights and other electric devices operate by converting electrical energy into other forms of energy.

Science Journal   For five electric devices, list the form of energy electrical energy is converted into by each device.

Start-Up Activities

Electric Circuits

No lights! No CD players! No computers, video games or TVs! Without electricity, many of the things that make your life enjoyable wouldn't exist. For these devices to operate, electric current must flow in the electric circuits that are part of the device. Under what conditions does electric current flow in an electric circuit?

1. Obtain a battery, a flashlight bulb, and some wire.

2. Connect the materials so that the lightbulb lights.

3. Draw diagrams of all the ways that you were able to light the bulb.

4. Record a few of the ways that didn't work.

5. Can you light the bulb using only one wire and one battery?

6. **Think Critically** Write a paragraph describing the requirements to light the bulb. Write out a procedure for lighting the bulb and have a classmate follow your procedure.

Electricity Make the following Foldable to help you organize information about electricity.

STEP 1 **Fold** a vertical sheet of paper from top to bottom. Make the top edge about 2 cm shorter than the bottom edge.

STEP 2 **Turn** lengthwise and fold into thirds.

STEP 3 **Unfold and cut** only the top layer along both folds to make three tabs.

STEP 4 **Label** the Foldable as shown.

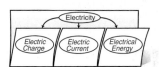

Organize Information As you read Chapter 7, organize the information you find about electric charge, electric current, and electrical energy under the appropriate tab.

Preview this chapter's content and activities at gpscience.com

Electric Charge

What You'll Learn

- **Describe** how electric charges exert forces on each other.
- **Compare** the strengths of electric and gravitational forces.
- **Distinguish** between conductors and insulators
- **Explain** how objects become electrically charged.

Why It's Important

The electrical energy that all electrical devices use comes from the forces electric charges exert on each other.

Review Vocabulary

atom: the smallest particle of an element

New Vocabulary

- static electricity
- law of conservation of charge
- conductor
- insulator
- charging by contact
- charging by induction

Positive and Negative Charge

Why does walking across a carpeted floor and then touching something sometimes result in a shock? The answer has to do with electric charge. Atoms contain particles called protons, neutrons, and electrons, as shown in **Figure 1.** Protons and electrons have electric charge, and neutrons have no electric charge.

There are two types of electric charge. Protons have positive electric charge and electrons have negative electric charge. The amount of positive charge on a proton equals the amount of negative charge on an electron. An atom contains equal numbers of protons and electrons, so the positive and negative charges cancel out and an atom has no net electric charge. Objects with no net charge are said to be electrically neutral.

Transferring Charge Electrons are bound more tightly to some atoms and molecules. For example, compared to the electrons in carpet atoms, electrons are bound more tightly to the atoms in the soles of your shoes. **Figure 2** shows that when you walk on the carpet, electrons are transferred from the carpet to the soles of your shoes. The soles of your shoes have an excess of electrons and become negatively charged. The carpet has lost electrons and has an excess of positive charge. The carpet has become positively charged. The accumulation of excess electric charge on an object is called **static electricity.**

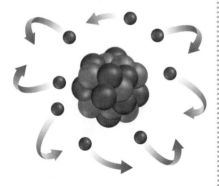

Figure 1 The center of an atom contains protons (orange) and neutrons (blue). Electrons (red) swarm around the atom's center.

Before the shoe scuffs against the carpet, both the sole of the shoe and the carpet are electrically neutral.

As the shoes scuff against the carpet, electrons are transferred from the carpet to the soles of the shoes.

Conservation of Charge When an object becomes charged, charge is neither created nor destroyed. Usually it is electrons that have moved from one object to another. According to the **law of conservation of charge,** charge can be transferred from object to object, but it cannot be created or destroyed. Whenever an object becomes charged, electric charges have moved from one place to another.

Figure 2 Atoms in the shoe's sole hold their electrons more tightly than atoms in the carpet hold their electrons.

✔ **Reading Check** *How does an object become charged?*

Charges Exert Forces Have you noticed how clothes sometimes cling together when removed from the dryer? These clothes cling together because of the forces electric charges exert on each other. **Figure 3** shows that unlike charges attract other, and like charges repel each other. The force between electric charges also depends on the distance between charges. The force decreases as the charges get farther apart.

Just as for two electric charges, the force between any two objects that are electrically charged decreases as the objects get farther apart. This force also depends on the amount of charge on each object. As the amount of charge on either object increases, the electrical force also increases.

As clothes tumble in a dryer, the atoms in some clothes gain electrons and become negatively charged. Meanwhile the atoms in other clothes lose electrons and become positively charged. Clothes that are oppositely charged attract each other and stick together.

Opposite charges attract

Like charges repel

Figure 3 Positive and negative charges exert forces on each other.

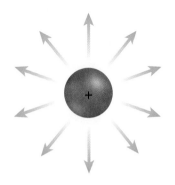

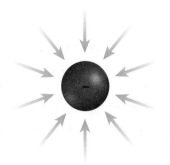

Figure 4 Surrounding every electric charge is an electric field that exerts forces on other electric charges. The arrows point in the direction a positive charge would move.

Electric Fields You might have seen bits of paper fly up and stick to a charged balloon. The bits of paper do not need to touch the charged balloon for an electric force to act on them. If the balloon and the paper are not touching, what causes the paper to move?

An electric field surrounds every electric charge, as shown in **Figure 4,** and exerts the force that causes other electric charges to be attracted or repelled. Any charge that is placed in an electric field will be pushed or pulled by the field. Electric fields are represented by arrows that show how the electric field would make a positive charge move.

Comparing Electric and Gravitational Forces The force of gravity between you and Earth seems to be strong. Yet, compared with electric forces, the force of gravity is much weaker. For example, the attractive electric force between a proton and an electron in a hydrogen atom is about a thousand trillion trillion trillion times larger, or 10^{36} times larger, than the attractive gravitational force between the two particles.

In fact, all atoms are held together by electric forces between protons and electrons that are tremendously larger than the gravitational forces between the same particles. The chemical bonds that form between atoms in molecules also are due to the electric forces between the atoms. These electric forces are much larger than the gravitational forces between the atoms.

 *Compare the strength of electric and gravitational forces between protons and electrons.*

Figure 5 As you walk across a carpeted floor, excess electrons can accumulate on your body. When you reach for a metal doorknob, electrons flow from your hand to the doorknob and you see a spark.

However, the electric forces between the objects around you are much less than the gravitational forces between them. Most objects that you see are nearly electrically neutral and have almost no net electric charge. As a result, there is usually no noticeable electric force between these objects. But even if a small amount of charge is transferred from one object to another, the electric force between the objects can be noticeable.

For example, you probably have noticed your hair being attracted to a rubber comb after you comb your hair. Transferring about one trillionth of the electrons in a single hair to the comb results in an electric force strong enough to overcome the force of gravity on the strand of hair.

Conductors and Insulators

If you reach for a metal doorknob after walking across a carpet, you might see a spark. The spark is caused by electrons moving from your hand to the doorknob, as shown in **Figure 5.** Recall that electrons were transferred from the carpet to your shoes. How did these electrons move from your shoes to your hand?

Conductors A material in which electrons are able to move easily is a **conductor.** Electrons on your shoes repel each other and some are pushed onto your skin. Because your skin is a better conductor than your shoes, the electrons spread over your skin, including your hand.

The best electrical conductors are metals. The atoms in metals have electrons that are able to move easily through the material. Electric wires usually are made of copper because copper metal is one of the best conductors.

Insulators A material in which electrons are not able to move easily is an **insulator.** Electrons are held tightly to atoms in insulators. Most plastics are insulators. The plastic coating around electric wires, shown in **Figure 6,** prevents a dangerous electric shock when you touch the wire. Other good insulators are wood, rubber, and glass.

Charging Objects

You might have noticed socks clinging to each other after they have been tumbling in a clothes dryer. Rubbing two materials together can result in a transfer of electrons. Then one material is left with a positive charge and the other with an equal amount of negative charge. The process of transferring charge by touching or rubbing is called **charging by contact.**

Figure 6 The plastic coating around wires is an insulator. A damaged electrical cord is hazardous when the conducting wire is exposed.

195

Figure 7 The balloon on the left is neutral. The balloon on the right is negatively charged. It produces a positively charged area on the sleeve by repelling electrons.
Determine *the direction of the force acting on the balloon.*

Charging at a Distance

Because electrical forces act at a distance, charged objects brought near a neutral object will cause electrons to rearrange their positions on the neutral object. Suppose you charge a balloon by rubbing it with a cloth. If you bring the negatively charged balloon near your sleeve, the extra electrons on the balloon repel the electrons in the sleeve. The electrons near the sleeve's surface move away from the balloon, leaving a positively charged area on the surface of the sleeve, as shown in **Figure 7.** As a result, the negatively charged balloon attracts the positively charged area of the sleeve. The rearrangement of electrons on a neutral object caused by a nearby charged object is called **charging by induction.** The sweater was charged by induction. The balloon will now cling to the sweater, being held there by an electrical force.

Lightning Have you ever seen lightning strike Earth? Lightning is a large static discharge. A static discharge is a transfer of charge between two objects because of a buildup of static electricity. A thundercloud is a mighty generator of static electricity. As air masses move and swirl in the cloud, areas of positive and negative charge build up. Eventually, enough charge builds up to cause a static discharge between the cloud and the ground. As the electric charges move through air, they collide with atoms and molecules. These collisions cause the atoms and molecules in air to emit light. You see this light as a spark, as shown in **Figure 8.**

Thunder Not only does lightning produce a brilliant flash of light, it also generates powerful sound waves. The electrical energy in a lightning bolt rips electrons off atoms in the atmosphere and produces great amounts of heat. The surrounding air temperature can rise to about 30,000°C—several times hotter than the Sun's surface. The heat causes air in the bolt's path to expand rapidly, producing sound waves that you hear as thunder.

The sudden discharge of so much energy can be dangerous. It is estimated that Earth is struck by lightning about 100 times every second. Lightning strikes can cause power outages, injury, loss of life, and fires.

Science Online

Topic: Lightning
Visit gpscience.com for Web links to information about lightning strikes.

Activity Make a table listing tips on how people can protect themselves from lightning.

Figure 8

Storm clouds can form when humid, sun-warmed air rises to meet a colder air layer. As these air masses churn together, the stage is set for the explosive electrical display we call lightning. Lightning strikes when negative charges at the bottom of a storm cloud are attracted to positive charges on the ground.

A Convection currents in the storm cloud cause charge separation. The top of the cloud becomes positively charged, the bottom negatively charged.

B Negative charges on the bottom of the cloud induce a positive charge on the ground below the cloud by repelling negative charges in the ground.

C When the bottom of the cloud has accumulated enough negative charges, the attraction of the positive charges below causes electrons in the bottom of the cloud to move toward the ground.

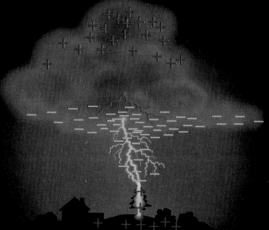

D When the electrons get close to the ground, they attract positive charges that surge upward, completing the connection between cloud and ground. This is the spark you see as a lightning flash.

INTRA-CLOUD LIGHTNING never strikes Earth and can occur ten times more often in a storm than cloud-to-ground lightning.

Figure 9 A lightning rod directs the charge from a lightning bolt safely to the ground.

Grounding The sensitive electronics in a computer can be harmed by large static discharges. A discharge can occur any time that charge builds up in one area. Providing a path for charge to reach Earth prevents any charge from building up. Earth is a large, neutral object that is also a conductor of charge. Any object connected to Earth by a good conductor will transfer any excess electric charge to Earth. Connecting an object to Earth with a conductor is called grounding. For example, buildings often have a metal lightning rod that provides a conducting path from the highest point on the building to the ground to prevent damage by lightning, as shown in **Figure 9.**

Plumbing fixtures, such as metal faucets, sinks, and pipes, often provide a convenient ground connection. Look around. Do you see anything that might act as a path to the ground?

Detecting Electric Charge

The presence of electric charges can be detected by an electroscope. One kind of electroscope is made of two thin, metal leaves attached to a metal rod with a knob at the top. The leaves are allowed to hang freely from the metal rod. When the device is not charged, the leaves hang straight down, as shown in **Figure 10A.**

Suppose a negatively charged balloon touches the knob. Because the metal is a good conductor, electrons travel down the rod into the leaves. Both leaves become negatively charged as they gain electrons, as shown in **Figure 10B.** Because the leaves have similar charges, they repel each other.

If a glass rod is rubbed with silk, electrons move away from the atoms in the glass rod and build up on the silk. The glass rod becomes positively charged.

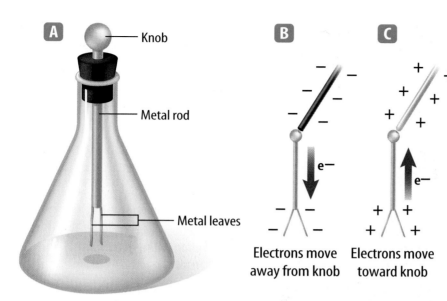

A Knob

Metal rod

Metal leaves

B Electrons move away from knob

C Electrons move toward knob

Figure 10 Notice the position of the leaves on the electroscope when they are **A** uncharged, **B** negatively charged, and **C** positively charged.
Infer *How can you tell whether an electroscope is positively or negatively charged?*

When the positively charged glass rod is brought into contact with the metal knob of an uncharged electroscope, electrons flow out of the metal leaves and onto the rod. The leaves repel each other because each leaf becomes positively charged as it loses electrons, as shown in **Figure 10C.**

section 1 review

Summary

Positive and Negative Charge

- There are two types of electric charge— positive charge and negative charge.
- Electric charges can be transferred between objects, but cannot be created or destroyed.
- Like charges repel and unlike charges attract.
- An electric charge is surrounded by an electric field that exerts forces on other charges.

Electrical Conductors and Insulators

- A conductor contains electrons that can move easily. The best conductors are metals.
- The electrons in an electrical insulator do not move easily. Rubber, glass, and most plastics are examples of insulators.

Charging Objects

- Electric charge can be transferred between objects by bringing them into contact.
- Charging by induction occurs when the electric field around a charged object rearranges electrons in a nearby neutral object.

Checking Concepts

1. **Define** static electricity.
2. **Describe** how lightning is produced.
3. **Explain** why if charge cannot be created or destroyed, electrically neutral objects can become electrically charged.
4. **Predict** what would happen if you touched the knob of a positively charged electroscope with another positively charged object.
5. **Think Critically** Humid air is a better electrical conductor than dry air. Explain why you're more likely to receive a shock after walking across a carpet when the air is dry than when the air is humid.

Applying Math

6. **Determine Lightning Strikes** Suppose Earth is struck by 100 lighting strikes each second. How many times is Earth struck by lightning in one day?
7. **Calculate Electric Force** A balloon with a mass of 0.020 kg is charged by rubbing and then is stuck to the ceiling. If the acceleration of gravity is 9.8 m/s^2, what is the electrical force on the balloon?

Electric Current

Reading Guide

What You'll Learn
- **Describe** how voltage difference causes current to flow.
- **Explain** how batteries produce a voltage difference in a circuit.
- **List** the factors that affect an object's electrical resistance.
- **Define** Ohm's law.

Why It's Important
You control electric current every time you change the volume on a TV, stereo, or CD player.

Review Vocabulary
pressure: amount of force exerted per unit area

New Vocabulary
- electric current
- voltage difference
- circuit
- resistance
- Ohm's law

Figure 11 Electric forces in a material cause electric current to flow, just as forces in the water cause water to flow.

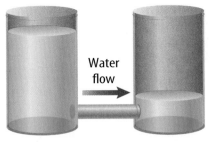

High pressure   Low pressure

The force that causes water to flow is related to a pressure difference.

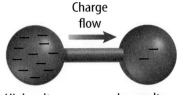

Charge flow

High voltage   Low voltage

The force that causes a current to flow is related to a voltage difference.

Current and Voltage Difference

When a spark jumps between your hand and a metal doorknob, electric charges move quickly from one place to another. The net movement of electric charges in a single direction is an **electric current.** In a metal wire, or any material, electrons are in constant motion in all directions. As a result, there is no net movement of electrons in one direction. However, when an electric current flows in the wire, electrons continue their random movement, but they also drift in the direction that the current flows.

Electric current is measured in amperes. One ampere is equal to 6,250 million billion electrons flowing past a point every second.

✔ **Reading Check** *What is electric current?*

Voltage Difference The movement of an electron in an electric current is similar to a ball bouncing down a flight of stairs. Even though the ball changes direction when it strikes a stair, the net motion of the ball is downward. The downward motion of the ball is caused by the force of gravity. When a current flows, the net movement of electric charges is caused by an electric force acting on the charges.

In some ways, the electric force that causes charges to flow is similar to the force acting on the water in a pipe. Water flows from higher pressure to lower pressure, as shown in **Figure 11.** In a similar way, electric charge flows from higher voltage to lower voltage. A **voltage difference** is related to the force that causes electric charges to flow. Voltage difference is measured in volts.

Figure 12 Water or electric current will flow continually only through a closed loop. If any part of the loop is broken or disconnected, the flow stops.

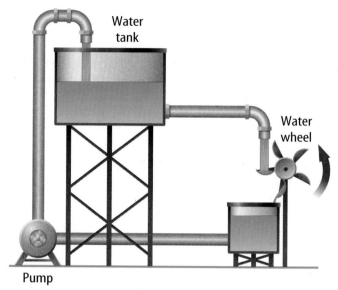

A pump provides the pressure difference that keeps water flowing.

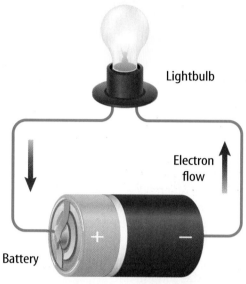

A battery provides the voltage difference that keeps electric current flowing.

Electric Circuits A way to have flowing water perform work is shown in **Figure 12.** Water flows out of the tank and falls on a paddle wheel, causing it to rotate. A pump then provides a pressure difference that lifts the water back up into the tank. The constant flow of water would stop if the pump stopped working. The flow of water also would stop if one of the pipes broke. Then water no longer could flow in a closed loop, and the paddle wheel would stop rotating.

Figure 12 also shows an electric current doing work by lighting a lightbulb. Just as the water current stops flowing if there is no longer a closed loop to flow through, the electric current stops if there is no longer a closed path to follow. A closed path that electric current follows is a **circuit.** If the circuit in Figure 12 is broken by removing the battery, or the light bulb, or one of the wires, current will not flow.

Batteries

In order to keep water flowing continually in the water circuit in **Figure 12,** a pump is used to provide a pressure difference. In a similar way, to keep an electric current continually flowing in the electric circuit in **Figure 12,** a voltage difference needs to be maintained in the circuit. A battery can provide the voltage difference that is needed to keep current flowing in a circuit. Current flows as long as there is a closed path that connects one battery terminal to the other battery terminal.

Investigating Battery Addition

Procedure

1. Make a circuit by using wire to link two **bulbs** and one **D-cell battery** in a loop. Observe the brightness of the bulbs.
2. Assemble a new circuit by linking two bulbs and two D-cell batteries in a loop. Observe the brightness of the bulbs.

Analysis

1. What is the voltage difference of each D cell? Add them together to find the total voltage difference for the circuit you tested in step 2.
2. Assuming that a brighter bulb indicates a greater current, what can you conclude about the relationship between the voltage difference and current?

Dry-Cell Batteries You probably are most familiar with dry-cell batteries. A cell consists of two electrodes surrounded by a material called an electrolyte. The electrolyte enables charges to move from one electrode to the other. Look at the dry cell shown in **Figure 13.** One electrode is the carbon rod, and the other is the zinc container. The electrolyte is a moist paste containing several chemicals. The cell is called a dry cell because the electrolyte is a moist paste, and not a liquid solution.

INTEGRATE Chemistry When the two terminals of a dry-cell battery are connected in a circuit, such as in a flashlight, a reaction involving zinc and several chemicals in the paste occurs. Electrons are transferred between some of the compounds in this chemical reaction. As a result, the carbon rod becomes positive, forming the positive (+) terminal. Electrons accumulate on the zinc, making it the negative (−) terminal.

The voltage difference between these two terminals causes current to flow through a closed circuit. You make a battery when you connect two or more cells together to produce a higher voltage difference.

Wet-Cell Batteries Another commonly used type of battery is the wet-cell battery. A wet cell, like the one shown in **Figure 13,** contains two connected plates made of different metals or metallic compounds in a conducting solution. A wet-cell battery contains several wet cells connected together.

Figure 13 Chemical reactions in batteries produce a voltage difference between the positive and negative terminals. **Identify** *when these chemical reactions occur.*

Positive terminal
Plastic insulator
Moist paste
Carbon rod
Zinc container
Negative terminal

Dry cell

In this dry cell, chemical reactions in the moist paste transfer electrons to the zinc container.

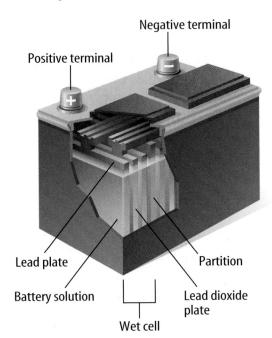

Negative terminal
Positive terminal
Lead plate
Battery solution
Partition
Lead dioxide plate
Wet cell

In this wet cell, chemical reactions transfer electrons from the lead plates to the lead dioxide plates.

Lead-Acid Batteries Most car batteries are lead-acid batteries, like the wet-cell battery shown in **Figure 13.** A lead-acid battery contains a series of six wet cells made up of lead and lead dioxide plates in a sulfuric acid solution. The chemical reaction in each cell provides a voltage difference of about 2 V, giving a total voltage difference of 12 V. As a car is driven, the alternator recharges the battery by sending current through the battery in the opposite direction to reverse the chemical reaction.

A voltage difference is provided at electrical outlets, such as a wall socket. This voltage difference usually is higher than the voltage difference provided by batteries. Most types of household devices are designed to use the voltage difference supplied by a wall socket. In the United States, the voltage difference across the two holes in a wall socket is usually 120 V. Some wall sockets supply 240 V, which is required by appliances such as electric ranges and electric clothes dryers.

Resistance

Flashlights use dry-cell batteries to provide the electric current that lights a lightbulb. What makes a lightbulb glow? Look at the lightbulb in **Figure 14.** Part of the circuit through the bulb is a thin wire called a filament. As the electrons flow through the filament, they bump into the metal atoms that make up the filament. In these collisions, some of the electrical energy of the electrons is converted into thermal energy. Eventually, the metal filament becomes hot enough to glow, producing radiant energy that can light up a dark room.

Resisting the Flow of Current Electric current loses energy as it moves through the filament because the filament resists the flow of electrons. **Resistance** is the tendency for a material to oppose the flow of electrons, changing electrical energy into thermal energy and light. With the exception of some substances that become superconductors at low temperatures, all materials have some electrical resistance. Electrical conductors have much less resistance than insulators. Resistance is measured in ohms (Ω).

Copper is an excellent conductor and has low resistance to the flow of electrons. Copper is used in household wiring because only a small amount of electrical energy is converted to thermal energy as current flows in copper wires.

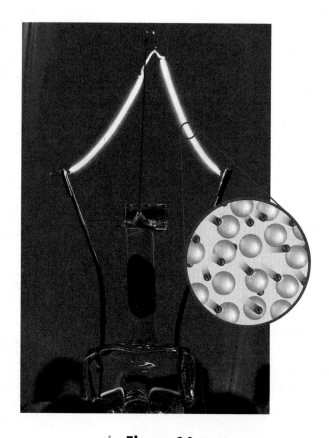

Figure 14 As electrons move through the filament in a lightbulb, they bump into metal atoms. Due to the collisions, the metal heats up and starts to glow.
Describe *the energy conversions that occur in a lightbulb filament.*

Temperature, Length, and Thickness The electric resistance of most materials usually increases as the temperature of the material increases. The resistance of an object such as a wire also depends on the length and diameter of the wire. The resistance of a wire, or any conductor, increases as the wire becomes longer. The resistance also increases as the wire becomes thinner.

In a 60 watt lightbulb, the filament is a piece of tungsten wire made into a short coil a few cm long. The uncoiled wire is about 2 m long and only about 0.25 mm thick. Even though tungsten metal is a good conductor, by making the wire thin and long, the resistance of the filament is made large enough to cause the bulb to glow.

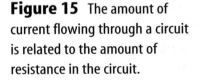 **Reading Check** *How does changing the length and thickness of a wire affect its resistance?*

The Current in a Simple Circuit

A simple electric circuit contains a source of voltage difference, such as a battery, a device, such as lightbulb, that has resistance, and conductors that connect the device to the battery terminals. When the wires are connected to the battery terminals, current flows in the closed path. An example of a simple circuit is shown in **Figure 15.**

The voltage difference, current, and resistance in a circuit are related. If the voltage difference doesn't change, decreasing the resistance increases the current in the circuit, as shown in **Figure 15.** Also, if the resistance doesn't change, increasing the voltage difference increases the current.

Figure 15 The amount of current flowing through a circuit is related to the amount of resistance in the circuit.

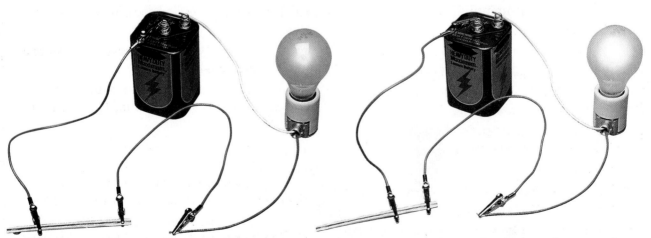

When the clips on the graphite rod are farther apart, the resistance of the rod in the circuit is larger. As a result, less current flows in the circuit and the lightbulb is dim.

When the clips on the graphite rod are closer together, the resistance of the rod in the circuit is less. As a result, more current flows in the circuit and the lightbulb is brighter.

Ohm's Law The relationship between voltage difference, current and resistance in a circuit is known as Ohm's law. According to **Ohm's law,** the current in a circuit equals the voltage difference divided by the resistance. If I stands for electric current, Ohm's law can be written as the following equation.

> ### Ohm's Law
>
> $$\text{current (in amperes)} = \frac{\text{voltage difference (in volts)}}{\text{resistance (in ohms)}}$$
>
> $$I = \frac{V}{R}$$

Ohm's law provides a way to measure the resistance of objects and materials. First the equation above is written as:

$$R = \frac{V}{I}$$

An object is connected to a source of voltage difference and the current flowing in the circuit is measured. The object's resistance then equals the voltage difference divided by the measured current.

Current and the Human Body When an electric shock occurs, an electric current moves through some part of the body. The damage caused by an electric shock depends on how large the current is. Research the effects of current on the human body. Make a table showing the effects on the body at different amounts of current.

section 2 review

Summary

Current and Voltage Difference

- Electric current is the net movement of electric charge in a single direction.
- A voltage difference is related to the force that causes charges to flow.
- A circuit is a closed, conducting path.

Batteries

- Chemical reactions in a battery produce a voltage difference between the positive and negative battery terminals.
- Two commonly used types of batteries are dry-cell batteries and wet-cell batteries.

Resistance and Ohm's Law

- Resistance is the tendency of a material to oppose the flow of electrons.
- Ohm's law relates the current, I, resistance, R, and voltage difference, V, in a circuit:

$$I = \frac{V}{R}$$

Self Check

1. **Compare and contrast** a current traveling through a circuit with a static discharge.
2. **Explain** how a carbon-zinc dry cell produces a voltage difference between the positive and negative terminals.
3. **Identify** two ways to increase the current in a simple circuit.
4. **Compare and contrast** the flow of water in a pipe and the flow of electrons in a wire.
5. **Think Critically** Explain how the resistance of a lightbulb filament changes after the light has been turned on.

Applying Math

6. **Calculate** the voltage difference in a circuit with a resistance of 25 Ω if the current in the circuit is 0.5 A.
7. **Calculate Resistance** A current of 0.5 A flows in a 60-W lightbulb when the voltage difference between the ends of the filament is 120 V. What is the resistance of the filament?

Identifying Conductors and Insulators

The resistance of an insulator is so large that only a small current flows when it is connected in a circuit. As a result, a lightbulb connected in a circuit with an insulator usually will not glow. In this lab, you will use the brightness of a lightbulb to identify conductors and insulators.

● Real-World Question

What materials are conductors and what materials are insulators?

Goals
- ■ **Identify** conductors and insulators.
- ■ **Describe** the common characteristics of conductors and insulators.

Materials
battery	bulb holder
flashlight bulb	insulated wire

Safety Precautions

● Procedure

1. Set up an incomplete circuit as pictured in the photograph.
2. Touch the free bare ends of the wires to various objects around the room. Test at least 12 items.
3. Copy the table below. In your table, record which materials make the lightbulb glow and which don't.

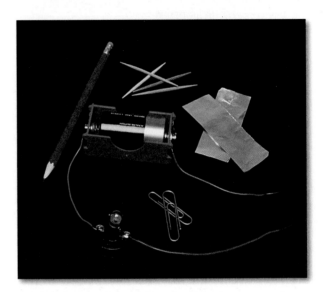

Material Tested with Lightbulb Circuit	
Lightbulb Glows	Lightbulb Doesn't Glow
Do not write in this book.	

● Conclude and Apply

1. Is there a pattern to your data?
2. Do all or most of the materials that light the lightbulb have something in common?
3. Do all or most of the materials that don't light the lightbulb have something in common?
4. **Explain** why one material may allow the lightbulb to light and another prevent the lightbulb from lighting.
5. **Predict** what other materials will allow the lightbulb to light and what will prevent the lightbulb from lighting.
6. **Classify** all the materials you have tested as conductors or insulators.

𝒞ommunicating Your Data

Compare your conclusions with those of other students in your class. **For more help, refer to the** Science Skill Handbook.

Electrical Energy

Series and Parallel Circuits

Look around. How many electrical devices such as lights, clocks, stereos, and televisions do you see that are plugged into electrical outlets? Circuits usually include three components. One is a source of voltage difference that can be provided by a battery or an electrical outlet. Another is one or more devices that use electrical energy. Circuits also include conductors such as wires that connect the devices to the source of voltage difference to form a closed path.

Think about using a hair dryer. The dryer must be plugged into an electrical outlet to operate. A generator at a power plant produces a voltage difference across the outlet, causing charges to move when the circuit is complete. The dryer and the circuit in the house contain conducting wires to carry current. The hair dryer turns the electrical energy into thermal energy and mechanical energy. When you unplug the hair dryer or turn off its switch, you open the circuit and break the path of the current. To use electrical energy, a complete circuit must be made. There are two kinds of circuits.

Series Circuits One kind of circuit is called a series circuit. In a **series circuit,** the current has only one loop to flow through, as shown in **Figure 16.** Series circuits are used in flashlights and some holiday lights.

✔ Reading Check *How many loops are in a series circuit?*

Figure 16 A series circuit provides only one path for the current to follow.
Infer *What happens to the brightness of each bulb as more bulbs are added?*

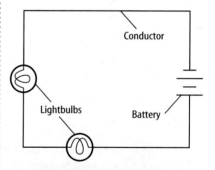

Conductor

Lightbulbs   Battery

Open Circuit If you have ever decorated a window or a tree with a string of lights, you might have had the frustrating experience of trying to find one burned-out bulb. How can one faulty bulb cause the whole string to go out? Because the parts of a series circuit are wired one after another, the amount of current is the same through every part. When any part of a series circuit is disconnected, no current flows through the circuit. This is called an open circuit. The burned-out bulb causes an open circuit in the string of lights.

Parallel Circuits What would happen if your home were wired in a series circuit and you turned off one light? This would cause an open circuit, and all the other lights and appliances in your home would go out, too. This is why houses are wired with parallel circuits. **Parallel circuits** contain two or more branches for current to move through. Look at the parallel circuit in **Figure 17.** The current can flow through both or either of the branches. Because all branches connect the same two points of the circuit, the voltage difference is the same in each branch. Then, according to Ohm's law, more current flows through the branches that have lower resistance.

Parallel circuits have several advantages. When one branch of the circuit is opened, such as when you turn a light off, the current continues to flow through the other branches. Houses, automobiles, and most electrical systems use parallel wiring so individual parts can be turned off without affecting the entire circuit.

Figure 17 In parallel circuits, the current follows more than one path. **Describe** *how the voltage difference will compare in each branch.*

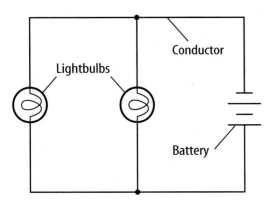

Figure 18 The wiring in a house must allow for the individual use of various appliances and fixtures. **Identify** *the type of circuit that is most common in household wiring.*

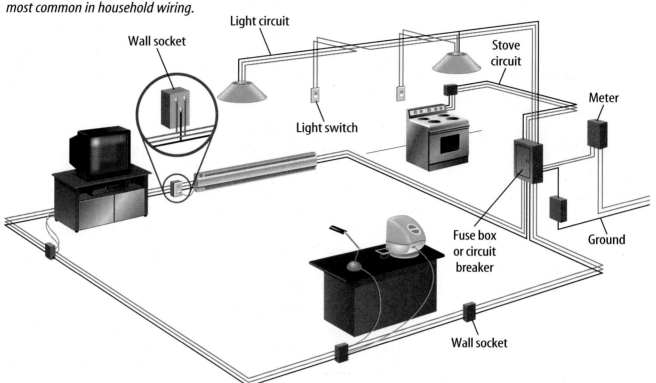

Light circuit

Wall socket

Stove circuit

Meter

Light switch

Fuse box or circuit breaker

Ground

Wall socket

Household Circuits

Count how many different things in your home require electrical energy. You don't see the wires because most of them are hidden behind the walls, ceilings, and floors. This wiring is mostly a combination of parallel circuits connected in an organized and logical network. **Figure 18** shows how electrical energy enters a home and is distributed. In the United States, the voltage difference in most of the branches is 120 V. In some branches that are used for electric stoves or electric clothes dryers, the voltage difference is 240 V. The main switch and circuit breaker or fuse box serve as an electrical headquarters for your home. Parallel circuits branch out from the breaker or fuse box to wall sockets, major appliances, and lights.

In a house, many appliances draw current from the same circuit. If more appliances are connected, more current will flow through the wires. As the amount of current increases, so does the amount of heat produced in the wires. If the wires get too hot, the insulation can melt and the bare wires can cause a fire. To protect against overheating of the wires, all household circuits contain either a fuse or a circuit breaker.

A

B

Figure 19 Two useful devices to prevent electric circuits from overheating are **A** fuses and **B** circuit breakers.
Evaluate *which device, a fuse or a circuit breaker, would be more convenient to have in the home.*

Fuses When you hear that somebody has "blown a fuse," it means that the person has lost his or her temper. This expression comes from the function of an electrical fuse, shown in **Figure 19A,** which contains a small piece of metal that melts if the current becomes too high. When it melts, it causes a break in the circuit, stopping the flow of current through the overloaded circuit. To enable current to flow again in the circuit, you must replace the blown fuse with a new one. However, before you replace the blown fuse, you should turn off or unplug some of the appliances. Too many appliances in use at the same time is the most likely cause for the overheating of the circuit.

Circuit Breaker A circuit breaker, shown in **Figure 19B,** is another device that prevents a circuit from overheating and causing a fire. A circuit breaker contains a piece of metal that bends when the current in it is so large that it gets hot. The bending causes a switch to flip and open the circuit, stopping the flow of current. Circuit breakers usually can be reset by pushing the switch to its "on" position. Again, before you reset a circuit breaker, you should turn off or unplug some of the appliances from the overloaded circuit. Otherwise, the circuit breaker will switch off again.

✓ **Reading Check** *What is the purpose of fuses and circuit breakers in household circuits?*

Electric Power

The reason that electricity is so useful is that electrical energy is converted easily to other types of energy. For example, electrical energy is converted to mechanical energy as the blades of a fan rotate to cool you. Electrical energy is converted to light energy in lightbulbs. A hair dryer changes electrical energy into thermal energy. The rate at which electrical energy is converted to another form of energy is the **electric power.**

The electric power used by appliances varies. Appliances often are labeled with a power rating that describes how much power the appliance uses, as shown in **Figure 20.** Appliances that have electric heating elements, such as ovens and hair dryers, usually use more electric power than other appliances.

Figure 20 All appliances come with a power rating.

Calculating Electric Power The electric power used depends on the voltage difference and the current. Electric power can be calculated from the following equation.

Electric Power Equation

electric power (in watts) = current (in amperes) ×
voltage difference (in volts)

$$P = IV$$

The unit for power is the watt (W). Because the watt is a small unit of power, electric power is often expressed in kilowatts (kW). One kilowatt equals 1,000 watts.

Applying Math   Solve a Simple Equation

POWER USED BY A CLOTHES DRYER The current in an electric clothes dryer is 15 A when it is plugged into a 240-volt outlet. How much power does the clothes dryer use?

IDENTIFY known values and the unknown value

Identify the known values:

current in the clothes dryer is 15 A   means→ $I = 15$ A

plugged into a 240-volt outlet   means→ $V = 240$ V

Identify the unknown value:

how much power does the clothes dryer use   means→ $P = ?$ W

SOLVE the problem

Substitute the known values $I = 15$ A and $V = 240$ V into the electric power equation:
$$P = IV = (15 \text{ A})(240 \text{ V}) = 3,600 \text{ W} = 3.6 \text{ kW}$$

CHECK the answer

Does your answer seem reasonable? Check your answer by dividing the power you calculated by the current given in the problem. The result should be the voltage difference given in the problem.

Practice Problems

1. A toaster oven is plugged into an outlet that provides a voltage difference of 120 V. What power does the oven use if the current is 10 A?

2. A VCR that is not playing still uses 10.0 W of power. What is the current if the VCR is plugged into a 120-V electrical outlet?

3. A flashlight bulb uses 2.4 W of power when the current in the bulb is 0.8 A. What is the voltage difference?

For more practice problems go to page 834, and visit gpscience.com/extra_problems.

Electrical Energy Using electric power costs money. However, electric companies charge by the amount of electrical energy used, rather than by the electric power used. Electrical energy usually is measured in units of kilowatt hours (kWh) and can be calculated from this equation:

Electric Energy Equation

electrical energy (in kWh) = electric power (in kW) × time (in hours)

$$E = Pt$$

In the above equation, electric power is in units of kW and the time is the number of hours that the electric power is used.

Applying Math   Solve a Simple Equation

ELECTRICAL ENERGY USED BY A MICROWAVE OVEN A microwave oven with a power rating of 1,200 W is used for 0.25 h. How much electrical energy is used by the microwave?

IDENTIFY known values and the unknown value

Identify the known values:

power rating of 1,200 W means $P = 1{,}200 \text{ W} = 1.2 \text{ kW}$

is used for 0.25 hours means $t = 0.25 \text{ h}$

Identify the unknown value:

how much electrical energy is used? means $E = ? \text{ kWh}$

SOLVE the problem

Substitute the known values $p = 1.2$ kW and $t = 0.25$ h into the electrical energy equation:

$$E = Pt = (1.2 \text{ kW})(0.25 \text{ h}) = 0.30 \text{ kWh}$$

CHECK your answer

Does your answer seem reasonable? Check your answer by dividing the electrical energy you calculated by the power given in the problem. The results should be the time given in the problem.

Practice Problems

1. A refrigerator operates on average for 10.0 h a day. If the power rating of the refrigerator is 700 W, how much electrical energy does the refrigerator use in one day?

2. A TV with a power rating of 200 W uses 0.8 kWh in one day. For how many hours was the TV on during this day?

For more practice problems go to page 834, and visit gpscience.com/extra_problems.

The Cost of Using Electrical Energy

The cost of using the appliance can be computed by multiplying the electrical energy used by the amount the power company charges for each kWh. For example, if a 100-W lightbulb is left on for 5 h, the amount of electrical energy used is

$$E = Pt = (0.1 \text{ kW}) (5 \text{ h}) = 0.5 \text{ kWh}$$

If the power company charges $0.10 per kWh, the cost of using the bulb for 5 h is

$$\text{cost} = (\text{kWh used}) (\text{cost per kWh})$$
$$= (0.5 \text{ kWh}) (\$0.10/\text{kWh}) = \$0.05$$

The cost of using some household appliances is given in **Table 1,** where the cost per kWh is assumed to be $0.09/kWh.

Table 1 Cost of Using Home Appliances

Appliance	Hair Dryer	Stereo	Color Television
Power rating	1,000	100	200
Hours used daily	0.25	2.0	4.0
kWh used monthly	7.5	6.0	24.0
Cost per kWh	$0.09	$0.09	$0.09
Monthly cost	$0.68	$0.54	$2.16

section 3 review

Summary

Series and Parallel Circuits

- A series circuit has only one path that current can flow in.
- A parallel circuit has two or more branches that current can flow in.
- Household wiring usually consists of a number of connected parallel circuits.
- Fuses and circuit breakers are used to prevent wires from overheating when the current flowing in the wires becomes too large.

Electric Power

- Electric power is the rate at which electrical energy is converted into other forms of energy.
- Electric power can be calculated by multiplying the current by the voltage difference:
$$P = IV$$

Electrical Energy

- The electrical energy used can be calculated by multiplying the power by the time:
$$E = Pt$$
- Electric power companies charge customers for the amount of electrical energy they use.

Checking Concepts

1. **Explain** how electric power and electrical energy are related.
2. **Discuss** why fuses and circuit breakers are used in household circuits.
3. **Explain** what determines the current in each branch of a parallel circuit.
4. **Explain** whether or not a fuse or circuit breaker should be connected in parallel to the circuit it is protecting.
5. **Think Critically** A parallel circuit consisting of four branches is connected to a battery. Explain how the amount of current that flows out of the battery is related to amount of current in the branches of the circuit.

Applying Math

6. **Calculate** the current flowing into a desktop computer plugged into a 120-V outlet if the power used is 180 W.
7. **Calculate Electric Power** A circuit breaker is tripped when the current in the circuit is greater than 15 A. If the voltage difference is 120 V, what is the power being used when the circuit breaker is tripped?
8. **Calculate** the monthly cost of using a 700-W refrigerator that runs for 10 h a day if the cost per kWh is $0.09.

Design Your Own

Comparing Series and Parallel Circuits

Real-World Question

Imagine what a bedroom might be like if it were wired in series. For an alarm clock to keep time and wake you in the morning, your lights and anything else that uses electricity would have to be on. Fortunately, most outlets in homes are wired in parallel circuits on separate branches of the main circuit. How do the behaviors of series and parallel circuits compare?

Form a Hypothesis

Predict what will happen to the other bulbs when one bulb is unscrewed from a series circuit and from a parallel circuit. Explain your prediction. Also, form a hypothesis to explain in which circuit the lights shine the brightest.

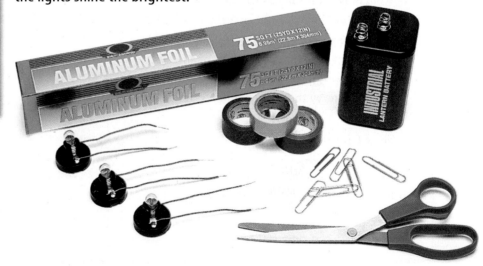

▶ Test Your Hypothesis

Make a Plan

1. As a group, agree upon and write the hypothesis statement.

2. Work together determining and writing the steps you will take t test your hypothesis. Include a list of the materials you will need

3. How will your circuits be arranged? On a piece of paper, draw a large parallel circuit of three lights and the dry-cell battery as shown. On the other side, draw another circuit with the three bulbs arranged in series.

4. Make conducting wires by taping a 30-cm piece of transparent tape to a sheet of aluminum foil and folding the foil over twice to cover the tape. Cut these to any length that works in you design.

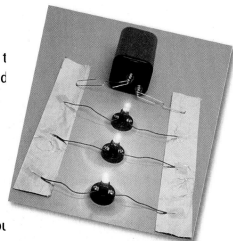

Follow Your Plan

1. Make sure your teacher approves your plan before you start.

2. Carry out the experiment. **WARNING:** *Leave the circuit on for only a few seconds at a time to avoid overheating.*

3. As you do the experiment, record your predictions and your observations in your Science Journal.

▶ Analyze Your Data

1. **Predict** what will happen in the series circuit when a bulb is unscrewed at one end. What will happen in the parallel circuit?

2. **Compare** the brightness of the lights in the different circuits. Explain.

3. **Predict** what happens to the brightness of the bulbs in the series circuit if you complete it with two bulbs instead of three bulbs. Test it. How does this demonstrate Ohm's law?

▶ Conclude and Apply

1. Did the results support your hypothesis? Explain by using your observations.

2. Where in the parallel circuit would you place a switch to control all three lights? Where would you place a switch to control only one light? Test it.

Communicating Your Data

Prepare a poster to highlight the differences between a parallel and a series circuit. Include possible practical applications of both types of circuits. **For more help, refer to the Science Skill Handbook.**

Invisible Man
by Ralph Ellison

I am an invisible man. No, I am not a spook like those who haunted Edgar Allen Poe; nor am I one of your Hollywood-movie ectoplasms.[1] A am a man of substance, of flesh and bone, fiber and liquids—and I might even be said to possess a mind. I am invisible, understand, simply because people refuse to see me.... Nor is my invisibility exactly a matter of biochemical accident to my epidermis.[2] That invisibility to which I refer occurs because of a peculiar disposition ... of those with whom I come in contact....

.... Now don't jump to the conclusion that because I call my home a "hole" it is damp and cold like a grave.... Mine is a warm hole.

My hole is warm and full of light. Yes, *full* of light. I doubt if there is a brighter spot in all New York than this hole of mine.... Perhaps you'll think it strange that an invisible man should need light, desire light, love light. Because maybe it is exactly because I *am invisible.* Light confirms my reality, gives birth to my form.... I myself, after existing some twenty years, did not become alive until I discovered my invisibility.

... In my hole in the basement there are exactly 1,369 lights. I've wired the entire ceiling, every inch of it.... Though invisible, I am in the great American tradition of tinkers. That makes me kin to Ford, Edison and Franklin.

1 The outer layer of a part of the cell.

2 The outer layer of skin.

Ralph Ellison

Understanding Literature

Prologue A prologue is an introduction to a novel, play, or other work of literature. Often a prologue contains useful information about events to come in the story. Foreshadowing is the use of clues by the author to prepare readers for future events or recurring themes.

Respond to the Reading

1. What clues does the narrator give that he is not really invisible?
2. Why does the narrator believe he is in the "great American tradition of tinkers"?
3. **Linking Science and Writing** Write a prologue to a make-believe book describing Edison's invention of the lightbulb. Recall that a prologue is not a summary of the book. Rather, it can state general themes that the work of literature will address or set the stage or describe the setting of the story.

INTEGRATE Physics

If all 1,369 lightbulbs were all wired together in a series circuit, the electrical resistance in the circuit would be high. By Ohm's law, the current in the circuit would be low and the bulbs wouldn't glow. If all the bulbs were wired in a parallel circuit, so much current would flow in the circuit that the connecting wires would melt. For the bulbs to light, the narrator must have wired them in many independent circuits.

Reviewing Main Ideas

Section 1 Electric Charge

1. There are two types of electric charge—positive charge and negative charge.

2. Electric charges exert forces on each other. Like charges repel and unlike charges attract.

3. Electric charges can be transferred from one object to another, but cannot be created or destroyed.

4. Electrons can move easily in an electrical conductor. Electrons do not move easily in an insulator.

5. Objects can be charged by contact or by induction. Charging by induction occurs when a charged object is brought near an electrically neutral object.

Section 2 Electric Current

1. Electric current is the net movement of electric charges in a single direction. A voltage difference causes an electric current to flow.

2. A circuit is a closed path along which charges can move. Current will flow continually only along a circuit that is unbroken.

3. Chemical reactions in a battery produce a voltage difference between the positive and negative terminals of the battery.

4. Electrical resistance is the tendency of a material to oppose the flow of electric current.

5. In an electric circuit, the voltage difference, current, and resistance are related by Ohm's law:

$$I = \frac{V}{R}$$

Section 3 Electrical Energy

1. Current has only one path in a series circuit and more than one path in a parallel circuit.

2. Circuit breakers and fuses prevent excessive current from flowing in a circuit.

3. Electrical power is the rate at which electrical energy is used, and can be calculated from $P = IV$.

4. The electrical energy used by a device can be calculated from the equation $E = Pt$.

FOLDABLES Use the Foldable you made at the beginning of the chapter to help you review electric charge, electric current, and electrical energy.

Using Vocabulary

charging by contact p. 195	law of conservation of
charging by induction	charge p. 193
p. 196	Ohm's law p. 205
circuit p. 201	parallel circuit p. 208
conductor p. 195	resistance p. 203
electric current p. 200	series circuit p. 207
electrical power p. 210	static electricity p. 192
insulator p. 195	voltage difference p. 200

Complete each statement using a word(s) from the vocabulary list above.

1. A(n) _____ is a circuit with only one path for current to follow.

2. An accumulation of excess electric charge is _____.

3. The electric force that makes current flow in a circuit is related to the _____.

4. According to _____, electric charge cannot be created or destroyed.

5. _____ is the result of electrons colliding with atoms as current flows in a material.

6. Charging a balloon by rubbing it on wool is an example of _____.

Checking Concepts

Choose the word or phrase that best answers the question.

7. Which of the following is a conductor?
   A) glass
   B) wood
   C) tungsten
   D) plastic

8. Resistance in wires causes electrical energy to be converted into which form of energy?
   A) chemical energy
   B) nuclear energy
   C) thermal energy
   D) sound

9. The electric force between two charged objects depends on which of the following?
   A) their masses and their separation
   B) their speeds
   C) their charge and their separation
   D) their masses and their charge

10. An object becomes positively charged when which of the following occurs?
    A) loses electrons
    C) gains electrons
    B) loses protons
    D) gains neutrons

11. Which of the following does NOT provide a voltage difference in a circuit?
    A) wet cell
    C) electrical outlet
    B) wires
    D) dry cell

12. A commonly used unit for electrical energy is which of the following?
    A) kilowatt-hour
    C) ohm
    B) ampere
    D) newton

13. Which of the following is the rate at which appliances use electrical energy?
    A) power
    C) resistance
    B) current
    D) speed

Interpreting Graphics

Use the table below to answer question 14.

Current in Electric Circuits	
Circuit	Current (A)
A	2.3
B	0.6
C	0.2
D	1.8

14. The table shows the current in circuits that were each connected to a 6-V dry cell. Calculate the resistance of each circuit. Graph the current versus the resistance of each circuit. Describe the shape of the line on your graph.

Science Online gpscience.com/vocabulary_puzzlemaker

15. Copy and complete the following concept map on electric current.

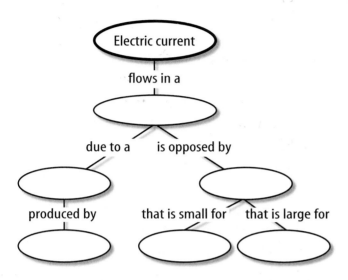

Thinking Critically

16. Identify and Manipulate Variables Design an experiment to test the effect on current and voltage differences in a circuit when two identical batteries are connected in series. What is your hypothesis? What are the variables and controls?

17. Explain A metal rod is charged by induction when a negatively-charged plastic rod is brought nearby. Explain how the net charge on the metal rod has changed.

18. Predict You walk across a carpet on a dry day and touch a glass doorknob. Predict whether or not you would receive an electric shock. Explain your reasoning.

19. Explain The electric force between electric charges is much larger than the gravitational force between the charges. Why then is the gravitational force between Earth and the Moon much larger than the electric force between Earth and the Moon?

20. Diagram Draw a circuit diagram showing how a stereo, a TV, and a computer can be connected to a single source of voltage difference, such that turning off one appliance does not turn off all the others. Include a circuit breaker in your diagram that will protect all the appliances.

Applying Math

21. Calculate Current Using the information in the circuit diagram below, compute the current flowing in the circuit.

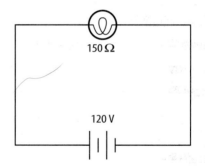

22. Calculate Current A toy car with a resistance of 20 Ω is connected to a 3-V battery. How much current flows in the car?

23. Calculate Electrical Energy The current flowing in an appliance connected to a 120-V source is 2 A. How many kilowatt-hours of electrical energy does the appliance use in 4 h?

24. Calculate Electrical Energy Cost A self-cleaning oven uses 5,400 W when cleaning the oven. If it takes 1.5 h to clean, how many kilowatt-hours of electricity are used? At a cost of $0.09 per kWh, what does it cost to clean the oven?

25. Calculate Power A calculator uses a 9-V battery and draws 0.1 A of current. How much power does it use?

Part 1 **Multiple Choice**

Record your answers on the answer sheet provided by your teacher or on a sheet of paper.

1. Which of the following is true about two adjacent electric charges?
   A. If both are positive, they attract.
   B. If both are negative, they attract.
   C. If one is positive and one is negative, they attract.
   D. If one is positive and one is negative, they repel.

The figure below shows a negatively charged electroscope. Use the figure to answer questions 2 and 3.

2. If a negatively-charged rod is brought close to, but not touching, the knob, the two leaves will
   A. move closer together.
   B. move farther apart.
   C. not move at all.
   D. become positively charged.

3. If a positively charged rod touches the knob, the two leaves will
   A. move closer together.
   B. move farther apart.
   C. not move at all.
   D. become positively charged.

4. Which of these is the SI unit of current?
   A. volt   **C.** ampere
   B. ohm   **D.** watt

5. A kilowatt-hour is a unit of
   A. power.   **C.** current.
   B. electric energy.   **D.** resistance.

6. When two objects become charged by contact, which of the following is true?
   A. The net charge on each object doesn't change.
   B. Both become negatively charged.
   C. Both become positively charged.
   D. Electrons are transferred.

7. When an object becomes charged by induction, which of the following best describes the net charge on the object?
   A. The net charge increases.
   B. The net charge decreases.
   C. The object is electrically neutral.
   D. The net charge is negative.

Use the figure below to answer questions 8, 9, and 10.

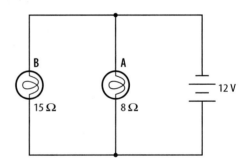

8. Which of the following is the same for each lightbulb?
   A. current in the filament
   B. voltage difference
   C. electric resistance
   D. charging by induction

9. Which of the following is the current that flows through lightbulb B?
   A. 1.25 A   **C.** 0.8 A
   B. 0.67 A   **D.** 1.5 A

10. Which of the following is the electric power used by lightbulb A?
   A. 8 W   **C.** 12 W
   B. 18 W   **D.** 15 W

Part 2 | **Short Response/Grid In**

Record your answers on the answer sheet provided by your teacher or on a sheet of paper.

11. The current flowing through a lightbulb is 2.5 A. The lamp is connected to a battery supplying a voltage difference of 12 V. What is the power used by the lightbulb?

12. A person spends four hours a day in a room, but leaves a 100-W lightbulb burning 24 hours a day. How much energy would be saved if the bulb burned for only the four hours the person was in the room?

Use the figure below to answer questions 13, 14, and 15.

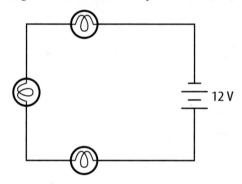

13. If the current in the circuit is 1.0 A, what is the total resistance of the circuit?

14. If the current in the circuit is 1.0 A, how much electrical energy is used by the circuit in 2.5 h?

15. How does the current in the circuit change if one of the lightbulbs is removed and the circuit is reconnected?

16. Two balloons are rubbed with a piece of wool. Describe what happens as the balloons are brought close together.

17. Calculate the cost of the electrical energy used by a TV for 30 days if the TV is on for four hours a day and the TV uses 250 W. Assume that the cost of electrical energy is $0.10 per kWh.

Part 3 | **Open Ended**

Record your answers on a sheet of paper.

18. Explain how a refrigerator with a power rating of 650 W can use more electrical energy in a day than a hair dryer with a power rating of 1,000 W.

19. Two copper wires have the same length and the same temperature. However, the electric resistance of wire A is twice as large as the resistance of wire B. Explain how wire A and wire B are different.

20. A rubber rod rubbed on hair becomes negatively charged. A glass rod rubbed with a piece of silk becomes positively charged. Suppose the charged rubber rod is suspended so it is free to rotate. Describe how the rubber rod moves as the piece of silk is brought close to it.

Use the table below to answer question 21.

Resistance of Copper Wire	
Length (m)	Resistance (Ω)
10	0.8
20	1.6
25	2.0
30	2.4

21. Suppose you were asked to estimate the resistance of a 5-m length of wire based on data in the table. What additional information would be needed to make your estimate more accurate?

22. The filament in a 75-W lightbulb has a smaller electric resistance than the filament in a 40-W lightbulb. Describe two ways the filament in the 75-W bulb could be different from the filament in the 40-W lightbulb.

Magnetism and Its Uses

A Natural Light Show

Have you ever seen an aurora? This light display occurs when blasts of charged particles from the Sun are captured by Earth's magnetic field. Atoms in the upper atmosphere emit the aurora light when they collide with charged particles that result from the Sun's blast.

Science Journal   List three things you know about magnets.

The Strength of Magnets

Did you know that magnets are used in TV sets, computers, stereo speakers, electric motors, and many other devices? Magnets also help create images of the inside of the human body. Even Earth acts like a giant bar magnet. How do magnets work?

1. Hold a bar magnet horizontally and put a paper clip on one end. Touch a second paper clip to the end of the first one. Continue adding paper clips until none will stick to one end of the chain. Copy the data table below and record the number of paper clips the magnet held. Remove the paper clips from the magnet.

2. Repeat step 1 three more times. First, start the chain about 2 cm from the end of the magnet. Second, start the chain near the center of the magnet. Third, start the chain at the other end of the magnet.

3. **Think Critically** Infer which part of the magnet exerts the strongest attraction. Compare the attraction at the center of the magnet with the attraction at the ends.

Magnet/Paper Clip Data	
	Paper Clip Chain (number of clips)
Trial 1 (end)	
Trial 2 (2 cm)	
Trial 3 (center)	
Trial 4 (other end)	

Using Magnets Many devices you use contain magnets that help convert one form of energy to another. Make the following Foldable to help you understand how magnets are used to transform electrical and mechanical energy.

STEP 1 Fold a sheet of paper in half lengthwise.

STEP 2 Fold the paper down about 2 cm from the top.

STEP 3 Open and draw lines along the top fold. Label as shown.

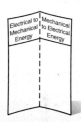

Summarize As you read the chapter, summarize how magnets are used to convert electrical energy to mechanical energy in the left column, and how magnets are used to convert mechanical energy to electrical energy in the right column.

Preview this chapter's content and activities at gpscience.com

section 1

Magnetism

Reading Guide

What You'll Learn

- **Explain** how a magnet exerts a force.
- **Describe** the properties of temporary and permanent magnets.
- **Explain** why some materials are magnetic and some are not.
- **Model** magnetic behavior using magnetic domains.

Why It's Important

Without the forces exerted by magnets, you could not use televisions, computers, CD players, or even refrigerators.

Review Vocabulary

electric field: surrounds an electric charge and exerts a force on other electric charges

New Vocabulary

- magnetism
- magnetic field
- magnetic pole
- magnetic domain

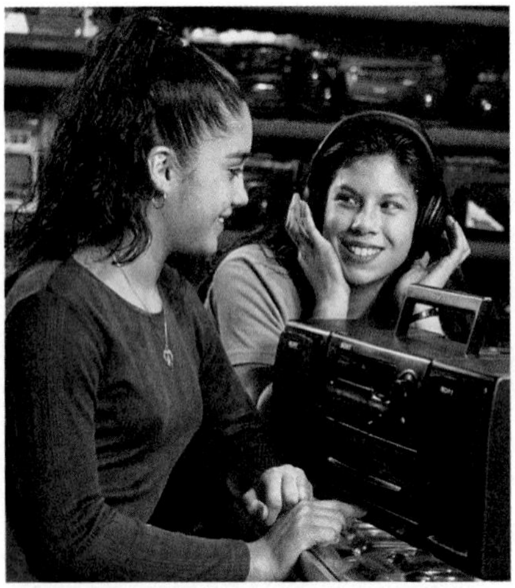

Figure 1 Magnets can be found in many devices you use everyday, such as TVs, video games, telephones. Headphones and CD players also contain magnets.

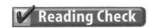

Magnets

More than 2,000 years ago Greeks discovered deposits of a mineral that was a natural magnet. They noticed that chunks of this mineral could attract pieces of iron. This mineral was found in a region of Turkey that then was known as Magnesia, so the Greeks named the mineral magnetic. The mineral is now called magnetite. In the twelfth century Chinese sailors used magnetite to make compasses that improved navigation. Since then many devices have been developed that rely on magnets to operate. Today, the word **magnetism** refers to the properties and interactions of magnets. **Figure 1** shows a device you might be familiar with that uses magnets and magnetism.

Magnetic Force You probably have played with magnets and might have noticed that two magnets exert a force on each other. Depending on which ends of the magnets are close together, the magnets either repel or attract each other. You might have noticed that the interaction between two magnets can be felt even before the magnets touch. The strength of the force between two magnets increases as magnets move closer together and decreases as the the magnets move farther apart.

Reading Check *What does the force between two magnets depend on?*

Figure 2 A magnet is surounded by a magnetic field.

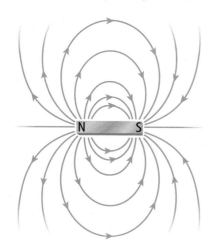

A magnet's magnetic field is represented by magnetic field lines.

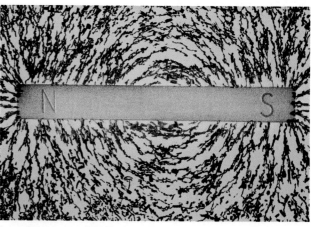

Iron filings sprinkled around a magnet line up along the magnetic field lines.

Magnetic Field A magnet is surrounded by a magnetic field. A **magnetic field** exerts a force on other magnets and objects made of magnetic materials. The magnetic field is strongest close to the magnet and weaker far away. The magnetic field can be represented by lines of force, or magnetic field lines. **Figure 2** shows the magnetic field lines surrounding a bar magnet. A magnetic field also has a direction. The direction of the magnetic field around a bar magnet is shown by the arrows of the left side of **Figure 2.**

Magnetic Poles Look again at **Figure 2.** Do you notice that the magnetic field lines are closest together at the ends of the bar magnet? These regions, called the **magnetic poles,** are where the magnetic force exerted by the magnet is strongest. All magnets have a north pole and a south pole. For a bar magnet, the north and south poles are at the opposite ends.

   Figure 3 shows the north and south poles of magnets with more complicated shapes. The two ends of a horseshoe-shaped magnet are the north and south poles. A magnet shaped like a disk has opposite poles on the top and bottom of the disk. Magnetic field lines always connect the north pole and the south pole of a magnet.

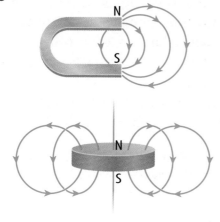

Figure 3 The magnetic field lines around horseshoe and disk magnets begin at each magnet's north pole and end at the south pole.
Identify *where the magnetic field is strongest.*

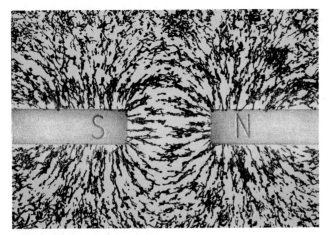

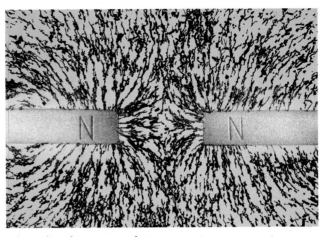

Unlike poles closest together

Like poles closest together

Figure 4 Two magnets can attract or repel each other, depending on which poles are closest together.

How Magnets Interact Two magnets can either attract or repel each other. Two north poles or two south poles of two magnets repel each other. However, north poles and south poles always attract each other. Like magnetic poles repel each other and unlike poles attract each other. When two magnets are brought close to each other, their magnetic fields combine to produce a new magnetic field. **Figure 4** shows the magnetic field that results when like poles and unlike poles of bar magnets are brought close to each other.

✔ **Reading Check** *How do magnetic poles interact with each other?*

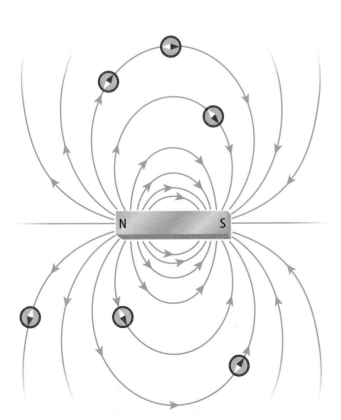

Magnetic Field Direction When a compass is brought near a bar magnet, the compass needle rotates. The compass needle is a small bar magnet with a north pole and a south pole. The force exerted on the compass needle by the magnetic field causes the needle to rotate. The compass needle rotates until it lines up with the magnetic field lines, as shown in **Figure 5.** The north pole of a compass points in the direction of the magnetic field. This direction is always away from a north magnetic pole and toward a south magnetic pole.

Figure 5 Compass needles placed around a bar magnet line up along magnetic field lines. The north poles of the compass needles are shaded red.

Earth's Magnetic Field A compass can help determine direction because the north pole of the compass needle points north. This is because Earth acts like a giant bar magnet and is surrounded by a magnetic field that extends into space. Just as with a bar magnet, the compass needle aligns with Earth's magnetic field lines, as shown in **Figure 6.**

Earth's Magnetic Poles The north pole of a magnet is defined as the end of the magnet that points toward the geographic north. Sometimes the north pole and south pole of magnets are called the north-seeking pole and the south-seeking pole. Because opposite magnetic poles attract, the north pole of a compass is being attracted by a south magnetic pole. So Earth is like a bar magnet with its south magnetic pole near its geographic north pole.

Currently, Earth's south magnetic pole is located in northern Canada about 1,500 km from the geographic north pole. However, Earth's magnetic poles move slowly with time. Sometimes Earth's magnetic poles switch places so that Earth's south magnetic pole is the southern hemisphere near the geographic south pole. Measurements of magnetism in rocks show that Earth's magnetic poles have changed places over 150 times in the past seventy million years.

No one is sure what produces Earth's magnetic field. Earth's inner core is made of a solid ball of iron and nickel, surrounded by a liquid layer of molten iron and nickel. According to one theory, circulation of the molten iron and nickel in Earth's outer core produces Earth's magnetic field.

Mini LAB

Observing Magnetic Interference

Procedure 👓

1. Clamp a **bar magnet** to a **ring stand.** Tie a **thread** around one end of a **paper clip** and stick the paper clip to one pole of the magnet.
2. Anchor the other end of the thread under a **book** on the table. Slowly pull the thread until the paper clip is suspended below the magnet but not touching the magnet.
3. Without touching the paper clip, slip a piece of paper between the magnet and the paper clip. Does the paper clip fall?
4. Try other materials, such as **aluminum foil, fabric,** or a **butter knife.**

Analysis

1. Which materials caused the paper clip to fall? Why do you think these materials interfered with the magnetic field?
2. Which materials did not cause the paper clip to fall? Why do you think these materials did not interfere with the magnetic field?

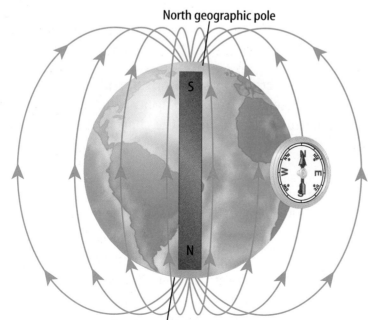

North geographic pole

South geographic pole

Figure 6 A compass needle aligns with the magnetic field lines of Earth's magnetic field. **Predict** *Which way would a compass needle point if Earth's magnetic poles switched places?*

Magnets in Organisms
Some organisms may use Earth's magnetic field to help find their way around. Some species of birds, insects, and bacteria have been shown to contain small amounts of the mineral magnetite. Research how one species uses Earth's magnetic field, and report your findings to your class.

Magnetic Materials

You might have noticed that a magnet will not attract all metal objects. For example, a magnet will not attract pieces of aluminum foil. Only a few metals, such as iron, cobalt, or nickel, are attracted to magnets or can be made into permanent magnets. What makes these elements magnetic? Remember that every atom contains electrons. Electrons have magnetic properties. In the atoms of most elements, the magnetic properties of the electrons cancel out. But in the atoms of iron, cobalt, and nickel, these magnetic properties don't cancel out. Each atom in these elements behaves like a small magnet and has its own magnetic field.

Even though these atoms have their own magnetic fields, objects made from these metals are not always magnets. For example, if you hold an iron nail close to a refrigerator door and let go, it falls to the floor. However, you can make the nail behave like a magnet temporarily.

Applying Science

How can magnetic parts of a junk car be salvaged?

Every year over 10 million cars containing plastics, glass, rubber, and various metals are scrapped. Magnets are often used to help retrieve some of these materials from scrapped cars. The materials can then be reused, saving both natural resources and energy. Once the junk car has been fed into a shredder, big magnets can easily separate many of its metal parts from its nonmetal parts. How much of the car does a magnet actually help separate? Use your ability to interpret a circle graph to find out.

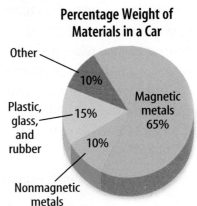

Percentage Weight of Materials in a Car

Other — 10%
Plastic, glass, and rubber — 15%
Nonmagnetic metals — 10%
Magnetic metals — 65%

Identifying the Problem

The graph at the right shows the average percent by weight of the different materials in a car. Included in the magnetic metals are steel and iron. The nonmagnetic metals refer to aluminum, copper, lead, zinc, and magnesium. According to the chart, how much of the car can a magnet separate for recycling?

Solving the Problem

1. What percent of the car's weight will a magnet recover?
2. A certain scrapped car has a mass of 1,500 kg. What is the mass of the materials in this car that cannot be recovered using a magnet?
3. If the average mass of a scrapped car is 1,500 kg, and 10 million cars are scrapped each year, what is the total mass of iron and steel that could be recovered from scrapped cars each year?

Magnetic Domains—A Model for Magnetism In iron, cobalt, nickel, and some other magnetic materials, the magnetic field created by each atom exerts a force on the other nearby atoms. Because of these forces, large groups of atoms align their magnetic poles so that almost all like poles point in the same direction. The groups of atoms with aligned magnetic poles are called **magnetic domains.** Each domain contains an enormous number of atoms, yet the domains are too small to be seen with the unaided eye. Because the magnetic poles of the individual atoms in a domain are aligned, the domain itself behaves like a magnet with a north pole and a south pole.

Lining Up Domains An iron nail contains an enormous number of these magnetic domains, so why doesn't the nail behave like a magnet? Even though each domain behaves like a magnet, the poles of the domains are arranged randomly and point in different directions, as shown in **Figure 7.** As a result, the magnetic fields from all the domains cancel each other out.

If you place a magnet against the same nail, the atoms in the domains orient themselves in the direction of the nearby magnetic field, as shown on the right in **Figure 7.** The like poles of the domains point in the same direction and no longer cancel each other out. The nail itself now acts as a magnet. But when the external magnetic field is removed, the constant motion and vibration of the atoms bump the magnetic domains out of their alignment. The magnetic domains in the nail return to random arrangement. For this reason, the nail is only a temporary magnet. Paper clips and other objects containing iron also can become temporary magnets.

Mini LAB

Making Your Own Compass

Procedure

WARNING: *Use care when handling sharp objects.*

1. Cut off the bottom of a **plastic foam cup** to make a polystyrene disk.
2. Magnetize a **sewing needle** by continuously stroking the needle in the same direction with a magnet for 1 min.
3. **Tape** the needle to the center of the foam disk.
4. Fill a **plate** with **water** and float the disk, needle-side up, in the water.
5. Bring the magnet close to the foam disk.

Analysis

1. How did the needle and disk move when you placed them in the water? Explain.
2. How did the needle and disk move when the magnet was brought near it? Explain.

Try at Home

Figure 7 Magnetic materials contain magnetic domains.

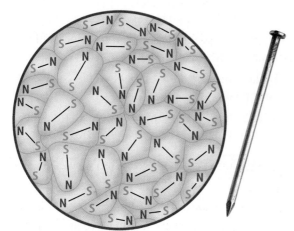

A normal iron nail is made up of microscopic domains that are arranged randomly.

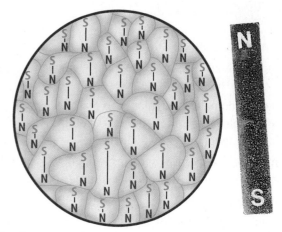

The domains will align themselves along the magnetic field lines of a nearby magnet.

Figure 8 Each piece of a broken magnet still has a north and a south pole.

Permanent Magnets A permanent magnet can be made by placing a magnetic material, such as iron, in a strong magnetic field. The strong magnetic field causes the magnetic domains in the material to line up. The magnetic fields of these aligned domains add together and create a strong magnetic field inside the material. This field prevents the constant motion of the atoms from bumping the domains out of alignment. The material is then a permanent magnet.

But even permanent magnets can lose their magnetic behavior if they are heated. Heating causes atoms in the magnet to move faster. If the permanent magnet is heated enough, its atoms may be moving fast enough to jostle the domains out of alignment. Then the permanent magnet loses its magnetic field and is no longer a magnet.

Can a pole be isolated? What happens when a magnet is broken in two? Can one piece be a north pole and one piece be a south pole? Look at the domain model of the broken magnet in **Figure 8.** Recall that even individual atoms of magnetic materials act as tiny magnets. Because every magnet is made of many aligned smaller magnets, even the smallest pieces have both a north pole and a south pole.

section 1 review

Summary

Magnets
- Magnets are surrounded by a magnetic field that exerts a force on magnetic materials.
- Magnets have a north pole and a south pole.
- Like magnetic poles repel and unlike poles attract.

Magnetic Materials
- Iron, cobalt, and nickel are magnetic elements because their atoms behave like magnets.
- Magnetic domains are regions in a material that contain an enormous number of atoms with their magnetic poles aligned.
- A magnetic field causes domains to align. In a temporary magnet, the domains return to random alignment when the field is removed.
- In a permanent magnet, a stong magnetic field aligns domains and they remain aligned when the field is removed.

Self Check

1. **Describe** what happens when you move two unlike magnetic poles closer together. Draw a diagram to illustrate your answer.
2. **Describe** how a compass needle moves when it is placed in a magnetic field.
3. **Explain** why only certain materials are magnetic.
4. **Predict** how the properties of a bar magnet would change if it were broken in half.
5. **Explain** how heating a bar magnet would change its magnetic field.
6. **Think Critically** Use the magnetic domain model to explain why a magnet sticks to a refrigerator door.

Applying Math

7. **Calculate Number of Domains** The magnetic domains in a magnet have an average volume of 0.0001 mm^3. If the magnet has dimensions 50 mm by 10 mm by 4 mm, how many domains does the magnet contain?

Science Online gpscience.com/self_check_quiz

Electricity and Magnetism

Reading Guide

What You'll Learn
- **Describe** the magnetic field produced by an electric current.
- **Explain** how an electromagnet produces a magnetic field.
- **Describe** how electromagnets are used.
- **Explain** how an electric motor operates.

Why It's Important
Electric motors contained in many of the devices you use every day operate because electric currents produce magnetic fields.

⊙ Review Vocabulary
electric current: the flow of electric charges in a wire or any conductor

New Vocabulary
- electromagnet
- solenoid
- galvanometer
- electric motor

Electric Current and Magnetism

In 1820, Hans Christian Oersted, a Danish physics teacher, found that electricity and magnetism are related. While doing a demonstration involving electric current, he happened to have a compass near an electric circuit. He noticed that the flow of the electric current affected the direction the compass needle pointed. Oersted hypothesized that the electric current must produce a magnetic field around the wire, and the direction of the field changes with the direction of the current.

Moving Charges and Magnetic Fields Oersted's hypothesis that an electric current creates a magnetic field was correct. It is now known that moving charges, like those in an electric current, produce magnetic fields. Around a current-carrying wire the magnetic field lines form circles, as shown in **Figure 8.** The direction of the magnetic field around the wire reverses when the direction of the current in the wire reverses. As the current in the wire increases the strength of the magnetic field increases. As you move farther from the wire the strength of the magnetic field decreases.

Figure 8 When electric current flows through a wire, a magnetic field forms around the wire. The direction of the magnetic field depends on the direction of the current in the wire.

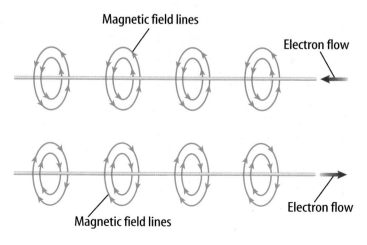

Magnetic field lines

Electron flow

Magnetic field lines

Electron flow

Electromagnets

The magnetic field that surrounds a current-carrying wire can be made much stronger in an electromagnet. An **electromagnet** is a temporary magnet made by wrapping a wire coil carrying a current around an iron core. When a current flows through a wire loop, such as the one shown in **Figure 9A,** the magnetic field inside the loop is stronger than the field around a straight wire. A single wire wrapped into a cylindrical wire coil is called a **solenoid.** The magnetic field inside a solenoid is stronger than the field in a single loop. The magnetic field around each loop in the solenoid combines to form the field shown in **Figure 9B.**

If the solenoid is wrapped around an iron core, an electromagnet is formed, as shown in **Figure 9C.** The solenoid's magnetic field magnetizes the iron core. As a result, the field inside the solenoid with the iron core can be more than 1,000 times greater than the field inside the solenoid without the iron core.

Properties of Electromagnets Electromagnets are temporary magnets because the magnetic field is present only when current is flowing in the solenoid. The strength of the magnetic field can be increased by adding more turns of wire to the solenoid or by increasing the current passing through the wire.

An electromagnet behaves like any other magnet when current flows through the solenoid. One end of the electromagnet is a north pole and the other end is a south pole. If placed in a magnetic field, an electromagnet will align itself along the magnetic field lines, just as a compass needle will. An electromagnet also will attract magnetic materials and be attracted or repelled by other magnets. What makes electromagnets so useful is that their magnetic properties can be controlled by changing the electric current flowing through the solenoid.

When current flows in the electromagnet and it moves toward or away from another magnet, electric energy is converted into mechanical energy to do work. Electromagnets do work in various devices such as stereo speakers and electric motors.

Figure 9 An electromagnet is made from a current-carrying wire.

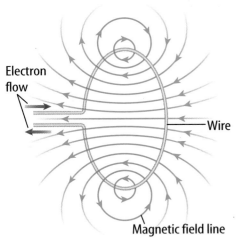

A The magnetic fields around different parts of the wire loop combine to form the field inside the loop.

B When many loops of current-carrying wire are formed into a solenoid, the magnetic field is increased inside the solenoid. The solenoid has a north pole and a south pole.
Predict *how the field would change if the current reversed direction.*

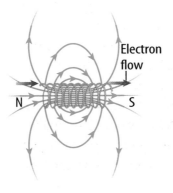

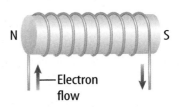

C A solenoid wrapped around an iron core forms an electromagnet.

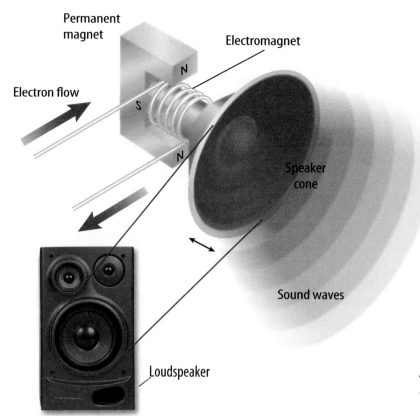

Permanent magnet

Electromagnet

Electron flow

N

S

N

Speaker cone

Sound waves

Loudspeaker

Figure 10 The electromagnet in a speaker converts electrical energy into mechanical energy to produce sound.
Explain *why a speaker needs a permanent magnet to produce sound.*

Using Electromagnets to Make Sound
How does musical information stored on a CD become sound you can hear? The sound is produced by a loudspeaker that contains an electromagnet connected to a flexible speaker cone that is usually made from paper, plastic, or metal. The electromagnet changes electrical energy to mechanical energy that vibrates the speaker cone to produce sound, as shown on **Figure 10.**

✔ Reading Check *How does a stereo speaker use an electromagnet to produce sound?*

When you listen to a CD, the CD player produces a voltage that changes according to the musical information on the CD. This varying voltage produces a varying electric current in the electromagnet connected to the speaker cone. Both the amount and the direction of the electric current change, depending on the information on the CD. The varying electric current causes both the strength and the direction of the magnetic field in the electromagnet to change. The electromagnet is surrounded by a permanent, fixed magnet. The changing direction of the magnetic field in the electromagnet causes the electromagnet to be attracted to or repelled by the permanent magnet. This makes the electromagnet move back and forth, causing the speaker cone to vibrate and reproduce the sound that was recorded on the CD.

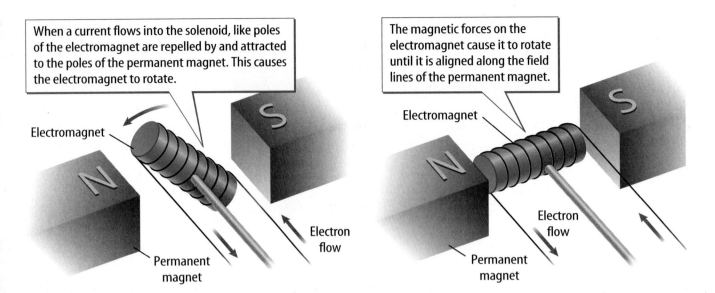

When a current flows into the solenoid, like poles of the electromagnet are repelled by and attracted to the poles of the permanent magnet. This causes the electromagnet to rotate.

Electromagnet

S

N

Permanent magnet

Electron flow

The magnetic forces on the electromagnet cause it to rotate until it is aligned along the field lines of the permanent magnet.

Electromagnet

N

S

Electron flow

Permanent magnet

Figure 11 An electromagnet can be made to rotate in a magnetic field.

Making an Electromagnet Rotate
The forces exerted on an electromagnet by another magnet can be used to make the electromagnet rotate. **Figure 11** shows an electromagnet suspended between the poles of a permanent magnet. The poles of the electromagnet are repelled by the like poles and attracted by the unlike poles of the permanent magnet. When the electromagnet is in the position shown on the left side of **Figure 11,** there is a downward force on the left side and an upward force on the right side of the electromagnet forces. These forces cause the electromagnet to rotate as shown.

Reading Check *How can a permanent magnet cause an electromagnet to rotate?*

The electromagnet continues to rotate until its poles are next to the opposite poles of the permanent magnet, as shown on the right side of **Figure 11.** In this position, the forces on the north and south poles of the electromagnet are in opposite directions. Then the net force on the electromagnet is zero, and the electromagnet stops rotating.

One way to change the forces that make the electromagnet rotate is to change the current in the electromagnet. Increasing the current increases the strength of the forces between the two magnets.

Galvanometers
You've probably noticed the gauges in the dashboard of a car. One gauge shows the amount of gasoline left in the tank, and another shows the engine temperature. How does a change in the amount of gasoline in a tank or the water temperature in the engine make a needle move in a gauge on the dashboard? These gauges are **galvanometers,** which are devices that use an electromagnet to measure electric current.

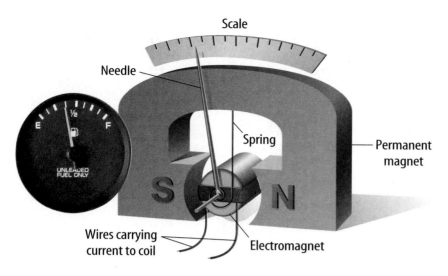

Scale

Needle

Spring

Permanent magnet

S N

Wires carrying current to coil

Electromagnet

Figure 12 The rotation of the needle in a galvanometer depends on the amount of current flowing in the electromagnet. The current flowing into the galvanometer in a car's fuel gauge changes as the amount of fuel changes.

Using Galvanometers An example of a galvanometer is shown in **Figure 12.** In a galvanometer, the electromagnet is connected to a small spring. Then the electromagnet rotates until the force exerted by the spring is balanced by the magnetic forces on the electromagnet. Changing the current in the electromagnet causes the needle to rotate to different positions on the scale.

For example, a car's fuel gauge uses a galvanometer. A float in the fuel tank is attached to a sensor that sends a current to the fuel gauge galvanometer. As the level of the float in the tank changes, the current sent by the sensor changes. The changing current in the galvanometer causes the needle to rotate by different amounts. The gauge is calibrated so that the current sent when the tank is full causes the needle to rotate to the full mark on the scale.

Electric Motors

On sizzling summer days, do you ever use an electric fan to keep cool? A fan uses an **electric motor,** which is a device that changes electrical energy into mechanical energy. The motor in a fan turns the fan blades, moving air past your skin to make you feel cooler.

Electric motors are used in all types of industry, agriculture, and transportation, including airplanes and automobiles. If you were to look carefully, you probably could find electric motors in every room of your house. Almost every appliance in which something moves contains an electric motor. Electric motors are used in devices such as in VCRs, CD players, computers, hair dryers, and other appliances shown in **Figure 13.**

Figure 13 All the devices shown here contain electric motors. **List** *three additional devices that contain electric motors.*

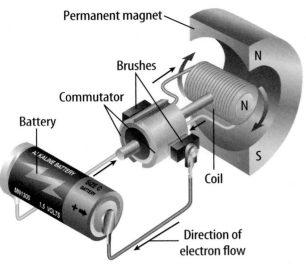

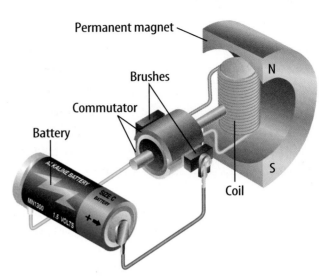

Step 1 When a current flows in the coil, the magnetic forces between the permanent magnet and the coil cause the coil to rotate.

Step 2 In this position, the brushes are not in contact with the commutator and no current flows in the coil. The inertia of the coil keeps it rotating.

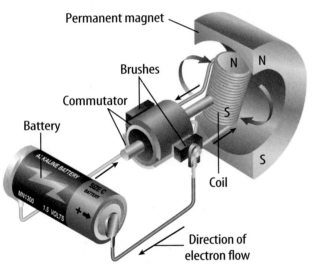

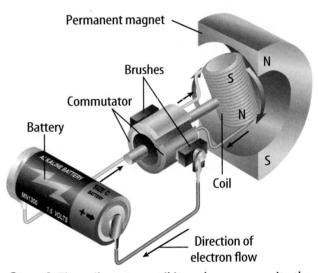

Step 3 The commutator reverses the direction of the current in the coil. This flips the north and south poles of the magnetic field around the coil.

Step 4 The coil rotates until its poles are opposite the poles of the permanent magnet. The commutator reverses the current, and the coil keeps rotating.

Figure 14 In a simple electric motor, a coil rotates between the poles of a permanent magnet. To keep the coil rotating, the current must change direction twice during each rotation.

A Simple Electric Motor A diagram of the simplest type of electric motor is shown in **Figure 14**. The main parts of a simple electric motor include a wire coil, a permanent magnet, and a source of electric current, such as a battery. The battery produces the current that makes the coil an electromagnet. A simple electric motor also includes components called brushes and a commutator. The brushes are conducting pads connected to the battery. The brushes make contact with the commutator, which is a conducting metal ring that is split. Each half of the commutator is connected to one end of the coil so that the commutator rotates with the coil. The brushes and the commutator form a closed electric circuit between the battery and the coil.

Making the Motor Spin When current flows in the coil, the forces between the coil and the permanent magnet cause the coil to rotate, as shown in step 1 of **Figure 14.** The coil continues to rotate until it reaches the position shown in step 2. Then the brushes no longer make contact with the commutator, and no current flows in the coil. As a result, there are no magnetic forces exerted on the coil. However, the inertia of the coil causes it to continue rotating.

In step 3 the coil has rotated so that the brushes again are in contact with the commutator. However, the halves of the commutator that are in contact with the positive and negative battery terminals have switched. This causes the current in the commutator to reverse direction. Now the top of the electromagnet is a north magnetic pole and the bottom is a south pole. These poles are repelled by the nearby like poles of the permanent magnet, and the magnet continues to rotate.

In step 4, the coil rotates until its poles are next to the opposite poles of the permanent magnet. Then the commutator again reverses the direction of the current, enabling the coil to keep rotating. In this way, the coil is kept rotating as long the battery remains connected to the commutator.

Science Online

Topic: Electric Motors
Visit gpscience.com for Web links to information about electric motors.

Activity Using the information provided at these links, construct a simple electric motor.

section 2 review

Summary

Electric Current and Magnetic Fields

- A magnetic field surrounds a moving electric charge.
- The strength of the magnetic field surrounding a current-carrying wire depends on the amount of current.

Electromagnets

- An electromagnet is a temporary magnet consisting of a current-carrying wire wrapped around an iron core.
- The magnetic properties of an electromagnet can be controlled by changing the current in the coil.
- A galvanometer uses an electromagnet to measure electric current.

Electric Motors

- In a simple electric motor, an electromagnet rotates between the poles of a permanent magnet.

Self Check

1. **Explain** why, if the same current flows in a wire coil and a single wire loop, the magnetic field inside the coil is stronger than the field inside the loop.

2. **Describe** two ways you could change the strength of the magnetic field produced by an electromagnet.

3. **Predict** how the magnetic field produced by an electromagnet would change if the iron core were replaced by an aluminum core.

4. **Explain** why it is necessary to continually reverse the direction of current flow in the coil of an electric motor.

5. **Think Critically** A bar magnet is repelled when an electromagnet is brought close to it. Describe how the bar magnet would have moved if the current in the electromagnet had been reversed.

Applying Math

6. **Use a Ratio** The magnetic field strength around a wire at a distance of 1 cm is twice as large as at a distance of 2 cm. How does the field strength at 0.5 cm compare to the field strength at 1 cm?

section 3

Producing Electric Current

Reading Guide

What You'll Learn

- **Define** electromagnetic induction.
- **Describe** how a generator produces an electric current.
- **Distinguish** between alternating current and direct current.
- **Explain** how a transformer can change the voltage of an alternating current.

Why It's Important

Electromagnetic induction enables power plants to generate the electric current an appliance uses when you plug it into an electric outlet.

🔊 Review Vocabulary

voltage difference: a measure of the electrical energy provided by charges as they flow in a circuit

New Vocabulary

- electromagnetic induction
- generator
- turbine
- direct current (DC)
- alternating current (AC)
- transformer

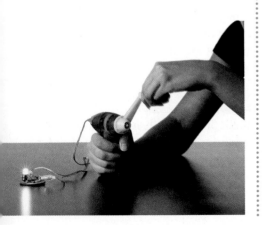

Figure 15 The coil in a generator is rotated by an outside source of mechanical energy. Here the student supplies the mechanical energy that is converted into electrical energy by the generator.

From Mechanical to Electrical Energy

Working independently in 1831, Michael Faraday in Britain and Joseph Henry in the United States both found that moving a loop of wire through a magnetic field caused an electric current to flow in the wire. They also found that moving a magnet through a loop of wire produces a current. In both cases, the mechanical energy associated with the motion of the wire loop or the magnet is converted into electrical energy associated with the current in the wire. The magnet and wire loop must be moving relative to each other for an electric current to be produced. This causes the magnetic field inside the loop to change with time. In addition, if the current in a wire changes with time, the changing magnetic field around the wire can also induce a current in a nearby coil. The generation of a current by a changing magnetic field is **electromagnetic induction.**

Generators Most of the electrical energy you use every day is provided by generators. A **generator** uses electromagnetic induction to transform mechanical energy into electrical energy. **Figure 15** shows one way a generator converts mechanical energy to electrical energy. The mechanical energy is provided by turning the handle on the generator.

An example of a simple generator is shown in **Figure 16.** In this type of generator, a current is produced in the coil as the coil rotates between the poles of a permanent magnet.

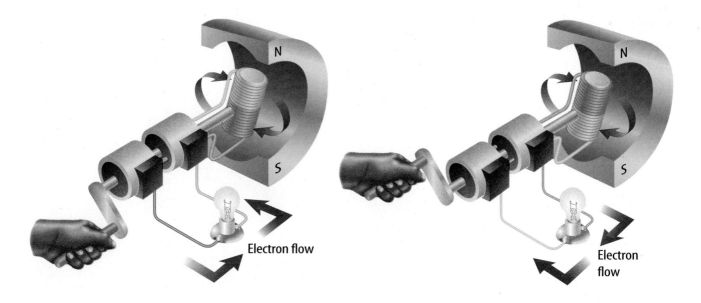

Electron flow

Electron flow

Switching Direction

As the generator's wire coil rotates through the magnetic field of the permanent magnet, current flows through the coil. After the wire coil makes one-half of a revolution, the ends of the coil are moving past the opposite poles of the permanent magnet. This causes the current to change direction. In a generator, as the coil keeps rotating, the current that is produced periodically changes direction. The direction of the current in the coil changes twice with each revolution, as **Figure 16** shows. The frequency with which the current changes direction can be controlled by regulating the rotation rate of the generator. In the United States, current is produced by generators that rotate 60 times a second, or 3,600 revolutions per minute.

Reading Check *For each revolution of the coil, how many times does the current change direction?*

Using Electric Generators

The type of generator shown in **Figure 16** is used in a car, where it is called an alternator. The alternator provides electrical energy to operate lights and other accessories. Spark plugs in the car's engine also use this electrical energy to ignite the fuel in the cylinders of the engine. Once the engine is running, it provides the mechanical energy that is used to turn the coil in the alternator.

Suppose that instead of using mechanical energy to rotate the coil in a generator, the coil was fixed, and the permanent magnet rotated instead. In fact, the current generated would be the same as when the coil rotates and the magnet doesn't move. The huge generators used in electric power plants are made this way. The current is produced in the stationary coil, and mechanical energy is used to rotate the magnet.

Figure 16 The current in the coil changes direction each time the ends of the coil move past the poles of the permanent magnet. **Explain** *how the frequency of the changing current can be controlled.*

Figure 17 Each of these generators at Hoover Dam can produce over 100,000 kW of electric power. In these generators, a rotating magnet induces an electric current in a stationary wire coil.

INTEGRATE
Career

Power Plant Operator
Many daily activities require electricity. Power plant operators control the machinery that generates electricity. Operators must have a high school diploma. College-level courses may be helpful. Research to find employers in your area that hire power plant operators.

Generating Electricity for Your Home You probably do not have a generator in your home that supplies the electrical energy you need to watch television or wash your clothes. This electrical energy comes from a power plant with huge generators like the one in **Figure 17.** The coils in these generators have many coils of wire wrapped around huge iron cores. The rotating magnets are connected to a **turbine** (TUR bine)—a large wheel that rotates when pushed by water, wind, or steam.

For example, some power plants first produce thermal energy by burning fossil fuels or using the heat produced by nuclear reactions. This thermal energy is used to heat water and produce steam. Thermal energy is then converted to mechanical energy as the steam pushes the turbine blades. The generator then changes the mechanical energy of the rotating turbine into the electrical energy you use. In some areas, fields of windmills, like those in **Figure 18,** can be used to capture the mechanical energy in wind to turn generators. Other power plants use the mechanical energy in falling water to drive the turbine. Both generators and electric motors use magnets to produce energy conversions between electrical and mechanical energy. **Figure 19** summarizes the differences between electric motors and generators.

Figure 18 The propeller on each of these windmills is connected to an electric generator. The rotating propeller rotates a coil or a permanent magnet.

Figure 19

Electric motors power many everyday machines, from CD players to vacuum cleaners. Generators produce the electricity those motors need to run. Both motors and generators use electromagnets, but in different ways. The table below compares motors and generators.

Permanent magnet

Permanent magnet

Coil

Coil

Coil

	Electric Motor	Generator
What does it do?	Changes electricity into movement 	Changes movement into electricity 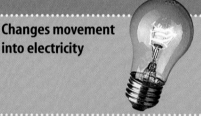
What makes its electromagnetic coil rotate?	Attractive and repulsive forces between the coil and the permanent magnet 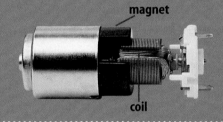 magnet coil	An outside source of mechanical energy 
What is the source of the current that flows in its coil?	An outside power source 	Electromagnetic induction from moving the coil through the field of the permanent magnet 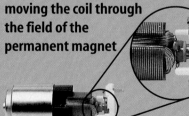
How often does the current in the coil change direction?	Twice during each rotation of the coil	Twice during each rotation of the coil

Topic: Transformers

Visit gpscience.com for links to information about how transformers transmit electric current.

Activity In England, voltage from an outlet is greater than required for U.S. appliances. Using information found at these links, diagram a transformer that would allow a U.S. appliance to operate in England.

Direct and Alternating Currents

Modern society relies heavily on electricity. Just how much you rely on electricity becomes obvious during a power outage. Out of habit, you might walk into a room and flip on the light switch. You might try to turn on a radio or television or check the clock to see what time it is. Because power outages sometimes occur, some electrical devices, like the radio in **Figure 20,** use batteries as a backup source of electrical energy. However, the current produced by a battery is different than the current from an electric generator.

A battery produces a direct current. **Direct current** (DC) flows only in one direction through a wire. When you plug your CD player or any other appliance into a wall outlet, you are using alternating current. **Alternating current** (AC) reverses the direction of the current in a regular pattern. In North America, generators produce alternating current at a frequency of 60 cycles per second, or 60 Hz. The electric current produced by a generator changes direction twice during each cycle, or each rotation, of the coil. So a 60-Hz alternating current changes direction 120 times each second.

Electronic devices that use batteries as a backup energy source usually require direct current to operate. When the device is plugged into a wall outlet, electronic components inside the device convert the alternating current to direct current and also reduce the voltage of the alternating current.

Transmitting Electrical Energy

The alternating current produced by an electric power plant carries electrical energy that is transmitted along electric transmission lines. However, when the electric energy is transmitted along power lines, some of the electrical energy is converted into heat due to the electrical resistance of the wires. The heat produced in the power lines warms the wires and the surrounding air and can't be used to power electrical devices. Also, the electrical resistance and heat production increases as the wires get longer. As a result, large amounts of heat can be produced when electrical energy is transmitted over long distances.

One way to reduce the heat produced in a power line is to transmit the electrical energy at high voltages, typically around 150,000 V. However, electrical energy at such high voltage cannot enter your home safely, nor can it be used in home appliances. Instead, a transformer is used to decrease the voltage.

Figure 20 Some devices, like this radio, can use either direct or alternating current. Electronic components in these devices change alternating current from an electric outlet to direct current.

Transformers

A **transformer** is a device that increases or decreases the voltage of an alternating current. A transformer is made of a primary coil and a secondary coil. These wire coils are wrapped around the same iron core, as shown in **Figure 21.** As an alternating current passes through the primary coil, the coil's magnetic field magnetizes the iron core. The magnetic field in the primary coil changes direction as the current in the primary coil changes direction. This produces a magnetic field in the iron core that changes direction at the same frequency. The changing magnetic field in the iron core then induces an alternating current with the same frequency in the secondary coil.

The voltage in the primary coil is the input voltage and the voltage in the secondary coil is the output voltage. The output voltage divided by the input voltage equals the number of turns in the secondary coil divided by the number of turns in the primary coil.

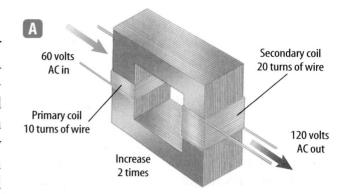

60 volts
AC in

Secondary coil
20 turns of wire

Primary coil
10 turns of wire

120 volts
AC out

Increase
2 times

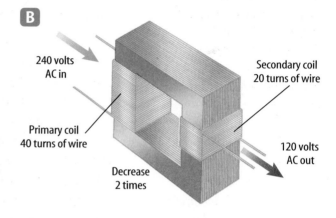

240 volts
AC in

Secondary coil
20 turns of wire

Primary coil
40 turns of wire

120 volts
AC out

Decrease
2 times

Reading Check *How does a transformer produce an alternating current in the secondary coil?*

Step-Up Transformer A transformer that increases the voltage so that the output voltage is greater than the input voltage is a step-up transformer. In a step-up transformer the number of wire turns on the secondary coil is greater than the number of turns on the primary coil. For example, the secondary coil of the step-up transformer in **Figure 21A** has twice as many turns as the primary coil has. So the ratio of the output voltage to the input voltage is two, and the output voltage is twice as large as the input voltage. For this transformer an input voltage of 60 V in the primary coil would be increased to 120 V in the secondary coil.

Step-Down Transformer A transformer that decreases the voltage so that the output voltage is less than the input voltage is a step-down transformer. In a step-down transformer the number of wire turns on the secondary coil is less than the number of turns on the primary coil. In **Figure 21B** the secondary coil has half as many turns as the primary coil has, so the ratio of the output voltage to the input voltage is one-half. The input voltage of 240 V in the primary coil is reduced to a voltage of 120 V in the secondary coil.

Figure 21 Transformers can increase or decrease voltage.
A A step-up transformer increases voltage. The secondary coil has more turns than the primary coil does. **B** A step-down transformer decreases voltage. The secondary coil has fewer turns than the primary coil does.
Infer *whether a transformer could change the voltage of a direct current.*

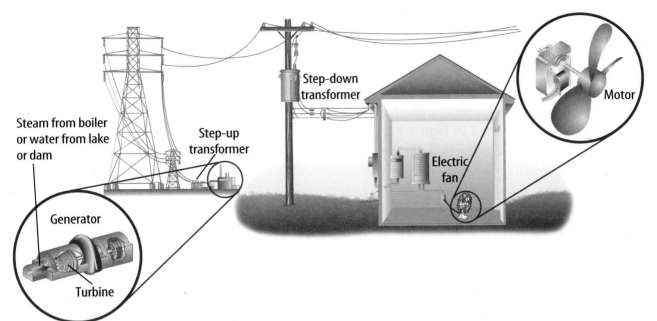

Step-down transformer

Step-up transformer

Steam from boiler or water from lake or dam

Generator

Turbine

Motor

Electric fan

Figure 22 Many steps are involved in the creation, transportation, and use of the electric current in your home.
Identify *the steps that involve electromagnetic induction.*

Transmitting Alternating Current Power plants commonly produce alternating current because the voltage can be increased or decreased with transformers. Although step-up transformers and step-down transformers change the voltage at which electrical energy is transmitted, they do not change the amount of electrical energy transmitted. **Figure 22** shows how step-up and step-down transformers are used in transmitting electrical energy from power plants to your home.

section 3 review

Summary

From Mechanical to Electrical Energy

- An electric current is produced by moving a wire loop through a magnetic field or a magnet through a wire loop.
- A generator can produce an electric current by rotating a wire coil in a magnetic field.

Direct and Alternating Currents

- A direct current flows in one direction. An alternating current changes direction in a regular pattern.

Transformers

- A transformer changes the voltage of an alternating current. The voltage can be increased or decreased.
- The changing magnetic field in the primary coil of a transformer induces an alternating current in the secondary coil.

Self Check

1. **Describe** the energy conversions that occur when water falls on a paddle wheel connected to a generator that is connected to electric lights.
2. **Compare and contrast** a generator with an electric motor.
3. **Explain** why the output voltage from a transformer is zero if a direct current flows through the primary coil.
4. **Explain** why electric current produced by power plants is transmitted as alternating current.
5. **Think Critically** A magnet is pushed into the center of a wire loop, and then stops. What is the current in the wire loop after the magnet stops moving? Explain.

Applying Math

6. **Use a Ratio** A transformer has 1,000 turns of wire in the primary coil and 50 turns in the secondary coil. If the input voltage is 2400 V, what is the output voltage?

Science Online gpscience.com/self_check_quiz

Magnets, C⦿ils, and Currents

Huge generators in power plants produce electricity by moving magnets past coils of wire. How does that produce an electric current?

⊙ Real-World Question

How can a magnet and a wire coil be used to produce an electric current?

Goals

- ■ **Observe** how a magnet and a wire coil can produce an electric current in a wire.
- ■ **Compare** the currents created by moving the magnet and the wire coil in different ways.

Materials

cardboard tube	scissors
bar magnet	galvanometer or ammeter
insulated wire	

Safety Precautions 🔪 🥽 ⚡

WARNING: *Do not touch bare wires when current is running through them.*

⊙ Procedure

1. Wrap the wire around the cardboard tube to make a coil of about 20 turns. Remove the tube from the coil.

2. Use the scissors to cut and remove 2 cm of insulation from each end of the wire.

3. Connect the ends of the wire to a galvanometer or ammeter. Record the reading on your meter.

4. Insert one end of the magnet into the coil and then pull it out. Record the current. Move the magnet at different speeds inside the coil and record the current.

5. Watch the meter and move the bar magnet in different ways around the outside of the coil. Record your observations.

6. Repeat steps 3 through 4, keeping the magnet stationary and moving the wire coil.

⊙ Conclude and Apply

1. How was the largest current generated?

2. Does the current generated always flow in the same direction? How do you know?

3. **Predict** what would happen if you used a coil made with fewer turns of wire.

4. **Infer** whether a current would have been generated if the cardboard tube were left in the coil. Why or why not? Try it.

𝒞ommunicating
Your Data

Compare the currents generated by different members of the class. What was the value of the largest current that was generated? How was this current generated?

Design Your Own

CONTROLLING ELECTROMAGNETS

Goals

■ **Measure** relative strengths of electromagnets.

■ **Determine** which factors affect the strength of an electromagnet.

Possible Materials

22-gauge insulated wire
16-penny iron nail
aluminum rod or nail
0–6 V DC power supply
three 1.5-V "D" cells
steel paper clips
magnetic compass
duct tape (to hold "D" cells together)

Safety Precautions

WARNING: *Do not leave the electromagnet connected for long periods of time because the battery will run down. Magnets will get hot with only a few turns of wire. Use caution in handling them when current is flowing through the coil. Do not apply voltages higher than 6 V to your electromagnets.*

⊙ *Real-World Question*

You use electromagnets every day when you use stereo speakers, power door locks, and many other devices. To make these devices work properly, the strength of the magnetic field surrounding an electromagnet must be controlled. How can the magnetic field produced by an electromagnet be made stronger or weaker? Think about the components that form an electromagnet. Make a hypothesis about how changing these components would affect the strength of the electromagnet's magnetic field.

⊙ *Form a Hypothesis*

As a group, write down the components of an electromagnet that might affect the strength of its magnetic field.

Make a Plan

1. Write your hypothesis for the best way to control the magnetic field strength of an electromagnet.

2. **Decide** how you will assemble and test the electromagnets. Which features will you change to determine the effect on the strength of the magnetic fields? How many changes will you need to try? How many electromagnets do you need to build?

3. **Decide** how you are going to test the strength of your electromagnets. Several ways are possible with the materials listed. Which way would be the most sensitive? Be prepared to change test methods if necessary.

4. Write your plan of investigation. Make sure your plan tests only one variable at a time.

Follow Your Plan

1. Before you begin to build and test the electromagnets, make sure your teacher approves of your plan.

2. Carry out your planned investigation.

3. Record your results.

Analyze Your Data

1. **Make a table** showing how the strength of your electromagnet depends on changes you made in its construction or operation.

2. **Examine** the trends shown by your data. Are there any data points which seem out of line? How can you account for them?

Testing Electromagnets		
Trial	Electromagnet Construction Features	Strength of of Electromagnet
	Do not write in this book.	

Conclude and Apply

1. **Describe** how the electromagnet's magnetic-field strength depended on its construction or operation.

2. **Identify** the features of the electromagnet's construction that had the greatest effect on its magnetic-field strength. Which do you think would be easiest to control?

3. **Explain** how you could use your electromagnet to make a switch. Would it work with both AC and DC?

4. **Evaluate** whether or not your results support your hypothesis. Why or why not?

Communicating

Your Data

Compare your group's result with those of other groups. Did any other group use a different method to test the strength of the magnet? Did you get the same results?

Using Vocabulary

alternating current (AC) p. 242	generator p. 238
	magnetic domains p. 229
direct current (DC) p. 242	magnetic field p. 225
electric motor p. 235	magnetic pole p. 225
electromagnet p. 232	magnetism p. 224
electromagnetic induction p. 238	solenoid p. 232
	transformer p. 243
galvanometer p. 234	turbine p. 240

Complete each statement with the correct vocabulary word or words.

1. A(n) _____ can be used to change the voltage of an alternating current.

2. A(n) _____ is the region where the magnetic field of a magnet is strongest.

3. _____ does not change direction.

4. The properties and interactions of magnets are called _____.

5. A(n) _____ can rotate in a magnetic field when a current passes through it.

6. The magnetic poles of atoms are aligned in a(n) _____.

7. A device that uses an electromagnet to measure electric current is a(n) _____.

Checking Concepts

Choose the word or phrase that best answers the question.

8. Where is the magnetic force exerted by a magnet strongest?
   A) both poles   C) north poles
   B) south poles   D) center

9. Which change occurs in an electric motor?
   A) electrical energy to mechanical energy
   B) thermal energy to wind energy
   C) mechanical energy to electrical energy
   D) wind energy to electrical energy

10. What happens to the magnetic force as the distance between two magnetic poles decreases?
    A) remains constant   C) increases
    B) decreases   D) decreases then increases

11. Which of the following best describes what type of magnetic poles the domains at the north pole of a bar magnet have?
    A) north magnetic poles only
    B) south magnetic poles only
    C) no magnetic poles
    D) north and south magnetic poles

12. Which of the following would not change the strength of an electromagnet?
    A) increasing the amount of current
    B) changing the current's direction
    C) inserting an iron core inside the coil
    D) increasing the number of loops

13. Which of the following would NOT be part of a generator?
    A) turbine   C) electromagnet
    B) battery   D) permanent magnet

14. Which of the following describes the direction of the electric current in AC?
    A) is constant   C) changes regularly
    B) is direct   D) changes irregularly

Interpreting Graphics

15. Copy and complete this Venn diagram. Include the functions, part names, and power sources for these devices.

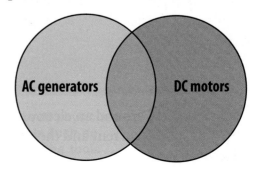

AC generators   DC motors

Science Online gpscience.com/vocabulary_puzzlemaker

Use the diagram below to answer questions 16 and 17.

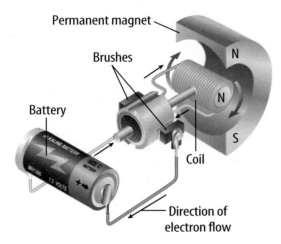

Permanent magnet

Brushes

Battery

N

N

S

Coil

Direction of electron flow

16. Using the diagram, describe the function of the permanent magnet, the electromagnet, and the current source in a simple electric motor.

17. Describe the sequence of steps that occur in an electric motor that forces the coil to spin. Include the role of the commutator in your description.

Use the graph below to answer questions 18–20.

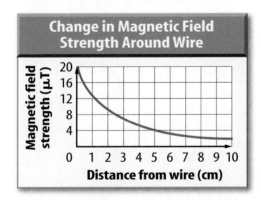

Change in Magnetic Field Strength Around Wire

Magnetic field strength (μT)

20
16
12
8
4

0 1 2 3 4 5 6 7 8 9 10

Distance from wire (cm)

18. How much larger is the magnetic field strength 1 cm from the wire compared to 5 cm from the wire?

19. Does the magnetic field strength decrease more rapidly with distance closer to the wire or farther from the wire? Explain.

20. Using the graph, estimate the magnetic field strength 11 cm from the wire.

Thinking Critically

21. **Infer** how you could you use a horseshoe magnet to find the direction north.

22. **Explain** In Europe, generators produce alternating current at a frequency of 50 Hz. Would the electric appliances you use in North America work if you plugged them into an outlet in Europe? Why or why not?

23. **Predict** Two generators are identical except for the loops of wire that rotate through their magnetic fields. One has twice as many turns of wire as the other one does. Which generator would produce the most electric current? Why?

24. **Explain** why a bar magnet will attract an iron nail to either its north pole or its south pole, but attract another magnet to only one of its poles.

25. **Compare and contrast** electromagnetic induction and the formation of electromagnets.

Applying Math

26. **Calculate** A step-down transformer reduces a 2,400-V current to 120 V. If the primary coil has 500 turns of wire, how many turns of wire are there on the secondary coil?

27. **Use a Ratio** To produce a spark, a spark plug requires a current at about 12,000 V. A car's engine uses a type of transformer called an induction coil to change the input voltage from 12 V to 12,000 V. In the induction coil, what is the ratio of the number of wire turns on the primary coil to the number of turns on the secondary coil?

Part 1 | Multiple Choice

Record your answers on the answer sheet provided by your teacher or on a sheet of paper.

A group of students built a transformer by wrapping 50 turns of wire on one side of an iron ring to form the primary coil. They then wrapped 10 turns of wire around the opposite side to form the secondary coil. Their results are shown in the table below.

Use the table below to answer questions 1–3.

Voltage and Current in a Transformer				
Trial	Input Voltage (V)	Primary Coil Current (A)	Output Voltage (V)	Secondary Coil Current (A)
1	5	0.1	1	0.5
2	10	0.2	2	1.0
3	20	0.2	4	1.0
4	50	0.5	10	2.5

1. What is the ratio of the input voltage to the output voltage for this transformer?
   A. 1:2.5
   B. 4:1
   C. 5:1
   D. 1:5

2. What is the ratio of the primary coil current to the secondary coil current?
   A. 1:2.5
   B. 4:1
   C. 5:1
   D. 1:5

3. The ratio of the secondary coil current to the primary coil current equals which of the following?
   A. the ratio of the secondary coil wire turns to the primary coil wire turns
   B. the ratio of the output voltage to the input voltage
   C. the ratio of the primary coil wire turns to the secondary wire turns
   D. It always equals one.

Use the figure below to answer questions 4 and 5.

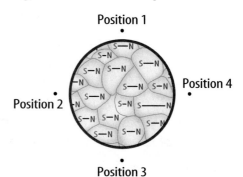

4. A steel paper clip is sitting on a desk. The figure above shows the magnetic domains in a section of the paper clip after the north pole of a magnet has been moved close to it. According to the diagram, the magnet's north pole is most likely at which of the following positions?
   A. position 1
   B. position 2
   C. position 3
   D. position 4

5. Which of the following diagrams shows the orientation of the needle of a compass that is placed at position 2?

   A.
   B.
   C.
   D.

Test-Taking Tip

Read All the Information If a question includes a text passage and a graphic, carefully read the information in the text passage and the graphic before answering the question.

Question 3 Review the information in the text above the table and the information in the table.

Part 2 | Short Response/Grid In

*Record your answers on the answer sheet
provided by your teacher or on a sheet of paper.*

6. A hydroelectric power plant uses water to spin a turbine attached to a generator. The generator produces 30,000 kW of electric power. If the turbine and generator are 85 percent efficient, how much power does the falling water supply to the turbine?

7. A bicycle has a small electric generator that is used to light a headlight. The generator is made to spin by rubbing against a wheel. Will the bicycle coast farther on a level surface if the light is turned on or turned off?

8. An electric motor rotates 60 times per second if the current source is 60 Hz alternating current. How many times will an electric motor rotate in one hour if the current source is changed to 50 Hz alternating current?

Use the figure below to answer questions 9 and 10.

9. A step-down transformer is plugged into a 120-V electric outlet and a light is plugged into the transformer. The transformer has 20 turns on the primary coil and 2 turns on the secondary coil. What is the voltage at the output coil?

10. If the light has a resistance of 8 Ω, what is the current in the light?

Part 3 | Open Ended

Record your answers on a sheet of paper.

11. A bar magnet is placed inside a wire coil. The bar magnet and coil are then carried across a room. Explain whether an electric current will flow in the coil as the magnet and coil are moving.

12. A student connects a battery to a step-up transformer in order to boost the voltage. Explain why a small electric motor does not spin when it is connected to the secondary coil of the transformer.

Use the figure below to answer question 13.

Effect of Rotation on Generator Voltage

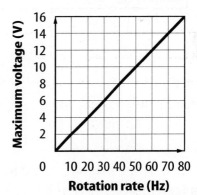

13. The graphic above shows how the voltage produced by a generator depends on the rotation rate of the coil. Explain whether this generator could produce household AC current which is 120 V at 60 Hz.

14. Describe how a permanent magnet is similar to and different from a piece of unmagnetized iron.

15. Compare and contrast the behavior and properties of positive and negative electric charges with north and south magnetic poles.

Energy Sources

That's a lot of kilowatts!

Look closely at this photo. Can you identify at least three items that require energy to operate? The welding torches require energy to join the pieces of the car together. The robots require energy to assemble the pieces. When the car is finished, energy will be required to operate the car.

Science Journal Describe how your day would be different if the electric power were off all day.

Start-Up Activities

Heating with Solar Energy

The Sun constantly bathes our planet with enormous amounts of energy. This energy can be captured and used to make electricity, heat homes, and provide hot water. How can the Sun's energy be used to heat water?

1. Use scissors to poke a small hole in the center of two coffee can lids.

2. Fill a coffee can that has been painted black with water at room temperature. Snap on the lid and push a thermometer through the hole in the lid. Record the temperature.

3. Repeat step 2 using the coffee can that has been painted white.

4. Place both cans in direct sunlight. After 15 min, record the temperature of the water in both cans again.

5. **Think Critically** Write a paragraph explaining why the temperature change differed between the two cans.

Preview this chapter's content and activities at gpscience.com

Energy Sources There are many sources of energy. Make the following Foldable to help you organize information about various types of energy sources.

STEP 1 **Fold** a sheet of paper in half lengthwise. Make the back edge about 5 cm longer than the front edge.

STEP 2 **Turn** the paper so the fold is on the bottom. Then **fold** it into thirds.

STEP 3 **Unfold and cut** only the top layer along both folds to make three tabs.

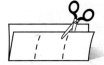

STEP 4 **Label** the Foldable as shown.

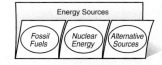

Energy Sources
Fossil Fuels | Nuclear Energy | Alternative Sources

Summarize As you read this chapter, summarize important information about each type of energy source under the appropriate tab.

Fossil Fuels

Reading Guide

What You'll Learn
- **Discuss** properties and uses of fossil fuels.
- **Explain** how fossil fuels are formed.
- **Describe** how the chemical energy in fossil fuels is converted into electrical energy.

Why It's Important
Fossil fuels are used to generate most of the energy you use every day.

Review Vocabulary
chemical potential energy: the energy stored in the chemical bonds between atoms in molecules

New Vocabulary
- fossil fuel
- petroleum
- nonrenewable resource

Figure 1 Energy is used in many ways.

Automobiles burn gasoline to provide energy.

Power lines like these carry the electrical energy you use every day.

Using Energy

How many different ways have you used energy today? You can see energy being used in many ways, throughout the day, such as those shown in **Figure 1.** Furnaces and stoves use thermal energy to heat buildings and cook food. Air conditioners use electrical energy to move thermal energy outdoors. Cars and other vehicles use mechanical energy to carry people and materials from one part of the country to another.

Transforming Energy According to the law of conservation of energy, energy cannot be created or destroyed. Energy can only be transformed, or converted, from one form to another. To use energy means to transform one form of energy to another form of energy that can perform a useful function. For example, energy is used when the chemical energy in fuels is transformed into thermal energy that is used to heat your home.

Sometimes energy is transformed into a form that isn't useful. For example, when an electric current flows through power lines, about 10 percent of the electrical energy is changed to thermal energy. This reduces the amount of useful electrical energy that is delivered to homes, schools, and businesses.

Energy Use in the United States

More energy is used in the United States than in any other country in the world. **Figure 2** shows energy usage in the United States. About 20 percent of the energy is used in homes for heating and cooling, to run appliances, and to provide lighting and hot water. About 27 percent is used for transportation, powering vehicles such as cars, trucks, and aircraft.

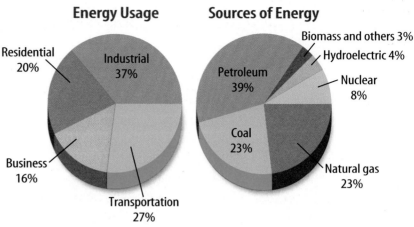

Another 16 percent is used by businesses to heat, cool, and light stores, shops, and office buildings. Finally, about 37 percent of this energy is used by industry and agriculture to manufacture products and produce food. **Figure 2** also shows the main sources of the energy used in the United States. Almost 85 percent of the energy used in the United States comes from burning petroleum, natural gas, and coal. Nuclear power plants provide about eight percent of the energy used in the United States.

Figure 2 These circle graphs show where energy is used in the United States and sources of this energy.

Making Fossil Fuels

In one hour of freeway driving a car might use several gallons of gasoline. It may be hard to believe that it took millions of years to make the fuels that are used to produce electricity, provide heat, and transport people and materials. **Figure 4** on the next page shows how coal, petroleum, and natural gas are formed by the decay of ancient plants and animals. Fuels such as petroleum, or oil, natural gas, and coal are called **fossil fuels** because they are formed from the decaying remains of ancient plants and animals.

Concentrated Energy Sources When fossil fuels are burned, carbon and hydrogen atoms combine with oxygen molecules in the air to form carbon dioxide and water molecules. This process converts the chemical potential energy that is stored in the chemical bonds between atoms to heat and light. Compared to other fuels such as wood, the chemical energy that is stored in fossil fuels is more concentrated. For example, burning 1 kg of coal releases two to three times as much energy as burning 1 kg of wood. **Figure 3** shows the amount of energy that is produced by burning different fossil fuels.

Figure 3 The bar graph shows the amount of energy released by burning one gram of four different fuels. **Determine** the ratio of the energy content of natural gas to the energy content of wood.

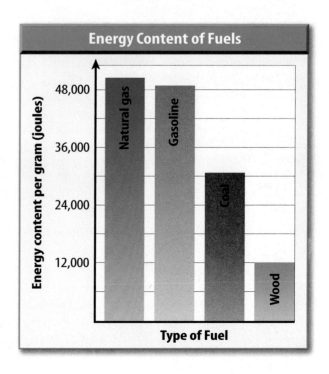

Figure 4

O il and natural gas form when organic matter on the ocean floor, gradually buried under additional layers of sediment, is chemically changed by heat and crushing pressure. The oil and gas may bubble to the surface or become trapped beneath a dense rock layer. Coal forms when peat—partially decomposed vegetation—is compressed by overlying sediments and transformed first into lignite (soft brown coal) and then into harder, bituminous (buh TYEW muh nus) coal. These two processes are shown below.

HOW OIL AND NATURAL GAS ARE FORMED

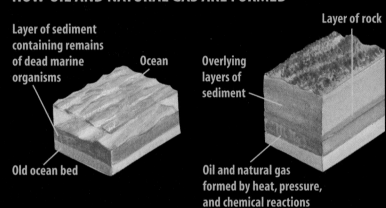

Layer of sediment containing remains of dead marine organisms

Ocean

Old ocean bed

Overlying layers of sediment

Layer of rock

Oil and natural gas formed by heat, pressure, and chemical reactions

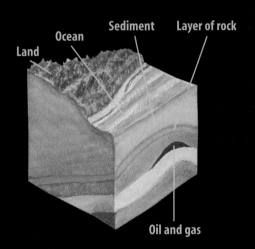

Land

Ocean

Sediment

Layer of rock

Oil and gas

HOW COAL IS FORMED

Vegetation

Peat

New layers of overlying sediment

Increasing pressure and temperature

Lignite

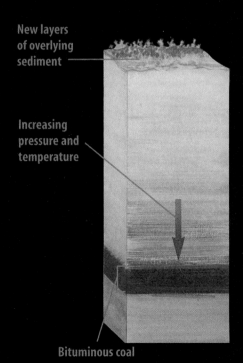

New layers of overlying sediment

Increasing pressure and temperature

Bituminous coal

Petroleum

Millions of gallons of petroleum, or crude oil, are pumped every day from wells deep in Earth's crust. **Petroleum** is a highly flammable liquid formed by decayed ancient organisms, such as microscopic plankton and algae. Petroleum is a mixture of thousands of chemical compounds. Most of these compounds are hydrocarbons, which means their molecules contain only carbon atoms and hydrogen atoms.

Separating Hydrocarbons The different hydrocarbon molecules found in petroleum have different numbers and arrangements of carbon and hydrogen atoms. The composition and structure of hydrocarbons determines their properties.

The many different compounds that are found in petroleum are separated in a process called fractional distillation. This separation occurs in the tall towers of oil-refinery plants. First, crude oil is pumped into the bottom of the tower and heated. The chemical compounds in the crude oil boil and vaporize according to their individual boiling points. Materials with the lowest boiling points rise to the top of the tower as vapor and are collected. Hydrocarbons with high boiling points, such as asphalt and some types of waxes, remain liquid and are drained off through the bottom of the tower.

✔ Reading Check *What is fractional distillation used for?*

Other Uses for Petroleum Not all of the products obtained from petroleum are burned to produce energy. About 15 percent of the petroleum-based substances that are used in the United States go toward nonfuel uses. Look around at the materials in your home or classroom. Do you see any plastics? In addition to fuels, plastics and synthetic fabrics are made from the hydrocarbons found in crude petroleum. Also, lubricants such as grease and motor oil, as well as the asphalt used in surfacing roads, are obtained from petroleum. Some synthetic materials produced from petroleum are shown in **Figure 5.**

Mini LAB

Designing an Efficient Water Heater

Procedure

1. Measure and record the mass of a **candle.**
2. Measure 50 mL of **water** into a **beaker.** Record the temperature of the water.
3. Use the lighted candle to increase the temperature of the water by 10°C. Put out the candle and measure its mass again.
4. Repeat steps 1 to 3 with an **aluminum chimney** surrounding the candle to help direct the heat upward.

Analysis

1. Compare the mass change in the two trials. Does a smaller or larger mass change in the candle show greater efficiency?
2. Gas burners are used to heat hot-water tanks. What must be considered in the design of these heaters?

Figure 5 The objects shown here are made from chemical compounds found in petroleum.
Identify *four objects in your classroom that are made from petroleum.*

Natural Gas

The chemical processes that produce petroleum as ancient organisms decay also produce gaseous compounds called natural gas. These compounds rise to the top of the petroleum deposit and are trapped there. Natural gas is composed mostly of methane, CH_4, but it also contains other hydrocarbon gases such as propane, C_3H_8, and butane, C_4H_{10}. Natural gas is burned to provide energy for cooking, heating, and manufacturing. About one fourth of the energy consumed in the United States comes from burning natural gas. There's a good chance that your home has a stove, furnace, hot-water heater, or clothes drier that uses natural gas.

Natural gas contains more energy per kilogram than petroleum or coal does. It also burns more cleanly than other fossil fuels, produces fewer pollutants, and leaves no residue such as ash.

Coal

Coal is a solid fossil fuel that is found in mines underground, such as the one shown in **Figure 6.** In the first half of the twentieth century, most houses in the United States were heated by burning coal. In fact, during this time, coal provided more than half of the energy that was used in the United States. Now, almost two-thirds of the energy used comes from petroleum and natural gas, and only about one-fourth comes from coal. About 90 percent of all the coal that is used in the United States is burned by power plants to generate electricity.

Figure 6 Coal mines usually are located deep underground.

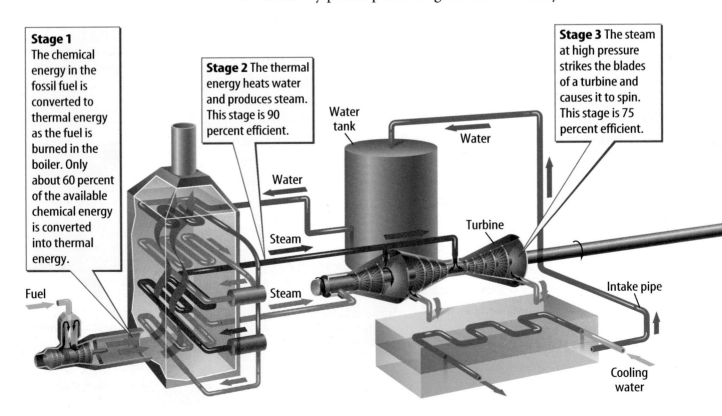

Stage 1 The chemical energy in the fossil fuel is converted to thermal energy as the fuel is burned in the boiler. Only about 60 percent of the available chemical energy is converted into thermal energy.

Stage 2 The thermal energy heats water and produces steam. This stage is 90 percent efficient.

Stage 3 The steam at high pressure strikes the blades of a turbine and causes it to spin. This stage is 75 percent efficient.

Water tank

Water

Water

Fuel

Steam

Steam

Turbine

Intake pipe

Cooling water

Origin of Coal Coal mines were once the sites of ancient swamps. Coal formed from the organic material that was deposited as the plants that lived in these swamps died. Worldwide, the amount of coal that is potentially available is estimated to be 20 to 40 times greater than the supply of petroleum.

Coal also is a complex mixture of hydrocarbons and other chemical compounds. Compared to petroleum and natural gas, coal contains more impurities, such as sulfur and nitrogen compounds. As a result, more pollutants, such as sulfur dioxide and nitrogen oxides, are produced when coal is burned.

Generating Electricity

Figure 7 shows that almost 70 percent of the electrical energy used in the United States is produced by burning fossil fuels. How is the chemical energy contained in fossil fuels converted to electrical energy in an electric power station?

The process is shown in **Figure 8.** In the first stage, fuel is burned in a boiler or combustion chamber, and it releases thermal energy. In the second stage, this thermal energy heats water and produces steam under high pressure. In the third stage, the steam strikes the blades of a turbine, causing it to spin. The shaft of the turbine is connected to an electric generator. In the fourth stage, electric current is produced when the spinning turbine shaft rotates magnets inside the generator. In the final stage, the electric current is transmitted to homes, schools, and businesses through power lines.

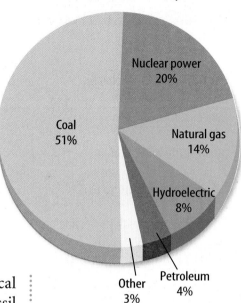

Sources of Electricity

Nuclear power 20%

Coal 51%

Natural gas 14%

Hydroelectric 8%

Other 3%

Petroleum 4%

Figure 7 This circle graph shows the percentage of electricity generated in the United States that comes from various energy sources.

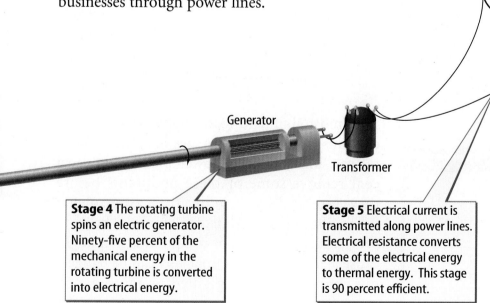

Generator

Transformer

Power lines

Stage 4 The rotating turbine spins an electric generator. Ninety-five percent of the mechanical energy in the rotating turbine is converted into electrical energy.

Stage 5 Electrical current is transmitted along power lines. Electrical resistance converts some of the electrical energy to thermal energy. This stage is 90 percent efficient.

Figure 8 Fossil fuels are burned to generate electricity in a power plant. **Determine** *which stage in this process is the most inefficient.*

Table 1 Efficiency of Fossil Fuel Conversion	
Process	**Efficiency (%)**
Chemical to thermal energy	60
Conversion of water to steam	90
Steam-turning turbine	75
Turbine spins electric generator	95
Transmission through power lines	90
Overall efficiency	35

Efficiency of Power Plants

When fossil fuels are burned to produce electricity, not all the chemical energy in the fuel is converted to electrical energy. In every stage of the process, some energy is converted into forms of energy that can't be used.

The overall efficiency of the entire process is given by multiplying the efficiencies of each stage of the process shown in **Table 1.** If you were to do this, you'd find that the resulting overall efficiency is only about 35 percent. This means that only about 35 percent of the energy contained in the fossil fuels is delivered to homes, schools, and businesses as electrical energy. The other 65 percent is converted mainly into thermal energy when the chemical energy in fuel is transformed into electrical energy that is delivered to energy users.

The Costs of Using Fossil Fuels

Although fossil fuels are a useful source of energy for generating electricity and providing the power for transportation, their use has some undesirable side effects. When petroleum products and coal are burned, smoke is given off that contains small particles called particulates. These particulates cause breathing problems for some people. Burning fossil fuels also releases carbon dioxide. **Figure 9** shows how the carbon dioxide concentration in the atmosphere has increased from 1960 to 2000. One consequence of increasing the atmospheric carbon dioxide concentration could be to cause Earth's surface temperature to increase.

Using Coal The most abundant fossil fuel is coal, but coal contains even more impurities than oil or natural gas. Many electric power plants that burn coal remove some of these pollutants before they are released into the atmosphere. Removing sulfur dioxide, for example, helps to prevent the formation of compounds that might cause acid rain. Mining coal also can be dangerous. Miners risk being killed or injured, and some suffer from lung diseases caused by breathing coal dust over long periods of time.

Figure 9 The carbon dioxide concentration in Earth's atmosphere has been measured at Mauna Loa in Hawaii. From 1960 to 2000, the carbon dioxide concentration has increased by about 16 percent.

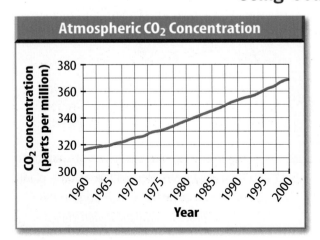

Nonrenewable Resources

All fossil fuels are **nonrenewable resources,** which means they are resources that cannot be replaced by natural processes as quickly as they are used. Therefore, fossil fuel reserves are decreasing at the same time that population and industrial demands are increasing. **Figure 10** shows how the production of oil might decline over the next 50 years as oil reserves are used up. As the production of energy from fossil fuels continues, the remaining reserves of fossil fuels will decrease. Fossil fuels will become more difficult to obtain, causing them to become more costly in the future.

Conserving Fossil Fuels

Even as reserves of fossil fuels decrease and they become more costly, the demand for energy continues to increase as the world's population increases. One way to meet these energy demands would be to reduce the use of fossil fuels and obtain energy from other sources.

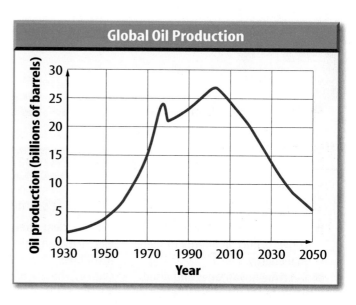

Figure 10 Some predictions show that worldwide oil production will peak by 2005 and then decline rapidly over the following 50 years.

---

section 1 review

Summary

Using Energy

- Energy cannot be created or destroyed, but can only be transformed from one form to another.

Fossil Fuels

- Petroleum, natural gas, and coal are fossil fuels formed by the decay of ancient plants and animals.
- Petroleum is a mixture of thousands of chemical compounds, most of which are hydrocarbons.
- About 90 percent of all coal used in the United States is burned by power plants to produce electricity.

Generating Electricity

- Power plants burn fossil fuels to produce steam that spins turbines attached to electric generators.

Self Check

1. **Describe** the advantages and disadvantages of using fossil fuels to generate electricity.
2. **Explain** how the different chemical compounds in crude oil are separated.
3. **Describe** how fossil fuels are formed.
4. **Name** three materials that are derived from the chemical compounds in petroleum.
5. **Think Critically** If fossil fuels are still forming, why are they considered to be a nonrenewable resource?

Applying Math

6. **Interpret a Graph** According to the graph in **Figure 9,** by how many parts per million did the concentration of atmospheric carbon dioxide increase from 1960 to 2000?
7. **Use a Table** In **Table 1,** if the efficiency of converting chemical to thermal energy was 90 percent, what would be the overall efficiency be?

Nuclear Energy

Reading Guide

What You'll Learn

- **Explain** how a nuclear reactor converts nuclear energy to thermal energy.
- **Describe** advantages and disadvantages of using nuclear energy to produce electricity.
- **Discuss** nuclear fusion as a possible energy source.

Why It's Important

Using nuclear energy to produce electricity can help reduce the use of fossil fuels. However, like all energy sources, the use of nuclear energy has advantages and disadvantages.

✪ Review Vocabulary

nuclear fission: the process of splitting an atomic nucleus into two or more nuclei with smaller masses

New Vocabulary

- nuclear reactor
- nuclear waste

Using Nuclear Energy

Over the past several decades, electric power plants have been developed that generate electricity without burning fossil fuels. Some of these power plants, such as the one shown in **Figure 11,** convert nuclear energy to electrical energy. Energy is released when the nucleus of an atom breaks apart. In this process, called nuclear fission, an extremely small amount of mass is converted into an enormous amount of energy. Today almost 20 percent of all the electricity produced in the United States comes from nuclear power plants. Overall, nuclear power plants produce about eight percent of all the energy consumed in the United States. In 2003, there were 104 nuclear reactors producing electricity at 65 nuclear power plants in the United States.

Figure 11 A nuclear power plant generates electricity using the energy released in nuclear fission. Each of the domes contain a nuclear reactor. A cooling tower is on the left.

Nuclear Reactors

A **nuclear reactor** uses the energy from controlled nuclear reactions to generate electricity. Although nuclear reactors vary in design, all have some parts in common, as shown in **Figure 12.** They contain a fuel that can be made to undergo nuclear fission; they contain control rods that are used to control the nuclear reactions; and they have a cooling system that keeps the reactor from being damaged by the heat produced. The actual fission of the radioactive fuel occurs in a relatively small part of the reactor known as the core.

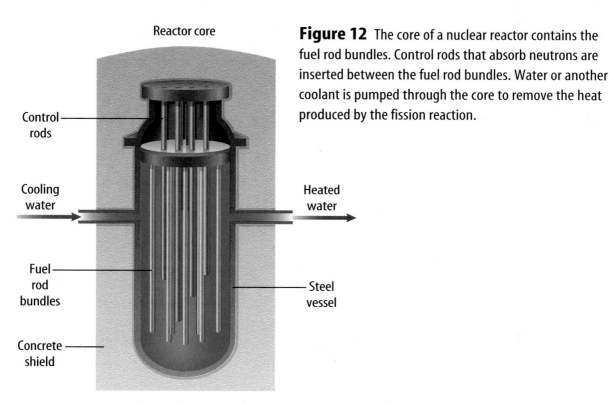

Reactor core

Control rods

Cooling water

Heated water

Fuel rod bundles

Steel vessel

Concrete shield

Figure 12 The core of a nuclear reactor contains the fuel rod bundles. Control rods that absorb neutrons are inserted between the fuel rod bundles. Water or another coolant is pumped through the core to remove the heat produced by the fission reaction.

Nuclear Fuel Only certain elements have nuclei that can undergo fission. Naturally occurring uranium contains an isotope, U-235, whose nucleus can split apart. As a result, the fuel that is used in a nuclear reactor is usually uranium dioxide. Naturally occurring uranium contains only about 0.7 percent of the U-235 isotope. In a reactor, the uranium usually is enriched so that it contains three percent to five percent U-235.

The Reactor Core The reactor core contains uranium dioxide fuel in the form of tiny pellets like the ones in **Figure 13.** The pellets are about the size of a pencil eraser and are placed end to end in a tube. The tubes are then bundled and covered with a metal alloy, as shown in **Figure 13.** The core of a typical reactor contains about a hundred thousand kilograms of uranium in hundreds of fuel rods. For every kilogram of uranium that undergoes fission in the core, 1 g of matter is converted into energy. The energy released by this gram of matter is equivalent to the energy released by burning more than 3 million kg of coal.

Figure 13 Nuclear fuel pellets are stacked together to form fuel rods. The fuel rods are bundled together, and the bundle is covered with a metal alloy.

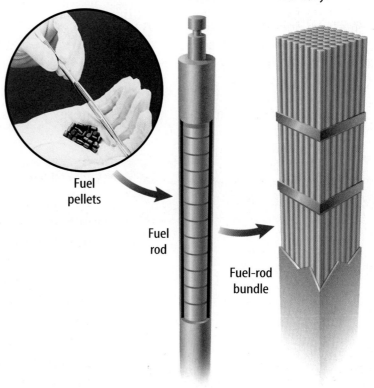

Fuel pellets

Fuel rod

Fuel-rod bundle

Figure 14 When a neutron strikes the nucleus of a U-235 atom, the nucleus splits apart into two smaller nuclei. In the process two or three neutrons also are emitted. The smaller nuclei are called fission products. **Explain** *what happens to the neutrons that are released in this reaction.*

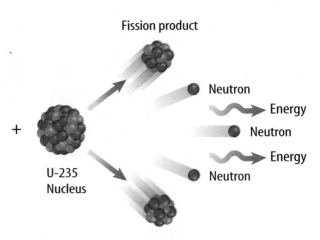

Fission product

Neutron

Energy

Neutron

Energy

Neutron

U-235 Nucleus

+

Fission product

Nuclear Fission How does the nuclear reaction proceed in the reactor core? Neutrons that are produced by the decay of U-235 nuclei are absorbed by other U-235 nuclei. When a U-235 nucleus absorbs a neutron, it splits into two smaller nuclei and two or three additional neutrons, as shown in **Figure 14.** These neutrons strike other U-235 nuclei, causing them to release two or three more neutrons each when they split apart.

Because every uranium atom that splits apart releases neutrons that cause other uranium atoms to split apart, this process is called a nuclear chain reaction. In the chain reaction involving the fission of uranium nuclei, the number of nuclei that are split can more than double at each stage of the process. As a result, an enormous number of nuclei can be split after only a small number of stages. For example, if the number of nuclei involved doubles at each stage, after only 50 stages more than a quadrillion nuclei might be split.

Nuclear chain reactions take place in a matter of milliseconds. If the process isn't controlled, the chain reaction will release energy explosively rather than releasing energy at a constant rate.

 **Reading Check** *What is a nuclear chain reaction?*

Controlling the Chain Reaction To control the chain reaction, some of the neutrons that are released when U-235 splits apart must be prevented from striking other U-235 nuclei. These neutrons are absorbed by rods containing boron or cadmium that are inserted into the reactor core. Moving these control rods deeper into the reactor causes them to absorb more neutrons and slow down the chain reaction. Eventually, only one of the neutrons released in the fission of each of the U-235 nuclei strikes another U-235 nucleus, and energy is released at a constant rate.

INTEGRATE Earth Science

Uranium-Lead Dating
Uranium is used to determine the age of rocks. As uranium decays into lead at a constant rate, the age of a rock can be found by comparing the amount of uranium to the amount of lead produced. Uranium-lead dating is used by scientists to date rocks as old as 4.6 billion years. Research other methods used to determine the age of rocks.

Nuclear Power Plants

Nuclear fission reactors produce electricity in much the same way that conventional power plants do. **Figure 15** shows how a nuclear reactor produces electricity. The thermal energy released in nuclear fission is used to heat water and produce steam. This steam then is used to drive a turbine that rotates an electric generator. To transfer thermal energy from the reactor core to heat water and produce steam, the core is immersed in a fluid coolant. The coolant absorbs heat from the core and is pumped through a heat exchanger. There thermal energy is transferred from the coolant and boils water to produce steam. The overall efficiency of nuclear power plants is about 35 percent, similar to that of fossil fuel power plants.

The Risks of Nuclear Power

Producing energy from nuclear fission has some disadvantages. Nuclear power plants do not produce the air pollutants that are released by fossil-fuel burning power plants. Also, nuclear power plants don't produce carbon dioxide.

The nuclear generation of electricity, however, has its problems. The mining of the uranium can cause environmental damage. Water that is used as a coolant in the reactor core must cool before it is released into streams and rivers. Otherwise, the excess heat could harm fish and other animals and plants in the water.

Ukraine The worst nuclear accident in history occurred at the Chernobyl nuclear power plant in the Ukraine in 1986. Many people in the area suffered from radiation sickness. Use a map or atlas to find the location of the Ukraine. Write a description of the location in your Science Journal.

Figure 15 A nuclear power plant uses the heat produced by nuclear fission in its core to produce steam. The steam turns an electric generator.

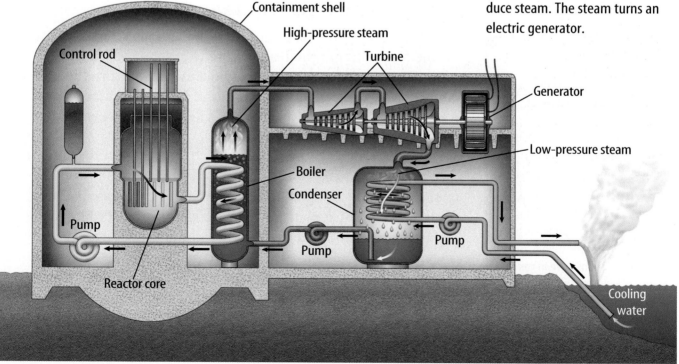

Figure 16 An explosion occurred at the Chernobyl reactor in the Ukraine after graphite control rods caught fire. The explosion shattered the reactor's roof.

Topic: Storing Nuclear Wastes

Visit gpscience.com for Web links to information about storing nuclear wastes.

Activity Obtain a map or sketch an outline of the United States. Mark the locations of the nuclear waste sites that you found. What do these locations have in common? Why do you think these locations were chosen over other sites that were closer to the nuclear waste generating sites?

The Release of Radioactivity One of the most serious risks of nuclear power is the escape of harmful radiation from power plants. The fuel rods contain radioactive elements with various half-lives. Some of these elements could cause damage to living organisms if they were released from the reactor core. Nuclear reactors have elaborate systems of safeguards, strict safety precautions, and highly trained workers in order to prevent accidents. In spite of this, accidents have occurred.

For example, in 1986 in Chernobyl, Ukraine, an accident occurred when a reactor core overheated during a safety test. Materials in the core caught fire and caused a chemical explosion that blew a hole in the reactor, as shown in **Figure 16.** This resulted in the release of radioactive materials that were carried by winds and deposited over a large area. As a result of the accident, 28 people died of acute radiation sickness. It is possible that 260,000 people might have been exposed to levels of radiation that could affect their health.

In the United States, power plants are designed to prevent accidents such as the one that occurred at Chernobyl. But many people still are concerned that similar accidents are possible.

The Disposal of Nuclear Waste

After about three years, not enough fissionable U-235 is left in the fuel pellets in the reactor core to sustain the chain reaction. The spent fuel contains radioactive fission products in addition to the remaining uranium. **Nuclear waste** is any radioactive by-product that results when radioactive materials are used.

Low-Level Waste Low-level nuclear wastes usually contain a small amount of radioactive material. They usually do not contain radioactive materials with long half-lives. Products of some medical and industrial processes are low-level wastes, including items of clothing used in handling radioactive materials. Low-level wastes also include used air filters from nuclear power plants and discarded smoke detectors. Low-level wastes usually are sealed in containers and buried in trenches 30 m deep at special locations. When dilute enough, low-level waste sometimes is released into the air or water.

High-Level Waste High-level nuclear waste is generated in nuclear power plants and by nuclear weapons programs. After spent fuel is removed from a reactor, it is stored in a deep pool of water, as shown in **Figure 17.** Many of the radioactive materials in high-level nuclear waste have short half-lives. However, the spent fuel also contains materials that will remain radioactive for tens of thousands of years. For this reason, the waste must be disposed of in extremely durable and stable containers.

 **Reading Check** *What is the difference between low-level and high-level nuclear wastes?*

One method proposed for the disposal of high-level waste is to seal the waste in ceramic glass, which is placed in protective metal-alloy containers. The containers then are buried hundreds of meters below ground in stable rock formations or salt deposits. It is hoped that this will keep the material from contaminating the environment for thousands of years.

Figure 17 Spent nuclear fuel rods are placed underwater after they are removed from the reactor core. The water absorbs the nuclear radiation and prevents it from escaping into the environment.

Applying Science

Can a contaminated radioactive site be reclaimed?

In the early 1900s, with the discovery of radium, extensive mining for the element began in the Denver, Colorado, area. Radium is a radioactive element that was used to make watch dials and instrument panels that glowed in the dark. After World War I, the radium industry collapsed. The area was left contaminated with 97,000 tons of radioactive soil and debris containing heavy metals and radium, which is now known to cause cancer. The soil was used as fill, foundation material, left in place, or mishandled.

> Radium
> 88
> **Ra**
> (226)

Identifying the Problem

In the 1980s, one area became known as the Denver Radium Superfund Site and was cleaned up by the Environmental Protection Agency. The land then was reclaimed by a local commercial establishment.

Solving the Problem

1. The contaminated soil was placed in one area and a protective cap was placed over it. This area also was restricted from being used for residential homes. Explain why it is important for the protective cap to be maintained and why homes could not be built in this area.

2. The advantages of cleaning up this site are economical, environmental, and social. Give an example of each.

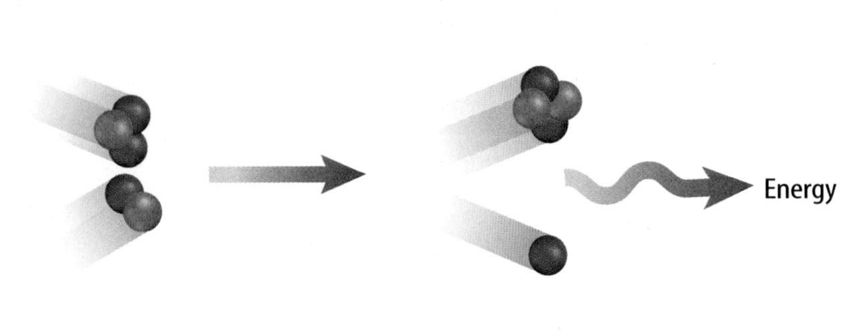

H-3 nucleus   He-4 nucleus

H-2 nucleus   Neutron

Energy

Figure 18 In nuclear fusion, two smaller nuclei join together to form a larger nucleus. Energy is released in the process. In the reaction shown here, two isotopes of hydrogen come together to form a helium nucleus.
Identify *the source of the energy released in a fusion reaction.*

Nuclear Fusion

The Sun gives off a tremendous amount of energy through a process called thermonuclear fusion. Thermonuclear fusion is the joining together of small nuclei at high temperatures, as shown in **Figure 18.** In this process, a small amount of mass is converted into energy. Fusion is the most concentrated energy source known.

An advantage of producing energy using nuclear fusion is that the process uses hydrogen as fuel. Hydrogen is abundant on Earth. Another advantage is that the product of the reaction is helium. Helium is not radioactive and is chemically nonreactive.

One disadvantage of fusion is that it occurs only at temperatures of millions of degrees Celsius. Research reactors often consume more energy to reach and maintain these temperatures than they produce. Another problem is how to contain a reaction that occurs at such extreme conditions. Until solutions to these and other problems are found, the use of nuclear fusion as an energy source is not practical.

section 2 review

Summary

Using Nuclear Energy
- Nuclear power plants produce about eight percent of the energy used each year in the United States.

Nuclear Power Plants
- Nuclear reactors use the energy released in the fission of U-235 to produce electricity.
- The energy released in the fission reaction is used to make steam. The steam drives a turbine that rotates an electric generator.

The Risks of Nuclear Energy
- Nuclear power generation produces high-level nuclear wastes.
- Organisms could be damaged if radiation is released from the reactor.
- Nuclear waste is the radioactive by-product produced by using radioactive materials.

Self Check

1. **Explain** why a chain reaction occurs when uranium-235 undergoes fission.
2. **Describe** how the chain reaction in a nuclear reactor is controlled.
3. **Compare** the advantages and disadvantages of nuclear power plants and those that burn fossil fuels.
4. **Describe** the advantages and disadvantages of using nuclear fusion reactions as a source of energy.
5. **Think Critically** A research project produced 10 g of nuclear waste with a short half-life. How would you classify this waste and how would it be disposed of?

Applying Math

6. **Use Percentages** Naturally occurring uranium contains 0.72 percent of the isotope uranium-235. What is the mass of uranium-235 in 2,000 kg of naturally-occurring uranium?

Renewable Energy Sources

Reading Guide

***What* You'll Learn**
- ■ **Analyze** the need for alternate energy sources.
- ■ **Describe** alternate methods for generating electricity.
- ■ **Compare** the advantages and disadvantages of various alternate energy sources.

***Why* It's Important**
The primary sources of energy in the United States are nonrenewable, so alternative energy sources need to be explored.

⊙ Review Vocabulary
radiant energy: the energy carried by an electromagnetic wave

New Vocabulary
- ● renewable resource
- ● photovoltaic cell
- ● hydroelectricity
- ● geothermal energy
- ● biomass

Energy Options

The demand for energy increases continually, but supplies of fossil fuels are decreasing. Using more nuclear reactors to produce electricity will produce more high-level nuclear waste that has to be disposed of safely. As a result, other sources of energy that can meet Earth's increasing energy demands are being developed. Some alternative energy sources are renewable resources. A **renewable resource** is an energy source that is replaced nearly as quickly as it is used.

Energy from the Sun

The average amount of solar energy that falls on the United States in one day is more than the total amount of energy used in the United States in one year. Because only about one billionth of the Sun's energy falls on Earth, and because the Sun is expected to continue producing energy for several billion years, solar energy cannot be used up. Solar energy is a renewable resource.

Many devices use solar energy for power including solar-powered calculators similar to the one in **Figure 19.** These devices use a **photovoltaic cell** that converts radiant energy from the Sun directly into electrical energy. Photovoltaic cells also are called solar cells.

Solar cell

Figure 19 This calculator uses a solar cell to produce the electricity it needs to operate.

Figure 20 Solar cells convert radiant energy from the Sun to electricity.

Identify *two devices that use solar cells for power.*

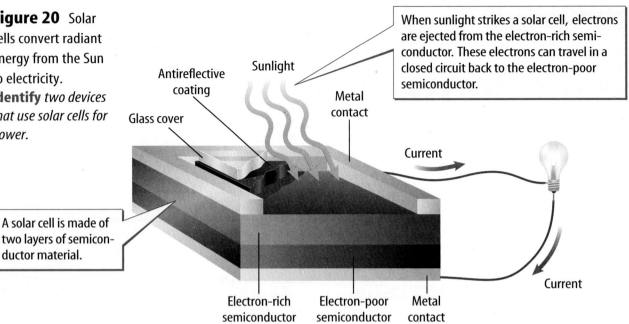

When sunlight strikes a solar cell, electrons are ejected from the electron-rich semiconductor. These electrons can travel in a closed circuit back to the electron-poor semiconductor.

Antireflective coating

Sunlight

Metal contact

Glass cover

Current

A solar cell is made of two layers of semiconductor material.

Current

Electron-rich semiconductor

Electron-poor semiconductor

Metal contact

How Solar Cells Work Solar cells are made of two layers of semiconductor materials sandwiched between two layers of conducting metal, as shown in **Figure 20.** One layer of semiconductor is rich in electrons, while the other layer is electron poor. When sunlight strikes the surface of the solar cell, electrons flow through an electrical circuit from the electron-rich semiconductor to the electron-poor material. This process of converting radiant energy from the Sun directly to electrical energy is only about 7 percent to 11 percent efficient.

Using Solar Energy Producing large amounts of electrical energy using solar cells is more expensive than producing electrical energy using fossil fuels. However, in remote areas where electric distribution lines are not available, the use of solar cells is a practical way of providing electrical power.

Currently, the most promising solar technologies are those that concentrate the solar power into a receiver. One such system is called the parabolic trough. The trough focuses the sunlight on a tube that contains a heat-absorbing fluid such as synthetic oil or liquid salt. The heated fluid is circulated through a boiler where it generates steam to turn a turbine, generating electricity.

The worlds' largest concentrating solar power plant is located in the Mojave Desert in California. This facility consists of nine units that generate over 350 megawatts of power. These nine units can generate enough electrical power to meet the needs of approximately 500,000 people. These units use natural gas as a backup power source for generating electricity at night and on cloudy days when solar energy is unavailable.

Energy from Water

Just as the expansion of steam can turn an electric generator, rapidly moving water can as well. The gravitational potential energy of the water can be increased if the water is retained by a high dam. This potential energy is released when the water flows through tunnels near the base of the dam. **Figure 21** shows how the rushing water spins a turbine, which rotates the shaft of an electric generator to produce electricity. Dams built for this purpose are called hydroelectric dams.

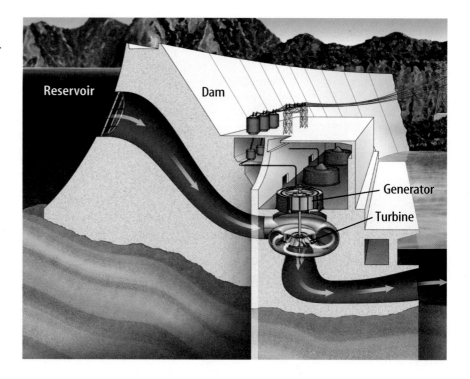

Using Hydroelectricity Electricity produced from the energy of moving water is called **hydroelectricity.** Currently about 8 percent of the electrical energy used in the United States is produced by hydroelectric power plants. Hydroelectric power plants are an efficient way to produce electricity with almost no pollution. Because no exchange of heat is involved in producing steam to spin a turbine, hydroelectric power plants are almost twice as efficient as fossil fuel or nuclear power plants.

Reading Check *Why are hydroelectric power plants more efficient than fossil fuel power plants?*

Another advantage is that the bodies of water held back by dams can form lakes that can provide water for drinking and crop irrigation. These lakes also can be used for boating and swimming. Also, after the initial cost of building a dam and a power plant, the electricity is relatively cheap.

However, artificial dams can disturb the balance of natural ecosystems. Some species of fish that live in the ocean migrate back to the rivers in which they were hatched to breed. This migration can be blocked by dams, which causes a decline in the fish population. Fish ladders, such as those shown in **Figure 22,** have been designed to enable fish to migrate upstream past some dams. Also, some water sources suitable for a hydroelectric power plant are located far from the regions needing power.

Figure 21 The potential energy in water stored behind the dam is converted to electrical energy in a hydroelectric power plant.
Diagram *the energy conversions that occur as a hydroelectric dam produces electrical energy.*

Figure 22 Fish ladders enable fish to migrate upstream past dams.

Energy from the Tides

The gravity of the Moon and Sun causes bulges in Earth's oceans. As Earth rotates, the two bulges of ocean water move westward. Each day, the level of the ocean on a coast rises and falls continually. Hydroelectric power can be generated by these ocean tides. As the tide comes in, the moving water spins a turbine that generates electricity. The water is then trapped behind a dam. At low tide the water behind the dam flows back out to the ocean, spinning the turbines and generating electric power.

Tidal energy is nearly pollution free. The efficiency of a tidal power plant is similar to that of a conventional hydroelectric power plant. However, only a few places on Earth have large enough differences between high and low tides for tidal energy to be a useful energy source. The only tidal power station in use in North America is at Annapolis Royal, Nova Scotia, shown in **Figure 23.** Tidal energy probably will be a limited source of energy in the future.

Figure 23 This tidal energy plant at Annapolis Royal, Nova Scotia, generates 20 megawatts of electric power.

Harnessing the Wind

You might have seen a windmill on a farm or pictures of windmills in a book. These windmills use the energy of the wind to pump water. Windmills also can use the energy of the wind to generate electricity. Wind spins a propeller that is connected to an electric generator. Windmill farms, like the one shown in **Figure 24,** may contain several hundred windmills.

However, only a few places on Earth consistently have enough wind to rely on wind power to meet energy needs. Also, windmills are only about 20 percent efficient on average. Research is underway to improve the design of wind generators and increase their efficiency. Other disadvantages of wind energy are that windmills can be noisy and change the appearance of a landscape. Also, they can disrupt the migration patterns of some birds. However, wind generators do not consume any nonrenewable natural resources, and they do not pollute the atmosphere or water.

Figure 24 Wind energy is converted to electricity as the spinning propeller turns a generator.

Energy from Inside Earth

Earth is not completely solid. Heat is generated within Earth by the decay of radioactive elements. This heat is called geothermal heat. Geothermal heat causes the rock beneath Earth's crust to soften and melt. This hot molten rock is called magma. The thermal energy that is contained in hot magma is called **geothermal energy.**

In some places, Earth's crust has cracks or thin spots that allow magma to rise near the surface. Active volcanoes, for example, permit hot gases and magma from deep within Earth to escape. Perhaps you have seen a geyser, like Old Faithful in Yellowstone National Park, shooting steam and hot water. The water that shoots from the geyser is heated by magma close to Earth's surface. In some areas, this hot water can be pumped into houses to provide heat.

 *What two natural phenomena are caused by geothermal heat?*

Geothermal Power Plants

Geothermal energy also can be used to generate electricity, as shown in **Figure 25.** Where magma is close to the surface, the surrounding rocks are also hot. A well is drilled and water is pumped into the ground, where it makes contact with the hot rock and changes into steam. The steam then returns to the surface, where it is used to rotate turbines that spin electric generators.

The efficiency of geothermal power plants is about 16 percent. Although geothermal power plants can release some gases containing sulfur compounds, pumping the water created by the condensed steam back into Earth can help reduce this pollution. However, the use of geothermal energy is limited to areas where magma is relatively close to the surface.

Figure 25 A geothermal power plant converts geothermal energy to electrical energy. Water is changed to steam by the hot rock. The steam is pumped to the surface where it turns a turbine attached to an electric generator.

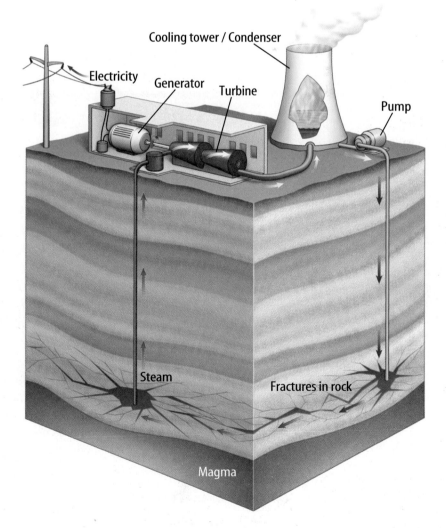

Cooling tower / Condenser

Electricity

Generator

Turbine

Pump

Steam

Fractures in rock

Magma

Alternative Fuels

The use of fossil fuels would be greatly reduced if cars could run on other fuels or sources of energy. For example, cars have been developed that use electrical energy supplied by batteries as a power source. Hybrid cars use both electric motors and gasoline engines. Hydrogen gas is another possible alternative fuel. It produces only water vapor when it burns and creates no pollution. **Figure 26** shows a car that is equipped to use hydrogen as fuel.

Biomass Fuels Could any other materials be used to heat water and produce electricity like fossil fuels and nuclear fission? Biomass can be burned in the presence of oxygen to convert the stored chemical energy to thermal energy. **Biomass** is renewable organic matter, such as wood, sugarcane fibers, rice hulls, and animal manure. Converting biomass is probably the oldest use of natural resources for meeting human energy needs.

Figure 26 Hydrogen may one day replace gasoline as a fuel for automobiles. Burning hydrogen produces water vapor, instead of carbon dioxide.

section 3 review

Summary

Energy Options

- The development of alternative energy sources can help reduce the use of fossil fuels.

Solar Energy

- Photovoltaic cells, or solar cells, convert radiant energy from the Sun into electrical energy.
- Producing large amounts of energy from solar cells is more expensive than using fossil fuels.

Other Renewable Energy Sources

- Hydroelectric power plants convert the potential energy in water to electrical energy.
- Tidal energy, wind energy, and geothermal energy can be converted into electrical energy, but are useable only in certain locations.
- Alternative fuels such as hydrogen could be used to power cars, and biomass can be burned to provide heat.

Self Check

1. **Explain** the need to develop and use alternative energy sources.
2. **Describe** three ways that solar energy can be used.
3. **Explain** how the generation of electricity by hydroelectric, tidal, and wind sources are similar to each other.
4. **Explain** why geothermal energy is unlikely to become a major energy source.
5. **Think Critically** What single energy source do most energy alternatives depend on, either directly or indirectly?

Applying Math

6. **Use Percentages** A house uses solar cells that generate 6.0 kW of electrical power to supply some of its energy needs. If the solar panels supply the house with 40 percent of the power it needs, how much power does the house use?

Science Online gpscience.com/self_check_quiz

S☀lar Heating

Energy from the Sun is absorbed by Earth and makes its temperature warmer. In a similar way, solar energy also is absorbed by solar collectors to heat water and buildings.

Temperature Due to Different Colors					
Color	2 min	4 min	6 min	8 min	10 min
Black					
White	Do not write in this book.				
Other color					

▶ Real-World Question

Does the rate at which an object absorbs solar energy depend on the color of the object?

Goals
■ **Demonstrate** solar heating.
■ **Compare** the effectiveness of heating items of different colors.
■ **Graph** your results.

Materials
small cardboard boxes
black, white, and colored paper
tape or glue
thermometer
watch with a second hand

▶ Procedure

1. Cover at least three small boxes with colored paper. The colors should include black and white as well as at least one other color.

2. Copy the data table into your Science Journal. Replace *Other color* with whatever color you are using.

3. Place the three objects on a windowsill or other sunny spot and note the starting time.

4. **Measure and record** the temperature inside each box at 2-min intervals for at least 10 min.

▶ Conclude and Apply

1. **Graph** your data using a line graph.

2. **Describe** the shapes of the lines on your graph. What color heated up the fastest? Which heated up the slowest?

3. **Explain** why the colored boxes heated at different rates.

4. **Infer** Suppose you wanted to heat a tub of water using solar energy. Based on the results of this activity, what color would you want the tub to be? Explain.

5. **Explain** why you might want to wear a white or light-colored shirt on a hot, sunny, summer day.

𝒞ommunicating Your Data

Compare your results with those of other students in your class. Discuss any differences found in your graphs, particularly if different colors were used by different groups.

Use the Internet

How much does energy really co$t?

Goals

■ **Identify** three energy sources that people use.
■ **Determine** the cost of the energy produced by each source.
■ **Describe** the environmental impact of each source.

Data Source

Visit **gpscience.com/ internet_lab** for more information about energy sources and for data collected by other students.

▶ Real-World Question

You know that it costs money to produce energy. Using energy also can have an impact on the environment. For example, coal costs less than some other fuels. However, combustion is a chemical reaction that can produce pollutants, and burning coal produces more pollution than burning other fossil fuels, such as natural gas. Even energy sources, such as hydroelectric power, that don't produce pollution can have an impact on the environment. What are some of the environmental impacts of the evergy sources used in the United States? How can these environmental impacts be compared to the cost of the energy produced?

▶ Make a Plan

1. **Research** the various sources of energy used in different areas of the United States and choose three energy sources to investigate.
2. **Research** the cost of the consumer of 1 kWh of electrical energy generated by energy sources you have choosen.
3. **Determine** the effects each of the three energy sources has on the environment.
4. **Use** your data to create a table showing the energy sources, and the energy cost and environmental impact of each energy source.
5. **Decide** how you will evaluate the environmental impact of each of your energy sources.
6. **Write** a summary describing which of your three energy sources is the most cost-effective for producing energy. Consider the cost of the energy and your evaluation of the environmental impact in making your decision. Use information from your research to support your conclusions.

Follow Your Plan

1. Make sure your teacher approves your plan before you start.

2. **Record** your data in your Science Journal.

Analyze Your Data

1. Of the energy sources you investigated, which is the most expensive to use? The least expensive?

2. Which energy source do you think has the most impact on the environment? The least impact?

Energy Sources		
Energy Source	Cost per kWh	Environmental Impacts
Energy source 1		
Energy source 2	Do not write in this book.	
Energy source 3		

Conclude and Apply

1. **Explain** Of the energy sources you investigated, which is the least expensive energy source? Which is the best choice to use? Why or why not?

2. **Explain** Of the energy sources you investigated, how did the environmental impact of using that energy source influence your choice of the best energy solution?

3. **Evaluate** Which data support your decision?

Communicating
Your Data

Find this lab using the link below. Post your data in the table provided. **Compare** your data to those of other students.

gpscience.com/internet_lab

Reacting to Nuclear Energy

Most people agree that thanks to energy sources, we have many things that make our quality of life better. Energy runs our cars, lights our homes, and powers our appliances. What many people don't agree on is where that energy should come from.

Almost all of the world's electric energy is produced by thermal power plants. Most of these plants burn fossil fuels—such as coal, oil, and natural gas—to produce energy. Nuclear energy is produced by fission, which is the splitting of an atom's nucleus. People in favor of nuclear energy argue that, unlike fossil fuels, nuclear energy is nonpolluting.

Opponents counter, though, that the poisonous radioactive waste created in nuclear reactors qualifies as pollution—and will be lingering in the ground and water for hundreds of thousands of years.

Supporters of nuclear energy also cite the spectacular efficiency of nuclear energy—one metric ton of nuclear fuel produces the same amount of energy as up to 3 million tons of coal. Opponents point out that uranium is in very short supply and, like fossil fuels, is likely to run out in the next 100 years.

Opponents worry that as utilities come under less government regulation, safety standards will be ignored in the interest of profit.

This could result in more accidents like the one that occurred at Chernobyl in the Ukraine. There, an explosion in the reactor core released radiation over a wide area.

Supporters counter that it will never be in the best interests of those running nuclear plants to relax safety standards since those safety standards are the best safeguard of workers' health. They cite the overall good safety record of nuclear power plants.

This site at Yucca Mountain, Nevada, is the location of a proposed high-level nuclear waste storage facility. Here radioactive materials would be buried for tens of thousands of years.

Debate Form three teams and have each team defend one of the views presented here. If you need more information, go to the Glencoe Science Web site. "Debrief" after the debate. Did the arguments change your understanding of the issues?

Science online

For more information, visit gpscience.com/time

Reviewing Main Ideas

Section 1 Fossil Fuels

1. Fossil fuels include oil, natural gas, and coal. They formed from the buried remains of plants and animals.

2. Fossil fuels can be burned to supply energy for generating electricity. Petroleum also is used to make plastics and synthetic fabrics.

3. Fossil fuels are nonrenewable energy resources. They can be replaced, but it takes millions of years.

Section 2 Nuclear Energy

1. A nuclear reactor transforms the energy from a controlled nuclear chain reaction to electrical energy.

2. Nuclear wastes must be contained and disposed of carefully so radiation from nuclear decay will not leak into the environment. These low-level nuclear wastes are buried to protect living organisms.

3. Nuclear fusion releases energy when two nuclei combine. Fusion only occurs at high temperatures that are difficult to produce in a laboratory.

Section 3 Renewable Energy Sources

1. Alternative energy resources can be used to supplement or replace nonrenewable energy resources.

2. Other sources of energy for generating electricity include hydroelectricity and solar, wind, tidal, and geothermal energy. Each source has its advantages and disadvantages. Also, some of these sources can damage the environment.

3. Although some alternative energy sources produce less pollution than fossil fuels do and are renewable, their use often is limited to the regions where the energy source is available. For example, tides can be used to generate electricity in coastal regions only.

4. It may be possible to use hydrogen as a fuel for automobiles and other vehicles. Biomass, such as wood and other renewable organic matter, has been used as fuel for thousands of years.

FOLDABLES Use the Foldable that you made at the beginning of the chapter to help you review energy sources.

Using Vocabulary

biomass p. 276
fossil fuel p. 257
geothermal energy p. 275
hydroelectricity p. 273
nonrenewable resource p. 263

nuclear reactor p. 264
nuclear waste p. 268
petroleum p. 259
photovoltaic cell p. 271
renewable resource p. 271

Complete each statement using a term from the vocabulary list above.

1. A(n) _____ uses the Sun to generate electricity.

2. _____ makes use of thermal energy inside the Earth.

3. Energy produced by the rise and fall of ocean levels is a(n) _____.

4. _____ includes the following: oil, natural gas, and coal.

5. Fossil fuels are a(n) _____ because they are being used up faster than they are being made.

6. A special caution should be taken in disposing of _____.

Checking Concepts

Choose the word or phrase that best answers the question.

7. Why are fossil fuels considered to be nonrenewable resources?
   A) They are no longer being produced.
   B) They are in short supply.
   C) They are not being produced as fast as they're being used.
   D) They contain hydrocarbons.

8. To generate electricity, nuclear power plants produce which of the following?
   A) steam   **C)** plutonium
   B) carbon dioxide   **D)** water

9. What is a major disadvantage of using nuclear fusion reactors?
   A) use of hydrogen as fuel
   B) less radioactivity produced
   C) extremely high temperatures required
   D) use of only small nuclei

10. How are spent nuclear fuel rods usually disposed of?
    A) burying them in a community landfill
    B) storing them in a deep pool of water
    C) burying them at the reactor site
    D) releasing them into the air

11. How much energy in the United States comes from burning petroleum, natural gas, and coal?
    A) 85%   **C)** 65%
    B) 35%   **D)** 25%

12. Solar cells would be more practical to use if they were which of the following?
    A) pollution free   **C)** less expensive
    B) nonrenewable   **D)** larger

13. Which energy source uses water that is heated naturally by Earth's internal heat?
    A) hydroelectricity   **C)** tidal energy
    B) nuclear fission   **D)** geothermal energy

14. What do hydrocarbons react with when fossil fuels are burned?
    A) carbon dioxide   **C)** oxygen
    B) carbon monoxide **D)** water

15. Which of the following is NOT a source of nuclear waste?
    A) products of fission reactors
    B) materials with short half-lives
    C) some medical and industrial products
    D) products of coal-burning power plants

16. Which of the following is the source of almost all of Earth's energy resources?
    A) plants   **C)** magma
    B) the Sun   **D)** fossil fuels

Science Online gpscience.com/vocabulary_puzzlemaker

Interpreting Graphics

17. Copy and complete the table below describing possible effects of changes in the normal operation of a nuclear reactor.

Reactor Problems	
Cause	**Effect**
The cooling water is released hot.	Do not write in this book.
The control rods are removed.	
	The reactor core overheats and meltdown occurs.

18. Copy and complete this concept map.

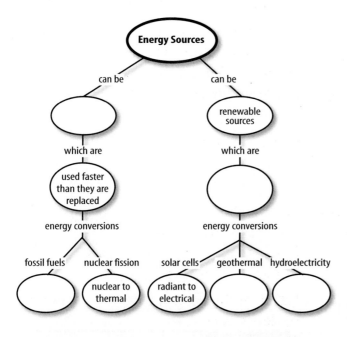

Thinking Critically

19. **Infer** why alternative energy resources aren't more widely used.

20. **Infer** whether fossil fuels should be conserved if renewable energy sources are being developed.

21. **Infer** Suppose new reserves of fossil fuels were found and a way to burn these fuels was developed that did not release pollutants and carbon dioxide into the atmosphere. Should fossil fuels still be conserved? Explain

22. **Explain** why coal is considered a nonrenewable energy source, but biomass, such as wood, is considered a renewable energy source.

23. **Make a table** listing two advantages and two disadvantages for each of the following energy sources: fossil fuels, hydroelectricity, wind turbines, nuclear fission, solar cells, and geothermal energy.

Applying Math

24. **Convert Units** Crude oil is sold on the world market in units called barrels. A barrel of crude oil contains 42 gallons. If 1 gallon is 3.8 liters, how many liters are there in a barrel of crude oil?

Use the table below to answer question 25.

High-Production Coal Mines	
Coal Mine	**Metric tons/year**
North Antelope Rochelle	6.78×10^7
Black Thunder	6.13×10^7

25. **Use Percentages** Nine of the top coal producing mines are located in Wyoming. Production information on two of the mines is in the table above. A total of about 1.02×10^9 metric tons is produced per year in the United States. What percentage do these two coal mines contribute to the total yearly coal production in the U.S.?

Part 1 Multiple Choice

Record your answers on the answer sheet provided by your teacher or on a sheet of paper.

Use the graph below to answer questions 1 and 2.

Sources of Electricity

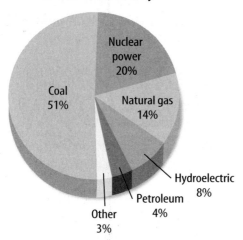

Nuclear power 20%

Coal 51%

Natural gas 14%

Hydroelectric 8%

Petroleum 4%

Other 3%

1. The graph above shows the percentage of electricity generated in the United States that comes from various energy sources. According to the graph, about what percentage comes from fossil fuels?
   A. 51%   **C.** 65%
   B. 55%   **D.** 69%

2. The graph shows that approximately what percentage of electricity comes from nonrenewable energy sources?
   A. 97%   **C.** 69%
   B. 89%   **D.** 55%

3. Which of the following is a typical efficiency for a solar cell?
   A. 10%   **C.** 75%
   B. 50%   **D.** 95%

4. Which of the following best describes wind mills used for the production of electricity?
   A. They are quiet.
   B. They can be used anywhere.
   C. They are 90 percent efficient.
   D. They are nonpolluting.

5. Which of the following forms only from ancient plant material, not from ancient animal remains?
   A. coal   **C.** natural gas
   B. crude oil   **D.** petroleum

Use the table below to answer questions 6 and 7.

Efficiency of Fossil Fuel Conversion	
Process	**Efficiency (%)**
Chemical to thermal energy	60
Conversion of water to steam	90
Steam spins turbine	75
Turbine spins electric generator	95
Transmission through power lines	90

6. The table above shows the efficiency of different steps in the conversion of fossil fuels to electricity at a power plant. According to the table, what is the efficiency for converting chemical energy in the fossil fuels to heat, and then converting water to steam?
   A. 30%   **C.** 75%
   B. 54%   **D.** 90%

7. What is the overall efficiency shown in the table for converting chemical energy in fossil fuels to electricity?
   A. 35%   **C.** 90%
   B. 82%   **D.** 95%

Test-Taking Tip

Determine the Information Needed Concentrate on what the question is asking about a table, instead of all the information in the table.

Question 7 Read the question carefully to determine which rows in the table contain the information needed to answer the question.

Part 2 | Short Response/Grid In

Record your answers on the answer sheet provided by your teacher or on a sheet of paper.

8. Explain why hydroelectric power plants are almost twice as efficient as fossil fuel or nuclear power plants.

9. About 90 percent of the coal that is used in the United States is used for what purpose?

10. What is the most inefficient stage in the production of electrical energy at a fossil-fuel burning power plant?

11. Describe the typical disposal method for low-level nuclear wastes.

Use the illustration below to answer questions 12 and 13.

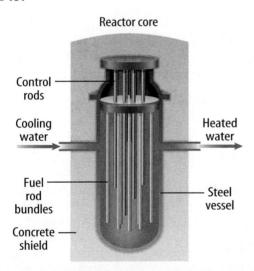

Reactor core

Control rods

Cooling water

Heated water

Fuel rod bundles

Steel vessel

Concrete shield

12. The core of a nuclear reactor might contain hundreds of fuel rods. Describe the composition of a fuel rod.

13. Describe the purpose of the control rods and explain how their placement in the reactor affects the nuclear chain reaction.

14. Fusion is the most concentrated energy source known. Why, then, is it not used at nuclear plants to make electricity?

Part 3 | Open Ended

Record your answers on a sheet of paper.

Use the photograph below to answer questions 15 and 16.

15. The photograph above shows a nuclear power plant that generates electricity using the energy released in nuclear fission of uranium-235. Draw a sketch showing this fission process. Describe your sketch and explain how the process results in a chain reaction.

16. Explain how a nuclear reactor at a nuclear power plant produces electricity. What is the purpose of the large tower shown in the photograph?

17. Explain how the steam that is used to run turbines is produced at a geothermal power plant.

18. Describe two advantages and three disadvantages of using solar energy to generate electricity.

19. Explain why biomass is considered a renewable energy source.

20. Describe the processes that form oil, natural gas, and coal.

21. What is the difference between low-level and high-level nuclear waste? Describe an example of each type.

How Are Glassblowing & X Rays Connected?

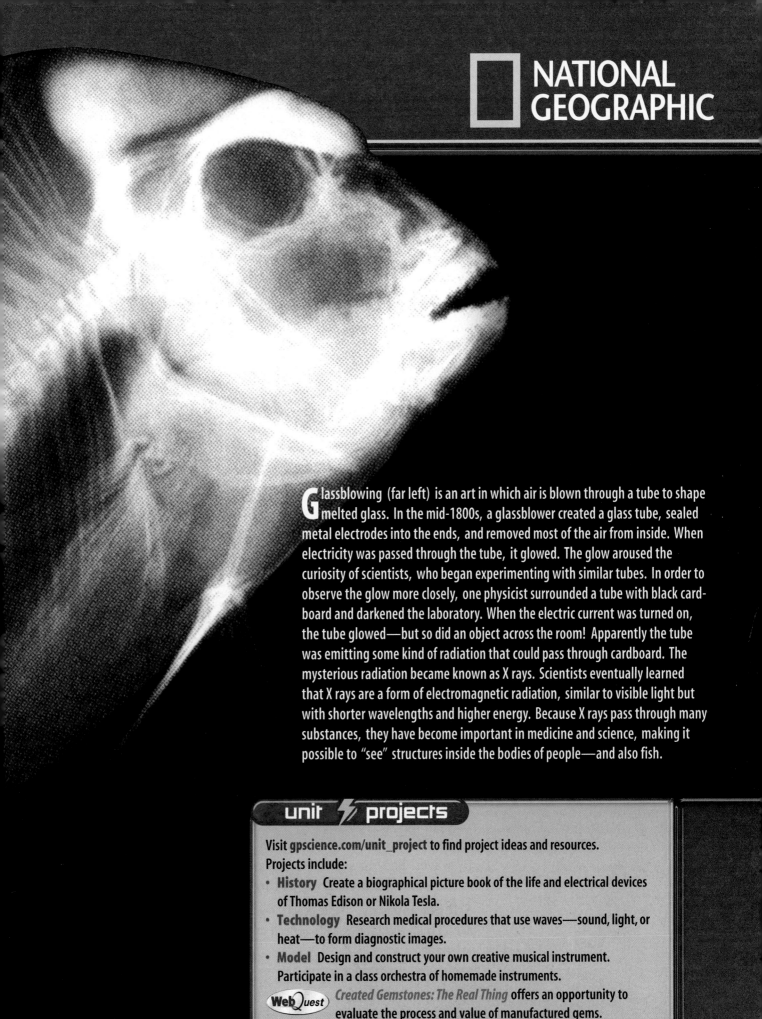

Glassblowing (far left) is an art in which air is blown through a tube to shape melted glass. In the mid-1800s, a glassblower created a glass tube, sealed metal electrodes into the ends, and removed most of the air from inside. When electricity was passed through the tube, it glowed. The glow aroused the curiosity of scientists, who began experimenting with similar tubes. In order to observe the glow more closely, one physicist surrounded a tube with black cardboard and darkened the laboratory. When the electric current was turned on, the tube glowed—but so did an object across the room! Apparently the tube was emitting some kind of radiation that could pass through cardboard. The mysterious radiation became known as X rays. Scientists eventually learned that X rays are a form of electromagnetic radiation, similar to visible light but with shorter wavelengths and higher energy. Because X rays pass through many substances, they have become important in medicine and science, making it possible to "see" structures inside the bodies of people—and also fish.

unit ⚡ projects

Visit **gpscience.com/unit_project** to find project ideas and resources.
Projects include:

- **History** Create a biographical picture book of the life and electrical devices of Thomas Edison or Nikola Tesla.
- **Technology** Research medical procedures that use waves—sound, light, or heat—to form diagnostic images.
- **Model** Design and construct your own creative musical instrument. Participate in a class orchestra of homemade instruments.

WebQuest *Created Gemstones: The Real Thing* offers an opportunity to evaluate the process and value of manufactured gems.

Waves

Waves anyone?

This surfer in Hawaii is surrounded by an ocean wave that forms a huge wall of water. But you are surrounded by waves, too. Everything you see or hear is brought to you by waves. Easy to see—like this ocean wave—or invisible, all waves carry energy.

Science Journal Write down three things you already know about waves, and one thing you would like to learn about waves.

Start-Up Activities

How do waves transfer energy?

Light enters your eyes and sound strikes your ears, enabling you to sense the world around you. Light and sound are waves that carry energy from one place to another. Do waves carry anything else along with their energy? Does a wave transfer matter too? In this activity you'll observe one way that waves can transfer energy. 👓

1. Place your textbook flat on your desk. Line up four marbles on the groove at the edge of the textbook so that the marbles are touching each other.
2. Hold the first three marbles in place using three fingers of one hand.
3. Use your other hand to tap the first marble with a pen or pencil.
4. Observe the behavior of the fourth marble.
5. **Think Critically** Write a paragraph explaining how the fourth marble reacted to the pen tap. Draw a diagram showing how energy was transferred through the marbles.

 **Preview this chapter's content and activities at** gpscience.com

 **Types of Waves** Make the following Foldable to compare and contrast two types of waves.

STEP 1 **Fold** one sheet of paper lengthwise.

STEP 2 **Fold** into thirds.

STEP 3 **Unfold and draw** overlapping ovals. Cut the top sheet along the folds.

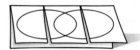

STEP 4 **Label** the ovals as shown.

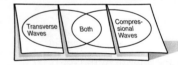

Construct a Venn Diagram As you read this chapter, list properties and characteristics unique to transverse waves under the left tab, those unique to compressional waves under the right tab, and those common to both under the middle tab.

The Nature of Waves

Reading Guide

What You'll Learn
- **Recognize** that waves carry energy but not matter.
- **Define** mechanical waves.
- **Compare and contrast** transverse waves and compressional waves.

Why It's Important
You can see and hear the world around you because of the energy carried by waves.

⦾ Review Vocabulary
energy: the ability to cause change

New Vocabulary
- ● wave
- ● medium
- ● transverse wave
- ● compressional wave

What's in a wave?

A surfer bobs in the ocean waiting for the perfect wave, microwaves warm up your leftover pizza, and sound waves from your CD player bring music to your ears. Do these and other types of waves have anything in common with one another?

A **wave** is a repeating disturbance or movement that transfers energy through matter or space. For example, ocean waves disturb the water and transfer energy through it. During earthquakes, energy is transferred in powerful waves that travel through Earth. Light is a type of wave that can travel through empty space to transfer energy from one place to another, such as from the Sun to Earth.

Figure 1 Falling pebbles transfer their kinetic energy to the particles of water in a pond, forming waves.

Waves and Energy

Kerplop! A pebble falls into a pool of water and ripples form. As **Figure 1** shows, the pebble causes a disturbance that moves outward in the form of a wave. Because it is moving, the falling pebble has energy. As it splashes into the pool, the pebble transfers some of its energy to nearby water molecules, causing them to move. Those molecules then pass the energy along to neighboring water molecules, which, in turn, transfer it to their neighbors. The energy moves farther and farther from the source of the disturbance. What you see is energy traveling in the form of a wave on the surface of the water.

Waves and Matter Imagine you're in a boat on a lake. Approaching waves bump against your boat, but they don't carry it along with them as they pass. The boat does move up and down and maybe even a short distance back and forth because the waves transfer some of their energy to it. But after the waves have moved on, the boat is still in nearly the same place. The waves don't even carry the water along with them. Only the energy carried by the waves moves forward. All waves have this property—they carry energy without transporting matter from place to place.

> **Reading Check** *What do waves carry?*

Making Waves A wave will travel only as long as it has energy to carry. For example, when you drop a pebble into a puddle, the ripples soon die out and the surface of the water becomes still again.

Suppose you are holding a rope at one end, and you give it a shake. You would create a pulse that would travel along the rope to the other end, and then the rope would be still again, as **Figure 2** shows. Now suppose you shake your end of the rope up and down for a while. You would make a wave that would travel along the rope. When you stop shaking your hand up and down, the rope will be still again. It is the up-and-down motion of your hand that creates the wave.

Anything that moves up and down or back and forth in a rhythmic way is vibrating. The vibrating movement of your hand at the end of the rope created the wave. In fact, all waves are produced by something that vibrates.

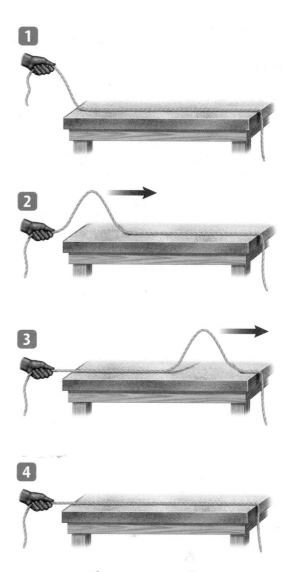

Figure 2 A wave will exist only as long as it has energy to carry. **Explain** *what happened to the energy that was carried by the wave in this rope.*

Mechanical Waves

Sound waves travel through the air to reach your ears. Ocean waves move through water to reach the shore. In both cases, the matter the waves travel through is called a **medium.** The medium can be a solid, a liquid, a gas, or a combination of these. For sound waves the medium is air, and for ocean waves the medium is water. Not all waves need a medium. Some waves, such as light and radio waves, can travel through space. Waves that can travel only through matter are called mechanical waves. The two types of mechanical waves are transverse waves and compressional waves.

Transverse Waves In a **transverse wave,** matter in the medium moves back and forth at right angles to the direction that the wave travels. For example, **Figure 3** shows how a wave in the ocean moves horizontally, but the water that the wave passes through moves up and down. When you shake one end of a rope while your friend holds the other end, you are making transverse waves. The wave and its energy travel from you to your friend as the rope moves up and down.

Figure 3 A water wave travels horizontally as the water moves vertically up and down.

Compressional Waves In a **compressional wave,** matter in the medium moves back and forth along the same direction that the wave travels. You can model compressional waves with a coiled spring toy, as shown in **Figure 4.** Squeeze several coils together at one end of the spring. Then let go of the coils, still holding onto coils at both ends of the spring. A wave will travel along the spring. As the wave moves, it looks as if the whole spring is moving toward one end. Suppose you watched the coil with yarn tied to it as in **Figure 4.** You would see that the yarn moves back and forth as the wave passes, and then stops moving after the wave has passed. The wave carries energy, but not matter, forward along the spring. Compressional waves also are called longitudinal waves.

Sound Waves Sound waves are compressional waves. When a noise is made, such as when a locker door slams shut and vibrates, nearby air molecules are pushed together by the vibrations. The air molecules are squeezed together like the coils in a coiled spring toy are when you make a compressional wave with it. The compressions travel through the air to make a wave.

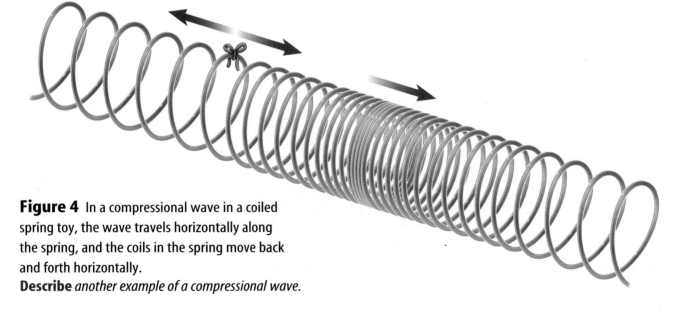

Figure 4 In a compressional wave in a coiled spring toy, the wave travels horizontally along the spring, and the coils in the spring move back and forth horizontally.
Describe *another example of a compressional wave.*

Sound in Other Materials Sound waves also can travel through other mediums, such as water and wood. Particles in these mediums also are pushed together and move apart as the sound waves travel through them. When a sound wave reaches your ear, it causes your eardrum to vibrate. Your inner ear then sends signals to your brain, and your brain interprets the signals as sound.

Reading Check *How do sound waves travel in solids?*

Water Waves Water waves are not purely transverse waves. The water moves up and down as the waves go by. But the water also moves a short distance back and forth along the direction the wave is moving. This movement happens because the low part of the wave can be formed only by pushing water forward or backward toward the high part of the wave, as in **Figure 5A.** Then as the wave passes, the water that was pushed aside moves back to its initial position, as in **Figure 5B.** In fact, if you looked closely, you would see that the combination of this up-and-down and back-and-forth motion causes water to move in circles. Anything floating on the surface of the water absorbs some of the waves' energy and bobs in a circular motion.

Ocean waves are formed most often by wind blowing across the ocean surface. As the wind blows faster and slower, the changing wind speed is like a vibration. The size of the waves that are formed depends on the wind speed, the distance over which the wind blows, and how long the wind blows. **Figure 6** on the next page shows how ocean waves are formed.

Figure 5 A water wave causes water to move back and forth, as well as up and down. Water is pushed back and forth to form the crests and troughs.

A The low point of a water wave is formed when water is pushed aside and up to the high point of the wave.

B The water that is pushed aside returns to its initial position.

Figure 6

When wind blows across an ocean, friction between the moving air and the water causes the water to move. As a result, energy is transferred from the wind to the surface of the water. The waves that are produced depend on the length of time and the distance over which the wind blows, as well as the wind speed.

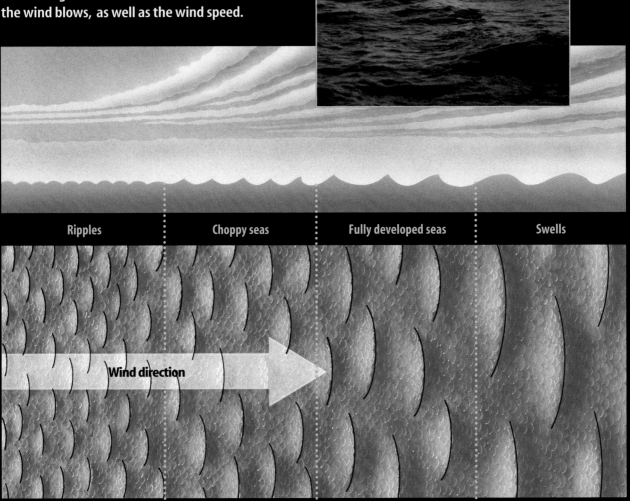

Ripples | Choppy seas | Fully developed seas | Swells

Wind direction

▲ Wind causes ripples to form on the surface of the water. As ripples form, they provide an even larger surface area for the wind to strike, and the ripples increase in size.

▲ Waves that are higher and have longer wavelengths grow faster as the wind continues to blow, but the steepest waves break up, forming whitecaps. The surface is said to be choppy.

▲ The shortest-wavelength waves break up, while the longest-wavelength waves continue to grow. When these waves have reached their maximum height, they form fully developed seas.

▲ After the wind dies down, the waves lose energy and become lower and smoother. These smooth, long-wavelength ocean waves are called swells.

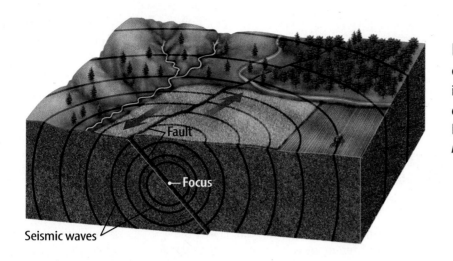

Figure 7 When Earth's crust shifts or breaks, the energy that is released is transmitted outward, causing an earthquake.
Explain *why earthquakes are mechanical waves.*

Seismic Waves A guitar string makes a sound when it breaks. The string vibrates for a short time after it breaks and produces sound waves. In a similar way, forces in Earth's crust can cause regions of the crust to shift, bend, or even break. The breaking crust vibrates, creating seismic (SIZE mihk) waves that carry energy outward, as shown in **Figure 7.** Seismic waves are a combination of compressional and transverse waves. They can travel through Earth and along Earth's surface. When objects on Earth's surface absorb some of the energy carried by seismic waves, they move and shake. The more the crust moves during an earthquake, the more energy is released.

Science Online

Topic: Seismic Waves
Visit gpscience.com for Web links to information about seismic waves.

Activity Write a summary of how seismic waves are used to map Earth's interior.

section 1 review

Summary

Waves and Energy

- A wave is a repeating disturbance or movement that transfers energy through matter or space.
- Waves carry energy without transporting matter.
- Waves are produced by something that is vibrating.

Mechanical Waves

- Mechanical waves must travel in matter.
- Mechanical waves can be transverse waves or compressional waves.
- In a transverse wave, matter in the medium moves at right angles to the wave motion.
- In a compressional wave, matter in the medium moves back and forth along the direction of the wave motion.

Self Check

1. **Compare and contrast** a transverse wave and a compressional wave. Give an example of each type.
2. **Describe** the motion of a buoy when a water wave passes. Does it move the buoy forward?
3. **Explain** how you could model a compressional wave using a coiled spring toy.
4. **List** the characteristics of a mechanical wave.
5. **Think Critically** Why do boats need anchors if ocean waves do not carry matter forward?

Applying Math

6. **Calculate Time** The average speed of sound in water is 1,500 m/s. How long would it take a sound wave to travel 9,000 m?

Wave Properties

What You'll Learn

- **Define** wavelength, frequency, period, and amplitude.
- **Describe** the relationship between frequency and wavelength.
- **Explain** how a wave's energy and amplitude are related.
- **Calculate** wave speed.

Why It's Important

Waves with different properties can be used in different ways.

Review Vocabulary

vibration: a back and forth movement

New Vocabulary

- crests
- troughs
- rarefaction
- wavelength
- frequency
- period
- amplitude

The Parts of a Wave

What makes sound waves, water waves, and seismic waves different from each other? Waves can differ in how much energy they carry and in how fast they travel. Waves also have other characteristics that make them different from each other.

Suppose you shake the end of a rope and make a transverse wave. The transverse wave in **Figure 8** has alternating high points, called **crests,** and low points, called **troughs.**

On the other hand, a compressional wave has no crests and troughs. When a compressional wave passes through a medium, it creates regions where the medium becomes crowded together and more dense, as in **Figure 8.** These regions are compressions. When you make compressional waves in a coiled spring, a compression is a region where the coils are close together. **Figure 8** also shows that the coils in the region next to a compression are spread apart, or less dense. This less-dense region of a compressional wave is called a **rarefaction.**

Figure 8 Transverse and compressional waves have different features that travel through a medium and form the wave.

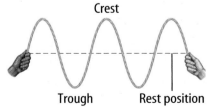

A transverse wave is made of crests and troughs that travel through the medium.

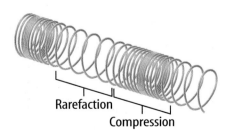

A compressional wave is made of compressions and rarefactions that travel through the medium.

Figure 9 One wavelength starts at any point on a wave and ends at the nearest point just like it.

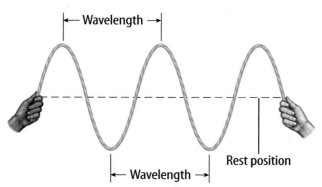

For transverse waves, a wavelength can be measured from crest to crest or trough to trough.

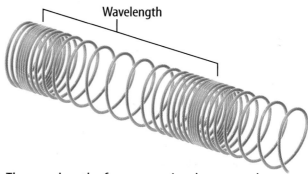

The wavelength of a compressional wave can be measured from compression to compression or from rarefaction to rarefaction.

Wavelength

Waves also have a property called wavelength. A **wavelength** is the distance between one point on a wave and the nearest point just like it. **Figure 9** shows that for transverse waves the wavelength is the distance from crest to crest or trough to trough.

A wavelength in a compressional wave is the distance between two neighboring compressions or two neighboring rarefactions, as shown in **Figure 9.** You can measure from the start of one compression to the start of the next compression or from the start of one rarefaction to the start of the next rarefaction. The wavelengths of sound waves that you can hear range from a few centimeters for the highest-pitched sounds to about 15 m for the deepest sounds.

 *How is wavelength measured in transverse and compressional waves?*

Frequency and Period

When you tune your radio to a station, you are choosing radio waves of a certain frequency. The **frequency** of a wave is the number of wavelengths that pass a fixed point each second. You can find the frequency of a transverse wave by counting the number of crests or troughs that pass by a point each second. The frequency of a compressional wave is the number of compressions or rarefactions that pass a point every second. Frequency is expressed in hertz (Hz). A frequency of 1 Hz means that one wavelength passes by in 1 s. In SI units, 1 Hz is the same as 1/s. The **period** of a wave is the amount of time it takes one wavelength to pass a point. As the frequency of a wave increases, the period decreases. Period has units of seconds.

Observing Wavelength

Procedure
1. Fill a **pie plate or other wide pan** with **water** about 2 cm deep.
2. Lightly tap your finger once per second on the surface of the water and observe the spacing of the water waves.
3. Increase the rate of your tapping, and observe the spacing of the water waves.

Analysis
1. How is the spacing of the water waves related to their wavelength?
2. How does the spacing of the water waves change when the rate of tapping increases?

Wavelength Is Related to Frequency If you make transverse waves with a rope, you increase the frequency by moving the rope up and down faster. Moving the rope faster also makes the wavelength shorter. This relationship is always true—as frequency increases, wavelength decreases. **Figure 10** compares the wavelengths and frequencies of two different waves.

The frequency of a wave is always equal to the rate of vibration of the source that creates it. If you move the rope up, down, and back up in 1 s, the frequency of the wave you generate is 1 Hz. If you move the rope up, down, and back up five times in 1 s, the resulting wave has a frequency of 5 Hz.

Reading Check *How are the wavelength and frequency of a wave related?*

Wave Speed

You're at a large stadium watching a baseball game, but you're high up in the bleachers, far from the action. The batter swings and you see the ball rise. An instant later you hear the crack of the bat hitting the ball. You see the impact before you hear it because light waves travel much faster than sound waves do. Therefore, the light waves reflected from the flying ball reach your eyes before the sound waves created by the crack of the bat reach your ears.

The speed of a wave depends on the medium it is traveling through. Sound waves usually travel faster in liquids and solids than they do in gases. However, light waves travel more slowly in liquids and solids than they do in gases or in empty space. Also, sound waves usually travel faster in a material if the temperature of the material is increased. For example, sound waves travel faster in air at 20°C than in air at 0°C.

Figure 10 The wavelength of a wave decreases as the frequency increases.

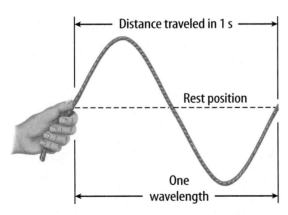

The rope is moved down, up, and down again one time in 1 s. One wavelength is created on the rope.

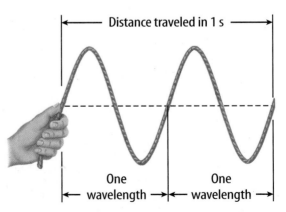

The rope is shaken down, up, and down again twice in 1 s. Two wavelengths are created on the rope.

Calculating Wave Speed You can calculate the speed of a wave represented by v by multiplying its frequency times its wavelength. Wavelength is represented by the Greek letter lambda (λ), and frequency is represented by f.

INTEGRATE
Social Studies

Deadly Ocean Waves
Tsunamis can cause serious damage when they hit land. These waves can measure up to 30 m tall and can travel faster than 700 km/h. Research to find which areas of the world are most vulnerable to tsunamis. Describe the effects of a tsunami that has occurred in these areas.

> **Wave Speed Equation**
>
> speed (in m/s) = frequency (in Hz) × wavelength (in m)
> $$v = f\lambda$$

Why does multiplying the frequency unit Hz by the distance unit m give the unit for speed, m/s? Recall the SI unit Hz is the same as 1/s. So multiplying m × Hz equals m × 1/s, which equals m/s.

Applying Math Solve a Simple Equation

THE SPEED OF SOUND What is the speed of a sound wave that has a wavelength of 2.00 m and a frequency of 170.5 Hz?

IDENTIFY known values and the unknown value

Identify the known values:

has a wavelength of 2.0 m means⟩ $\lambda = 2.00$ m

frequency of 170.5 Hz means⟩ $f = 170.5$ Hz

Identify the unknown value:

What is the speed of a sound wave? means⟩ $v = ?$ m/s

SOLVE the problem

Substitute the known values $\lambda = 2.00$ m and $f = 170.5$ Hz into the wave speed equation:

$$v = f\lambda = (170.5 \text{ Hz}) (2.00 \text{ m}) = 341 \text{ m/s}$$

CHECK the answer

Does your answer seem reasonable? Check your answer by dividing the wave speed you calculated by the wavelength given in the problem. The result should be the frequency given in the problem.

Practice Problems

1. A wave traveling in water has a frequency of 500.0 Hz and a wavelength of 3.0 m. What is the speed of the wave?

2. The lowest-pitched sounds humans can hear have a frequency of 20.0 Hz. What is the wavelength of these sound waves if their wave speed is 340.0 m/s?

3. The highest-pitched sound humans can hear have a wavelength of 0.017 m in air. What is the frequency of these sound waves if their wave speed is 340.0 m/s?

For more practice problems, go to page 834, and visit gpscience.com/extra_problems.

Amplitude and Energy

Why do some earthquakes cause terrible damage, while others are hardly felt? This is because the amount of energy a wave carries can vary. **Amplitude** is related to the energy carried by a wave. The greater the wave's amplitude is, the more energy the wave carries. Amplitude is measured differently for compressional and transverse waves.

Amplitude of Compressional Waves The amplitude of a compressional wave is related to how tightly the medium is pushed together at the compressions. The denser the medium is at the compressions, the larger its amplitude is and the more energy the wave carries. For example, it takes more energy to push the coils in a coiled spring toy tightly together than to barely move them. The closer the coils are in a compression, the farther apart they are in a rarefaction. So the less dense the medium is at the rarefactions, the more energy the wave carries. **Figure 11** shows compressional waves with different amplitudes.

Figure 11 The amplitude of a compressional wave depends on the density of the medium in the compressions and rarefactions.

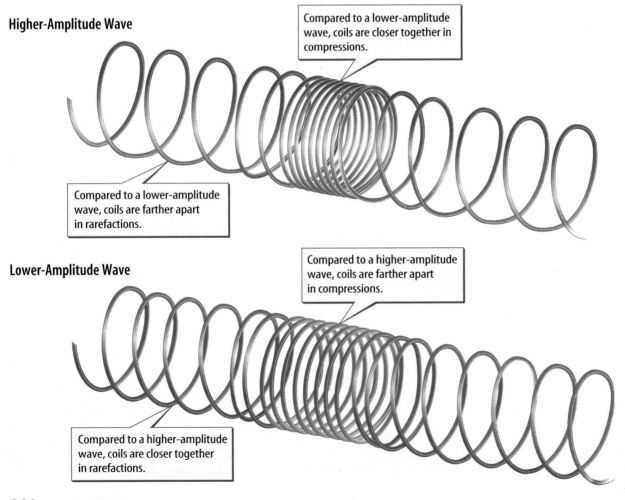

Higher-Amplitude Wave

Compared to a lower-amplitude wave, coils are closer together in compressions.

Compared to a lower-amplitude wave, coils are farther apart in rarefactions.

Lower-Amplitude Wave

Compared to a higher-amplitude wave, coils are farther apart in compressions.

Compared to a higher-amplitude wave, coils are closer together in rarefactions.

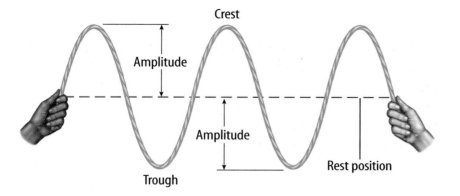

Figure 12 The amplitude of a transverse wave is the distance between a crest or a trough and the position of the medium at rest. **Describe** *how you could create waves with different amplitudes in a piece of rope.*

Amplitude of Transverse Waves If you've ever been knocked over by an ocean wave, you know that the higher the wave, the more energy it carries. Remember that the amplitude of a wave increases as the energy carried by the wave increases. So a tall ocean wave has a greater amplitude than a short ocean wave does. The amplitude of any transverse wave is the distance from the crest or trough of the wave to the rest position of the medium, as shown in **Figure 12.**

section 2 review

Summary

The Parts of a Wave

- Transverse waves have repeating high points called crests and low points called troughs.
- Compressional waves have repeating high-density regions called compressions, and low-density regions called rarefactions.

Wavelength, Frequency, and Period

- Wavelength is the distance between a point on a wave and the nearest point just like it.
- Wave frequency is the number of wavelengths passing a fixed point each second.
- Wave period is the amount of time it takes one wavelength to pass a fixed point.

Wave Speed and Amplitude

- The speed of a wave depends on the material it is traveling in, and the temperature.
- The speed of a wave is the product of its frequency and its wavelength:

$$v = f\lambda$$

- As the amplitude of a wave increases, the energy carried by the wave increases.

Self Check

1. **Describe** the difference between a compressional wave with a large amplitude and one with a small amplitude.

2. **Describe** how the wavelength of a wave changes if the wave slows down and its frequency doesn't change.

3. **Explain** how the frequency of a wave changes when the period of the wave increases.

4. **Form a hypothesis** to explain why a sound wave travels faster in a solid than in a gas.

5. **Think Critically** You make a transverse wave by shaking the end of a long rope up and down. Explain how you would shake the end of the rope to make the wavelength shorter. How would you shake the end of the rope to increase the energy carried by the wave?

Applying Math

6. **Calculate** the frequency of a water wave that has a wavelength of 0.5 m and a speed of 4.0 m/s.

7. **Calculate Wavelength** An FM radio station broadcasts radio waves with a frequency of 100,000,000 Hz. What is the wavelength of these radio waves if they travel at a speed of 300,000 km/s?

Waves in Different Mediums

Have you ever swum underwater? If so, even with your head underwater, you probably still heard some sounds. The noises probably sounded different underwater than they do in air. Waves can change properties when they travel from one medium into another.

● Real-World Question

How is the speed of a wave affected by the type of material it is traveling through?

Goals

- **Demonstrate** transverse and compressional waves.
- **Compare** the speed of waves traveling through different mediums.

Possible Materials

coiled spring toys (made out of both metal and plastic)
rope, both heavy and light
string
long rubber band, such as those used for exercising
strip of heavy cloth, such as a towel
strip of light cloth, such as nylon panty hose
stopwatch

Safety Precautions

● Procedure

1. Use pieces of each material that are about the same length. For each material, have a partner hold one end of the material still while you shake the material back and forth. Shake each material in the same way.

2. Have someone time how long a pulse takes to reach the opposite end of the material.

3. Tie two different types of rope together or tie a heavy piece of cloth to a lighter piece. Observe how the wave changes when it moves from one material to the other.

4. **Observe** compressional waves using coiled spring toys. You can connect two different types of coiled spring toys together to see how a compressional wave changes in different mediums.

● Conclude and Apply

1. **Describe** how the amplitude of the waves changed as they traveled from one material to a different material.

2. **Determine** if the waves travel at the same speed through the different mediums.

3. **Explain** how the waves changed when they moved from one material to another.

4. **Describe** how the waves created in this activity got their energy.

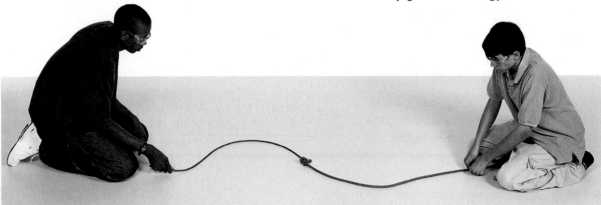

The Behavior of Waves

Reading Guide

What You'll Learn

- **State** the law of reflection.
- **Explain** why waves change direction when they travel from one material to another.
- **Compare and contrast** refraction and diffraction.
- **Describe** how waves interfere with each other.

Why It's Important

You can hear an echo, see shadows, and check your reflection in a mirror because of how waves behave.

🔍 Review Vocabulary

perpendicular: a line that forms a 90-degree angle with another line

New Vocabulary

- refraction
- diffraction
- interference
- standing wave
- resonance

Reflection

If you are one of the last people to leave your school building at the end of the day, you'll probably find the hallways quiet and empty. When you close your locker door, the sound echoes down the empty hall. Your footsteps also make a hollow sound. Thinking you're all alone, you may be startled by your own reflection in a classroom window. The echoes and your image looking back at you from the window are caused by wave reflection.

Reflection occurs when a wave strikes an object and bounces off of it. All types of waves—including sound, water, and light waves—can be reflected. How does the reflection of light allow the boy in **Figure 13** to see himself in the mirror? It happens in two steps. First, light strikes his face and bounces off. Then, the light reflected off his face strikes the mirror and is reflected into his eyes.

Echoes A similar thing happens to sound waves when your footsteps echo. Sound waves form when your foot hits the floor and the waves travel through the air to both your ears and other objects. Sometimes when the sound waves hit another object, they reflect off it and come back to you. Your ears hear the sound again, a few seconds after you first heard your footstep.

Bats and dolphins use echoes to learn about their surroundings. A dolphin makes a clicking sound and listens to the echoes. These echoes enable the dolphin to locate nearby objects.

Figure 13 The light that strikes the boy's face is reflected into the mirror. The light then reflects off the mirror into his eyes.
List *examples of waves that can be reflected.*

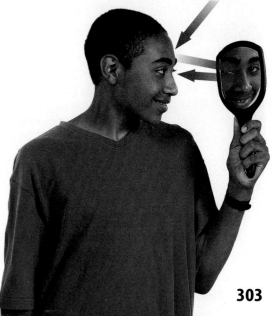

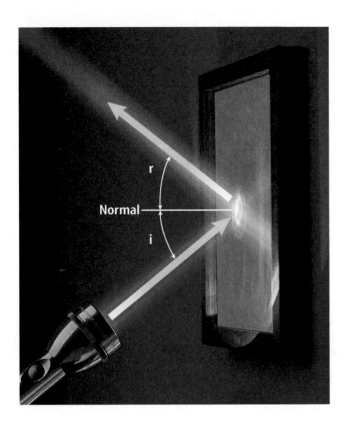

The Law of Reflection Look at the two light beams in **Figure 14.** The beam striking the mirror is called the incident beam. The beam that bounces off the mirror is called the reflected beam. The line drawn perpendicular to the surface of the mirror is called the normal. The angle formed by the incident beam and the normal is the angle of incidence, labeled i. The angle formed by the reflected beam and the normal is the angle of reflection, labeled r. According to the law of reflection, the angle of incidence is equal to the angle of reflection. All reflected waves obey this law. Objects that bounce from a surface sometimes behave like waves that are reflected from a surface. For example, suppose you throw a bounce pass while playing basketball. The angle between the ball's direction and the normal to the floor is the same before and after it bounces.

Figure 14 A flashlight beam is made of light waves. When any wave is reflected, the angle of incidence, i, equals the angle of reflection, r.

Refraction

Do you notice anything unusual in **Figure 15?** The pencil looks as if it is broken into two pieces. But if you pulled the pencil out of the water, you would see that it is unbroken. This illusion is caused by refraction. How does it work?

Remember that a wave's speed depends on the medium it is moving through. When a wave passes from one medium to another—such as when a light wave passes from air to water—it changes speed. If the wave is traveling at an angle when it passes from one medium to another, it changes direction, or bends, as it changes speed. **Refraction** is the bending of a wave caused by a change in its speed as it moves from one medium to another. The greater the change in speed is, the more the wave bends.

✔ **Reading Check** *When does refraction occur?*

Figure 16A on the next page shows what happens when a wave passes into a material in which it slows down. The wave is refracted (bent) toward the normal. **Figure 16B** shows what happens when a wave passes into a medium in which it speeds up. Then the wave is refracted away from the normal.

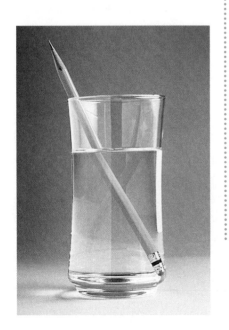

Figure 15 The pencil looks like it is broken at the surface of the water because of refraction.

Figure 16 Light waves travel slower in water than in air. This causes light waves to change direction when they move from water to air or air to water.
Predict *how the beam would bend if the speed were the same in both air and water.*

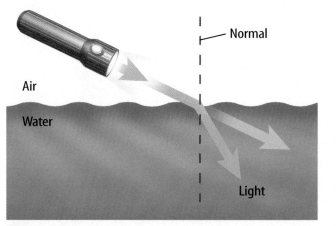

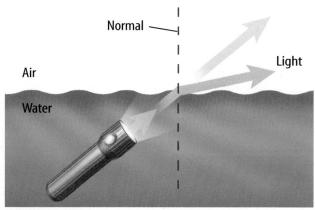

A When light waves travel from air to water, they slow down and bend toward the normal.

B When light waves travel from water to air, they speed up and bend away from the normal.

Refraction of Light in Water You may have noticed that objects that are underwater seem closer to the surface than they really are. **Figure 17** shows how refraction causes this illusion. In the figure, the light waves reflected from the swimmer's foot are refracted away from the normal and enter your eyes. However, your brain assumes that all light waves have traveled in a straight line. The light waves that enter your eyes seem to have come from a foot that was higher in the water. This is also why the pencil in **Figure 15** seems broken. The light waves coming from the part of the pencil that is underwater are refracted, but your brain interprets them as if they have traveled in a straight line. However, the light waves coming from the part of the pencil above the water are not refracted. So, the part of the pencil that is underwater looks as if it has shifted.

Figure 17 Light waves from the boy's foot bend away from the normal as they pass from water to air. This makes the foot look closer to the surface than it really is.
Infer *whether the boy's knee would seem closer to the surface than it is.*

Figure 18 Diffraction causes ocean waves to change direction as they pass a group of islands.

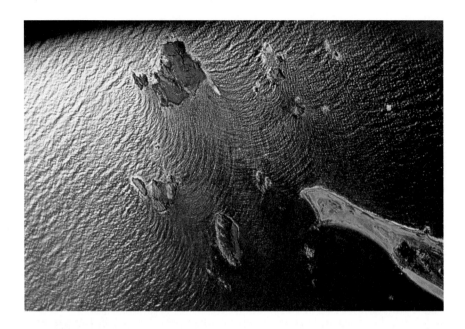

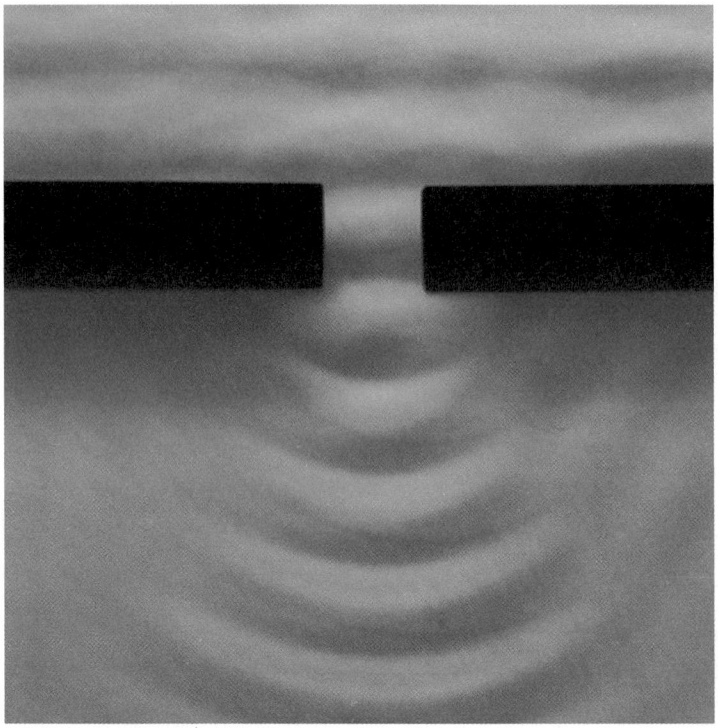

Figure 19 When water waves pass through a small opening in a barrier, they diffract and spread out after they pass through the hole.

Diffraction

When waves strike an object, several things can happen. The waves can bounce off, or be reflected. If the object is transparent, light waves can be refracted as they pass through it. Sometimes the waves may be both reflected and refracted. If you look into a glass window, sometimes you can see your reflection in the window, as well as objects behind it. Light is passing through the window and is also being reflected at its surface.

Waves also can behave another way when they strike an object. The waves can bend around the object. **Figure 18** shows how ocean waves change direction and bend after they strike an island. **Diffraction** occurs when an object causes a wave to change direction and bend around it. Diffraction and refraction both cause waves to bend. The difference is that refraction occurs when waves pass through an object, while diffraction occurs when waves pass around an object.

Reading Check *How do diffraction and refraction differ?*

Waves also can be diffracted when they pass through a narrow opening, as shown in **Figure 19.** After they pass through the opening, the waves spread out. In this case the waves are bending around the corners of the opening.

Diffraction and Wavelength How much does a wave bend when it strikes an object or an opening? The amount of diffraction that occurs depends on how big the obstacle or opening is compared to the wavelength, as shown in **Figure 20.** When an obstacle is smaller than the wavelength, the waves bend around it. But if the obstacle is larger than the wavelength, the waves do not diffract as much. In fact, if the obstacle is much larger than the wavelength, almost no diffraction occurs. The obstacle casts a shadow because almost no waves bend around it.

Hearing Around Corners For example, you're walking down the hallway and you can hear sounds coming from the lunchroom before you reach the open lunchroom door. However, you can't see into the room until you reach the doorway. Why can you hear the sound waves but not see the light waves while you're still in the hallway? The wavelengths of sound waves are similar in size to a door opening. Sound waves diffract around the door and spread out down the hallway. Light waves have a much shorter wavelength. They are hardly diffracted at all by the door. So you can't see into the room until you get close to the door.

Diffraction of Radio Waves Diffraction also affects your radio's reception. AM radio waves have longer wavelengths than FM radio waves do. Because of their longer wavelengths, AM radio waves diffract around obstacles like buildings and mountains. The FM waves with their short wavelengths do not diffract as much. As a result, AM radio reception is often better than FM reception around tall buildings and natural barriers such as hills.

ScienceOnline

Topic: Diffraction
Visit gpscience.com for Web links to information about diffraction.

Activity Research the wave lengths of several types of waves. For each wave type, give an example of an object that could cause diffraction to occur.

Figure 20 The diffraction of waves around an obstacle depends on the wavelength and the size of the obstacle.

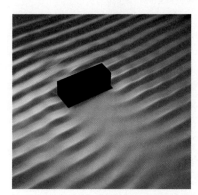

Less diffraction occurs if the wavelength is smaller than the obstacle.

More diffraction occurs if the wavelength is the same size as the obstacle.

Interference

Suppose two waves are traveling toward each other on a long rope as in **Figure 21A.** What will happen when the two waves meet? If you did this experiment, you would find that the two waves pass right through each other, and each one continues to travel in its original direction, as shown in **Figure 21B** and **Figure 21C.** If you look closely at the waves when they meet each other in **Figure 21B,** you see a wave that looks different than either of the two original waves. When the two waves arrive at the same place at the same time, they combine to form a new wave. When two or more waves overlap and combine to form a new wave, the process is called **interference.** This new wave exists only while the two original waves continue to overlap. The two ways that the waves can combine are called constructive interference and destructive interference.

Figure 21 Interference occurs while two waves are overlapping. Then the waves combine to form a new wave. Two waves traveling on a rope can interfere with each other.

A Two waves move toward each other on a rope.

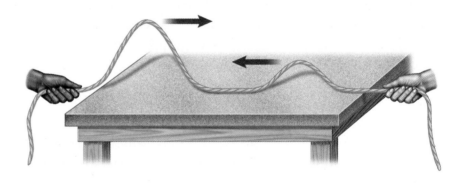

B As the waves overlap, they interfere to form a new wave. **Identify** *What is the amplitude of the new wave?*

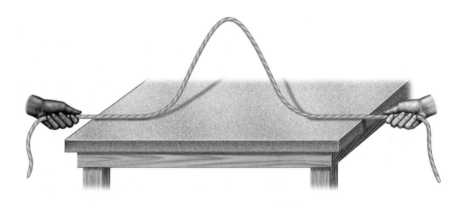

C While the two waves overlap, they continue to move right through each other. Afterward, they continue moving unchanged, as if they had never met.

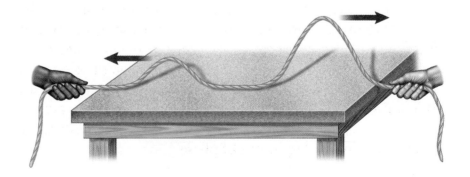

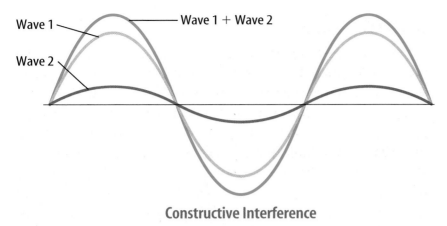

Constructive Interference

A If Wave 1 and Wave 2 overlap, they constructively interfere and form the green wave. Wave 1 and Wave 2 are in phase.

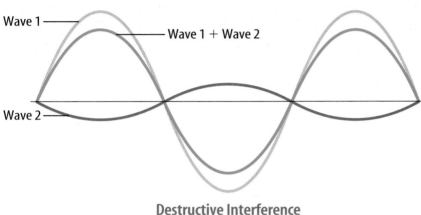

Destructive Interference

B If Wave 1 and Wave 2 overlap, they destructively interfere and form the green wave. Wave 1 and Wave 2 are out of phase.

Figure 22 When waves interfere with each other, constructive and destructive interference can occur. **Infer** *how the energy carried by each wave changes when interference occurs.*

Constructive Interference In constructive interference, shown in **Figure 22A,** the waves add together. This happens when the crests of two or more transverse waves arrive at the same place at the same time and overlap. The amplitude of the new wave that forms is equal to the sum of the amplitudes of the original waves. Constructive interference also occurs when the compressions of different compressional waves overlap. If the waves are sound waves, for example, constructive interference produces a louder sound. Waves undergoing constructive interference are said to be in phase.

Destructive Interference In destructive interference, the waves subtract from each other as they overlap. This happens when the crests of one transverse wave meet the troughs of another transverse wave, as shown in **Figure 22B.** The amplitude of the new wave is the difference between the amplitudes of the waves that overlapped. With compressional waves, destructive interference occurs when the compression of one wave overlaps with the rarefaction of another wave. The compressions and rarefactions combine and form a wave with reduced amplitude. When destructive interference happens with sound waves, it causes a decrease in loudness. Waves undergoing destructive interference are said to be out of phase.

INTEGRATE
Health

Noise Damage People who are exposed to constant loud noises, such as those made by airplane engines, can suffer hearing damage. Special ear protectors have been developed that use destructive interference to cancel damaging noise. With a classmate, list all the jobs you can think of that require ear protectors.

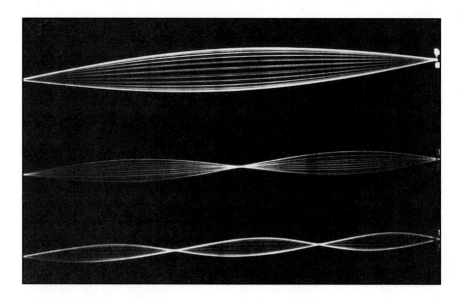

Figure 23 Standing waves form a wave pattern that seems to stay in the same place.

Explain *how nodes form in a standing wave.*

Standing Waves

When you make transverse waves with a rope, you might shake one end while your friend holds the other end still. What would happen if you both shook the rope continuously to create identical waves moving toward each other? As the two waves travel in opposite directions down the rope, they continually pass through each other. Interference takes place as the waves from each end overlap along the rope. At any point where a crest meets a crest, a new wave with a larger amplitude forms. But at points where crests meet troughs, the waves cancel each other and no motion occurs.

The interference of the two identical waves makes the rope vibrate in a special way, as shown in **Figure 23.** The waves create a pattern of crests and troughs that do not seem to be moving. Because the wave pattern stays in one place, it is called a standing wave. A **standing wave** is a special type of wave pattern that forms when waves equal in wavelength and amplitude, but traveling in opposite directions, continuously interfere with each other. The places where the two waves always cancel are called nodes. The nodes always stay in the same place on the rope. Meanwhile, the wave pattern vibrates between the nodes.

Standing Waves in Music When the string of a violin is played with a bow, it vibrates and creates standing waves. The standing waves in the string help produce a rich, musical tone. Other instruments also rely on standing waves to produce music. Some instruments, like flutes, create standing waves in a column of air. In other instruments, like drums, a tightly stretched piece of material vibrates in a special way to create standing waves. As the material in a drum vibrates, nodes are created on the surface of the drum.

Resonance

You might have noticed that bells of different sizes and shapes create different notes. When you strike a bell, the bell vibrates at certain frequencies called the natural frequencies. All objects have their own natural frequencies of vibration that depend on the object's size, shape, and the material it is made from.

There is another way to make something vibrate at its natural frequencies. Suppose you have a tuning fork that has a single natural frequency of 440 Hz. Imagine that a sound wave of the same frequency strikes the tuning fork. Because the sound wave has the same frequency as the natural frequency of the tuning fork, the tuning fork will vibrate. The process by which an object is made to vibrate by absorbing energy at its natural frequencies is called **resonance.**

Sometimes resonance can cause an object to absorb a large amount of energy. Remember that the amplitude of a wave increases as the energy it carries increases. In the same way, an object vibrates more strongly as it continues to absorb energy at its natural frequencies. If enough energy is absorbed, the object can vibrate so strongly that it breaks apart.

Mini LAB

Experimenting with Resonance

Procedure
1. Strike a **tuning fork** with a **mallet.**
2. Hold the vibrating tuning fork near a **second tuning fork** that has the same frequency.
3. Strike the tuning fork again. Hold it near a **third tuning fork** that has a different frequency.

Analysis
What happened when you held the vibrating tuning fork near each of the other two? Explain.

section 3 review

Summary

Reflection and Refraction

- When reflection of a wave occurs, the angle of incidence equals the angle of reflection.
- Refraction occurs when a wave changes direction as it moves from one medium to another.

Diffraction

- Diffraction occurs when a wave changes direction by bending around an obstacle.
- The effects of diffraction are greatest when the wavelength is nearly the obstacle size.

Interference and Resonance

- Interference occurs when two or more waves overlap and form a new wave.
- Interference between two waves with the same wavelength and amplitude, but moving in opposite directions, produces a standing wave.
- Resonance occurs when an object is made to vibrate by absorbing energy from vibrations at its natural frequencies.

Self Check

1. **Compare** the loudness of sound waves that are in phase when they interfere with the loudness of sound waves that are out of phase when they interfere.
2. **Describe** how the reflection of light waves enables you to see your image in a mirror.
3. **Describe** the energy transformations that occur when one tuning fork makes another tuning fork resonate.
4. **Think Critically** Suppose the speed of light was greater in water than in air. Draw a diagram to show whether an object underwater would seem deeper or closer to the surface than it really is.

Applying Math

5. **Use Percentages** You aim a flashlight at a window. The radiant energy in the reflected beam is two fifths of the energy in the incident beam. What percentage of the incident beam energy passed through the window?
6. **Calculate Angle of Incidence** The angle between a flashlight beam that strikes a mirror and the reflected beam is 80 degrees. What is the angle of incidence?

Measuring Wave Properties

Goals

■ **Measure** the speed of a transverse wave.

■ **Create** waves with different amplitudes.

■ **Measure** the wavelength of a transverse wave.

Materials

long spring, rope, or hose
meterstick
stopwatch

Safety Precautions

◉ Real-World Question

Some waves travel through space; others pass through a medium such as air, water, or earth. Each wave has a wavelength, speed, frequency, and amplitude. How can the speed of a wave be measured? How can the wavelength be determined from the frequency?

◉ Procedure

1. With a partner, stretch your spring across an open floor and measure the length of the spring. Record this measurement in the data table. Make sure the spring is stretched to the same length for each step.

2. Have your partner hold one end of the spring. Create a single wave pulse by shaking the other end of the spring back and forth.

3. Have a third person use a stopwatch to measure the time needed for the pulse to travel the length of the spring. Record this measurement in the *Wave Time* column of your data table.

4. Repeat steps 2 and 3 two more times.

5. **Calculate** the speed of waves 1, 2, and 3 in your data table by using the formula:

$$\text{speed} = \text{distance/time}$$

Average the speeds of waves 1, 2, and 3 to find the speed of waves on your spring.

6. **Create** a wave with several wavelengths. You make one wavelength when your hand moves left, right, and left again. Count the number of wavelengths that you generate in 10 s. Record this measurement for wave 4 in the *Wavelength Count* column in your data table.

7. **Repeat** step 6 two more times. Each time, create a wave with a different wavelength by shaking the spring faster or slower.

Analyze Your Data

1. **Calculate** the frequency of waves 4, 5, and 6 by dividing the number of wavelengths by 10 s.

2. Calculate the wavelength of waves 4, 5, and 6 using the formula
   wavelength = wave speed/frequency
   Use the average speed calculated in step 5 for the wave speed.

Wave Property Measurement			
	Spring Length	**Wave Time**	**Wave Speed**
Wave 1			
Wave 2			
Wave 3	Do not write in this book.		
	Wavelength Count	**Frequency**	**Wavelength**
Wave 4			
Wave 5			
Wave 6			

Conclude and Apply

1. Was the wave speed different for the three different pulses you created? Why or why not?

2. Why would you average the speeds of the three different pulses to calculate the speed of waves on your spring?

3. How did the wavelength of the waves you created depend on the frequency of the waves?

Communicating Your Data

Ask your teacher to set up a contest between the groups in your class. Have each group compete to determine who can create waves with the longest wavelength, the highest frequency, and the largest wave speed. Record the measurements of each group's efforts on the board.

MAKING WAVES

Sonar Helps Create Deep-Sea Pictures and Save Lives

This machine houses side-scan sonar. It was used to help locate the wreck of the *Titanic*.

What is sonar?

Sonar is a device that uses sound waves to locate and measure the distance to underwater objects. Its name is a shortened version of SOund NAvigation and Ranging.

How does sonar work?

Sonar sends out a ping sound that reflects back when it hits an underwater object. Since sound travels through water at a known speed (about 1,500 m/s), scientists measure how long the sound takes to return, then calculate the distance.

Why was it invented?

Sonar was developed by scientists in the early twentieth century as a way to detect icebergs and prevent boating disasters. Its technical advancement was hurried by the Allies' need to detect German submarines in World War I. By 1918, the United States and Britain had developed an active sonar system placed in submarines sent to attack other subs.

By World War II, sonar allowed ships to defend themselves effectively from enemy subs. Their strategy was to use sonar to find subs and then fire depth charges at them from a safe distance. After the war, sonar-absorbing hulls and quiet engines and machinery ensured that subs could partly shield themselves from sonar.

Sonar is now used to help fishermen and scientists find schools of fish. Oceanographers also use it to map ocean and lake floors. Sonar has been vital, too, in the discovery of downed airplanes and ships, including the *Titanic*—the passenger liner that sank in 1912.

In 1985, a French and American team used a new type of sonar device called the side-scan sonar to locate the *Titanic*. This kind of sonar projects a tight beam of sound to create detailed images of the sea bed. Members of the expedition towed this sonar device about 170 m above the seabed across a section of the Atlantic Ocean where the *Titanic* went down. Weeks later, video cameras finally spotted the wreck.

The *Titanic* was found thanks to sonar.

Report Research how sonar was used by navies in World War I and World War II. Did sonar affect each war's outcome? How did it save lives? What uses can you think of for sonar if it could be used in everyday life?

online

For more information, visit gpscience.com/time

Reviewing Main Ideas

Section 1 The Nature of Waves

1. Waves are rhythmic disturbances that transfer energy through matter or space.

2. Waves transfer only energy, not matter.

3. Mechanical waves need matter to travel through. Mechanical waves can be compressional or transverse.

4. When a transverse wave travels in a medium, matter in the medium moves at right angles to the direction the wave travels.

5. When a compressional wave travels in a medium, matter moves back and forth along the same direction as the wave travels.

Section 2 Wave Properties

1. The movement of high points in a medium called crests and low points called troughs forms a transverse wave.

2. The movement of more-dense regions called compressions and less-dense regions called rarefactions forms a compressional wave.

3. Transverse and compressional waves can be described by their wavelengths, frequencies, periods, and amplitudes. As frequency increases, wavelength always decreases.

4. The greater a wave's amplitude is, the more energy it carries.

5. A wave's velocity can be calculated by multiplying its frequency times its wavelength.

Section 3 The Behavior of Waves

1. For all waves, the angle of incidence equals the angle of reflection.

2. A wave is bent, or refracted, when it changes speed as it enters a new medium.

3. When two or more waves overlap, they combine to form a new wave. This process is called interference.

FOLDABLES Use the Foldable that you made at the beginning of this chapter to help you review transverse and compressional waves.

Using Vocabulary

amplitude p. 300
compressional wave p. 292
crest p. 296
diffraction p. 306
frequency p. 297
interference p. 308
medium p. 291
period p. 297

rarefaction p. 296
refraction p. 304
resonance p. 311
standing wave p. 310
transverse wave p. 292
trough p. 296
wave p. 290
wavelength p. 297

Answer the following questions using complete sentences.

1. Compare and contrast reflection and refraction.

2. Which type of wave has points called nodes that do not move?

3. Which part of a compressional wave has the lowest density?

4. Find two words in the vocabulary list that describe the bending of a wave.

5. What occurs when waves overlap?

6. What is the relationship among amplitude, crest, and trough?

7. What does frequency measure?

8. What does a mechanical wave always travel through?

Checking Concepts

Choose the word or phrase that best answers the question.

9. Which of the following do waves carry?
   A) matter
   B) energy
   C) matter and energy
   D) the medium

10. What is the formula for calculating wave speed?
   A) $v = \lambda f$
   B) $v = f - \lambda$
   C) $v = \lambda / f$
   D) $v = \lambda + f$

11. When a compressional wave travels through a medium, which way does matter in the medium move?
   A) backward
   B) forward
   C) perpendicular to the rest position
   D) along the same direction the wave travels

12. What is the highest point of a transverse wave called?
   A) crest
   B) compression
   C) wavelength
   D) trough

13. If the frequency of the waves produced by a vibrating object increases, how does the wavelength of the waves produced change?
   A) It stays the same.
   B) It decreases.
   C) It vibrates.
   D) It increases.

14. If the amplitude of a wave changes, which of the following changes?
   A) wave energy
   B) frequency
   C) wave speed
   D) refraction

15. Which term describes the bending of a wave around an object?
   A) resonance
   B) interference
   C) diffraction
   D) reflection

16. What is equal to the angle of reflection?
   A) refraction angle
   B) normal angle
   C) bouncing angle
   D) angle of incidence

Use the table below to answer question 17.

Speed of Sound in Air	
Temperature (°C)	Sound Speed (m/s)
0	331.4
10	337.4
20	343.4

17. Based on the data in the table above, which of the following would be the speed of sound in air at 30°C?
   A) 340.4 m/s
   B) 346.4 m/s
   C) 353.4 m/s
   D) 349.4 m/s

Science online gpscience.com/vocabulary_puzzlemaker

Interpreting Graphics

18. Copy and complete the following concept map on waves.

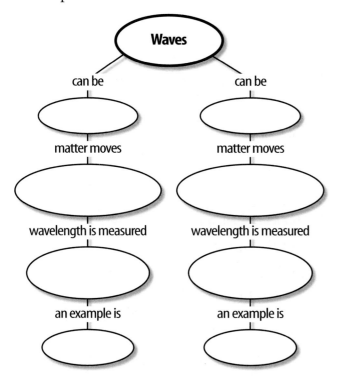

Thinking Critically

19. **Explain** An earthquake on the ocean floor produces a tsunami that hits a remote island. Is the water that hits the island the same water that was above the earthquake on the ocean floor?

20. **Compare** Suppose waves with different amplitudes are produced by a vibrating object. How do the frequencies of the waves with different amplitudes compare?

21. **Explain** Use the law of reflection to explain why you see only a portion of the area behind you when you look in a mirror.

22. **Explain** why you can hear a fire engine coming around a street corner before you can see it.

23. **Describe** the objects or materials that vibrated to produce three of the sounds you've heard today.

24. **Form a Hypothesis** In 1981, people dancing on the balconies of a Kansas City, Missouri, hotel caused the balconies to collapse. Use what have you learned about wave behavior to form a hypothesis that explains why this happened.

25. **Make and Use Tables** Find information in newspaper articles or magazines describing five recent earthquakes. Construct a table that shows for each earthquake the date, location, magnitude, and whether the damage caused by each earthquake was light, moderate, or heavy.

26. **Concept Map** Design a concept map that shows the characteristics of transverse waves. Include the terms *crest, trough, medium, wavelength, frequency, period,* and *amplitude.*

Applying Math

27. **Calculate Wavelength** Calculate the wavelength of a wave traveling on a spring if the wave moves at 0.2 m/s and has a period of 0.5 s.

28. **Calculate Wavespeed** The microwaves produced inside a microwave oven have a wavelength of 12.0 cm and a frequency of 2,500,000,000 Hz. At what speed do the microwaves travel in units of m/s?

29. **Calculate Frequency** Water waves on a lake travel toward a dock with a speed of 2.0 m/s and a wavelength of 0.5 m. How many wave crests strike the dock each second?

Part 1 Multiple Choice

Record your answers on the answer sheet provided by your teacher or on a sheet of paper.

1. When a transverse wave travels through a medium, which way does matter in the medium move?
   A. backward
   B. all directions
   C. at right angles to the direction the wave travels
   D. in the same direction the wave travels

Use the illustration below to answer questions 2 and 3.

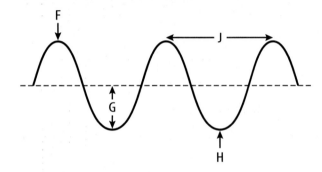

2. What wave property is shown at G?
   A. amplitude   C. crest
   B. wavelength   D. trough

3. What property of the wave is shown at H?
   A. amplitude   C. crest
   B. wavelength   D. trough

4. What is the number of waves that pass a point in a certain time called?
   A. wavelength   B. wave amplitude
   C. wave intensity   D. wave frequency

5. The period of a wave can be directly calculated from which of the following?
   A. troughs   C. frequency
   B. amplitude   D. wavelength

6. What is the energy of a wave related to?
   A. frequency   C. amplitude
   B. wave speed   D. refraction

7. What is the bending of a wave as it enters a new material called?
   A. refraction   C. reflection
   B. diffraction   D. interference

8. When the crests of two identical waves meet, what is the amplitude of the resulting wave?
   A. three times the amplitude of each wave
   B. half the amplitude of each wave
   C. twice the amplitude of each wave
   D. four times the amplitude of one of the original waves

Use the illustration below to answer questions 9 and 10.

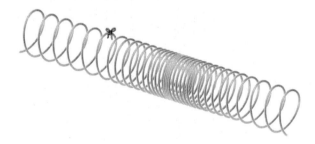

9. What kind of wave is shown?
   A. mechanical   C. transverse
   B. compressional   D. both A and B

10. What happens to the yarn tied to the coil?
    A. It moves back and forth as the wave passes.
    B. It moves up and down as the wave passes.
    C. It does not move as the wave passes.
    D. It moves to the next coil as the wave passes.

11. Through which of the following can sound waves NOT travel?
    A. water   C. outer space
    B. wood   D. air

12. What property of a wave is measured in hertz?
    A. amplitude   C. speed
    B. wavelength   D. frequency

Part 2 | Short Response/Grid In

Record your answers on the answer sheet provided by your teacher or on a sheet of paper.

13. Explain why water waves traveling toward a swimmer on a float do not move the float forward.

Use the illustration below to answer questions 14 and 15.

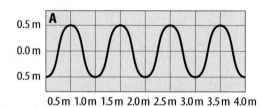

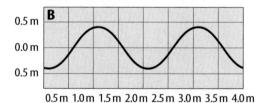

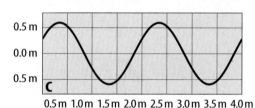

14. Determine the amplitudes and the wavelengths of each of the three waves.

15. If the length of the *x*-axis on each diagram represents 2 s of time, what is the frequency of each wave?

16. A tuning fork vibrates at a frequency of 256 Hz. The wavelength of the sound produced by the tuning fork is 1.32 m. What is the speed of the wave?

17. A sound wave has a speed in air of 330 m/s. If it has a wavelength of 15 m, what is the frequency of the wave?

18. A wave has a speed of 345 m/s and its frequency is 2050 Hz. What is its wavelength?

Part 3 | Open Ended

Record your answers on a sheet of paper.

19. Describe how a standing wave forms and why it has nodes.

20. Explain why objects that are underwater seem to be closer to the surface than they really are.

21. Compare and contrast refraction and diffraction of waves.

22. In a science fiction movie, a huge explosion occurs on the surface of a planet. People in a spaceship heading toward the planet see and hear the explosion. Is this realistic? Explain.

23. Would you expect better reception from AM or FM radio stations on your car radio when the car was traveling in a mountainous area? Explain.

Use the illustration below to answer questions 24 and 25.

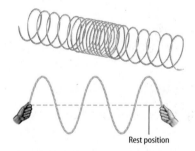

Rest position

24. Describe how the amplitude of each of the two waves shown is defined.

25. Describe how you would change both drawings to show waves that carry more energy.

Test-Taking Tip

Answer every question Never leave any open ended answer blank. Answer each question as best you can. You can receive partial credit for partially correct answers.

Question 19 Before you answer the question, list what you know about standing waves.

Sound

Cracking the Sound Barrier

Have you ever heard the crack of a sonic boom? A sonic boom occurs when a plane exceeds the speed of sound. Sound waves create a cone-shaped shock wave coming from the aircraft. Behind the cone are low-pressure regions that can cause water vapor to condense, forming the cloud you see here.

Science Journal Write three things that you would like to learn about sound.

Start-Up Activities

What sound does a ruler make?

Think of the musical instruments you've seen and heard. Some have strings, some have hollow tubes, and others have keys or pedals. Musical instruments come in many shapes and sizes and are played with various techniques. These differences give each instrument a unique sound. What would an instrument made from a ruler sound like?

1. Hold one end of a thin ruler firmly down on a desk, allowing the free end to extend beyond the edge of the desk.

2. Gently pull up on and release the end of the ruler. What do you see and hear?

3. Vary the length of the overhanging portion and repeat the experiment several times.

4. **Think Critically** In your Science Journal, write instructions for playing a song with the ruler. Explain how the length of the overhanging part of the ruler affects the sound.

 Preview this chapter's content and activities at gpscience.com

---

 **Sound** Make the following Foldable to help identify what you already know, what you want to know, and what you learned about sound.

STEP 1 Fold a vertical sheet of paper from side to side. Make the front edge about 1.25 cm shorter than the back edge.

STEP 2 Turn lengthwise and fold into thirds.

STEP 3 Unfold and cut only the top layer along both folds to make three tabs.

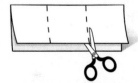

STEP 4 Label each tab as shown.

Identify Questions As you read the chapter, write what you learn about sound traveling through solids under the left tab of your Foldable, what you learn about sound traveling through liquids under the center tab, and what you learn about sound traveling through gases under the right tab.

The Nature of Sound

Reading Guide

What You'll Learn
- **Explain** how sound travels through different mediums.
- **Identify** what influences the speed of sound.
- **Describe** how the ear enables you to hear.

Why It's Important
The nature of sound affects how you hear and interpret sounds.

⚲ Review Vocabulary
vibration: rhythmic back-and-forth motion

New Vocabulary
- eardrum
- cochlea

What causes sound?

An amusement park can be a noisy place. With all the racket of carousel music and booming loudspeakers, it can be hard to hear what your friends say. These sounds are all different, but they do have something in common—each sound is produced by an object that vibrates. For example, your friends' voices are produced by the vibrations of their vocal cords, and music from a carousel and voices from a loudspeaker are produced by vibrating speakers. All sounds are created by something that vibrates.

Sound Waves

When an object like a radio speaker vibrates, it collides with nearby molecules in the air, transferring some of its energy to them. These molecules then collide with other molecules in the air and pass the energy on to them. The energy originally transferred by the vibrating object continues to pass from one molecule to another. This process of collisions and energy transfer forms a sound wave.

Sound waves are compressional waves. Remember that a compressional wave is made up of two types of regions called compressions and rarefactions. If you look at **Figure 1,** you'll see that when a radio speaker vibrates outward, the nearby molecules in the air are pushed together to form compressions. As the figure shows, when the speaker moves inward, the nearby molecules in the air have room to spread out, and a rarefaction forms. As long as the speaker continues to vibrate back and forth, compressions and rarefactions are formed.

Figure 1 The vibration of a speaker produces compressional waves.

Compression

When the speaker vibrates outward, molecules in the air next to it are pushed together to form a compression.

Rarefaction

When the speaker vibrates inward, the molecules spread apart to form a rarefaction.

Traveling as a Wave Compressions and rarefactions move away from the speaker as molecules in the air collide with their neighbors. As the speaker continues to vibrate, more molecules in the air are alternately pushed together and spread apart. A series of compressions and rarefactions forms that travels from the speaker to your ear. This sound wave is what you hear.

The Speed Of Sound

Most sounds you hear travel through air to reach your ears. However, if you've ever been swimming underwater and heard garbled voices, you know that sound also travels through water. In fact, sound waves can travel through any type of matter—solid, liquid, or gas. The matter that a wave travels through is called a medium. Sound waves create compressions and rarefactions in any medium they travel through.

What would happen if no matter existed to form a medium? Could sound be transmitted without particles of matter to compress, expand, and collide? On the Moon, which has no atmosphere, the energy in sound waves cannot be transmitted from particle to particle because no particles exist. Sound waves cannot travel through empty space. Astronauts must talk to each other using electronic communication equipment.

The Speed of Sound in Different Materials The speed of a sound wave through a medium depends on the substance the medium is made of and whether it is solid, liquid, or gas. For example, **Table 1** shows that at room temperature, sound travels at 347 m/s through air, at 1,498 m/s through water, and at 4,877 m/s through aluminum. In general, sound travels the slowest through gases, faster through liquids, and even faster through solids.

 **Reading Check** *What are two things that affect the speed of sound?*

Sound travels faster in liquids and solids than in gases because the individual molecules in a liquid or solid are closer together than the molecules in a gas. When molecules are close together, they can transmit energy from one to another more rapidly. However, the speed of sound doesn't depend on the loudness of the sound. Loud sounds travel through a medium at the same speed as soft sounds.

Mini LAB

Listening to Sound Through Different Materials

Procedure
1. Tie the middle of a length of **string** onto a **metal object,** such as a wire hanger or a spoon, so that the string has two long ends.
2. Wrap each string end around a finger on each hand.
3. Gently placing your fingers in your ears, swing the object until it bumps against the edge of a **chair** or **table.** Listen to the sound.
4. Take your fingers out of your ears and listen to the sound made by the collisions.

Analysis
1. Compare and contrast the sounds you hear when your fingers are and are not in your ears.
2. Do sounds travel better through air or the string?

Table 1 Speed of Sound in Different Mediums	
Medium	**Speed of Sound (in m/s)**
Air	347
Cork	500
Water	1,498
Brick	3,650
Aluminum	4,877

Figure 2 A line of people passing a bucket is a model for molecules transferring the energy of a sound wave.

When the people are far away from each other, like the molecules in a gas, it takes longer to transfer the bucket of water from person to person.

The bucket travels quickly down the line when the people stand close together.

Explain *why sound would travel more slowly in cork than in steel.*

A Model for Transmitting Sound

You can understand why solids and liquids transmit sound well by picturing a large group of people standing in a line. Imagine that they are passing a bucket of water from person to person. If everyone stands far apart, each person has to walk a long distance to transfer the bucket, as in the top photo of **Figure 2.** However, if everyone stands close together, as in the bottom photo of **Figure 2,** the bucket quickly moves down the line.

The people standing close to each other are like particles in solids and liquids. Those standing far apart are like gas particles. The closer the particles, the faster they can transfer energy from particle to particle.

Temperature and the Speed of Sound

The speed of sound waves also depends on the temperature of a medium. As the temperature of a substance increases, its molecules move faster. This makes them more likely to collide with each other. Remember that sound waves depend on the collisions of particles to transfer energy through a medium. If the particles in a medium are colliding with each other more often, more energy can be transferred in a shorter amount of time. Then sound waves move faster. For example, when the temperature is 0°C, sound travels through the air at only 331 m/s, but at a temperature of 20°C, it speeds up to 343 m/s.

Human Hearing

INTEGRATE
Life Science

Think of the last conversation you had. Vocal cords and mouths move in many different ways to produce various kinds of compressional waves, but you were somehow able to make sense of these different sound waves. Your ears and brain work together to turn the compressional waves caused by speech, music, and other sources into something that has meaning. Making sense of these waves involves four stages. First, the ear gathers the compressional waves. Next, the ear amplifies the waves. In the ear, the amplified waves are converted to nerve impulses that travel to the brain. Finally, the brain decodes and interprets the nerve impulses.

Gathering Sound Waves—The Outer Ear When you think of your ear, you probably picture just the fleshy, visible, outer part. But, as shown in **Figure 3,** the human ear has three sections called the outer ear, the middle ear, and the inner ear.

The visible part of your ear, the ear canal, and the eardrum make up the outer ear. The outer ear is where sound waves are gathered. The gathering process starts with the outer part of your ear, which is shaped to help capture and direct sound waves into the ear canal. The ear canal is a passageway that is 2-cm to 3-cm long and is a little narrower than your index finger. The sound waves travel along this passageway, which leads to the eardrum. The **eardrum** is a tough membrane about 0.1 mm thick. When incoming sound waves reach the eardrum, they transfer their energy to it and it vibrates.

> ✔ **Reading Check** *What makes the eardrum vibrate?*

Amplifying Sound Waves—The Middle Ear When the eardrum vibrates, it passes the sound vibrations into the middle ear, where three tiny bones start to vibrate. These bones are called the hammer, the anvil, and the stirrup. They make a lever system that multiplies the force and pressure exerted by the sound wave. The bones amplify the sound wave. The stirrup is connected to a membrane on a structure called the oval window, which vibrates as the stirrup vibrates.

Audiology Some types of hearing loss involve damage to the inner ear. The use of a hearing aid often can improve hearing. Audiologists diagnose and treat people with hearing loss. Research the field of audiology. Find out what education is required for people to gain certification as audiologists, and the settings in which they work.

Figure 3 The ear has three regions that perform specific functions in hearing.

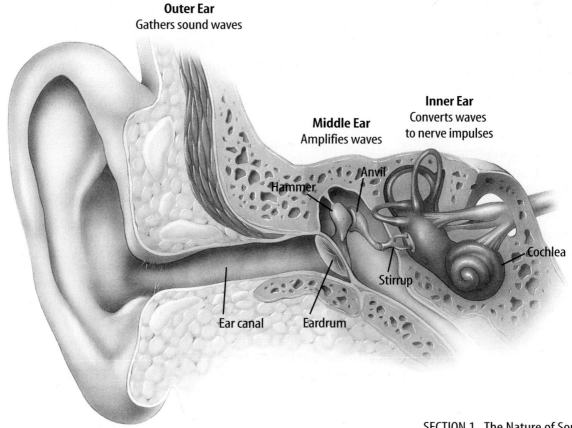

Outer Ear
Gathers sound waves

Middle Ear
Amplifies waves

Inner Ear
Converts waves
to nerve impulses

Anvil

Hammer

Stirrup

Cochlea

Ear canal   Eardrum

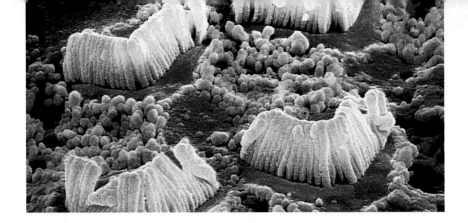

Figure 4 These hair cells in the human ear send nerve impulses to the brain when sound waves cause them to vibrate. In this photo the hair cells are magnified 3,900 times.

Converting Sound Waves—The Inner Ear When the membrane in the oval window vibrates, the sound vibrations are transmitted into the inner ear. The inner ear contains the **cochlea** (KOH klee uh), which is a spiral-shaped structure that is filled with liquid and contains tiny hair cells like those shown in **Figure 4.** When these tiny hair cells in the cochlea begin to vibrate, nerve impulses are sent through the auditory nerve to the brain. It is the cochlea that converts sound waves to nerve impulses.

When someone's hearing is damaged, it's usually because the tiny hair cells in the cochlea are damaged or destroyed, often by loud sounds. New research suggests that these hair cells may be able to repair themselves.

section 1 review

Summary

Sound Waves and Their Causes

- Sound results from compressional waves emanating from vibrating objects.
- The compressions and rarefactions of sound waves transfer energy.

Moving Through Materials

- Sound waves can travel through any matter.
- Sound travels fastest through solids, slower through liquids, and slowest through gases.
- Sound travels faster in warmer mediums.

Human Hearing

- The outer ear gathers sound waves and directs them to the eardrum.
- The middle ear contains three tiny bones that multiply the force and pressure of the vibrating eardrum.
- The inner ear's cochlea converts sound waves to nerve impulses.

Self Check

1. **Explain** how sound travels from your vocal cords to your friend's ears when you talk.
2. **Summarize** the physical reasons that sound waves travel at different speeds through liquids and gases.
3. **Explain** why sound speeds up when the temperature rises.
4. **Describe** in detail each section of the human ear and its role in hearing.
5. **Think Critically** Some people hear ringing in their ears, called tinnitus, even in the absence of sound. Form a hypothesis to explain why this occurs.

Applying Math

6. **Calculate Time** How long would it take a sound wave from a car alarm to travel 1 km if the temperature were 0°C?
7. **Calculate Time** How long would it take the same wave to travel 1 km if the temperature were 20°C?

Science online gpscience.com/self_check_quiz

Properties of Sound

Reading Guide

What You'll Learn

- **Recognize** how amplitude, intensity, and loudness are related.
- **Describe** how sound intensity is measured and what levels can damage hearing.
- **Explain** the relationship between frequency and pitch.
- **Discuss** the Doppler effect.

Why It's Important

Each property of a sound wave affects how things sound to you—from your blaring CD player to someone's whisper.

Review Vocabulary

potential energy: energy that is stored in an object's position

New Vocabulary

- intensity
- loudness
- decibel
- pitch
- ultrasonic
- Doppler effect

Intensity and Loudness

If the phone rings while you're listening to a stereo, you might have to turn down the volume on the stereo to be able to hear the person on the phone. What happens to the sound waves from your radio when you adjust the volume? The notes sound the same as when the volume was higher, but something about the sound changes. The difference is that quieter sound waves do not carry as much energy as louder sound waves do.

Recall that the amount of energy a wave carries corresponds to its amplitude. For a compressional wave, amplitude is related to the density of the particles in the compressions and rarefactions. Look at **Figure 5.** For a sound wave that carries less energy and has a lower amplitude, particles in the medium are less compressed in the compressions and less spread out in the rarefactions. For a sound wave that has a higher amplitude, particles in the medium are closer together, or more compressed, in the compressions and more spread out in the rarefactions.

To produce a wave that carries more energy, more energy is transferred from the vibrating object to the medium. More energy is transferred to the medium when the particles of the medium are forced closer together in the compressions and spread farther apart in the rarefactions.

Figure 5 The amplitude of a sound wave depends on how tightly packed molecules are in the compressions and rarefactions.

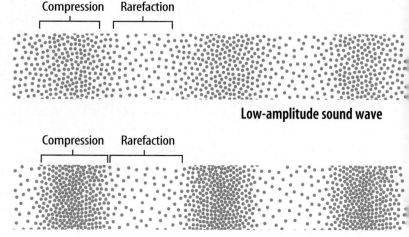

Low-amplitude sound wave

High-amplitude sound wave

Intensity Imagine sound waves moving through the air from your radio to your ear. If you held a square loop between you and the radio, as in **Figure 6,** and could measure how much energy passed through the loop in 1 s, you would measure intensity. **Intensity** is the amount of energy that flows through a certain area in a specific amount of time. When you turn down the volume of your radio, you reduce the energy carried by the sound waves, so you also reduce their intensity.

Intensity influences how far away a sound can be heard. If you and a friend whisper a conversation, the sound waves you create have low intensity and do not travel far. You have to sit close together to hear each other. However, when you shout to each other, you can be much farther apart. The sound waves made by your shouts have high intensity and can travel far.

Intensity Decreases With Distance Intensity influences how far a wave will travel because some of a wave's energy is converted to other forms of energy when it is passed from particle to particle. Think about what happens when you drop a basketball. The ball has potential energy as you hold it above the ground. This potential energy is converted into energy of motion as the ball falls. When the ball hits the ground and bounces up, a small amount of that energy has been transferred to the ground. The ball no longer has enough energy to bounce back to the original level. The ball transfers a small amount of energy with each bounce, until finally the ball has no more energy. If you held the ball higher above the ground, it would have more energy and would bounce for a longer time before it came to a stop. In a similar way, a sound wave of low intensity loses its energy more quickly, and travels a shorter distance than a sound wave of higher intensity.

Figure 6 The intensity of the sound waves from the CD player is related to the amount of energy that passes through the loop in a certain amount of time. **Describe** *how the intensity would change if the loop were 10 m away from the radio.*

Loudness When you hear different sounds, you do not need special equipment to know which sounds have greater intensity. Your ears and brain can tell the difference. **Loudness** is the human perception of sound intensity. Sound waves with high intensity carry more energy. When sound waves of high intensity reach your ear, they cause your eardrum to move back and forth a greater distance than sound waves of low intensity do. The bones of the middle ear convert the increased movement of the eardrum into increased movement of the hair cells in the inner ear. As a result, you hear a loud sound. As the intensity of a sound wave increases, the loudness of the sound you hear increases.

✔ **Reading Check** *How are intensity and loudness related?*

The Decibel Scale It's hard to say how loud too loud is. Two people are unlikely to agree on what is too loud, because people vary in their perception of loudness. A sound that seems fine to you may seem earsplitting to your teacher. Even so, the intensity of sound can be described using a measurement scale. Each unit on the scale for sound intensity is called a **decibel** (DE suh bel), abbreviated dB. On this scale, the faintest sound that most people can hear is 0 dB. Sounds with intensity levels above 120 dB may cause pain and permanent hearing loss. During some rock concerts, sounds reach this damaging intensity level. Wearing ear protection, such as earplugs, around loud sounds can help protect against hearing loss. **Figure 7** shows some sounds and their intensity levels in decibels.

Science nline

Topic: Sound Intensity
Visit gpscience.com for Web links to information about sound as well as a list of sounds and their intensities in decibels.

Activity Make a table that lists some sounds you heard today, in order from loudest to quietest. In another column, write the intensity level of each sound.

Figure 7 The decibel scale measures the intensity of sound. **Identify** *where a normal speaking voice would fall on the scale.*

Loudness in Decibels

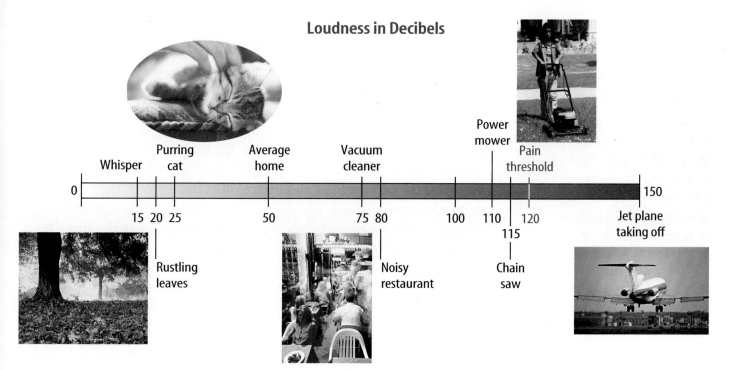

Whisper — Purring cat — Average home — Vacuum cleaner — Power mower — Pain threshold

0 — 15 20 25 — 50 — 75 80 — 100 110 — 120 — 150

115

Rustling leaves — Noisy restaurant — Chain saw — Jet plane taking off

C	D	E	F	G	A	B	C
do	re	mi	fa	so	la	ti	do
262 Hz	294 Hz	330 Hz	349 Hz	393 Hz	440 Hz	494 Hz	524 Hz

Figure 8 Every note has a different frequency, which gives it a distinct pitch.

Describe *how pitch changes when frequency increases.*

Simulating Hearing Loss

Procedure

1. Tune a **radio** to a news station. Turn the volume down to the lowest level you can hear and understand.
2. Turn the bass to maximum and the treble to minimum. If the radio does not have these controls, hold thick wads of **cloth** over your ears.
3. Observe which sounds are hardest and easiest to hear.

Analysis

1. Are high or low pitches harder to hear? Are vowel or consonant sounds harder to hear?
2. How could you help a person with hearing loss understand what you say?

Try at Home

Pitch

If you have ever taken a music class, you are probably familiar with the musical scale do, re, mi, fa, so, la, ti, do. If you were to sing this scale, your voice would start low and become higher with each note. You would hear a change in **pitch,** which is how high or low a sound seems to be. The pitch of a sound is related to the frequency of the sound waves.

Frequency and Pitch Frequency is a measure of how many wavelengths pass a particular point each second. For a compressional wave, such as sound, the frequency is the number of compressions or the number of rarefactions that pass by each second. Frequency is measured in hertz (Hz)—1 Hz means that one wavelength passes by in 1 s.

When a sound wave with high frequency hits your ear, many compressions hit your eardrum each second. The wave causes your eardrum and all the other parts of your ear to vibrate more quickly than if a sound wave with a low frequency hit your ear. Your brain interprets these fast vibrations caused by high-frequency waves as a sound with a high pitch. As the frequency of a sound wave decreases, the pitch becomes lower. **Figure 8** shows different notes and their frequencies. For example, a whistle with a frequency of 1,000 Hz has a high pitch, but low-pitched thunder has a frequency of less than 50 Hz.

A healthy human ear can hear sound waves with frequencies from about 20 Hz to 20,000 Hz. The human ear is most sensitive to sounds in the range of 440 Hz to about 7,000 Hz. In this range, most people can hear much fainter sounds than at higher or lower frequencies.

Ultrasonic and Infrasonic Waves Most people can't hear sound frequencies above 20,000 Hz, which are called **ultrasonic** waves. Dogs can hear sounds with frequencies up to about 35,000 Hz, and bats can detect frequencies higher than 100,000 Hz. Even though humans can't hear ultrasonic waves, they use them for many things. Ultrasonic waves are used in medical diagnosis and treatment. They also are used to estimate the size, shape, and depth of underwater objects.

Infrasonic, or subsonic, waves have frequencies below 20 Hz—too low for most people to hear. These waves are produced by sources that vibrate slowly, such as wind, heavy machinery, and earthquakes. Although you can't hear infrasonic waves, you may feel them as a rumble inside your body.

The Doppler Effect

Imagine that you are standing at the side of a racetrack with race cars zooming past. As they move toward you, the different pitches of their engines become higher. As they move away from you, the pitches become lower. The change in pitch or wave frequency due to a moving wave source is called the **Doppler effect. Figure 9** shows how the Doppler effect occurs.

✔ Reading Check *What is the Doppler effect?*

Moving Sound As a race car moves, it sends out sound waves in the form of compressions and rarefactions. In **Figure 9A,** the race car creates a compression, labeled A. Compression A moves through the air toward the flagger standing at the finish line. By the time compression B leaves the race car in **Figure 9B,** the car has moved forward. Because the car has moved since the time it created compression A, compressions A and B are closer together than they would be if the car had stayed still. Because the compressions are closer together, more compressions pass by the flagger each second than if the car were at rest. As a result, the flagger hears a higher pitch. You also can see from **Figure 9B** that the compressions behind the moving car are farther apart, resulting in a lower frequency and a lower pitch after the car passes and moves away from the flagger.

INTEGRATE Astronomy

Red Shift The Doppler effect can also be observed in light waves emanating from moving sources—although the sources must be moving at tremendous speeds. Astronomers have learned that the universe is expanding by observing the Doppler effect in light waves. Research the phenomenon known as red shift and explain in your Science Journal how it relates to the Doppler effect.

Figure 9 The Doppler effect occurs when the source of a sound wave is moving relative to a listener.

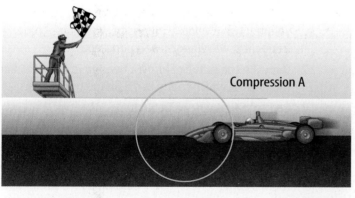

A The race car creates compression A.

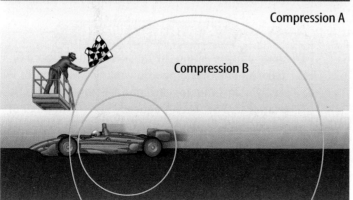

B The car is closer to the flagger when it creates compression B. Compressions A and B are closer together in front of the car, so the flagger hears a higher-pitched sound.

Figure 10 Doppler radar can show the movement of winds in storms, and, in some cases, can detect the wind rotation that leads to the formation of tornadoes, like the one shown here. This can help provide early warning and reduce the injuries and loss of life caused by tornadoes.

A Moving Observer You also can observe the Doppler effect when you are moving past a sound source that is standing still. Suppose you were riding in a school bus and passed a building with a ringing bell. The pitch would sound higher as you approached the building and lower as you rode away from it. The Doppler effect happens any time the source of a sound is changing position compared with the observer. It occurs no matter whether it is the sound source or the observer that is moving. The faster the change in position, the greater the change in frequency and pitch.

Using the Doppler Effect The Doppler effect also occurs for other waves besides sound waves. For example, the frequency of electromagnetic waves, such as radar waves, changes if an observer and wave source are moving relative to each other. Radar guns use the Doppler effect to measure the speed of cars. The radar gun sends radar waves toward a moving car. The waves are reflected from the car and their frequency is shifted, depending on the speed and direction of the car. From the Doppler shift of the reflected waves, the radar gun determines the car's speed. Weather radar also uses the Doppler shift to show the movement of winds in storms, such as the tornado in **Figure 10.**

section 2 review

Summary

Intensity and Loudness

- Tight, dense compressions in a sound wave mean high intensity, loudness, and more energy.
- Sound intensity is measured in decibels.

Pitch

- High frequency sound waves have closer compressions and rarefactions and higher pitch than those of low frequency.
- Ultrasonic waves are too high for people to hear, but are useful for medical purposes.

The Doppler Effect

- The Doppler effect occurs when a moving object emits sounds that change pitch as the object moves past you.
- Police radar uses the Doppler effect to detect speeding cars.

Self Check

1. **Determine** which will change if you turn up a radio's volume: wave velocity, intensity, pitch, amplitude, frequency, wavelength, or loudness.
2. **Describe** the range of human hearing in decibels, and the level at which sound can damage human ears.
3. **Contrast** frequency and pitch.
4. **Draw and label** a diagram that explains the Doppler effect.
5. **Think Critically** Why would a passing race car display more Doppler effect than a passing police siren?

Applying Math

6. **Make a Table** Using the musical scale in **Figure 8,** make a table that shows how many wavelengths will pass you in 1 min for each musical note. What is the relationship between frequency and the number of wavelengths that pass you in 1 min?

Science Online gpscience.com/self_check_quiz

Music

Reading Guide

What You'll Learn

- **Distinguish** between noise and music.
- **Describe** why different instruments have different sound qualities.
- **Explain** how string, wind, and percussion instruments produce music.
- **Describe** the formation of beats.

Why It's Important

Music makes life more enjoyable, but noise pollution is unpleasant.

Review Vocabulary

frequency: the number of vibrations occurring in 1 s

New Vocabulary

- music
- sound quality
- overtone
- resonator

What is music?

To someone else, your favorite music might sound like a jumble of noise. Music and noise are caused by vibrations—with some important differences, as shown in **Figure 11.** Noise has random patterns and pitches. **Music** is made of sounds that are deliberately used in a regular pattern.

Natural Frequencies Every material or object has a particular set of frequencies at which it vibrates. This set of frequencies are its natural frequencies. When you pluck a guitar string the pitch you hear depends on the string's natural frequencies. Each string on a guitar has a different set of natural frequencies. The natural frequencies of a guitar string depend on the string's thickness, its length, the material it is made from, and how tightly it is stretched. Musical instruments contain strings, membranes, or columns of air that vibrate at their natural frequencies to produce notes with different pitches.

Resonance The sound produced by musical instruments is amplified by resonance. Recall that resonance occurs when a material or an object is made to vibrate at its natural frequencies by absorbing energy from something that is also vibrating at those frequencies. The vibrations of the mouthpiece or the reed in a wind instrument cause the air inside the instrument to absorb energy and vibrate at its natural frequencies. The vibrating air makes the sound of the instrument louder.

Figure 11 These wave patterns represent the sound waves created by the piano and the scraping fingernails.
Determine *which of these has a regularly repeating pattern.*

Piano

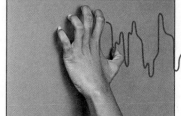

Scraping fingernails

Sound Quality

Suppose your classmate played a note on a flute and then a note of the same pitch and loudness on a piano. Even if you closed your eyes, you could tell the difference between the two instruments. Their sounds wouldn't be the same. Each of these instruments has a unique sound quality. **Sound quality** describes the differences among sounds of the same pitch and loudness. Objects can be made to vibrate at other frequencies besides their natural frequency. This produces sound waves with more than one frequency. The specific combination of frequencies produced by a musical instrument is what gives it a distinctive quality of sound.

 **Reading Check** *What does sound quality describe and how is it created?*

Overtones Even though an instrument vibrates at many different frequencies at once, you still hear just one note. All of the frequencies are not at the same intensity. The main tone that is played and heard is called the fundamental frequency. On a guitar, for example, the fundamental frequency is produced by the entire string vibrating back and forth, as in **Figure 12.** In addition to vibrating at the fundamental frequency, the string also vibrates to produce overtones. An **overtone** is a vibration whose frequency is a multiple of the fundamental frequency. The first two guitar-string overtones also are shown in **Figure 12.** These overtones create the rich sounds of a guitar. The number and intensity of overtones vary with each instrument. These overtones produce an instrument's distinct sound quality.

Musical Instruments

A musical instrument is any device used to produce a musical sound. Violins, cello, oboes, bassoons, horns, and kettledrums are musical instruments that you might have seen and heard in your school orchestra. These familiar examples are just a small sample of the diverse assortment of instruments people play throughout the world. For example, Australian Aborigines accompany their songs with a woodwind instrument called the didgeridoo (DIH juh ree dew). Caribbean musicians use rubber-tipped mallets to play steel drums, and a flutelike instrument called the nay is played throughout the Arab world.

Figure 12 A guitar string can vibrate at more than one frequency at the same time. Here the guitar string vibrations produce the fundamental frequency, and first and second overtones are shown.
Infer *how the string would vibrate to produce the third overtone.*

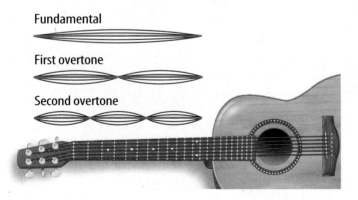

Fundamental

First overtone

Second overtone

Strings Soft violins, screaming electric guitars, and elegant harps are types of string instruments. In string instruments, sound is produced by plucking, striking, or drawing a bow across tightly stretched strings. Because the sound of a vibrating string is soft, string instruments usually have a resonator, like the violin in **Figure 13.** A **resonator** (RE zuh nay tur) is a hollow chamber filled with air that amplifies sound when the air inside of it vibrates. For example, if you pluck a guitar string that is stretched tightly between two nails on a board, the sound is much quieter than if the string were on a guitar. When the string is attached to a guitar, the guitar frame and the air inside the instrument begin to vibrate as they absorb energy from the vibrating string. The vibration of the guitar body and the air inside the resonator makes the sound of the string louder and also affects the quality of the sound.

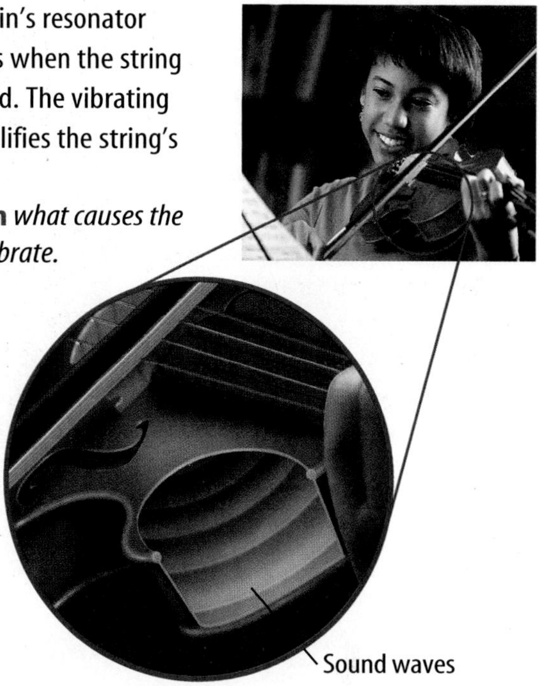

Brass and Woodwinds Brass and woodwind instruments rely on the vibration of air to make music. The many different brass and wind instruments—such as horns, oboes, and flutes—use various methods to make air vibrate inside the instrument. For example, brass instruments have cone-shaped mouthpieces like the one in **Figure 14.** This mouthpiece is inserted into metal tubing, which is the resonator in a brass instrument. As the player blows into the instrument, his or her lips vibrate against the mouthpiece. The air in the resonator also starts to vibrate, producing a pitch. On the other hand, to play a flute, a musician blows a stream of air against the edge of the flute's mouth hole. This causes the air inside the flute to vibrate.

In brass and wind instruments, the length of the vibrating tube of air determines the pitch of the sound produced. For example, in flutes and trumpets, the musician changes the length of the resonator by opening and closing finger holes or valves. In a trombone, however, the tubing slides in and out to become shorter or longer.

Figure 13 The air inside the violin's resonator vibrates when the string is played. The vibrating air amplifies the string's sound.
Explain *what causes the air to vibrate.*

Sound waves

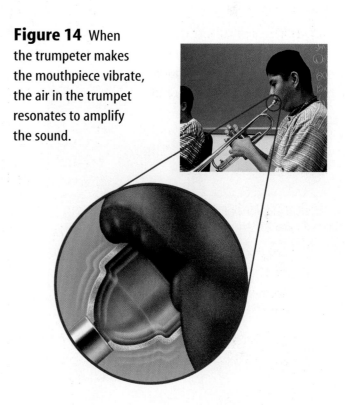

Figure 14 When the trumpeter makes the mouthpiece vibrate, the air in the trumpet resonates to amplify the sound.

Figure 15 The air inside the resonator of the drum amplifies the sound created when the musician strikes the membrane's surface. **Describe** *how the natural frequency of the air in the drum affects the sound it creates.*

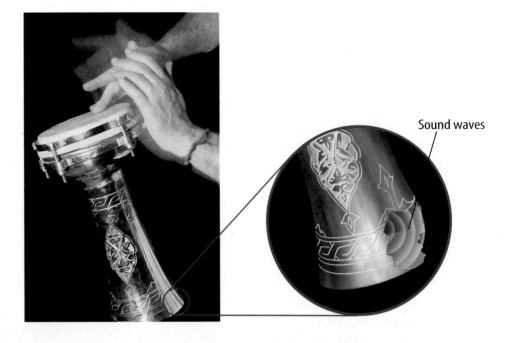

Sound waves

Figure 16 Xylophones are made with many wooden bars that each have their own resonator tubes. **Explain** *why the resonators and bars on a xylophone are different sizes.*

Percussion Does the sound of a bass drum make your heart start to pound? Since ancient times, people have used drums and other percussion instruments to send signals, accompany important rituals, and entertain one another. Percussion instruments are struck, shaken, rubbed, or brushed to produce sound. Some, such as kettledrums or the drum shown in **Figure 15,** have a membrane stretched over a resonator. When the drummer strikes the membrane with sticks or hands, the membrane vibrates and causes the air inside the resonator to vibrate. The resonator amplifies the sound made when the membrane is struck. Some drums have a fixed pitch, but others have a pitch that can be changed by tightening or loosening the membrane.

Caribbean steel drums were developed in the 1940s in Trinidad. As many as 32 different striking surfaces hammered from the ends of 55-gallon oil barrels create different pitches of sound. The side of a drum acts as the resonator.

✔ **Reading Check** *How have people used drums?*

The xylophone shown in **Figure 16** is another type of percussion instrument. It has a series of wooden bars, each with its own tube-shaped resonator. The musician strikes the bars with mallets, and the type of mallet affects the sound quality. Hard mallets make crisp sounds, while softer rubber mallets make duller sounds. Other types of percussion instruments include cymbals, rattles, and even old-fashioned washboards.

Beats

Have you ever heard two flutes play the same note when they weren't properly tuned? The sounds they produce have slightly different frequencies. You may have heard a pulsing variation in loudness, called beats.

When two instruments play at the same time, the sound waves produced by each instrument interfere. The amplitudes of the waves add together when compressions overlap and rarefactions overlap, causing an increase in loudness. When compressions and rarefactions overlap each other, the loudness decreases. Look at **Figure 17.** If two waves of different frequencies interfere, a new wave is produced that has a different frequency. The frequency of this wave is actually the difference in the frequencies of the two waves. The frequency of the beats that you hear decreases as the two waves become closer in frequency. If two flutes that are in tune play the same note, no beats are heard.

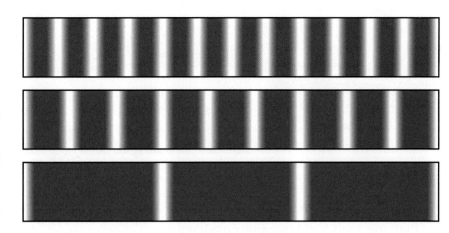

Figure 17 Beats can occur when sound waves of different frequencies, shown in the top two panels, combine. These sound waves interfere with each other, forming a wave with a lower frequency, shown in the bottom panel. This wave causes a listener to hear beats.

Reading Check *How are beats produced?*

section 3 review

Summary

Nature of Music

- Music is sound used in regular patterns.

- Every material has a set of natural frequencies at which it will vibrate.

- Resonance helps amplify the sound produced by musical instruments.

Sound Quality

- Quality of sound describes the differences among sounds of the same pitch and loudness.

- Sound quality results from specific combinations of frequencies produced in various musical instruments.

Beats

- The interference of two waves with different frequencies produces beats.

Self Check

1. **Compare and contrast** music and noise.

2. **Explain** how an instrument's overtones contribute to its sound quality.

3. **Explain** how a flute, violin, and kettledrum produce sound.

4. **Describe** what occurs when two out-of-tune instruments play the same note.

5. **Think Critically** Why do musical instruments vary in different regions of the world?

Applying Math

6. **Calculate Frequencies** A string on a guitar vibrates with a frequency of 440 Hz. Two beats are heard when this string and a string on another guitar are played at the same time. What are the possible frequencies of vibration of the second string?

Making Music

There are many different types of musical instruments. Early instruments were made from materials that were easily obtained such as clay, shells, skins, wood, and reeds. These materials were fashioned into various instruments that produced pleasing sounds. In this lab, you are going to create a musical instrument using materials that are available to you—just as your ancestors did.

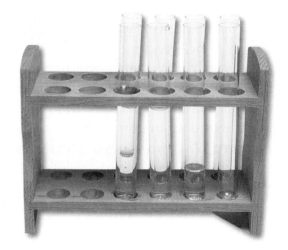

⊙ Real-World Question

How can you make different tones using only test tubes and water?

Goals

■ **Demonstrate** how to make music using water and test tubes.

■ **Predict** how the tones will change when there is more or less water in a test tube.

Materials

test tubes
test-tube rack

Safety Precautions

⊙ Procedure

1. Put different amounts of water into each of the test tubes.

2. **Predict** any differences you expect in how the tones from the different test tubes will sound.

3. Blow across the top of each test tube.

4. **Record** any differences that you notice in the tones that you hear from each test tube.

⊙ Conclude and Apply

1. **Describe** how the tones change depending on the amount of water in the test tube.

2. **Explain** why the pitch depends on the height of the water.

3. **Summarize** why each test tube produces a different tone.

4. **Explain** how resonance amplifies the sound from a test tube.

5. **Explain** how the natural frequencies of the air columns in each of the tubes differ.

6. **Compare and contrast** the way the test tubes make music with the way a flute makes music.

⊙ *C*ommunicating Your Data

When you are listening to music with family or friends, describe to them what you have learned about how musical instruments produce sound.

Using Sound

Acoustics

When an orchestra stops playing, does it seem as if the sound of its music lingers for a couple of seconds? The sounds and their reflections reach your ears at different times, so you hear echoes. This echoing effect produced by many reflections of sound is called reverberation (rih vur buh RAY shun).

During an orchestra performance, reverberation can ruin the sound of the music. To prevent this problem, scientists and engineers who design concert halls must understand how the size, shape, and furnishings of the room affect the reflection of sound waves. These scientists and engineers specialize in **acoustics** (uh KEW stihks), which is the study of sound. They know that soft, porous materials can reduce excess reverberation, so they might recommend that the walls of concert halls be lined with carpets and draperies. **Figure 18** shows a concert hall that has been designed to create a good listening environment.

Echolocation

At night, bats swoop around in darkness without bumping into anything. They even manage to find insects and other prey in the dark. Their senses of sight and smell help them navigate. Many species of bats also depend on echolocation. **Echolocation** is the process of locating objects by emitting sounds and interpreting the sound waves that are reflected back. Look at **Figure 19** to learn how echolocation works.

Figure 18 This concert hall uses cloth drapes to help reduce reverberations.
Explain *how the drapes absorb or reflect sound waves.*

Figure 19

Many bats emit ultrasonic—very high-frequency—sounds. The sound waves bounce off objects, and bats locate prey by using the returning echoes. Known as echolocation, this technique is also used by dolphins, which produce clicking sounds as they hunt. The diagrams below show how a bat uses echolocation to capture a flying insect.

A Sound waves from a bat's ultrasonic cries spread out in front of it.

B Some of the waves strike a moth and bounce back to the bat.

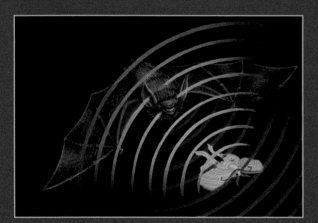

C The bat determines the moth's location by continuing to emit cries, then changes its course to catch the moth.

D By emitting a continuous stream of ultrasonic cries, the bat homes in on the moth and captures its prey.

Sonar

More than 140 years ago, a ship named the *Central America* disappeared in a hurricane off the coast of South Carolina. In its hold lay 21 tons of newly minted gold coins and bars that would be worth $1 billion or more in today's market. When the shipwreck occurred, there was no way to search for the ship in the deep water where it sank. The *Central America* and its treasures lay at the bottom of the ocean until 1988, when crews used sonar to locate the wreck under 2,400 m of water. **Sonar** is a system that uses the reflection of underwater sound waves to detect objects. First, a sound pulse is emitted toward the bottom of the ocean. The sound travels through the water and is reflected when it hits something solid, as shown in **Figure 20.** A sensitive underwater microphone called a hydrophone picks up the reflected signal. Because the speed of sound in water is known, the distance to the object can be calculated by measuring how much time passes between emitting the sound pulse and receiving the reflected signal.

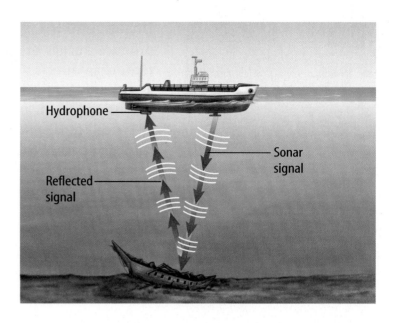

Figure 20 Sonar uses sound waves to find objects that are underwater.
Describe *how sonar is like echolocation.*

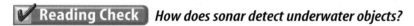

Reading Check *How does sonar detect underwater objects?*

The idea of using sonar to detect underwater objects was first suggested as a way of avoiding icebergs, but many other uses have been developed for it. Navy ships use sonar for detecting, identifying, and locating submarines. Fishing crews also use sonar to find schools of fish, and scientists use it to map the ocean floor. More detail can be revealed by using sound waves of high frequency. As a result, most sonar systems use ultrasonic frequencies.

Ultrasound in Medicine

High-frequency sound waves are used in more than just echolocation and sonar. Ultrasonic waves also are used to break up and remove dirt buildup from jewelry. Chemists sometimes use ultrasonic waves to clean glassware. One of the important uses of ultrasonic waves, though, is in medicine. Using special instruments, medical professionals can send ultrasonic waves into a specific part of a patient's body. Reflected ultrasonic waves are used to detect and monitor conditions such as pregnancy, certain types of heart disease, and cancer.

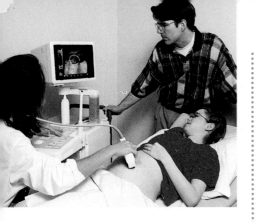

Figure 21 Ultrasonic waves are directed into a pregnant woman's uterus to form images of her fetus.

Ultrasound Imaging Like X rays, ultrasound can be used to produce images of internal structures. A medical ultrasound technician directs the ultrasound waves toward a target area of a patient's body. The sound waves reflect off the targeted organs or tissues, and the reflected waves are used to produce electrical signals. A computer program converts these electrical signals into video images, called sonograms. Physicians trained to interpret sonograms can use them to detect a variety of medical problems.

✅ **Reading Check** *How does ultrasound imaging use reflected waves?*

Medical professionals use ultrasound to examine many parts of the body, including the heart, liver, gallbladder, pancreas, spleen, kidneys, breast, and eye. Ultrasound also is used to monitor the development of a fetus, as shown in **Figure 21.** However, ultrasound does not produce good images of the bones and lungs, because hard tissues and air absorb the ultrasonic waves instead of reflecting them.

Applying Math — Solve a Simple Equation

USING SONAR A sonar pulse returns in 3 s from a sunken ship directly below. Find the depth of the ship if the speed of the pulse is 1,439 m/s. Hint: the sonar pulse travels a distance equal to twice the depth of the ship, so use the equation $d = st/2$ to find the depth.

IDENTIFY known values and the unknown value

Identify the known values:

speed of sound in water ⟶means⟶ $v = 1{,}439$ m/s

time for round trip ⟶means⟶ $t = 3$ s

Identify the unknown value:

The depth of the ship ⟶means⟶ $d = ?$ m

SOLVE the problem

Substitute the known values, $t = 3$ s and $s = 1{,}439$ m/s, into the equation for the depth:

$$d = \frac{st}{2} = \frac{(1{,}439 \text{ m/s})(3 \text{ s})}{2} = \frac{1}{2}\,4{,}317 \text{ m} = 2{,}158.5 \text{ m}$$

CHECK your answer

Divide the distance by the time and multiply by 2. You should get the same speed of sound.

Practice Problem

Find the speed of a sonar pulse that returns in 2.0 s from a sunken ship at 1,500 m depth.

For more practice problems go to page 834, and visit gpscience.com/extra_problems.

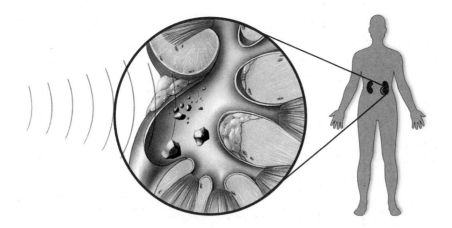

Figure 22 Ultrasonic waves can be used to break up kidney stones. **Explain** *why ultrasound therapy is beneficial for treating kidney stones.*

Treating with Ultrasound High-frequency sound waves can be used to treat certain medical problems. For example, sometimes small, hard deposits of calcium compounds or other minerals form in the kidneys, making kidney stones. In the past, physicians had to perform surgery to remove kidney stones. But now ultrasonic treatments are commonly used to break them up instead. Bursts of ultrasound create vibrations that cause the stones to break into small pieces, as shown in **Figure 22.** These fragments then pass out of the body with the urine. A similar treatment is available for gallstones. Patients who are treated successfully with ultrasound recover more quickly than those who must have surgery.

INTEGRATE Health

Doppler Waves Physicians can measure blood flow by studying the Doppler effect in ultrasonic waves. Brainstorm how the Doppler-shifted waves could help doctors diagnose diseases of the arteries and monitor their healing.

section 4 review

Summary

Acoustics

- Acoustics is a field in which scientists and engineers work to control the quality of sound in spaces.

Echolocation

- Bats locate objects by emitting sounds and then interpreting their reflected sound waves.

Sonar

- Humans using sonar can interpret reflected sound to locate objects underwater.

Ultrasound in Medicine

- High-frequency sound waves are useful for detecting and monitoring certain medical conditions.
- Medical ultrasound is also used to cure problems like kidney stones and gallstones.

Self Check

1. **Describe** some differences between a gym and a concert hall that might affect the amount of reverberation in each.

2. **Explain** how echolocation helps bats to find food and avoid obstacles.

3. **Explain** how sound waves can be used to find underwater objects.

4. **Describe** at least three uses of ultrasonic technology in medicine.

5. **Think Critically** How is sonar technology useful in locating deposits of oil and minerals?

Applying Math

6. **Calculate Distance** Sound travels at about 1,500 m/s in seawater. How far will a sonar pulse travel in 45 s?

7. **Calculate Time** How long will it take for an undersea sonar pulse to travel 3 km?

Blocking N⏰ise Pollution

Goals

- **Design** an experiment that tests the effectiveness of various types of barriers and materials for blocking out noise pollution.
- **Test** different types of materials and barriers to determine the best noise blockers.

Possible Materials

radio, CD player, horn, drum, or other loud noise source

shrubs, trees, concrete walls, brick walls, stone walls, wooden fences, parked cars, or hanging laundry

sound meter

meterstick or metric tape measure

⊙ Real-World Question

What loud noises do you enjoy, and which ones do you find annoying? Most people enjoy a music concert performed by their favorite artist, booming displays of fireworks on the Fourth of July, and the roar of a crowd when their team scores a goal or touchdown. Although these are loud noises, most people enjoy them for short periods of time. Most people find certain loud noises, such as traffic, sirens, and loud talking, annoying. Constant, annoying noises are called noise pollution. What can be done to reduce noise pollution? What types of barriers best block out loud noises? What types of barriers will best block out noise pollution?

⊙ Form a Hypothesis

Based on your experiences with loud noises, form a hypothesis that predicts the effectiveness of different types of barriers at blocking out noise pollution.

⊙ Test Hypothesis

Make a Plan

1. **Decide** what type of barriers or materials you will test.

2. **Describe** exactly how you will use these materials.

3. **Identify** the controls and variables you will use in your experiment.

4. **List** the steps you will use and describe each step precisely.

5. **Prepare** a data table in your Science Journal to record your measurements.

6. **Organize** the steps of your experiment in logical order.

Follow Your Plan

1. Ask your teacher to approve your plan and data table before you start.

2. **Conduct** your experiment as planned.

3. **Test** each barrier two or three times.

4. **Record** your test results in your data table in your Science Journal.

◉ *Analyze Your Data*

1. **Identify** the barriers that most effectively reduced noise pollution.

2. **Identify** the barriers that least effectively reduced noise pollution.

3. **Compare** the effective barriers and identify common characteristics that might explain why they reduced noise pollution.

4. **Compare** the natural barriers you tested with the artificial barriers. Which type of barrier best reduced noise pollution?

5. **Compare** the different types of materials the barriers were made of. Which type of material best reduced noise pollution?

◉ *Conclude and Apply*

1. **Evaluate** whether your results support your hypothesis.

2. **Predict** how your results would differ if you use a louder source of noise such as a siren.

3. **Infer** from your results how people living near a busy street could reduce noise pollution.

4. **Identify** major sources of noise pollution in or near your home. How could this be reduced?

5. **Research** how noise pollution can be unhealthy.

Communicating Your Data

Draw a poster illustrating how builders and landscapers could use certain materials to better insulate a home or office from excess noise pollution.

Noise Pollution
AND HEARING LOSS

Now hear this: More than 28 million Americans have hearing loss. Twelve million more people have a condition called tinnitus (TIN uh tus), or ringing in the ears. In at least 10 million of the 28 million cases mentioned above, hearing loss could have been prevented, because it was caused by noise pollution.

People take their music very seriously in the United States. And a lot of people like it loud. So loud, in fact, that it can damage their hearing. The medical term is *auditory overstimulation.* You may have experienced a high-pitched ringing in your ears for days after standing too close to a loudspeaker at a concert. That's how hearing loss starts.

Music isn't the only cause of hearing loss caused by noise pollution. Other kinds of environmental noise can be strong enough to damage the ears, too. Intense, short-duration noise, like the sound of a gunshot or even a thunderclap, can cause some hearing loss. All of the structures of the inner ear can be damaged this way.

A less intense but longer duration noise, like the sound of a lawn mower, a low-flying plane, or a drill blasting away at cement, can possibly damage the ear, as well.

All the Better to Hear You

There are 20,000 to 30,000 sensory receptors, or hair cells, located in the inner ear, or the cochlea. When vibrations reach these hair cells, electrical impulses are triggered.

The impulses send messages to the auditory center of the brain. But the human ear was not made to withstand all the very loud sounds of the modern world. Once hair cells are damaged, they don't grow back.

What can you do to avoid hearing damage? Well, the first thing is to turn the volume down on the stereo and TV. And keep the volume low when you've got your earphones on, no matter how tempting it is to blast it. Also, if you go to a rock concert, you can wear earplugs to muffle the sound. You'll still hear everything, but you won't damage your ears. And earplugs are now small enough that they're pretty much undetectable. So enjoy all that music and the street sounds of urban life, but mind your ears.

Test Put on a blindfold and have a friend test your hearing. Have your friend choose several noise-making objects. (Not too loud, please, and make sure you have your teacher's permission.) Guess what object is making each sound.

nline

For more information, visit **gpscience.com/time**

Reviewing Main Ideas

Section 1 ▸ The Nature of Sound

1. Sound is a compressional wave created by something that is vibrating.

2. Sound travels fastest through solids and slowest through gases. You see the flash of lightning before you hear the clap of thunder because light travels faster than sound.

3. The human ear can be divided into three sections—the outer ear, the middle ear, and the inner ear. Each section plays a specific role in hearing.

4. Hearing involves four stages: gathering sound waves, amplifying them, converting them to nerve impulses, and interpreting the signals in the brain.

Section 2 ▸ Properties of Sound

1. Intensity is a measure of how much energy a wave carries. Humans interpret the intensity of sound waves as loudness.

2. The pitch of a sound becomes higher as the frequency increases.

3. The Doppler effect is a change in frequency that occurs when a source of sound is moving relative to a listener.

Section 3 ▸ Music

1. Music is made of sounds used deliberately in a regular pattern.

2. Instruments, such as this flute, use a variety of methods to produce and amplify sound waves.

3. When sound waves of similar frequencies overlap, they interfere with each other to form beats.

Section 4 ▸ Using Sound

1. Acoustics is the study of sound. This gym is great for playing basketball, but the sound quality would be poor for a concert.

2. Sonar uses reflected sound waves to detect objects.

3. Ultrasound waves can be used for imaging body tissues or treating medical conditions.

FOLDABLES Use the Foldable that you made at the beginning of this chapter to help you review sound.

chapter 11 Review

Using Vocabulary

acoustics p. 339	music p. 333
cochlea p. 326	overtone p. 334
decibel p. 329	pitch p. 330
Doppler effect p. 331	resonator p. 335
eardrum p. 325	sonar p. 341
echolocation p. 339	sound quality p. 334
intensity p. 328	ultrasonic p. 330
loudness p. 329	

Fill in the blanks with the correct vocabulary word or words.

1. The _____ is filled with fluid and contains tiny hair cells that vibrate.

2. _____ is the study of sound.

3. A change in pitch or wave frequency due to a moving wave source is called _____.

4. _____ is a combination of sounds and pitches that follows a specified pattern.

5. Differences among sounds of the same pitch and loudness are described as _____.

6. _____ is how humans perceive the intensity of sound.

7. _____ is a scale for sound intensity.

Checking Concepts

Choose the word or phrase that best answers the question.

8. For a sound with a low pitch, what else is always low?
   A) amplitude   C) wavelength
   B) frequency   D) wave velocity

9. Sound intensity decreases when which of the following decreases?
   A) wave velocity   C) quality
   B) wavelength   D) amplitude

10. When specific pitches and sounds are put together in a pattern, what are they called?
   A) overtones   C) white noise
   B) music   D) resonance

11. Sound can travel through all but which of the following?
   A) solids   C) gases
   B) liquids   D) empty space

12. What is the term for variations in the loudness of sound caused by wave interference?
   A) beats
   B) standing waves
   C) pitch
   D) forced vibrations

13. What does the outer ear do to sound waves?
   A) scatter them   C) gather them
   B) amplify them   D) convert them

14. Which of the following occurs when a sound source moves away from you?
   A) The sound's velocity decreases.
   B) The sound's loudness increases.
   C) The sound's frequency decreases.
   D) The sound's frequency increases.

15. Sounds with the same pitch and loudness traveling in the same medium may differ in which of these properties?
   A) frequency   C) quality
   B) amplitude   D) wavelength

16. What part of a musical instrument amplifies sound waves?
   A) resonator   C) mallet
   B) string   D) finger hole

17. What is the name of the method used to find objects that are underwater?
   A) sonogram
   B) ultrasonic bath
   C) sonar
   D) percussion

Science Online gpscience.com/vocabulary_puzzlemaker

Interpreting Graphics

18. Copy and complete the following table on musical instruments.

Characteristics of Musical Instruments	Guitar	Flute	Bongo Drum
How Played	plucked	blown into	
Role of Resonator	amplifies sound		amplifies sound
Type of Instrument		wind	percussion

Use the table below to answer question 19.

Federally Recommended Noise Exposure Limits	
Sound Level (dB)	Time Permitted (hours per day)
90	8
95	4
100	2
105	1
110	0.5

19. You use a lawn mower with a sound level of 100 dB. Using the table above, determine the maximum number of hours a week you can safely work mowing lawns without ear protection.

Thinking Critically

20. Infer A car comes to a railroad crossing. The driver hears a train's whistle and its pitch becomes lower. What can be assumed about how the train is moving?

21. Form Hypotheses Sound travels slower in air at high altitudes than at low altitudes. Form a hypothesis to explain this.

22. Apply Acoustic scientists sometimes do research in rooms that absorb all sound waves. How could such a room be used to study how bats find their food?

23. Explain why windows might begin to rattle when an airplane flies overhead.

24. Communicate Some people enjoy using snowmobiles. Others object to the noise that they make. Write a proposal for a policy that seems fair to both groups for the use of snowmobiles in a state park.

Applying Math

Use the wave speed equation $v = f\lambda$ to answer questions 25–27.

25. Calculate Frequency A sonar pulse has a wavelength of 3.0 cm. If the pulse has a speed in water of 1,500 m/s, calculate its frequency.

26. Calculate Wavelength What is the wavelength of a sound wave with a frequency of 440 Hz if the speed of sound in air is 340 m/s?

27. Calculate Wave Speed An earthquake produces a seismic wave that has a wavelength of 650 m and a frequency of 10 Hz. How fast does the wave travel?

28. Calculate Distance The sound of thunder travels at a speed of 340 m/s and reaches you in 2.6 s. How far away is the storm?

29. Calculate Time The shipwrecked Central America was discovered lying beneath 2,400 m of water. If the speed of sound in seawater is 1,500 m/s, how long will it take a sonar pulse to travel to the shipwreck and return?

Part 1 Multiple Choice

Record your answers on the answer sheet provided by your teacher or on a sheet of paper.

1. What does a sound's frequency determine?
   A. pitch   **C.** intensity
   B. amplitude   **D.** energy

2. Which medium does sound travel fastest through?
   A. empty space   **C.** gases
   B. liquids   **D.** solids

Use the table below to answer questions 3 and 4.

Sound	Loudness (dB)
Jet taking off	150
Pain threshold	120
Chain saw	115
Power mower	110
Vacuum cleaner	75
Average home	50
Purring cat	25

3. Which of the following would sound the loudest?
   A. vacuum cleaner
   B. chain saw
   C. power mower
   D. purring cat

4. Which of the following statements is true about a sound of 65 decibels?
   A. It causes intense pain.
   B. It can cause permanent hearing loss.
   C. It cannot be heard by anyone.
   D. It can be heard without discomfort or damage.

5. To which group of instruments does a clarinet belong?
   A. electronic   **C.** stringed
   B. percussion   **D.** wind

6. Which of the following will lower the pitch of the sound made by a guitar string?
   A. shortening the vibrating string
   B. tightening the string
   C. plucking the string harder
   D. loosening the string

Use the illustration below to answer questions 7 and 8.

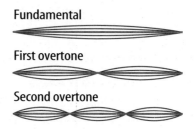

Fundamental

First overtone

Second overtone

7. If the fundamental frequency of a guitar string is 262 Hz, what is the frequency of the first overtone?
   A. 262 Hz   **C.** 786 Hz
   B. 524 Hz   **D.** 1048 Hz

8. What is the frequency of the second overtone if the fundamental frequency is 294 Hz?
   A. 294 Hz   **C.** 882 Hz
   B. 588 Hz   **D.** 1176 Hz

9. If you were on a moving train, what would happen to the pitch of the bell at a crossing as you approached and then passed by the crossing?
   A. pitch would increase and then decrease
   B. pitch would remain the same
   C. pitch would decrease and then increase
   D. pitch would keep decreasing

10. On what does a guitar string's natural frequency depend?
    A. thickness   **C.** length
    B. tightness   **D.** all of these

Part 2 Short Response/Grid In

Record your answers on the answer sheet provided by your teacher or on a sheet of paper.

Use the table below to answer questions 11–13.

Medium	Speed of Sound (m/s)
Air	347
Water	1,498
Iron	5,103

11. A "fishfinder" sends out a pulse of ultra-sound and measures the time needed for the sound to travel to a school of fish and back to the boat. If the fish are 16 m below the surface, how long would it take sound to make the round trip in the water?

12. Suppose you are sitting in the bleachers at a baseball game 150 m from home plate. How long after the batter hits the ball do you hear the "crack" of the ball and bat?

13. Suppose a friend is 500 m away along a railroad track while you have your ear to the track. He drops a stone on the tracks. How long will the sound take to reach your ear?

14. Explain why placing your hand on a bell that has just been rung will stop the sound immediately.

15. Why do different musical instruments have different sound qualities?

Test-Taking Tip

Make Sure the Units Match Read carefully and make note of the units used in any measurement.

Question 11 Compare the units given in the table with those in the problem. If the units do not match, you will have to do unit conversions.

Part 3 Open Ended

Record your answers on a sheet of paper.

16. Explain why people who work on the ground near jet runways wear big ear muffs filled with a sound insulator.

Use the illustration below to answer questions 17 and 18.

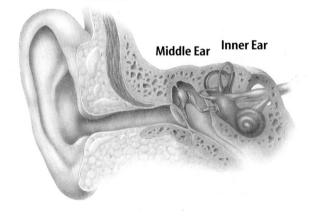

Outer Ear

Middle Ear   Inner Ear

17. Describe the path of sound from the time it enters the ear until a message reaches the brain.

18. Would sound waves traveling through the outer ear travel faster or slower that those traveling through the inner ear? Explain.

19. Science-fiction movies often show battles in space where an enemy spaceship explodes with a very loud sound. Explain whether or not this scenario is accurate.

20. Compare the way a violin, a flute, and drum produce sound waves. What acts as a resonator in each instrument?

21. How can interference produce a sound with decreased loudness? How can interference produce a sound with increased loudness?

22. Imagine you have been hired by a school to reduce the amount of reverberation that occurs in the classrooms. What recommendations would you make?

Electromagnetic Waves

How's the reception?

These giant 25-m dishes aren't picking up TV signals, unless they're coming from distant stars and galaxies. They are part of a group of 27 antennas that detect radio waves, which are electromagnetic waves. However, all objects emit electromagnetic waves, not just stars and galaxies.

Science Journal List six objects around you that emit light or feel warm.

Start-Up Activities

Can electromagnetic waves change materials?

You often hear about the danger of the Sun's ultraviolet rays, which can damage the cells of your skin. When the exposure isn't too great, your cells can repair themselves, but too much at one time can cause a painful sunburn. Repeated overexposure to the Sun over many years can damage cells and cause skin cancer. In the lab below, observe how energy carried by ultraviolet waves can cause changes in other materials.

1. Cut a sheet of red construction paper in half.
2. Place one piece outside in direct sunlight. Place the other in a shaded location.
3. Keep the construction paper in full sunlight for at least 45 min. If possible, allow it to stay there for 3 h or more before taking it down. Be sure the other piece remains in the shade.
4. **Think Critically** In your Science Journal, describe any differences you notice in the two pieces of construction paper. Comment on your results.

FOLDABLES™ Study Organizer

Electromagnetic Waves Make the following Foldable to help you understand electromagnetic waves.

STEP 1 Fold a vertical sheet of paper in half from top to bottom.

STEP 2 Fold in half from side to side with the fold at the top.

STEP 3 Unfold the paper once. Cut only the fold of the top flap to make two tabs.

STEP 3 Write on the front tabs as shown.

How do electromagnetic waves travel through space?

How do electromagnetic waves transfer energy to matter?

Identify Questions As you read the chapter, write answers to the questions on the back of the appropriate tabs.

Science Online

Preview this chapter's content and activities at gpscience.com

What are electromagnetic waves?

Reading Guide

What You'll Learn

- **Describe** how electric and magnetic fields form electromagnetic waves.
- **Explain** how vibrating charges produce electromagnetic waves.
- **Describe** properties of electromagnetic waves.

Why It's Important

You, and all the objects and materials around you, are radiating electromagnetic waves.

Review Vocabulary

Hertz: the SI unit of frequency, abbreviated Hz; 1 Hz equals one vibration per second

New Vocabulary

- electromagnetic wave
- radiant energy
- photon

Waves in Space

Stay calm. Do not panic. As you are reading this sentence, no matter where you are, you are surrounded by electromagnetic waves. Even though you can't feel them, some of these waves are traveling right through your body. They enable you to see. They make your skin feel warm. You use electromagnetic waves when you watch television, talk on a cordless phone, or prepare popcorn in a microwave oven.

Sound and Water Waves Waves are produced by something that vibrates, and they carry energy from one place to another. Look at the sound wave and the water wave in **Figure 1.** Both waves are moving through matter. The sound wave is moving through air and the water wave through water. These waves travel because energy is transferred from particle to particle. Without matter to transfer the energy, they cannot move.

Electromagnetic Waves However, electromagnetic waves do not require matter to transfer energy. **Electromagnetic waves** are made by vibrating electric charges and can travel through space where matter is not present. Instead of transferring energy from particle to particle, electromagnetic waves travel by transferring energy between vibrating electric and magnetic fields.

Figure 1 Water waves and sound waves require matter to move through. Energy is transferred from one particle to the next as the wave travels through the matter.

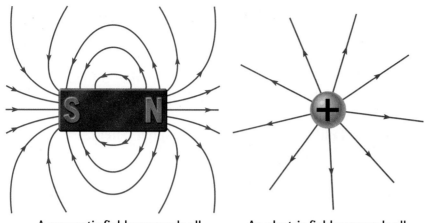

A magnetic field surrounds all magnets.

An electric field surrounds all charges.

Figure 2 Fields enable magnets and charges to exert forces at a distance. These fields extend throughout space.
Explain *how you could detect a magnetic field.*

Electric and Magnetic Fields

When you bring a magnet near a metal paper clip, the paper clip moves toward the magnet and sticks to it. The paper clip moved because the magnet exerted a force on it. The magnet exerted this force without having to touch the paper clip. The magnet exerts a force without touching the paper clip because all magnets are surrounded by a magnetic field, as shown in **Figure 2.** Magnetic fields exist around magnets even if the space around the magnet contains no matter.

Just as magnets are surrounded by magnetic fields, electric charges are surrounded by electric fields, also shown in **Figure 2.** An electric field enables charges to exert forces on each other even when they are far apart. Just as a magnetic field around a magnet can exist in empty space, an electric field exists around an electric charge even if the space around it contains no matter.

Magnetic Fields and Moving Charges

Electric charges also can be surrounded by magnetic fields. An electric current flowing through a wire is surrounded by a magnetic field, as shown in **Figure 3.** An electric current in a wire is the flow of electrons in a single direction. It is the motion of these electrons that creates the magnetic field around the wire. In fact, any moving electric charge is surrounded by a magnetic field, as well as an electric field.

Figure 3 Electrons moving in a wire are surrounded by a magnetic field.
Describe *how you would confirm that a magnetic field exists around a current-carrying wire.*

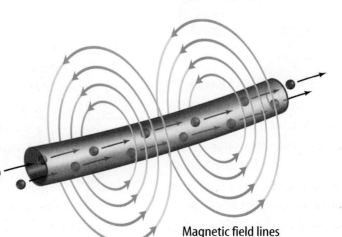

Magnetic field lines

Investigating Electromagnetic Waves

Procedure

1. Point your **television remote control** in different directions and observe whether it will still control the **television.**

2. Place various materials in front of the infrared receiver on the television and observe whether the remote still will control the television. Some materials you might try are **glass, a book, your hand, paper,** and a **metal pan.**

Analysis

1. Was it necessary for the remote to be pointing exactly toward the receiver to control the television? Explain.

2. Did the remote continue to work when the various materials were placed between it and the receiver? Explain why or why not.

Try at Home

Figure 4 A vibrating electric charge creates an electromagnetic wave that travels outward in all directions from the charge. The wave in only one direction is shown here.
Determine *whether an electromagnetic wave is a transverse wave or a compressional wave.*

Changing Electric and Magnetic Fields A changing magnetic field creates a changing electric field. For example, in a transformer, changing electric current in the primary coil produces a changing magnetic field. This changing magnetic field then creates a changing electric field in the secondary coil that produces current in the coil. The reverse is also true—a changing electric field creates a changing magnetic field.

Making Electromagnetic Waves

Waves such as sound waves are produced when something vibrates. Electromagnetic waves also are produced when something vibrates—an electric charge that moves back and forth.

Reading Check *What produces an electromagnetic wave?*

When an electric charge vibrates, the electric field around it changes. Because the electric charge is in motion, it also has a magnetic field around it. This magnetic field also changes as the charge vibrates. As a result, the vibrating electric charge is surrounded by changing electric and magnetic fields.

How do the vibrating electric and magnetic fields around the charge become a wave that travels through space? The changing electric field around the charge creates a changing magnetic field. This changing magnetic field then creates a changing electric field. This process continues, with the magnetic and electric fields continually creating each other. These vibrating electric and magnetic fields are perpendicular to each other and travel outward from the moving charge, as shown in **Figure 4.** Because the electric and magnetic fields vibrate at right angles to the direction the wave travels, an electromagnetic wave is a transverse wave.

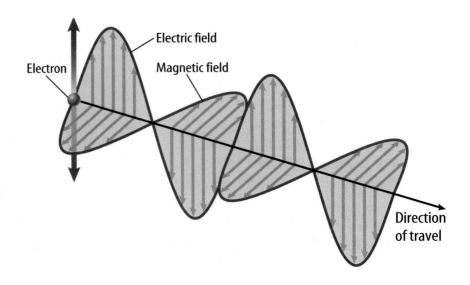

Properties of Electromagnetic Waves

All matter contains charged particles that are always in motion. As a result, all objects emit electromagnetic waves. The wavelengths of the emitted waves become shorter as the temperature of the material increases. As an electromagnetic wave moves, its electric and magnetic fields encounter objects. These vibrating fields can exert forces on charged particles and magnetic materials, causing them to move. For example, electromagnetic waves from the Sun cause electrons in your skin to vibrate and gain energy, as shown in **Figure 5.** The energy carried by an electromagnetic wave is called **radiant energy.** Radiant energy makes a fire feel warm and enables you to see.

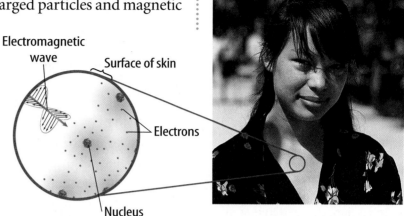

Figure 5 As an electromagnetic wave strikes your skin, electrons in your skin gain energy from the vibrating electric and magnetic fields.

Electromagnetic wave
Surface of skin
Electrons
Nucleus

Applying Math   Using Scientific Notation

THE SPEED OF LIGHT IN WATER The speed of light in water is 226,000 km/s. Write this number in scientific notation.

IDENTIFY known values and the unknown value

Identify the known values:
The speed of light in water is 226,000 km/s.

Identify the unknown value:
the number 226,000 written in scientific notation

SOLVE the problem

A number written in scientific notation has the form $M \times 10^N$. N is the number of places the decimal point in the number has to be moved so that the number M that results has only one digit to the left of the decimal point.

Write the number in scientific notation form:	$226,000. \times 10^N$
Move the decimal point five places to the left:	2.26000×10^N
The decimal point was moved 5 places, so N equals 5:	2.26000×10^5
Delete remaining zeroes at the end of the number.	2.26×10^5

CHECK your answer

Add zeroes to the end of the number and move the decimal point in the opposite direction five places. The result should be the original number.

Practice Problems

Write the following numbers in scientific notation: 433; 812,000,000; 73,000,000,000; 84,500.
For more practice problems go to page 834, and visit gpscience.com/extra_problems.

Table 1 Speed of Visible Light	
Material	Speed (km/s)
Vacuum	300,000
Air	slightly less than 300,000
Water	226,000
Glass	200,000
Diamond	124,000

Wave Speed All electromagnetic waves travel at 300,000 km/s in the vacuum of space. Because light is an electromagnetic wave, the speed of electromagnetic waves in space is usually called the "speed of light." The speed of light is nature's speed limit—nothing travels faster than the speed of light. In matter, the speed of electromagnetic waves depends on the material they travel through. Electromagnetic waves usually travel the slowest in solids and the fastest in gases. **Table 1** lists the speed of visible light in various materials.

Reading Check *What is the speed of light?*

Wavelength and Frequency Like all waves, electromagnetic waves can be described by their wavelength and frequency. The wavelength of an electromagnetic wave is the distance from one crest to another, as shown in **Figure 6.**

The frequency of any wave is the number of wavelengths that pass a point in 1 s. The frequency of an electromagnetic wave also equals the frequency of the vibrating charge that produces the wave. This frequency is the number of vibrations, or back and forth movements, of the charge in one second. The frequency and wavelength of electromagnetic waves are related. As the frequency increases, the wavelength becomes smaller.

Waves and Particles

The difference between a wave and a particle might seem obvious—a wave is a disturbance that carries energy, and a particle is a piece of matter. However, in reality the difference is not so clear.

Waves as Particles In 1887, Heinrich Hertz found that by shining light on a metal, electrons were ejected from the metal. Hertz found that whether or not electrons were ejected depended on the frequency of the light and not the amplitude. Because the energy carried by a wave depends on its amplitude and not its frequency, this result was mysterious. Years later, Albert Einstein provided an explanation—electromagnetic waves can behave as a particle, called a **photon,** whose energy depends on the frequency of the waves.

Figure 6 The wavelength of an electromagnetic wave is the distance between the crests of the vibrating electric field or magnetic field.

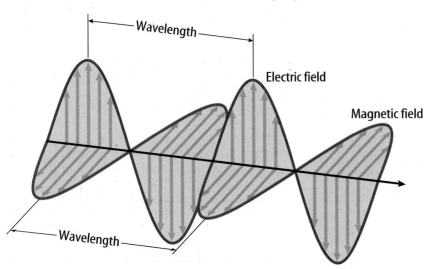

Wavelength

Electric field

Magnetic field

Wavelength

Particles of paint sprayed through two slits coat only the area behind the slits.

Electrons fired at two closely-spaced openings form a wave-like interference pattern.

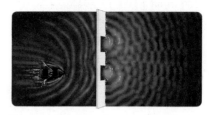

Water waves produce an interference pattern after passing through two openings.

Particles as Waves Because electromagnetic waves could behave as a particle, others wondered whether matter could behave as a wave. If a beam of electrons were sprayed at two tiny slits, you might expect that the electrons would strike only the area behind the slits, like the spray paint in **Figure 7.** Instead, it was found that the electrons formed an interference pattern. This type of pattern is produced by waves when they pass through two slits and interfere with each other, as the water waves do in **Figure 7.** This experiment showed that electrons can behave like waves. It is now known that all particles, not only electrons, can behave like waves.

Figure 7 When electrons are sent through two narrow slits, they behave as a wave.

section 1 review

Summary

Making Electromagnetic Waves

- Moving electric charges are surrounded by an electric field and a magnetic field.
- A vibrating electric charge produces an electromagnetic wave.
- An electromagnetic wave consists of vibrating electric and magnetic fields that are perpendicular to each other and travel outward from the vibrating electric charge.

Properties of Electromagnetic Waves

- Electromagnetic waves carry radiant energy.
- In empty space electromagnetic waves travel at 300,000 km/s—the speed of light.
- Electromagnetic waves travel slower in matter, with a speed that depends on the material.

Waves and Particles

- Electromagnetic waves can behave as particles that are called photons.
- In some circumstances, particles, such as electrons can behave as waves.

Self Check

1. **Explain** why an electromagnetic wave is a transverse wave and not a compressional wave.
2. **Compare** the frequency of an electromagnetic wave with the frequency of the vibrating charge that produces the wave.
3. **Describe** how electromagnetic waves transfer radiant energy to matter.
4. **Explain** why an electromagnetic wave can travel through empty space that contains no matter.
5. **Think Critically** Suppose a moving electric charge was surrounded only by an electric field. Infer whether or not a vibrating electric charge would produce an electromagnetic wave.

Applying Math

6. **Calculate Time** How many minutes does it take an electromagnetic wave to travel 150,000,000 km?
7. **Use Scientific Notation** Calculate the distance an electromagnetic wave in space would travel in one day. Express your answer in scientific notation.

The Electromagnetic Spectrum

What You'll Learn

- **Describe** the waves in the different regions of the electromagnetic spectrum.
- **Compare** the properties of different electromagnetic waves.
- **Identify** uses for different types of electromagnetic waves.

Why It's Important

Every day, waves in different regions of the electromagnetic spectrum are used in many ways.

Review Vocabulary

spectrum: a continuous sequence arranged by a particular property

New Vocabulary

- radio wave
- microwave
- infrared wave
- visible light
- ultraviolet wave
- X ray
- gamma ray

A Range of Frequencies

Electromagnetic waves can have a wide variety of frequencies. They might vibrate once each second or trillions of times each second. The entire range of electromagnetic wave frequencies is known as the electromagnetic spectrum, shown in **Figure 8.** Various portions of the electromagnetic spectrum interact with matter differently. As a result, they are given different names. The electromagnetic waves that humans can detect with their eyes, called visible light, are a small portion of the entire electromagnetic spectrum. However, various devices have been developed to detect the other frequencies. For example, the antenna of your radio detects radio waves.

Figure 8 Electromagnetic waves are described by different names depending on their frequency and wavelength.

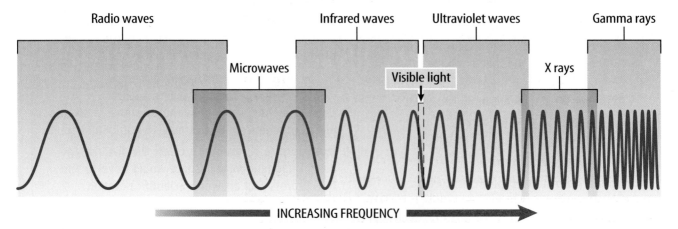

Radio waves Microwaves Infrared waves Visible light Ultraviolet waves X rays Gamma rays

INCREASING FREQUENCY

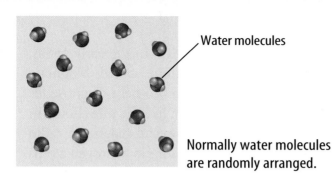

Water molecules

Normally water molecules are randomly arranged.

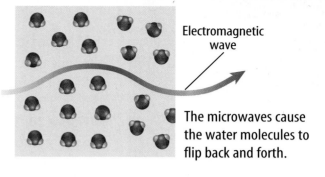

Electromagnetic wave

The microwaves cause the water molecules to flip back and forth.

Radio Waves

Stop and look around you. Even though you can't see them, radio waves are moving everywhere you look. Some radio waves carry an audio signal from a radio station to a radio. However, even though these radio waves carry information that a radio uses to create sound, you can't hear radio waves. You hear a sound wave when the compressions and rarefactions the sound wave produces reach your ears. A radio wave does not produce compressions and rarefactions as it travels through air.

Microwaves **Radio waves** are low-frequency electromagnetic waves with wavelengths longer than about 1 mm. Radio waves with wavelengths of less than about 30 cm are called **microwaves.** Microwaves with wavelengths of about 1 cm to 20 cm are widely used for communication, such as for cellular telephones and satellite signals. You are probably most familiar with microwaves because of their use in microwave ovens.

☑ Reading Check *What is the difference between a microwave and a radio wave?*

Microwave ovens heat food when microwaves interact with water molecules in food, as shown in **Figure 9.** Each water molecule is positively charged on one side and negatively charged on the other side. The vibrating electric field inside a microwave oven causes water molecules in food to rotate back and forth billions of times each second. This rotation causes a type of friction between water molecules that generates thermal energy. It is the thermal energy produced by the interactions between the water molecules that causes your food to cook.

Radar Another use for radio waves is to find the position and movement of objects by a method called radar. Radar stands for **RA**dio **D**etecting **A**nd **R**anging. With radar, radio waves are transmitted toward an object. By measuring the time required for the waves to bounce off the object and return to a receiving antenna, the location of the object can be found. Law enforcement officers use radar to measure how fast a vehicle is moving. Radar also is used for tracking the movement of aircraft, watercraft, and spacecraft.

Figure 9 Microwave ovens use electromagnetic waves to heat food.

Heating with Microwaves

Procedure

1. Obtain two small **beakers** or baby-food jars. Place 50 mL of **dry sand** into each. To one of the jars, add 20 mL of **room-temperature water** and stir well.
2. Record the temperature of the sand in each jar.
3. Together, **microwave** both jars of sand for 10 s and immediately record the temperature again.

Analysis

1. Compare the initial and final temperatures of the wet and dry sand.
2. Infer why there was a difference.

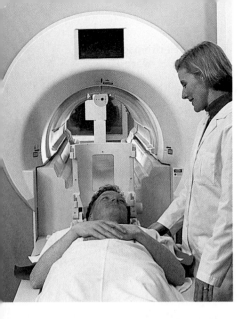

Figure 10 Magnetic resonance imaging technology uses radio waves as an alternative to X-ray imaging.

Magnetic Resonance Imaging (MRI) In the early 1980s, medical researchers developed a technique called Magnetic Resonance Imaging, which uses radio waves to help diagnose illness. The patient lies inside a large cylinder, like the one shown in **Figure 10.** Housed in the cylinder is a powerful magnet, a radio wave emitter, and a radio wave detector. Protons in hydrogen atoms in bones and soft tissue behave like magnets and align with the strong magnetic field. Energy from radio waves causes some of the protons to flip their alignment. As the protons flip, they release radiant energy. A radio receiver detects this released energy. The amount of energy a proton releases depends on the type of tissue it is part of. The released energy detected by the radio receiver is used to create a map of the different tissues. A picture of the inside of the patient's body is produced painlessly.

Infrared Waves

Most of the warm air in a fireplace moves up the chimney, yet when you stand in front of a fireplace, you feel the warmth of the blazing fire. Why do you feel the heat? The warmth you feel is thermal energy transmitted to you by **infrared waves,** which are a type of electromagnetic wave with wavelengths between about 1 mm and about 750 billionths of a meter.

You use infrared waves every day. A remote control emits infrared waves to control your television. A computer uses infrared waves to read CD-ROMs. In fact, every object emits infrared waves. Hotter objects emit more infrared waves than cooler objects emit. The wavelengths emitted also become shorter as the temperature increases. Infrared detectors can form images of objects from the infrared radiation they emit. Infrared sensors on satellites can produce infrared images that can help identify the vegetation over a region. **Figure 11** shows how cities appear different from surrounding vegetation in infrared satellite imagery.

Figure 11 Infrared images and visible light images can provide different types of information.

This visible light image of the region around San Francisco Bay in California was taken from an aircraft at an altitude of 20,000 m.

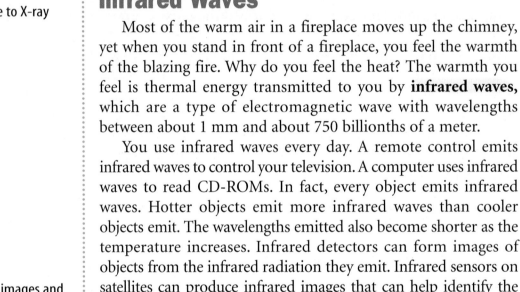

This infrared image of the same area was taken from a satellite. In this image, vegetation is red and buildings are gray.

Visible Light

Visible light is the range of electromagnetic waves that you can detect with your eyes. Light differs from radio waves and infrared waves only by its frequency and wavelength. Visible light has wavelengths around 750 billionths to 400 billionths of a meter. Your eyes contain substances that react differently to various wavelengths of visible light, so you see different colors. These colors range from short-wavelength blue to long-wavelength red. If all the colors are present, you see the light as white.

Ultraviolet Waves

Ultraviolet waves are electromagnetic waves with wavelengths from about 400 billionths to 10 billionths of a meter. Ultraviolet waves are energetic enough to enter skin cells. Overexposure to ultraviolet rays can cause skin damage and cancer. Most of the ultraviolet radiation that reaches Earth's surface are longer-wavelength UVA rays. The shorter-wavelength UVB rays cause sunburn, and both UVA and UVB rays can cause skin cancers and skin damage such as wrinkling. Although too much exposure to the Sun's ultraviolet waves is damaging, some exposure is healthy. Ultraviolet light striking the skin enables your body to make vitamin D which is needed for healthy bones and teeth.

Useful UVs A useful property of ultraviolet waves is their ability to kill bacteria on objects such as food or medical supplies. When ultraviolet light enters a cell, it damages protein and DNA molecules. For some single-celled organisms, damage can mean death, which can be a benefit to health. Ultraviolet waves are also useful because they make some materials fluoresce (floor ES). Fluorescent materials absorb ultraviolet waves and reemit the energy as visible light. As shown in **Figure 12,** police detectives sometimes use fluorescent powder to show fingerprints when solving crimes.

CT Scans In certain situations, doctors will perform a CT scan on a patient instead of a traditional X ray. Research to find out more about CT scans. Compare and contrast CT scans with X rays. What are the advantages and disadvantages of a CT scan? Write a paragraph about your findings in your Science Journal.

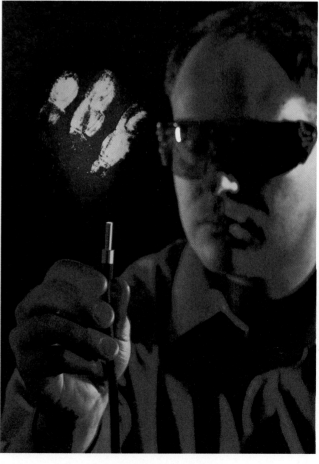

Figure 12 The police detective in this picture is shining ultraviolet light on a fingerprint dusted with fluorescent powder.

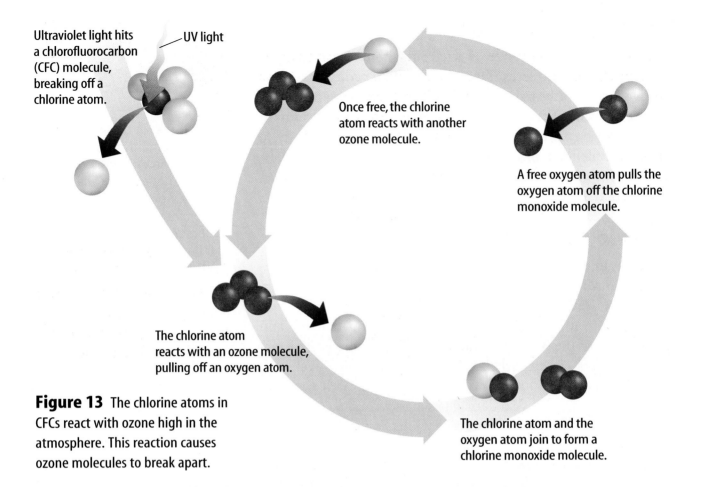

Ultraviolet light hits a chlorofluorocarbon (CFC) molecule, breaking off a chlorine atom.

UV light

Once free, the chlorine atom reacts with another ozone molecule.

A free oxygen atom pulls the oxygen atom off the chlorine monoxide molecule.

The chlorine atom reacts with an ozone molecule, pulling off an oxygen atom.

The chlorine atom and the oxygen atom join to form a chlorine monoxide molecule.

Figure 13 The chlorine atoms in CFCs react with ozone high in the atmosphere. This reaction causes ozone molecules to break apart.

The Ozone Layer About 20 to 50 km above Earth's surface in the stratosphere is a region called the ozone layer. Ozone is a molecule composed of three oxygen atoms. It is continually being formed and destroyed by ultraviolet waves high in the atmosphere. The ozone layer is vital to life on Earth because it absorbs most of the Sun's harmful ultraviolet waves. However, over the past few decades the amount of ozone in the ozone layer has decreased. Averaged globally, the decrease is about three percent, but is greater at higher latitudes.

✔ Reading Check *Why is the ozone layer vital to life on Earth?*

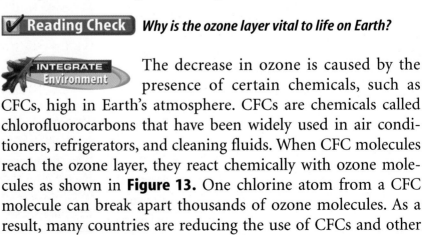

The decrease in ozone is caused by the presence of certain chemicals, such as CFCs, high in Earth's atmosphere. CFCs are chemicals called chlorofluorocarbons that have been widely used in air conditioners, refrigerators, and cleaning fluids. When CFC molecules reach the ozone layer, they react chemically with ozone molecules as shown in **Figure 13.** One chlorine atom from a CFC molecule can break apart thousands of ozone molecules. As a result, many countries are reducing the use of CFCs and other ozone-depleting chemicals.

X Rays and Gamma Rays

The electromagnetic waves with the shortest wavelengths and highest frequencies are X rays and gamma rays. Both X rays and gamma rays are high energy electromagnetic waves. **X rays** have wavelengths between about ten billionths of a meter and ten trillionths of a meter. Doctors and dentists use low doses of X rays to form images of internal organs, bones, and teeth, like the image shown in **Figure 14.** X rays also are used in airport screening devices to examine the contents of luggage.

Electromagnetic waves with wavelengths shorter than about 10 trillionths of a meter are gamma rays. These are the highest-energy electromagnetic waves and can penetrate through several centimeters of lead. Gamma rays are produced by processes that occur in atomic nuclei. Both X rays and gamma rays are used in a technique called radiation therapy to kill diseased cells in the human body. A beam of X rays or gamma rays can damage the biological molecules in living cells, causing both healthy and diseased cells to die. However, by carefully controlling the amount of X ray or gamma ray radiation received by the diseased area, the damage to healthy cells can be reduced.

Figure 14 Bones are more dense than surrounding tissues and absorb more X rays. The image of a bone on an X ray is the shadow cast by the bone as X rays pass through the soft tissue.

---

section 2 review

Summary

Radio Waves and Infrared Waves

- Radio waves are electromagnetic waves with wavelengths longer than about 1 mm.
- Microwaves are radio waves with wavelengths between about 1 mm and 1 m.
- Infrared waves have wavelengths between about 1 mm and 750 billionths of a meter.

Visible Light and Ultraviolet Waves

- Visible light waves have wavelengths between about 750 and 400 billionths of a meter.
- Ultraviolet waves have wavelengths between about 400 and 40 billionths of a meter.
- Most of the harmful ultraviolet waves emitted by the Sun are absorbed by the ozone layer.

X Rays and Gamma Rays

- X rays and gamma rays are the most energetic electromagnetic waves.
- Gamma rays have wavelengths less than 10 trillionths of a meter and are produced in the nuclei of atoms.

Self Check

1. **Explain** A mug of water is heated in a microwave oven. Explain why the water gets hotter than the mug.
2. **Describe** why you can see visible light waves, but not other electromagnetic waves.
3. **List** the beneficial effects and the harmful effects of human exposure to ultraviolet rays.
4. **Identify** three objects in a home that produce electromagnetic waves and describe how the electromagnetic waves are used.
5. **Think Critically** What could an infrared image of their house reveal to the homeowners?

Applying Math

6. **Use Scientific Notation** Express the range of wavelengths corresponding to visible light, ultraviolet waves, and X rays in scientific notation.
7. **Convert Units** A nanometer, abbreviated nm, equals one billionth of a meter, or 10^{-9} meters. Express the range of wavelengths corresponding to visible light, ultraviolet waves and X rays in nanometers.

The Shape of Satellite Dishes

Communications satellites transmit signals with a narrow beam pointed toward a particular area of Earth. To detect this signal, receivers are typically large, parabolic dishes.

▶ Real-World Question

How does the shape of a satellite dish improve reception?

Goals

- **Make** a model of a satellite reflecting dish.
- **Observe** how the shape of the dish affects reception.

Materials

flashlight	small bowl
several books	*large, metal spoon
aluminum foil	*Alternate materials

Safety Precautions

▶ Procedure

1. Cover one side of a book with aluminum foil. Be careful not to wrinkle the foil.

2. Line the inside of the bowl with foil, also keeping it as smooth as possible.

3. Place some of the books on a table. Put the flashlight on top of the books so that its beam of light will shine several centimeters above and across the table.

4. Hold the foil-covered book on its side at a right angle to the top of the table. The foil-covered side should face the beam of light.

5. **Observe** the intensity of the light on the foil.

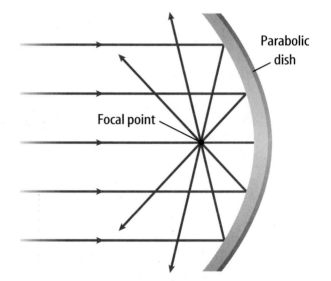

Parabolic dish

Focal point

6. Repeat steps 4 and 5, replacing the foil-covered book with the bowl.

▶ Conclude and Apply

1. **Compare** the brightness of the light reflected from the two surfaces.

2. **Explain** why the light you see from the curved surface is brighter.

3. **Infer** why bowl-shaped dishes are used to receive signals from satellites.

𝒞ommunicating
Your Data

Compare your conclusions with those observed by other students in your class. **For more help, refer to the** Science Skill Handbook.

Radio Communication

Radio Transmission

When you listen to the radio, you hear music and words that are produced at a distant location. The music and words are sent to your radio by radio waves. The metal antenna of your radio detects radio waves. As the electromagnetic waves pass by your radio's antenna, the electrons in the metal vibrate, as shown in **Figure 15.** These vibrating electrons produce a changing electric current that contains the information about the music and words. An amplifier boosts the current and sends it to speakers, causing them to vibrate. The vibrating speakers create sound waves that travel to your ears. Your brain interprets these sound waves as music and words

Dividing the Radio Spectrum Each radio station is assigned to broadcast at one particular radio frequency. Turning the tuning knob on your radio allows you to select a particular frequency to listen to. The specific frequency of the electromagnetic wave that a radio station is assigned is called the **carrier wave.**

The radio station must do more than simply transmit a carrier wave. The station has to send information about the sounds that you are to receive. This information is sent by modifying the carrier wave. The carrier wave is modified to carry information in one of two ways, as shown in **Figure 16.**

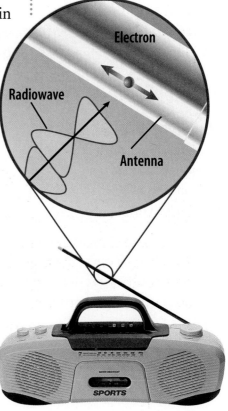

Figure 15 Radio waves exert a force on the electrons in an antenna, causing the electrons to vibrate.

Electron

Radiowave

Antenna

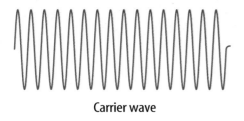

Carrier wave

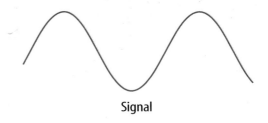

Signal

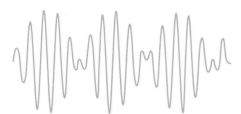

Amplitude modulation

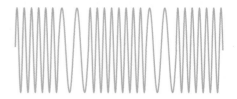

Frequency modulation

Figure 16 A carrier wave broadcast by a radio station can be altered in one of two ways to transmit a signal: amplitude modulation (AM) or frequency modulation (FM).

AM Radio An AM radio station broadcasts information by varying the amplitude of the carrier wave, as shown in **Figure 16.** Your radio detects the variations in amplitude of the carrier wave and produces a changing electric current from these variations. The changing electric current makes the speaker vibrate. AM carrier wave frequencies range from 540,000 to 1,600,000 Hz.

FM Radio Electronic signals are transmitted by FM radio stations by varying the frequency of the carrier wave, as in **Figure 16.** Your radio detects the changes in frequency of the carrier wave. Because the strength of the FM waves is kept fixed, FM signals tend to be more clear than AM signals. FM carrier frequencies range from 88 million to 108 million Hz. This is much higher than AM frequencies, as shown in **Figure 17. Figure 18** shows how radio signals are broadcast.

Figure 17 Cell phones, TVs, and radios broadcast at frequencies that range from more than 500,000 Hz to almost 1 billion Hz.

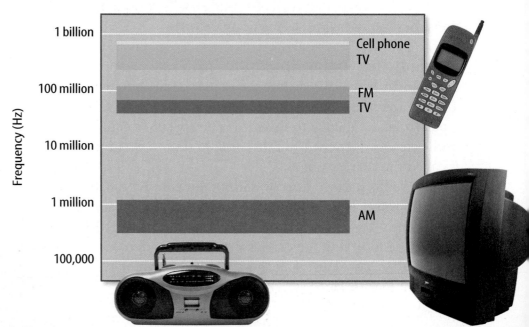

Figure 18

You flick a switch, turn the dial, and music from your favorite radio station fills the room. Although it seems like magic, sounds are transmitted over great distances by converting sound waves to electromagnetic waves and back again, as shown here.

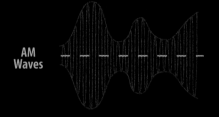

A At the radio station, musical instruments and voices create sound waves by causing air molecules to vibrate. Microphones convert these sound waves to a varying electric current, or electronic signal.

B This signal then is added to the station's carrier wave. If the station is an AM station, the electronic signal modifies the amplitude of the carrier wave. If the station is a FM station, the electronic signal modifies the frequency of the carrier wave.

AM Waves

FM Waves

C The modified carrier wave is used to vibrate electrons in the station's antenna. These vibrating electrons create a radio wave that travels out in all directions at the speed of light.

D The radio wave from the station makes electrons in your radio's antenna vibrate. This creates an electric current. If your radio is tuned to the station's frequency, the carrier wave is removed from the original electronic signal. This signal then makes the radio's speaker vibrate, creating sound waves that you hear as music.

INTEGRATE
Career

Astronomers Do you ever look up at the stars at night and wonder how they were formed? With so many stars and so many galaxies, life might be possible on other planets. Research ways that astronomers use electromagnetic waves to investigate the universe. Choose one project astronomers currently are working on that interests you, and write about it in your Science Journal. Discuss the benefits of a career in astronomy.

Television

What would people hundreds of years ago have thought if they had seen a television? They might seem like magic, but not if you know how they work. Television and radio transmissions are similar. At the television station, sound and images are changed into electronic signals. These signals are broadcast by carrier waves. The audio part of television is sent by FM radio waves. Information about the color and brightness is sent at the same time by AM signals.

Cathode-Ray Tubes In many television sets, images are displayed on a cathode-ray tube (CRT), as shown in **Figure 19.** A **cathode-ray tube** is a sealed vacuum tube in which one or more beams of electrons are produced. The CRT in a color TV produces three electron beams that are focused by a magnetic field and strike a coated screen. The screen is speckled with more than 100,000 rectangular spots that are of three types. One type glows red, another glows green, and the third type glows blue when electrons strike it. The spots are grouped together with a red, green, and blue spot in each group.

An image is created when the three electron beams of the CRT sweep back and forth across the screen. Each electron beam controls the brightness of each type of spot, according to the information in the video signal from the TV station. By varying the brightness of each spot in a group, the three spots together can form any color so that you see a full-color image.

Reading Check *What is a cathode-ray tube?*

Figure 19 Cathode-ray tubes produce the images you see on television. The inside surface of a television screen is covered by groups of spots that glow red, green, or blue when struck by an electron beam.

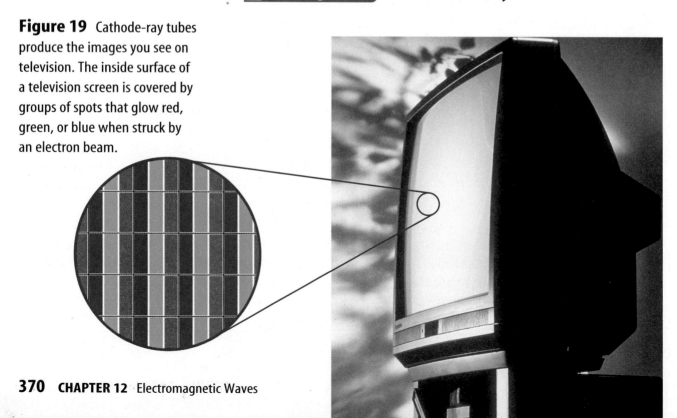

Telephones

Until about 1950, human operators were needed to connect many calls between people. Just 20 years ago you never would have seen someone walking down the street talking on a telephone. Today, cell phones are seen everywhere. When you speak into a telephone, a microphone converts sound waves into an electrical signal. In cell phones, this current is used to create radio waves that are transmitted to and from a microwave tower, as shown in **Figure 20.** A cell phone uses one radio signal for sending information to a tower at a base station. It uses another signal for receiving information from the base station. The base stations are several kilometers apart. The area each one covers is called a cell. If you move from one cell to another while using a cell phone, an automated control station transfers your signal to the new cell.

Reading Check *What are the cells in a cell phone system?*

Figure 20 The antenna at the top of a microwave tower receives signals from nearby cell phones. **Determine** *whether any microwave towers are located near your school or home. Describe their locations.*

Cordless Telephones Like a cellular telephone, a cordless telephone is a transceiver. A **transceiver** transmits one radio signal and receives another radio signal from a base unit. Having two signals at different frequencies allows you to talk and listen at the same time. Cordless telephones work much like cell phones. With a cordless telephone, however, you must be close to the base unit. Another drawback is that when someone nearby is using a cordless telephone, you could hear that conversation on your phone if the frequencies match. For this reason, many cordless phones have a channel button. This allows you to switch your call to another frequency.

Pagers Another method of transmitting signals is a pager, which allows messages to be sent to a small radio receiver. A caller leaves a message at a central terminal by entering a call-back number through a telephone keypad or by entering a text message from a computer. At the terminal, the message is changed into an electronic signal and transmitted by radio waves. Each pager is given a unique number for identification. This identification number is sent along with the message. Your pager receives all messages that are transmitted in the area at its assigned frequency. However, your pager responds only to messages with its particular identification number. Newer pagers can send data as well as receive them.

Science Online

Topic: Radio Wave Technology

Visit gpscience.com for Web links to information about advances in radio wave technology.

Activity List the advances you find, and write about the significance of each one in your Science Journal.

Figure 21 Communications satellites, like the one shown here, use solar panels to provide the electrical energy they need to communicate with receivers on Earth. The solar panels are the structures on either side of the central body of the satellite.

Topic: Satellite Communication

Visit gpscience.com for Web links to information about ways satellites are used for communication.

Activity Write a paragraph describing the advantages of placing a communications satellite in a geosynchronus orbit. Include a diagram.

Communications Satellites

Since satellites were first developed, thousands have been launched into Earth's orbit. Many of these, like the one in **Figure 21,** are used for communication. A station broadcasts a high-frequency microwave signal to the satellite. The satellite receives the signal, amplifies it, and transmits it to a particular region on Earth. To avoid interference, the frequency broadcast by the satellite is different than the frequency broadcast from Earth.

Satellite Telephone Systems If you have a mobile telephone, you can make a phone call when sailing across the ocean. To call on a mobile telephone, the telephone transmits radio waves directly to a satellite. The satellite relays the signal to a ground station, and the call is passed on to the telephone network. Satellite links work well for one-way transmissions, but two-way communications can have an annoying delay caused by the large distance the signals travel to and from the satellite.

Television Satellites The satellite-reception dishes that you sometimes see in yards or attached to houses are receivers for television satellite signals. Satellite television is used as an alternative to ground-based transmission. Communications satellites use microwaves rather than the longer-wavelength radio waves used for normal television broadcasts. Short-wavelength microwaves travel more easily through the atmosphere. The ground receiver dishes are rounded to help focus the microwaves onto an antenna.

The Global Positioning System

Getting lost while hiking is not uncommon, but if you are carrying a Global Positioning System receiver, it is much less likely to happen. The **Global Positioning System (GPS)** is a system of satellites, ground monitoring stations, and receivers that determine your exact location at or above Earth's surface. The 24 satellites necessary for 24-hour, around-the-world coverage became fully operational in 1995. GPS satellites are owned and operated by the United States Department of Defense, but the microwave signals they send out can be used by anyone. As shown in **Figure 22,** signals from four satellites are needed to determine the location of an object using a GPS receiver. Today GPS receivers are used in airplanes, ships, cars, and even by hikers.

Figure 22 A GPS receiver uses signals from orbiting satellites to determine the receiver's location.

section 3 review

Summary

Radio Transmission

- Radio stations transmit electromagnetic waves that receivers convert to sound waves.
- Each AM radio station is assigned a carrier wave frequency and varies the amplitude of the carrier waves to transmit a signal.
- Each FM radio station is assigned a carrier wave frequency and varies the frequency of the carrier waves to transmit a signal.

Television

- TV sets use cathode-ray tubes to convert electronic signals from TV stations into both sound and images.

Telephones

- Telephones contain transceivers that convert sound waves into electrical signals and also convert electrical signals into sound waves.
- Wires, microwave towers, and satellites are used to transmit and receive telephone signals.

Global Positioning System

- The Global Positioning System uses a system of satellites to determine your exact position.

Self Check

1. **Explain** the difference between AM and FM radio. Make a sketch of how a carrier wave is modulated in AM and FM radio.

2. **Define** a cathode-ray tube, and explain how it is used in a television.

3. **Describe** what happens if you are talking on a cell phone while riding in a car and you travel from one cell to another cell.

4. **Explain** some of the uses of a Global Positioning System. Why might emergency vehicles all be equipped with GPS receivers?

5. **Think Critically** Why do cordless telephones stop working if you move too far from the base unit?

Applying Math

6. **Calculate a Ratio** A group of red, green and blue spots on a TV screen is a pixel. A standard TV has 460 pixels horizontally and 360 pixels vertically. A high-definition TV has 1,920 horizontal and 1,080 vertical pixels. What is the ratio of the number of pixels in a high-definition TV to the number in a standard TV?

Rado Frequencies

Goals

- **Research** which frequencies are used by different radio stations.
- **Observe** the reception of your favorite radio station.
- **Make** a chart of your findings and communicate them to other students.

Data Source

ScienceOnline

Visit gpscience.com/ internet_lab for more information on radio frequencies, different frequencies of radio stations around the country, and the ranges of AM and FM broadcasts.

Real-World Question

The signals from many radio stations broadcasting at different frequencies are hitting your radio's antenna at the same time. When you tune to your favorite station, the electronics inside your radio amplify the signal at the frequency broadcast by the station. The signal from your favorite station is broadcast from a transmission site that may be several miles away.

You may have noticed that if you're listening to a radio station while driving in a car, sometimes the station gets fuzzy and you'll hear another station at the same time. Sometimes you lose the station completely. How far can you drive before that happens? Does the distance vary depending on the station you listen to? What are the ranges of radio stations? Form a hypothesis about how far you think a radio station can transmit? Which type of signal, AM or FM, has a greater range? Form a hypothesis about the range of your favorite radio station.

Make a Plan

1. **Research** what frequencies are used by AM and FM radio stations in your area and other areas around the country.
2. **Determine** these stations' broadcast locations.
3. **Determine** the broadcast range of radio stations in your area.
4. **Observe** how frequencies differ. What is the maximum difference between frequencies for FM stations in your area? AM stations?

▶ Follow Your Plan

1. **Make** sure your teacher approves your plan before you start.

2. Visit the link shown below for links to different radio stations.

3. **Compare** the different frequencies of the stations and the locations of the broadcasts.

4. **Determine** the range of radio stations in your area and the power of their broadcast signal in watts.

5. **Record** your data in your Science Journal.

▶ Analyze Your Data

1. **Make** a map of the radio stations in your area. Do the ranges of AM stations differ from FM stations?

2. **Make** a map of different radio stations around the country. Do you see any patterns in the frequencies for stations that are located near each other?

3. **Write** a description that compares how close the frequencies of AM stations are and how close the frequencies of FM stations are. Also compare the power of their broadcast signals and their ranges.

4. **Share** your data by posting it at the link shown below.

▶ Conclude and Apply

1. **Compare** your findings to those of your classmates and other data that was posted at the link shown below. Do all AM stations and FM stations have different ranges?

2. **Observe** your map of the country. How close can stations with similar frequencies be? Do AM and FM stations appear to be different in this respect?

3. **Infer** The power of the broadcast signal also determines its range. How does the power (wattage) of the signals affect your analysis of your data?

Communicating Your Data

Find this lab using the link below. Post your data in the table provided. **Compare** your data to that of other students. Then combine your data with theirs and make a map for your class that shows all of the data.

gpscience.com/internet_lab

Riding a Beam of Light

Einstein and the Special Theory of Relativity

Catch a Wave

At age sixteen, Albert Einstein wondered "What would it be like to ride a beam of light?" He imagined what might happen if he turned on a flashlight while riding a light beam. Because the flashlight was already traveling at the speed of light, would light from the flashlight travel at twice the speed of light?

What's so special?

Einstein thought about this problem and in 1905 he published the special theory of relativity. This theory stated that the speed of light measured by any observer that moves with a constant speed always would be the same. The measured speed of light would not depend on the speed of the observer or on how fast the source of light was moving. Einstein answered the question he asked himself when he was sixteen. He had found the universal speed limit that can't be broken.

It Doesn't Add Up

According to Einstein, electromagnetic waves like light waves behave very differently from other waves. For example, sound waves from the siren of an ambulance moving toward you move faster than they would if the ambulance were not moving. The speed of the ambulance adds to the speed of the sound waves. However, for light waves, the speed of a light source doesn't add to the speed of light.

Very Strange But True

Einstein's special theory of relativity makes other strange predictions. According to this theory, no object can travel faster than the speed of light. Another prediction is that the measured length of a moving object is shorter than when the object is at rest. Also, moving clocks should run slower than when they are at rest. These predictions have been confirmed by experiments. Measurements have shown, for example, that a moving clock does run slower.

Communicate Research the life of Albert Einstein and make a timeline showing important events in his life. Also include on your timeline major historical events that occurred during Einstein's lifetime.

Science Online

For more information, visit gpscience.com/time

Reviewing Main Ideas

Section 1 — What are electromagnetic waves?

1. Electromagnetic waves consist of vibrating electric and magnetic fields, and are produced by vibrating electric charges.

2. Electromagnetic waves carry radiant energy and can travel through a vacuum or through matter.

3. Electromagnetic waves sometimes behave like particles called photons.

Section 2 — The Electromagnetic Spectrum

1. Electromagnetic waves with the longest wavelengths are called radio waves. Radio waves have wavelengths greater than about 1 mm. Microwaves are radio waves with wavelengths between about 1 m and 1 mm.

2. Infrared waves have wavelengths between about 1 mm and 750 billionths of a meter. Warmer objects emit more infrared waves than cooler objects.

3. Visible light rays have wavelengths between about 750 and 400 billionths of a meter. Substances in your eyes react with visible light to enable you to see.

4. Ultraviolet waves have frequencies between about 400 and 10 billionths of a meter. Excessive exposure to ultraviolet waves can damage human skin.

5. X rays and gamma rays are high-energy electromagnetic waves with wavelengths less than 10 billionths of a meter. X rays are used in medical imaging.

Section 3 — Radio Communication

1. Modulated radio waves are used often for communication. AM and FM are two forms of carrier wave modulation.

2. Television signals are transmitted as a combination of AM and FM waves.

3. Cellular telephones, cordless telephones, and pagers use radio waves to transmit signals. Communications satellites are used to relay telephone and television signals over long distances.

4. The Global Positioning System enables an accurate position on Earth to be determined.

FOLDABLES Use the Foldable that you made at the beginning of this chapter to help you review electromagnetic waves.

Using Vocabulary

carrier wave p. 367
cathode-ray tube p. 370
electromagnetic wave p. 354
gamma rays p. 365
Global Positioning System p. 373
infrared waves p. 362

microwaves p. 361
photon p. 358
radiant energy p. 357
radio waves p. 361
transceiver p. 371
ultraviolet waves p. 363
visible light p. 363
X rays p. 365

Complete each statement using the correct word or words from the vocabulary list above.

1. _____ are the type of electromagnetic waves often used for communication.

2. A remote control uses _____ to communicate with a television set.

3. Electromagnetic waves carry _____ .

4. If you stay outdoors too long, your skin might be burned by exposure to _____ from the Sun.

5. A radio station broadcasts radio waves called _____ that have the specific frequency assigned to the station.

6. The image on a television screen is produced by a _____.

7. Transverse waves that are produced by vibrating electric charges and consist of vibrating electric and magnetic fields are _____.

Checking Concepts

Choose the word or phrase that best answers the question.

8. Which type of electromagnetic wave is the most energetic?
   A) gamma rays
   B) ultraviolet waves
   C) infrared waves
   D) microwaves

9. Electromagnetic waves can behave like what type of particle?
   A) electrons
   B) molecules
   C) photons
   D) atoms

10. Which type of electromagnetic wave enables skin cells to produce vitamin D?
    A) visible light
    B) ultraviolet waves
    C) infrared waves
    D) X rays

11. Which of the following describes X rays?
    A) short wavelength, high frequency
    B) short wavelength, low frequency
    C) long wavelength, high frequency
    D) long wavelength, low frequency

12. Which of the following is changing in an AM radio wave?
    A) speed
    B) frequency
    C) amplitude
    D) wavelength

13. Which type of electromagnetic wave has wavelengths greater than about 1 mm?
    A) X rays
    B) radio waves
    C) gamma rays
    D) ultraviolet waves

14. What is the name of the ability of some materials to absorb ultraviolet light and re-emit it as visible light?
    A) modulation
    B) handoff
    C) transmission
    D) fluorescence

15. Which of these colors of visible light has the shortest wavelength?
    A) blue
    B) green
    C) red
    D) white

16. Which type of electromagnetic wave has wavelengths slightly longer than humans can see?
    A) X rays
    B) ultraviolet waves
    C) infrared waves
    D) gamma rays

Science Online gpscience.com/vocabulary_puzzlemaker

Interpreting Graphics

17. Copy and complete the following table about the electromagnetic spectrum.

Uses of Electromagnetic Waves

Type of Electromagnetic Waves	Examples of How Electromagnetic Waves Are Used
	radio, TV transmission
Infrared waves	
Visible light	vision
X rays	
	destroying harmful cells

18. Copy and complete the following events chain about the destruction of ozone molecules in the ozone layer by CFC molecules.

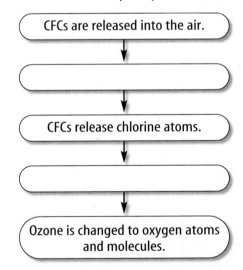

CFCs are released into the air.

⬇

⬇

CFCs release chlorine atoms.

⬇

⬇

Ozone is changed to oxygen atoms and molecules.

Thinking Critically

19. **Explain** why X rays are used in medical imaging.

20. **Predict** whether an electromagnetic wave would travel through space if its electric and magnetic fields were not changing with time. Explain your reasoning.

21. **Infer** Electromagnetic waves consist of vibrating electric and magnetic fields. A magnetic field can make a compass needle. Why doesn't a compass needle move when visible light strikes the compass?

22. **Classify** Look around your home, school, and community. Make a list of the different devices that use electromagnetic waves. Beside each device, write the type of electromagnetic wave the device uses.

23. **Form a hypothesis** to explain why communications satellites don't use ultraviolet waves to receive information and transmit signals to Earth's surface.

24. **Compare** the energy of photons corresponding to infrared waves with the energy of photons corresponding to ultraviolet waves.

25. **Determine** whether or not all electromagnetic waves always travel at the speed of light. Explain.

Applying Math

26. **Use Fractions** When visible light waves travel in ethyl alcohol, their speed is three fourths of the speed of light in air. What is the speed of light in ethyl alcohol?

27. **Use Scientific Notation** The speed of light in a vacuum has been determined to be 299, 792, 458 m/s. Express this number to four significant digits using scientific notation.

28. **Calculate Wavelength** A radio wave has a frequency of 540,000 Hz and travels at a speed of 300,000 km/s. Use the wave speed equation to calculate the wavelength of the radio wave. Express your answer in meters.

Part 1 Multiple Choice

Record your answers on the answer sheet provided by your teacher or on a sheet of paper.

1. Which of the following produces electromagnetic waves?
   A. vibrating charge
   B. direct current
   C. static charge
   D. constant magnetic field

Use the photograph below to answer questions 2 and 3.

2. A television image is produced by three electron beams. What device inside a television set produces the electron beams?
   A. transceiver   C. antenna
   B. transmitter   D. cathode-ray tube

3. What colors are the three types of glowing spots that are combined to form the different colors in the image on the screen?
   A. red, yellow, blue
   B. red, green, blue
   C. cyan, magenta, yellow
   D. cyan, magenta, blue

Test-Taking Tip

Marking on Tests Be sure to ask if it is okay to write on the test booklet when taking the test, but make sure you mark all answers on your answer sheet.

4. Which of the following explains how interference is avoided between the signals communications satellites receive and the signals they broadcast?
   A. The signals travel at different speeds.
   B. The signals have different amplitudes.
   C. The signals have different frequencies.
   D. The signals are only magnetic.

5. Which of the following people explained how light can behave as a particle, called a photon, whose energy depends on the frequency of light?
   A. Einstein   C. Newton
   B. Hertz   D. Galileo

Use the table below to answer questions 6 and 7.

Regions of the Electromagnetic Spectrum		
Infrared waves	Radio waves	Gamma rays
X rays	Visible light	Ultraviolet waves

6. If you arranged the list of electromagnetic waves shown above in order from shortest to longest wavelength, which would be first on the list?
   A. radio waves   C. gamma rays
   B. X rays   D. visible light

7. Which region of the electromagnetic spectrum listed in the table above includes microwaves?
   A. gamma rays   C. ultraviolet waves
   B. radio waves   D. infrared waves

8. The warmth you feel when you stand in front of a fire is thermal energy transmitted to you by what type of electromagnetic waves?
   A. X rays   C. ultraviolet waves
   B. microwaves   D. infrared waves

Part 2 | Short Response/Grid In

Record your answers on the answer sheet provided by your teacher or on a sheet of paper.

Use the illustrations below to answer questions 9 and 10.

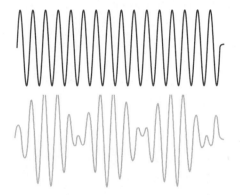

9. The illustration above shows two radio waves broadcast by a radio station. What is the upper, unmodulated wave called?

10. The lower figure shows the same wave that has been modulated to carry sound information. What type of modulation does it show?

11. The frequency of electromagnetic waves is measured in what units? What does this unit mean?

12. Even on a cloudy day, you can get sunburned outside. However, inside a glass greenhouse, you won't get sunburned. Which type of electromagnetic waves will pass through clouds, but not glass?

13. What term refers to the energy carried by an electromagnetic wave?

14. The following sentence is *not* true: A magnetic field creates an electric field, and an electric field creates a magnetic field. Rewrite the sentence so that it is true.

15. Which type of radio station transmits radio waves that have a higher frequency, AM stations or FM stations?

Part 3 | Open Ended

Record your answers on a sheet of paper.

16. A CD player converts the musical information on a CD to a varying electric current. Describe how the varying electric current produced by a CD player in a radio station is converted into radio waves.

17. Explain how an electromagnetic wave that strikes a material transfers radiant energy to the atoms in the material.

18. How would changing the amount of ozone in the ozone layer affect the amount of the different types of electromagnetic waves emitted by the Sun that reach Earth's surface?

19. Explain how the cathode-ray tube in a television is able to produce all the colors that you see in an image on a television screen, using just three electron beams.

20. If all atoms contain electric charges, and if all atoms are constantly in motion, explain why all objects should emit electromagnetic waves.

Use the illustration below to answer questions 21 and 22.

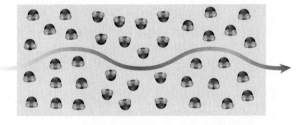

21. The illustration above shows microwaves interacting with water molecules in food. How does the electric field in microwaves affect water molecules?

22. Describe how thermal energy inside food is produced by microwaves interacting with water molecules.

Light

That Inner Glow

Some organisms, like this squid, can produce light. This process, called bioluminines-
cence, results from a chemical reaction, like
the one that causes fireflies to light up at
night. Bioluminescent organisms glow to
lure prey, attract a mate, coordinate group
movements, and evade predators.

Science Journal   Find some other examples of living
organisms that give off light.

Start-Up Activities

Rainbows of Light

Light passing through a prism can produce exciting patterns of color. Imagine what your surroundings would look like now if humans could see only shades of gray instead of distinct colors. The ability to see color depends on the cells in your eyes that are sensitive to different wavelengths of light. What color is the light produced by a flashlight or the Sun?

1. In a darkened room, shine a flashlight through a glass prism. Project the resulting colors onto a white wall or ceiling.

2. In a darkened room, shine a flashlight over the surface of some water with dishwashing liquid bubbles in it. What do you see?

3. Aim a flashlight at the surface of a compact disc.

4. **Think Critically** How did your observations in each case differ? Explain where you think the colors came from.

Preview this chapter's content and activities at gpscience.com

---

FOLDABLES™
Study Organizer

Light Transmission Make the following Foldable to help identify the characteristics of opaque, translucent, and transparent objects.

STEP 1 **Fold** a vertical sheet of paper from side to side. Make the front edge about 1.25 cm shorter than the back edge.

STEP 2 **Turn** lengthwise and fold into thirds.

STEP 3 **Unfold** and cut only the top layer along both folds to make three tabs.

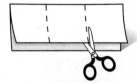

STEP 4 **Label** each tab as shown.

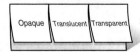

Find Main Ideas As you read this chapter, list the characteristics of opaque, translucent, and transparent objects.

The Behavior of Light

Reading Guide

What You'll Learn

- **Describe** how light waves interact with matter.
- **Explain** the difference between regular and diffuse reflection.
- **Define** the index of refraction of a material.
- **Explain** why a prism separates white light into different colors.

Why It's Important

The images you see every day are due to the behavior of light waves.

Review Vocabulary

visible light: an electromagnetic wave with wavelengths between about 400 and 750 billionths of a meter

Vocabulary

- opaque
- translucent
- transparent
- index of refraction
- mirage

Figure 1 These candleholders have different light-transmitting properties.

A Opaque

B Translucent

C Transparent

Light and Matter

Look around your darkened room at night. After your eyes adjust to the darkness, you begin to recognize some familiar objects. You know that some of the objects are brightly colored, but they look gray or black in the dim light. Turn on the light, and you clearly can see all the objects in the room, including their colors. What you see depends on the amount of light in the room and the color of the objects. For you to see an object, it must reflect some light back to your eyes.

Opaque, Transparent, and Translucent Objects can absorb light, reflect light, and transmit light—allow light to pass through them. The type of matter in an object determines the amount of light it absorbs, reflects, and transmits. For example, the **opaque** (oh PAYK) material in the candleholder in **Figure 1A** only absorbs and reflects light—no light passes through it. As a result, you cannot see the candle inside. Materials that allow some light to pass through them, like the material of the candleholder in **Figure 1B,** are described as **translucent** (trans LEW sunt). You cannot see clearly through translucent materials.

Transparent materials like the candleholder in **Figure 1C** transmit almost all the light striking them, so you can see objects clearly through them. Only a small amount of light is absorbed and reflected.

Reflection of Light

Just before you left for school this morning, did you take a glance in a mirror to check your appearance? For you to see your reflection in the mirror, light had to reflect off you, hit the mirror, and reflect off the mirror into your eye. Reflection occurs when a light wave strikes an object and bounces off.

The Law of Reflection

Because light behaves as a wave, it obeys the law of reflection, as shown in **Figure 2.** According to the law of reflection, the angle at which a light wave strikes a surface is the same as the angle at which it is reflected. Light reflected from any surface—a mirror or a sheet of paper—follows this law.

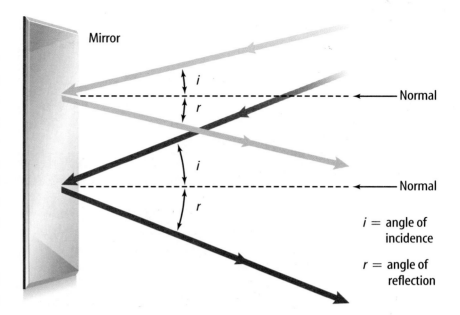

Mirror

i = angle of incidence

r = angle of reflection

Figure 2 According to the law of reflection, light is reflected so that the angle of incidence always equals the angle of reflection.

Regular and Diffuse Reflection

Why can you see your reflection in a store window but not in a brick wall? The answer has to do with the smoothness of the surfaces. A smooth, even surface like that of a pane of glass produces a sharp image by reflecting parallel light waves in only one direction. Reflection of light waves from a smooth surface is regular reflection. A brick wall has an uneven surface that causes incoming parallel light waves to be reflected in many directions, as shown in **Figure 3.** Reflection of light from a rough surface is diffuse reflection.

Reading Check *What are some more examples of objects that produce regular or diffuse reflection?*

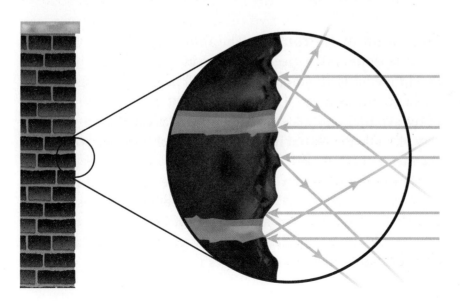

Figure 3 This brick wall has an uneven surface, so it produces a diffuse reflection.
Explain *Use the law of reflection to explain why a rough surface causes parallel light waves to be reflected in many directions.*

Figure 4 Although the surface of this pot may seem smooth, it produces a diffuse reflection. At high magnification, the surface is seen to be rough.

Roughness of Surfaces

Even a surface that appears to be smooth can be rough enough to cause diffuse reflection. For example, a metal pot might seem smooth, but at high magnification, the surface shows rough spots, as shown in **Figure 4.** To cause a regular reflection, the roughness of the surface must be less than the wavelengths it reflects.

Refraction of Light

What occurs when a light wave passes from one material to another—from air to water, for example? Refraction is caused by a change in the speed of a wave when it passes from one material to another. If the light wave is traveling at an angle and the speed that light travels is different in the two materials, the wave will be bent, or refracted.

Reading Check *How does refraction occur?*

The Index of Refraction The amount of bending that takes place depends on the speed of light in both materials. The greater the difference is, the more the light will be bent as it passes at an angle from one material to the other. **Figure 5** shows an example of refraction. Every material has an **index of refraction**—a property of the material that indicates how much the speed of light in the material is reduced.

The larger the index of refraction, the more light is slowed down in the material. For example, because glass has a larger index of refraction than air, light moves more slowly in glass than air. Many useful devices like eyeglasses, binoculars, cameras, and microscopes form images using refraction.

Figure 5 The spoon looks bent because light waves are refracted as they change speed when they pass from the water to the air.

Prisms A sparkling glass prism hangs in a sunny window, refracting the sunlight and projecting a colorful pattern onto the walls of the room. How does the bending of light create these colors? It occurs because the amount of bending usually depends on the wavelength of the light. Wavelengths of visible light range from the longer red waves to the shorter violet waves. White light, such as sunlight, is made up of this whole range of wavelengths.

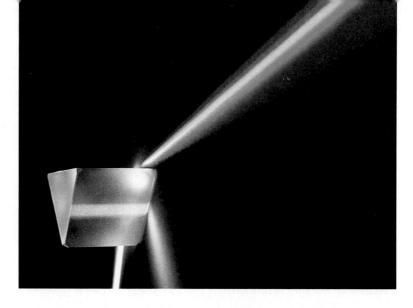

Figure 6 shows what occurs when white light passes through a prism. The triangular prism refracts the light twice—once when it enters the prism and again when it leaves the prism and reenters the air. Because the longer wavelengths of light are refracted less than the shorter wavelengths are, red light is bent the least. As a result of these different amounts of bending, the different colors are separated when they emerge from the prism. Which color of light would you expect to bend the most?

Rainbows Does the light leaving the prism in **Figure 6** remind you of a rainbow? Like prisms, rain droplets also refract light. The refraction of the different wavelengths can cause white light from the Sun to separate into the individual colors of visible light, as shown in **Figure 7**. In a rainbow, the human eye usually can distinguish only about seven colors clearly. In order of decreasing wavelength, these colors are red, orange, yellow, green, blue, indigo, and violet.

Figure 7 As white light passes through the water droplet, different wavelengths are refracted by different amounts. This produces the separate colors seen in a rainbow.

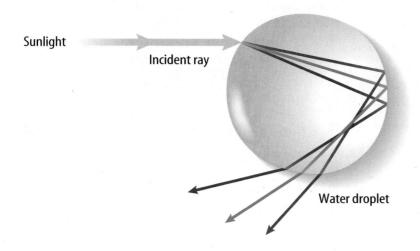

Figure 6 Refraction causes a prism to separate a beam of white light into different colors.

Observing Refraction in Water

Procedure
1. Place a **penny** at the bottom of a **short, opaque cup.** Set it on a **table** in front of you.
2. Have a partner slowly slide the cup away from you until you can't see the penny.
3. Without disturbing the penny or the cup and without moving your position, have your partner slowly pour **water** into the cup until you can see the penny.
4. Reverse roles and repeat the experiment.

Analysis
1. What did you observe? Explain how this is possible.
2. In your **Science Journal,** sketch the light path from the penny to your eye after the water was added.

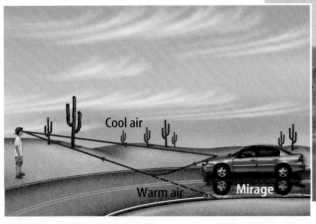

Cool air

Warm air   Mirage

Figure 8 Mirages result when air near the ground is much warmer or cooler than the air above. This causes some light-waves reflected from the object to refract, creating one or more additional images.

Mirages You might have seen what looks like a pool of water on the road ahead. As you get closer, the water seems to disappear. You saw a **mirage,** an image of a distant object produced by the refraction of light through air layers of different densities. Mirages result when the air at ground level is much warmer or cooler than the air above it, as **Figure 8** shows. The density of air increases as air cools. Light waves travel slower as the density of air increases, so that light travels slower in cooler air. As a result, light waves refract as they pass through air layers with different temperatures.

section 1 review

Summary

Light and Matter

- When light waves strike an object, the light can be absorbed, reflected, and transmitted.
- The amount of light that is absorbed, reflected, or transmitted depends on the material an object is made from.

Reflection of Light

- Light waves always obey the law of reflection—the angle of incidence equals the angle of reflection.
- Regular reflection occurs when the roughness of a surface is less than the wavelengths reflected.
- Diffuse reflection causes parallel light waves to be reflected in many directions.

Refraction of Light

- Refraction occurs if a light wave changes speed in moving from one material to another.
- The index of refraction of a material indicates how much light slows down in the material.
- In a material, different wavelengths of light can be refracted by different amounts.

Self Check

1. **Compare and contrast** opaque, transparent, and translucent materials. Give at least one example of each.

2. **Discuss** why you can see your reflection in a smooth piece of aluminum foil but not in a crumpled ball of foil.

3. **Explain** why you are more likely to see a mirage on a hot day than on a mild day.

4. **Infer** what happens to white light when it passes through a prism.

5. **Think Critically** Decide whether the lens of your eye, a fingernail, your skin, and your tooth are opaque, translucent, or transparent. Explain.

Applying Math

6. **Find an Angle** A light ray strikes a mirror at an angle of 42° from the surface of the mirror. What angle does the reflected ray make with the normal?

7. **Find an Angle** A ray of light hits a mirror at 27° from the normal. What is the angle between the reflected ray and the normal?

 Science Online   gpscience.com/self_check_quiz

Light and Color

What You'll Learn

- **Explain** how you see color.
- **Describe** the difference between light color and pigment color.
- **Predict** what happens when different colors are mixed.

Why It's Important

From traffic lights to great works of art, color plays an important role in your world.

Review Vocabulary

retina: inner layer of the eye containing cells that convert light images into electrical signals

New Vocabulary

- pigment

Colors

Why do some apples appear red, while others look green or yellow? An object's color depends on the wavelengths of light it reflects. You know that white light is a blend of all colors of visible light. When a red apple is struck by white light, it reflects red light back to your eyes and absorbs all of the other colors. **Figure 9** shows white light striking a green leaf. Only the green light is reflected to your eyes.

Although some objects appear to be black, black isn't a color that is present in visible light. Objects that appear black absorb all colors of light and reflect little or no light back to your eye. White objects appear to be white because they reflect all colors of visible light.

Reading Check *Why does a white object appear white?*

Colored Filters Wearing tinted glasses changes the color of almost everything you look at. If the lenses are yellow, the world takes on a golden glow. If they are rose colored, everything looks rosy. Something similar would occur if you placed a colored, clear plastic sheet over this white page. The paper would appear to be the same color as the plastic. The plastic sheet and the tinted lenses are filters. A filter is a transparent material that transmits one or more colors of light but absorbs all others. The color of a filter is the color of the light that it transmits.

Figure 9 This green leaf absorbs all wavelengths of visible light except green.

Figure 10 The color of this cooler seems to change under different lighting conditions.

A The blue cooler is shown in white light.

B The cooler appears blue when viewed through a blue filter.

C The cooler appears black through a red filter.

Looking Through Colored Filters **Figure 10** shows what happens when you look at a colored object through various colored filters. In the white light in **Figure 10A,** a blue cooler looks blue because it reflects only the blue light in the white light striking it. It absorbs the light of all other colors. If you look at the cooler through a blue filter as in **Figure 10B,** the cooler still looks blue because the filter transmits the reflected blue light. **Figure 10C** shows how the cooler looks when you examine it through a red filter.

 *Why does the blue cooler appear black through a red filter?*

Seeing Color

As you approach a busy intersection, the color of the traffic light changes from green to yellow to red. On the cross street, the color changes from red to green. At a busy intersection, traffic safety depends on your ability to detect immediate color changes. How do you see colors?

Light and the Eye In a healthy eye, light enters and is focused on the retina, an area on the inside of your eyeball, as shown in **Figure 11A.** The retina is made up of two types of cells that absorb light, as shown in **Figure 11B.** When these cells absorb light energy, chemical reactions convert light energy into nerve impulses that are transmitted to the brain. One type of cell in the retina, called a cone, allows you to distinguish colors and detailed shapes of objects. Cones are most effective in daytime vision.

Figure 11 Light enters the eye and focuses on the retina. The two types of light-detecting cells that make up the retina are called rods and cones.

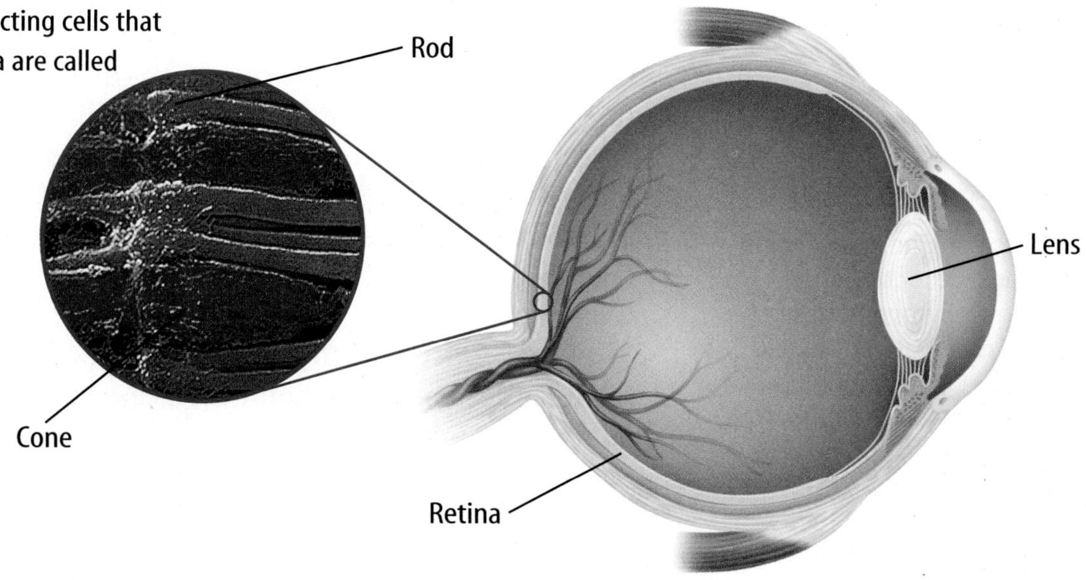

Rod

Lens

Cone

Retina

Cones and Rods

Your eyes have three types of cones, each of which responds to a different range of wavelengths. Red cones respond to mostly red and yellow, green cones respond to mostly yellow and green, and blue cones respond to mostly blue and violet. The second type of cell, called a rod, is sensitive to dim light and is useful for night vision.

INTEGRATE Life Science

Interpreting Color

Why does a banana look yellow? The light reflected by the banana causes the cone cells that are sensitive to red and green light to send signals to your brain. Your brain would get the same signal if a mixture of red light and green light reached your eye. Again, your red and green cones would respond, and you would see yellow light because your brain can't perceive the difference between incoming yellow light and yellow light produced by combining red and green light. The next time you are at a play or a concert, look at the lighting above the stage. Watch how the colored lights combine to produce effects onstage.

Color Blindness

If one or more of your sets of cones did not function properly, you would not be able to distinguish between certain colors. About eight percent of men and one-half percent of women have a form of color blindness. Most people who are said to be color-blind are not truly blind to color, but they have difficulty distinguishing between a few colors, most commonly red and green. **Figure 12** shows a page of a color blindness test. Because these two colors are used in traffic signals, drivers and pedestrians must be able to identify them.

Figure 12 Color blindness is an inherited sex-linked condition in which certain cones do not function properly.
Identify *what number you see in the dots.*

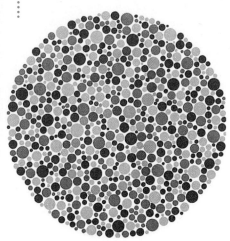

Color for Photosynthesis
Plant pigments determine the wavelengths of light for photosynthesis. Leaves usually look green due to the pigment chlorophyll. Chlorophyll absorbs most wavelengths of visible light except green, which it reflects. But not all plants are green. Research different plant pigments to find how they allow plant species to survive in diverse habitats.

Figure 13 White light is produced when the three primary colors of light are mixed.

Mixing Colors

If you have ever browsed through a paint store, you have probably seen displays where customers can select paint samples of almost every imaginable color. The colors are a result of mixtures of pigments. For example, you might have mixed blue and yellow paint to produce green paint. A **pigment** is a colored material that is used to change the color of other substances. The color of a pigment results from the different wavelengths of light that the pigment reflects.

Mixing Colored Lights From the glowing orange of a sunset to the deep blue of a mountain lake, all the colors you see can be made by mixing three colors of light. These three colors—red, green, and blue—are the primary colors of light. They correspond to the three different types of cones in the retina of your eye. When mixed together in equal amounts, they produce white light, as **Figure 13** shows. Mixing the primary colors in different proportions can produce the colors you see.

✔ Reading Check *What are primary colors?*

Paint Pigments If you were to mix equal amounts of red, green, and blue paint, would you get white paint? If mixing colors of paint were like mixing colors of light, you would, but mixing paint is different. Paints are made with pigments. Paint pigments usually are made of chemical compounds such as titanium oxide, a bright white pigment, and lead chromate, which is used for painting yellow lines on highways.

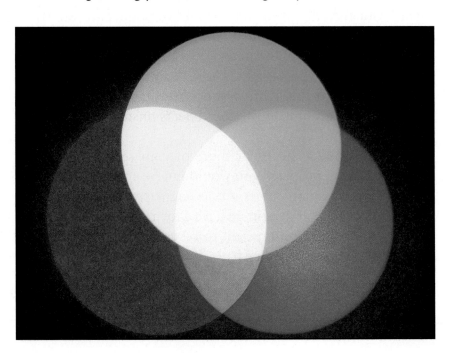

Mixing Pigments You can make any pigment color by mixing different amounts of the three primary pigments—magenta (bluish red), cyan (greenish blue), and yellow. In fact, color printers use those pigments to make full-color prints like the pages in this book. However, color printers also use black ink to produce a true black color. A primary pigment's color depends on the color of light it reflects. Actually, pigments both absorb and reflect a range of colors in sending a single color message to your eye. For example, in white light, the yellow pigment appears yellow because it reflects yellow, red, orange, and green light but absorbs blue and violet light. The color of a mixture of two primary pigments is determined by the primary colors of light that both pigments reflect.

Look at **Figure 14.** The area in the center where the colors all overlap appears to be black because the three blended primary pigments absorb all the primary colors of light. Recall that the primary colors of light combine to produce white light. They are called additive colors. However, the primary pigment colors combine to produce black. Because black results from the absence of reflected light, the primary pigments are called subtractive colors.

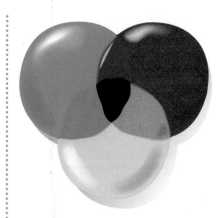

Figure 14 The three primary colors of pigment appear to be black when they are mixed.

---

section 2 review

Summary

Colors
- The color of an object is determined by the wavelengths of light it reflects.
- The color of a filter is the color of the light the filter transmits.

Seeing Color
- Rod and cone cells are light-sensitive cells found in the retina of the human eye.
- Rod cells are sensitive to dim light. Cone cells enable the human eye to see colors.
- There are three types of cone cells. One type responds to red light, another to green light, and another to blue light.

Mixing Colors
- Red, green and blue are the primary light colors. Any color can be created by mixing these primary light colors.
- Any pigment color can be formed by mixing the primary pigment colors—magenta, cyan, and yellow.

Self Check

1. **Identify** what colors are reflected and what colors are absorbed if a white light shines on a red shirt.
2. **Discuss** how the primary colors of light differ from the primary pigment colors.
3. **Explain** why a color-blind person can distinguish among some colors but not others.
4. **Determine** why a white fence appears to be white instead of multicolored if all colors are present in white light.
5. **Think Critically** Light reflected from an object passes through a green filter, then a red filter, and finally a blue filter. What color will the object seem to be?

Applying Math

6. **Use Percentages** In the human eye there are about 120,000,000 rods. If 90,000,000 rods trigger at once, what percentage of the total number of rods triggered?
7. **Convert Units** The wavelengths of a color are measured in nanometers (nm) which is 0.000001 mm. Find the wavelength in mm of a light wave that has a wavelength of 690 nm.

Producing Light

Incandescent Lights

Most of the lightbulbs in your house probably produce **incandescent light**, which is generated by heating a piece of metal until it glows. Inside an incandescent lightbulb is a small wire coil, called a filament, that usually is made of tungsten metal. When an electric current flows in the filament, the electric resistance of the metal causes the filament to become hot enough to give off light. However, about 90% of the energy given off by an incandescent bulb is in the form of thermal energy.

Reading Check *Why does an incandescent lightbulb get hot?*

Fluorescent Lights

Your house also may have fluorescent (floo RE sunt) lights. A fluorescent bulb, like the one shown in **Figure 15,** is filled with a gas at low pressure. The inside of the bulb is coated with phosphors that emit visible light when they absorb ultraviolet radiation. The tube also contains electrodes at each end. Electrons are given off when the electrodes are connected in a circuit. When these electrons collide with the gas atoms, ultraviolet radiation is emitted. The phosphors on the inside of the bulb absorb this radiation and give off visible light.

Figure 15 Fluorescent light-bulbs do not use filaments. **Determine** *what property of phosphors makes them useful in fluorescent bulbs.*

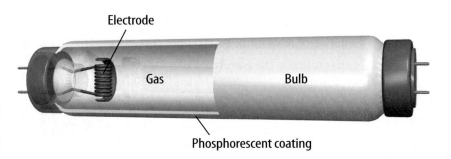

Electrode

Gas

Bulb

Phosphorescent coating

Efficient Lighting A **fluorescent light** uses phosphors to convert ultraviolet radiation to visible light. Fluorescent lights use as little as one fifth the electrical energy to produce the same amount of light as incandescent bulbs. Fluorescent bulbs also last much longer than incandescent bulbs. This higher efficiency can mean lower energy costs over the life of the bulb. Reduced energy usage could reduce the amount of fossil fuels burned to generate electricity, which also decreases the amount of carbon dioxide and pollutants released into Earth's atmosphere.

Fluorescent bulbs are used widely in hospitals, office buildings, schools, and factories. Compact fluorescent bulbs, which can be screwed into traditional lightbulb sockets, have been developed for use in homes.

Neon Lights

The vivid, glowing colors of neon lights, such as those shown in **Figure 16,** make them a popular choice for signs and eye-catching decorations on buildings. These lighting devices are glass tubes filled with gas, typically neon, and work similarly to fluorescent lights. When an electric current flows through the tube, electrons collide with the gas molecules. In this case, however, the collisions produce visible light. If the tube contains only neon, the light is bright red. Different colors can be produced by adding other gases to the tube.

✓ Reading Check *What causes the color in a neon light?*

Mini LAB

Discovering Energy Waste in Lightbulbs

Procedure
1. Obtain an **incandescent bulb** and a **fluorescent bulb** of identical wattage.
2. Make a heat collector by covering the top of a **foam cup** with a piece of **plastic food wrap** to make a window. Carefully make a small hole (diameter less than the thermometer's) in the side of the cup. Push a **thermometer** through the hole.
3. Measure the temperature of the air inside the cup. Then, hold the window of the tester 1 cm from one of the lights for 2 min and measure the temperature.
4. Cool the heat collector and thermometer. Repeat step 3 using the second bulb.

Analysis
1. What was the temperature for each bulb?
2. Which bulb appears to give off more heat? Explain why this occurs.

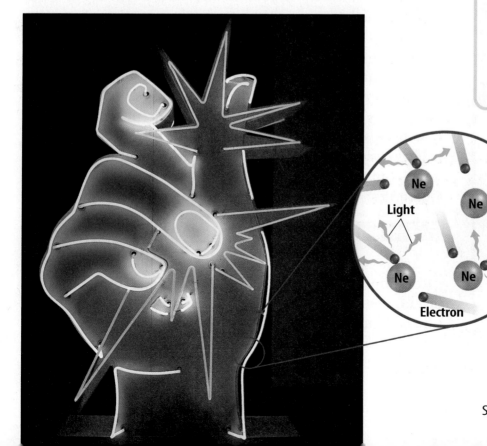

Figure 16 This neon light has vivid, glowing colors.

Figure 17 Sodium-vapor lights emit mostly yellow light. Half of this photo was taken under sunlight and half was taken under sodium-vapor lighting.

Sodium-Vapor Lights

Sodium-vapor lights often are used for streetlights and other outdoor lighting. Inside a sodium-vapor lamp is a tube that contains a mixture of neon gas, a small amount of argon gas, and a small amount of sodium metal. When the lamp is turned on, the gas mixture becomes hot. The hot gases cause the sodium metal to turn to vapor, and the hot sodium vapor emits a yellow-orange glow, as shown in **Figure 17.**

Tungsten-Halogen Lights

Tungsten-halogen lights sometimes are used to create intensely bright light. These lights have a tungsten filament inside a quartz bulb or tube. The tube is filled with a gas that contains one of the halogen elements, such as fluorine or chlorine. The presence of this gas enables the filament to become much hotter than the filament in an ordinary incandescent bulb. As a result, the light is much brighter and also lasts longer. Tungsten-halogen lights sometimes are used on movie sets and in underwater photography.

Lasers

From laser surgery to a laser light show, lasers have become a large part of the world you live in. A laser's light begins when a number of light waves are emitted at the same time. To achieve this, a number of identical atoms each must be given the same amount of energy. When they release their energy, each atom sends off an identical light wave. This light wave is reflected between two facing mirrors at opposite ends of the laser. One of the mirrors is coated only partially with reflective material, so it reflects most light but allows some to get through. Some emitted light waves travel back and forth between the mirrors many times, stimulating other atoms to emit identical light waves also. **Figure 18** shows how this process produces a beam of laser light.

✔ **Reading Check** *How do mirrors help in creating lasers?*

Lasers can be made with many different materials, including gases, liquids, and solids. One of the most common is the helium-neon laser, which produces a beam of red light. A mixture of helium and neon gases sealed in a tube with mirrors at both ends is excited by a flashtube. The excited atoms then lose their excess energy by emitting coherent light waves.

Science Online

Topic: Light Pollution
Visit gpscience.com for Web links to information about sodium vapor lights and light pollution.

Activity Research light pollution and how it affects astronomers. Determine which types of outdoor lights create more and which create less light pollution.

Figure 18

Lasers produce light waves that have the same wavelength. Almost all of these waves travel in the same direction and are in phase. As a result, beams of laser light can be made more intense than ordinary light. In modern eye surgery, shown at the right, lasers are often used instead of a traditional scalpel.

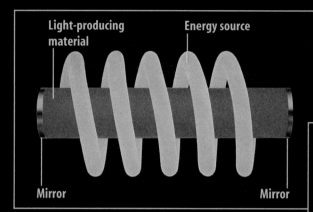

A The key parts of a laser include a material that can be stimulated to produce light, such as a ruby rod, and an energy source. In this example, the energy source is a lightbulb that spirals around the ruby rod and emits an intense light.

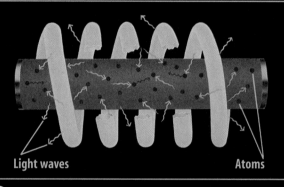

B When the lightbulb is turned on, energy is absorbed by the atoms in the rod. These atoms then re-emit that energy as light waves that are in phase and have the same wavelength.

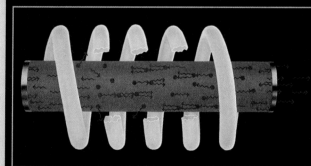

C Most of these waves are reflected between the mirrors located at each end of the laser. One of the mirrors, however, is only partially reflective, allowing one percent of the light waves to pass through it and form a beam.

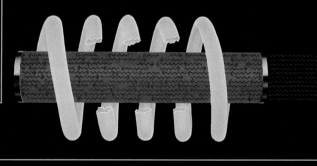

D As the waves travel back and forth between the mirrors, they stimulate other atoms in the ruby rod to emit light waves. In a fraction of a second, billions of identical waves are bouncing between the mirrors. The waves are emitted from the partially reflective mirror in a stream of laser light.

Figure 19 Light waves can be either coherent or incoherent.

A These waves are coherent because they have the same wavelength and travel with their crests and troughs aligned.

B Incoherent waves such as these can contain more than one wavelength, and do not travel with their crests and troughs aligned.

Coherent Light Lasers produce the narrow beams of light that zip across the stage and through the auditorium during some rock concerts. Beams of laser light do not spread out because laser light is coherent. **Coherent light** is light of only one wavelength that travels with its crests and troughs aligned. The beam does not spread out because all the waves travel in the same direction, as shown in **Figure 19A.** As a result, the energy carried by the beam remains concentrated over a small area.

Incoherent Light Light from an ordinary light-bulb is incoherent. **Incoherent light** can contain more than one wavelength, and its electromagnetic waves are not aligned, as in **Figure 19B.** The waves don't travel in the same direction, so the beam spreads out. The energy carried by the light waves is spread over a large area, so the intensity of the light is much less than that of the laser beam.

Using Lasers

Compact disc players, surgical tools, and many other useful devices take advantage of the unique properties of lasers. A laser beam is narrow and does not spread out as it travels over long distances. So lasers can apply large amounts of energy to small areas. In industry, powerful lasers are used for cutting and welding materials. Surveyors and builders use lasers for measuring and leveling. To measure the moon's orbit with great accuracy, scientists use laser light reflected from mirrors placed on the Moon's surface. Information also can be coded in pulses of light from lasers. This makes them useful for communications. In telephone systems, pulses of laser light transmit conversations through long glass fibers called optical fibers.

Lasers in Medicine Lasers are routinely used to remove cataracts, reshape the cornea, and repair the retina. In the eye and other parts of the body, surgeons can use lasers in place of scalpels to cut through body tissues. The energy from the laser seals off blood vessels in the incision and reduces bleeding. Because most lasers do not penetrate deeply through the skin, they can be used to remove small tumors or birthmarks on the surface without damaging deeper tissues. By sending laser light into the body through an optical fiber, physicians can also treat conditions such as blocked arteries.

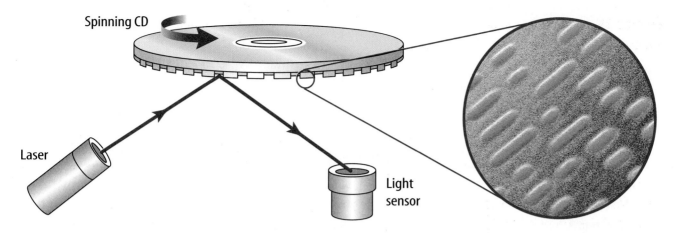

Spinning CD

Laser

Light sensor

Compact Discs Compact discs are plastic discs with reflective surfaces used to store sound, images, and text in digital form. When a CD is produced, the information is burned into the surface of the disc with a laser. The laser creates millions of tiny pits in a spiral pattern that starts at the center of the disc and moves out to the edge. A CD player, shown in **Figure 20,** also uses a laser to read the disc. As the laser beam strikes a pit or flat spot, different amounts of light are reflected to a light sensor. The reflected light is converted to an electric signal that the speakers use to create sound.

Figure 20 The blowup shows the pits (blue) on the bottom surface of a CD. A CD player uses a laser to convert the information on the CD to an electric signal.

section 3 review

Summary

Incandescent and Fluorescent Lights

- In an incandescent bulb, an electric current heats a tungsten filament so that it glows.
- The phosphorescent coating on the inside of a fluorescent bulb absorbs ultraviolet light and emits visible light.

Other Light Sources

- In a neon tube, red light is produced when electrons collide with atoms of neon gas.
- In a sodium-vapor light, sodium metal is vaporized and emits a yellow-orange glow.
- A tungsten-halogen bulb is brighter and hotter than an ordinary incandescent bulb.

Lasers

- Laser light is produced when identical atoms are stimulated to emit coherent light.
- A laser beam does not spread, so energy carried by the beam can be concentrated in a small area.

Self Check

1. **Explain** how light is produced in an ordinary incandescent bulb.

2. **Discuss** the advantages of using a fluorescent bulb instead of an incandescent bulb.

3. **Describe** the difference between coherent and incoherent light.

4. **Think Critically** Which type of lighting device would you use for each of the following needs: an economical light source in a manufacturing plant, an eye-catching sign that will be visible at night, and a baseball stadium? Explain.

Applying Math

5. **Calculate Efficiency** A 25-W fluorescent light emits 5.0 J of thermal energy each second. What is the efficiency of the fluorescent light?

6. **Use Percentages** If 90 percent of the energy emitted by an incandescent bulb is thermal energy, how much thermal energy is emitted by a 60-W bulb each second?

Using Light

What **You'll Learn**
- **Distinguish** polarized light from unpolarized light.
- **Explain** how a hologram is made.
- **Determine** when total internal reflection occurs.
- **Describe** the uses of optical fibers.

Why **It's Important**
Light waves were used to transmit a TV program you watched today, or to send a telephone call to a friend.

Review Vocabulary
interference: occurs when two or more waves overlap and form a new wave

New Vocabulary
- polarized light
- holography
- total internal reflection

Polarized Light

You may have a pair of sunglasses with a sticker on them that says "polarized." Do you know what makes them different from other sunglasses? The difference has to do with the vibration of light waves that pass through the lenses. You can make transverse waves in a rope vibrate in any direction—horizontal, vertical, or anywhere in between. Light also is a transverse wave and can vibrate in any direction. In **polarized light,** however, the waves vibrate in only one direction.

Polarizing Filters If the light passes through a special polarizing filter, the light becomes polarized. A polarizing filter acts like a group of parallel slits. Only light waves vibrating in the same direction as the slits can pass through. If a second polarizing filter is lined up with its slits at right angles to those of the first filter, no light can pass through, as **Figure 21** shows.

Polarized lenses are useful for reducing glare without interfering with your ability to see clearly. When light is reflected from a horizontal surface, such as a lake or a shiny car hood, it becomes partially horizontally polarized. The lenses of polarizing sunglasses have vertical polarizing filters that block out the reflected light that has been polarized horizontally.

Figure 21 Slats in a fence behave like a polarizing filter for a transverse wave on a rope.

If the slats are aligned at right angles to each other, the wave can't pass through.

If the slats are in the same direction, the wave passes through.

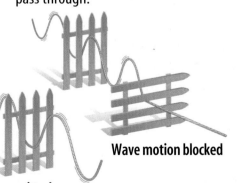

Wave motion blocked

Wave motion transmitted

Figure 22 Lasers can be used to make holograms like this one.

Holography

Science museums often have exhibits where a three-dimensional image seems to float in space, like the one shown in **Figure 22.** You can see the image from different angles, just as you would if you viewed the real object. Three-dimensional images on credit cards are produced by holography. **Holography** is a technique that produces a hologram—a complete three-dimensional photographic image of an object.

Making Holograms Illuminating objects with laser light produces holograms. Laser light reflects from the object onto photographic film. At the same time, a second beam split from the laser also is directed at the film. The light from the two beams creates an interference pattern on the film. The pattern looks nothing like the original object, but when laser light shines on the pattern on the film, a holographic image is produced.

Information in Light An ordinary photographic image captures only the brightness or intensity of light reflected from an object's surface, but a hologram records the intensity as well as the direction. As a result, it conveys more information to your eye than a conventional two-dimensional photograph does, but it also is more difficult to copy. Holographic images are used on credit cards, identification cards, and on the labels of some products to help prevent counterfeiting. Using X-ray lasers, scientists can produce holographic images of microscopic objects. It may be possible to create three-dimensional views of biological cells.

Topic: Holograms
Visit gpscience.com for Web links to information about holograms.

Activity Make an events-chain concept map of how a hologram is produced.

 *How are holographic images produced?*

Optical Fibers

When laser light must travel long distances or be sent into hard-to-reach places, optical fibers often are used. These transparent glass fibers can transmit light from one place to another. A process called total internal reflection makes this possible.

Total Internal Reflection Remember what happens when light speeds up as it travels from one medium to another. For example, when light travels from water to air the direction of the light ray is bent away from the normal, as shown in **Figure 23.** If the underwater light ray makes a larger angle with the normal, the light ray in the air bends closer the surface of the water. At a certain angle, called the critical angle, the refracted ray has been bent so that it is traveling along the surface of the water, as shown in **Figure 23.** For a light ray traveling from water into air, the critical angle is about 49°.

Figure 23 shows what happens if the underwater light ray strikes the boundary between the air and water at an angle larger than the critical angle. There is no longer any refraction, and the light ray does not travel in the air. Instead, the light ray is reflected at the boundary, just as if a mirror were there. This behavior of light is called total internal reflection. **Total internal reflection** occurs when light traveling from one medium to another is completely reflected at the boundary between the two materials. Then the light ray obeys the law of reflection. For total internal reflection to occur, light must travel slower in the first medium, and must strike the boundary at an angle greater than the critical angle.

Reading Check *How does total internal reflection occur?*

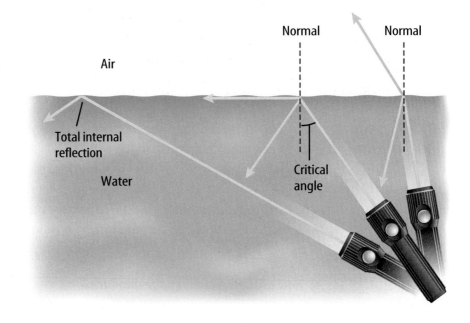

Figure 23 A light wave is bent away from the normal as it passes from water to air. At the critical angle, the refracted wave is traveling along the water surface. At angles greater than the critical angle, total internal reflection occurs.

Figure 24 Optical fibers make use of total internal reflection.

Light

Plastic covering

Glass core

Glass layer

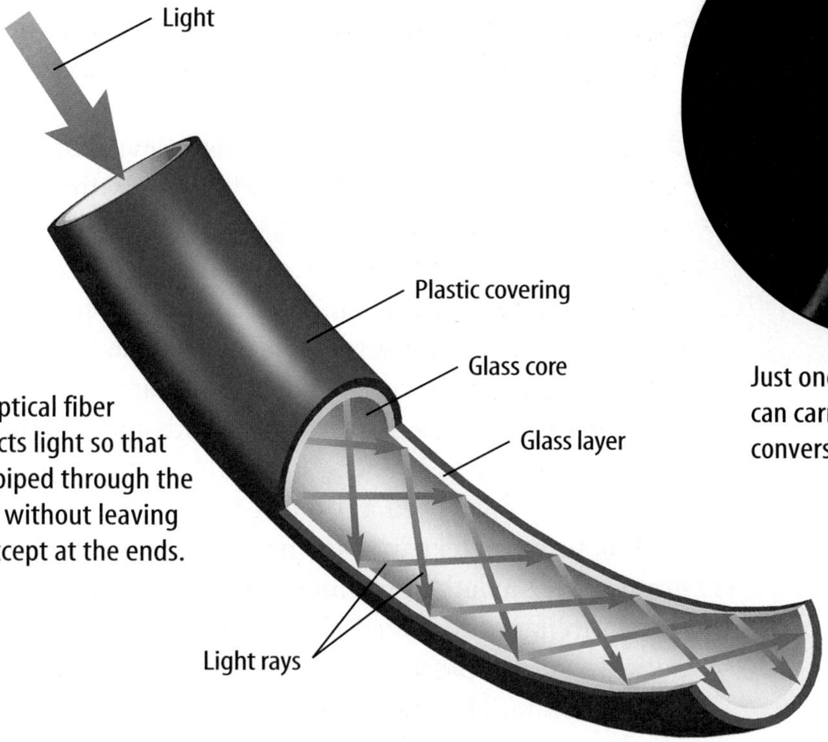

An optical fiber reflects light so that it is piped through the fiber without leaving it, except at the ends.

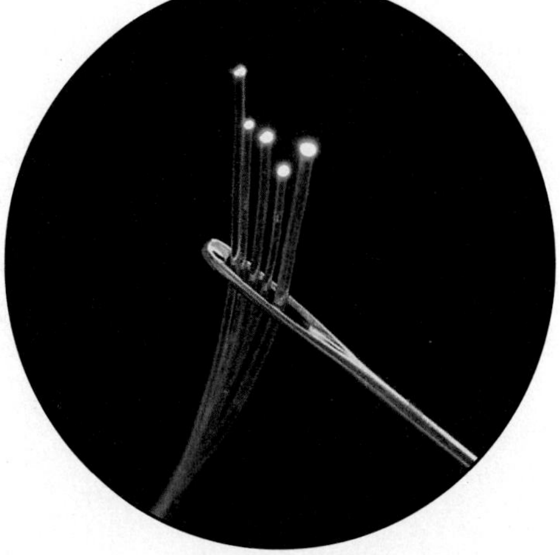

Light rays

Just one of these optical fibers can carry thousands of phone conversations at the same time.

Light Pipes Total internal reflection makes light transmission in optical fibers possible. As shown in **Figure 24,** light entering one end of the fiber is reflected continuously from the sides of the fiber until it emerges from the other end. Like water moves through a pipe, almost no light is lost or absorbed in optical fibers.

Using Optical Fibers Optical fibers are most often used in communications. Telephone conversations, television programs, and computer data can be coded in light beams. Signals can't leak from one fiber to another and interfere with other messages, so the signal is transmitted clearly. To send telephone conversations through an optical fiber, sound is converted into digital signals consisting of pulses of light by a light-emitting diode or a laser. Some systems use multiple lasers, each with its own wavelength, to fit multiple signals into the same fiber. You could send a million copies of the play *Romeo and Juliet* in one second on a single fiber. **Figure 24** shows the size of typical optical fibers.

Optical fibers also are used to explore the inside of the human body. One bundle of fibers transmits light, while another carries the reflected light back to the doctor.

INTEGRATE History

Wire Communication
At one time telegraph and telephone communications were transmitted only through wire lines. Today, optical fibers sometimes are used instead of wires to transmit communications. Research the history of wire communication. Create a time line of what you learn.

Optical Scanners

In supermarkets and many other kinds of stores, a cashier passes your purchases over a glass window in the checkout counter or holds a handheld device up to each item, like the one in **Figure 25.** In an instant, the optical scanner beeps and the price of the item appears on a screen. An optical scanner is a device that reads intensities of reflected light and converts the information to digital signals. You may have noticed that somewhere on each item the cashier scans is a pattern of thick and thin stripes called a bar code. An optical scanner detects the pattern and translates it into a digital signal, which goes to a computer. The computer searches its database for a matching item, finds its price, and sends the information to the cash register.

You may have used another type of optical scanner to convert pictures or text into forms you can use in computer programs. With a flatbed scanner, for example, you lay a document or picture facedown on a sheet of glass and close the cover. An optical scanner passes underneath the glass and reads the pattern of colors. The scanner converts the pattern to an electronic file that can be stored on a computer.

Figure 25 Optical scanners like this one use lasers to find the price of various products.

section 4 review

Summary

Polarized Light

- Polarized light has light waves that vibrate in only one direction.
- Polarizing filters block light waves that vibrate at right angles to the direction of the filter.

Holography

- A hologram is produced by interference between laser light reflected from an object and a second laser beam.

Optical Fibers

- Total internal reflection occurs when light is completely reflected at the boundary between two materials.
- Total internal reflection will occur when the angle between the light beam and the normal is equal to, or greater than, the critical angle.
- Optical fibers use total internal reflection to transmit light waves over long distances.

Self Check

1. **Explain** why polarized sunglasses can reduce the glare from light reflected from lakes and ponds.

2. **Describe** why a holographic image is considered to be three-dimensional.

3. **List** all the conditions that are necessary for total internal reflection to occur.

4. **Discuss** how optical fibers are used to transmit telephone conversations.

5. **Think Critically** On a sunny day you are looking at the surface of a lake through polarized sunglasses. How could you use your sunglasses to tell if the light reflected from the lake is polarized?

Applying Math

6. **Calculate Number of Fibers** An optical fiber has a diameter of 0.3 mm. How many fibers would be needed to form a cable with a square cross section, if the cross section was 1.5 cm on a side?

Make a Light Bender

From a hilltop you can see the reflection of pine trees and a cabin in the calm surface of a lake. This is possible because some of the light that reflects off these objects strikes the water's surface and reflects into your eyes. However, you don't see a clear, colorful image because much of the light enters the water rather than being reflected.

▶ Real-World Question

How does water affect the viewer's image of an object that is above the water's surface?

Goals
- ■ **Identify** reflection of an image in water.
- ■ **Identify** refraction of an image in water.

Materials
light source
unsharpened pencil
clear rectangular container
water
clay

Safety Precautions

▶ Procedure

1. Fill the container with water.

2. Place the container so that a light source—window or overhead light—reaches it.

3. Stand the pencil on end in the clay and place it by the container as shown in the figure above. The pencil must be taller than the level of the water. Also, place the pencil on the same side of the container as the light source.

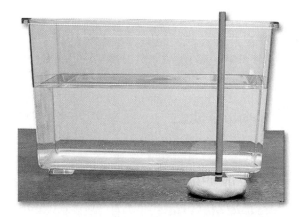

4. Looking down through the surface of the water from the side opposite the pencil, observe the reflection and refraction of the image of the pencil.

5. **Draw** a diagram of the image and label *Reflection* and *Refraction*.

6. Repeat steps 4 and 5 two more times, but position the pencil at two different angles.

▶ Conclude and Apply

1. **Discuss** how the image you see would change or be different if the surface of the water were a mirror.

2. **Predict** how the angles of reflection or refraction would change if the surface of the container were curved. Explain.

*C*ommunicating
Your Data

Make a poster of your diagrams and use it to explain reflection and refraction of light waves to your class. For more help, refer to your Science Skill Handbook.

Polarizing Filters

Goals

- **Demonstrate** when light does and does not shine through a pair of polarizing filters.
- **Predict** what will happen when you add a third polarizing filter.

Possible Materials

polarizing filters (3)
lamp or flashlight

Safety Precautions

WARNING: *Never look directly at the Sun, even with a polarizing filter.*

▶ Real-World Question

Polarizing filters cause light waves to vibrate only in one direction. Wearing polarized sunglasses can help reduce glare while allowing you to see clearly. If you have two polarizing filters on top of one another, when will light shine through and when will it not? What might happen if you added a third filter in between the first two?

▶ Form a Hypothesis

Form a hypothesis about how two polarizing filters that are placed on top of one another must be oriented for light to shine through and for no light to shine through.

▶ Test Your Hypothesis

Make a Plan

1. Using a pair of polarizing filters, choose at least three orientations of the filters to test your hypothesis.

2. When the two filters are oriented to allow the maximum amount of light to shine through, predict how a third filter placed between the two must be oriented for the same results.

3. Repeat step 2 but allow no light to shine through.

4. Make sure that your teacher approves your plan before you start.

Follow Your Plan

1. Using an appropriate light source, test when light does and does not shine through a pair of polarizing filters. Test each of the orientations you planned in step 1. Record the results.

2. Test three orientations of the third filter for allowing the maximum amount of light to pass through. Record the results.

3. Repeat step 2 but allow no light to shine through. Record the results.

⊙ *Analyze Your Data*

1. **Describe** how the pair of polarizing filters were oriented when light did and did not shine through. In cases where light did shine through, was it always the same amount of light? Or did the amount of light change in different orientations?

2. In each case, describe what happened when you added a third filter between the first two. How did the three orientations of the third filter change the amount of light that passed through? Explain.

⊙ *Conclude and Apply*

1. **Explain** why light did or did not shine through two polarizing filters in the various orientations.

2. **Evaluate** why or why not your hypothesis was supported.

3. **Infer** why light did or did not shine through various orientations of three polarizing filters.

4. **Analyze** whether or not your predictions were correct.

5. **Discuss** what you can conclude about the polarization of reflected light if a polarizing filter reduces the brightness of light reflected from the surface of a lake.

Communicating

Your Data

The next time you see a family member or friend wearing sunglasses, explain to them how polarizing lenses can reduce problems of glare.

A Haiku Garden:

The Four Seasons in Poems and Prints

by Stephen Addiss with Fumiko and Akira Yamamoto

Withered by winter
the sound of the wind—
one-color world

Basho

Lingering
in every pool of water—
spring sunlight

Issa

Understanding Literature

Japanese Haiku A haiku is a verse that consists of three lines and 17 syllables in the Japanese language. The first and third lines have five syllables each, and the middle line has seven syllables. Why is imagination important in reading Haiku?

Respond to the Reading

1. How do the illustrations help the reader better understand the poems?
2. What do you think is meant by the word *lingering* in the Haiku about spring sunlight?
3. **Linking Science and Writing** Write one haiku about summer and another about fall. In one poem, use color to help you describe the season. In the other, use light or some property of light to help describe the season.

INTEGRATE
Physics

Research has determined that there is a connection between color and mood. Warm colors have longer wavelengths, and can be more stimulating. Cool colors, which have shorter wavelengths, tend to have a calming or soothing effect on people. Light and color have long been used as literary symbols. Does the use of color change what you imagine when you read the haiku?

Reviewing Main Ideas

Section 1 · The Behavior of Light

1. When light interacts with matter, some light can be absorbed, some can be transmitted, and some can be reflected.

2. When light waves are reflected, they obey the law of reflection—the angle of incidence equals the angle of reflection.

3. Light waves are refracted, or bent, when a light wave changes speed as it travels from one material to another.

Section 2 · Light and Color

1. You see color when light is reflected off objects and into your eyes.

2. Specialized cells in your eyes called cones allow you to distinguish colors and shapes of objects. Other cells, called rods, allow you to see in dim light.

3. Red, blue, and green are the three primary colors of light and can be mixed to form all other colors.

4. The color of a pigment is due to the wavelengths of the light reflected from the pigment. The primary pigment colors are magenta, cyan, and yellow.

Section 3 · Producing Light

1. Incandescent bulbs produce light by heating a tungsten filament until it glows brightly.

2. Fluorescent bulbs give off light when ultraviolet radiation produced inside the bulb causes the phosphor coating inside the bulb to glow.

3. Neon lights contain a gas that glows when electric current passes through it.

4. A laser produces coherent light by emitting a beam of light waves that have only one wavelength, have their crests and troughs aligned, and are moving in a single direction.

Section 4 · Using Light

1. Polarized light consists of transverse waves that vibrate along only one plane.

2. Total internal reflection occurs when a light wave strikes the boundary between two materials at an angle greater than the critical angle.

3. Optical scanners sense reflected light and convert the information to digital signals.

FOLDABLES Use the Foldable that you made at the beginning of this chapter to help you review light transmission.

Using Vocabulary

coherent light p.398	opaque p.384
fluorescent light p.395	pigment p.392
holography p.401	polarized light p.400
incandescent light p.394	total internal reflection
incoherent light p.398	p.402
index of refraction p.386	translucent p.384
mirage p.388	transparent p.384

Answer the following questions using complete sentences.

1. What type of light does heating a filament until it glows produce?

2. What process would you use to produce a complete three-dimensional image of an object?

3. How would you describe an object that you can see through?

4. What process makes it possible for optical fibers to transmit telephone conversations over long distances?

5. What is a false image of a distant object?

6. What type of light has light waves that vibrate in only one direction?

Checking Concepts

Choose the word or phrase that best answers the question.

7. Which word describes materials that absorb or reflect all light?
   - **A)** translucent
   - **B)** opaque
   - **C)** ultraviolet
   - **D)** diffuse

8. What is the term for the property of a material that indicates how much light slows down when traveling in the material?
   - **A)** pigment
   - **B)** filter
   - **C)** index of refraction
   - **D)** mirage

9. Which of the following explains why a prism separates white light into the colors of the rainbow?
   - **A)** interference
   - **B)** fluorescence
   - **C)** diffraction
   - **D)** refraction

10. What do you see when noting the color of an object?
    - **A)** the light it reflects
    - **B)** the light it absorbs
    - **C)** polarization
    - **D)** diffuse reflection

11. What do the phosphors inside fluorescent bulbs absorb to create a glow?
    - **A)** incandescent light
    - **B)** ultraviolet radiation
    - **C)** halogens
    - **D)** argon

12. What term describes objects that allow some light, but not all light to pass through them?
    - **A)** translucent
    - **B)** reflective
    - **C)** transparent
    - **D)** opaque

13. Which light waves are bent most when passing through a prism?
    - **A)** red waves
    - **B)** yellow waves
    - **C)** blue waves
    - **D)** violet waves

14. Which type of cells in your eyes allows you to see the color violet?
    - **A)** red cones
    - **B)** green cones
    - **C)** blue cones
    - **D)** rods

15. What color of light is produced when the three primary colors of light are combined in equal amounts?
    - **A)** black
    - **B)** yellow
    - **C)** white
    - **D)** cyan

16. Which of the following terms best describes laser light?
    - **A)** incoherent
    - **B)** coherent
    - **C)** incandescent
    - **D)** fluorescent

Interpreting Graphics

17. **Concept Map** Copy and complete this concept map to show the steps in the production of fluorescent light.

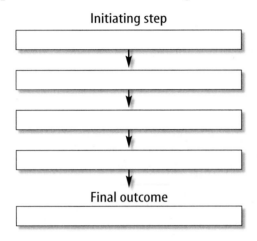

Initiating step

↓

↓

↓

Final outcome

18. **Concept Map** Copy and complete the following concept map about producing light.

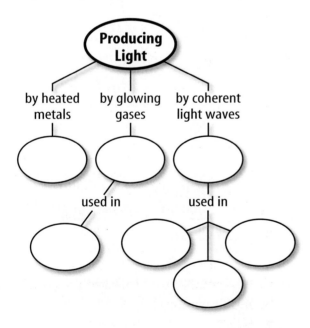

Thinking Critically

19. **Explain** how light is produced by an incandescent bulb. What is a disadvantage of these bulbs?

20. **Compare and contrast** the reflection of light from a white wall with a rough surface with the reflection of light from a mirror.

21. **Predict** what color a white shirt would appear to be if the light reflected from the shirt passed through a red filter and then through a green filter.

22. **Identify** the color of light that changes speed the most as it passes through a prism. Explain your reasoning.

23. **Infer** Most mammals, including dogs and cats, can't see colors. Infer how the retina of a cat's eye might be different from the retina of a human eye.

24. **Compare** White light passes through a translucent pane of glass, and shines on a shirt. Both the translucent glass and the shirt appear green. Compare the colors of light that are absorbed and transmitted by the glass and the shirt.

25. **Explain** The speed of light is greater in air than in glass. Explain whether or not internal reflection could occur when a light wave traveling in air strikes the glass.

Applying Math

26. **Calculate Angle of Incidence** A light ray is reflected from a mirror. If the angle between the incident ray and the reflected ray is 136 degrees, what is the angle of incidence?

27. **Calculate Speed** The index of refraction of a material equals the speed of light in a vacuum, which is 300,000 km/s, divided by the speed of light in the material. What is the speed of light in water if the index of refraction for water is 1.33?

Part 1 | Multiple Choice

Record your answer on the answer sheet provided by your teacher or on a sheet of paper.

1. Which word describes materials that transmit almost all of the light that strikes them?
   - **A.** translucent
   - **B.** transparent
   - **C.** opaque
   - **D.** diffuse

Use the diagram below to answer questions 2 and 3.

2. How are the light waves shown described?
   - **A.** coherent
   - **B.** incoherent
   - **C.** opaque
   - **D.** translucent

3. What device produces these light waves?
   - **A.** fluorescent light
   - **B.** sodium-vapor light
   - **C.** laser
   - **D.** incandescent light

4. What happens when light traveling at an angle passes from one material into another?
   - **A.** The light is reflected.
   - **B.** The light is refracted.
   - **C.** The light always speeds up.
   - **D.** The light changes color.

5. Why does an apple look red?
   - **A.** It reflects red light.
   - **B.** It absorbs red light.
   - **C.** It reflects all colors of light.
   - **D.** It reflects all colors of light except red.

6. Which of the following processes is used in optical fibers to transmit light?
   - **A.** diffuse reflection
   - **B.** total internal reflection
   - **C.** polarization
   - **D.** incandescence

7. What part of the eye enables you see color in bright light?
   - **A.** rods
   - **B.** retina
   - **C.** cones
   - **D.** lens

8. What do the phosphors inside a fluorescent bulb absorb to make the bulb emit visible light?
   - **A.** infrared radiation
   - **B.** ultraviolet radiation
   - **C.** electrons
   - **D.** sodium vapor

Use the illustration below to answer questions 9–11.

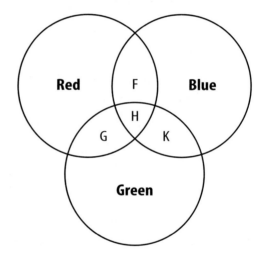

9. When the primary colors of light are added together, what color appears in area H?
   - **A.** magenta
   - **B.** yellow
   - **C.** cyan
   - **D.** white

10. What color appears in area G?
    - **A.** magenta
    - **B.** yellow
    - **C.** cyan
    - **D.** white

11. Which of the following light sources produces light by making a piece of metal hot enough to glow?
    - **A.** fluorescent light
    - **B.** incandescent light
    - **C.** neon light
    - **D.** laser light

Part 2 | Short Response/Grid In

Record your answers on the answer sheet provided by your teacher or on a sheet of paper.

Use the illustration below to answer questions 12 and 13.

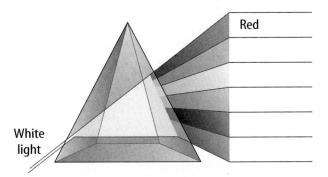

12. Identify the object shown in the figure and explain how it affects white light that passes through it.

13. Identify the colors that would be seen on the right side of the object, in order of decreasing wavelength.

14. Why would a green leaf appear to be black when you look at the leaf through a red filter?

15. Describe the effect that diffuse reflection would have on a beam of parallel light rays.

16. Describe the difference between an object that appears black and an object that appears white.

17. If green light shines on a black-and-white striped shirt, what colors will the stripes appear to be?

18. Why is an electric current needed to produce light in fluorescent lights and in neon lights?

19. For a regular reflection to occur, how must the roughness of a surface compare to the wavelengths of light it reflects?

Part 3 | Open Ended

Record your answers on a sheet of paper.

20. Why can the human eye see colors better in bright light than in dim light?

21. Explain why, compared to the light from a lightbulb, a beam of laser light can deliver a large amount of energy to a small area.

22. Explain whether or not a large bottle made of green glass would be a suitable container in which to grow small plants.

Use the illustration below to answer questions 23 and 24.

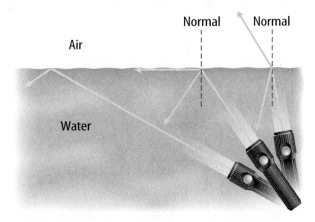

23. When a light wave traveling in water reaches the water's surface, under what circumstances will total internal reflection occur?

24. If the flashlight were in air and the direction of the light beams shown in the figure were reversed. Under what circumstances would total internal reflection occur? Explain.

Test-Taking Tip

Answer All Parts Make sure each part of a question is answered when listing discussion points. For example, if the question asks you to compare and contrast, make sure you list similarities and differences

Question 17 Be sure to describe the change in color of both the black and the white stripes.

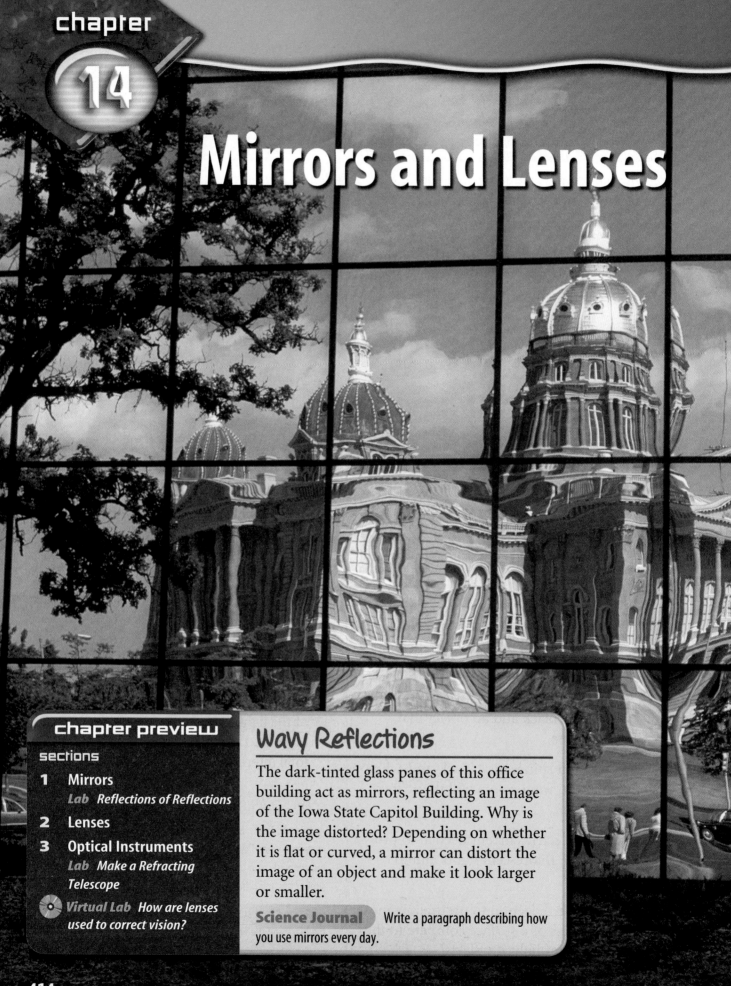

Mirrors and Lenses

Wavy Reflections

The dark-tinted glass panes of this office building act as mirrors, reflecting an image of the Iowa State Capitol Building. Why is the image distorted? Depending on whether it is flat or curved, a mirror can distort the image of an object and make it look larger or smaller.

Science Journal   Write a paragraph describing how you use mirrors every day.

Start-Up Activities

Making a Water Lens

Have you ever used a magnifying glass, a camera, a microscope, or a telescope? If so, you were using a lens to create an image. A lens is a transparent material that bends rays of light and forms an image. In this activity, you will use water to create a lens.

1. Cut a 10-cm × 10-cm piece of plastic wrap. Set it on a page of printed text.

2. Place a small water drop on the plastic. Look at the text through the drop. What do you observe?

3. Make your water drop larger and observe the text through it again.

4. Carefully lift the piece of plastic wrap a few centimeters above the text and look at the text through the water drop again.

5. **Think Critically** Describe how the text looked in steps 2, 3, and 4. Why do you think water affects the way the text looks? What other materials might you use to change the appearance of the text?

  Preview this chapter's content and activities at gpscience.com

Types of Mirrors Make the following Foldable to help identify the three different types of mirrors and their characteristics.

STEP 1 Fold a vertical sheet of paper from side to side. Make the front edge about 1.25 cm shorter than the back edge.

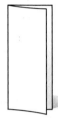

STEP 2 Turn lengthwise and fold into thirds.

STEP 3 Unfold and cut only the top layer along both folds to make three tabs.

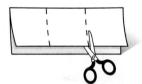

STEP 4 Label each tab as shown.

Plane Mirror | Concave Mirror | Convex Mirror

Read and Write As you read Section 1 of this chapter, write down important information under the appropriate tab about each type of mirror.

Reading Guide

What You'll Learn

- **Describe** how an image is formed in three types of mirrors.
- **Explain** the difference between real and virtual images.
- **Identify** examples and uses of plane, concave, and convex mirrors.

Why It's Important

Mirrors enable you to check your appearance, see objects behind you, and produce beams of light.

Review Vocabulary

reflection: occurs when waves change direction after striking a surface

New Vocabulary

- plane mirror
- virtual image
- concave mirror
- optical axis
- focal point
- focal length
- real image
- convex mirror

How do you use light to see?

Have you tried to read a book under the covers with only a small flashlight? Or have you ever tried to find an address number on a house or an apartment at night on a poorly lit street? It's harder to do those activities in the dark than it is when there is plenty of light. Your eyes see by detecting light, so anytime you see something, it is because light has come from that object to your eyes. Light is emitted from a light source, such as the Sun or a lightbulb, and then reflects off an object, such as the page of a book or someone's face. When light travels from an object to your eye, you see the object. Light can reflect more than once. For example, light can reflect off of an object into a mirror and then reflect into your eyes. When no light is available to reflect off of objects and into your eye, your eyes cannot see anything. This is why it is hard to read a book or see an address in the dark.

Light Rays Light sources send out light waves that travel in all directions. These waves spread out from the light source just as ripples on the surface of water spread out from the point of impact of a pebble.

You also could think of the light coming from the source as being many narrow beams of light. Each narrow beam of light travels in a straight line and is called a light ray. **Figure 1** shows how a light source, such as a candle, gives off light rays that travel away from the source in all directions. Even though light rays can change direction when they are reflected or refracted, your brain interprets images as if light rays travel in a single direction.

Figure 1 A light source, like a candle, sends out light rays in all directions.

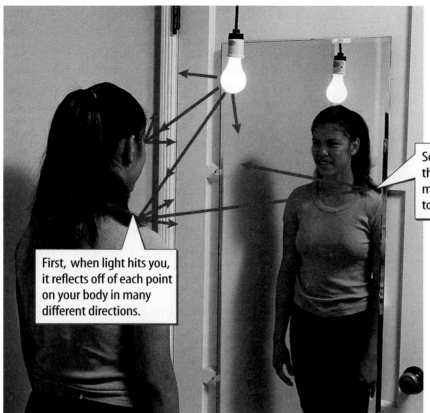

Figure 2 Seeing an image of yourself in a mirror involves two sets of reflections.

First, when light hits you, it reflects off of each point on your body in many different directions.

Some of the light rays then travel toward the mirror and reflect back toward your eyes.

Seeing Reflections with Plane Mirrors

Greek mythology tells the story of a handsome young man named Narcissus who noticed his image in a pond and fell in love with himself. Like pools of water, mirrors are smooth surfaces that reflect light to form images. Just as Narcissus did, you can see yourself as you glance into a quiet pool of water or walk past a shop window. Most of the time, however, you probably look for your image in a flat, smooth mirror called a **plane mirror.**

Reading Check *What is a plane mirror?*

Reflection from Plane Mirrors What do you see when you look into a plane mirror? Your reflection appears upright. If you were 1 m from the mirror, your image would appear to be 1 m behind the mirror, or 2 m from you. In fact, your image is what someone standing 2 m from you would see. **Figure 2** shows how your image is formed by a plane mirror. First, light rays from a light source strike you. Every point that is struck by the light rays reflects these rays so they travel outward in all directions. If your friend were looking at you, these reflected light rays coming from you would enter her eyes so she could see you. However, if a mirror is placed between you and your friend, the light rays are reflected from the mirror back to your eyes.

INTEGRATE
Life Science

Mirror Images Your left hand and right hand are mirror images of each other. Some of the molecules in your body exist in two forms that are mirror images. However, your body uses some molecules only in the left-handed form and other molecules only in the right-handed form. Using different colors of gumdrops and toothpicks, make a model of a molecule that has a mirror image.

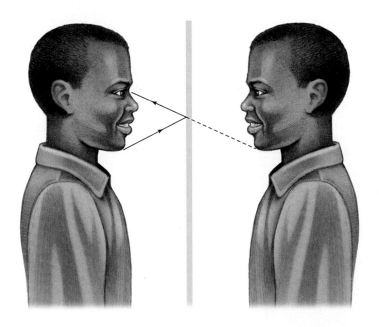

Figure 3 Your brain thinks that the light rays that reflect off of the mirror come from a point behind the mirror.

Infer *how the size of your image in a plane mirror depends on your distance from the mirror.*

Virtual Images You can understand how your brain interprets your reflection in a mirror by looking at **Figure 3.** The light waves that are reflected off of you travel in all directions. Light rays reflected from your chin strike the mirror at different places. Then, they reflect off of the mirror in different directions. Recall that your brain always interprets light rays as if they have traveled in a straight line. It doesn't realize that the light rays have been reflected and that they changed direction. If the reflected light rays were extended back behind the mirror, they would meet at a single point. Your brain interprets the rays that enter your eye as coming from this point behind the mirror. You seem to see the reflected image of your chin at this point. An image like this, which your brain perceives even though no light rays pass through it, is called a **virtual image.** The virtual image formed by a plane mirror is always upright and appears to be as far behind the mirror as the object is in front of it.

Concave Mirrors

Not all mirrors are flat like plane mirrors are. If the surface of a mirror is curved inward, it is called a **concave mirror.** Concave mirrors, like plane mirrors, reflect light waves to form images. The difference is that the curved surface of a concave mirror reflects light in a unique way.

Features of Concave Mirrors A concave mirror has an optical axis. The **optical axis** is an imaginary straight line drawn perpendicular to the surface of the mirror at its center. Every light ray traveling parallel to the optical axis as it approaches the mirror is reflected through a point on the optical axis called the **focal point.** Using the focal point and the optical axis, you can diagram how some of the light rays that travel to a concave mirror are reflected, as shown in **Figure 4.** On the other hand, if a light ray passes through the focal point before it hits the mirror, it is reflected parallel to the optical axis. The distance from the center of the mirror to the focal point is called the **focal length.**

Figure 4 A concave mirror has an optical axis and a focal point. When light rays travel toward the mirror parallel to the optical axis, they reflect through the focal point.

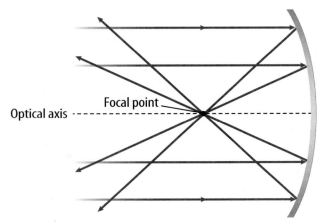

Optical axis

Focal point

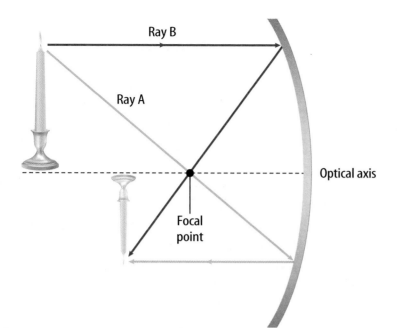

Ray B

Ray A

Optical axis

Focal
point

Figure 5 Rays A and B start from the same place on the candle, travel in different directions, and meet again on the reflected image.

Diagram *how other points on the image of the candle are formed.*

How a Concave Mirror Works The image that is formed by a concave mirror changes depending on where the object is located relative to the focal point of the mirror. You can diagram how an image is formed. For example, suppose that the distance between the object, such as the candle in **Figure 5,** and the mirror is a little greater than the focal length. Light rays bounce off of each point on the candle in all directions. One light ray, labeled Ray A, starts from a point on the flame of the candle and passes through the focal point on its way to the mirror. Ray A is then reflected so it travels parallel to the optical axis. Another ray, Ray B, starts from the same point on the candle's flame but travels parallel to the optical axis as it moves toward the mirror. When Ray B is reflected by the mirror, it passes through the focal point. The place where Ray A and Ray B meet after they are reflected forms a point on the flame of the reflected image.

More points on the reflected image can be located in this way. From each point on the candle, one ray can be drawn that passes through the focal point and is reflected parallel to the optical axis. Another ray can be drawn that travels parallel to the optical axis and passes through the focal point after it is reflected. The point where the two rays meet is on the reflected image.

Real Images The image that is formed by the concave mirror is not virtual. Rays of light pass through the location of the image. A **real image** is formed when light rays converge to form the image. You could hold a sheet of paper at the location of a real image and see the image projected on the paper. When an object is farther from a concave mirror than twice the focal length, the image that is formed is real, smaller, and upside down, or inverted.

Mini LAB

Observing Images in a Spoon

Procedure
1. Look at the inside of a shiny **spoon**. Move it close to your face and then far away. The place where your image changes is the focal point.
2. Hold the inside of the spoon facing a bright **light,** a little farther away than the focal length of the spoon.
3. Place a piece of **poster board** between the light and the spoon without blocking all of the light.
4. Move the poster board between the spoon and the light until you see the reflected light on it.

Analysis
Which of the images you observed were real and which were virtual?

Try at Home

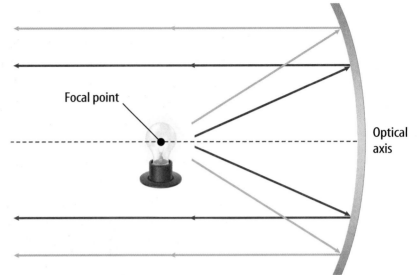

Figure 6 A flashlight uses a concave mirror to create a beam of light.
Explain *why the reflected rays of light in the diagram are parallel to each other.*

Focal point

Optical axis

Creating Light Beams What happens if you place an object exactly at the focal point of the concave mirror? **Figure 6** shows that if the object is at the focal point, the mirror reflects all light rays parallel to the optical axis. No image forms because the rays never meet—not even if the rays are extended back behind the mirror. Therefore, a light placed at the focal point is reflected in a beam. Car headlights, flashlights, lighthouses, spotlights, and other devices use concave mirrors in this way to create concentrated light beams of nearly parallel rays.

Mirrors That Magnify The image formed by a concave mirror changes again when you place an object between it and its focal point. The location of the reflected image again can be found by drawing two rays from each point. **Figure 7** shows that in this case, these rays never meet after they are reflected. Instead, the reflected rays diverge. Just as it does with a plane mirror, your brain interprets the diverging rays as if they came from one point behind the mirror. You can find this point by extending the rays behind the mirror until they meet. Because no light rays are behind the mirror where the image seems to be, the image formed is virtual. The image also is upright and enlarged.

Shaving mirrors and makeup mirrors are concave mirrors. They form an enlarged, upright image of a person's face so it's easier to see small details. The bowl of a shiny spoon also forms an enlarged, upright image of your face when it is placed close to your face.

Figure 7 If the candle is between the mirror and its focal point, the reflected image is enlarged and virtual.
Infer *why this image couldn't be projected on a screen.*

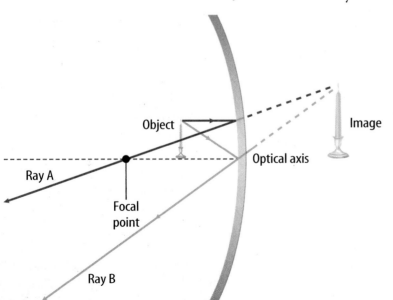

Object

Image

Optical axis

Ray A

Focal point

Ray B

Convex Mirrors

Why do you think the security mirrors in banks and stores are shaped the way they are? The next time you are in a store, look up to one of the back corners or at the end of an aisle to see if a large, rounded mirror is mounted there. You can see a large area of the store in the mirror. A mirror that curves outward like the back of a spoon is called a **convex mirror.** Light rays that hit a convex mirror diverge, or spread apart, after they are reflected. Look at **Figure 8** to see how the rays from an object are reflected to form an image. The reflected rays diverge and never meet, so the image formed by a convex mirror is a virtual image. The image also is always upright and smaller than the actual object is.

✔ **Reading Check** *Describe the image formed by a convex mirror.*

Uses of Convex Mirrors Because convex mirrors cause light rays to diverge, they allow large areas to be viewed. As a result, a convex mirror is said to have a wide field of view. In addition to increasing the field of view in places like grocery stores and factories, convex mirrors can widen the view of traffic that can be seen in rearview or side-view mirrors of automobiles. However, because the image created by a convex mirror is smaller than the actual object, your perception of distance can be distorted. Objects look farther away than they truly are in a convex mirror. Distances and sizes seen in a convex mirror are not realistic, so most convex side mirrors carry a printed warning that says "Objects in mirror are closer than they appear."

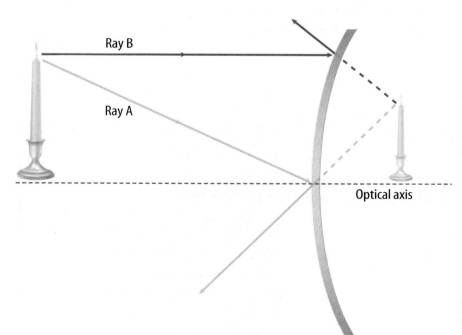

Ray B

Ray A

Optical axis

Figure 8 A convex mirror forms a reduced, upright, virtual image.

Table 1 Images Formed by Mirrors

Mirror Shape	Position of Object	Virtual/Real	Image Created Upright/Upside Down	Size
Plane		virtual	upright	same as object
Concave	Object more than two focal lengths from mirror	real	upside down	smaller than object
	Object between one and two focal lengths	real	upside down	larger than object
	Object at focal point	none	none	none
	Object within focal length	virtual	upright	larger than object
Convex		virtual	upright	smaller than object

Mirror Images The different shapes of plane, concave, and convex mirrors cause them to reflect light in distinct ways. Each type of mirror has different uses. **Table 1** summarizes the images formed by plane, concave, and convex mirrors.

section 1 review

Summary

How do you use light to see?
- You see an object because your eyes detect the light reflected from that object.

Seeing with Plane Mirrors
- Plane mirrors are smooth and flat.
- A plane mirror forms upright, virtual images.
- No light rays pass through the location of a virtual image.

Concave Mirrors
- A concave mirror curves inward.
- The image formed by a concave mirror depends on the location of an object.

Convex Mirrors
- A convex mirror curves outward.
- Convex mirrors produce virtual, upright images that are smaller than the object.

Self Check

1. **Describe** how your image in a plane mirror changes as you move closer to the mirror.
2. **Diagram** how light rays from an object are reflected by a convex mirror to form an image.
3. **Describe** the image of an object that is 38 cm from a concave mirror that has a focal length of 10 cm.
4. **Infer** An object is less than one focal length from a concave mirror. How does the size of the image change as the object gets closer to the mirror?
5. **Think Critically** Determine whether or not a virtual image can be photographed.

Applying Math

6. **Calculate Angle of Reflection** A light ray from a flashlight strikes a plane mirror so that the angle between the mirror's surface and the light ray is 60°. What is the angle of reflection?

Science Online gpscience.com/self_check_quiz

REFLECTIONS OF REFLECTIONS

How can you see the back of your head? You can use two mirrors to view a reflection of a reflection of the back of your head.

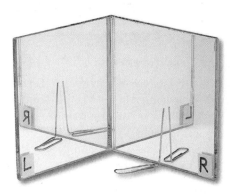

▶ Real-World Question

How many reflections can you see with two mirrors?

Goals

■ **Infer** how the number of reflections depends on the angle between mirrors.

Materials

plane mirrors (2)   protractor
masking tape   paper clip

Safety Precautions

Handle glass mirrors and paper clips carefully.

▶ Procedure

1. Lay one mirror on top of the other with the mirror surfaces inward. Tape them together so they will open and close. Use tape to label them *L* and *R*.

2. Stand the mirrors up on a sheet of paper. Using the protractor, close the mirrors to an angle of 72°.

3. Bend one leg of a paper clip up 90° and place it close to the front of the R mirror.

4. Count the number of images of the clip you see in the *R* and *L* mirrors. Record these numbers in the data table.

5. The mirror arrangement creates an image of a circle divided into wedges by the mirrors. Record the number of wedges.

6. Hold the *R* mirror still and slowly open the *L* mirror to 90°. Count and record the images of the clip and the wedges in the circle. Repeat, this time opening the mirrors to 120°.

▶ Conclude and Apply

1. **Infer** the relationship between the number of wedges and paper clip images you can see.

2. **Determine** the angle that would divide a circle into six wedges. Hypothesize how many images would be produced.

𝒞ommunicating Your Data

Demonstrate for younger students the relationship between the angle of the mirrors and the number of reflections.

Images and Wedges Seen in the Mirrors

Angle of Mirrors	Number of Paper Clip Images		Number of Wedges
	R	L	
72°			
90°	Do not write in this book.		
120°			

Lenses

Reading Guide

***What* You'll Learn**

- **Describe** the shapes of convex and concave lenses.
- **Explain** how convex and concave lenses form images.
- **Explain** how lenses are used to correct vision problems.

***Why* It's Important**

Even if you don't wear eyeglasses or contacts, you still use lenses to see.

Review Vocabulary

transparent: a material that transmits almost all the light that strikes it

New Vocabulary

- convex lens
- concave lens
- cornea
- retina

What is a lens?

What do your eyes have in common with cameras, eyeglasses, and microscopes? Each of these things contains at least one lens. A lens is a transparent material with at least one curved surface that causes light rays to bend, or refract, as they pass through. The image that a lens forms depends on the shape of the lens. Like curved mirrors, a lens can be convex or concave.

Convex Lenses

A **convex lens** is thicker in the middle than at the edges. Its optical axis is an imaginary straight line that is perpendicular to the surface of the lens at its thickest point. When light rays approach a convex lens traveling parallel to its optical axis, the rays are refracted toward the center of the lens, as in **Figure 9.** All light rays traveling parallel to the optical axis are refracted so they pass through a single point, which is the focal point of the lens. The focal length of the lens depends on the shape of the lens. If the sides of a convex lens are less curved, light rays are bent less. As a result, lenses with flatter sides have longer focal lengths. **Figure 9** also shows that light rays traveling along the optical axis are not bent at all.

Figure 9 Convex lenses are thicker in the middle than at the edges. A convex lens focuses light rays at a focal point. A light ray that passes straight through the center of the lens is not refracted.

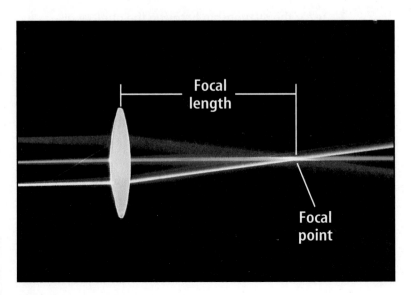

Focal length

Focal point

Figure 10 The image formed by a convex lens depends on the positions of the lens and the object.

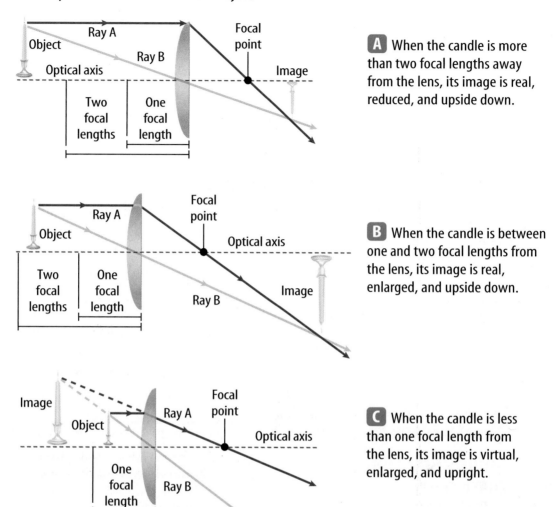

A When the candle is more than two focal lengths away from the lens, its image is real, reduced, and upside down.

B When the candle is between one and two focal lengths from the lens, its image is real, enlarged, and upside down.

C When the candle is less than one focal length from the lens, its image is virtual, enlarged, and upright.

Forming Images with a Convex Lens The type of image a convex lens forms depends on where the object is relative to the focal point of the lens. If an object is more than two focal lengths from the lens, as in **Figure 10A,** the image is real, reduced, and inverted, and on the opposite side of the lens from the object.

As the object moves closer to the lens, the image gets larger. **Figure 10B** shows the image formed when the object is between one and two focal lengths from the lens. Now the image is larger than the object, but is still inverted.

When an object is less than one focal length from the lens, as in **Figure 10C,** the image becomes an enlarged, virtual image. The image is virtual because light rays from the object diverge after they pass through the lens. When you use a magnifying glass, you move a convex lens so that it is less than one focal length from an object. This causes the image of the object to be magnified.

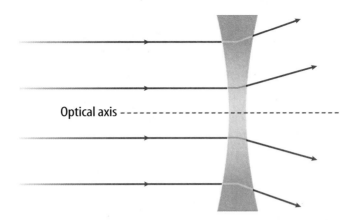

Figure 11 A concave lens refracts light rays so they spread out.
Classify *Is a concave lens most like a concave mirror or a convex mirror?*

Optical axis

Concave Lenses

A **concave lens** is thinner in the middle and thicker at the edges. As shown in **Figure 11,** light rays that pass through a concave lens bend outward away from the optical axis. The rays spread out and never meet at a focal point, so they never form a real image. The image is always virtual, upright, and smaller than the actual object is. Concave lenses are used in some types of eyeglasses and some telescopes. Concave lenses usually are used in combination with other lenses. A summary of the images formed by concave and convex lenses is shown in **Table 2** on the next page.

Applying Science

Comparing Object and Image Distances

The size and orientation of an image formed by a convex lens depends on the location of the object. What happens to the location of the image formed by a convex lens as the object moves closer to or farther from the lens? The distance from the lens to the object is the object distance, and the distance from the lens to the image is the image distance. How are the focal length, object distance, and image distance related to each other?

Identifying the Problem

A 5-cm-tall object is placed at different lengths from a double convex lens with a focal length of 15 cm. The table above lists the different object and image distances. How are these two measurements related?

Object and Image Distances

Focal Length	Object Distance	Image Distance
15.0 cm	45.0 cm	22.5 cm
15.0 cm	30.0 cm	30.0 cm
15.0 cm	20.0 cm	60.0 cm

Solving the Problem

1. What is the relationship between the object distance and the image distance?
2. The lens equation describes the relationship between the focal length and the image and object distances.
1/focal length = 1/object distance + 1/image distance
Using this equation, calculate the image distance of an object placed at a distance of 60.0 cm from the lens.

Table 2 Images Formed by Lenses

Lens Shape	Location of Object	Type of Image Virtual/Real	Type of Image Upright/Inverted	Type of Image Size
Convex	Object beyond 2 focal lengths from lens	real	inverted	smaller than object
	Object between 1 and 2 focal lengths	real	inverted	larger than object
	Object within 1 focal length	virtual	upright	larger than object
Concave	Object at any position	virtual	upright	smaller than object

Lenses and Eyesight

INTEGRATE Life Science

What determines how well you can see the words on this page? If you don't need eyeglasses, the structure of your eye gives you the ability to focus on these words and other objects around you. Look at **Figure 12.** Light enters your eye through a transparent covering on your eyeball called the **cornea** (KOR nee uh). The cornea causes light rays to bend so that they converge. The light then passes through an opening called the pupil. Behind the pupil is a flexible convex lens. The lens helps focus light rays so that a sharp image is formed on your retina. The **retina** is the inner lining of your eye. It has cells that convert the light image into electrical signals, which are then carried along the optic nerve to your brain to be interpreted.

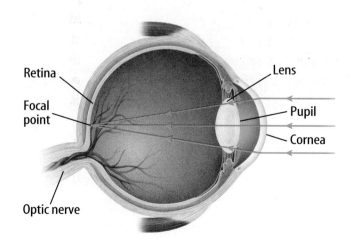

Retina
Focal point
Optic nerve
Lens
Pupil
Cornea

Figure 12 The cornea and lens in your eye focus light rays so that a sharp image is formed on the retina.

Focusing on Near and Far How can your eyes focus both on close objects, like the watch on your wrist, and distant objects, like a clock across the room? For you to see an object clearly, its image must be focused sharply on your retina. However, the retina is always a fixed distance from the lens. Remember that the location of an image formed by a convex lens depends on the focal length of the lens and the location of the object. For example, look back at **Figure 10.** As an object moves farther from a convex lens, the position of the image moves closer to the lens.

For an image to be formed on the retina, the focal length of the lens needs to be able to change as the distance of the object changes. The lens in your eye is flexible, and muscles attached to it change its shape and its focal length. This is why you can see objects that are near and far away.

Look at **Figure 13.** As an object gets farther from your eye, the focal length of the lens has to increase. The muscles around the lens stretch it so it has a less convex shape. But when you focus on a nearby object, these muscles make the lens more curved, causing the focal length to decrease.

 *How does the shape of the lens in your eye change when you focus on a nearby object?*

When an object is far away, your lens is less convex.

Lens less convex

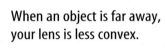

Figure 13 The lens in your eye changes shape so you can focus on objects at different distances.

To focus on a close object, your lens becomes more convex.

Lens more convex

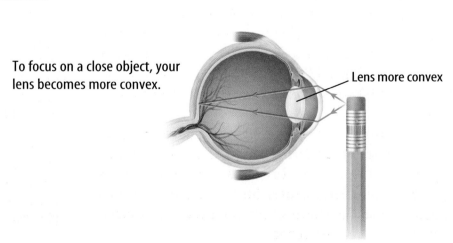

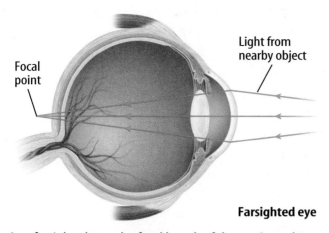

Farsighted eye

In a farsighted eye, the focal length of the eye is too long to form a sharp image of nearby objects on the retina.

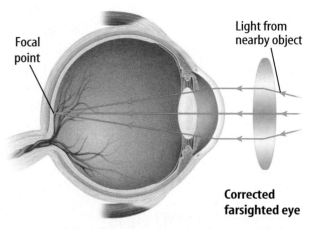

Corrected farsighted eye

A convex lens in front of a farsighted eye enables a sharp image to be focused on the retina.

Figure 14 Farsightedness can be corrected by convex lenses.

Vision Problems

People that have good vision can see objects clearly that are about 25 cm or farther away from their eyes. However, people with the most common vision problems see objects clearly only at some distances, or see all objects as being blurry.

Farsightedness A person who is farsighted can see distant objects clearly, but can't bring nearby objects into focus. Light rays from nearby objects do not converge enough after passing through the cornea and the lens to form a sharp image on the retina, as shown in **Figure 14.** The problem can be corrected by using convex lenses that cause incoming light rays to converge before they enter the eye, as in **Figure 14.**

As many people age, their eyes develop a condition that makes them unable to focus on close objects. The lenses in their eyes become less flexible. The muscles around the lenses still contract as they try to change the shape of the lens. However, the lenses have become more rigid, and cannot be made curved enough to form an image on the retina. People who are more than 40 years old might not be able to focus on objects closer than 1 m from their eyes. Some vision problems are caused by diseases of the retina. **Figure 15** shows how using new technology allows people with diseased retinas to recover some vision.

Astigmatism Another vision problem, called astigmatism occurs when the surface of the cornea is curved unevenly. When people have astigmatism, their corneas are more oval than round in shape. Astigmatism causes blurry vision at all distances. Corrective lenses also have an uneven curvature, canceling out the effect of an uneven cornea.

Eyeglasses and the Printing Press
Eyeglasses were first developed in Italy in the thirteenth century and were used mainly by nobles and the clergy. However, in 1456 the printing press was invented in Germany by Johannes Gutenberg. As books became more available, the demand for eyeglasses increased. In turn, the increasing availability of eyeglasses enabled more people to read. This helped increase the demand for books. Research the development of eyeglasses.

Figure 15

Millions of people worldwide suffer from vision problems associated with diseases of the retina. Until recently, such people had little hope of improving their eyesight. Now, however, scientists are developing specialized silicon chips that convert light into electrical pulses, mimicking the function of the retina. When implanted in the eye, these artificial silicon retinas may restore sight.

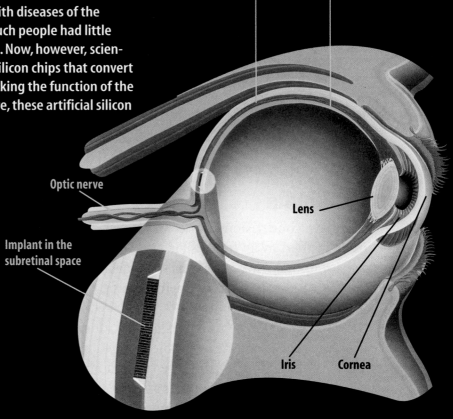

Outer retina
Inner retina
Optic nerve
Implant in the subretinal space
Lens
Iris
Cornea

Viewed with normal vision

Viewed with retinitis pigmentosa

Viewed with macular degeneration

▲ After making a number of incisions, surgeons implant the artificial silicon retina between the outer and inner retinal layers. Then they reseal the retina over the silicon chip.

▲ These three photos show how normal vision can deteriorate as a result of diseases that attack the retina. Retinitis pigmentosa (ret uh NYE tis pig men TOE suh) causes a lack of peripheral vision. Macular degeneration can lead to total blindness.

▲ The artificial silicon retina, above right, is thinner than a human hair and only 2 mm in diameter—the same diameter as the white dot on this penny.

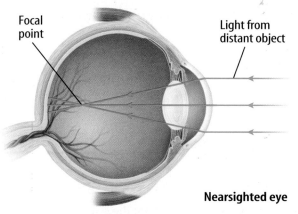

Focal point

Light from distant object

Nearsighted eye

When a nearsighted person looks at distant objects, the light rays from the objects are focused in front of the retina.

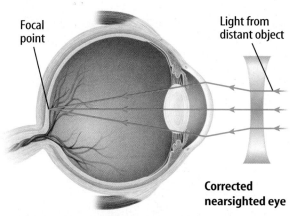

Focal point

Light from distant object

Corrected nearsighted eye

A concave lens in front of a nearsighted eye will diverge the light rays so they are focused on the retina.

Nearsightedness A person who is nearsighted can see objects clearly only when they are nearby. Objects that are far away appear blurred. In a nearsighted eye, the cornea and the lens bring light rays from distant objects to a focus in front of the retina, as shown in **Figure 16.** To correct this problem, a nearsighted person can wear concave lenses. **Figure 16** shows how a concave lens causes incoming light rays to diverge before they enter the eye. Then the light rays from distant objects can be focused by the eye to form a sharp image on the retina.

Figure 16 Nearsightedness can be corrected with concave lenses.

section 2 review

Summary

Convex Lenses

- A convex lens is thicker in the middle than at the edges. Light rays are refracted toward the optical axis.
- The image formed by a convex lens depends on the distance of the object from the lens.

Concave Lenses

- A concave lens is thinner in the middle and thicker at the edges. Light rays are refracted away from the optical axis.

The Eye and Vision Problems

- The eye contains a lens that changes shape to produce sharp images on the retina of objects that are at different distances.
- Farsightedness results when the lens of the eye cannot form a sharp image of nearby objects on the retina. Nearsightedness results when a sharp image of distant objects isn't formed.

Self Check

1. **Explain** how the focal length of a convex lens changes as the sides of the lens become less curved.
2. **Compare** the image of an object less than one focal length from a convex lens with the image of an object more than two focal lengths from the lens.
3. **Describe** the image formed by a concave lens.
4. **Explain** how the focal length of the lens in the eye changes to focus on a nearby object.
5. **Think Critically** If image formation by a convex lens is similar to image formation by a concave mirror, describe the image formed by a light source placed at the focal point of a convex lens.

Applying Math

6. **Calculate Object Distance** If you looked through a convex lens with a focal length of 15 cm and saw a real, inverted, enlarged image, what is the maximum distance between the lens and the object?

Optical Instruments

What You'll Learn

■ **Compare** refracting and reflecting telescopes.
■ **Explain** why a telescope in space is useful.
■ **Describe** how a microscope uses lenses to magnify small objects.
■ **Explain** how a camera creates an image.

Why It's Important

Optical instruments, such as microscopes and telescopes, enable your eyes to see objects that otherwise would be too small or far away to see.

⊙ Review Vocabulary

refraction: the change in direction of a wave when it changes speed as it moves from one medium to another

New Vocabulary

● refracting telescope
● reflecting telescope
● microscope

Telescopes

You know from your experience that it's hard to see faraway objects clearly. When you look at an object, only some of the light reflected from its surface enters your eye. As the object moves farther away, the amount of light entering your eye decreases, as shown in **Figure 17.** As a result, the object appears dimmer and less detailed.

A telescope uses a lens or a concave mirror that is much larger than your eye to gather more of the light from distant objects. The largest telescopes can gather more than a million times more light than the human eye. As a result, objects such as distant galaxies appear much brighter. Because the image formed by a telescope is so much brighter, more detail can be seen when the image is magnified.

Figure 17 As the cup gets farther away, fewer light rays from any point on the cup enter the viewer's eye. The amount of light from an object that enters the eye decreases as the object gets farther away.

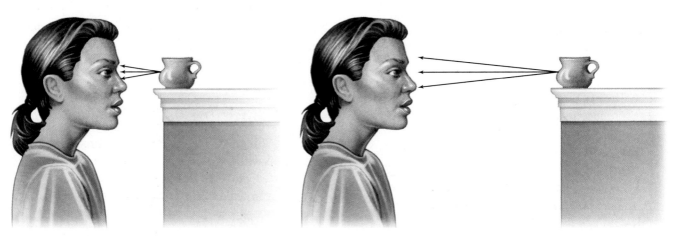

Refracting Telescopes One common type of telescope is the refracting telescope. A simple **refracting telescope,** shown in **Figure 18,** uses two convex lenses to gather and focus light from distant objects. Incoming light from distant objects passes through the first lens, called the objective lens. Because they are so far away, light rays from distant objects are nearly parallel to the optical axis of the lens. As a result, the objective lens forms a real image at the focal point of the lens, within the body of the telescope. The second convex lens, called the eyepiece lens, acts like a magnifying glass and magnifies this real image. When you look through the eyepiece lens, you see an enlarged, inverted, virtual image of the real image formed by the objective lens.

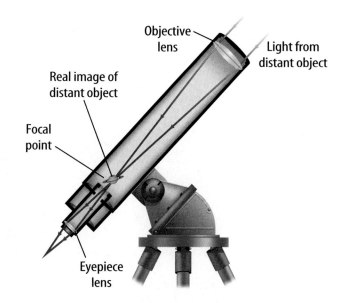

Figure 18 Light from a distant object passes through an objective lens and an eyepiece lens in a refracting telescope. The two lenses produce a large virtual image.

> ✔ **Reading Check** *What type of image is formed by the objective lens in a refracting telescope?*

Several problems are associated with refracting telescopes. In order to form a detailed image of distant objects, such as planets and galaxies, the objective lens must be as large as possible. A large lens is heavy and can be supported in the telescope tube only around its edge. The lens can sag or flex due to its own weight, distorting the image it forms. Also, these heavy glass lenses are costly and difficult to make.

Reflecting Telescopes Due to the problems with making large lenses, most large telescopes today are reflecting telescopes. A **reflecting telescope** uses a concave mirror, a plane mirror, and a convex lens to collect and focus light from distant objects. **Figure 19** shows a reflecting telescope. Light from a distant object enters one end of the telescope and strikes a concave mirror at the opposite end. The light reflects off of this mirror and converges. Before it converges at a focal point, the light hits a plane mirror that is placed at an angle within the telescope tube. The light is reflected from the plane mirror toward the telescope's eyepiece. The light rays converge at the focal point, creating a real image of the distant object. Just as in a refracting telescope, a convex lens in the eyepiece then magnifies this image.

Figure 19 Reflecting telescopes use two mirrors to create a real image, which then is magnified by a convex lens.
Infer *whether the image produced by the eyepiece lens is real or virtual.*

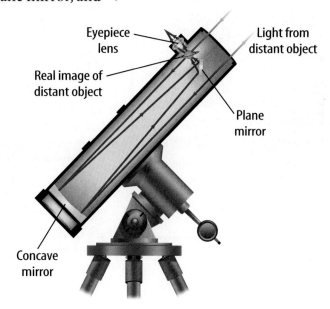

Figure 20 The view from telescopes on Earth is different from the view from telescopes in space.

The distorting effects of Earth's atmosphere can cause telescopes on Earth to form blurry images.

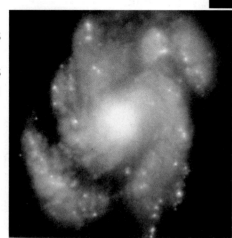

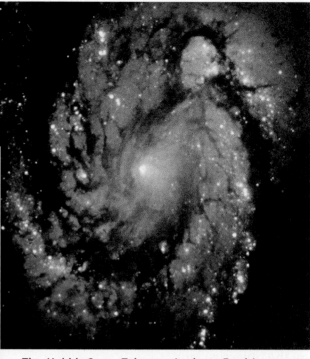

The *Hubble Space Telescope* is above Earth's atmosphere and forms clearer images of objects in space.

Sciencenline

Topic: *Hubble Space Telescope*

Visit gpscience.com for Web links to information and data about the *Hubble Space Telescope*.

Activity Prepare a speech to defend your opinion on whether or not the *Hubble Space Telescope* is useful and important. Hold a class debate.

Telescopes In Space Imagine being at the bottom of a swimming pool and trying to read a sign by the pool's edge. The water in the pool would distort your view of any object beyond the water's surface. In a similar way, Earth's atmosphere blurs the view of objects in space. To overcome the blurriness of humans' view into space, the National Aeronautics and Space Administration (NASA) built a telescope called the *Hubble Space Telescope* to be placed into space high above Earth's atmosphere. On April 25, 1990, NASA used the space shuttle *Discovery* to launch this telescope into an orbit about 600 km above Earth. The *Hubble Space Telescope* has produced images much sharper and more detailed than the largest telescopes on Earth can. **Figure 20** shows the difference in the images produced by telescopes on Earth and the *Hubble* telescope. With the *Hubble Space Telescope*, scientists can detect visible light—as well as other types of radiation—that is affected by Earth's atmosphere from the planets, stars, and distant galaxies.

✔ **Reading Check** *Why is the* **Hubble Space Telescope** *able to produce clearer images than telescopes on Earth?*

The *Hubble* telescope is a type of reflecting telescope that uses two mirrors to collect and focus light to form an image. The primary mirror in the telescope is 2.4 m across. When the *Hubble* was first launched, a defect in this primary mirror caused the telescope to create blurry images. The telescope was repaired by astronauts in December 1993.

Microscopes

A telescope would be useless if you were trying to study the cells in a butterfly wing, a sample of pond scum, or the differences between a human hair and a horse hair. You would need a microscope to look at such small objects. A **microscope** uses two convex lenses with relatively short focal lengths to magnify small, close objects. A microscope, like a telescope, has an objective lens and an eyepiece lens. However, it is designed differently because the objects viewed are close to the lens.

Figure 21 shows a simple microscope. The object to be viewed is placed on a transparent slide and illuminated from below. The light passes by or through the object on the slide and then travels through the objective lens. The objective lens is a convex lens. It forms a real, enlarged image of the object, because the distance from the object to the lens is between one and two focal lengths. The real image is then magnified again by the eyepiece lens (another convex lens) to create a virtual, enlarged image. This final image can be hundreds of times larger than the actual object, depending on the focal lengths of the two lenses.

Figure 21 A microscope uses two convex lenses to magnify small objects. **Explain** *where the object must be placed in relation to the objective lens's focal point.*

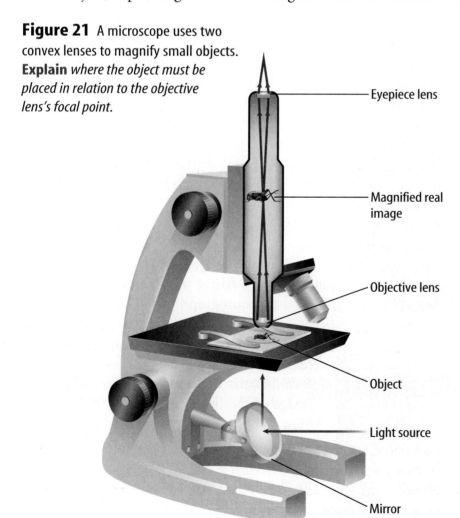

- Eyepiece lens
- Magnified real image
- Objective lens
- Object
- Light source
- Mirror

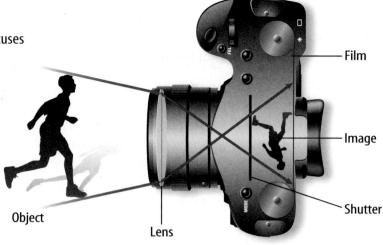

Figure 22 A camera's lens focuses an image on photographic film.

Film

Image

Shutter

Object

Lens

Cameras

Imagine swirls of lavender, gold, and magenta clouds sweeping across the sky at sunset. With the click of a button, you can capture the beautiful scene in a photo. How does a camera make a reduced image of a life-sized scene on film? A camera works by gathering and bending light with a lens. This lens then projects an image onto light-sensitive film to record a scene.

When you take a picture with a camera, a shutter opens to allow light to enter the camera for a specific length of time. The light reflected off your subject enters the camera through an opening called the aperture. It passes through the camera lens, which focuses the image on the film, as in **Figure 22.** The image is real, inverted, and smaller than the actual object. The size of the image depends upon the focal length of the lens and how close the lens is to the film.

Figure 23 Each object in the image produced by a wide-angle lens is small. This allows more of the surroundings to be seen.

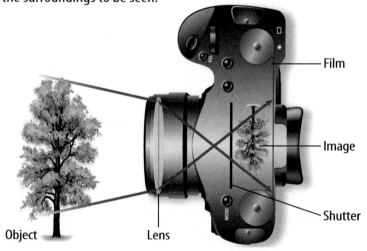

Film

Image

Shutter

Object

Lens

Wide-Angle Lenses Suppose you and a friend use two different cameras to photograph the same object at the same distance. If the cameras have different lenses, your pictures might look different. For example, some lenses have short focal lengths that produce a relatively small image of the object but have a wide field of view. These lenses are called wide-angle lenses, and they must be placed close to the film to form a sharp image with their short focal length. **Figure 23** shows how a wide-angle lens works. The photo in **Figure 23** was taken with a wide-angle lens.

Figure 24 A telephoto lens creates a larger image of an object than a wide-angle lens does.

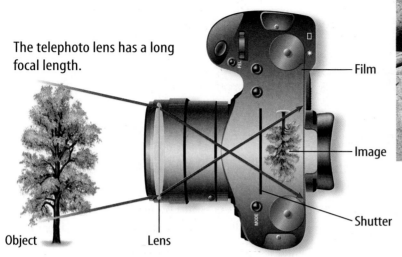

The telephoto lens has a long focal length.

Film

Image

Shutter

Object

Lens

Less of the surroundings can be seen, though a close-up of one of the objects can be photographed.

Telephoto Lenses Telephoto lenses have longer focal lengths. **Figure 24** shows how a telephoto lens forms an image. The image through a telephoto lens seems enlarged and closer than it actually is. Telephoto lenses are easy to recognize because they usually protrude from the camera to increase the distance between the lens and the film.

section 3 review

Summary

Telescopes

- Refracting telescopes use two convex lenses to gather and focus light.
- Reflecting telescopes use a concave mirror, a plane mirror, and a convex lens to collect, reflect, and focus light.
- Placing a telescope in orbit avoids the distorting effects of Earth's atmosphere.

Microscopes

- A microscope uses two convex lenses with short focal lengths to magnify small, close objects.

Cameras

- A wide-angle lens has a short focal length that produces a wide field of view.
- Telephoto lenses have longer focal lengths and are located farther from the film than wide-angle lenses are.

Self Check

1. **Describe** the image formed by the objective lens in a microscope.
2. **Infer** how the amount of light that enters the eye from an object changes as the object moves closer.
3. **Identify** the advantage to making the objective lens larger in a refracting telescope.
4. **Explain** why the largest telescopes are reflecting telescopes instead of refracting telescopes.
5. **Think Critically** Which optical instrument—a telescope, a microscope, or a camera—forms images in a way most like your eye? Explain.

Applying Math

6. **Calculate Magnification** Suppose the objective lens in a microscope forms an image that is 100 times the size of an object. The eyepiece lens magnifies this image ten times. What is the total magnification?

Model & Invent

Make a *Refracting* Telescope

Goals
- **Build** a simple telescope.
- **Estimate** the magnification of the telescope.
- **Compare** convex and concave eyepieces.

Possible Materials
objective lens—convex, 25 cm to 30 cm focal length, about 4 cm diameter

eyepiece lenses—one each convex and concave, 2 cm to 3 cm focal length, about 2.5 cm to 3 cm diameter

cardboard tubes—one with inside diameter of about 4 cm; one with inside diameter of about 3 cm. (The smaller tube should slide inside the larger one with a snug fit.)

clay to hold the lenses in place

*cellophane tape or duct tape

scissors

*Alternate materials

Safety Precautions

WARNING: *Do not look directly at the Sun through a telescope. Permanent eye damage can result.*

⊙ Real-World Question

Galileo used the telescope to enhance his eyesight. It enabled him to see planets and stars beyond the range of his eyes alone. By combining two lenses, distant objects can be magnified. A simple refracting telescope uses a small convex eyepiece lens and a larger convex objective lens at the other end. How do the lenses in a simple telescope form an image?

⊙ Procedure

1. Check that the smaller-diameter tube can slide in and out of the larger-diameter tube.

2. Hold the small concave eyepiece lens near your eye. Hold the objective lens in front of the eyepiece lens and move the objective lens away from you until a distant object is in focus. Estimate the distance between the two lenses.

3. Subtract half the length of the larger-diameter tube from the distance you estimated in step 2 to get the length needed for the smaller tube.

4. Cut the smaller-diameter tube to the length determined in step 3. Make two pieces this length.

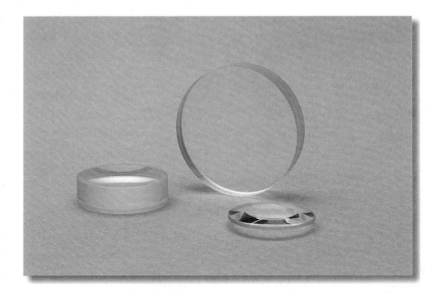

5. Attach the objective lens with clay or tape to the end of the larger tube. Make sure that the lens is perpendicular to the sides of the tube.

6. Attach the convex eyepiece lens with clay or tape to the end of one of the smaller tubes. Make sure the lens is perpendicular to the sides of the tube.

7. Slide the smaller tube into the larger one and look through the eyepiece.

8. Move the smaller tube in and out of the larger tube until a distant object is focused clearly.

Analyze Your Data

1. Estimate how much larger the image seen through the eyepiece is than the image you see with your unaided eye. Describe the appearance of the image.

2. Attach the concave eyepiece to the second smaller tube that you cut.

3. Repeat your observations using the concave eyepiece. Describe the appearance of an object seen through the concave eyepiece.

4. How does the image produced using the convex and concave eyepiece lenses change when you look through the objective lens instead of the eyepiece lens?

Conclude and Apply

1. **Infer** the estimated magnification of your telescope.

2. **Discuss** how you could change the magnification of your telescope.

3. **Diagram** the path of light rays that pass through the telescope and then into your eye.

4. **Explain** how you could build a telescope with higher magnification than the one you constructed here.

Communicating Your Data

Compare your telescope and its operation with those of other members of your class. Try reading numbers or letters on a distant sign. Which telescope helps you see more detail?

Sight Lines

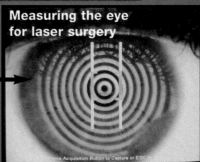

Measuring the eye for laser surgery

Lasers make it possible to throw away eyeglasses

Back in the 1970s, scientists developed a special kind of laser to make microscopic notches in computer chips. This laser is also perfect for eye surgery. It does not generate much heat, so it doesn't damage the delicate tissues of the eye. With this technology, most of the 160 million Americans who wear eyeglasses or contact lenses can kiss them goodbye forever.

The most common type of laser surgery used to correct poor vision is LASIK. This painless procedure takes only about five minutes per eye. The patient is awake the entire time and usually sees well immediately after the surgery.

The Cornea and Vision

The eyeball has two structures, the cornea and a flexible lens, that cause light to be focused on the retina. The cornea is a transparent structure at the front of the eye. Most of the bending of light rays occurs when they pass through the cornea. The lens fine-tunes the focus of light from objects by adjusting its shape so that a sharp image is formed on the retina. Unlike the lens, the shape of the cornea doesn't change.

How It Works:

The LASIK procedure fixes vision problems by reshaping the cornea. For farsighted eyes, the laser vaporizes a ring of tissue from the cornea. This makes the cornea more curved so that light rays are bent more. For nearsighted eyes, the laser vaporizes tissue from the center of the cornea, making it flatter.

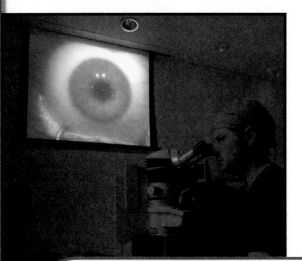

A microscope mounted on the laser gives the doctor a detailed view during the surgery. The screen allows others to view the surgery.

Interview Ophthalmologists are medical doctors who specialize in healing eyes. Optometrists are doctors who specialize in correcting poor vision by prescribing glasses and contact lenses. Interview an optometrist or ophthalmologist to find out how he or she detects eye problems and how these problems can be corrected.

Science online

For more information, visit gpscience.com/time

Reviewing Main Ideas

Section 1 Mirrors

1. Plane mirrors reflect light to form upright, virtual images.

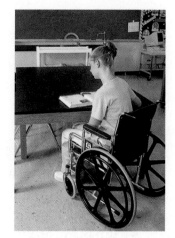

2. Concave mirrors can form various types of images, depending on where an object is relative to the focal point of the mirror. Concave mirrors can be used to magnify objects or create beams of light.

3. Convex mirrors spread out reflected light to form a reduced image. Convex mirrors allow you to see large areas.

Section 2 Lenses

1. Convex lenses converge light rays. Convex lenses can form real or virtual images, depending on the distance from the object to the lens.

2. Concave lenses diverge light rays to form virtual smaller, upright images. They often are used in combination with other lenses.

3. The human eye has a flexible lens that changes shape to focus an image on the retina.

4. People with imperfect vision can use corrective lenses to improve their vision. Farsighted people wear convex lenses, and nearsighted people wear concave lenses.

Section 3 Optical Instruments

1. A refracting telescope uses convex lenses to magnify distant objects.

2. A reflecting telescope uses concave and plane mirrors and a convex lens to magnify distant objects.

3. By avoiding atmospheric distortion, the *Hubble Space Telescope* produces sharper images than telescopes on Earth are able to produce.

4. A simple microscope uses a convex objective lens and eyepiece lens with short focal lengths to magnify small objects.

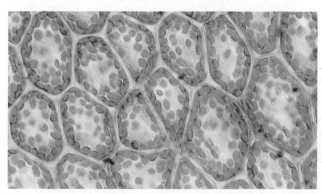

5. Light passing through the lens of a camera is focused on light-sensitive film inside the camera. The image on the film is inverted and reduced.

FOLDABLES Use the Foldable that you made at the beginning of the chapter to help you review image formation by mirrors.

Using Vocabulary

concave lens p. 426
concave mirror p. 418
convex lens p. 424
convex mirror p. 421
cornea p. 427
focal length p. 418
focal point p. 418
microscope p. 435

optical axis p. 418
plane mirror p. 417
real image p. 419
reflecting telescope p. 433
refracting telescope p. 433
retina p. 427
virtual image p. 418

Complete each sentence with the correct vocabulary word.

1. A flat, smooth surface that reflects light and forms an image is a(n) _____.

2. A(n) _____ uses two convex lenses to magnify small, close objects.

3. Every light ray that travels parallel to the optical axis before hitting a concave mirror is reflected such that it passes through the _____.

4. A(n) _____ is thicker in the middle than at the edges.

5. The inner lining of the eye that converts light images into electrical signals is called the _____.

Checking Concepts

Choose the word or phrase that best answers the question.

6. Which of the following best describes image formation by a plane mirror?
   A) A real image is formed in front of the mirror.
   B) A real image is formed behind the mirror.
   C) A virtual image is formed in front of the mirror.
   D) A virtual image is formed behind the mirror.

7. Which mirror can form an enlarged image?
   A) convex   **C)** concave
   B) plane   **D)** transparent

8. Which of the following is used in a headlight or a flashlight to create a beam of light?
   A) concave lens   **C)** concave mirror
   B) convex lens   **D)** convex mirror

9. What do lenses do?
   A) reflect light   **C)** diffract light
   B) refract light   **D)** interfere with light

10. Which way does a concave lens bend light?
    A) toward its optical axis
    B) toward its center
    C) toward its edges
    D) toward its focal point

11. What type of lens is used to correct farsightedness?
    A) flat lens   **C)** concave lens
    B) convex lens   **D)** plane lens

12. Which of the following is NOT part of a reflecting telescope?
    A) plane mirror   **C)** convex lens
    B) concave mirror **D)** concave lens

13. Which of the following images do light rays never pass through?
    A) real   **C)** enlarged
    B) virtual   **D)** reduced

14. The image formed by a camera lens must always be which of the following?
    A) real   **C)** virtual
    B) upright   **D)** enlarged

15. What happens to a light ray traveling parallel to the optical axis of a convex lens that passes through the lens?
    A) It travels parallel to the optical axis.
    B) It passes through the focal point.
    C) It is bent away from the optical axis.
    D) It forms a virtual image.

Science online gpscience.com/vocabulary_puzzlemaker

Interpreting Graphics

Use the illustration below to answer question 16.

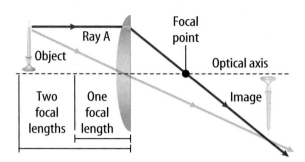

16. Suppose the image of the candle moves away from the focal point. How did the position of the candle change?

17. Copy and complete the following table on the image formation by lenses and mirrors.

Image Formation by Lenses and Mirrors		
Type of Lens or Mirror	Position of Object	Type of Image
Concave lens	All positions of object	virtual, upright, reduced
Convex lens	closer than one focal length	
	between one and two focal lengths	
	farther than two focal lengths	real, inverted reduced
Concave mirror	closer than one focal length	
	object placed at focal point	
	farther than two focal lengths	
Convex mirror	All positions of object	

Thinking Critically

18. **Describe** how the shape of the lens in your eye changes when you look at a nearby object, and then a distant object.

19. **Compare and contrast** a refracting telescope and a microscope.

20. **Infer** why a convex mirror and a conccave lens can never produce a real image.

21. **Explain** why people often become far-sighted as they grow older.

22. **Infer** why it would be easier to make a concave mirror for a reflecting telescope than an objective lens of the same size for a refracting telescope.

23. **Determine** whether a convex lens could form an image that is enlarged, real, and upright.

24. **Compare** A concave lens made of plastic is placed in a liquid. Light rays traveling in the liquid are not refracted when they pass through the lens. Comare the speed of light in the plastic and in the liquid.

Applying Math

25. **Calculate Magnification** The magnification of a refracting telescope can be calculated by dividing the focal length of the objective lens by the focal length of the eyepiece lens. If an objective lens has a focal length of 1 m and the eyepiece has a focal length of 1 cm, what is the magnification of the telescope?

26. **Determine Object Distance** You hold an object in front of a concave mirror with a focal length of 30 cm. If you do not see a reflected image, how far from the mirror is the object?

Part 1 | Multiple Choice

Record your answers on the answer sheet provided by your teacher or on a sheet of paper.

1. Which of the following describes the image formed by a convex mirror?
   A. real
   B. enlarged
   C. inverted
   D. virtual

Use the illustration below to answer questions 2 and 3.

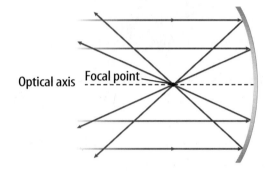

2. Which of the following describes a light ray that passes through the focal point and then is reflected by the mirror?
   A. It travels parallel to the optical axis.
   B. It forms a real image.
   C. It is reflected back through the focal point.
   D. It forms a virtual image

3. If the mirror becomes flatter and the focal point moves farther from the mirror, which of the following best describes the reflection of the parallel rays shown in the figure?
   A. They pass through the old focal point.
   B. They do not pass through either the old or the new focal point.
   C. They pass through the new focal point.
   D. They reverse direction.

Test-Taking Tip

Be Prepared Bring at least two sharpened No. 2 pencils and a good eraser to the test. Make sure the eraser erases completely.

4. How far is an object from a concave mirror if the image formed is upright?
   A. one focal length
   B. less than one focal length
   C. more than two focal lengths
   D. two focal lengths

5. What is an advantage to increasing the diameter of the concave mirror in a reflecting telescope?
   A. Brighter images are formed.
   B. The mirror forms larger images.
   C. The mirror forms sharper images.
   D. The focal length increases.

Use the table below to answer questions 6–8.

Image Magnification by a Convex Lens		
Object Distance (cm)	Image Distance (cm)	Magnification
250.0	62.5	0.25
200.0	66.7	0.33
150.0	75.0	0.50
100.0	100.0	1.00
75.0	150.0	2.00

6. How does the image change as the object gets closer to the lens?
   A. It gets larger.
   B. It gets smaller
   C. It gets closer.
   D. It becomes real.

7. Which of the following is the best estimate of the magnification if the image is 225 cm from the lens?
   A. 0.40
   B. 1.25
   C. 1.5
   D. 0.3

8. What should the object distance be if the lens is to be used as a magnifying glass?
   A. 150 cm
   B. 100 cm
   C. greater than 250 cm
   D. less than 100 cm

Part 2 **Short Response/Grid In**

Record your answers on the answer sheet provided by your teacher or on a sheet of paper.

9. Describe how you could determine whether the image formed by a lens or mirror is a real image or a virtual image.

10. The largest refracting telescope has an objective lens with a diameter of 1.0 m. Calculate the area of this lens.

11. The objective lens in a microscope has a magnification of 30. What is the magnification of the microscope if the eyepiece lens has a magnification of 20?

12. Describe how the focal length of a convex lens changes as the lens becomes more curved.

Use the graph below to answer questions 13 and 14.

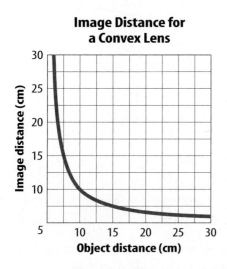

Image Distance for a Convex Lens

13. Determine how far the image is from the lens when the object is 15 cm from the lens.

14. At what object distance are the image distance and object distance equal?

15. Explain why the largest telescopes are reflecting telescopes instead of refracting telescopes.

Part 3 **Open Ended**

Record your answers on a sheet of paper.

Use the illustration below to answer questions 16 and 17.

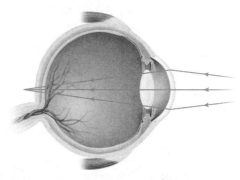

16. Describe the vision problem shown by the illustration. Why does this vision problem become more prevalent as people age?

17. Explain how the vision problem shown by the illustration can be corrected.

18. A convex lens is formed out of a transparent substance. Light travels with the same speed in this substance as in air. Explain why this lens would not cause light rays to converge.

19. Some cameras have zoom lenses that have a focal length that varies between 35 mm and 155 mm. Determine which focal length would correspond to a wide-angle lens and which would correspond to a telephoto lens.

20. Explain why side-view convex mirrors on the right side of cars have the printed warning "Objects in mirror are closer than they appear."

21. Describe the change in the lens in each of your eyes when you look at this book and then look out the window at a distant object.

22. Explain why objects become dimmer and less detailed as they move farther away.

How Are Playing Cards & the Periodic Table Connected?

By 1860, scientists knew of about 60 elements. However, they had yet to clearly organize their knowledge. A Russian scientist named Dmitri Mendeleev changed that. Mendeleev loved to play solitaire, a type of card game in which playing cards are arranged into patterns according to their properties. One day, Mendeleev decided to make a set of cards on which he wrote the names and properties of the known elements. Then he began to arrange the cards into rows. The result was a table in which certain chemical properties could be seen to occur periodically—that is, to occur in a repeating pattern. In 1869, Mendeleev published his periodic table (seen here in a more advanced version). He left blank spaces in the table where the pattern seemed to call for elements that were not yet known. Over the next several decades, other scientists refined the table, and new elements were added. Modern periodic tables—like the one probably hanging in your classroom—still follow the basic pattern laid out by Mendeleev.

unit ⚡ projects

Visit gpscience.com/unit_project to find project ideas and resources.
Projects include:

- **History** Discover the diverse uses of lasers. Compile a class spider map of laser use in different professions.
- **Technology** Research elements and create a 3-dimensional periodic table with characteristics of each element on element cubes.
- **Model** Design and construct a unit review game to include questions, answers, directions, playing pieces, and a creative box.

WebQuest *Art of Neon* is an investigation of the noble gases and how they are inserted into glass tubes to be used in art and signage.

Classification of Matter

The Art of Mixtures

When painting, colors can be skillfully mixed to achieve a certain shade, or the artist can show several individual pigments in a single brush stroke. Paint undergoes physical and chemical changes with time that can make the artist's work endure.

Science Journal Describe the changes that take place as the paint dries, particularly which changes are physical and which ones are chemical.

Start-Up Activities

Demonstrate the Distillation of Water

Matter is classified according to the different properties that it exhibits. These differences in properties allow drinking water to be obtained from seawater. These properties could be very important to you if you were stranded on a desert island and needed drinking water. Purified water can be obtained through a process called distillation.

1. Place 75 mL of water in a 200-mL beaker and add 20 drops of red food coloring.

2. Place the beaker on a hot plate.

3. Add ice to an evaporating dish until the dish is half full. Add boiling chips. Place the evaporating dish on the beaker as shown in the photo.

4. Turn on the hot plate and slowly bring the water and food coloring solution to a boil.

5. After boiling the solution for five minutes, carefully remove the evaporating dish using heat-resistant gloves. Touch the drops of liquid on the bottom of the dish to a piece of white paper.

6. Observe the liquid on the paper.

7. **Think Critically** In your Science Journal, write a paragraph explaining where the liquid came from. What was in the beaker that is not in the liquid on the paper?

FOLDABLES ™
Study Organizer

Classification of Matter Make the following Foldable to ensure that you have understood the content by defining the vocabulary terms from this chapter.

STEP 1 **Fold** a vertical sheet of paper from side to side. Make the front edge about 1.25 cm shorter than the back edge.

STEP 2 **Turn** lengthwise, with the fold on top, and **fold** into thirds.

STEP 3 **Unfold and cut** only the top layer along both folds to make three tabs.

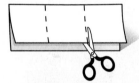

STEP 4 **Label** the tabs *Elements, Compounds,* and *Mixtures* as shown.

Define As you read the chapter, define each term and list examples of each under the appropriate tab.

Preview this chapter's content and activities at gpscience.com

Composition of Matter

What You'll Learn
- **Define** substances and mixtures.
- **Identify** elements and compounds.
- **Compare and contrast** solutions, colloids, and suspensions.

Why It's Important
You can form a better picture of your world when you understand the concepts of elements and compounds.

⊙ **Review Vocabulary**
property: characteristic or essential quality

New Vocabulary
- substance
- element
- compound
- heterogeneous mixture
- homogeneous mixture
- solution
- colloid
- Tyndall effect
- suspension

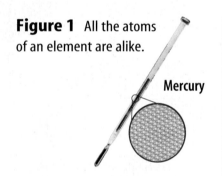

Figure 1 All the atoms of an element are alike.

Mercury

Copper

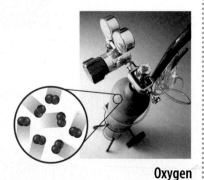

Oxygen

Pure Substances

Have you ever seen a picture hanging on a wall that looked just like a real painting? Did you have to touch it to find out? If so, the rough or smooth surface told you which it was. Each material has its own properties. The properties of materials can be used to classify them into general categories.

Materials are made of a pure substance or a mixture of substances. A pure **substance,** or simply a substance, is a type of matter with a fixed composition. A substance can be either an element or a compound. Some substances you might recognize are helium, aluminum, water, and salt.

Elements All substances are built from atoms. If all the atoms in a substance have the same identity, that substance is an **element.** The graphite in your pencil point and the copper coating of most pennies are examples of elements. In graphite all the atoms are carbon atoms, and in a copper sample, all the atoms are copper atoms. The metal substance beneath the copper in the penny is another element—zinc. About 90 elements are found on Earth. More than 20 others have been made in laboratories, but most of these are unstable and exist only for short periods of time. Some elements you might recognize are shown in **Figure 1.** Some less common elements and their properties are shown in **Figure 2.**

Figure 2

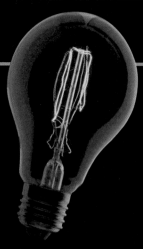

Most of us think of gold as a shiny yellow metal used to make jewelry. However, it is an element that is also used in more unexpected ways, such as in spacecraft parts. On the other hand, some less common elements, such as americium (am uh REE see um), are used in everyday objects. Some elements and their uses are shown here.

▲ **ALUMINUM** Aluminum is an excellent reflector of heat. Here, an aluminum plastic laminate is used to retain the body heat of a newborn baby.

▲ **TUNGSTEN** Although tungsten can be combined with steel to form a very durable metal, in its pure form it is soft enough to be stretched to form the filament of a lightbulb. Tungsten has the highest melting point of any metal.

▲ **TITANIUM** (tie TAY nee um) Parts of the exterior of the Guggenheim Museum in Bilbao, Spain, are made of titanium panels. Strong and lightweight, titanium is also used for body implants.

▲ **GOLD** Gold's resistance to corrosion and its ability to reflect infrared radiation make it an excellent coating for space vehicles. The electronic box on the six-wheel Sojourner Rover, above, part of NASA's Pathfinder 1997 mission to Mars, is coated with gold.

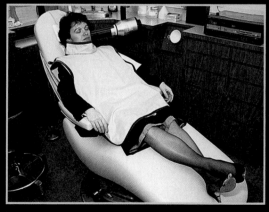

▲ **LEAD** Because lead has a high density, it is a good barrier to radiation. Dentists drape lead aprons on patients before taking X rays of the patient's teeth to reduce radiation exposure.

◄ **AMERICIUM** Named after America, where it was first produced, americium is a component of this smoke detector. It is a radioactive metal that must be handled with care to avoid contact.

Compounds Two or more elements can combine to form substances called compounds. A **compound** is a substance in which the atoms of two or more elements are combined in a fixed proportion. For example, water is a compound in which two atoms of the element hydrogen combine with one atom of the element oxygen. Chalk contains calcium, carbon, and oxygen in the proportion of one atom of calcium and carbon to three atoms of oxygen.

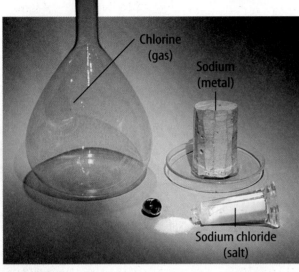

Figure 3 Chlorine gas and sodium metal combine dramatically in the ratio of one to one to form sodium chloride.

Labels in Figure 3: Chlorine (gas), Sodium (metal), Sodium chloride (salt)

✓ **Reading Check** *How are elements and compounds related?*

Can you imagine yourself putting something made from a silvery metal and a greenish-yellow, poisonous gas on your food? You might have shaken some on your food today—table salt is a chemical compound that fits this description. Even though it looks like white crystals and adds flavor to food, its components—sodium and chlorine—are neither white nor salty, as shown in **Figure 3.** Like salt, compounds usually look different from the elements in them.

Mixtures

Are pizza and a soft drink one of your favorite lunches? If so, you enjoy two foods that are classified as mixtures—but two different kinds of mixtures. A mixture, such as the pizza or soft drink shown in **Figure 4,** is a material made up of two or more substances that can be easily separated by physical means.

Figure 4 Pizza and soft drinks, like most foods, are mixtures.

Heterogeneous Mixtures Unlike compounds, mixtures do not always contain the same proportions of the substances that make them up—the pizza chef doesn't measure precisely how much of each topping is sprinkled on. You easily can see most of the toppings on a pizza. A mixture in which different materials can be distinguished easily is called a **heterogeneous** (he tuh ruh JEE nee us) **mixture.** Granite, concrete, and dry soup mixes are other heterogeneous mixtures you can recognize.

You might be wearing another heterogeneous mixture—clothing made of permanent-press fabric like that seen in **Figure 5A.** Such fabric contains fibers of two materials—polyester and cotton. The amounts of polyester and cotton can vary from one article of clothing to another, as shown by the label. Though you might not be able to distinguish the two fibers just by looking at them with your naked eye, you probably could tell using a microscope, as shown in **Figure 5B.** Therefore, a permanent-press fabric is also a heterogeneous mixture.

Most of the substances you come in contact with every day are heterogeneous mixtures. Some components are easy to see, like the ingredients in pizza, but others are not. In fact, the component you see can be a mixture itself. For example, the cheese in pizza is also a mixture, but you cannot see the individual components. Cheese contains many compounds, such as milk proteins, butterfat, colorings, and other food additives.

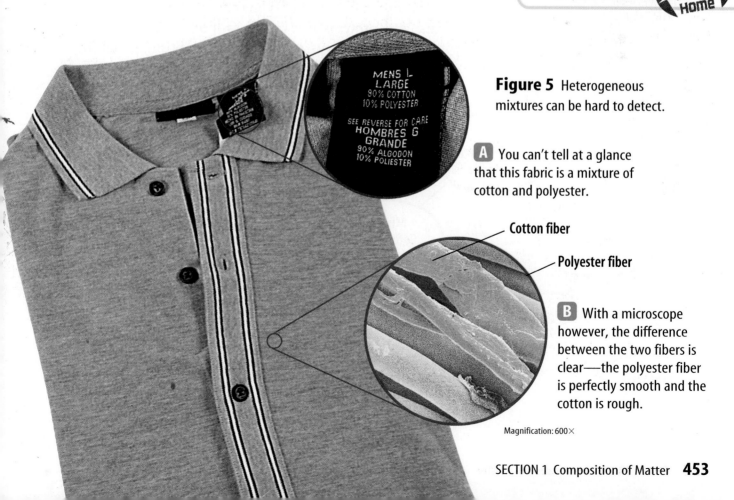

Figure 5 Heterogeneous mixtures can be hard to detect.

A You can't tell at a glance that this fabric is a mixture of cotton and polyester.

Cotton fiber

Polyester fiber

B With a microscope however, the difference between the two fibers is clear—the polyester fiber is perfectly smooth and the cotton is rough.

Magnification: 600×

Figure 6 A soft drink can be either heterogeneous or homogeneous. As carbon dioxide fizzes out it is a heterogeneous mixture.

The resulting flat soft drink is a homogeneous mixture of water, sugar, flavor, color, and some remaining carbon dioxide.

Homogeneous Mixtures Remember that soft drink you had with your pizza? Soft drinks contain water, sugar, flavoring, coloring, and carbon dioxide gas. **Figure 6** will help you to visualize these particles in a liquid soft drink.

Soft drinks in sealed bottles are examples of homogeneous mixtures. A **homogeneous** (hoh muh JEE nee us) **mixture** contains two or more gaseous, liquid, or solid substances blended evenly throughout. However, a soft drink in which you can see bubbles of carbon dioxide gas and ice cubes is a heterogeneous mixture.

Vinegar is another homogeneous mixture. It appears clear even though it is made up of particles of acetic acid mixed with water. Another name for homogeneous mixtures like vinegar and a cold soft drink is solution. A **solution** is a homogeneous mixture of particles so small that they cannot be seen with a microscope and will never settle to the bottom of their container. Solutions remain constantly and uniformly mixed. The differences between substances and mixtures are summarized in **Figure 7.**

Reading Check *What kind of mixture is a solution?*

Colloids Milk is an example of a specific kind of mixture called a colloid. Like a heterogeneous mixture, it contains water, fats, and proteins in varying proportions. Like a solution, its components won't settle if left standing. A **colloid** (KAH loyd) is a type of mixture with particles that are larger than those in solutions but not heavy enough to settle out. The word *colloid* comes from a Greek word for glue. The first colloids studied were in gelatin, a source of some types of glue.

Paint is an example of a liquid with suspended colloid particles. Gases and solids can contain colloidal particles, too. For example, fog consists of particles of liquid water suspended in air, and smoke contains solids suspended in air.

Figure 7 All matter can be divided into substances and mixtures.

Matter
Has mass and takes up space

Substance
Composition definite

Mixture
Composition variable

Compound
Two or more kinds of atoms

Heterogeneous
Unevenly mixed

Element
One kind of atom

Homogeneous
Evenly mixed; a solution

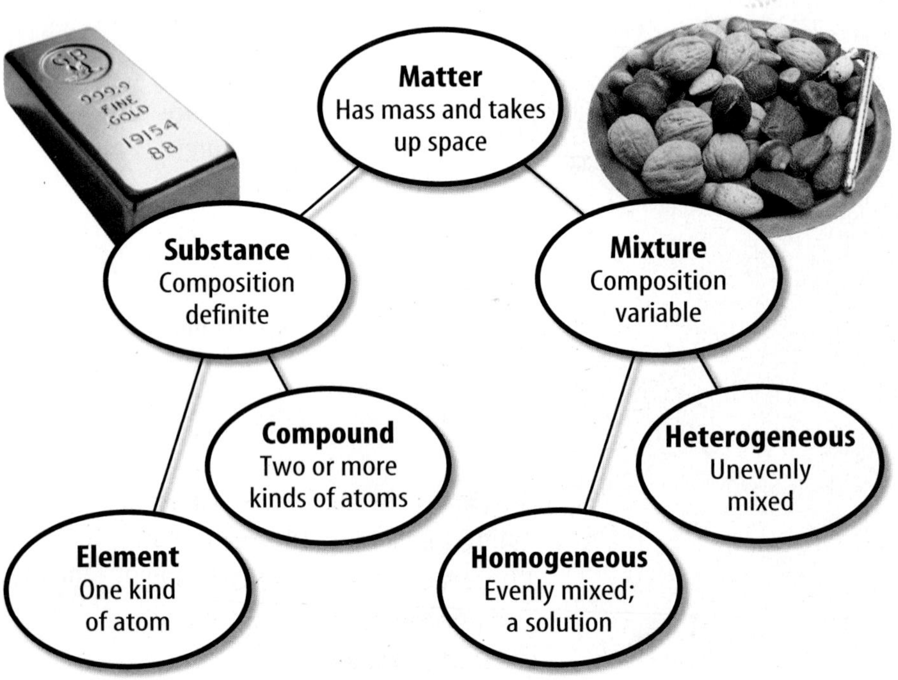

Figure 8 Fog is a colloid composed of water droplets suspended in air.

The light from the headlights is scattered by fog.

The same colloid allows you to see the sunlight as it streams through the trees.

Detecting Colloids One way to distinguish a colloid from a solution is by its appearance. Fog appears white because its particles are large enough to scatter light as shown in **Figure 8.** Sometimes it is not so obvious that a liquid is a colloid. For example, some shampoos and gelatins are colloids called gels that appear almost clear. You can tell for certain if a liquid is a colloid by passing a beam of light through it, as shown in **Figure 9.** A light beam is invisible as it passes through a solution, but can be seen readily as it passes through a colloid. This occurs because the particles in the colloid are large enough to scatter light, but those in the solution are not. This scattering of light by colloidal particles is called the **Tyndall effect.**

 **Reading Check** *How can you distinguish a colloid from a solution?*

Figure 9 Because of the Tyndall effect, a light beam is scattered by the colloid suspension on the right, but passes invisibly through the solution on the left.

Figure 10 The mud deposited by the Mississippi River is said to be more than 10,000 m thick.

Table 1 Comparing Solutions, Colloids, and Suspensions

Description	Solutions	Colloids	Suspensions
Settle upon standing?	no	no	yes
Separate using filter paper?	no	no	yes
Particle size	0.1–1 nm	1–1,000 nm	>1,000 nm
Scatter light?	no	yes	yes

Suspensions Some mixtures are neither solutions nor colloids. One example is muddy pond water. If pond water stands long enough, some mud particles will fall to the bottom, and the water clears. Pond water is a **suspension,** which is a heterogeneous mixture containing a liquid in which visible particles settle. **Table 1** summarizes the properties of different types of mixtures.

INTEGRATE Earth Science River deltas are a large scale example of how a suspension settles. Rivers flow swiftly through narrow channels, picking up soil and debris along the way. As the river widens, it flows more slowly. Suspended particles settle forming deltas at the mouth, as shown in **Figure 10.**

section ① review

Summary

Pure Substances

- An element is a substance in which all atoms have the same identity.
- A compound is a substance that has two or more elements combined in a fixed proportion.

Mixtures

- Heterogeneous mixtures are mixtures in which different materials can be distinguished easily.
- A homogeneous mixture contains two or more gaseous, liquid, or solid substances that are blended evenly throughout.
- Mixtures can be heterogeneous, homogeneous, colloids, or suspensions.

Self Check

1. **Describe** How is a compound similar to a homogeneous mixture? How is it different?
2. **Distinguish** between a substance and a mixture. Give two examples of each.
3. **Describe** the differences between colloids and suspensions.
4. **Think Critically** Why do the words "Shake well before using" indicate that the fruit juice is a suspension?

Applying Skills

5. **Compare and Contrast** In terms of suspensions and colloids, compare and contrast a glass of milk and a glass of fresh-squeezed orange juice.

Science online gpscience.com/self_check

Elements, Compounds, and Mixtures

Elements, compounds, and mixtures all contain atoms. In elements, the atoms all have the same identity. In compounds, two or more elements have been combined in a fixed ratio. In mixtures, the ratio of substances can vary.

Real-World Question

What are some differences among elements, compounds, and mixtures?

Goals

■ **Determine** whether several materials are elements, compounds, or mixtures.

Materials

plastic freezer bag containing the following labeled items:

copper wire
small package of salt
pencil
aluminum foil

chalk (calcium carbonate)
piece of granite
sugar water in a vial

Safety Precautions

Procedure

1. Copy the data table into your Science Journal and use it to record your observations.

2. Obtain a bag of objects. Identify each object and classify it as an element, compound, heterogeneous mixture, or homogeneous mixture. The elements appear in the periodic table. Compounds are named as examples in Section 1.

Conclude and Apply

1. If you know the name of a substance, how can you find out whether or not it is an element?

2. **Explain** how the appearance of the items is different. Which would you classify as a compound? Why? Which would you classify as a mixture? Why?

3. **Discuss** how chalk is different from the granite in relation to compounds and mixtures.

Classification of Objects		
Object	Identity	Classification
1		
2	Do not write in this book.	
3		
4		
5		
6		
7		

Communicating Your Data

Enter your data in the data table and compare your findings with those of your classmates. **For more help, refer to the Science Skill Handbook.**

Properties of Matter

What You'll Learn

- **Identify** substances using physical properties.
- **Compare and contrast** physical and chemical changes.
- **Identify** chemical changes.
- **Determine** how the law of conservation of mass applies to chemical changes.

Why It's Important

Understanding chemical and physical properties can help you use materials properly.

🔍 Review Vocabulary

state of matter: one of three physical forms of matter—solid, liquid, or gas

New Vocabulary

- physical property
- physical change
- distillation
- chemical property
- chemical change
- law of conservation of mass

Physical Properties

You can stretch a rubber band, but you can't stretch a piece of string very much, if at all. You can bend a piece of wire, but you can't easily bend a matchstick. In each case, the materials change shape, but the identity of the substances—rubber, string, wire, wood—does not change. The abilities to stretch and bend are physical properties. Any characteristic of a material that you can observe without changing the identity of the substances that make up the material is a **physical property.** Examples of other physical properties are color, shape, size, density, melting point, and boiling point. What physical properties can you use to describe the items in **Figure 11?**

Appearance How would you describe a tennis ball? You could begin by describing its shape, color, and state of matter. For example, you might describe the tennis ball as a brightly colored, hollow sphere. You can measure some physical properties, too. For instance, you could measure the diameter of the ball. What physical property of the ball is measured with a balance?

To describe a soft drink in a cup, you could start by calling it a liquid with a brown color. You could measure its volume and temperature. Each of these characteristics is a physical property of that soft drink.

Figure 11 Appearance is the most obvious physical property. **Describe** *the appearance of these items.*

Figure 12 The best way to separate substances depends on their physical properties. Size is the property used to separate poppy seeds from sunflower seeds in this example.

Behavior Some physical properties describe the behavior of a material or a substance. As you might know, objects that contain iron, such as a safety pin, are attracted by a magnet. Attraction to a magnet is a physical property of the substance iron. Every substance has a specific combination of physical properties that make it useful for certain tasks. Some metals, such as copper, can be drawn out into wires. Others, such as gold, can be pounded into sheets as thin as 0.1 micrometers (μm), about 4-millionths of an inch. This property of gold makes it useful for decorating picture frames and other objects. Gold that has been beaten or flattened in this way is called gold leaf.

Think again about your soft drink. If you knock over the cup, the drink will spread out over the table or floor. If you knock over a jar of molasses, however, it does not flow as easily. The ability to flow is a physical property of liquids.

Using Physical Properties to Separate Removing the seeds from a watermelon can be easily done based on the physical properties of the seeds compared to the rest of the fruit. **Figure 12** shows a mixture of poppy seeds and sunflower seeds. You can identify the two kinds of seeds by differences in color, shape, and size. By sifting the mixture, you can separate the poppy seeds from the sunflower seeds quickly because their sizes differ.

Now look at the mixture of iron filings and sand shown in **Figure 12.** You probably won't be able to sift out the iron filings because they are similar in size to the sand particles. What you can do is pass a magnet through the mixture. The magnet attracts only the iron filings and pulls them from the sand. This is an example of how a physical property, such as magnetic attraction, can be used to separate substances in a mixture. Something like this is done to separate iron for recycling.

Magnetism easily separates iron from sand.

INTEGRATE Environment

Recycling and Physical Properties Recycling conserves natural resources. In some large recycling projects, aluminum metal must be separated from scrap iron. What physical properties of the two metals could be used to separate them?

Physical Change

If you break a piece of chewing gum, you change some of its physical properties—its size and shape. However, you have not changed the identity of the materials that make up the gum.

The Identity Remains the Same When a substance freezes, boils, evaporates, or condenses, it undergoes physical changes. A change in size, shape, or state of matter is called a **physical change.** These changes might involve energy changes, but the kind of substance—the identity of the element or compound—does not change. Because all substances have distinct properties like densities, specific heats, and boiling and melting points, which are constant, these properties can be used to help identify them when a particular mixture contains substances which are not yet identified.

Reading Check *Does a change in state mean that a new substance has formed? Explain.*

Iron is a substance that can change states if it absorbs or releases enough energy—at high temperatures, it melts. However, in both the solid and liquid state, iron has physical properties that identify it as iron. Color changes can accompany a physical change, too. For example, when iron is heated it first glows red. Then, if it is heated to a higher temperature, it turns white, as shown in **Figure 13.**

Using Physical Change to Separate A cool drink of water is something most people take for granted, but in some parts of the world, drinkable water is scarce. Not enough drinkable water can be obtained from wells. Many such areas that lie close to the sea obtain drinking water by using physical properties of water to separate it from the salt. One of these methods, which uses the property of boiling point, is a type of distillation.

Figure 13 Heating iron raises its energy level and changes its color. These energy changes are physical changes because it is still iron.

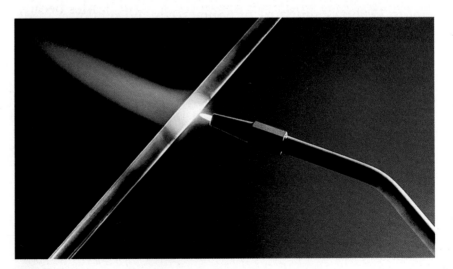

Distillation The process for separating substances in a mixture by evaporating a liquid and recondensing its vapor is **distillation.** It usually is done in the laboratory using an apparatus similar to that shown in **Figure 14.** As you can see, the liquid vaporizes and condenses, leaving the solid material behind.

Two liquids having different boiling points can be separated in a similar way. The mixture is heated slowly until it begins to boil. Vapors of the liquid with the lowest boiling point form first and are condensed and collected. Then, the temperature is increased until the second liquid boils, condenses, and is collected. Distillation is used often in industry. For instance, natural oils such as mint are distilled.

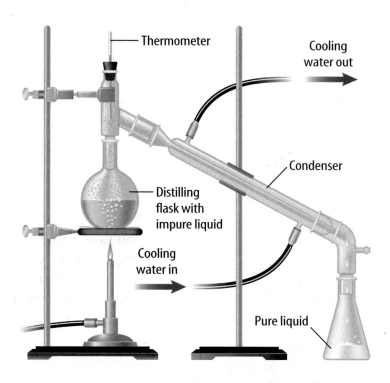

Figure 14 Distillation can easily separate liquids from solids dissolved in them. The liquid is heated until it vaporizes and moves up the column. Then, as it touches the water-cooled surface of the condenser, it becomes liquid again.

Chemical Properties and Changes

You probably have seen warnings on cans of paint thinners and lighter fluids for charcoal grills that say these liquids are flammable (FLA muh buhl). The tendency of a substance to burn, or its flammability, is an example of a chemical property because burning produces new substances during a chemical change. A **chemical property** is a characteristic of a substance that indicates whether it can undergo a certain chemical change. Many substances used around the home, such as lighter fluids, are flammable. Knowing which ones are flammable helps you to use them safely.

A less dramatic chemical change can affect some medicines. Look at **Figure 15.** You probably have seen bottles like this in a pharmacy. Many medicines are stored in dark bottles because they contain compounds that can change chemically if they are exposed to light.

Figure 15 The brown color of these bottles tells you that these vitamins may react to light. Reaction to light is a chemical property.

Alchemy In the Middle Ages, alchemy was an early form of chemistry devoted to the study of changing baser metals into gold and also to finding the elixir of perpetual youth. Based on what we know now about the properties of metals and biology, it is easy to understand why this field of study is no longer practiced.

Detecting Chemical Change

If you leave a pan of chili cooking unattended on the stove for too long, your nose soon tells you that something is wrong. Instead of a spicy aroma, you detect an unpleasant smell that alerts you that something is burning. This burnt odor is a clue telling you that a new substance has formed.

The Identity Changes The smell of rotten eggs and the formation of rust on bikes or car fenders are signs that a chemical change has taken place. A change of one substance to another is a **chemical change.** The foaming of an antacid tablet in a glass of water and the smell in the air after a thunderstorm are other signs of new substances being produced. In some chemical changes, a rapid release of energy—detected as heat, light, and sound—is a clue that changes are occurring.

✔ **Reading Check** *What is a chemical change?*

Clues such as heat, cooling, or the formation of bubbles or solids in a liquid are helpful indicators that a reaction is taking place. However, the only sure proof is that a new substance is produced. Consider the following example. The heat, light, and sound produced when hydrogen gas combines with oxygen in a rocket engine are clear evidence that a chemical reaction has taken place. But no clues announce the reaction that takes place when iron combines with oxygen to form rust because the reaction takes place so slowly. The only clue that iron has changed into a new substance is the presence of rust. Burning and rusting are chemical changes because new substances form. You sometimes can follow the progress of a chemical reaction visually. For example, you can see lead nitrate forming in **Figure 16.**

Figure 16 The solid forming from two liquids is another sign that a chemical reaction has taken place.

Using Chemical Change to Separate One case where you might separate substances using a chemical change is in cleaning tarnished silver. Tarnish is a chemical reaction between silver metal and sulfur compounds in the air which results in silver sulfide. It can be changed back into silver using a chemical reaction. This chemical reversal back to silver takes place when the tarnished item is placed in a warm water bath with baking soda and aluminum foil. You don't usually separate substances using chemical changes in the home. In industry and chemical laboratories, however, this kind of separation is common. For example, many metals are separated from their ores and then purified using chemical changes.

Applying Math — Calculate

CALCULATIONS WITH THE LAW OF CONSERVATION OF MASS When a chemical reaction takes place, the total mass of reactants equals the total mass of products. If 18 g of hydrogen react completely with 633 g of chlorine, how many grams of HCl are formed? $H_2 + Cl_2 \rightarrow 2HCl$

IDENTIFY known values and the unknown value

Identify the known values:

   mass of H_2 = 18 g

   mass of Cl_2 = 633 g

   mass of reactants = mass of products

Identify the unknown value:

   mass of HCl $\boxed{\text{means}}\!\!\Rightarrow$? g

SOLVE the problem

Solve for the mass of HCl. $g\ H_2 + g\ Cl_2 = g\ HCl$

Substitute the known values. $18\ g + 633\ g = 651\ g\ HCl$

CHECK your answer

Does your answer seem reasonable? Check your answer by subtracting the mass of H_2 from the mass of HCl. Do you obtain the mass of the Cl_2? If so, the answer is correct.

Practice Problems

1. In the following reaction, 24 g of CH_4 (methane) react completely with 96 g of O_2 to form 66 g of CO_2. How many grams of H_2O are formed? $CH_4 + 2O_2 \rightarrow CO_2 + 2H_2O$

2. In the following equation, 54.0 g of Al react completely with 409.2 g of $ZnCl_2$ to form 196.2 g of Zn metal. How many grams of $AlCl_3$ are formed? $2Al + 3ZnCl_2 \rightarrow 3Zn + 2AlCl_3$

For more practice problems, go to page 834, and visit gpscience.com/extra_problems.

Flowing water shaped and smoothed these rocks in a physical process.

Both chemical and physical changes shaped the famous White Cliffs of Dover lining the English Channel.

Figure 17 Weathering can involve physical or chemical change.

Topic: Caves
Visit gpscience.com for Web links to information about cave formations.

Activity Describe the formation of stalactites and stalagmites. Explain the differences between the two and whether they are the result of a physical and/or chemical process.

Weathering—Chemical or Physical Change?

The forces of nature continuously shape Earth's surface. Rocks split, deep canyons are carved out, sand dunes shift, and curious limestone formations decorate caves. Do you think these changes, often referred to as weathering, are physical or chemical? The answer is both. Geologists, who use the same criteria that you have learned in this chapter, say that some weathering changes are physical and some are chemical.

Physical Large rocks can split when water seeps into small cracks, freezes, and expands. However, the smaller pieces of newly exposed rock still have the same properties as the original sample. This is a physical change. Streams can cut through softer rock, forming canyons, and can smooth and sculpt harder rock, as shown at left in **Figure 17.** In each case, the stream carries rock particles far downstream before depositing them. Because the particles are unchanged, the change is a physical one.

Chemical In other cases, the change is chemical. For example, solid calcium carbonate, a compound found in limestone, does not dissolve easily in water. However, when the water is even slightly acidic, as it is when it contains some dissolved carbon dioxide, calcium carbonate reacts. It changes into a new substance, calcium hydrogen carbonate, which does dissolve in water. This change in limestone is a chemical change because the identity of the calcium carbonate changes. The White Cliffs of Dover, shown at right in **Figure 17,** are made of limestone and undergo such chemical changes, as well as physical changes. A similar chemical change produces caves and the icicle-shaped rock formations that often are found in them.

The Conservation of Mass

Wood is combustible, or burnable. As you just learned, this is a chemical property. Suppose you burn a large log in the fireplace, as shown in **Figure 18,** until nothing is left but a small pile of ashes. Smoke, heat, and light are given off and the changes in the appearance of the log confirm that a chemical change took place. At first, you might think that matter was lost during this change because the pile of ashes looks much smaller than the log did. In fact, the mass of the ashes is less than that of the log. However, suppose that you could collect all the oxygen in the air that was combined with the log during the burning and all the smoke and gases that escaped from the burning log and measure their masses, too. Then you would find that no mass was lost after all.

Not only is no mass lost during burning, mass is not gained or lost during any chemical change. In other words, matter is neither created nor destroyed during a chemical change. According to the **law of conservation of mass,** the mass of all substances that are present before a chemical change equals the mass of all the substances that remain after the change.

Figure 18 This reaction appears to be destroying these logs. When it is over, only ashes will remain. Yet you know that no mass is lost in a chemical reaction. **Explain** why this is so.

 *Explain what is meant by the law of conservation of mass.*

---

section 2 review

Summary

Physical Properties

- You can observe physical properties without changing the identity of a substance.

Physical Change

- Change in the size, shape, or state of matter is a physical change.

Chemical Properties and Changes

- A chemical property is a characteristic of a substance that indicates whether it can undergo a certain chemical change.
- A change of one substance to another is a chemical change.
- Many metals are separated from their ores and purified using chemical changes.

Self Check

1. **Explain** why evaporation of water is a physical change and not a chemical change.
2. **List** four physical properties you could use to describe a liquid.
3. **Describe** why flammability is a chemical property rather than a physical property.
4. **Explain** how the law of conservation of mass applies to chemical changes.
5. **Think Critically** How might you demonstrate this law of conservation of mass for melting ice and distillation of water?

Applying Math

6. **Calculate** In the following equation, 417.96 g of Bi (bismuth) react completely with 200 g of F (fluorine). How many grams of BiF_3 (bismuth fluoride) are formed? $2\,Bi + 3\,F_2 \longrightarrow 2\,BiF_3$

Checking Out Chemical Changes

Goals

- **Observe** the results of adding dilute hydrochloric acid to baking soda.
- **Infer** that the production of new substances indicates that a chemical change has occurred.
- **Design** an experiment that allows you to compare the activity of baking soda with that of a product formed when baking soda reacts.

Possible Materials

baking soda
small evaporating dish
magnifying lens
1*M* hydrochloric acid (HCl)
10-mL graduated cylinder
electric hot plate

Safety Precautions

● *Real-World Question*

Mixing materials together does not always produce a chemical change. You must find evidence of a new substance with new properties being produced before you can conclude that a chemical change has taken place. Try this lab and use your observation skills to deduce what kind of change has occurred. What evidence indicates a chemical change?

● *Form a Hypothesis*

Think about what happens when small pieces of limestone are mixed with sand. What happens when limestone is mixed with an acid? Based on these thoughts, form a hypothesis about how to determine when mixing substances together produces a chemical change.

● *Test Your Hypothesis*

Make a Plan

1. As a group, agree upon a hypothesis and decide how to test it. Write the hypothesis statement.

2. To test your hypothesis, devise a plan to compare two different mixtures. The first mixture consists of 3 mL of hydrochloric acid and 0.5 g of baking soda. The second mixture is 3 mL of hydrochloric acid and the solid product of the first mixture. Describe exactly what you will do at each step.

Limestone

Sand

3. Make a list of the materials needed to complete your experiment.

4. **Design** a table for data and observations in your Science Journal so that it is ready to use as your group observes what happens.

Follow Your Plan

1. Make sure your teacher approves your plan before you start.

2. Read over your entire experiment to make sure that all steps are in logical order.

3. **Identify** any constants and the variables of the experiment.

4. Should you run any test more than once? How will observations be summarized?

5. Assemble your materials and carry out the experiment according to your plan. Be sure to record your results as you work.

◉ *Analyze Your Data*

1. **Observe** what happened to the baking soda. Did anything happen to the product formed from the first mixture? Explain why this occurred.

2. **Describe** What different properties of any new substances did you observe after adding hydrochloric acid to the baking soda?

◉ *Conclude and Apply*

1. Did the results support your hypothesis? Explain.

2. If you had used vinegar, which contains acetic acid, as the acid, do you think a new susbstance would have formed? How could you test this?

Write a description of your observations in your Science Journal. **Compare** your results with those of other groups. **Discuss** your conclusions.

SCIENCE Stats

Intriguing Elements

Did you know...

... Silver-white cobalt, which usually is combined with other elements in nature, is used to create rich paint pigments. It can be used to form powerful magnets, treat cancer patients, build jet engines, and prevent disease in sheep.

... Gold is the most ductile (stretchable) of all the elements. Just 29 g of gold—about ten wedding bands—can be pulled into a wire 100 km long. That's long enough to stretch from Toledo, Ohio, to Detroit, Michigan, and beyond.

Percent of Elements in the Human Body

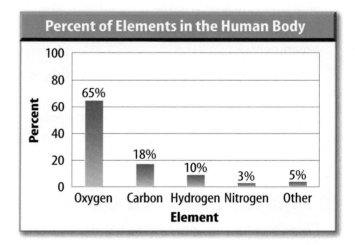

... Zinc makes chewing gum taste better. Up to 0.3 mg of zinc acetate can be added per 1,000 mg of chewing gum to provide a tart, zingy flavor.

Applying Math

1. Zinc acetate is approximately 35% zinc. How many grams of chewing gum would be needed to provide a total of 10.0 mg of zinc?

2. If you wanted to produce a gold wire as mentioned in our example, how many grams would be needed to make a wire one kilometer in length?

3. Table sugar has the chemical formula $C_{12}H_{22}O_{11}$. If you were going to build a scale model of sucrose (table sugar) and you had 36 carbon model atoms, how many of each of the others (hydrogen and oxygen) would you need to build an accurate, complete sucrose model?

Reviewing Main Ideas

Section 1 **Composition of Matter**

1. Elements and compounds are substances. A mixture is composed of two or more substances.

2. You can distinguish between the different materials in a heterogeneous mixture using either your unaided eye or a microscope.

3. Colloids and suspensions are two types of heterogeneous mixtures. The particles in a suspension will settle eventually. Particles of a colloid will not. Milk is an example of a colloid.

4. In a homogeneous mixture, the particles are distributed evenly and are not visible, even when using a microscope. Homogeneous mixtures can be composed of solids, liquids, or gases.

5. A solution is another name for a homogeneous mixture that remains constantly and uniformly mixed.

This is a clear indication that this is a suspension.

Section 2 **Properties of Matter**

1. Physical properties are characteristics of materials that you can observe without changing the identity of the substance.

2. Chemical properties indicate what chemical changes substances can undergo. Many medicines are stored in dark bottles because they react with light.

3. In physical changes, the identities of substances remain unchanged.

4. In chemical changes, the identities of substances change—new substances are formed. There is a visible chemical change that takes place when rust is cleaned with bleach.

5. The law of conservation of mass states that during any chemical change, matter is neither created nor destroyed.

FOLDABLES Use the Foldable that you made at the beginning of the chapter to help you review the classifications of matter.

Using Vocabulary

chemical change p. 462
chemical property p. 461
colloid p. 454
compound p. 452
distillation p. 461
element p. 450
heterogeneous mixture
    p. 453
homogeneous mixture
    p. 454

law of conservation of
    mass p. 465
physical change p. 460
physical property p. 458
solution p. 454
substance p. 450
suspension p. 456
Tyndall effect p. 455

Complete each sentence with the correct vocabulary word or words.

1. Substances formed from atoms of two or more elements are called _____.

2. A(n) _____ is a heterogeneous mixture in which visible particles settle.

3. Freezing, boiling, and evaporation are all examples of _____.

4. According to the _____, matter is neither created nor destroyed during a chemical change.

5. A mixture in which different materials are easily identified is _____.

6. Compounds are made from the atoms of two or more _____.

7. Distillation is a process that can separate two liquids using _____.

Checking Concepts

Choose the word or phrase that best answers the question.

8. Bending a copper wire is an example of what type of property?
   A) chemical
   B) physical
   C) conservation
   D) element

9. Which of the following is NOT an element?
   A) water
   B) carbon
   C) oxygen
   D) hydrogen

10. Which of the following is an example of a chemical change?
    A) boiling
    B) burning
    C) evaporation
    D) melting

11. What type of substance is gelatin?
    A) colloid
    B) compound
    C) substance
    D) suspension

12. A visible sunbeam is an example of which of the following?
    A) an element
    B) a solution
    C) a compound
    D) the Tyndall effect

13. You start to eat some potato chips from an open bag you found in your locker and notice that they taste unpleasant. What do you think might cause this unpleasant taste?
    A) combustion
    B) melting
    C) physical change
    D) chemical change

14. How would you classify the color of a rose?
    A) chemical change
    B) physical change
    C) chemical property
    D) physical property

15. How would you describe the process of evaporating water from seawater?
    A) chemical change
    B) physical change
    C) chemical property
    D) physical property

16. Which of these warnings refers to a chemical property of the material?
    A) Fragile
    B) Flammable
    C) Handle with Care
    D) Shake Well

17. Which of the following is a substance?
    A) colloid
    B) element
    C) mixture
    D) solution

18. Which of these properties can be used to help identify an unknown substance?
    A) specific heat
    B) combustion
    C) temperature
    D) Tyndall effect

Science Online gpscience.com/vocabulary_puzzlemaker

Interpreting Graphics

19. Copy and complete the concept map below about matter.

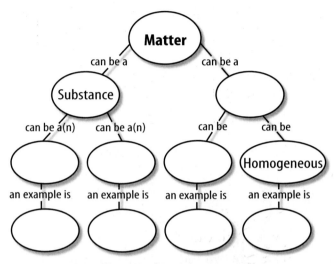

Use the table below to answer question 20.

Common Colloids	
Colloid	**Example**
Solid in a liquid	Gelatin
Solid in a gas	
Gas in a solid	
Solid in a liquid	
Liquid in a gas	

20. Different colloids can involve different states. For example, gelatin is formed from solid particles in a liquid. Complete this table using these colloids: *smoke, marshmallow, fog,* and *paint.*

Thinking Critically

21. Describe the contents of a carton of milk using at least four physical properties.

22. Explain Carbon and the gases hydrogen and oxygen combine to form sugar. How do you know sugar is a compound?

23. Explain The word *colloid* means "gluelike." Why was this term chosen to name certain mixtures?

24. Use a nail rusting in air to explain the law of conservation of mass.

25. Explain Mai says that ocean water is a solution. Tom says that it's a suspension. Can they both be correct? Explain.

26. Use Variables, Constants, and Controls Marcos took a 100-cm^3 sample of a suspension, shook it well, and poured equal amounts into four different test tubes. He placed one test tube in a rack, one in hot water, one in warm water, and the fourth in ice water. He then observed the time it took for each suspension to settle. What was the variable in the experiment? What was one constant?

27. Concept Map Make a network tree to show types of liquid mixtures. Include these terms: *homogeneous mixtures, heterogeneous mixtures, solutions, colloids,* and *suspensions.*

Applying Math

28. Interpret Data Hannah started with a 25-mL sample of pond water. Without shaking the sample, she poured 5 mL through a piece of filter paper. She repeated this with four more pieces of filter paper. She dried each piece of filter paper and measured the mass of the sediment. Why did the last sample have a higher mass than did the first sample?

29. Use Numbers In the following equation, 243.5 g of Sb (antimony) react completely with 1000 g of I_2 (iodine) to form 1004.9 g of SbI_3 (antimony triiodide). How many grams of I_2 were consumed in the reaction? $2\ Sb + 3\ I_2 \rightarrow 2\ SbI_3$

Part 1 | Multiple Choice

Record your answers on the answer sheet provided by your teacher or on a sheet of paper.

1. Which statement about elements is FALSE?
   A. All atoms in an element are alike.
   B. There are about 1,000 elements found in nature.
   C. Some elements have been made in laboratories.
   D. Zinc, copper, and iron are elements.

2. $CaCO_3$ is an example of which type of material?
   A. element   C. compound
   B. mixture   D. colloid

Use the graph below to answer questions 3 and 4.

Elements in the Universe

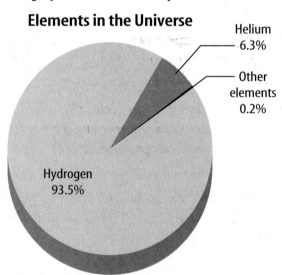

Helium — 6.3%

Other elements 0.2%

Hydrogen 93.5%

3. What percentage do the elements hydrogen and helium account for in the universe?
   A. 100%   C. 98%
   B. 99.9%   D. 99.8%

Test-Taking Tip

Directions and Instructions Listen carefully to the instructions from the teacher and read the directions and each question carefully.

4. The most plentiful element in the universe readily burns in air. What is this chemical property called?
   A. flammability
   B. ductility
   C. density
   D. boiling point

Use the graph below to answer questions 5 and 6.

Elements in Earth's Crust

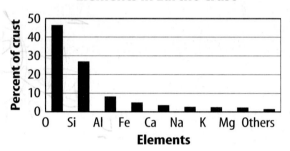

Percent of crust — O, Si, Al, Fe, Ca, Na, K, Mg, Others — Elements

5. Which element makes up 8 percent of Earth's crust?
   A. iron   C. silicon
   B. aluminum   D. oxygen

6. Which element has the physical property of magnetism?
   A. sodium   C. oxygen
   B. iron   D. calcium

7. Which statement best decribes the law of conservation of mass?
   A. The mass of the products is always greater than the mass of the materials which react in a chemical change.
   B. The mass of the products is always less than the mass of the materials which react in a chemical change.
   C. A certain mass of material must be present for a reaction to occur.
   D. Matter is neither lost nor gained during a chemical change.

Part 2 | Short Response/Grid In

Record your answers on the answer sheet provided by your teacher or on a sheet of paper.

Use the illustrations below to answer questions 8 and 9.

8. What physical properties could be used to identify these minerals?

9. Describe some ways these minerals, or rocks which contain these minerals, are weathered in nature.

10. Why are some medicines stored in dark bottles?

11. Describe the properties of tungsten which make it a useful material. How do you use this element daily?

12. What physical and chemical properties of gold make it useful as a coating for space vehicles?

13. Compare and contrast the properties of heterogeneous and homogeneous mixtures. What is another name for a homogeneous mixture?

14. You are given a mixture of iron filings, sand, and salt. Describe how to separate this mixture.

Part 3 | Open Ended

Record your answers on a sheet of paper.

Use the illustration below to answer questions 15 and 16.

15. What indicates that more than just a physical change is taking place as the wood burns?

16. Design an experiment which shows that this type of chemical change is governed by the law of conservation of mass.

17. Illustrate and describe the process of distillation of sea water using a multi-step process. On what physical property is this process based?

18. Explain the statement "Everything on Earth is made from 90 elements." Does this surprise you? Why or why not?

19. What type of mixture is a salad dressing made with oil, vinegar, and herbs? How do you know?

20. Choose an object in the room and describe its physical properties as completely as possible. Use available tools to describe measurable properties of the object.

21. Describe how a compound is a combination of elements in a fixed proportion. Give two examples.

Solids, Liquids, and Gases

Surrounded by Science

Driving down this road, you can't help but notice the scenery. Now look at this photo through a scientist's eyes. Can you find three states of water? The solid is present as snow on the mountaintops. The liquid is found as water in the lake. The gas is present in the atmosphere as water vapor.

Science Journal Identify examples of a solid, a liquid, and a gas in your classroom.

Start-Up Activities

The Expansion of a Gas

Why does the mercury in a thermometer rise? Why do sidewalks, streets, and bridges have cracks? Many substances expand when heated and contract when cooled, as you will see during this lab.

Observe the Expansion and Contraction of Air

1. Blow up a balloon until it is half filled. Use a tape measure to measure the circumference of the balloon.

2. Pour water into a large beaker until it is half full. Place the beaker on a hot plate and wait for the water to boil.

3. Set the balloon on the mouth of the beaker and observe for five minutes. Be careful not to allow the balloon to touch the hot plate. Measure the circumference of the balloon.

4. **Think Critically** Write a paragraph in your Science Journal describing the changing size of the balloon's circumference. Infer why the balloon's circumference changed.

Solids, Liquids, and Gases The matter that surrounds you is either a solid, liquid, or gas. Make the following Foldable to help you organize information about solids, liquids, and gases.

STEP 1 Fold a sheet of paper in half lengthwise. Make the back edge about 5 cm longer than the front edge.

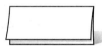

STEP 2 Turn the paper so the fold is on the bottom. Then fold it into thirds.

STEP 3 Unfold and cut only the top layer along both folds to make three tabs.

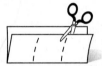

STEP 4 Label the Foldable as shown.

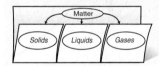

Read for Main Ideas As you read the chapter, list the characteristics of solids, liquids, and gases under the appropriate tab. List some examples of each under the tab also.

Preview this chapter's content and activities at gpscience.com

Kinetic Theory

Reading Guide

What You'll Learn
- **Explain** the kinetic theory of matter.
- **Describe** particle movement in the four states of matter.
- **Explain** particle behavior at the melting and boiling points.

Why It's Important
You can use energy that is lost or gained when a substance changes from one state to another.

🔍 Review Vocabulary
kinetic energy: energy in the form of motion

New Vocabulary
- kinetic theory
- melting point
- heat of fusion
- boiling point
- heat of vaporization
- diffusion
- plasma
- thermal expansion

States of Matter

You probably do not think of the states of matter as you do everyday activities. An everyday activity such as eating lunch may include solids, liquids, and gases. Look at **Figure 1.** Can you identify the states of matter present? The boiling soup on the stove and the visible steam above the boiling soup is in the liquid state. The ice cube dropped into the soup to cool it, is in the solid state. How are these states alike and different?

Kinetic Theory The **kinetic theory** is an explanation of how particles in matter behave. To explain the behavior of particles, it is necessary to make some basic assumptions. The three assumptions of the kinetic theory are as follows:

1. All matter is composed of small particles (atoms, molecules, and ions).

2. These particles are in constant, random motion.

3. These particles are colliding with each other and the walls of their container.

Particles lose some energy during collisions with other particles. But the amount of energy lost is very small and can be neglected in most cases.

To visualize the kinetic theory, think of each particle as a tiny table-tennis ball in constant motion. These balls are bouncing and colliding with each other. Mentally visualizing matter in this way can help you understand the movement of particles in matter.

Figure 1 Two states of water are visible in this photograph.
Identify *Can you identify the solid and liquid states of water?*

Thermal Energy Think about the ice cube in the soup. Does the ice cube appear to be moving? How can a frozen, solid ice cube have motion? Remember to focus on the particles. Atoms in solids are held tightly in place by the attraction between the particles. This attraction between the particles gives solids a definite shape and volume. However, the thermal energy in the particles causes them to vibrate in place. Thermal energy is the total energy of a material's particles, including kinetic—vibrations and movement within and between the particles—and potential—resulting from forces that act within or between particles. When the temperature of the substance is lowered, the particles will have less thermal energy and will vibrate more slowly.

 **Reading Check** *What is thermal energy?*

Average Kinetic Energy Temperature is the term used to explain how hot or cold an object is. In science, temperature means the average kinetic energy of particles in the substance, or how fast the particles are moving. On average, molecules of frozen water at 0°C will move slower than molecules of water at 100°C. Therefore, water molecules at 0°C have lower average kinetic energy than the molecules at 100°C. Molecules will have kinetic energy at all temperatures, including absolute zero. Scientists theorize that at absolute zero, or –273.15°C, particle motion is so slow that no additional thermal energy can be removed from a substance.

Reading Check *How are kinetic energy and temperature related?*

Solid State An ice cube is an example of a solid. The particles of a solid are closely packed together, as shown in **Figure 2.** Most solid materials have a specific type of geometric arrangement in which they form when cooled. The type of geometric arrangement formed by a solid is important. Chemical and physical properties of solids often can be attributed to the type of geometric arrangement that the solid forms. **Figure 3** shows the geometric arrangement of solid water. Notice that the hydrogen and oxygen atoms are alternately spaced in the arrangement.

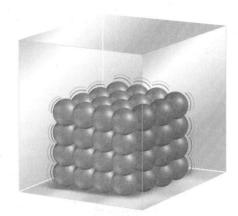

Solid

Figure 2 The particles in a solid are packed together tightly and are constantly vibrating in place.

Figure 3 The particles in solid water align themselves in an ordered geometric pattern. Even though a solid ice cube doesn't look like it is moving, its molecules are vibrating in place.

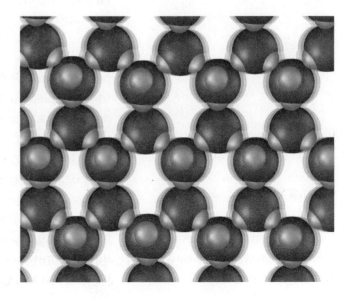

Liquid

Figure 4 The particles in a liquid are moving more freely than the particles in a solid. They have enough kinetic energy to slip out of the ordered arrangement of a solid.

Liquid State What happens to a solid when thermal energy or heat is added to it? Think about the ice cube in the hot soup. The particles in the hot soup are moving fast and colliding with the vibrating particles in the ice cube. The collisions of the particles transfer energy from the soup to the ice cube. The particles on the surface of the ice cube vibrate faster. These particles collide with and transfer energy to other ice particles. Soon the particles of ice have enough kinetic energy to overcome the attractive forces. The particles of ice gain enough kinetic energy to slip out of their ordered arrangement and the ice melts. This is known as the **melting point,** or the temperature at which a solid begins to liquefy. Energy is required for the particles to slip out of the ordered arrangement. The amount of energy required to change a substance from the solid phase to the liquid phase at its melting point is known as the **heat of fusion.**

Reading Check *What is heat of fusion?*

Liquids Flow Particles in a liquid, shown in **Figure 4,** have more kinetic energy than particles in a solid. This extra kinetic energy allows particles to partially overcome the attractions to other particles. Thus, the particles can slide past each other, allowing liquids to flow and take the shape of their container. However, the particles in a liquid have not completely overcome the attractive forces between them. This causes the particles to cling together, giving liquids a definite volume.

Reading Check *Why do liquids flow?*

Gas State Particles in the gas state are shown in **Figure 5.** Gas particles have enough kinetic energy to overcome the attractions between them. Gases do not have a fixed volume or shape. Instead, gas particles spread out so that they fill whatever container they are in. How does a liquid become a gas? The particles in a liquid are constantly moving. Some particles are moving faster and have more kinetic energy than others. The particles that are moving fast enough can escape the attractive forces of other particles and enter the gas state. This process is called vaporization. Vaporization can occur in two ways— evaporation and boiling. Evaporation is vaporization that occurs at the surface of a liquid and can occur at temperatures below the liquid's boiling point. To evaporate, particles must have enough kinetic energy to escape the attractive forces of the liquid. They must be at the liquid's surface and traveling away from the liquid.

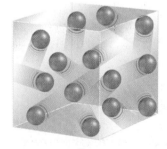

Gas

Figure 5 In gases, the particles are far apart and the attractive forces between the particles are overcome. Gases do not have a definite volume or shape.

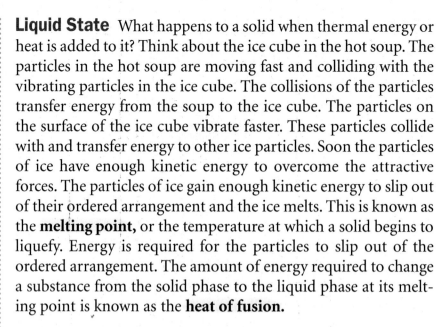

Figure 6 Boiling occurs throughout a liquid when the pressure of the vapor in the liquid equals the pressure of the vapor on the surface of the liquid. **Explain** *the difference between boiling and evaporation.*

Boiling Point A second way that a liquid can vaporize is by boiling. Unlike evaporation, boiling occurs throughout a liquid at a specific temperature depending on the pressure on the surface of the liquid. Boiling is shown in **Figure 6.** The **boiling point** of a liquid is the temperature at which the pressure of the vapor in the liquid is equal to the external pressure acting on the surface of the liquid. This external pressure is a force pushing down upon a liquid, keeping particles from escaping. Particles require energy to overcome this force. **Heat of vaporization** is the amount of energy required for the liquid at its boiling point to become a gas.

 *How does external pressure affect the boiling point of a liquid?*

Gases Fill Their Container What happens to the attractive forces between the particles in a gas? The gas particles are moving so quickly and are so far apart that they have overcome the attractive forces between them. Because the attractive forces between them are overcome, gases do not have a definite shape or a definite volume. The movement of particles and the collisions between them cause gases to diffuse. **Diffusion** is the spreading of particles throughout a given volume until they are uniformly distributed. Diffusion occurs in solids and liquids but occurs most rapidly in gases. For example, if you spray air freshener in one corner of a room, it's not long before you smell the scent all over the room. The particles of gas have moved, collided, and "filled" their container—the room. The particles have diffused. Gases will fill the container that they are in even if the container is a room. The particles continue to move and collide in a random motion within their container.

Topic: States of Matter
Visit gpscience.com for Web links to information about states of matter.

Activity Create a slide presentation about the states of matter using the computer and presentation software.

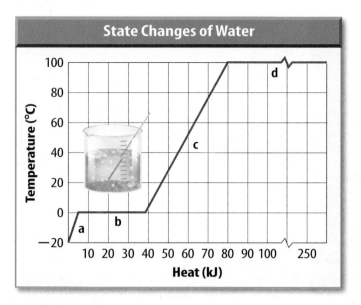

State Changes of Water

Temperature (°C) vs Heat (kJ)

Figure 7 This graph shows the heating curve of water. At **a** and **c** the water is increasing in kinetic energy. At **b** and **d** the added energy is used to overcome the bonds between the particles.

Figure 8 Stars including the Sun contain matter that is in the plasma phase. Plasma exists where the temperature is extremely high. **Describe** *the plasma phase.*

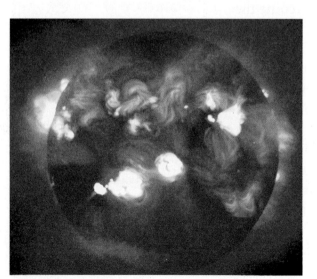

Heating Curve of a Liquid A graph of water being heated from −20°C to 100°C is shown in **Figure 7.** This type of graph is called a heating curve because it shows the temperature change of water as thermal energy, or heat, is added. Notice the two areas on the graph where the temperature does not change. At 0°C, ice is melting. All of the energy put into the ice at this temperature is used to overcome the attractive forces between the particles in the solid. The temperature remains constant during melting. After the attractive forces are overcome, particles move more freely and their average kinetic energy, or temperature, increases. At 100°C, water is boiling or vaporizing and the temperature remains constant again. All of the energy that is put into the water goes to overcoming the remaining attractive forces between the water particles. When all of the attractive forces in the water are overcome, the energy goes to increasing the temperature of the particles.

 **Reading Check** *What is occurring at the two temperatures on the heat curve where the graph is a flat line?*

 **INTEGRATE Astronomy**

Plasma State So far, you've learned about the three familiar states of matter—solids, liquids, and gases. But none of these is the most common state of matter in the universe. Scientists estimate that much of the matter in the universe is plasma. **Plasma** is matter consisting of positively and negatively charged particles. Although this matter contains positive and negative particles, its overall charge is neutral because equal numbers of both charges are present. Recall that on average, particles of matter move faster as the matter is heated to higher temperatures. The faster the particles move the greater the force is with which they collide. The forces produced from high-energy collisions are so great that electrons from the atom are stripped off. This state of matter is called plasma. All of the observed stars including the Sun, shown in **Figure 8,** consist of plasma. Plasma also is found in lightning bolts, neon and fluorescent tubes, and auroras.

Reading Check *What is plasma?*

Thermal Expansion

You have learned how the kinetic theory is used to explain the behavior of particles in different states of matter. The kinetic theory also explains other characteristics of matter in the world around you. Have you noticed the seams in a concrete driveway or sidewalk? A gap often is left between the sections to clearly separate them. These separation lines are called expansion joints. When concrete absorbs heat, it expands. Then when it cools, it contracts. If expansion joints are not used, the concrete will crack when the temperature changes.

Expansion of Matter The kinetic theory can be used to explain this behavior in concrete. Recall that particles move faster and separate as the temperature rises. This separation of particles results in an expansion of the entire object, known as thermal expansion. **Thermal expansion** is an increase in the size of a substance when the temperature is increased. The kinetic theory can be used to explain the contraction in objects, too. When the temperature of an object is lowered, particles slow down. The attraction between the particles increases and the particles move closer together. The movements of the particles closer together result in an overall shrinking of the object, known as contraction.

Expansion in Liquids Expansion and contraction occur in most solids, liquids, and gases. A common example of expansion in liquids occurs in thermometers, as shown in **Figure 9.** The addition of energy causes the particles of the liquid in the thermometer to move faster. The particles in the liquid in the narrow thermometer tube start to move farther apart as their motion increases. The liquid has to expand only slightly to show a large change on the temperature scale.

Expansion in Gases An example of thermal expansion in gases is shown in **Figure 10.** Hot-air balloons are able to rise due to thermal expansion of air. The air in the balloon is heated, causing the distance between the particles in the air to increase. As the hot-air balloon expands, the number of particles per cubic centimeter decreases. This expansion results in a decreased density of the hot air. Because the density of the air in the hot-air balloon is lower than the density of the cooler air outside, the balloon will rise.

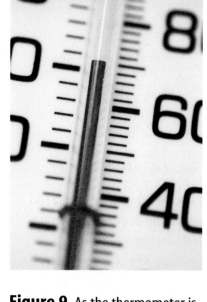

Figure 9 As the thermometer is heated, the column of liquid in the thermometer expands. As the temperature cools, the liquid in the thermometer contracts.

Figure 10 Heating the air in this hot-air balloon causes the particles in the air to move apart, creating a lower density inside the balloon.

Partial negative charge

Partial positive charge

Figure 11 The positively and negatively charged regions on a water molecule interact to create empty spaces in the crystal lattice. These interactions cause water to expand when it is in the solid phase. **Explain** *why ice floats on water.*

Understanding Unusual Behavior Pierre-Gilles de Gennes, a French physicist, was awarded the 1991 Nobel Prize for Physics for his discoveries about the behavior of molecules in liquid crystals and polymers (plastics). His discoveries led to a better understanding about the behavior and control of these substances.

The Strange Behavior of Water Normally, substances expand as the temperature rises, because the particles move farther apart. An exception to this rule, however, is water. Water molecules are unusual in that they have highly positive and highly negative areas. **Figure 11** is a diagram of the water molecule showing these charged regions. These charged regions affect the behavior of water. As the temperature of water drops, the particles move closer together. The unlike charges will be attracted to each other and line up so that only positive and negative zones are near each other. Because the water molecules orient themselves according to charge, empty spaces occur in the structure. These empty spaces are larger in ice than in liquid water, so water expands when going from a liquid to a solid state. Solid ice is less dense than liquid water. That is why ice floats on the top of lakes in the winter.

Solid or a Liquid?

Other substances also have unusual behavior when changing states. Amorphous solids and liquid crystals are two classes of materials that do not react as you would expect when they are changing states.

Amorphous Solids Ice melts at 0°C, gold melts at 1,064°C, and lead melts at 327°C. But not all solids have a definite temperature at which they change from solid to liquid. Some solids merely soften and gradually turn into a liquid over a temperature range. There is not an exact temperature like a boiling point where the phase change occurs. These solids lack the highly ordered structure found in crystals. They are known as amorphous solids from the Greek word for "without form."

You are familiar with two amorphous solids—glass and plastics. The particles that make up amorphous solids are typically long, chainlike structures that can get jumbled and twisted instead of being neatly stacked into geometric arrangements. Interactions between the particles occur along the chain, which gives amorphous solids some properties that are very different from crystalline solids.

Liquids do not have an orderly arrangement of particles. Some amorphous solids form when liquid matter changes to solid matter too quickly for an orderly structure to form. One example of this is obsidian—a volcanic glass. Obsidian forms when lava, made of molten rock, cools quickly, such as when it spills into water.

 **Reading Check** *What are two examples of amorphous solids?*

Liquid Crystals Liquid crystals are another group of materials that do not change states in the usual manner. Normally, the ordered geometric arrangement of a solid is lost when the substance goes from the solid state to the liquid state. Liquid crystals start to flow during the melting phase similar to a liquid, but they do not lose their ordered arrangement completely, as most substances do. Liquid crystals will retain their geometric order in specific directions.

Liquid crystals are placed in classes depending upon the type of order they maintain when they liquefy. They are highly responsive to temperature changes and electric fields. Scientists use these unique properties of liquid crystals to make liquid crystal displays (LCD) in the displays of watches, clocks, and calculators, as shown in **Figure 12.**

Figure 12 Liquid crystals are used in the displays of watches, clocks, calculators, and some notebook computers because they respond to electric fields.

section 1 review

Summary

States of Matter

- The kinetic theory is an explanation of how particles in matter move.
- Thermal energy is the total energy of a material's particles, including kinetic and potential energy.
- Temperature is the average kinetic energy of a substance.
- In most substances, as temperature increases the kinetic energy and disorder of the particles increase.

Thermal Expansion

- Some materials undergo thermal expansion when heated.
- Water expands when it changes from a liquid to a solid.

Solid or a Liquid?

- Amorphous solids have no definite melting point. They liquefy over a temperature range.
- Liquid crystals maintain some geometric order in the liquid state.

Self Check

1. **List** the three basic assumptions of the kinetic theory.
2. **Describe** the movement of the particles in solids, liquids, and gases.
3. **Describe** the movement of the particles at the melting point of a substance.
4. **Describe** the movement of the particles at the boiling point of a substance.
5. **Think Critically** Would the boiling point of water be higher or lower on the top of a mountain peak? How would the boiling point be affected in a pressurized boiler system? Explain.

Applying Math

6. **Interpret Data** Using the graph in **Figure 7,** describe the energy changes that are occurring when water goes from $-15°C$ to $100°C$.
7. **Make and Use Graphs** The melting point of acetic acid is $16.6°C$ and the boiling point is $117.9°C$. Draw a graph similar to the graph in **Figure 7** showing the phase changes for acetic acid. Clearly mark the three phases, the boiling point, and the melting point on the graph.

Thermal Energy Changes In Matter

Thermal energy changes in matter are important in your home, but you may not realize it. Your refrigerator removes thermal energy from warm food and releases it into the room. This process keeps food from spoiling by decreasing the temperature of the food.

▶ Real-World Question

Can a study of thermal energy changes lead to better understanding of matter and energy?

Goals

■ **Explain** the thermal energy changes that occur as matter goes from the solid to gas state.

Materials

beakers (2)	wire mesh	hot plate
ring clamp	ice	
ring stand	thermometer	

Safety Precautions

▶ Procedure

1. Set up the equipment as pictured. Prepare a data table in your Science Journal.

2. Gently heat the ice in the lower beaker. Every 3 min record your observations and the temperature of the water in the bottom container. Do not touch the thermometer to the bottom or sides of the container.

3. After the ice in the beaker melts and the water begins to boil, observe the system for several more minutes and record your observations.

4. Turn off the heat and let your system completely cool before you clean up.

▶ Conclude and Apply

1. **Draw** a picture of the system used in this lab in your Science Journal. Label the state the water started at in the lower beaker, the state it changed into in the lower beaker, the state above the lower beaker, and the state on the outside of the upper beaker.

2. **Find** the location on the diagram that has the greatest thermal energy and which has the least amount of thermal energy.

3. **Draw** a time-temperature graph using your data from your Science Journal.

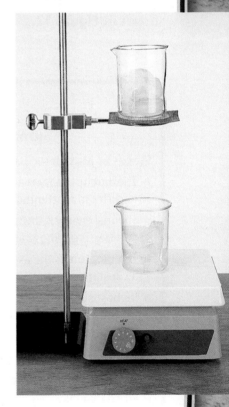

*C*ommunicating Your Data

Compare your results with other groups in the lab. **For more help, refer to the** Science Skill Handbook.

section 2

Properties of Fluids

Reading Guide

What You'll Learn
- **Explain** Archimedes' principle.
- **Explain** Pascal's principle.
- **Explain** Bernoulli's principle and explain how we use it.

Why It's Important
Properties of fluids determine the design of ships, airplanes, and hydraulic machines.

Review Vocabulary
density: mass per unit volume of a material

New Vocabulary
- buoyancy
- pressure
- viscosity

How do ships float?

Some ships are so huge that they are like floating cities. For example, aircraft carriers are large enough to allow airplanes to take off and land on their decks. Despite their weight, these ships are able to float. This is because a greater force pushing up on the ship opposes the weight—or force—of the ship pushing down. What is this force? This supporting force is called the buoyant force. **Buoyancy** is the ability of a fluid—a liquid or a gas—to exert an upward force on an object immersed in it. If the buoyant force is equal to the object's weight, the object will float. If the buoyant force is less than the object's weight, the object will sink.

Archimedes' Principle In the third century B.C., a Greek mathematician named Archimedes made a discovery about buoyancy. Archimedes found that the buoyant force on an object is equal to the weight of the fluid displaced by the object. For example, if you place a block of wood in water, it will push water out of the way as it begins to sink—but only until the weight of the water displaced equals the block's weight. When the weight of water displaced—the buoyant force—becomes equal to the weight of the block, it floats. If the weight of the water displaced is less than the weight of the block, the object sinks. **Figure 13** shows the forces that affect an object in a fluid.

Reading Check *Why do rocks sink and rubber balls float in a swimming pool?*

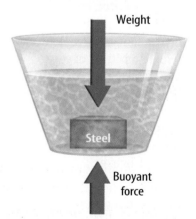

Figure 13 If the buoyant force of the fluid is equal to the weight of the object, the object floats. If the buoyant force of the fluid is less than the weight of the object, the object sinks.

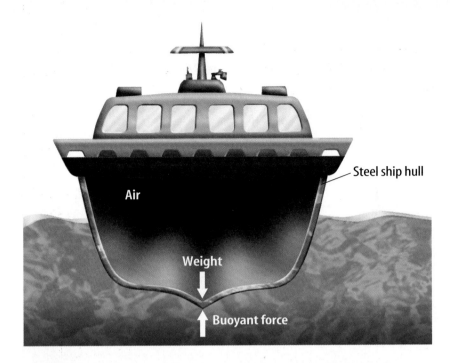

Figure 14 An empty hull of a ship contains mostly air. Its density is much lower than the density of a solid-steel hull. The lower density of the steel and air combination is what allows the ship to float in water. **Explain** *why a boat that takes in water will sink.*

Steel ship hull

Air

Weight

Buoyant force

Density Would a steel block the same size as a wood block float in water? They both displace the same volume and weight of water when submerged. Therefore, the buoyant force on the blocks is equal. Yet the steel block sinks and the wood block floats. What is different? The volume of the blocks and the volume of the water displaced each have different masses. If the three equal volumes have different masses, they must have different densities. Remember that density is mass per unit volume. The density of the steel block is greater than the density of water. The density of the wood block is less than the density of water. An object will float if its density is less than the density of the fluid it is placed in.

Suppose you formed the steel block into the shape of a hull filled with air, as in **Figure 14.** Now the same mass takes up a larger volume. The overall density of the steel boat and air is less than the density of water. The boat will now float.

Pascal's Principle

If you are underwater, you can feel the pressure of the water all around you. **Pressure** is force exerted per unit area, or $P = F/A$. Do you realize that Earth's atmosphere is a fluid? Earth's atmosphere exerts pressure all around you.

Blaise Pascal (1623–1662), a French scientist, discovered a useful property of fluids. According to Pascal's principle, pressure applied to a fluid is transmitted throughout the fluid. For example, when you squeeze one end of a balloon, the balloon expands out on the other end. When you squeeze one end of a toothpaste tube, toothpaste emerges from the other end. The pressure has been transmitted through the fluid toothpaste.

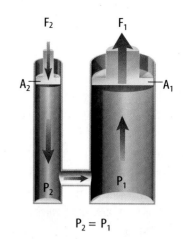

Applying the Principle Hydraulic machines are machines that move heavy loads in accordance with Pascal's principle. Maybe you've seen a car raised using a hydraulic lift in an auto repair shop. A pipe that is filled with fluid connects small and large cylinders as shown in **Figure 15.** Pressure applied to the small cylinder is transferred through the fluid to the large cylinder. Because pressure remains constant throughout the fluid, according to Pascal's principle, more force is available to lift a heavy load by increasing the surface area. With a hydraulic machine, you could use your weight to lift something much heavier than you are. Do the following activity to see how force, pressure, and area are related.

Figure 15 The pressure remains the same throughout the fluid in a hydraulic lift.

Applying Math — Solve a One-Step Equation

CALCULATING FORCES USING PASCAL'S PRINCIPLE A hydraulic lift is used to lift a heavy machine that is pushing down on a 2.8 m² piston (A_1) with a force (F_1) of 3,700 N. What force (F_2) needs to be exerted on a 0.072 m² piston (A_2) to lift the machine?

IDENTIFY known values and the unknown value

Identify the known values:

The force on piston one is 3,700 N　means　$F_1 = 3,700$ N

The area of piston one is 2.8 m²　means　$A_1 = 2.8$ m²

The area of piston two is 0.072 m²　means　$A_2 = 0.072$ m²

Identify the unknown value:

What force needs to be exerted?　means　$F_2 = ?$

SOLVE the problem

Substitute the known values $F_1 = 3,700$ N, $A_1 = 2.8$ m², and $A_2 = 0.072$ m²:

$$P_1 = P_2 = \frac{F_1}{A_1} = \frac{F_2}{A_2}, \text{ then } F_2 = \frac{F_1 A_2}{A_1} = \frac{3,700\text{N} \times 0.072\text{m}^2}{2.8\text{m}^2} = 95 \text{ N}$$

In this equation, m² is divided by m²:

$$\frac{\text{N} \cdot m^2}{m^2} = \text{N}$$

the result is N, which is the unit for force.

CHECK your answer

Does your answer seem reasonable? Substitute the values for F_1, A_1, F_2, and A_2 back into the original equation. If both sides are equal, the answer is correct.

The value you calculate for F_2 can be used to recalculate one of the known values, F_1, A_1, or A_2. If F_2 is correct, the value you calculate for the known value will be the same as the value given in the problem.

Practice Problems

1. A heavy crate applied a force of 1,500 N on a 25-m² piston. What force needs to be exerted on the 0.80-m² piston to lift the crate?

For more practice problems, go to page 834, and visit gpscience.com/extra_problems.

Figure 16 The air above the sheet of paper is moving faster than the air under the paper, creating a low-pressure area above the paper, so the paper rises.

Bernoulli's Principle

Daniel Bernoulli (1700–1782) was a Swiss scientist who studied the properties of moving fluids such as water and air. He published his discovery in 1738. According to Bernoulli's principle, as the velocity of a fluid increases, the pressure exerted by the fluid decreases. One way to demonstrate Bernoulli's principle is to blow across the top surface of a sheet of paper, as in **Figure 16.** The paper will rise. The velocity of the air you blew over the top surface of the paper is greater than that of the quiet air below it. As a result, the air pressure pushing down on the top of the paper is lower than the air pressure pushing up on the paper. The net force below the paper pushes the paper upward. This principle is used today when designing aircraft wings and fluid-transporting piping systems.

Another application of Bernoulli's principle is the hose-end sprayer. This sprayer is used to apply fertilizers, herbicides, and insecticides to yards and gardens. To use this sprayer, a concentrated solution of the chemical that is to be applied is placed in the sprayer. The sprayer is attached to a garden hose, as shown in **Figure 17.** A strawlike tube is attached to the lid of the unit. The end of the tube is submerged into the concentrated chemical. The water to the garden hose is turned to a high flow rate. When you are ready to apply the chemicals to the lawn or plant area, you must push a trigger on the sprayer attachment. This allows the water in the hose to flow at a high rate of speed, creating a low pressure area above the strawlike tube. The concentrated chemical solution is sucked up through the straw and into the stream of water. The concentrated solution is mixed with water, reducing the concentration to the appropriate level and creating a spray that is easy to apply.

Figure 17 Bernoulli's principle was used in designing the hose-end sprayer.
Define *Bernoulli's principle.*

 **Reading Check** *How does pressure change as the velocity of a fluid increases?*

Water moves through the sprayer at high speed.

The fast-moving water creates a low-pressure area, pulling chemicals up the tube.

The water-chemical mixture sprays out of the tip.

Strawlike tube

Concentrated chemical solution (atmospheric pressure)

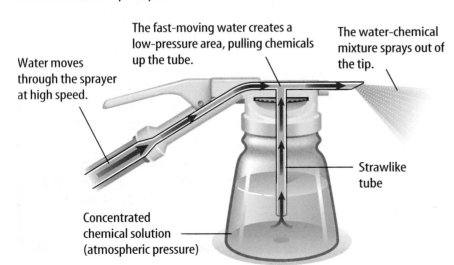

Fluid Flow

Another property exhibited by fluid is its tendency to flow. The resistance to flow by a fluid is called **viscosity**. Fluids vary in their tendency to flow. For example, when you take syrup out of the refrigerator and pour it, the flow of syrup is slow. But if this syrup were heated, it would flow much faster. Water has a low viscosity because it flows easily. Cold syrup has a high viscosity because it flows slowly.

When a container of liquid is tilted to allow flow to begin, the flowing particles will transfer energy to the particles that are stationary. In effect, the flowing particles are pulling the other particles, causing them to flow, too. If the flowing particles do not effectively pull the other particles into motion, then the liquid has a high viscosity, or a high resistance to flow. If the flowing particles pull the other particles into motion easily, then the liquid has low viscosity, or a low resistance to flow.

INTEGRATE Earth Science

Magma and Viscosity
Magma, or liquefied rock from a volcano, is an example of a liquid with varying viscosity. The viscosity of magma depends upon its composition. The viscosity of the magma flow determines the shape of the volcanic cone. In your Science Journal, infer the type of volcano cone that is created with high- and low-viscosity lava flows.

☑ Reading Check *How does temperature affect viscosity?*

section 2 review

Summary

How do ships float?

- Buoyancy is the ability of a liquid or gas to exert an upward force on an object immersed in it.
- If the buoyant force is equal to the object's weight, the object will float. If the buoyant force is less than the object's weight, the object will sink.

Pascal's Principle

- Pascal's principle says that pressure applied to a fluid is transmitted throughout the fluid.

Bernoulli's Principle

- Bernoulli's principle says that as the velocity of a fluid increases, the pressure exerted by the fluid decreases.

Fluid Flow

- The resistance to flow by a fluid is called viscosity.
- Increasing the temperature increases the rate of energy transfer between the particles. This decreases the resistance to flow, or the viscosity.

Self Check

1. **Describe** the two opposing forces that are acting on an object floating in water.

2. **Explain** how a heavy boat floats on water.

3. **Explain** Use Pascal's principle to explain why squeezing a plastic mustard bottle forces mustard out the top.

4. **Describe**, using Bernoulli's principle, how roofs are lifted off buildings in tornados.

5. **Think Critically** If you fill a balloon with air, tie it off, and release it, it will fall to the floor. Why does it fall instead of float? What would happen if the balloon contained helium?

Applying Math

6. **Find Force** The density of water is 1.0 g/cm^3. How many kilograms of water does a submerged 120-cm^3 block displace? One kilogram has a force of 9.8 N. What is the buoyant force on the block?

7. **Solve an Equation** To lift an object weighing 20,000 N, how much force is needed on a small piston with an area of 0.072 m^2 if the large piston has an area of 2.8 m^2?

Behavior of Gases

Pressure

You learned from the kinetic theory that gas particles are constantly moving and colliding with anything in their path. The collisions of these particles in the air result in pressure. Pressure is the amount of force exerted per unit of area.

Often, gases are confined within containers. A balloon and a bicycle tire are considered to be containers. They remain inflated because of collisions the air particles have with the walls of their container, as shown in **Figure 18.** This collection of forces, caused by the collisions of the particles, pushes the walls of the container outward. If more air is pumped into the balloon, the number of air particles is increased. This causes more collisions with the walls of the container, which causes it to expand. Since the bicycle tire can't expand much, its pressure increases.

Pressure is measured in a unit called **pascal** (Pa), the SI unit of pressure. Because pressure is the amount of force divided by area, one pascal of pressure is one Newton per square meter or 1 N/m^2. This is a small pressure unit, so most pressures are given in kilopascals (kPa), or 1,000 pascals. At sea level, atmospheric pressure is 101.3 kPa. This means that at Earth's surface, the atmosphere exerts a force of about 101,300 N on every square meter—about the weight of a large truck. More information about the atmosphere is shown in **Figure 19.** Notice the temperature and pressure differences in the atmosphere as the distance from the surface of Earth increases.

Figure 18 The force created by the many particles in air striking the balloon's walls pushes the wall outward, keeping the balloon inflated.

Explain why the term pressure can be used to describe these forces.

✔ **Reading Check** *How are force, area, and pressure related?*

VISUALIZING ATMOSPHERIC LAYERS

Figure 19

Earth's atmosphere is divided into five layers. The air gets thinner as distance from Earth's surface increases. Temperature is variable, however, due to differences in the way the layers absorb incoming solar energy.

500 km

The space shuttle crosses all the atmosphere's layers.

Auroras

Meteors

Jets and weather balloons fly in the atmosphere's lowest layers.

85 km

50 km

Ozone Layer

10 km
0 km

The Hubble Space Telescope

Exosphere (on average, 1,100°C; pressure negligible)

Gas molecules are sparse in the exosphere (beyond 500 km). The *Landsat 7* satellite and the *Hubble Space Telescope* orbit in this layer, at an altitude of about 700 km and 600 km respectively. Beyond the exosphere there is nothing but the vacuum of interplanetary space.

Thermosphere (−80°C to 1,000°C; pressure negligible)

Compared to the exosphere, gas molecules are slightly more concentrated in the thermosphere (85–500 km). Air pressure is still very low, however, and temperatures range widely. Light displays called auroras form in this layer over polar regions.

The temperature drops dramatically in the mesosphere (50–85 km), the coldest layer. The stratosphere (10–50 km) contains a belt of ozone, a gas that absorbs most of the Sun's harmful ultraviolet rays. Clouds and weather systems form in the troposphere (1–10 km), the only layer in which air-breathing organisms typically can survive.

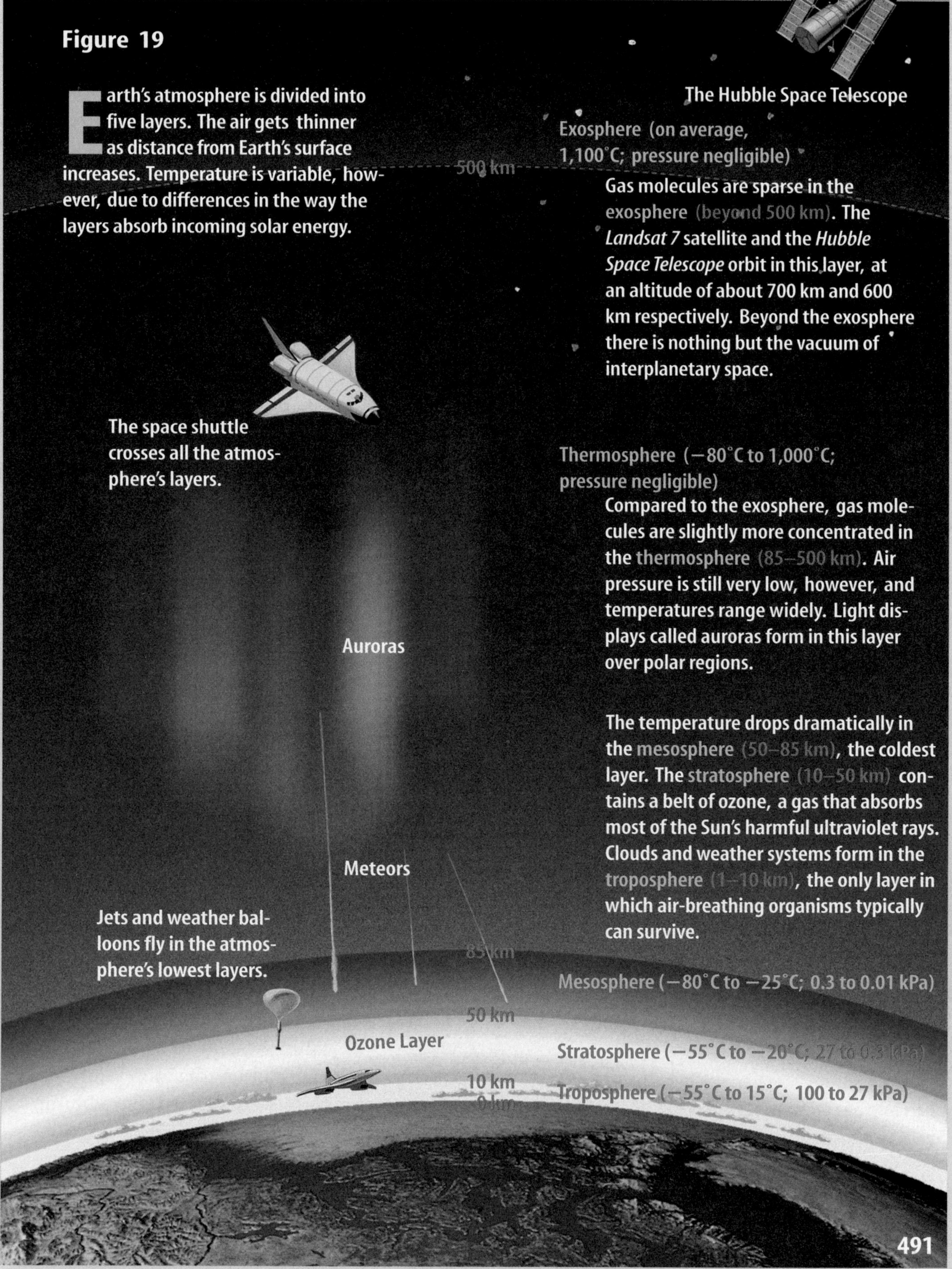

Mesosphere (−80°C to −25°C; 0.3 to 0.01 kPa)

Stratosphere (−55°C to −20°C; 27 to 0.3 kPa)

Troposphere (−55°C to 15°C; 100 to 27 kPa)

Figure 20
Balloons are used to measure the weather conditions at high altitudes. These balloons expand as they rise due to decreased pressure. **Describe** *what eventually happens to the balloon.*

Boyle's Law

You now know how gas creates pressure in a container. What happens to the gas pressure if you decrease the size of the container? You know that the pressure of a gas depends on how often its particles strike the walls of the container. If you squeeze gas into a smaller space, its particles will strike the walls more often—giving an increased pressure. The opposite is true, too. If you give the gas particles more space, they will hit the walls less often—gas pressure will be reduced. Robert Boyle (1627–1691), a British scientist, described this property of gases. According to Boyle's law, if you decrease the volume of a container of gas and hold the temperature constant, the pressure of the gas will increase. An increase in the volume of the container causes the pressure to drop, if the temperature remains constant.

The behavior of weather balloons, as shown in **Figure 20,** can be explained using Boyle's law. Rubber or neoprene weather balloons are used to carry sensing instruments to high altitudes to detect weather information. The balloons are inflated near Earth's surface with a low-density gas. As the balloon rises, the atmospheric pressure decreases. The balloon gradually expands to a volume of 30 to 200 times its original size. At some point the expanding balloon ruptures. Boyle's law states that as pressure is decreased the volume increases, as demonstrated by the weather balloon. The opposite also is true, as shown by the graph in **Figure 21.** As the pressure is increased, the volume will decrease.

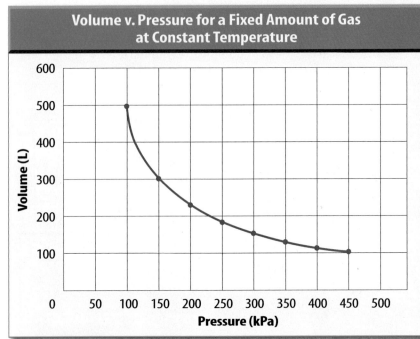

Volume v. Pressure for a Fixed Amount of Gas at Constant Temperature

Figure 21 The graph shows that, as pressure increases, volume decreases; as pressure decreases, volume increases.
Use Graphs *What is the volume of the gas at 100 kPa?*

Boyle's Law in Action When Boyle's law is applied to a real life situation, we find that the pressure multiplied by the volume is always equal to a constant if the temperature is constant. As the pressure and volume change indirectly, the constant will remain the same. You can use the equations $P_1V_1 = \text{constant} = P_2V_2$ to express this mathematically. This shows us that the product of the initial pressure and volume—designated with the subscript 1—is equal to the product of the final pressure and volume—designated with the subscript 2. Using this equation, you can find one unknown value, as shown in the example problem below.

✔ **Reading Check** *What is $P_1V_1 = P_2V_2$ known as?*

Applying Math — Solve a One-Step Equation

USING BOYLE'S LAW A balloon has a volume of 10.0 L at a pressure of 101 kPa. What will be the new volume when the pressure drops to 43.0 kPa?

IDENTIFY known values and the unknown value

Identify the known values:

The initial pressure is 101 kPa. `means` ➤ $P_1 = 101$ kPa

The initial volume is 10.0 L. `means` ➤ $V_1 = 10.0$ L

The final pressure is 43.0 kPa. `means` ➤ $P_2 = 430$ kPa

Identify the unknown value:

What is the new volume? `means` ➤ $V_2 = ?$

SOLVE the problem

Substitute the known values $P_1 = 101$ kPa, $V_1 = 10.0$ L, and $P_2 = 43.0$ kPa into the equation:

$$P_1V_1 = P_2V_2, \text{ then } V_2 = \frac{P_1V_1}{P_2} = \frac{(101 \text{ kPa})(10.0 \text{ L})}{43.0 \text{ kPa}} = 23.5 \text{ L}$$

> In this equation, kPa is divided by kPa:
> $$\frac{\cancel{kPa} \, L}{\cancel{kPa}} = L$$
> the result is L, which is the unit for volume.

CHECK your answer

Does your answer seem reasonable? Multiplying the values for P_1 and V_1 gives a pressure of 1010 kPa. Multiplying the values for P_2 and V_2 gives a pressure of 1010.5 kPa. These values are reasonably close and differ because of round-off error. So the calculated volume is correct.

> The value you calculate for V_2 can be used to recalculate one of the known values, P_1, V_1, or P_2. If V_2 is correct, the value you calculate for the known value will be the same as the value given in the problem.

Practice Problems

1. A volume of helium occupies 11.0 L at 98.0 kPa. What is the new volume if the pressure drops to 86.2 kPa?

For more practice problems, go to page 834, and visit gpscience.com/extra_problems.

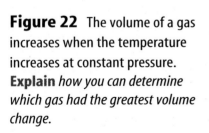
The Pressure-Temperature Relationship

Have you ever read the words "keep away from heat" on a pressurized spray canister? What happens if you heat an enclosed gas? The particles of gas will strike the walls of the canister more often. Because this canister is rigid, its volume cannot increase. Instead, its pressure increases. If the pressure becomes greater than the canister can hold, it will explode. At a constant volume, an increase in temperature results in an increase in pressure.

Charles's Law

If you've watched a hot-air balloon being inflated, you know that gases expand when they are heated. Because particles in the hot air are farther apart than particles in the cool air, the hot air is less dense than the cool air. This difference in density allows the hot air balloon to rise. Jacques Charles (1746–1823) was a French scientist who studied gases. According to Charles's law, the volume of a gas increases with increasing temperature, as long as pressure does not change. As with Boyle's law, the reverse is true, also. The volume of a gas shrinks with decreasing temperature, as shown in **Figure 22.**

Charles's law can be explained using the kinetic theory of matter. As a gas is heated, its particles move faster and faster and its temperature increases. Because the gas particles move faster, they begin to strike the walls of their container more often and with more force. In the hot-air balloon, the walls have room to expand so instead of increased pressure, the volume increases.

Figure 22 The volume of a gas increases when the temperature increases at constant pressure. **Explain** *how you can determine which gas had the greatest volume change.*

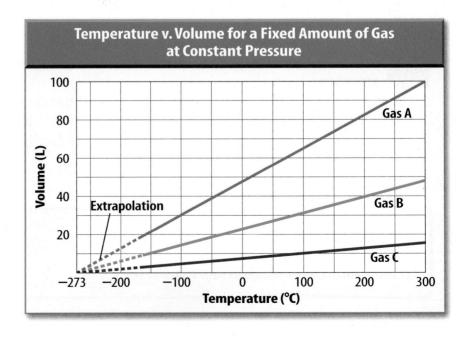

Using Charles's Law The formula that relates the variables of temperature to volume shows a direct relationship, $V_1/T_1 = V_2/T_2$, when temperature is given in kelvin. When using Charles's law, the pressure must be kept constant. What would be the resulting volume of a 2.0-L balloon at 25.0°C that was placed in a container of ice water at 3.0°C, as shown in **Figure 23?**

$$V_1 = 2.0 \text{ L} \qquad T_1 = 25.0°C + 273 = 298 \text{ K}$$

$$V_2 = ? \qquad T_2 = 3.0°C + 273 = 276 \text{ K}$$

$$\frac{V_1}{T_1} = \frac{V_2}{T_2} = \frac{2.0 \text{ L}}{298 \text{ K}} = \frac{V_2}{276 \text{ K}}$$

$$V_2 = \frac{(2.0 \text{ L})(276 \text{ K})}{298 \text{ K}} = 1.9 \text{ L}$$

As Charles's law predicts, the volume decreased as the temperature of the trapped gas decreased. This assumed no changes in pressure.

Figure 23 Charles's law states that as the temperature of a gas is lowered, the volume decreases. **Calculate** *If the balloon in the text was placed in a freezer at 5°C, what would be the new volume?*

 **Reading Check** *According to Charles's law, what happens to the volume of a gas if the temperature increases?*

---

section 3 review

Summary

Pressure

- Pressure is the amount of force exerted per unit area.
- Pressure is measured in pascals.

Boyle's Law

- Boyle's law states that if the temperature is constant, as the volume of a gas decreases the pressure increases. It states also that at constant temperature, as the volume of a gas increases the pressure decreases.

The Pressure-Temperature Relationship

- This relationship describes how, at a constant volume, the pressure increases with increasing temperature.

Charles's Law

- Charles's law states that at constant pressure, the volume of a gas increases with increasing temperature.

Self Check

1. **Explain** why a gas has pressure.
2. **Describe** Earth's atmosphere at sea level. How does the pressure change as the distance from Earth increases?
3. **Explain,** using Boyle's law, the volume change of an inflated balloon that a diver takes to a pressure of 2 atm.
4. **Explain,** using Charles's law, the purpose of a gas burner on a hot-air balloon.
5. **Think Critically** Labels on cylinders of compressed gases state the highest temperature to which the cylinder may be exposed. Give a reason for this warning.

Applying Math

6. **Find Volume** A helium balloon has a volume of 2.00 L at 101 kPa. As the balloon rises the pressure drops to 97.0 kPa. What is the new volume?
7. **Solve One-Step Equations** If a 5-L balloon at 25°C was gently heated to 30°C, what new volume would the balloon have?

LAB

Testing the Viscosity of Common Liquids

Goals
■ Observe and compare the viscosity of common liquids.

Materials
room temperature household liquids such as:
- dish detergent
- corn syrup
- pancake syrup
- shampoo
- vegetable oil
- vinegar
- molasses
- water

spheres such as glass marbles or steel balls
100-mL graduated cylinders
150-mL beaker
ruler
stopwatch

Safety Precautions

Dispose of wastes as directed by your teacher.

● Real-World Question

The resistance to flow of a liquid is called viscosity, and it can be measured and compared. One example of the importance of a liquid's viscosity is motor oil in car engines. The viscosity of motor oil in your family car is important because it keeps the engine lubricated. It must cling to the moving parts and not run off, leaving the parts dry and unlubricated. If the engine is not properly lubricated, it will be damaged eventually. The motor oil must maintain its viscosity in all types of weather, from extreme heat in the summer to freezing cold in the winter. Can the study of viscosity lead to a better understanding of the properties of matter?

● Procedure

1. Measure equal amounts of the liquids to be tested into the graduated cylinders.
2. Measure the depth of the liquid.
3. Copy the data chart into your Science Journal.
4. Place the sphere on the surface of the liquid. Using a stopwatch, measure and record how long it takes for it to travel to the bottom of the liquid.
5. Remove the sphere and repeat step 4 two more times for the same liquid.
6. Rinse and dry the sphere.
7. Repeat steps 4, 5, and 6 for two more liquids.

Viscosity of Common Liquids				
Substance	Trial	Depth of Liquid (cm)	Time (s)	Speed (cm/s)
	Do not write in this book.			

▶ Analyze Your Data

1. **Graph** the average speed of the sphere for each liquid on a bar graph.

2. **Interpret Data** In which liquid did the sphere move the fastest? Would that liquid have a high or low viscosity? Explain.

▶ Conclude and Apply

1. **Infer** Would it matter if you dropped or threw the sphere into the liquid instead of placing it there? Explain your answer.

2. **Analyze Results** What effect does temperature play in the viscosity of a liquid? What would happen to the viscosity of your slowest liquid if you made it colder? Explain.

3. **Infer** If the temperature of the liquids is dropped to 10°C, would all of the liquids have an equivalent change in viscosity? Explain your answer.

4. **Explain** Would corn syrup, molasses, or pancake syrup make a good lubricant in a car engine? Explain your answer.

*C*ommunicating
Your Data

Compare your results with other groups and discuss differences noted. Why might these differences have occurred? **For more help, refer to the** Science Skill Handbook.

SCIENCE Stats

Hot and Cold

Did you know...

... The world's coldest substance, liquid helium, is about −269°C. It's used in cryogenics research, which is the study of extremely low temperatures. Cryogenics has enabled physicians to freeze and preserve body parts, such as corneas from human eyes. The freezing keeps cells alive until they are needed.

Cryogenics laboratory

Oxyacetylene torch

... The hottest known flame is made by burning a mixture of oxygen and acetylene. The flame of an oxyacetylene torch can become as hot as 3,300°C. That's more than two times hotter than the melting point of steel.

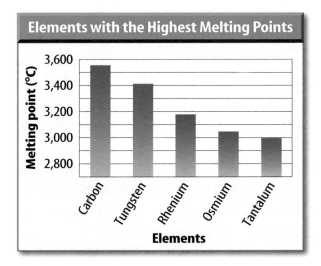

Elements with the Highest Melting Points

Melting point (°C) vs. Elements (Carbon, Tungsten, Rhenium, Osmium, Tantalum)

Applying Math

1. In 1983, the temperature dropped to −89°C in Vostok, Antarctica. How many more degrees Celsius would the temperature need to drop for the air to become a liquid?

2. Look at the graph above. List the elements that could be melted by an oxyacetylene torch.

Reviewing Main Ideas

Section 1   Kinetic Theory

1. Four states of matter exist: solid, liquid, gas, and plasma.

2. According to the kinetic theory, all matter is made of constantly moving particles that collide without losing energy.

3. Most matter expands when heated and contracts when cooled. This expansion joint allows the concrete to expand and contract without damage.

4. Changes of state can be interpreted in terms of the kinetic theory of matter.

Section 2   Properties of Fluids

1. Archimedes' principle states that the buoyant force of an object in a fluid is equal to the weight of the fluid displaced. The buoyant force on this penny was less than its weight, so the penny sank.

2. Pascal's principle states that pressure applied to a fluid is transmitted unchanged throughout the fluid.

3. Bernoulli's principle states that the pressure exerted by a fluid decreases as its velocity increases.

Section 3   Behavior of Gases

1. Gas pressure results from moving particles colliding with the inside walls of the container.

2. The SI unit of pressure is the pascal (Pa). Because this is a small pressure unit, pressures often are given in kilopascals.

3. Boyle's law states that the volume of a gas decreases when the pressure increases at constant temperature.

4. Charles's law states that the volume of a gas increases when the temperature increases at constant pressure.

5. At constant volume, as the temperature of a gas increases, so does the pressure of a gas. The pressure in this cylinder will increase as the sun increases the temperature.

FOLDABLES Use the Foldable that you made at the beginning of this chapter to help you review solids, liquids, and gases.

Using Vocabulary

boiling point p. 479
buoyancy p. 485
diffusion p. 479
heat of fusion p. 478
heat of vaporization p. 479
kinetic theory p. 476

melting point p. 478
pascal p. 490
plasma p. 480
pressure p. 486
thermal expansion p. 481
viscosity p. 489

Answer the following questions using complete sentences.

1. What is the property of a fluid that represents its resistance to flow?

2. What is the SI unit of pressure?

3. What term is used to describe the amount of force exerted per unit of area?

4. What is the temperature when a solid begins to liquefy?

5. What theory is used to explain the behavior of particles in matter?

6. What is the ability of a fluid to exert an upward force on an object?

Checking Concepts

Choose the word or phrase that best answers the question.

7. What is the temperature at which all particle motion of matter ceases?
   A) absolute zero
   B) melting point
   C) boiling point
   D) heat of fusion

8. What is the most common state of matter in the universe?
   A) solid
   B) liquid
   C) gas
   D) plasma

9. Which of the following would be used to measure pressure?
   A) gram
   B) newtons
   C) kilopascals
   D) kilograms

10. Which of the following uses Pascal's principle?
    A) aerodynamics
    B) hydraulics
    C) buoyancy
    D) changes of state

11. Which of the following uses Bernoulli's principle?
    A) airplane
    B) piston
    C) skateboard
    D) snowboard

12. The particles in which of the following are farthest apart from each other?
    A) gas
    B) solid
    C) liquid
    D) plasma

13. What is the upward force in a liquid?
    A) pressure
    B) kinetic theory
    C) buoyancy
    D) diffusion

14. What is the amount of energy needed to change a solid to a liquid at its melting point called?
    A) heat of fusion
    B) heat of vaporization
    C) temperature
    D) absolute zero

Interpreting Graphics

Use the graph below to answer question 15.

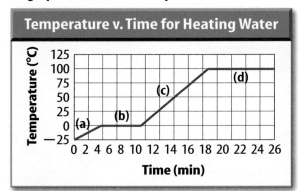

Temperature v. Time for Heating Water

15. A group of students heated ice until it turned to steam. They measured the temperature each minute. Their graph is provided above. Explain what is happening at each letter (a, b, c, d) in the graph.

16. Copy and complete this concept map.

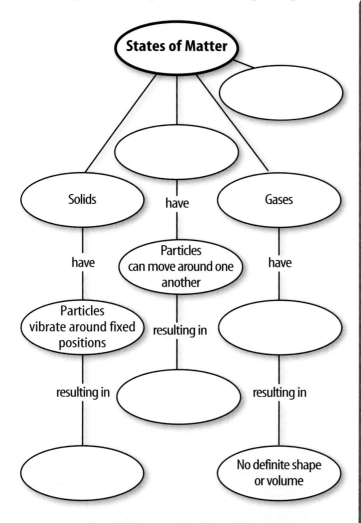

21. Use Numbers As elevation increases, boiling point decreases. List each of the following locations as *at sea level, above sea level,* or *below sea level.* (Boiling point of water is given in parenthesis.)

Death Valley (100.3°C), Denver (94°C), Madison (99°C), Mt. Everest (76.5°C), Mt. McKinley (79°C), New York City (100°C), Salt Lake City (95.6°C)

Use the illustration below to answer question 22.

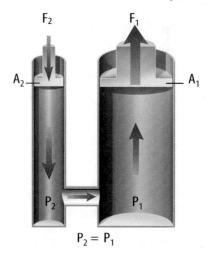

22. Solve One-Step Equations A hydraulic lift is used to lift a heavy box that is pushing down on a 3.0 m² piston (A_1) with a force (F_1) of 1,500 N. What force needs to be exerted on a 0.08 m² piston (A_2) to lift the machine?

23. Calculate What would be the resulting volume of a 1.5 L balloon at 25.0°C that was placed in a container of hot water at 90.0°C?

24. Use Numbers A balloon has a volume of 25.0 L at a pressure of 98.7 kPa. What will be the new volume when the pressure is 51.2 kPa?

Thinking Critically

17. Explain Use the temperature-pressure relationship to explain why you should check your tire pressure when the temperature changes.

18. Describe the changes that occur inside a helium balloon as it rises from sea level.

19. Explain why aerosol cans have a "do not incinerate" warning.

20. Explain The Dead Sea is a solution that is so dense you float on it easily. Explain why you are able to float easily, using the terms *density* and *buoyant force.*

Part 1 Multiple Choice

Record your answers on the answer sheet provided by your teacher or on a sheet of paper.

1. Which state of matter would you expect to find water, at −25°C and 1 atm on Earth?
   A. solid
   B. liquid
   C. gas
   D. plasma

Use the graph below to answer questions 2 and 3.

State Changes of Water

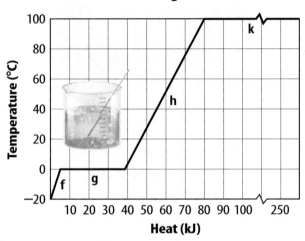

2. Which points on the graph is water increasing in kinetic energy?
   A. F and G
   B. G and K
   C. F and H
   D. H and K

3. On which points on the graph is the added energy used to overcome the bonds between particles?
   A. F and G
   B. G and K
   C. F and H
   D. H and K

4. Which of the following is unlikely to contain plasma?
   A. stars
   B. neon lights
   C. lightning
   D. water

Test-Taking Tip

Read Carefully Read all choices before answering the questions.

5. Which term is the amount of energy required for a liquid at its boiling point to become a gas?
   A. heat of vaporization
   B. diffusion
   C. heat of fusion
   D. thermal energy

6. In which state of matter do particles stay close together, yet are able to slide past each other?
   A. solid
   B. liquid
   C. gas
   D. plasma

Use the graph below to answer questions 7 and 8

Gas Characteristics

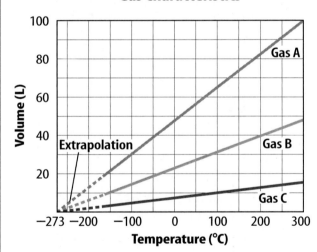

7. Which of the following statements is true?
   A. Gas A had the greatest increase in volume.
   B. Gas B had the greatest increase in volume.
   C. Gas C had the greatest increase in volume.
   D. The gases had the same increase in volume.

8. Approximately what temperature is the volume of Gas B about 40 L^3?
   A. 100°C
   B. 150°C
   C. 200°C
   D. 300°C

Part 2 | Short Response/Grid In

Record your answers on the answer sheet provided by your teacher or on a sheet of paper.

9. If you place two wood blocks in water and one sinks while the other floats, what do you know about the densities of the blocks?

Use the illustration below to answer question 10.

10. A weather balloon is inflated near Earth's surface with a low-density gas. Explain why the balloon rises when it is released.

11. Motor oil with a lower viscosity grade flows more easily than oil with a higher viscosity grade. Which grade of oil would be better for cold-weather driving? Explain.

12. The air in a scuba tank is under more than 200 times normal air pressure. Why should a filled scuba tank never be left in a hot car for an extended period of time?

13. A large crate applies a force of 2,500 N to a piston with an area of 25 m². What force must be applied to a piston with an area of 5.0 m² in order to lift the crate?

Part 3 | Open Ended

Record your answers on a sheet of paper.

Use the illustration below to answer questions 14 and 15.

14. Explain why the hot-air balloon can still float in the air when it is carrying a basket filled with people.

15. Explain what will happen when the burner on the hot-air balloon is turned off when the balloon is in the air.

16. Scuba divers often add or remove air from their buoyancy vests to maintain neutral buoyancy. This means they will neither sink to the bottom, nor float to the surface. If the diver takes a deep breath, the diver will rise slightly. When the diver exhales, the diver will sink slightly. Explain why this happens.

17. A penny will sink in a beaker of water, but it will float in a beaker of mercury. Explain how this is possible.

18. Explain the difference between evaporation and boiling.

19. Use the kinetic theory to explain how the temperature and the pressure of a given amount of gas are related.

Properties of Atoms and the Periodic Table

Atoms Compose All Things—Great and Small

Everything in this photo and the universe is composed of tiny particles called atoms. You will learn about atoms and their components—protons, neutrons, electrons, and quarks.

Science Journal In your Science Journal, write a few paragraphs about what you know about atoms.

Start-Up Activities

Inferring What You Can't Observe

How do detectives solve a crime when no witnesses saw it happen? How do scientists study atoms when they cannot see them? In situations such as these, techniques must be developed to find clues to answer the question. Do the lab below to see how clues might be gathered.

1. Take an envelope from your teacher.
2. Place an assortment of dried beans in the envelope and seal it. **WARNING:** *Do not eat any lab materials.*
3. Trade envelopes with another group.
4. Without opening the envelope, try to figure out the types and number of beans that are in the envelope. Record a hypothesis about the contents of the envelope in your Science Journal.
5. After you record your hypothesis, open the envelope and see what is inside.
6. **Think Critically** Describe the contents of your envelope. Was your hypothesis correct?

Preview this chapter's content and activities at gpscience.com

FOLDABLES™
Study Organizer

Atoms You have probably studied atoms before. Make the following Foldable to help identify what you already know, what you want to know, and what you learned about atoms.

STEP 1 Fold a vertical sheet of paper from side to side. Make the front edge about 1.25 cm shorter than the back edge.

STEP 2 Turn lengthwise and **fold** into thirds.

STEP 3 Unfold and cut only the top layer along both folds to make three tabs.

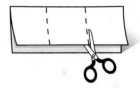

STEP 4 Label each tab as shown.

Identify Questions Before you read the chapter, write what you already know about atoms under the left tab of your Foldable, and write questions about what you'd like to know under the center tab. After you read the chapter, list what you learned under the right tab.

Structure of the Atom

Reading Guide

What You'll Learn

- **Identify** the names and symbols of common elements.
- **Identify** quarks as subatomic particles of matter.
- **Describe** the electron cloud model of the atom.
- **Explain** how electrons are arranged in an atom.

Why It's Important

Everything that you see, touch, and breathe is composed of tiny atoms.

Review Vocabulary

element: substance with atoms that are all alike

New Vocabulary

- atom
- nucleus
- proton
- neutron
- electron
- quark
- electron cloud

Scientific Shorthand

Do you have a nickname? Do you use abbreviations for long words or the names of states? Scientists also do this. In fact, scientists have developed their own shorthand for dealing with long, complicated names.

Do the letters C, Al, Ne, and Ag mean anything to you? Each letter or pair of letters is a chemical symbol, which is a short or abbreviated way to write the name of an element. Chemical symbols, such as those in **Table 1,** consist of one capital letter or a capital letter plus one or two small letters. For some elements, the symbol is the first letter of the element's name. For other elements, the symbol is the first letter of the name plus another letter from its name. Some symbols are derived from Latin. For instance, *Argentum* is Latin for "silver." Elements have been named in a variety of ways. Some elements are named to honor scientists, for places, or for their properties. Other elements are named using rules established by an international committee. Regardless of the origin of the name, scientists derived this international system for convenience. It is much easier to write H for hydrogen, O for oxygen, and H_2O for dihydrogen oxide (water). Because scientists worldwide use this system, everyone understands what the symbols mean.

Table 1 Symbols of Some Elements			
Element	Symbol	Element	Symbol
Aluminum	Al	Iron	Fe
Calcium	Ca	Mercury	Hg
Carbon	C	Nitrogen	N
Chlorine	Cl	Oxygen	O
Gold	Au	Potassium	K
Hydrogen	H	Sodium	Na

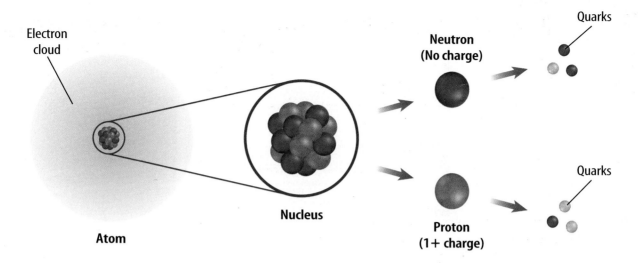

Figure 1 The nucleus of the atom contains protons and neutrons that are composed of quarks. The proton has a positive charge and the neutron has no charge. A cloud of negatively charged electrons surrounds the nucleus of the atom.

Atomic Components

An element is matter that is composed of one type of **atom,** which is the smallest piece of matter that still retains the property of the element. For example, the element silver is composed of only silver atoms and the element hydrogen is composed of only hydrogen atoms. Atoms are composed of particles called protons, neutrons, and electrons, as shown in **Figure 1.** Protons and neutrons are found in a small, positively-charged center of the atom called the **nucleus** that is surrounded by a cloud containing electrons. **Protons** are particles with an electrical charge of 1+. **Neutrons** are neutral particles that do not have an electrical charge. **Electrons** are particles with an electrical charge of 1−. Atoms of different elements differ in the number of protons they contain.

 *What are the particles that make up the atom and where are they located?*

Quarks—Even Smaller Particles

Are the protons, electrons, and neutrons that make up atoms the smallest particles that exist? Scientists hypothesize that electrons are not composed of smaller particles and are one of the most basic types of particles. Protons and neutrons, however, are made up of smaller particles called **quarks.** So far, scientists have confirmed the existence of six uniquely different quarks. Scientists theorize that an arrangement of three quarks held together with the strong nuclear force produces a proton. Another arrangement of three quarks produces a neutron. The search for the composition of protons and neutrons is an ongoing effort.

Figure 2 The Tevatron is a huge machine. The aerial photograph of Fermi National Accelerator Laboratory shows the circular outline of the Tevatron particle accelerator. The close-up photograph of the Tevatron gives you a better view of the tunnel.
Infer *Why is such a long tunnel needed?*

Figure 3 Bubble chambers can be used by scientists to study the tracks left by subatomic particles.

Finding Quarks To study quarks, scientists accelerate charged particles to tremendous speeds and then force them to collide with—or smash into—protons. This collision causes the proton to break apart. The Fermi National Accelerator Laboratory, a research laboratory in Batavia, Illinois, houses a machine that can generate the forces that are required to collide protons. This machine, the Tevatron, shown in **Figure 2,** is approximately 6.4 km in circumference. Electric and magnetic fields are used to accelerate, focus and collide the fast-moving particles.

The particles that result from the collision can be detected by various collection devises. Often, scientists use multiple collection devices to collect the most possible information about the particles created in a collision. Just as police investigators can reconstruct traffic accidents from tire marks and other clues at the scene, scientists are able to examine and gather information about the particles, as shown in **Figure 3.** Scientists use inference to identify the subatomic particles and to reveal information about each particle's inner structure.

The Sixth Quark Finding evidence for the existence of the quarks was not an easy task. Scientists found five quarks and hypothesized that a sixth quark existed. However, it took a team of nearly 450 scientists from around the world several years to find the sixth quark. The tracks of the sixth quark were hard to detect because only about one billionth of a percent of the proton collisions performed showed the presence of a sixth quark—typically referred to as the *top* quark.

Models—Tools for Scientists

Scientists and engineers use models to represent things that are difficult to visualize—or picture in your mind. You might have seen models of buildings, the solar system, or airplanes. These are scaled-down models. Scaled-down models allow you to see either something too large to see all at once, or something that has not been built yet. Scaled-up models are often used to visualize things that are too small to see. To give you an idea of how small the atom is, it would take about 24,400 atoms stacked one on top of the other to equal the thickness of a sheet of aluminum foil. To study the atom, scientists have developed scaled-up models that they can use to visualize how the atom is constructed. For the model to be useful, it must support all of the information that is known about matter and the behavior of atoms. As more information about the atom is collected, scientists change their models to include the new information.

✔ Reading Check *Explain how models can simplify science.*

The Changing Atomic Model You know now that all matter is composed of atoms, but this was not always known. Around 400 B.C., Democritus proposed the idea that atoms make up all substances. However, another famous Greek philosopher, Aristotle, disputed Democritus's theory and proposed that matter was uniform throughout and was not composed of smaller particles. Aristotle's incorrect theory was accepted for about 2,000 years. In the 1800s, John Dalton, an English scientist, was able to offer proof that atoms exist.

Dalton's model of the atom, a solid sphere shown in **Figure 4,** was an early model of the atom. As you can see in **Figure 5,** the model has changed somewhat over time. Dalton's modernization of Aristotle's idea of the atom provided a physical explanation for chemical reactions. Scientists could then express these reactions in quantitative terms using chemical symbols and equations.

Mini LAB

Modeling an Aluminum Atom

Procedure

1. Arrange thirteen 3-cm circles cut from **orange paper** and fourteen 3-cm circles cut from **blue paper** on a **flat surface** to represent the nucleus of an atom. Each orange circle represents one proton, and each blue circle represents one neutron.
2. Position two holes punched from **red paper** about 20 cm from your nucleus.
3. Position eight punched holes about 40 cm from your nucleus.
4. Position three punched holes about 60 cm from your nucleus.

Analysis

1. How many protons, neutrons, and electrons does an aluminum atom have?
2. Explain how your circles model an aluminum atom.
3. Explain why your model does not accurately represent the true size and distances in an aluminum atom.

Figure 4 John Dalton's atomic model was a simple sphere.

Figure 5

The ancient Greek philosopher Democritus proposed that elements consisted of tiny, solid particles that could not be subdivided (A). He called these particles *atomos,* meaning "uncuttable." This concept of the atom's structure remained largely unchallenged until the 1900s, when researchers began to discover through experiments that atoms were composed of still smaller particles. In the early 1900s, a number of models for atomic structure were proposed (B-D). The currently accepted model (E) evolved from these ideas and the work of many other scientists.

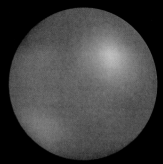

A DEMOCRITUS'S UNCUTTABLE ATOM

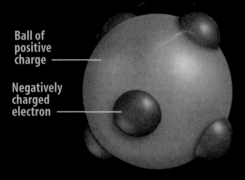

Ball of positive charge

Negatively charged electron

B THOMSON MODEL, 1904 English physicist Joseph John Thomson inferred from his experiments that atoms contained small, negatively charged particles. He thought these "electrons" (in red) were evenly embedded throughout a positively charged sphere, much like chocolate chips in a ball of cookie dough.

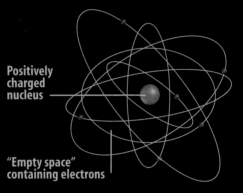

Positively charged nucleus

"Empty space" containing electrons

C RUTHERFORD MODEL, 1911 Another British physicist, Ernest Rutherford, proposed that almost all the mass of an atom—and all its positive charges—were concentrated in a central atomic nucleus surrounded by electrons.

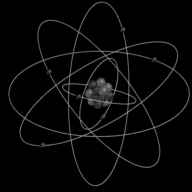

D BOHR MODEL, 1913 Danish physicist Niels Bohr hypothesized that electrons traveled in fixed orbits around the atom's nucleus. James Chadwick, a student of Rutherford, concluded that the nucleus contained positive protons and neutral neutrons.

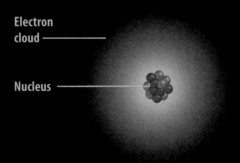

Electron cloud

Nucleus

E ELECTRON CLOUD MODEL, CURRENT According to the currently accepted model of atomic structure, electrons do not follow fixed orbits but tend to occur more frequently in certain areas around the nucleus at any given time.

The Electron Cloud Model By 1926, scientists had developed the electron cloud model of the atom that is in use today. An **electron cloud** is the area around the nucleus of an atom where its electrons are most likely found. The electron cloud is 100,000 times larger than the diameter of the nucleus. In contrast, each electron in the cloud is much smaller than a single proton.

Because an electron's mass is small and the electron is moving so quickly around the nucleus, it is impossible to describe its exact location in an atom. Picture the spokes on a moving bicycle wheel. They are moving so quickly that you can't pinpoint any single spoke. All you see is a blur that contains all of the spokes somewhere within it. In the same way, an electron cloud is a blur containing all of the electrons of the atom somewhere within it. **Figure 6** illustrates what the electron cloud might look like.

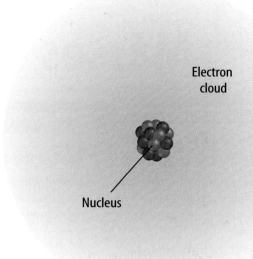

Electron cloud

Nucleus

Figure 6 The electrons are located in an electron cloud surrounding the nucleus of the atom.

---

section 1 review

Summary

Scientific Shorthand
- Scientists use chemical symbols as shorthand when naming elements.

Atomic Components
- Atoms are composed of small particles that have known charges.
- The particles that make up the atom are located in predictable locations within the atom.

Quarks—Even Smaller Particles
- So far, scientists have confirmed the existence of six different quarks.

Models—Tools for Scientists
- Models are used by scientists to simplify the study of concepts and things.
- The current atomic model is an accumulation of over two hundred years of knowledge.
- The electron cloud model is the current atomic model.

Self Check

1. **List** the chemical symbols for the elements carbon, aluminum, hydrogen, oxygen, and sodium.
2. **Identify** the names, charges, and locations of three kinds of particles that make up an atom.
3. **Identify** the smallest particle of matter. How were they discovered?
4. **Describe** the electron cloud model of the atom.
5. **Think Critically** Explain how a rotating electric fan might be used to model the atom. Explain how the rotating fan is unlike an atom.

Applying Math

6. **Use Numbers** The mass of a proton is estimated to be 1.6726×10^{-24} g and the mass of an electron is estimated to be 9.1093×10^{-28} g. How many times larger is the mass of a proton compared to the mass of an electron?
7. **Calculate** What is the difference between the mass of a proton and the mass of an electron?

Masses of Atoms

Reading Guide

What You'll Learn
- **Compute** the atomic mass and mass number of an atom.
- **Identify** the components of isotopes.
- **Interpret** the average atomic mass of an element.

Why It's Important
Some elements naturally exist in more than one form—radioactive and nonradioactive.

Review Vocabulary
mass: amount of matter in an object

New Vocabulary
- atomic number
- mass number
- isotope
- average atomic mass

Table 2 Subatomic Particle Masses

Particle	Mass (g)
Proton	1.6726×10^{-24}
Neutron	1.6749×10^{-24}
Electron	9.1093×10^{-28}

Figure 7 If you held a textbook and placed a paper clip on it, you wouldn't notice the added mass because the mass of a paper clip is small compared to the mass of the book. In a similar way, the masses of an atom's electrons are negligible compared to an atom's mass.

Atomic Mass

The nucleus contains most of the mass of the atom because protons and neutrons are far more massive than electrons. The mass of a proton is about the same as that of a neutron—approximately 1.6726×10^{-24} g, as shown in **Table 2.** The mass of each is approximately 1,836 times greater than the mass of the electron. The electron's mass is so small that it is considered negligible when finding the mass of an atom, as shown in **Figure 7.**

If you were asked to estimate the height of your school building, you probably wouldn't give an answer in kilometers. The number would be too cumbersome to use. Considering the scale of the building, you would more likely give the height in a smaller unit, meters. When thinking about the small masses of atoms, scientists found that even grams were not small enough to use for measurement. Scientists need a unit that results in more manageable numbers. The unit of measurement used for atomic particles is the atomic mass unit (amu). The mass of a proton or a neutron is almost equal to 1 amu. This is not coincidence—the unit was defined that way. The atomic mass unit is defined as one-twelfth the mass of a carbon atom containing six protons and six neutrons. Remember that the mass of the carbon atom is contained almost entirely in the mass of the protons and neutrons that are located in the nucleus. Therefore, each of the 12 particles in the nucleus must have a mass nearly equal to one.

Reading Check *Where is the majority of the mass of an atom located?*

Table 3 Mass Numbers of Some Atoms

Element	Symbol	Atomic Number	Protons	Neutrons	Mass Number	Average Atomic Mass*
Boron	B	5	5	6	11	10.81 amu
Carbon	C	6	6	6	12	12.01 amu
Oxygen	O	8	8	8	16	16.00 amu
Sodium	Na	11	11	12	23	22.99 amu
Copper	Cu	29	29	34	63	63.55 amu

*The atomic mass units are rounded to two decimal places.

Protons Identify the Element You learned earlier that atoms of different elements are different because they have different numbers of protons. In fact, the number of protons tells you what type of atom you have and vice versa. For example, every carbon atom has six protons. Also, all atoms with six protons are carbon atoms. Atoms with eight protons are oxygen atoms. The number of protons in an atom is equal to a number called the **atomic number.** The atomic number of carbon is six. Therefore, if you are given any one of the following—the name of the element, the number of protons in the element, or the atomic number of the element, you can determine the other two.

✔ Reading Check *Which element is an atom with six protons in the nucleus?*

Mass Number The **mass number** of an atom is the sum of the number of protons and the number of neutrons in the nucleus of an atom. Look at **Table 3** and see if this is true.

If you know the mass number and the atomic number of an atom, you can calculate the number of neutrons. The number of neutrons is equal to the atomic number subtracted from the mass number.

number of neutrons = mass number − atomic number

Atoms of the same element with different numbers of neutrons can have different properties. For example, carbon with a mass number equal to 12, or carbon-12, is the most common form of carbon. Carbon-14 is present on Earth in much smaller quantities. Carbon-14 is radioactive and carbon-12 is not.

Carbon Dating Living organisms on Earth contain carbon. Carbon-12 makes up 99 percent of this carbon. Carbon-13 and carbon-14 make up the other one percent. Which isotopes are archaeologists most interested in when they determine the age of carbon-containing remains? Explain your answer in your Science Journal.

Isotopes

Not all the atoms of an element have the same number of neutrons. Atoms of the same element that have different numbers of neutrons are called **isotopes.** Suppose you have a sample of the element boron. Naturally occurring atoms of boron have mass numbers of 10 or 11. How many neutrons are in a boron atom? It depends upon the isotope of boron to which you are referring. Obtain the number of protons in boron from the periodic table. Then use the formula on the previous page to calculate the number of neutrons in each boron isotope. You can determine that boron can have five or six neutrons.

 **Reading Check** *Uranium-238 has 92 protons. How many neutrons does it have?*

Applying Science

Radioactive Isotopes Help Tell Time

Atoms can be used to measure the age of bones or rock formations that are millions of years old. The time it takes for half of the radioactive atoms in a piece of rock or bone to change into another element is called its half-life. Scientists use the half-lives of radioactive isotopes to measure geologic time.

Half-Lives of Radioactive Isotopes		
Radioactive Element	**Changes to This Element**	**Half-Life**
uranium-238	lead-206	4,460 million years
potassium-40	argon-40, calcium-40	1,260 million years
rubidium-87	strontium-87	48,800 million years
carbon-14	nitrogen-14	5,715 years

Identifying the Problem

The table above lists the half-lives of a sample of radioactive isotopes and into which elements they change. For example, it would take 5,715 years for half of the carbon-14 atoms in a rock to change into atoms of nitrogen-14. After another 5,715 years, half of the remaining carbon-14 atoms will change, and so on. You can use these radioactive clocks to measure different periods of time.

Solving the Problem

1. How many years would it take half of the rubidium-87 atoms in a piece of rock to change into strontium-87? How many years would it take for 75% of the atoms to change?
2. After a long period, only 25% of the atoms in a rock remained uranium-238. How many years old would you predict the rock to be? The other 75% of the atoms are now which radioactive element?

Identifying Isotopes Models of two isotopes of boron are shown in **Figure 8.** Because the numbers of neutrons in the isotopes are different, the mass numbers are also different. You use the name of the element followed by the mass number of the isotope to identify each isotope: boron-10 and boron-11. Because most elements have more than one isotope, each element has an average atomic mass. The **average atomic mass** of an element is the weighted-average mass of the mixture of its isotopes. For example, four out of five atoms of boron are boron-11, and one out of five is boron-10. To find the weighted-average or the average atomic mass of boron, you would solve the following equation:

$$\frac{4}{5}(11 \text{ amu}) + \frac{1}{5}(10 \text{ amu}) = 10.8 \text{ amu}$$

The average atomic mass of the element boron is 10.8 amu. Note that the average atomic mass of boron is close to the mass of its most abundant isotope, boron-11.

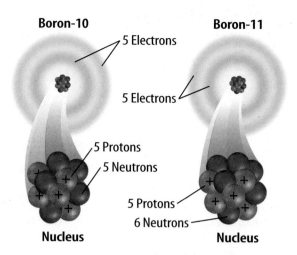

Figure 8 Boron-10 and boron-11 are two isotopes of boron. These two isotopes differ by one neutron.
Explain why these atoms are isotopes.

---

section 2 review

Summary

Atomic Mass

- The nucleus contains most of the mass of an atom.
- The mass of a proton and neutron are approximately equal.
- The mass of an electron is considered negligible when finding the mass of an atom.
- The unit of measurement for atomic particles is the atomic mass unit.
- The carbon-12 isotope was used to define the atomic mass unit.
- The number of protons identifies the element.

Isotopes

- Atoms of the same element with different numbers of neutrons are called isotopes.
- The average atomic mass of an element is the weighted-average mass of the mixture of isotopes.

Self Check

1. **Identify** the mass number and atomic number of a chlorine atom that has 17 protons and 18 neutrons.
2. **Explain** how the isotopes of an element are alike and how are they different.
3. **Explain** why the atomic mass of an element is an average mass.
4. **Explain** how you would calculate the number of neutrons in potassium-40.
5. **Think Critically** Chlorine has an average atomic mass of 35.45 amu. The two naturally occurring isotopes of chlorine are chlorine-35 and chlorine-37. Why does this indicate that most chlorine atoms contain 18 neutrons?

Applying Math

6. **Use Numbers** If a hydrogen atom has 2 neutrons and 1 proton, what is its mass number?
7. **Use Tables** Use the information in **Table 2** to find the mass in kilograms of each subatomic particle.

The Periodic Table

Reading Guide

What You'll Learn
- **Explain** the composition of the periodic table.
- **Use** the periodic table to obtain information.
- **Explain** what the terms *metal, nonmetal,* and *metalloid* mean.

Why It's Important
The periodic table is an organized list of the elements that compose all living and nonliving things that are known to exist in the universe.

Review Vocabulary
chemical property: any characteristic of a substance that indicates whether it can undergo a certain chemical change

New Vocabulary
- periodic table
- group
- electron dot diagram
- period

Figure 9 Mendeleev discovered that the elements had a periodic pattern in their chemical properties. Notice the question marks in his chart. These were elements that had not been discovered at that time.

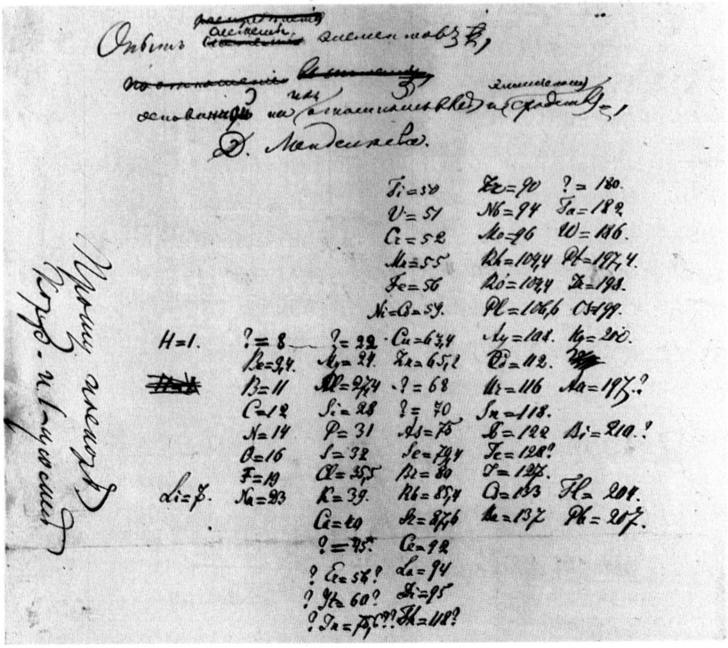

Organizing the Elements

On a clear evening, you can see one of the various phases of the Moon. Each month, the Moon seems to grow larger, then smaller, in a repeating pattern. This type of change is periodic. *Periodic* means "repeated in a pattern." The days of the week are periodic because they repeat themselves every seven days. The calendar is a periodic table of days and months.

In the late 1800s, Dmitri Mendeleev, a Russian chemist, searched for a way to organize the elements. When he arranged all the elements known at that time in order of increasing atomic masses, he discovered a pattern. **Figure 9** shows Mendeleev's early periodic chart. Chemical properties found in lighter elements could be shown to repeat in heavier elements. Because the pattern repeated, it was considered to be periodic. Today, this arrangement is called a periodic table of elements. In the **periodic table,** the elements are arranged by increasing atomic number and by changes in physical and chemical properties.

Table 4 Mendeleev's Predictions

Predicted Properties of Ekasilicon (Es)	Actual Properties of Germanium (Ge)
Existence Predicted—1871	*Actual Discovery—1886*
Atomic mass = 72	Atomic mass = 72.61
High melting point	Melting point = 938°C
Density = 5.5 g/cm^3	Density = 5.323 g/cm^3
Dark gray metal	Gray metal
Density of EsO$_2$ = 4.7 g/cm^3	Density of GeO$_2$ = 4.23 g/cm^3

Mendeleev's Predictions Mendeleev had to leave blank spaces in his periodic table to keep the elements properly lined up according to their chemical properties. He looked at the properties and atomic masses of the elements surrounding these blank spaces. From this information, he was able to predict the properties and the mass numbers of new elements that had not yet been discovered. **Table 4** shows Mendeleev's predicted properties for germanium, which he called ekasilicon. His predictions proved to be accurate. Scientists later discovered these missing elements and found that their properties were extremely close to what Mendeleev had predicted.

Reading Check *How did Mendeleev organize his periodic chart?*

Improving the Periodic Table Although Mendeleev's arrangement of elements was successful, it did need some changes. On Mendeleev's table, the atomic mass gradually increased from left to right. If you look at the modern periodic table, shown in **Table 5,** you will see several examples, such as cobalt and nickel, where the mass decreases from left to right. You also might notice that the atomic number always increases from left to right. In 1913, the work of Henry G.J. Moseley, a young English scientist, led to the arrangement of elements based on their increasing atomic numbers instead of an arrangement based on atomic masses. This new arrangement seemed to correct the problems that had occurred in the old table. The current periodic table uses Moseley's arrangement of the elements.

Reading Check *How is the modern periodic table arranged?*

Organizing a Personal Periodic Table

Procedure

1. Collect as many of the following items as you can find: **feather, penny, container of water, pencil, dime, strand of hair, container of milk, container of orange juice, square of cotton cloth, nickel, crayon, quarter, container of soda, golf ball, sheet of paper, baseball, marble, leaf, paper clip.**
2. Organize these items into several columns based on their similarities to create your own periodic table.

Analysis

1. Explain the system you used to group your items.
2. Were there any items on the list that did not fit into any of your columns?
3. Infer how your activity modeled Mendeleev's work in developing the periodic table of the elements.

Try at Home

PERIODIC TABLE OF THE ELEMENTS

Columns of elements are called groups. Elements in the same group have similar chemical properties.

Gas

Liquid

Solid

Synthetic

Element	—	Hydrogen
Atomic number	—	1
Symbol	—	H
Atomic mass	—	1.008

State of matter

The first three symbols tell you the state of matter of the element at room temperature. The fourth symbol identifies elements that are not present in significant amounts on Earth. Useful amounts are made synthetically.

1		2		3	4	5	6	7	8	9
1 Hydrogen 1 **H** 1.008										
2 Lithium 3 **Li** 6.941		Beryllium 4 **Be** 9.012								
3 Sodium 11 **Na** 22.990		Magnesium 12 **Mg** 24.305								
4 Potassium 19 **K** 39.098		Calcium 20 **Ca** 40.078		Scandium 21 **Sc** 44.956	Titanium 22 **Ti** 47.867	Vanadium 23 **V** 50.942	Chromium 24 **Cr** 51.996	Manganese 25 **Mn** 54.938	Iron 26 **Fe** 55.845	Cobalt 27 **Co** 58.933
5 Rubidium 37 **Rb** 85.468		Strontium 38 **Sr** 87.62		Yttrium 39 **Y** 88.906	Zirconium 40 **Zr** 91.224	Niobium 41 **Nb** 92.906	Molybdenum 42 **Mo** 95.94	Technetium 43 **Tc** (98)	Ruthenium 44 **Ru** 101.07	Rhodium 45 **Rh** 102.906
6 Cesium 55 **Cs** 132.905		Barium 56 **Ba** 137.327		Lanthanum 57 **La** 138.906	Hafnium 72 **Hf** 178.49	Tantalum 73 **Ta** 180.948	Tungsten 74 **W** 183.84	Rhenium 75 **Re** 186.207	Osmium 76 **Os** 190.23	Iridium 77 **Ir** 192.217
7 Francium 87 **Fr** (223)		Radium 88 **Ra** (226)		Actinium 89 **Ac** (227)	Rutherfordium 104 **Rf** (261)	Dubnium 105 **Db** (262)	Seaborgium 106 **Sg** (266)	Bohrium 107 **Bh** (264)	Hassium 108 **Hs** (277)	Meitnerium 109 **Mt** (268)

The number in parentheses is the mass number of the longest-lived isotope for that element.

Rows of elements are called periods. Atomic number increases across a period.

The arrow shows where these elements would fit into the periodic table. They are moved to the bottom of the table to save space.

	Cerium 58 **Ce** 140.116	Praseodymium 59 **Pr** 140.908	Neodymium 60 **Nd** 144.24	Promethium 61 **Pm** (145)	Samarium 62 **Sm** 150.36
Lanthanide series					
Actinide series	Thorium 90 **Th** 232.038	Protactinium 91 **Pa** 231.036	Uranium 92 **U** 238.029	Neptunium 93 **Np** (237)	Plutonium 94 **Pu** (244)

Science Online

Topic: Periodic Table Updates

Visit gpscience.com for updates to the periodic table.

				13	14	15	16	17	18
									Helium 2 **He** 4.003
				Boron 5 **B** 10.811	Carbon 6 **C** 12.011	Nitrogen 7 **N** 14.007	Oxygen 8 **O** 15.999	Fluorine 9 **F** 18.998	Neon 10 **Ne** 20.180
	10	11	12	Aluminum 13 **Al** 26.982	Silicon 14 **Si** 28.086	Phosphorus 15 **P** 30.974	Sulfur 16 **S** 32.065	Chlorine 17 **Cl** 35.453	Argon 18 **Ar** 39.948
	Nickel 28 **Ni** 58.693	Copper 29 **Cu** 63.546	Zinc 30 **Zn** 65.409	Gallium 31 **Ga** 69.723	Germanium 32 **Ge** 72.64	Arsenic 33 **As** 74.922	Selenium 34 **Se** 78.96	Bromine 35 **Br** 79.904	Krypton 36 **Kr** 83.798
	Palladium 46 **Pd** 106.42	Silver 47 **Ag** 107.868	Cadmium 48 **Cd** 112.411	Indium 49 **In** 114.818	Tin 50 **Sn** 118.710	Antimony 51 **Sb** 121.760	Tellurium 52 **Te** 127.60	Iodine 53 **I** 126.904	Xenon 54 **Xe** 131.293
	Platinum 78 **Pt** 195.078	Gold 79 **Au** 196.967	Mercury 80 **Hg** 200.59	Thallium 81 **Tl** 204.383	Lead 82 **Pb** 207.2	Bismuth 83 **Bi** 208.980	Polonium 84 **Po** (209)	Astatine 85 **At** (210)	Radon 86 **Rn** (222)
	Darmstadtium 110 **Ds** (281)	Roentgenium 111 **Rg** (272)	Ununbium * 112 **Uub** (285)		Ununquadium * 114 **Uuq** (289)				

* The names and symbols for elements 112 and 114 are temporary. Final names will be selected when the elements' discoveries are verified.

Europium 63 **Eu** 151.964	Gadolinium 64 **Gd** 157.25	Terbium 65 **Tb** 158.925	Dysprosium 66 **Dy** 162.500	Holmium 67 **Ho** 164.930	Erbium 68 **Er** 167.259	Thulium 69 **Tm** 168.934	Ytterbium 70 **Yb** 173.04	Lutetium 71 **Lu** 174.967
Americium 95 **Am** (243)	Curium 96 **Cm** (247)	Berkelium 97 **Bk** (247)	Californium 98 **Cf** (251)	Einsteinium 99 **Es** (252)	Fermium 100 **Fm** (257)	Mendelevium 101 **Md** (258)	Nobelium 102 **No** (259)	Lawrencium 103 **Lr** (262)

Research Physicist The study of nuclear interactions is shared by chemists and physicists. Research physicists use their knowledge of the physical laws of nature to explain the behavior of the atom and its composition. Explain in your Science Journal why physicists would study the amount of energy that electrons contain.

The Atom and the Periodic Table

Objects often are sorted or grouped according to the properties they have in common. This also is done in the periodic table. The vertical columns in the periodic table are called **groups,** or families, and are numbered 1 through 18. Elements in each group have similar properties. For example, in Group 11, copper, silver, and gold have similar properties. Each is a shiny metal and a good conductor of electricity and heat. What is responsible for the similar properties? To answer this question, look at the structure of the atom.

Electron Cloud Structure You have learned about the number and location of protons and neutrons in an atom. But where are the electrons located? How many are there? In a neutral atom, the number of electrons is equal to the number of protons. Therefore, a carbon atom, with an atomic number of six, has six protons and six electrons. These electrons are located in the electron cloud surrounding the nucleus.

Scientists have found that electrons within the electron cloud have different amounts of energy. Scientists model the energy differences of the electrons by placing the electrons in energy levels, as in **Figure 10.** Energy levels nearer the nucleus have lower energy than those levels that are farther away. Electrons fill these energy levels from the inner levels (closer to the nucleus) to the outer levels (farther from the nucleus).

Elements that are in the same group have the same number of electrons in their outer energy level. It is the number of electrons in the outer energy level that determines the chemical properties of the element. It is important to understand the link between the location on the periodic table, chemical properties, and the structure of the atom.

Figure 10 Energy levels in atoms can be represented by a flight of stairs. Each stair step away from the nucleus represents an increase in the amount of energy within the electrons. The higher energy levels contain more electrons.

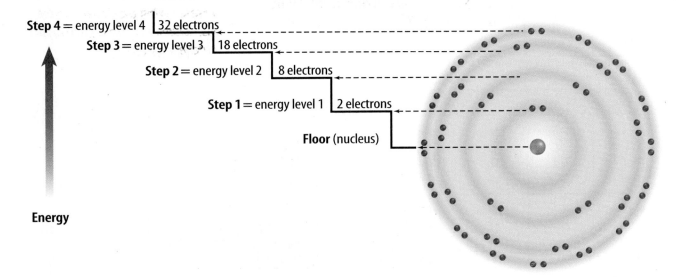

Step 4 = energy level 4 | 32 electrons
Step 3 = energy level 3 | 18 electrons
Step 2 = energy level 2 | 8 electrons
Step 1 = energy level 1 | 2 electrons
Floor (nucleus)

Energy

Energy Levels These energy levels are named using numbers one to seven. The maximum number of electrons that can be contained in each of the first four levels is shown in **Figure 10.** For example, energy level one can contain a maximum of two electrons. Energy level two can contain a maximum of eight electrons. Notice that energy levels three and four contain several electrons. A complete and stable outer energy level will contain eight electrons. In elements in periods three and higher, additional electrons can be added to inner energy levels although the outer energy level contains only eight electrons.

Rows on the Table Remember that the atomic number found on the periodic table is equal to the number of electrons in an atom. Look at **Figure 11.** The first row has hydrogen with one electron and helium with two electrons both in energy level one. Because energy level one is the outermost level containing an electron, hydrogen has one outer electron. Helium has two outer electrons. Recall from **Figure 10** that energy level one can hold only two electrons. Therefore, helium has a full or complete outer energy level.

The second row begins with lithium, which has three electrons—two in energy level one and one in energy level two. Lithium has one outer electron. Lithium is followed by beryllium with two outer electrons, boron with three, and so on until you reach neon with eight outer electrons. Again, looking at **Figure 10,** energy level two can only hold eight electrons. Therefore, neon has a complete outer energy level. Do you notice how the row in the periodic table ends when an outer energy level is filled? In the third row of elements, the electrons begin filling energy level three. The row ends with argon, which has a full outer energy level of eight electrons.

 *How many electrons are needed to fill the outer energy level of sulfur?*

Topic: Atomic Energy Level Structure
Visit gpscience.com for Web links to information about the structure of atomic energy levels.

Activity Draw a diagram that details how the energy levels are structured.

Figure 11 One proton and one electron are added to each element as you go across a period in the periodic table.
Explain *what the elements in the last column share in relation to their outer energy levels.*

Hydrogen 1 H								Helium 2 He
Lithium 3 Li	Beryllium 4 Be		Boron 5 B	Carbon 6 C	Nitrogen 7 N	Oxygen 8 O	Fluorine 9 F	Neon 10 Ne
Sodium 11 Na	Magnesium 12 Mg		Aluminum 13 Al	Silicon 14 Si	Phosphorus 15 P	Sulfur 16 S	Chlorine 17 Cl	Argon 18 Ar

Figure 12 The elements in Group 1 have one electron in their outer energy level. This electron dot diagram represents that one electron.

Electron Dot Diagrams Did you notice that hydrogen, lithium, and sodium have one electron in their outer energy level? Elements that are in the same group have the same number of electrons in their outer energy level. These outer electrons are so important in determining the chemical properties of an element that a special way to represent them has been developed. American chemist G. N. Lewis created this method while teaching a college chemistry class. An **electron dot diagram** uses the symbol of the element and dots to represent the electrons in the outer energy level. **Figure 12** shows the electron dot diagram for Group 1 elements. Electron dot diagrams are used also to show how the electrons in the outer energy level are bonded when elements combine to form compounds.

Same Group—Similar Properties The elements in Group 17, the halogens, have electron dot diagrams similar to chlorine, shown in **Figure 13.** All halogens have seven electrons in their outer energy levels. Since all of the members of a group on the periodic table have the same number of electrons in their outer energy level, group members will undergo chemical reactions in similar ways.

A common property of the halogens is the ability to form compounds readily with elements in Group 1. Group 1 elements have only one electron in their outer energy level. **Figure 13** shows an example of a compound formed by one such reaction. The Group 1 element, sodium, reacts easily with the Group 17 element, chlorine. The result is the compound sodium chloride, or NaCl—ordinary table salt.

Not all elements will combine readily with other elements. The elements in Group 18 have complete outer energy levels. This special configuration makes Group 18 elements relatively unreactive. You will learn more about why and how bonds form between elements in the later chapters.

✔ Reading Check *Why do elements in a group undergo similar chemical reactions?*

Figure 13 Electron dot diagrams show the electrons in an element's outer energy level.

The electron dot diagram for Group 17 consists of three sets of paired dots and one single dot.

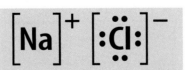

Sodium combines with chlorine to give each element a complete outer energy level in the resulting compound.

Neon, a member of Group 18, has a full outer energy level. Neon has eight electrons in its outer energy level, making it unreactive.

Regions on the Periodic Table

The periodic table has several regions with specific names. The horizontal rows of elements on the periodic table are called **periods.** The elements increase by one proton and one electron as you go from left to right in a period.

All of the elements in the blue squares in **Figure 14** are metals. Iron, zinc, and copper are examples of metals. Most metals exist as solids at room temperature. They are shiny, can be drawn into wires, can be pounded into sheets, and are good conductors of heat and electricity.

Those elements on the right side of the periodic table, in yellow, are classified as nonmetals. Oxygen, bromine, and carbon are examples of nonmetals. Most nonmetals are gases, are brittle, and are poor conductors of heat and electricity at room temperature. The elements in green are metalloids or semimetals. They have some properties of both metals and nonmetals. Boron and silicon are examples of metalloids.

 **Reading Check** *What are the properties of the elements located on the left side of the periodic table?*

A Growing Family Scientists around the world are continuing their research into the synthesis of elements. In 1994, scientists at the Heavy-Ion Research Laboratory in Darmstadt, Germany, discovered element 111. As of 1998, only one isotope of element 111 has been found. This isotope had a life span of 0.002 s. In 1996, element 112 was discovered at the same laboratory. As of 1998, only one isotope of element 112 has been found. The life span of this isotope was 0.00048 s. Both of these elements are produced in the laboratory by joining smaller atoms into a single atom. The search for elements with higher atomic numbers continues. Scientists think they have synthesized elements 114 and 116. However, the discovery of these elements has not yet been confirmed.

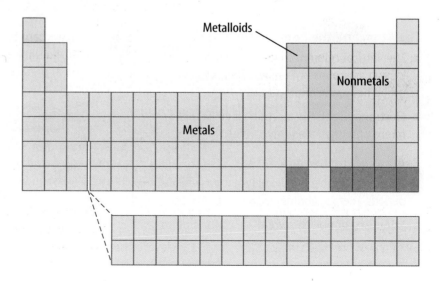

Figure 14 Metalloids are located along the green stair-step line. Metals are located to the left of the metalloids. Nonmetals are located to the right of the metalloids.

Science Online

Topic: New Elements
Visit gpscience.com for Web links to information about newly synthesized elements.

Activity Write a paragraph explaining how several new elements were synthesized and who synthesized them.

Elements in the Universe

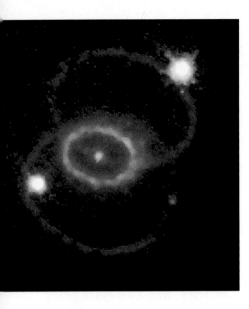

Figure 15 Scientists think that some elements are found in nature only within stars.

INTEGRATE Astronomy Using the technology that is available today, scientists are finding the same elements throughout the universe. They have been able to study only a small portion of the universe, though, because it is so vast. Many scientists believe that hydrogen and helium are the building blocks of other elements. Atoms join together within stars to produce elements with atomic numbers greater than 1 or 2—the atomic numbers of hydrogen and helium. Exploding stars, or supernovas, shown in **Figure 15,** give scientists evidence to support this theory. When stars go supernova a mixture of elements, including the heavy elements such as iron, are flung into the galaxy. Many scientists believe that supernovas have spread the elements that are found throughout the universe. Promethium, technetium, and elements with an atomic number above 92 are rare or are not found on Earth. Some of these elements are found only in trace amounts in Earth's crust as a result of uranium decay. Others have been found only in stars.

section 3 review

Summary

Organizing the Elements

- Mendeleev organized the elements using increasing atomic mass and chemical and physical properties.

- Mendeleev left blank spaces in his table to allow for elements that were yet undiscovered.

- Moseley corrected the problems in the periodic table by arranging the elements in order of increasing atomic number.

The Atom and the Periodic Table

- The vertical columns in the periodic table are known as groups or families. Elements in a group have similar properties.

- Electrons within the electron cloud have different amounts of energy.

Regions of the Periodic Table

- The periodic table is divided into these regions: periods, metals, nonmetals, and metalloids.

- Scientists around the world are continuing to try to synthesize new elements.

Self Check

1. **Identify** Use the periodic table to find the name, atomic number, and average atomic mass of the following elements: N, Ca, Kr, and W.

2. **List** the period and group in which each of these elements is found: nitrogen, sodium, iodine, and mercury.

3. **Classify** each of these elements as a metal, a nonmetal, or a metalloid and give the full name of each: K, Si, Ba, and S.

4. **Think Critically** The Mendeleev and Mosely periodic charts have gaps for the as-then-undiscovered elements. Why do you think the chart used by Mosely was more accurate at predicting where new elements would be placed?

Applying Math

5. **Make a Graph** Construct a circle graph showing the percentage of elements classified as metals, metalloids, and nonmetals. Use markers or colored pencils to distinguish clearly between each section on the graph. Record your calculations in your Science Journal.

A Periodic Table of Fds

▶ Real-World Question

Mendeleev's task of organizing a collection of loosely related items probably seemed daunting at first. How will using your favorite foods to create your own periodic table be similar to the task that Mendeleev had?

Goals

■ **Organize** 20 of your favorite foods into a periodic table of foods.

■ **Analyze** and **evaluate** your periodic table for similar characteristics among groups or family members on your table.

■ **Infer** where new foods added to your table would be placed.

Materials

11 × 17 paper
12- or 18-inch ruler
colored pencils or markers

▶ Procedure

1. **List** 20 of your favorite foods and drinks.

2. **Describe** basic characteristics of each of your food and drink items. For example, you might describe the primary ingredient, nutritional value, taste, and color of each item. You also could identify the food group of each item such as fruits/vegetables, grains, dairy products, meat, and sweets.

3. **Create** a data table to organize the information that you collect.

4. Using your data table, construct a periodic table of foods on your 11 × 17 sheet of paper. Determine which characteristics you will use to group your items. Create families (columns) of food and drink items that share similar characteristics on your table.

For example, potato chips, pretzels, and cheese-flavored crackers could be combined into a family of salty tasting foods. Create as many groups as you need, and you do not need to have the same number of items in every family.

▶ Conclude and Apply

1. **Evaluate** the characteristics you used to make the groups on your periodic table. Do the characteristics of each group adequately describe all the family members? Do the characteristics of each group distinguish its family members from the family members of the other groups?

2. **Analyze** the reasons why some items did not fit easily into a group.

3. **Infer** why chemists have not created a periodic table of compounds.

*C*ommunicating

Your Data

Construct a bulletin board of the periodic tables of foods created by the class. How are the tables similar?

Use the Internet

What's in a name?

Real-World Question

The symbols used for different elements sometimes are easy to figure out. After all, it makes sense for the symbol for carbon to be C and the symbol for nitrogen to be N. However, some symbols aren't as easy to figure out. For example, the element silver has the symbol Ag. This symbol comes from the Latin word for silver, *Argentum*. How are symbols and names chosen for elements?

Make a Plan

1. Make a list of particular elements you wish to study.
2. **Compare and contrast** these elements' names to their symbols.
3. **Research** the discovery of these elements. Do their names match their symbols? Were they named after a property of the element, a person, their place of discovery, or a system of nomenclature? What was that system?

Goals

- **Research** the names and symbols of various elements.
- **Study** the methods that are used to name elements and how they have changed through time.
- **Organize** your data by making your own periodic table.
- **Study** the history of certain elements and their discoveries.
- **Create** a table of your findings and communicate them to other students.

Data Source

Science Online

Visit **gpscience.com/ internet_lab** for more information on naming elements, elements' symbols, and the discovery of new elements, and for data from other students.

Follow Your Plan

1. Make sure your teacher approves your plan before you start.

2. Visit the Web site provided for links to different sites about elements, their history, and how they were named.

3. **Research** these elements.

4. Carefully record your data in your Science Journal.

Analyze Your Data

1. **Record** in your Science Journal how the symbols for your elements were chosen. What were your elements named after?

2. Make a periodic table that includes the research information on your elements that you found.

3. Make a chart of your class's findings. Sort the chart by year of discovery for each element.

4. How are the names and symbols for newly discovered elements chosen? Make a chart that shows how the newly discovered elements will be named.

Conclude and Apply

1. **Compare** your findings to those of your classmates. Did anyone's data differ for the same element? Were all the elements in the periodic table covered?

2. **Explain** the system that is used to name the newly discovered elements today.

3. **Explain** Some elements were assigned symbols based on their name in another language. Do these examples occur for elements discovered today or long ago?

Communicating Your Data

Find this lab using the link below. Post your data in the table provided. **Compare** your data to those of other students. Combine your data with those of other students to complete your periodic table with all of the elements.

 **Science**nline

gpscience.com/internet_lab

A CHILLING STORY

A scientist inspects an ice core sample from the Greenland Ice Sheet. The samples are stored in a freezer at −36°C.

Picture this: It's 1361. A ship from Norway arrives at a Norwegian settlement in Greenland. The ship's crew hopes to trade its cargo with the people living there. The crew gets off the ship. They look around. The settlement is deserted. More than 1,000 people had vanished!

New evidence has shed some light on the mysterious disappearance of the Norse settlers. The evidence came from a place on the Greenland Ice Sheet over 600 km away from the settlement. This part of Greenland is so cold that snow never melts. As new snow falls, the existing snow is buried and turns to ice.

By drilling deep into this ice, scientists can recover an ice core.

Air bubbles and dirt trapped in ice provide clues to Earth's past climate.

The core is made up of ice formed from snowfalls going way, way back in time.

By measuring the ratio of oxygen isotopes in the ice core, scientists can estimate Greenland's past air temperatures. The cores provide a detailed climate history going back over 80,000 years. Individual ice layers can be dated much like tree rings to determine their age, and the air bubbles trapped within each layer are used to learn about climate variations. Dust and pollen trapped in the ice also yield clues to ancient climates.

A Little Ice Age

Based on their analysis, scientists think the Norse moved to Greenland during an unusually warm period. Then in the 1300s, the climate started to cool and a period known as the Little Ice Age began. The ways the Norse hunted and farmed were inadequate for survival in this long chill. Since they couldn't adapt to their colder surroundings, the settlers died out.

Research Report Evidence seems to show that Earth is warming. Rising temperatures could affect our lives. Research global warming to find out how Earth may change. Share your report with the class.

Science online

For more information, visit gpscience.com/time

Reviewing Main Ideas

Section 1 Structure of the Atom

1. A chemical symbol is a shorthand way of writing the name of an element.

2. An atom consists of a nucleus made of protons and neutrons surrounded by an electron cloud as shown in the figure to the right.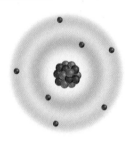

3. Quarks are particles of matter that make up protons and neutrons.

4. The model of the atom changes over time. As new information is discovered, scientists incorporate it into the model.

Section 2 Masses of Atoms

1. The number of neutrons in an atom can be computed by subtracting the atomic number from the mass number.

2. The isotopes of an element are atoms of that same element that have different numbers of neutrons. The figure below shows the isotopes of hydrogen.

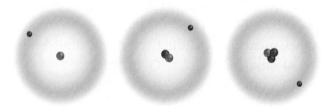

3. The average atomic mass of an element is the weighted-average mass of the mixture of its isotopes. Isotopes are named by using the element name, followed by a dash, and its mass number.

Section 3 The Periodic Table

1. In the periodic table, the elements are arranged by increasing atomic number resulting in periodic changes in properties. Knowing that the number of protons, electrons, and atomic number are equal gives you partial composition of the atom.

2. In the periodic table, the elements are arranged in 18 vertical columns, or groups, and seven horizontal rows, or periods.

3. Metals are found at the left of the periodic table, nonmetals at the right, and metalloids along the line that separates the metals from the nonmetals as shown below.

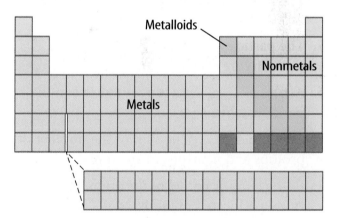

4. Elements are placed on the periodic table in order of increasing atomic number. A new row on the periodic table begins when the outer energy level of the element is filled.

FOLDABLES Use the Foldable that you made at the beginning of the chapter to help you review properties of atoms and the periodic table.

Using Vocabulary

atom p.507	mass number p.513
atomic number p.513	neutron p.507
average atomic mass p.515	nucleus p.507
electron p.507	period p.523
electron cloud p.511	periodic table p.516
electron dot diagram p.522	proton p.507
group p.520	quark p.507
isotope p.514	

Fill in the blanks with the correct word or words.

1. Mendeleev created an organized table of elements called the _____.

2. Two elements with the same number of protons but a different number of neutrons are called _____.

3. _____ is the weighted-average mass of all the known isotopes for an element.

4. The positively charged center of an atom is called the _____.

5. The particles that make up protons and neutrons are called _____.

6. A(n) _____ is a horizontal row in the periodic table.

7. The _____ is the sum of the number of protons and neutrons in an atom.

8. In the current model of the atom, the electrons are located in the _____.

Checking Concepts

Choose the word or phrase that best answers the question.

9. In which state of matter are most of the elements to the left of the stair-step line in the periodic table?
   A) gas   C) plasma
   B) liquid   D) solid

10. Which is a term for a pattern that repeats?
    A) isotopic   C) periodic
    B) metallic   D) transition

11. Which of the following is an element that would have similar properties to those of neon?
    A) aluminum   C) arsenic
    B) argon   D) silver

12. Which of the following terms describes boron?
    A) metal   C) noble gas
    B) metalloid   D) nonmetal

13. How many outer-level electrons do lithium and potassium have?
    A) 1   C) 3
    B) 2   D) 4

14. Which of the following is NOT found in the nucleus of an atom?
    A) proton   C) electron
    B) neutron   D) quark

15. The halogens are located in which group?
    A) 1   C) 15
    B) 11   D) 17

16. In which of the following states is nitrogen found at room temperature?
    A) gas   C) metal
    B) metalloid   D) liquid

17. Which of the elements below is a shiny element that conducts electricity and heat?
    A) chlorine   C) hydrogen
    B) sulfur   D) magnesium

18. The atomic number of Re is 75. The atomic mass of one of its isotopes is 186. How many neutrons are in an atom of this isotope?
    A) 75   C) 186
    B) 111   D) 261

Science Online gpscience.com/vocabulary_puzzlemaker

Interpreting Graphics

19. As a star dies, it becomes more dense. Its temperature rises to a point where He nuclei are combined with other nuclei. When this happens, the atomic numbers of the other nuclei are increased by 2 because each gains the two protons contained in the He nucleus. For example, Cr fuses with He to become Fe. Copy and complete the concept map showing the first four steps in He fusion.

He
↓ +He
Be
↓ +He
↓ +He
↓ +He

20. Copy and complete the concept map below.

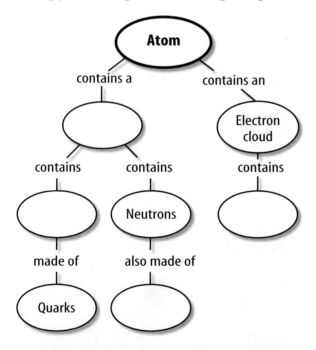

Atom
contains a   contains an
Electron cloud
contains   contains   contains
Neutrons
made of   also made of
Quarks

Thinking Critically

21. **Infer** Lead and mercury are two pollutants in the environment. From information about them in the periodic table, determine why they are called heavy metals.

22. **Explain** why it is necessary to change models as new information becomes available.

23. **Infer** Why did scientists choose carbon to base the atomic mass unit? Which isotope of carbon did they use?

24. **Infer** Ge and Si are used in making semiconductors. Are these two elements in the same group or the same period?

25. **Explain** Using the periodic table, predict how many outer level electrons will be in elements 114, 116, and 118. Explain your answer.

26. **Infer** Ca is used by the body to make bones and teeth. Sr-90 is radioactive. Ca is safe for people and Sr-90 is hazardous. Why is Sr-90 hazardous to people?

Applying Math

27. **Solve One-Step Equations** The atomic number of Yttrium is 39. The atomic mass of one of its isotopes is 89. How many neutrons are in an atom of this isotope?

Use the table below for question 28.

Electrons per Energy Level	
Energy Level	Maximum Number of Electrons
1	2
2	
3	
4	

28. **Use Tables** Use the information in **Figure 10** to determine how many electrons should be in the 2nd, 3rd, and 4th energy levels for Argon, atomic number 18. Copy and complete the table above with the number of electrons for each energy level.

Part 1 | Multiple Choice

Record your answers on the answer sheet provided by your teacher or on a sheet of paper.

1. Atoms of different elements are different because they have different numbers of what type of particle?
   A. electrons
   B. photons
   C. protons
   D. neutrons

2. Which group of elements on the periodic table do not combine readily with other elements?
   A. Group 1
   B. Group 2
   C. Group 17
   D. Group 18

Use the illustration below to answer questions 3 and 4.

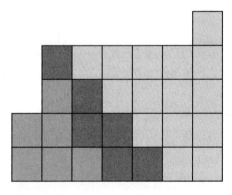

3. What is the name by which the elements along the stair-step line, darkly shaded on the periodic table shown above, are known?
   A. lanthanides
   B. metalloids
   C. metals
   D. nonmetals

4. Which of the regions shown on the periodic table contains mostly elements that are gases at room temperature?
   A. region 1
   B. region 2
   C. region 3
   D. region 4

5. Which scientist proposed the idea that atoms make up all substances?
   A. Aristotle
   B. Dalton
   C. Democritus
   D. Galileo

Use the table below to answer questions 6 and 7.

Element	Electrons in a Neutral Atom	Electrons in Outer Energy Level
Carbon	6	4
Oxygen	8	6
Neon	10	8
Sodium	11	1
Chlorine	17	7

6. The table above lists properties of some elements. Which element in the table has a complete outer energy level?
   A. carbon
   B. oxygen
   C. neon
   D. sodium

7. Which element would you expect to be located in Group 1 of the periodic table?
   A. oxygen
   B. neon
   C. sodium
   D. chlorine

8. How many quarks have been found to exist?
   A. six
   B. eight
   C. ten
   D. twelve

9. The element nickel has five naturally occurring isotopes. Which of the following describes the relationship of these isotopes?
   A. same mass, same atomic number
   B. same mass, different atomic number
   C. different mass, same atomic number
   D. different mass, different atomic number

Test-Taking Tip

Partial Credit Never leave any open ended answer blank. Answer each question as best as you can. Often, you can receive partial credit for partially correct answers.

Question 20 If you have some of the particles mixed up, try to write as much as you remember.

Part 2 | Short Response/Grid In

*Record your answers on the answer sheet
provided by your teacher or on a sheet of paper.*

10. According to the periodic table, an atom of lead has an atomic number of 82. How many neutrons does lead-207 have?

11. About three out of four chlorine atoms are chlorine-35, and about one out of four are chlorine-37. What is the average atomic mass of chlorine?

Use the illustration below to answer questions 12 and 13.

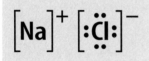

12. The electron dot diagram above shows how a sodium atom, Na, combines with a chlorine atom, Cl, to form sodium chloride. What do the + and the − symbols indicate in the diagram?

13. What do the dots around the chlorine atom indicate?

14. Why isn't the mass of the electron included in the mass of an atom on the periodic table?

15. What determines the chemical properties of an element?

16. What property of radioactive isotopes can scientists use to determine the age of bones or rock formations?

17. The atomic mass for silicon is listed as 28.09 amu on the periodic table. A student claims that no silicon atom has this atomic mass. Is this true? Explain why or why not.

Part 3 | Open Ended

Record your answers on a sheet of paper.

Use the illustration below to answer questions 18 and 19.

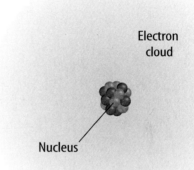

Electron
cloud

Nucleus

18. The illustration above shows the currently accepted model of atomic structure. Describe this model.

19. Compare and contrast the model shown above with Bohr's model of an atom.

20. Describe the composition of protons, neutrons, and electrons.

21. How can you use the periodic table to determine the average number of neutrons an element has, even though the number of neutrons is not listed?

22. Describe the concept of energy levels and how they relate to the placement of elements on the periodic table.

23. Explain the importance of the rows in the periodic table's organization of elements.

24. Describe how Dalton's modernization of the ancient Greek's ideas of element, atom, and compound provided a basis for understanding chemical reactions. Give an example.

Radioactivity and Nuclear Reactions

Planet Power

Although less than one billionth of the energy emitted by the Sun falls on Earth, the energy Earth receives from the Sun powers the entire planet. Without the Sun's energy, life on Earth could not exist. This energy is produced inside the Sun by nuclear fusion—a nuclear reaction in which atomic nuclei join together.

Science Journal In your Science Journal, write a paragraph describing your impressions of the Sun.

Start-Up Activities

The Size of a Nucleus

Do you realize you are made up mostly of empty space? Your body is made of atoms, and atoms are made of electrons whizzing around a small nucleus of protons and neutrons. The size of an atom is the size of the space in which the electrons move around the nucleus. In this lab, you'll find out how the size of an atom compares with the size of a nucleus.

1. Go outside and pour several grains of sugar onto a sheet of paper.

2. Choose one of the grains of sugar to represent the nuculeus of an atom.

3. Brush the rest of the sugar off the paper and place the sugar grain in the center of the paper.

4. Use a meterstick to measure a distance of 10 m from the sugar grain. This distance represents the radius of the electron cloud around an atom.

5. **Think Critically** In your Science Journal, explain why an atom contains mostly empty space. Use the fact that an electron is much smaller than the nucleus.

Radioactivity and Nuclear Reactions Make the following Foldable to help you understand radioactivity and nuclear reactions.

STEP 1 Fold a sheet of paper in half lengthwise.

STEP 2 Fold paper down 2.5 cm from the top. (Hint: From the tip of your index finger to your middle knuckle is about 2.5 cm.)

STEP 3 Open and draw lines along the 2.5-cm fold. Label as shown.

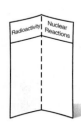

Summarize in a Table As you read the chapter, write what you learn about radioactivity in the left column, and what you learn about nuclear reactions in the right column.

Science Online
Preview this chapter's content and activities at gpscience.com

Radioactivity

Reading Guide

What You'll Learn
- **Describe** the structure of an atom and its nucleus.
- **Explain** what radioactivity is.
- **Contrast** properties of radioactive and stable nuclei.
- **Discuss** the discovery of radioactivity.

Why It's Important
Radioactivity is everywhere because every element on the periodic table has some atomic nuclei that are radioactive.

⊙ Review Vocabulary
long-range force: a force that becomes weaker with distance, but never vanishes

New Vocabulary
- strong force
- radioactivity

Figure 1 The size of a nucleus in an atom can be compared to a marble sitting in the middle of an empty football stadium.

The Nucleus

Every second you are being bombarded by energetic particles. Some of these particles come from unstable atoms in soil, rocks, and the atmosphere. What types of atoms are unstable? What type of particles do unstable atoms emit? The answers to these questions begin with the nucleus of an atom.

Recall that atoms are composed of protons, neutrons, and electrons. The nucleus of an atom contains the protons, which have a positive charge, and neutrons, which have no electric charge. The total amount of charge in a nucleus is determined by the number of protons, which also is called the atomic number. You might remember that an electron has a charge that is equal but opposite to a proton's charge. Atoms usually contain the same number of protons as electrons. Negatively charged electrons are electrically attracted to the positively charged nucleus and swarm around it.

Protons and Neutrons in the Nucleus Protons and neutrons are packed together tightly in a nucleus. The region outside the nucleus in which the electrons are located is large compared to the size of the nucleus. As **Figure 1** shows, the nucleus occupies only a tiny fraction of the space in the atom. If an atom were enlarged so that it was 1 km in diameter, its nucleus would have a diameter of only a few centimeters. But the nucleus contains almost all the mass of the atom, because the mass of one proton or neutron is almost 2,000 times greater than the mass of an electron.

Figure 2 The particles in the nucleus are attracted to each other by the strong force.

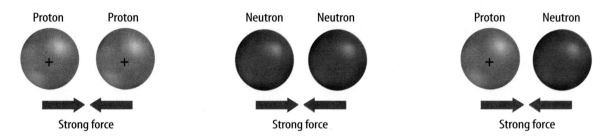

Proton   Proton
Strong force

Neutron   Neutron
Strong force

Proton   Neutron
Strong force

The Strong Force

How do you suppose protons and neutrons are held together so tightly in the nucleus? Positive electric charges repel each other, so why don't the protons in a nucleus push each other away? Another force, called the **strong force,** causes protons and neutrons to be attracted to each other, as shown in **Figure 2.**

The strong force is one of the four basic forces in nature and is about 100 times stronger than the electric force. The attractive forces between all the protons and neutrons in a nucleus keep the nucleus together. However, protons and neutrons have to be close together, like they are in the nucleus, to be attracted by the strong force. The strong force is a short-range force that quickly becomes extremely weak as protons and neutrons get farther apart. The electric force is a long-range force, so protons that are far apart still are repelled by the electric force, as shown in **Figure 3.**

✓ **Reading Check** *What causes the attraction between protons and neutrons?*

Figure 3 The total force between two protons depends on how far apart they are. **Infer** *whether the total force between two protons could become zero.*

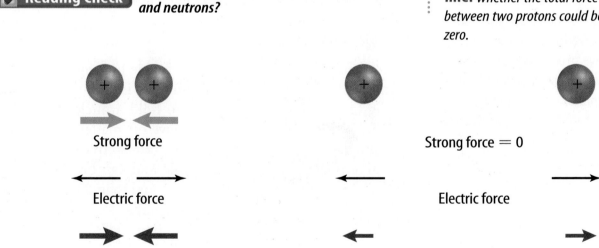

Strong force

Electric force

Total force

When protons are close together, they are attracted to each other. The attraction due to the short-range strong force is much stronger than the repulsion due to the long-range electric force.

Strong force = 0

Electric force

Total force

When protons are too far apart to be attracted by the strong force, they still are repelled by the electric force between them. Then the total force between them is repulsive.

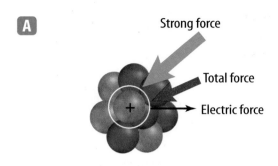

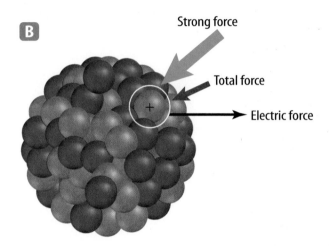

Figure 4 Protons and neutrons are held together less tightly in large nuclei. The circle shows the range of the attractive strong force. **A** Small nuclei have few protons, so the repulsive force on a proton due to the other protons is small. **B** In large nuclei, the attractive strong force is exerted only by the nearest neighbors, but all the protons exert repulsive forces. The total repulsive force is large.

Attraction and Repulsion Some atoms, such as uranium, have many protons and neutrons in their nuclei. These nuclei are held together less tightly than nuclei containing only a few protons and neutrons. To understand this, look at **Figure 4A.** If a nucleus has only a few protons and neutrons, they are all close enough together to be attracted to each other by the strong force. Because only a few protons are in the nucleus, the total electric force causing protons to repel each other is small. As a result, the overall force between the protons and the neutrons attracts the particles to each other.

Forces in a Large Nucleus However, if nuclei have many protons and neutrons, each proton or neutron is attracted to only a few neighbors by the strong force, as shown in **Figure 4B.** The other protons and neutrons are too far away. Because only the closest protons and neutrons attract each other in a large nucleus, the strong force holding them together is about the same as in a small nucleus. However, all the protons in a large nucleus exert a repulsive electric force on each other. Thus, the electric repulsive force on a proton in a large nucleus is larger than it would be in a small nucleus. Because the repulsive force increases in a large nucleus while the attractive force on each proton or neutron remains about the same, protons and neutrons are held together less tightly in a large nucleus.

Radioactivity

In many nuclei the strong force is able to keep the nucleus permanently together, and the nucleus is stable. When the strong force is not large enough to hold a nucleus together tightly, the nucleus can decay and give off matter and energy. This process of nuclear decay is called **radioactivity.**

Large nuclei tend to be unstable and can break apart or decay. In fact, all nuclei that contain more than 83 protons are radioactive. However, many other nuclei that contain fewer than 83 protons also are radioactive. Even some nuclei with only one or a few protons are radioactive.

Almost all elements with more than 92 protons don't exist naturally on Earth. They have been produced only in laboratories and are called synthetic elements. These synthetic elements are unstable, and decay soon after they are created.

Isotopes The atoms of an element all have the same number of protons in their nuclei. For example, the nuclei of all carbon atoms contains six protons. However, naturally occurring carbon nuclei can have six, seven, or eight neutrons. Nuclei that have the same number of protons but different numbers of neutrons are called isotopes. The element carbon has three isotopes that occur naturally. The atoms of all isotopes of an element have the same number of electrons, and have the same chemical properties. **Figure 5** shows two isotopes of helium.

Stable and Unstable Nuclei The ratio of neutrons to protons is related to the stability of the nucleus. In less massive elements, an isotope is stable if the ratio is about 1 to 1. Isotopes of the heavier elements are stable when the ratio of neutrons to protons is about 3 to 2. However, the nuclei of any isotopes that differ much from these ratios are unstable, whether the elements are light or heavy. In other words, nuclei with too many or too few neutrons compared to the number of protons are radioactive.

Nucleus Numbers A nucleus can be described by the number of protons and neutrons it contains. The number of protons in a nucleus is called the atomic number. Because the mass of all the protons and neutrons in a nucleus is nearly the same as the mass of the atom, the number of protons and neutrons is called the mass number.

> ✓ **Reading Check** *What is the atomic number of a nucleus?*

A nucleus can be represented by a symbol that includes its atomic number, mass number, and the symbol of the element it belongs to. The symbol for the nucleus of the stable isotope of carbon is shown below as an example.

$$\text{mass number} \rightarrow \ \ \ ^{12}_{\ 6}\text{C} \leftarrow \text{element symbol}$$
$$\text{atomic number} \rightarrow$$

This isotope is called carbon-12. The number of neutrons in the nucleus is the mass number minus the atomic number. So the number of neutrons in the carbon-12 nucleus is $12 - 6 = 6$. Carbon-12 has six protons and six neutrons. Now, compare the isotope carbon-12 to this radioactive isotope of carbon:

$$\text{mass number} \rightarrow \ \ \ ^{14}_{\ 6}\text{C} \leftarrow \text{element symbol}$$
$$\text{atomic number} \rightarrow$$

The radioactive isotope is carbon-14. How many neutrons does carbon-14 have?

Helium-3   Helium-4

Figure 5 These two isotopes of helium each have the same number of protons, but different numbers of neutrons.
Identify *the ratio of protons to neutrons in each of these isotopes of helium.*

Modeling the Strong Force

Procedure
1. Gather **15 yellow candies** to represent neutrons and **13 red** and **2 green candies** to represent protons.
2. Model a small nucleus by placing 2 red protons and 3 neutrons around a green proton so they touch.
3. Model a larger nucleus by arranging the remaining candies around the other green proton so they are touching.

Analysis
1. Compare the number of protons and neutrons touching a green proton in both models.
2. Suppose the strong force on a green proton is due to protons and neutrons that touch it. Compare the strong force on a green proton in both models.

Try at Home

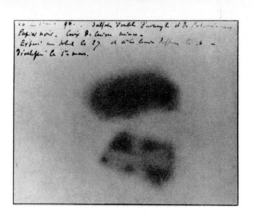

Figure 7 The dark spots on this photographic plate were made by the radiation emitted by radioactive uranium atoms. Uranium salt had been placed next to the plate by Henri Becquerel in 1896.

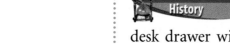

Topic: Marie Curie

Visit gpscience.com for Web links to information about the life of Marie Curie.

Activity Create a timeline showing important events in the life of Marie Curie.

INTEGRATE History

The Discovery of Radioactivity In 1896, Henri Becquerel left uranium salt in a desk drawer with a photographic plate. Later, when he developed the plate, shown in **Figure 6,** he found an outline of the clumps of the uranium salt. He hypothesized that the uranium salt had emitted some unknown invisible rays, or radiation, that had darkened the film.

Two years after Becquerel's discovery, Marie and Pierre Curie discovered two new elements, polonium and radium, that also were radioactive. To obtain a sample of radium large enough to be studied, they developed a process to extract radium from the mineral pitchblende. After more than three years, they were able to obtain about 0.1 g of radium from several tons of pitchblende. Years of additional processing gradually produced more radium that was made available to other researchers all over the world.

section 1 review

Summary

The Strong Force

- The short-ranged strong force causes neutrons and protons to be attracted to each other.
- The long-ranged electric force causes protons to repel each other.
- The combination of the strong and electric forces causes protons and neutrons in a large nucleus to be held together less tightly than in a small nucleus.

Radioactivity and Isotopes

- Radioactivity is the process of nuclear decay.
- Isotopes of an element have the same number of protons, but different numbers of neutrons.
- The atomic number is the number of protons in a nucleus. The mass number is the number of protons and neutrons in a nucleus.

Self Check

1. **Describe** the properties of the strong force.
2. **Compare** the strong force between protons and neutrons in a small nucleus and a large nucleus.
3. **Explain** why large nuclei are unstable.
4. **Identify** the contributions of the three scientists who discovered the first radioactive elements.
5. **Think Critically** What is the ratio of protons to neutrons in lead-214? Explain whether you would expect this isotope to be radioactive or stable.

Applying Math

6. **Calculate a Ratio** What is the ratio of neutrons to protons in a nucleus of radon-222?
7. **Use Percentages** A silicon rod contains 30.21 g of silicon-28, 1.53 g of silicon-29, and 1.02 g of silicon-30. Calculate the percentage of each isotope in the rod.

 gpscience.com/self_check_quiz

Nuclear Decay

Reading Guide

What You'll Learn
- **Compare and contrast** alpha, beta, and gamma radiation.
- **Define** the half-life of a radioactive material.
- **Describe** the process of radioactive dating.

Why It's Important
Nuclear decay produces nuclear radiation that can both harm people and be useful.

Review Vocabulary
electromagnetic wave: a transverse wave consisting of vibrating electric and magnetic fields

New Vocabulary
- alpha particle
- transmutation
- beta particle
- gamma ray
- half-life

Nuclear Radiation

When an unstable nucleus decays, particles and energy called nuclear radiation are emitted from it. The three types of nuclear radiation are alpha, beta (BAY tuh), and gamma radiation. Alpha and beta radiation are particles. Gamma radiation is an electromagnetic wave.

Alpha Particles

When alpha radiation occurs, an **alpha particle**—made of two protons and two neutrons, as shown in **Table 1**—is emitted from the decaying nucleus. An alpha particle is the same as the nucleus of a helium atom and has a charge of +2 and an atomic mass of 4. Its symbol is the same as the symbol of a helium nucleus, $^4_2$He.

✔ Reading Check *What does an alpha particle consist of?*

Compared to beta and gamma radiation, alpha particles are much more massive. They also have the most electric charge. As a result, alpha particles lose energy more quickly when they interact with matter than the other types of nuclear radiation do. When alpha particles pass through matter, they exert an electric force on the electrons in atoms in their path. This force pulls electrons away from atoms and leaves behind charged ions. Alpha particles lose energy quickly during this process. As a result, alpha particles are the least penetrating form of nuclear radiation. Alpha particles can be stopped by a sheet of paper.

Table 1 · Alpha Particles

Symbol	$^4_2$He
Mass	4
Charge	+2

Figure 7 When alpha particles collide with molecules in the air, positively-charged ions and electrons result. The ions and electrons move toward charged plates, creating a current in the smoke detector.

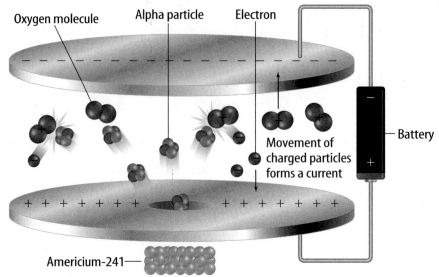

Oxygen molecule   Alpha particle   Electron

Movement of charged particles forms a current

Battery

Americium-241

Damage from Alpha Particles

Alpha particles can be dangerous if they are released by radioactive atoms inside the human body. Biological molecules inside your body are large and easily damaged. A single alpha particle can damage many fragile biological molecules. Damage from alpha particles can cause cells not to function properly, leading to illness and disease.

Smoke Detectors

Some smoke detectors give off alpha particles that ionize the surrounding air. Normally, an electric current flows through this ionized air to form a circuit, as in **Figure 7.** But if smoke particles enter the ionized air, they will absorb the ions and electrons. The circuit is broken and the alarm goes off.

Transmutation

When an atom emits an alpha particle, it has two fewer protons, so it is a different element. **Transmutation** is the process of changing one element to another through nuclear decay. In alpha decay, two protons and two neutrons are lost from the nucleus. The new element has an atomic number two less than that of the original element. The mass number of the new element is four less than the original element. **Figure 8** shows a nuclear transmutation caused by alpha decay. The charge of the original nucleus equals the sum of the charges of the nucleus and the alpha particle that are formed.

Figure 8 In this transmutation, polonium emits an alpha particle and changes into lead.
Determine *whether the charges and mass numbers of the products equal the charge and mass number of the polonium nucleus.*

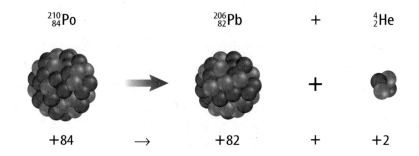

$^{210}_{84}\text{Po}$ $^{206}_{82}\text{Pb}$ + $^{4}_{2}\text{He}$

+84 $\rightarrow$ +82   +   +2

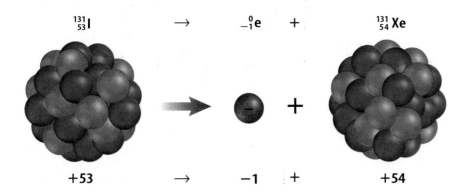

$$^{131}_{53}\text{I} \rightarrow \ ^{0}_{-1}\text{e} \ + \ ^{131}_{54}\text{Xe}$$

$$+53 \rightarrow -1 \ + \ +54$$

Figure 9 Nuclei that emit beta particles undergo transmutation. In beta decay shown here, iodine changes to xenon.
Compare *the total atomic number and mass number of the products with the atomic number and mass number of the iodine nucleus.*

Beta Particles

A second type of radioactive decay is called beta decay, which is summarized in **Table 2.** Sometimes in an unstable nucleus a neutron decays into a proton and emits an electron. The electron is emitted from the nucleus and is called a **beta particle.** Beta decay is caused by another basic force called the weak force.

Because the atom now has one more proton, it becomes the element with an atomic number one greater than that of the original element. Atoms that lose beta particles undergo transmutation. However, because the total number of protons and neutrons does not change during beta decay, the mass number of the new element is the same as that of the original element. **Figure 9** shows a transmutation caused by beta decay.

Damage from Beta Particles Beta particles are much faster and more penetrating than alpha particles. They can pass through paper but are stopped by a sheet of aluminum foil. Just like alpha particles, beta particles can damage cells when they are emitted by radioactive nuclei inside the human body.

Gamma Rays

The most penetrating form of nuclear radiation is gamma radiation. **Gamma rays** are electromagnetic waves with the highest frequencies and the shortest wavelengths in the electromagnetic spectrum. They have no mass and no charge and travel at the speed of light. They usually are emitted from a nucleus when alpha decay or beta decay occurs. The properties of gamma rays are summarized in **Table 3.**

Thick blocks of dense materials, such as lead and concrete, are required to stop gamma rays. However, gamma rays cause less damage to biological molecules as they pass through living tissue. Suppose an alpha particle and a gamma ray travel the same distance through matter. The gamma ray produces fewer ions because it has no electric charge.

Table 2 Beta Particles	
Symbol	$^{0}_{-1}\text{e}$
Mass	0.0005
Charge	-1

Table 3 Gamma Rays	
Symbol	γ
Mass	0
Charge	0

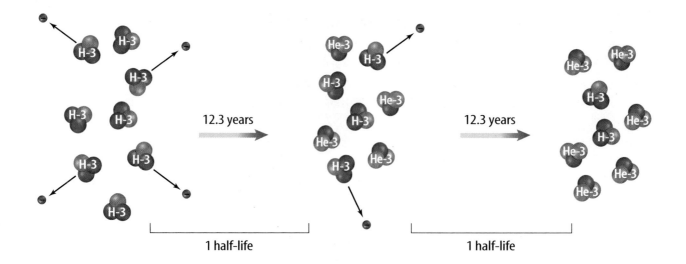

1 half-life | 1 half-life

Figure 10 The half-life of $_1^3$H is 12.3 years. During each half-life, half of the atoms in the sample decay into helium.

Infer *how many hydrogen atoms will be left in the sample after the next half-life.*

Table 4	Sample Half-Lives
Isotope	**Half-Life**
$_1^3$H	12.3 years
$_{82}^{212}$Pb	10.6 hr
$_6^{14}$C	5,730 years
$_{84}^{211}$Po	0.5 s
$_{92}^{235}$U	7.04 × 10⁸ years
$_{53}^{131}$I	8.04 days

Radioactive Half-Life

If an element is radioactive, how can you tell when its atoms are going to decay? Some radioisotopes decay to stable atoms in less than a second. However, the nuclei of certain radioactive isotopes require millions of years to decay. A measure of the time required by the nuclei of an isotope to decay is called the half-life. The **half-life** of a radioactive isotope is the amount of time it takes for half the nuclei in a sample of the isotope to decay. The nucleus left after the isotope decays is called the daughter nucleus. **Figure 10** shows how the number of decaying nuclei decreases after each half-life.

Half-lives vary widely among the radioactive isotopes. For example, polonium-214 has a half-life of less than a thousandth of a second, but uranium-238 has a half-life of 4.5 billion years. The half-lives of some other radioactive elements are listed in **Table 4.**

✔ **Reading Check** *What is a daughter nucleus?*

Radioactive Dating

Some geologists, biologists, and archaeologists, among others, are interested in the ages of rocks and fossils found on Earth. The ages of these materials can be determined using radioactive isotopes and their half-lives. First, the amounts of the radioactive isotope and its daughter nucleus in a sample of material are measured. Then, the number of half-lives that need to pass to give the measured amounts of the isotope and its daughter nucleus is calculated. The number of half-lives is the amount of time that has passed since the isotope began to decay. It is also usually the amount of time that has passed since the object was formed, or the age of the object. Different isotopes are useful in dating different types of materials.

Carbon Dating The radioactive isotope carbon-14 often is used to estimate the ages of plant and animal remains. Carbon-14 has a half-life of 5,730 years and is found in molecules such as carbon dioxide. Plants use carbon dioxide when they make food, so all plants contain carbon-14. When animals eat plants, carbon-14 is added to their bodies.

The decaying carbon-14 in a plant or animal is replaced when an animal eats or when a plant makes food. As a result, the ratio of the number of carbon-14 atoms to the number of carbon-12 atoms in the organism remains nearly constant. But when an organism dies, its carbon-14 atoms decay without being replaced. The ratio of carbon-14 to carbon 12 then decreases with time. By measuring this ratio, the age of an organism's remains can be estimated. However, only material from plants and animals that lived within the past 50,000 years contains enough carbon-14 to be measured.

Uranium Dating Radioactive dating also can be used to estimate the ages of rocks. Some rocks contain uranium, which has two radioactive isotopes with long half-lives. Each of these uranium isotopes decays into a different isotope of lead. The amount of these uranium isotopes and their daughter nuclei are measured. From the ratios of these amounts, the number of half-lives since the rock was formed can be calculated.

section 2 review

Summary

Nuclear Radiation

- When an unstable nucleus decays it emits nuclear radiation that can be alpha particles, beta particles, or gamma rays.
- An alpha particle consists of two protons and two neutrons.
- A beta particle is an electron and is emitted when a neutron decays into a proton.
- Gamma rays are electromagnetic waves of very high frequency that usually are emitted when alpha decay or beta decay occurs.

Half-Life and Radioactive Dating

- The half-life of a radioactive isotope is the amount of time for half the nuclei in a sample of the isotope to decay.
- The amounts of a radioactive isotope and its daughter nucleus are needed to date materials.

Self Check

1. **Infer** how the mass number and the atomic number of a nucleus change when it emits a beta particle.
2. **Determine** the daughter nucleus formed when a radon-222 nucleus emits an alpha particle.
3. **Describe** how each of the three types of radiation can be stopped.
4. **Think Critically** Sample 1 contains nuclei with a half-life of 10.6 hr and sample 2 contains an equal number of nuclei with a half-life of 0.5 s. After 3 half-lives pass for each sample, which sample contains more of the original nuclei?

Applying Math

5. **Use Percentages** What is the percentage of radioactive nuclei left after 3 half-lives pass?
6. **Use Fractions** If the half-life of iodine 131 is 8 days, how much of a 5-g sample is left after 32 days?

Detecting Radioactivity

What You'll Learn

- **Describe** how radioactivity can be detected in cloud and bubble chambers.
- **Explain** how an electroscope can be used to detect radiation.
- **Explain** how a Geiger counter can measure nuclear radiation.

Why It's Important

Devices that detect and measure radioactivity are used to monitor exposure to humans.

Review Vocabulary

ion: an atom that has gained or lost electrons

New Vocabulary

- cloud chamber
- bubble chamber
- Geiger counter

Radiation Detectors

Because you can't see or feel alpha particles, beta particles, or gamma rays, you must use instruments to detect their presence. Some tools that are used to detect radioactivity rely on the fact that radiation forms ions in the matter it passes through. The tools detect these newly formed ions in several ways.

Cloud Chambers A **cloud chamber,** shown in **Figure 11,** can be used to detect alpha or beta particle radiation. A cloud chamber is filled with water or ethanol vapor. When a radioactive sample is placed in the cloud chamber, it gives off charged alpha or beta particles that travel through the water or ethanol vapor. As each charged particle travels through the chamber, it knocks electrons off the atoms in the air, creating ions. It leaves a trail of ions in the chamber. The water or ethanol vapor condenses around these ions, creating a visible path of droplets along the track of the particle. Beta particles leave long, thin trails, and alpha particles leave shorter, thicker trails.

Reading Check *Why are trails produced by alpha and beta particles seen in cloud chambers?*

Figure 11 If a sample of radioactive material is placed in a cloud chamber, a trail of condensed vapor will form along the paths of the emitted particles.

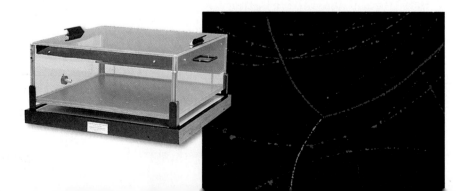

Bubble Chambers Another way to detect and monitor the paths of nuclear particles is by using a bubble chamber. A **bubble chamber** holds a superheated liquid, which doesn't boil because the pressure in the chamber is high. When a moving particle leaves ions behind, the liquid boils along the trail. The path shows up as tracks of bubbles, like the ones in **Figure 12.**

Electroscopes Do you remember how an electroscope can be used to detect electric charges? When an electroscope is given a negative charge, its leaves repel each other and spread apart, as in **Figure 13A.** They will remain apart until their extra electrons have somewhere to go and discharge the electroscope. The excess charge can be neutralized if it combines with positive charges. Nuclear radiation moving through the air can remove electrons from some molecules in air, as shown in **Figure 13B,** and cause other molecules in air to gain electrons. When this occurs near the leaves of the electroscope, some positively charged molecules in the air can come in contact with the electroscope and attract the electrons from the leaves, as **Figure 13C** shows. As these negatively charged leaves lose their charges, they move together. **Figure 13D** shows this last step in the process. The same process also will occur if the electroscope leaves are positively charged. Then the electrons move from negative ions in the air to the electroscope leaves.

Figure 12 Particles of nuclear radiation can be detected as they leave trails of bubbles in a bubble chamber.

Figure 13 Nuclear radiation can cause an electroscope to lose its charge.

A The electroscope leaves are charged with negative charge.

B Nuclear radiation, such as alpha particles, can create positive ions.

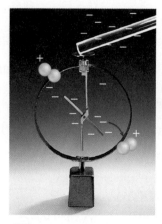

C Negative charges move from the leaves to positively charged ions.

D The electroscope leaves lose their negative charge and come together.

Measuring Radiation

It is important to monitor the amount of radiation a person is being exposed to because large doses of radiation can be harmful to living tissue. A **Geiger counter** is a device that measures the amount of radiation by producing an electric current when it detects a charged particle.

Applying Math   Use Logarithms

MEASURING THE AGE OF ROCKS The radioactive nucleus uranium-238 produces the daughter nucleus lead-206 with a half-life of 4.5 billion years. The age of a mineral sample can be calculated from the equation:

$$\text{age} = (1.44H)\ln\left[1 + \frac{N_L}{N_U}\right]$$

In this equation, H is the half-life of uranium 238, N_L is the number of lead-206 atoms in the sample, and N_U is the number of uranium-238 atoms in the sample. Find the age of a sample in which the ratio N_L/N_U is measured to be 0.55.

IDENTIFY   known values and unknown values

Identify the known values:

the ratio of the number of uranium-238 atoms   means⟩ $\dfrac{N_L}{N_U} = 0.55$
to the number of lead-206 atoms is 0.55

a half-life of 4.5 billion years   means⟩ $H = 4.5$ billion years

Identify the unknown value:

what is the age of the sample?   means⟩   age = ? years

SOLVE   the problem

Substitute the known values into the equation for the age:

$$\text{age} = (1.44H)\ln\left[1 + \frac{N_L}{N_U}\right] = (1.44 \times 4.5 \text{ billion yrs}) \ln[1 + 0.55]$$
$$= (6.48 \text{ billion yrs}) \ln(1.55)$$

To calculate $\ln(1.55)$, enter 1.55 on your calculator and press the "ln" button. The result is 0.438. Substitute this value in the above equation:

$$\text{age} = (6.48 \text{ billion yrs})(0.438) = 2.84 \text{ billion yrs}$$

CHECK   the answer

Does your answer seem reasonable? The ratio $N_L/N_U = 0.55$, means that less than one half-life has elapsed. The calculated age is less than one half-life.

Practice Problems

Find the age of a sample in which the ratio N_L/N_U has been measured to be 1.21.

For more practice problems go to page 834, and visit gpscience.com/extra_problems.

Figure 14 Electrons that are stripped off gas molecules in a Geiger counter move to a positively charged wire in the device. This causes current to flow in the wire. The current then is used to produce a click or a flash of light.

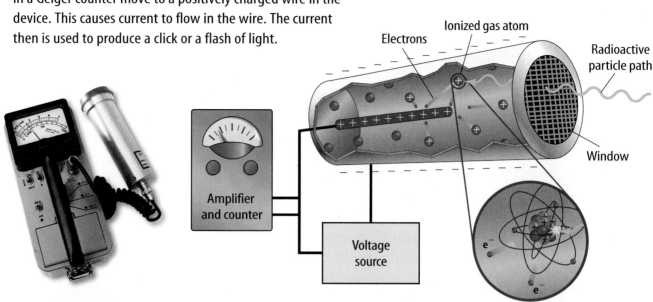

Electrons

Ionized gas atom

Radioactive particle path

Window

Amplifier and counter

Voltage source

e⁻

Geiger Counters A Geiger counter, shown in **Figure 14,** has a tube with a positively charged wire running through the center of a negatively charged copper cylinder. This tube is filled with gas at a low pressure. When radiation enters the tube at one end, it knocks electrons from the atoms of the gas. These electrons then knock more electrons off other atoms in the gas, and an "electron avalanche" is produced. The free electrons are attracted to the positive wire in the tube. When a large number of electrons reaches the wire, a short, intense current is produced in the wire. This current is amplified to produce a clicking sound or flashing light. The intensity of radiation present is determined by the number of clicks or flashes of light each second.

 *How does a Geiger counter indicate that radiation is present?*

Background Radiation

It might surprise you to know that you are bathed in radiation that comes from your environment. This radiation, called background radiation, is not produced by humans. Instead it is low-level radiation emitted mainly by naturally occurring radioactive isotopes found in Earth's rocks, soils, and atmosphere. Building materials such as bricks, wood, and stones contain traces of these radioactive materials. Traces of naturally occurring radioactive isotopes are found in the food, water, and air consumed by all animals and plants. As a result, animals and plants also contain small amounts of these isotopes.

Artificial Rainmaking It may be possible to ease the health and economic hardships caused by severe droughts by artificially making rain. The formation of raindrops in a cloud is similar to the formation of droplets in a cloud chamber. Rain forms when cold droplets freeze around microscopic particles of dust, and then melt as they fall through warmer air. Research artificial rainmaking and report you findings to your class.

Sources of Background Radiation

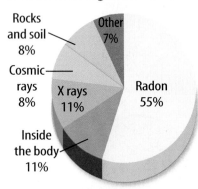

Sources of Background Radiation

Rocks and soil 8%
Other 7%
Cosmic rays 8%
X rays 11%
Radon 55%
Inside the body 11%

Figure 15 This circle graph shows the sources of background radiation received on average by a person living in the United States.

Sources of Background Radiation Background radiation comes from several sources, as shown in **Figure 15.** The largest source comes from the decay of radon gas. Radon is produced in Earth's crust by the decay of uranium-238 and emits an alpha particle when it decays. Radon gas can seep into houses and basements from the surrounding soil and rocks.

Some background radiation comes from high-speed nuclei, called cosmic rays, that strike Earth's atmosphere. They produce showers of particles, including alpha, beta, and gamma radiation. Most of this radiation is absorbed by the atmosphere. Higher up, there is less atmosphere to absorb this radiation, so the background radiation from cosmic rays increases with altitude.

Radiation in Your Body Some of the elements that are essential for life have naturally occurring radioactive isotopes. For example, about one out of every trillion carbon atoms is carbon-14, which emits a beta particle when it decays. With each breath, you inhale about 3 million carbon-14 atoms.

The amount of background radiation a person receives can vary greatly. The amount depends on the type of rocks underground, the type of materials used to construct the person's home, and the elevation at which the person lives, among other things. However, because it comes from naturally occurring processes, background radiation never can be eliminated.

section 3 review

Summary

Radiation Detectors

- Alpha and beta particles can be detected by the trail of ions they form when they pass through a cloud chamber or a bubble chamber.

- The presence of alpha or beta particles can cause an electroscope to become discharged.

- A Geiger counter produces a clicking sound or a flash of light when alpha or beta particles enter the Geiger counter tube, and is used to measure radiation levels.

Background Radiation

- Background radiation is low-level radiations emitted mainly by radioactive isotopes in Earth's rocks, soils, and atmosphere.

- The largest source of background radiation is from the alpha decay of radon gas.

Self Check

1. **Describe** why a charged electroscope will discharge when placed near a radioactive material.

2. **Compare and contrast** cloud and bubble chambers.

3. **Describe** that process that occurs in a Geiger counter when a click is produced.

4. **Explain** why background radiation never can be completely eliminated.

5. **Think Critically** If the radioactive isotope radon-222 has a half-life of only four days, how can radon gas be continually present inside houses?

Applying Math

6. **Use Percentages** The amount of radiation can be measured in units called millirems. If 25 millirems from cosmic rays is 8.0 percent of the average background radiation, what is the amount of the average background radiation in millirems?

Science Online gpscience.com/self_check_quiz

Nuclear Reactions

Reading Guide

What You'll Learn
- **Explain** nuclear fission and how it can begin a chain reaction.
- **Discuss** how nuclear fusion occurs in the Sun.
- **Describe** how radioactive tracers can be used to diagnose medical problems.
- **Discuss** how nuclear reactions can help treat cancer.

Why It's Important
Almost all of the different atoms that you are made of were formed by the nuclear reactions inside ancient, distant stars.

⊙ Review Vocabulary
kinetic energy: energy of motion; increases as the mass or speed of an object increases

New Vocabulary
- nuclear fission
- chain reaction
- critical mass
- nuclear fusion
- tracer

Nuclear Fission

In the 1930s the physicist Enrico Fermi thought that by bombarding nuclei with neutrons, nuclei would absorb neutrons and heavier nuclei would be produced. However, in 1938, Otto Hahn and Fritz Strassmann found that when a neutron strikes a uranium-235 nucleus, the nucleus splits apart into smaller nuclei.

In 1939 Lise Meitner was the first to offer a theory to explain these results. She proposed that the uranium-235 nucleus is so distorted when the neutron strikes it that it divides into two smaller nuclei, as shown in **Figure 16.** The process of splitting a nucleus into several smaller nuclei is **nuclear fission.** The word *fission* means "to divide."

✔ Reading Check *What initiates nuclear fission of a uranium-235 nucleus?*

Only large nuclei, such as the nuclei of uranium and plutonium atoms, can undergo nuclear fission. The products of a fission reaction usually include several individual neutrons in addition to the smaller nuclei. The total mass of the products is slightly less than the mass of the original nucleus and the neutron. This small amount of missing mass is converted to a tremendous amount of energy during the fission reaction.

Figure 16 When a neutron hits a uranium-235 nucleus, the uranium nucleus splits into two smaller nuclei and two or three free neutrons. Energy also is released.

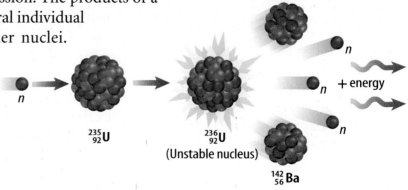

$^{91}_{36}$Kr

n

$+$ energy

n

n

$^{235}_{92}$U

$^{236}_{92}$U
(Unstable nucleus)

$^{142}_{56}$Ba

Figure 17 A chain reaction occurs when neutrons emitted from a split nucleus cause other nuclei to split and emit additional neutrons.

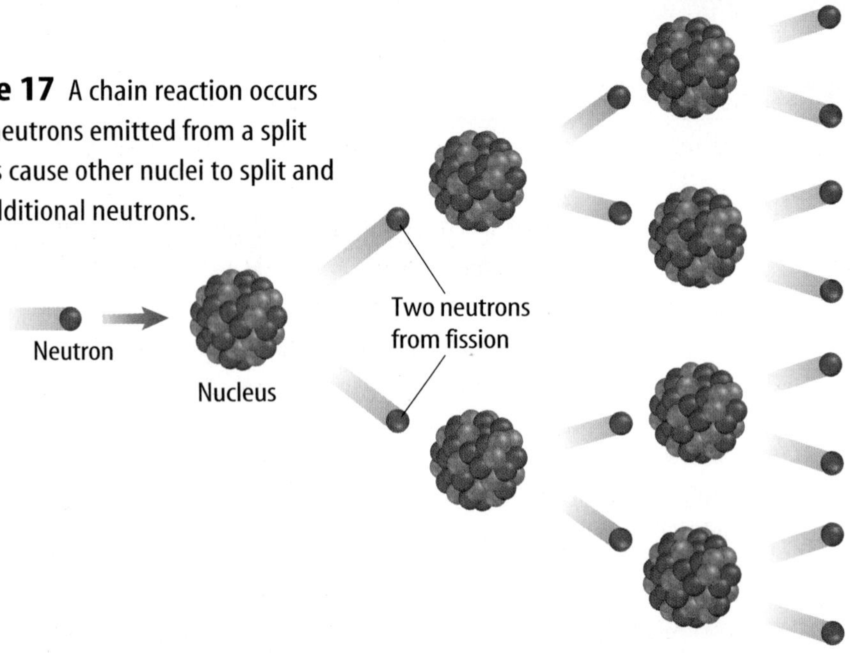

Neutron

Nucleus

Two neutrons from fission

Mini LAB

Modeling a Nuclear Reaction

Procedure

1. Put **32 marbles,** each with an attached lump of **clay,** into a large **beaker.** These marbles with clay represent unstable atoms.
2. During a 1-min period, remove half of the marbles and pull off the clay. Place the removed marbles into another beaker and place the lumps of clay into a pile. Marbles without clay represent stable atoms. The clay represents waste from the reaction—smaller atoms that still might decay and give off energy.
3. Repeat this procedure four more times.

Analysis

1. What is the half-life of this reaction?
2. Explain whether the waste products could undergo nuclear fission.

Mass and Energy Albert Einstein proposed that mass and energy were related in his special theory of relativity. According to this theory, mass can be converted to energy and energy can be converted to mass. The relation between mass and energy is given by this equation:

> **Mass-Energy Equation**
>
> Energy (joules) = mass (kg) × [speed of light (m/s)]^2
>
> $$E = mc^2$$

A small amount of mass can be converted into an enormous amount of energy. For example, if one gram of mass is converted to energy, about 100 trillion joules of energy are released.

Chain Reactions When a nuclear fission reaction occurs, the neutrons emitted can strike other nuclei in the sample, and cause them to split. These reactions then release more neutrons, causing additional nuclei to split, as shown in **Figure 17.** The series of repeated fission reactions caused by the release of neutrons in each reaction is a **chain reaction.**

If the chain reaction is uncontrolled, an enormous amount of energy is released in an instant. However, a chain reaction can be controlled by adding materials that absorb neutrons. If enough neutrons are absorbed, the reaction will continue at a constant rate.

For a chain reaction to occur, a critical mass of material that can undergo fission must be present. The **critical mass** is the amount of material required so that each fission reaction produces approximately one more fission reaction. If less than the critical mass of material is present, a chain reaction will not occur.

Nuclear Fusion

Tremendous amounts of energy can be released in nuclear fission. In fact, splitting one uranium-235 nucleus produces about 30 million times more energy than chemically reacting one molecule of dynamite. Even more energy can be released in another type of nuclear reaction, called nuclear fusion. In **nuclear fusion,** two nuclei with low masses are combined to form one nucleus of larger mass. Fusion fuses atomic nuclei together, and fission splits nuclei apart.

Temperature and Fusion For nuclear fusion to occur, positively charged nuclei must get close to each other. However, all nuclei repel each other because they have the same positive electric charge. If nuclei are moving fast, they can have enough kinetic energy to overcome the repulsive electrical force between them and get close to each other.

Remember that the kinetic energy of atoms or molecules increases as their temperature increases. Only at temperatures of millions of degrees Celsius are nuclei moving so fast that they can get close enough for fusion to occur. These extremely high temperatures are found in the center of stars, including the Sun.

Nuclear Fusion and the Sun The Sun is composed mainly of hydrogen. Most of the energy given off by the Sun is produced by a process involving the fusion of hydrogen nuclei. This process occurs in several stages, and one of the stages is shown in **Figure 18.** The net result of this process is that four hydrogen nuclei are converted into one helium nucleus. As this occurs, a small amount of mass is changed into an enormous amount of energy. Earth receives a small amount of this energy as heat and light.

As the Sun ages, the hydrogen nuclei are used up as they are converted into helium. So far, only about one percent of the Sun's mass has been converted into energy. Eventually, no hydrogen nuclei will be left, and the fusion reaction that changes hydrogen into helium will stop. However, it is estimated that the Sun has enough hydrogen to keep this reaction going for another 5 billion years.

Figure 18 The fusion of hydrogen to form helium takes place in several stages in the Sun. One of these stages is shown here. An isotope of helium is produced when a proton and the hydrogen isotope H-2 undergo fusion.

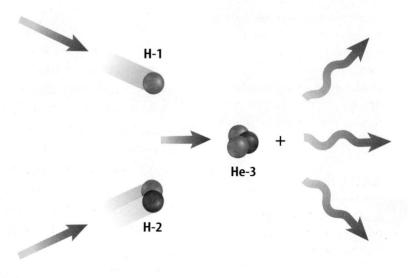

Radioactive Decay Equations A uranium-235 atom can fission, or break apart, to form barium and krypton. Use a periodic table to find the atomic numbers of barium and krypton. What do they add up to? A uranium-235 atom can fission in several other ways such as producing neodymium and another element. What is the other element?

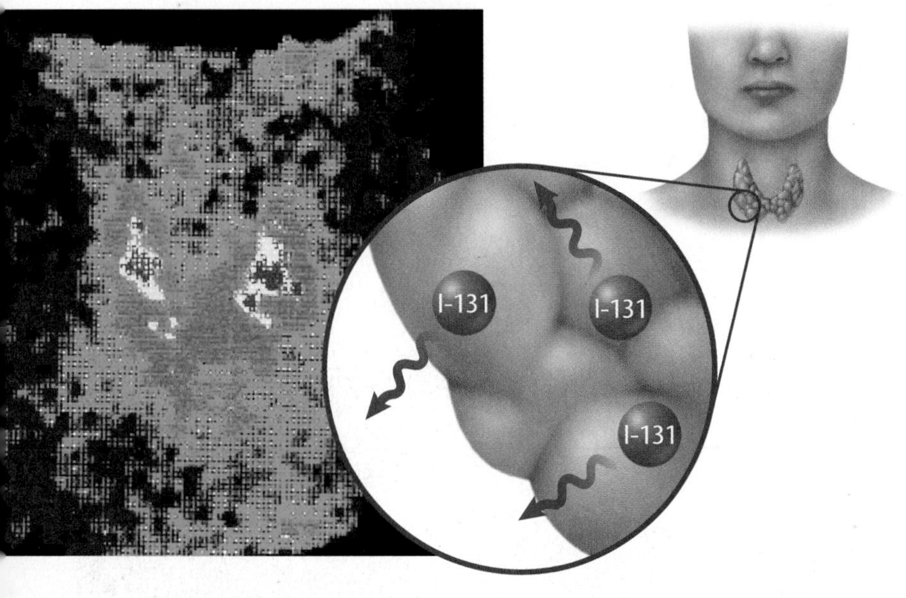

Figure 19 Radioactive iodine-131 accumulates in the thyroid gland and emits gamma rays, which can be detected to form an image of a patient's thyroid. **List** *some advantages of being able to use iodine-131 to form an image of a thyroid.*

Using Nuclear Reactions in Medicine

If you were going to meet a friend in a crowded area, it would be easier to find her if your friend told you that she would be wearing a red hat. In a similar way, scientists can find one molecule in a large group of molecules if they know that it is "wearing" something unique. Although a molecule can't wear a red hat, if it has a radioactive atom in it, it can be found easily in a large group of molecules, or even in a living organism. Radioactive isotopes can be located by detecting the radiation they emit.

When a radioisotope is used to find or keep track of molecules in an organism, it is called a **tracer.** Scientists can use tracers to follow where a particular molecule goes in your body or to study how a particular organ functions. Tracers also are used in agriculture to monitor the uptake of nutrients and fertilizers. Examples of tracers include carbon-11, iodine-131, and sodium-24. These three radioisotopes are useful tracers because they are important in certain body processes. As a result, they accumulate inside the organism being studied.

✓ Reading Check *How are tracers located inside the human body?*

Iodine Tracers in the Thyroid The thyroid gland is located in your neck and produces chemical compounds called hormones. These hormones help regulate several body processes, including growth. Because the element iodine accumulates in the thyroid, the radioisotope iodine-131 can be used to diagnose thyroid problems. As iodine-131 atoms are absorbed by the thyroid, their nuclei decay, emitting beta particles and gamma rays. The beta particles are absorbed by the surrounding tissues, but the gamma rays penetrate the skin. The emitted gamma rays can be detected and used to determine whether the thyroid is healthy, as shown in **Figure 19.** If the detected radiation is not intense, then the thyroid has not properly absorbed the iodine-131 and is not functioning properly. This could be due to the presence of a tumor. **Figure 20** shows how radioactive tracers are used to study the brain.

I-131 I-131 I-131

Figure 20

The diagram below shows an imaging technique known as Positron Emission Tomography, or PET. Positrons are emitted from the nuclei of certain radioactive isotopes when a proton changes to a neutron. PET can form images that show the level of activity in different areas of the brain. These images can reveal tumors and regions of abnormal brain activity.

A When positrons are emitted from the nucleus of an atom, they can hit electrons from other atoms and become transformed into gamma rays.

B The radioactive isotope fluorine-18 emits positrons when it decays. Fluorine-18 atoms are chemically attached to molecules that are absorbed by brain tissue. These compounds are injected into the patient and carried by blood to the brain.

C Inside the patient's brain, the decay of the radioactive fluorine-18 nuclei emits positrons that collide with electrons. The gamma rays that are released are sensed by the detectors.

D A computer uses the information collected by the detectors to generate an image of the activity level in the brain. This image shows normal activity in the right side of the brain (red, yellow, green) but below-normal activity in the left (purple).

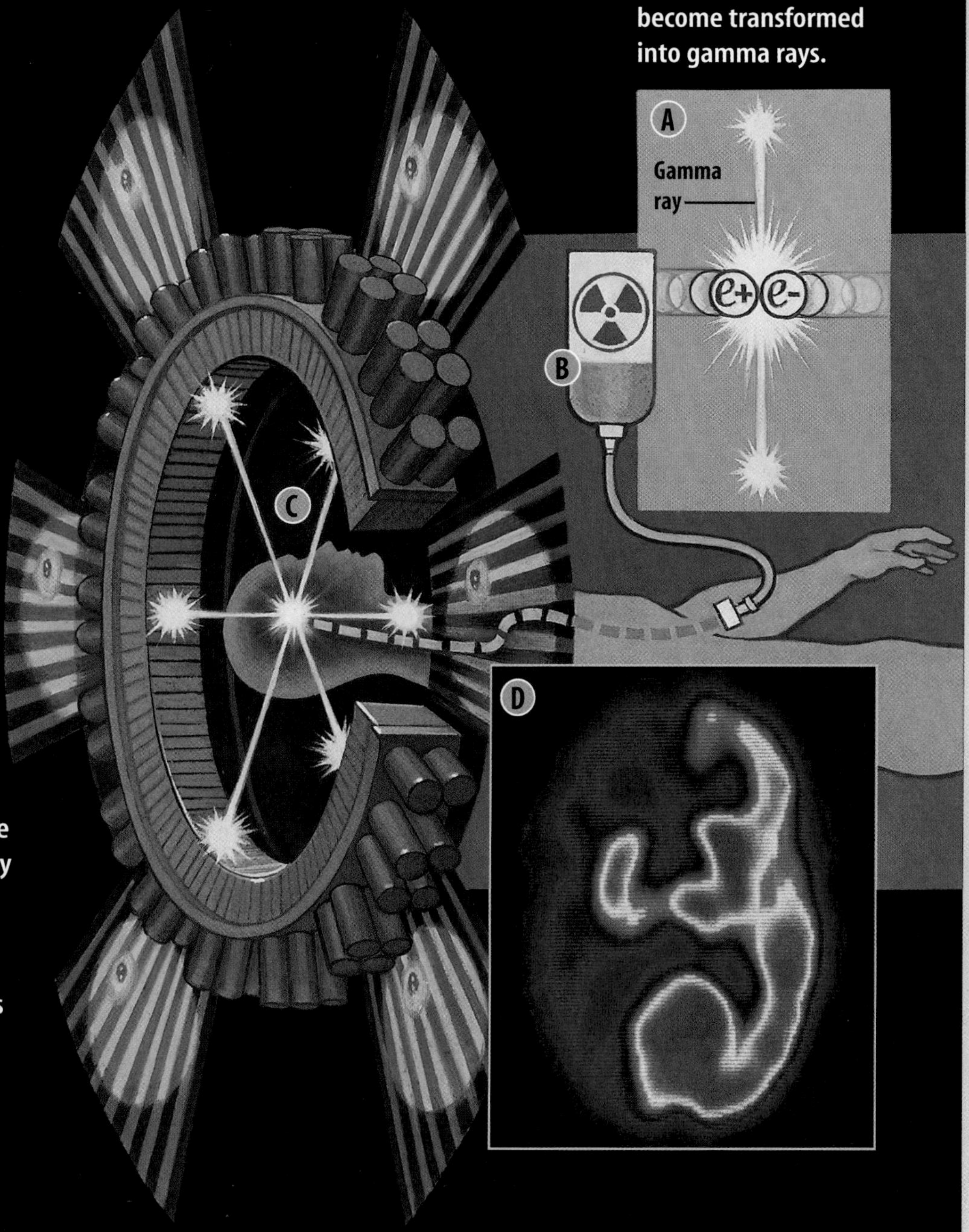

Gamma ray

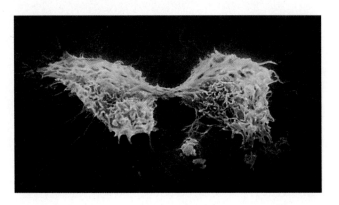

Figure 21 Cancer cells, such as the ones shown here, can be killed with carefully measured doses of radiation.

Treating Cancer with Radioactivity

When a person has cancer, a group of cells in that person's body grows out of control and can form a tumor. Radiation can be used to stop some types of cancerous cells from growing. Remember that the radiation that is given off during nuclear decay is strong enough to ionize nearby atoms. If a source of radiation is placed near cancer cells, such as those shown in **Figure 21,** atoms in the cells can be ionized. If the ionized atoms are in a critical molecule, such as the DNA or RNA of a cancer cell, then the molecule might no longer function properly. The cell then could die or stop growing.

When possible, a radioactive isotope such as gold-198 or iridium-192 is implanted within or near the tumor. Other times, tumors are treated from outside the body. Typically, an intense beam of gamma rays from the decay of cobalt-60 is focused on the tumor for a short period of time. The gamma rays pass through the body and into the tumor. How can physicians be sure that only the cancer cells will absorb radiation? Because cancer cells grow quickly, they are more susceptible to absorbing radiation and being damaged than healthy cells are. However, other cells in the body that grow quickly also are damaged, which is why cancer patients who have radiation therapy sometimes experience severe side effects.

section 4 review

Summary

Nuclear Fission

- Nuclear fission occurs when a neutron strikes a nucleus, causing it to split into smaller nuclei.
- A chain reaction requires a critical mass of fissionable material.

Nuclear Fusion

- Nuclear fusion occurs when two nuclei combine to form another nucleus.
- Nuclear fusion occurs at temperatures of millions of degrees, which occur inside the Sun.

Medical Uses of Radiation

- Radioactive isotopes are used as tracers to locate various atoms or molecules in organisms.
- Radiation emitted by radioactive isotopes is used to kill cancer cells.

Self Check

1. **Infer** whether mass is conserved in a nuclear reaction.
2. **Explain** why fusion reactions can occur inside stars.
3. **Explain** how a chain reaction can be controlled.
4. **Describe** two properties of a tracer isotope used for monitoring the functioning of an organ in the body.
5. **Think Critically** Explain why high temperatures are needed for fusion reactions to occur, but not for fission reactions to occur.

Applying Math

6. **Calculate Number of Nuclei** In a chain reaction, two neutrons are emitted by each nucleus that is split. If one nucleus is split in the first step of the reaction, how many nuclei will have been split after the fifth step?

Chain Reactions

In an uncontrolled nuclear chain reaction, the number of reactions increases as additional neutrons split more nuclei. In a controlled nuclear reaction, neutrons are absorbed, so the reaction continues at a constant rate. How could you model a controlled and an uncontrolled nuclear reaction in the classroom?

▶ Real-World Question

How can you use dominoes to model chain reactions?

Goals
- **Model** a controlled and uncontrolled chain reaction.
- **Compare** the two types of chain reactions.

Materials
dominoes   stopwatch

▶ Procedure

1. Set up a single line of dominoes standing on end so that when the first domino is pushed over, it will knock over the second and each domino will knock over the one following it.

2. Using the stopwatch, time how long it takes from the moment the first domino is pushed over until the last domino falls over. Record the time.

3. Using the same number of dominoes as in step 1, set up a series of dominoes in which at least one of the dominoes will knock down two others, so that two lines of dominoes will continue falling. In other words, the series should have at least one point that looks like the letter Y.

4. Repeat step 2.

▶ Conclude and Apply

1. **Compare** the amount of time it took for all of the dominoes to fall in each of your two arrangements.

2. **Determine** the average number of dominoes that fell per second in both domino arrangements.

3. **Identify** which of your domino arrangements represented a controlled chain reaction and which represented an uncontrolled chain reaction.

4. **Describe** how the concept of critical mass was represented in your model of a controlled chain reaction.

5. Assuming that they had equal amounts of material, which would finish faster—a controlled or an uncontrolled nuclear chain reaction? Explain.

✒ommunicating
Your Data

Explain to friends or members of your family how a controlled nuclear chain reaction can be used in nuclear power plants to generate electricity.

Modeling Transmutations

◉ Real-World Question

Imagine what would happen if the oxygen atoms around you began changing into nitrogen atoms. Without oxygen, most living organisms, including people, could not live. Fortunately, more than 99.9 percent of all oxygen atoms are stable and do not decay. Usually, when an unstable nucleus decays, an alpha or beta particle is thrown out of its nucleus, and the atom becomes a new element. A uranium-238 atom, for example, will undergo eight alpha decays and six beta decays to become lead. This process of one element changing into another element is called transmutation. How could you create a model of a uranium-238 atom and the decay process it undergoes during transmutation? What types of materials could you use to represent the protons and neutrons in a U-238 nucleus? How could you use these materials to model transmutation?

◉ Make a Model

1. **Choose** two materials of different colors or shapes for the protons and neutrons of your nucleus model. Choose a material for the negatively charged beta particle.

Possible Materials
brown rice
white rice
colored candies
dried beans
dried seeds
glue
poster board

Safety Precautions

WARNING: *Never eat foods used in the lab.*

Data Source
Refer to your textbook for general information about transmutation.

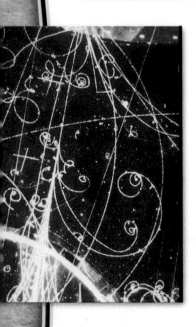

2. **Decide** how to model the transmutation process. Will you create a new nucleus model for each new element? How will you model an alpha or beta particle leaving the nucleus?

3. **Create** a transmutation chart to show the results of each transmutation step of a uranium-238 atom with the identity, atomic number, and mass number of each new element formed and the type of radiation particle emitted at each step. A uranium-238 atom will undergo the following decay steps before transmuting into a lead-206 atom: alpha decay, beta decay, beta decay, alpha decay, alpha decay, alpha decay, alpha decay, alpha decay, beta decay, beta decay, alpha decay, beta decay, beta decay, alpha decay.

4. **Describe** your model plan and transmutation chart to your teacher and ask how they can be improved.

5. **Present** your plan and chart to your class. Ask classmates to suggest improvements in both.

6. **Construct** your model of a uranium-238 nucleus showing the correct number of protons and neutrons.

▶ Test Your Model

1. Using your nucleus model, demonstrate the transmutation of a uranium-238 nucleus into a lead-206 nucleus by following the decay sequence outlined in the previous section.

2. Show the emission of an alpha particle or beta particle between each transmutation step.

▶ Analyze Your Data

1. **Compare** how alpha and beta decay change an atom's atomic number.

2. **Compare** how alpha and beta decay change the mass number of an atom.

▶ Conclude and Apply

1. **Calculate** the ratio of neutrons to protons in lead-206 and uranium-238. In which nucleus is the ratio closer to 1.5?

2. **Identify** Alchemists living during the Middle Ages spent much time trying to turn lead into gold. Identify the decay processes needed to accomplish this task.

*C*ommunicating Your Data

Show your model to the class and explain how your model represents the transmutation of U-238 into Pb-206.

The Nuclear Alchemists

The colored tracks are alpha particles emitted from a speck of radium salt placed on a special photographic plate.

For centuries, ancient alchemists tried in vain to convert common metals into gold. However, in the early 20th century, some scientists realized there was a way to convert atoms of some elements into other elements—nuclear fission.

A Startling Discovery

As the twentieth century dawned, most scientists thought atoms could not be broken apart. In 1902, a New Zealand physicist Ernest Rutherford and his colleague Frederick Soddy showed that heavy elements uranium and thorium decayed into slightly lighter elements, with the production of helium gas. "Don't call it transmutation. They'll have our heads off as alchemists!" Rutherford warned Soddy. In 1908, Rutherford showed that the alpha particles emitted in radioactive decay were the same as a helium nucleus.

Something's Missing

In 1938 in Germany, Otto Hahn and Fritz Strassmann found the uranium-235 nucleus would split if struck by a neutron. The process was called nuclear fission.

Enrico Fermi lead the development of the first nuclear reactor.

A year later, Austrian physicist Lise Meitner pointed out that the total mass of the particles produced when the uranium nucleus split was less than that of the original uranium nucleus. According to the special theory of relativity, this small amount of missing mass results in the release of a tremendous amount of energy when fission occurs. But is there any way this energy could be controlled?

Lise Meitner was the first to explain how nuclear fission occurs.

Controlling a Chain Reaction

Only a few years later, Italian physicist Enrico Fermi, working with colleagues in the United States, found the answer. Fermi realized that the neutrons released when fission occurs could lead to a chain reaction. However, materials that absorb neutrons could be used to control the chain reaction. In late 1942, Fermi and his colleagues built the first nuclear reactor by using cadmium rods to absorb neutrons and control the chain reaction. The tremendous energy released by nuclear fission could be controlled.

Research Find out more about the contributions these scientists made to understanding radioactivity and the nucleus. What other discoveries did Rutherford and Fermi make?

online

For more information, visit gpscience.com/time

Reviewing Main Ideas

Section 1 Radioactivity

Nucleus

1. The protons and neutrons in an atomic nucleus, like the one to the right, are held together by the strong force.

2. The ratio of protons to neutrons indicates whether a nucleus will be stable or unstable. Large nuclei tend to be unstable.

3. Radioactivity is the emission of energy or particles from an unstable nucleus.

4. Radioactivity was discovered accidentally by Henri Becquerel about 100 years ago.

Section 2 Nuclear Decay

1. Unstable nuclei can decay by emitting alpha particles, beta particles, and gamma rays.

2. Alpha particles consist of two protons and two neutrons. A beta particle is an electron.

3. Gamma rays are the highest frequency electromagnetic waves.

4. Half-life is the amount of time in which half of the nuclei of a radioactive isotope will decay.

5. Because all living things contain carbon, the radioactive isotope carbon-14 can be used to date the remains

of organisms that lived during the past 50,000 years, such as this skeleton.

6. Radioactive isotopes of uranium are used to date rocks.

Section 3 Detecting Radioactivity

1. Radioactivity can be detected with a cloud chamber, a bubble chamber, an electroscope, or a Geiger counter.

2. A Geiger counter measures the amount of radiation by producing electric current when it is struck by a charged particle.

3. Background radiation is low-level radiation emitted by naturally occurring isotopes found in Earth's rocks and soils, the atmosphere, and inside your body.

Section 4 Nuclear Reactions

1. When nuclear fission occurs, a nucleus splits into smaller nuclei. Neutrons and a large amount of energy are emitted.

2. Neutrons emitted when a nuclear fission reaction occurs can cause a chain reaction. A chain reaction can occur only if a critical mass of material is present.

3. Nuclear fusion occurs at high temperatures when light nuclei collide and form heavier nuclei, releasing a large amount of energy.

4. Radioactive tracers that are absorbed by specific organs can help diagnose health problems. Nuclear radiation is used to kill cancer cells.

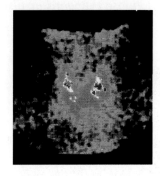

FOLDABLES Use the Foldable that you made at the beginning of this chapter to help you review advantages and disadvantages of using radioactive materials and nuclear reactions.

Using Vocabulary

alpha particle p. 541	half-life p. 544
beta particle p. 543	nuclear fission p. 551
bubble chamber p. 547	nuclear fusion p. 553
chain reaction p. 552	radioactivity p. 538
cloud chamber p. 546	strong force p. 537
critical mass p. 552	tracer p. 554
gamma ray p. 543	transmutation p. 542
Geiger counter p. 548	

Use what you know about the vocabulary words to explain the differences in the following sets of words. Then explain how the words are related.

1. cloud chamber—bubble chamber

2. chain reaction—critical mass

3. nuclear fission—nuclear fusion

4. radioactivity—half-life

5. alpha particle—beta particle—gamma ray

6. Geiger counter—tracer

7. nuclear fission—transmutation

8. electroscope—Geiger counter

9. strong force—radioactivity

Checking Concepts

Choose the word or phrase that best answers the question.

10. What keeps particles in a nucleus together?
    A) strong force   C) electrical force
    B) repulsion   D) atomic glue

11. Which device would be most useful for measuring the amount of radiation in a nuclear laboratory?
    A) a cloud chamber
    B) a Geiger counter
    C) an electroscope
    D) a bubble chamber

12. What is an electron that is produced when a neutron decays called?
    A) an alpha particle
    B) a beta particle
    C) gamma radiation
    D) a negatron

13. Which of the following describes an isotope's half-life?
    A) a constant time interval
    B) a varied time interval
    C) an increasing time interval
    D) a decreasing time interval

14. For which of the following could carbon-14 dating be used?
    A) a bone fragment
    B) a marble column
    C) dinosaur fossils
    D) rocks

15. Which term describes an ongoing series of fission reactions?
    A) chain reaction   C) positron emission
    B) decay reaction   D) fusion reaction

16. Which process is responsible for the tremendous energy released by the Sun?
    A) nuclear decay   C) nuclear fusion
    B) nuclear fission   D) combustion

17. Which radioisotope acts as an external source of ionizing radiation in the treatment of cancer?
    A) cobalt-60   C) gold-198
    B) carbon-14   D) technetium-99

18. Which of the following describes all nuclei with more than 83 protons?
    A) radioactive   C) synthetic
    B) repulsive   D) stable

19. Which of the following describes atoms with the same number of protons and a different number of neutrons?
    A) unstable   C) radioactive
    B) synthetic   D) isotopes

Interpreting Graphics

20. Copy and complete the following concept map on radioactivity.

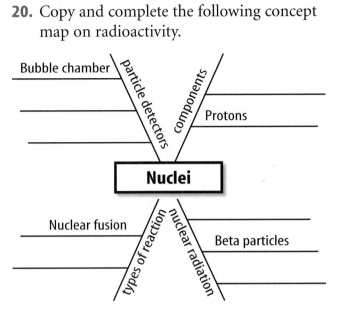

21. Make a table summarizing the use of radioactive isotopes or nuclear radiation in the following applications: radioactive dating, monitoring the thyroid gland, and treating cancer. Include a description of the radioactive isotope or radiation involved.

Use the data in the table below to answer question 22.

Isotope Half-Lives

Isotope	Mass Number	Half-Life
Radon-222	222	4 days
Thorium-234	234	24 days
Iodine-131	131	8 days
Bismuth-210	210	5 days
Polonium-210	210	138 days

22. Graph the data in the table above with the *x*-axis the mass number and the *y*-axis the half-life. Infer from your graph whether there is a relation between the half-life and the mass number. If so, how does half-life depend on mass number?

Thinking Critically

23. Explain why the amount of background radiation a person receives can vary greatly from place to place.

24. Infer how the atomic number of a nucleus changes when the nucleus emits only gamma radiation.

25. Identify the properties of alpha particles that make them harmful to living cells.

26. Determine the type of nuclear radiation that is emitted by each of the following nuclear reactions:
  a. uranium-238 to thorium-234
  b. boron-12 to carbon-12
  c. cesium-130 to cesium 130
  d. radium-226 to radon-222

27. Determine how the motion of an alpha particle is affected when it passes between a positively-charged electrode and a negatively charged electrode. How is the motion of a gamma ray affected?

28. Infer how the background radiation a person receives changes when they fly in a jet airliner.

Applying Math

29. Use a Ratio The mass of an alpha particle is 4.0026 mass units, and the mass of a beta particle is 0.000548 mass units. How many times larger is the mass of an alpha particle than the mass of a beta particle?

30. Calculate Number of Half-Lives How many half-lives have elapsed when the amount of a radioactive isotope in a sample is reduced to 3.125 percent of the original amount in the sample.

Part 1 Multiple Choice

Record your answers on the answer sheet provided by your teacher or on a sheet of paper.

1. If a radioactive material has a half-life of 10 y, what fraction of the material will remain after 30 y?
   A. one half
   B. one third
   C. one fourth
   D. one eighth

2. Which of the follow statements is true about all the isotopes of an element?
   A. They have the same mass number.
   B. They have different numbers of protons.
   C. They have different numbers of neutrons.
   D. They have the same number of neutrons.

3. How does the beta decay of a nucleus cause the nucleus to change?
   A. The number of protons increases.
   B. The number of neutrons increases.
   C. The number of protons decreases.
   D. The number of protons plus the number of neutrons decreases.

Use the illustration below to answer questions 4 and 5.

Helium-3   Helium-4

4. What does the illustration show?
   A. nuclear fusion
   B. nuclear decay
   C. isotopes
   D. half-lives

5. Which is a TRUE statement about the two nuclei?
   A. They have the same atomic number.
   B. They have the same mass number.
   C. They have different numbers of electrons.
   D. They have different numbers of protons.

6. What is the atomic number of a nucleus equal to?
   A. the number of neutrons
   B. the number of protons
   C. the number of neutrons and protons
   D. the number of neutrons minus the number of protons

Use the illustration below to answer questions 7 and 8.

$^{210}_{84}Po$ $^{206}_{82}Pb$ + $^{4}_{2}He$

 → + +

7. What process is shown by this illustration?
   A. nuclear fusion
   B. chain reaction
   C. transmutation
   D. beta decay

8. How do the total charge and total mass number of the products compare to the charge and mass number of the polonium nucleus.
   A. The charges are equal but the mass numbers are not equal.
   B. The mass numbers are equal but the charges are not equal.
   C. Neither the mass numbers or the charges are equal
   D. The mass numbers and charges are equal.

9. Radioactive isotopes of which element are used to study the brain?
   A. uranium
   B. fluorine
   C. carbon
   D. lead

Test-Taking Tip

Understand the Question Be sure you understand the question before you read the answer choices. Make special note of words like NOT or EXCEPT. Read and consider all the answer choices before you mark your answer sheet.

Part 2 | Short Response/Grid In

Record your answers on the answer sheet provided by your teacher or on a sheet of paper.

10. What process contributes the most to the background radiation received by a person in the United States?

11. Explain why alpha particles tend to produce more ions than beta particles or gamma rays when they pass through matter.

Use the table below to answer questions 12–14.

Half-Lives of Isotopes	
Isotope	**Half-life**
Carbon-14	5,730 years
Potassium-40	1.28 billion years
Iodine-131	8.04 days
Radon-222	4 days

12. Calculate how much of an 80 g sample of carbon-14 will be left after 17,190 years.

13. Potassium-40 decays to argon-40. What is the age of a rock in which 87.5 percent of the atoms are argon-40?

14. A sample containing which radioactive isotope will have one-eighth of the isotope left after 24 days?

15. Explain how control rods are able to control a chain reaction.

16. Describe the sequence of events that must occur for a nuclear chain reaction to occur.

17. A radioactive tracer with a half-life of 2 h is used to study the accumulation of a compound in the kidneys. Explain whether this study could be done over a 24-h period.

18. When the boron isotope boron-10 is bombarded with neutrons, it absorbs a neutron, and then emits an alpha particle. Identify the isotope that is formed in this process.

Part 3 | Open Ended

Record your answers on a sheet of paper.

19. Compare the strength of the strong force on a proton and the strength of the electric force on a proton in a small nucleus and a large nucleus.

20. Explain how the alpha particles emitted by the decay of the radioactive isotope americium-241 in a smoke detector produce an electric current between the charged plates of the smoke detector.

21. The geothermal heat that flows from inside Earth is produced by the decay of radioactive isotopes. Form a hypothesis about how the rate of geothermal heat production will change with time.

Use the illustration below to answer questions 22 and 23.

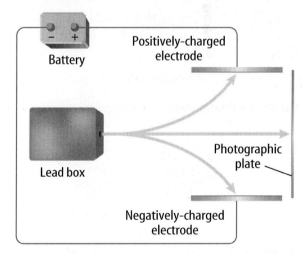

22. In the figure above, nuclear radiation is escaping from a small hole in the lead box. Which type of nuclear radiation is deflected toward the positively-charged electrode, and why is this radiation deflected toward this electrode?

23. Explain why the radiation that struck the photographic plate was not deflected by the electrodes.

How Are
Billiards & Bottles
Connected?

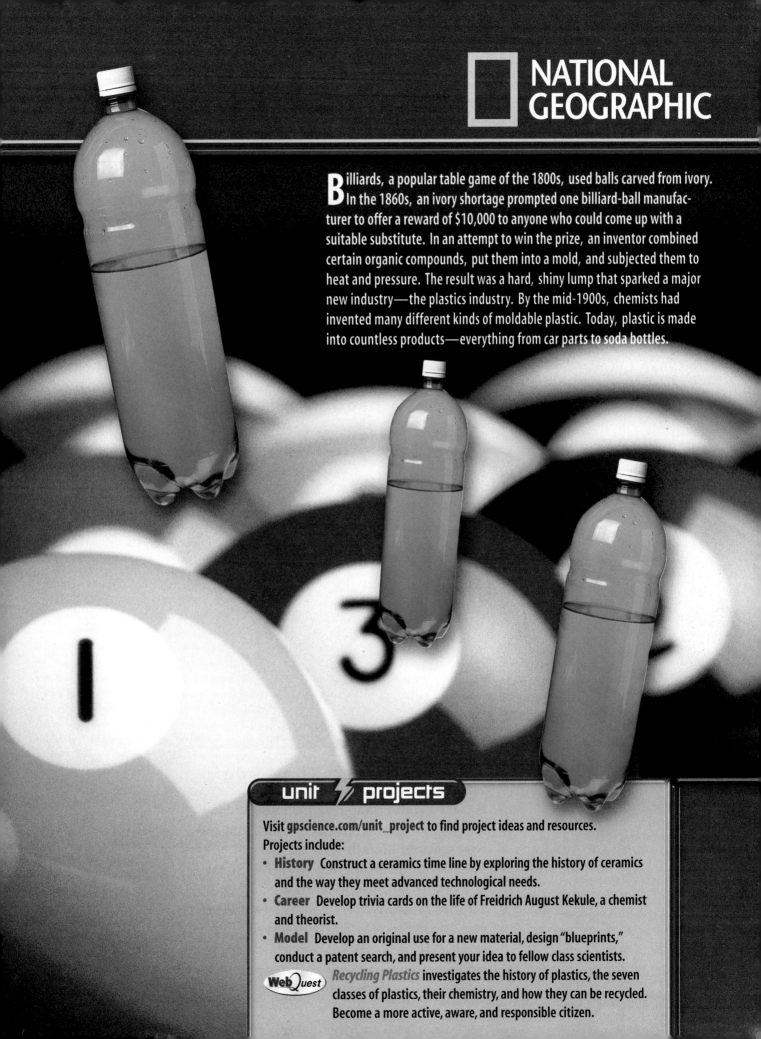

Billiards, a popular table game of the 1800s, used balls carved from ivory. In the 1860s, an ivory shortage prompted one billiard-ball manufacturer to offer a reward of $10,000 to anyone who could come up with a suitable substitute. In an attempt to win the prize, an inventor combined certain organic compounds, put them into a mold, and subjected them to heat and pressure. The result was a hard, shiny lump that sparked a major new industry—the plastics industry. By the mid-1900s, chemists had invented many different kinds of moldable plastic. Today, plastic is made into countless products—everything from car parts to soda bottles.

unit projects

Visit gpscience.com/unit_project to find project ideas and resources.
Projects include:

- **History** Construct a ceramics time line by exploring the history of ceramics and the way they meet advanced technological needs.
- **Career** Develop trivia cards on the life of Freidrich August Kekule, a chemist and theorist.
- **Model** Develop an original use for a new material, design "blueprints," conduct a patent search, and present your idea to fellow class scientists.

WebQuest *Recycling Plastics* investigates the history of plastics, the seven classes of plastics, their chemistry, and how they can be recycled. Become a more active, aware, and responsible citizen.

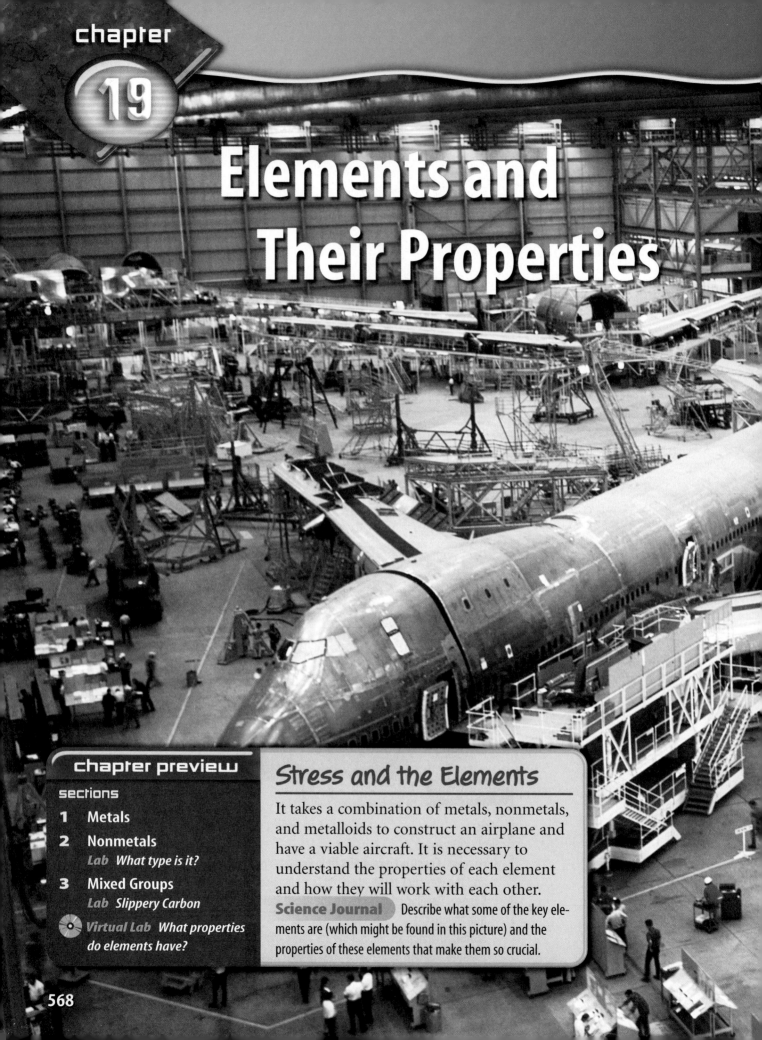

Elements and Their Properties

Stress and the Elements

It takes a combination of metals, nonmetals, and metalloids to construct an airplane and have a viable aircraft. It is necessary to understand the properties of each element and how they will work with each other.

Science Journal Describe what some of the key elements are (which might be found in this picture) and the properties of these elements that make them so crucial.

Start-Up Activities

Observe Colorful Clues

It is the distinct physical properties of each element that make it so that one element can be identified from another. In this lab, you will observe how the heated atoms of some elements absorb energy and then in a short time release the absorbed energy, which you see as colored light.

1. Wearing gloves and using tongs, carefully hold a clean paper clip in the hottest part of a lab burner flame for 45 seconds.

2. Dip the hot paper clip into a solution of copper(II) sulfate.

3. Using the tongs with the same paper clip, repeat step 1, observing any color change.

4. Repeat all three steps using solutions of strontium chloride and sodium chloride with clean tongs and new paper clips.

5. **Think Critically** Which element—chlorine or strontium—was responsible for the color observed when strontium chloride was placed in the flame? How do you know? Devise a plan to determine whether copper or sulfate was responsible for the color in step 2.

Groups Make the following Foldable to help classify and organize elements into groups based on their common features.

STEP 1 Fold a vertical sheet of paper in half from top to bottom.

STEP 2 Fold in half from side to side with the fold at the top.

STEP 3 Unfold the paper once. Cut only the fold of the top flap to make two tabs.

STEP 4 Turn the paper horizontally and label the tabs *Metals* and *Nonmetals* as shown.

Illustrate and Label Before reading the chapter, list all of the metal and nonmetal elements you know under the appropriate tab. As you read the chapter, check your list and make changes as needed.

 **Preview this chapter's content and activities at** gpscience.com

Metals

What You'll Learn
- **Describe** the properties of a typical metal.
- **Identify** the alkali metals and alkaline earth metals.
- **Differentiate** among three groups of transition elements.

Why It's Important
Metals are a part of your everyday life—from electric cords to the cars you ride in.

🔍 **Review Vocabulary**
element: substance with atoms that are all alike

New Vocabulary
- metal
- malleable
- ductile
- metallic bonding
- radioactive element
- transition element

Properties of Metals

Figure 1 The various properties of metals make them useful.

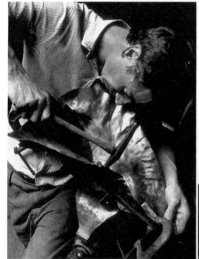

Metals, like the one shown, can be hammered into thin sheets.
Explain *one use for a sheet of metal.*

Metals can be drawn into wires, like the wire that is being used here.
Describe *what this property of metals is called.*

The first metal used about 6,000 years ago was gold. The use of copper and silver followed a few thousand years later. Then came tin and iron. Aluminum wasn't refined until the 1800s because it must go through a much more complicated refining process that earlier civilizations had not yet developed.

In the periodic table, metals are elements found to the left of the stair-step line. In the table on the inside back cover of your book, the metal element blocks are colored blue. **Metals** usually have common properties—they are good conductors of heat and electricity, and all but one are solid at room temperature. Mercury is the only metal that is not a solid at room temperature. Metals also reflect light. This is a property called luster. Metals are **malleable** (MAL yuh bul), which means they can be hammered or rolled into sheets, as shown in **Figure 1.** Metals are also **ductile,** which means they can be drawn into wires like the ones shown in **Figure 1.** These properties make metals suitable for use in objects ranging from eyeglass frames to computers to building structures.

Ionic Bonding in Metals The atoms of metals generally have one to three electrons in their outer energy levels. In chemical reactions, metals tend to give up electrons easily because of the strength of charge of the protons in the nucleus. When metals combine with nonmetals, the atoms of the metals tend to lose electrons to the atoms of nonmetals, forming ionic bonds, as shown in **Figure 2.** Both metals and nonmetals become more chemically stable when they form ions. They take on the electron structure of the nearest noble gas.

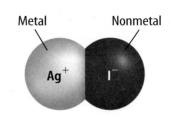

Figure 2 Metals can form ionic bonds with nonmetals.

Metallic Bonding Another type of bonding, neither ionic nor covalent, occurs among the atoms in a metal. In **metallic bonding,** positively charged metallic ions are surrounded by a cloud of electrons. Outer-level electrons are not held tightly to the nucleus of an atom. Rather, the electrons move freely among many positively charged ions. As shown in **Figure 3,** the electrons form a cloud around the ions of the metal.

The idea of metallic bonding explains many of the properties of metals. For example, when a metal is hammered into a sheet or drawn into a wire, it does not break because the ions are in layers that slide past one another without losing their attraction to the electron cloud. Metals are also good conductors of electricity because the outer-level electrons are weakly held.

Reading Check *Why do metals conduct electricity?*

Look at the periodic table inside the back cover of your book. How many of the elements in the table are classified as metals? All of the blue-shaded boxes represent metals. Except for hydrogen, all the elements in Groups 1 through 12 are metals, as well as the elements under the stair-step line in Groups 13 through 15. You will learn more about metals in some of these groups throughout this chapter.

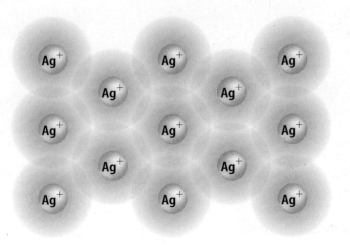

Figure 3 In metallic bonding, the electrons represented by the cloud are not attached to any one silver ion. This allows them to move and conduct electricity.

The Alkali Metals

The elements in Group 1 of the periodic table are the alkali (AL kuh li) metals. Like other metals, Group 1 metals are shiny, malleable, and ductile. They are also good conductors of heat and electricity. However, they are softer than most other metals. The alkali metals are the most reactive of all the metals. They react rapidly—sometimes violently—with oxygen and water, as shown in **Figure 4.** Because they combine so readily with other elements, alkali metals don't occur in nature in their elemental form and are stored in substances that are unreactive, such as an oil.

Each atom of an alkali metal has one electron in its outer energy level. This electron is given up when an alkali metal combines with another atom. As a result, the alkali metal becomes a positively charged ion in a compound such as sodium chloride, $NaCl$, or potassium bromide, KBr.

Alkali metals and their compounds have many uses. You and other living things need potassium and sodium compounds to stay healthy. Doctors sometimes use lithium compounds to treat bipolar disorder. The lithium keeps chemical levels that are important to mental health within a narrow range. The operation of some photocells depends upon rubidium or cesium compounds. Francium, the last element in Group 1, is extremely rare and radioactive. A **radioactive element** is one in which the nucleus breaks down and gives off particles and energy. Francium can be found in uranium minerals, but only 25 g to 30 g of francium are in all of Earth's crust at one time.

The Alkali Metals

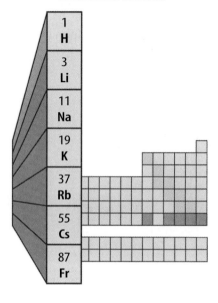

Figure 4 Alkali metals are very reactive.

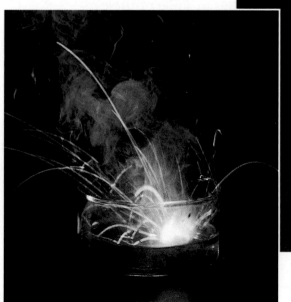

Potassium reacts strongly in water.

Sodium will burn in air if it is heated.

The Alkaline Earth Metals

The alkaline earth metals make up Group 2 of the periodic table. Like most metals, these metals are shiny, malleable, and ductile. They are also similar to alkali metals in that they combine so readily with other elements that they are not found as free elements in nature. Each atom of an alkaline earth metal has two electrons in its outer energy level. These electrons are given up when an alkaline earth metal combines with a nonmetal. As a result, the alkaline earth metal becomes a positively charged ion in a compound such as calcium fluoride, CaF_2.

Fireworks and Other Uses Magnesium metal is one of the metals used to produce the brilliant white color in fireworks like the ones in **Figure 5.** Compounds of strontium produce the bright red flashes. Magnesium's lightness and strength account for its use in cars, planes, and spacecraft. Magnesium also is used in compounds to make such things as household ladders and baseball and softball bats. Most life on Earth depends upon chlorophyll, a magnesium compound that enables plants to make food. Marble statues and some countertops are made of the calcium compound calcium carbonate.

The Alkaline Earth Metals and Your Body Calcium is seldom used as a free metal, but its compounds are needed for life. You may take a vitamin with calcium. Calcium phosphate in your bones helps make them strong.

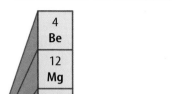

The Alkaline Earth Metals

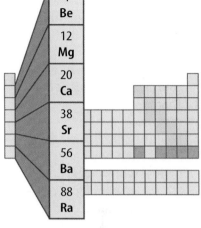

Figure 5 Alkaline earth metals make spectacular fireworks.

The barium compound $BaSO_4$ is used to diagnose some digestive disorders because it absorbs X-ray radiation well. First, the patient swallows a barium compound. Next, an X ray is taken while the barium compound is going through the digestive tract. A doctor can then see where the barium is in the body. In this way, doctors can diagnose internal abnormalities in the body.

Radium, the last element in Group 2, is radioactive and is found associated with uranium. It was once used to treat cancers. Today, other radioactive elements that are more readily available are replacing radium in cancer therapy.

Transition Elements

A titanium bike frame and a glowing tungsten lightbulb filament are examples of objects made from transition elements. **Transition elements** are those elements in Groups 3 through 12 in the periodic table. They are called transition elements because they are considered to be elements in transition between Groups 1 and 2 and Groups 13 through 18. Look at the periodic table inside the back cover of your book. Which elements do you think of as being typical metals? Transition elements are the most familiar because they often occur in nature as uncombined elements, unlike Group 1 and Group 2 metals which are less stable.

Transition elements often form colored compounds. The gems in **Figure 6** show brightly colored compounds containing chromium. Cadmium yellow and cobalt blue paints are made from compounds of transition elements. However, cadmium and cobalt paints are so toxic that their use is limited.

Iron, Cobalt, and Nickel The first elements in Groups 8, 9, and 10—iron, cobalt, and nickel—form a unique cluster of transition elements. These three sometimes are called the iron triad. All three elements are used in the process to create steel and other metal mixtures.

Iron—the main component of steel—is the most widely used of all metals. It is the second most abundant metallic element in Earth's crust after aluminum. Other metals are added to steel to give it various characteristics. Some steels contain cobalt or nickel. Nickel is added to some metals to give them strength. Also, nickel is used to give a shiny, protective coating to other metals.

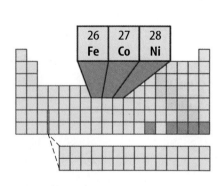

Figure 6 The colors of the ruby and emerald are due to the transition element chromium.

Figure 7 The coinage metals have many uses.

Because gold and silver are so expensive, copper is more common in coins.

Silver is used in compounds to make photographic materials.

Gold frequently is used in jewelry.

Copper, Silver, and Gold The main metals in the objects in **Figure 7** are copper, silver, and gold—the three elements in Group 11. Because they are so stable and malleable and can be found as free elements in nature, these metals were once used widely to make coins. For this reason, they are known as the coinage metals. Because they are so expensive, silver and gold rarely are used in coins anymore. The United States stopped using gold in the production of its coins in 1933 and silver in 1964. Most coins now are made of nickel and copper.

Copper often is used in electrical wiring because of its superior ability to conduct electricity and its relatively low cost. Can you imagine a world without photographs and movies? Silver iodide and silver bromide break down when exposed to light, producing an image on paper. Consequently, these compounds are used to make photographic film and paper. Silver and gold are used in jewelry because of their attractive color, relative softness, resistance to corrosion, and rarity.

 **Reading Check** *Why does gold's relative softness make it a good choice for jewelry?*

The Coinage Metals

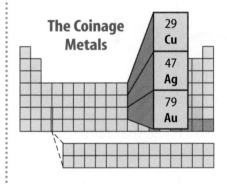

| 29 Cu |
| 47 Ag |
| 79 Au |

Zinc, Cadmium, and Mercury Zinc, cadmium, and mercury are found in Group 12 of the periodic table. Zinc combines with oxygen in the air to form a thin, protective coating of zinc oxide on its surface. Zinc and cadmium often are used to coat, or plate, other metals such as iron because of this protective quality. Cadmium is used also in rechargeable batteries.

Mercury is a silvery, liquid metal—the only metal that is a liquid at room temperature. It is used in thermometers, thermostats, switches, and batteries. Mercury is poisonous and can accumulate in the body. People have died of mercury poisoning after eating fish that lived in mercury-contaminated water.

Zinc, Cadmium, and Mercury

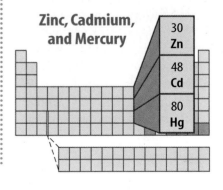

| 30 Zn |
| 48 Cd |
| 80 Hg |

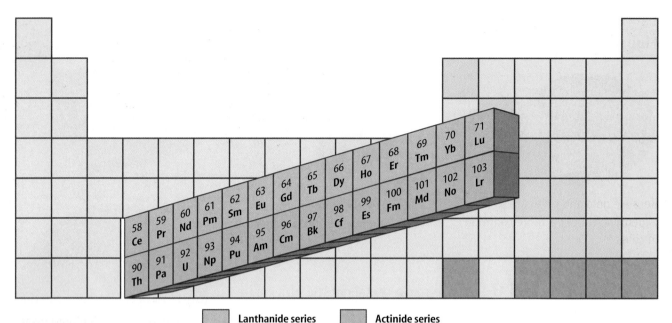

| Lanthanide series | Actinide series |

Figure 8 To save space, the periodic table usually isn't shown with the inner transition elements positioned where they should be. **Identify** *where the lanthanide and actinide series are found.*

Mining Engineer The processing of ores mined from Earth normally begins with metals in very low concentrations and generally not in the most desirable form. To refine the ores and concentrate the metal into the desired form, the mining engineer must develop a process that is economical and environmentally sound to be a viable process. Investigate some of the processes used to refine various metals and give examples of a process that is environmentally friendly but not cost effective and of a process that is just the opposite.

The Inner Transition Metals

The two rows of elements that seem to be disconnected from the rest on the periodic table are called the inner transition elements. They are called this because like the transition elements, they fit in the periodic table between Groups 3 and 4 in periods 6 and 7, as shown in **Figure 8.** To save room, they are listed below the table.

The Lanthanides The first row includes a series of elements with atomic numbers of 58 to 71. These elements are called the lanthanide series because they follow the element lanthanum.

Lanthanum, cerium, praseodymium, and samarium are used with carbon to make a compound that is used extensively by the motion picture industry. Europium, gadolinium, and terbium are used to produce the colors you see on your TV screen.

The Actinides The second row of inner transition metals includes elements with atomic numbers ranging from 90 to 103. These elements are called the actinide series because they follow the element actinium. All of the actinides are radioactive and unstable. Their unstable nature makes researching them difficult. Thorium and uranium are the actinides found in the Earth's crust in usable quantities. Thorium is used in making the glass for high-quality camera lenses because it bends light without much distortion. Uranium is best known for its use in nuclear reactors and in weapons applications, but one of its compounds has been used as photographic toner, as well.

Metals in the Crust

Earth's hardened outer layer, called the crust, contains many compounds and a few uncombined metals such as gold and copper. Metals must be mined and separated from their ores, as shown in **Figure 9.**

Most of the world's platinum is found in South Africa. Chromium is important because it is used to harden steel, to manufacture stainless steel, and to form other alloys. The United States imports most of its chromium from South Africa, the Philippines, and Turkey.

Ores: Minerals and Mixtures Metals in Earth's crust that combined with other elements are found as ores. Most ores consist of a metal compound, or mineral, within a mixture of clay or rock. After an ore is mined from Earth's crust, the rock is separated from the mineral. Then the mineral often is converted to another physical form. This step usually involves heat and is called roasting. Finally, the metal is refined into a pure form. Later it can be alloyed with other metals.

Removing the waste rock can be expensive. If the cost of removing the waste rock becomes greater than the value of the desired material, the mineral no longer is classified as an ore.

Figure 9 Copper is mined in the United States at the Bingham Canyon Copper Mine in Utah.

section 1 review

Summary

Properties of Metals

- Metals tend to form ionic and metallic bonds due to low numbers of electrons in their outer energy level.

Alkali and Alkaline Earth Metals

- Elements in Group 1 are called alkali metals.
- Elements in Group 2 are called alkaline earth metals.

Transition Elements and Inner Transition Metals

- Transition elements are elements in Groups 3–12 in the periodic table.
- Inner transition metals fit in the periodic table between Groups 3 and 4 in periods 6 and 7.

Self Check

1. **Describe** how to test palladium to see if it is a metal.
2. **Explain** how arrangement of the iron triad differs from arrangements of coinage metals.
3. **Identify** how metallic bonds differ from ionic and covalent bonds.
4. **Think Critically** If X stands for a metal, how can you tell from the following formulas—XCl and XCl$_2$—which compound contains an alkali metal and which contains an alkaline earth metal?

Applying Math

5. **Use Percentages** Pennies used to be made of copper and zinc, and weighed 3.11 g. Today, pennies are made of copper-plated zinc, and weighs 2.5 g. A new penny weighs what percent of an old penny?

Nonmetals

Properties of Nonmetals

Most of your body's mass is made of oxygen, carbon, hydrogen, and nitrogen, as shown in **Figure 10.** Calcium, a metal, and other elements make up the remaining four percent of your body's mass. Phosphorus, sulfur, and chlorine are among these other elements found in your body. These elements are classified as nonmetals. **Nonmetals** are elements that usually are gases or brittle solids at room temperature. Because solid nonmetals are brittle or powdery, they are not malleable or ductile. Most nonmetals do not conduct heat or electricity well, and generally they are not shiny.

In the periodic table, all nonmetals except hydrogen are found at the right of the stairstep line. On the table in the inside back cover of your book, the nonmetal element blocks are colored yellow. The noble gases, Group 18, make up the only group of elements that are all nonmetals. Group 17 elements, except for astatine, are also nonmetals. Other nonmetals, found in Groups 13 through 16, will be discussed later.

Figure 10 As a percentage of mass, humans are made up of mostly nonmetals.

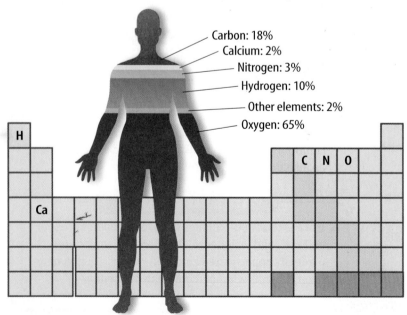

Elements in the Human Body

Carbon: 18%
Calcium: 2%
Nitrogen: 3%
Hydrogen: 10%
Other elements: 2%
Oxygen: 65%

Lead and sulfur bond ionically to form lead sulfide, PbS, also known as galena.

Carbon and oxygen can bond covalently to form carbon dioxide, CO_2.

Bonding in Nonmetals
The electrons in most nonmetals are strongly attracted to the nucleus of the atom. So, as a group, nonmetals are poor conductors of heat and electricity.

Most nonmetals can form ionic and covalent compounds. Examples of these two kinds of compounds are shown in **Figure 11.**

When nonmetals gain electrons from metals, the nonmetals become negative ions in ionic compounds. An example of such an ionic compound is potassium iodide, KI, which often is added to table salt. KI is formed from the nonmetal iodine and the metal potassium. When bonded with other nonmetals, atoms of nonmetals usually share electrons to form covalent compounds. An example is ammonia, NH_3, the strong, unpleasant-smelling compound you notice when you open a bottle of some household cleaners.

Hydrogen
If you could count all the atoms in the universe, you would find that about 90 percent of them are hydrogen. Most hydrogen on Earth is found in the compound water. The word *hydrogen* is derived from the Greek term for "water forming." When water is broken down into its elements, hydrogen becomes a gas made up of diatomic molecules. A **diatomic molecule** consists of two atoms of the same element in a covalent bond.

Hydrogen is highly reactive. A hydrogen atom has a single electron, which the atom shares when it combines with other nonmetals. For example, hydrogen burns in oxygen to form water, H_2O, in which hydrogen shares electrons with oxygen.

Hydrogen can gain an electron when it combines with alkali and alkaline earth metals. The compounds formed are hydrides, such as sodium hydride, NaH.

Reading Check *What is a diatomic molecule?*

Figure 11 Nonmetals form ionic bonds with metals and covalent bonds with other nonmetals.

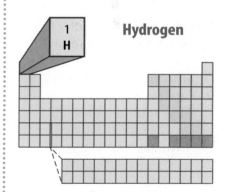

Hydrogen

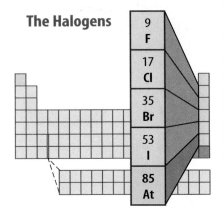

The Halogens

| 9
F |
| 17
Cl |
| 35
Br |
| 53
I |
| 85
At |

Identifying Chlorine Compounds in Your Water

Procedure

1. In three labeled **test tubes,** obtain 2 mL of **chlorine standard solution, distilled water,** and **drinking water.**
2. Carefully add five drops of **silver nitrate solution** to each and stir. **WARNING:** *Avoid contact with the silver nitrate solution. Silver nitrate is a corrosive liquid that can stain skin and clothes.*

Analysis

1. Which solution will definitely show a presence of chlorine? How did this result compare to the result with distilled water?
2. Which result most resembled your drinking water?

The Halogens

Halogen lights contain small amounts of bromine or iodine. These elements, as well as fluorine, chlorine, and astatine, are called halogens and are in Group 17. They are very reactive in their elemental form, and their compounds have many uses. As shown in **Figure 12,** fluorides are added to toothpastes and to city water systems to prevent tooth decay, and chlorine compounds are added to water to disinfect it.

Because an atom of a halogen has seven electrons in its outer energy level, only one electron is needed to complete this energy level. If a halogen gains an electron from a metal, an ionic compound, called a **salt,** is formed. An example of this is NaCl. In the gaseous state, the halogens form reactive diatomic covalent molecules and can be identified by their distinctive colors. Chlorine is greenish yellow, bromine is reddish orange, and iodine is violet.

Fluorine is the most chemically active of all elements. Hydrofluoric acid, a mixture of hydrogen fluoride and water, is used to etch glass and to frost the inner surfaces of lightbulbs and is also used in the fabrication of semiconductors.

Figure 12 The halogens have many uses.

Chlorine compounds are used in pools to disinfect the water.

Fluoride compounds are used in toothpaste to prevent tooth decay.

Figure 13 This ocean-salt recovery site uses evaporation to separate the halogen compounds from the water so the salts can be refined further.

Uses of Halogens The odor you sometimes smell near a swimming pool is chlorine. Chlorine compounds are used to disinfect water. Chlorine, the most abundant halogen, is obtained from seawater at ocean-salt recovery sites like the one in **Figure 13.** Household and industrial bleaches used to whiten flour, clothing, and paper also contain chlorine compounds.

Bromine, the only nonmetal that is a liquid at room temperature, also is extracted from compounds in seawater. Other bromine compounds are used as dyes in cosmetics.

Iodine, a shiny purple-gray solid at room temperature, is obtained from seawater. When heated, iodine changes directly to a purple vapor. The process of a solid changing directly to a vapor without forming a liquid is called **sublimation,** as shown in **Figure 14.** Iodine is essential in your diet for the production of the hormone thyroxin and to prevent goiter, an enlarging of the thyroid gland in the neck.

Chlorofluorocarbons Compounds called chlorofluorocarbons are used in refrigeration systems. If released, these compounds destroy ozone in the atmosphere. The ozone protects you from some of the harmful rays from the Sun. Find the advantages and disadvantages of these compounds. Write your answer in your Science Journal.

Reading Check *What is sublimation?*

Astatine is the last member of Group 17. It is radioactive and rare, but has many properties similar to those of the other halogens. There are no known uses due to its rarity.

Figure 14 Frozen carbon dioxide, or dry ice, is used to make inexpensive, visible gas for theatrical productions. The carbon dioxide is brought out as a solid, then it sublimes as shown here.

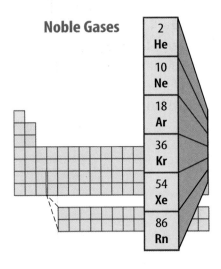

| 2 He |
| 10 Ne |
| 18 Ar |
| 36 Kr |
| 54 Xe |
| 86 Rn |

The Noble Gases

The noble gases exist as isolated atoms. They are stable because their outermost energy levels are full. No naturally occurring noble gas compounds are known, but several compounds of xenon and krypton with fluorine have been created in a laboratory.

The stability of noble gases is what makes them useful. In addition, the light weight of helium makes it useful in lighter-than-air blimps and balloons. Neon and argon are used in "neon lights" for advertising. Argon and krypton are used in electric lightbulbs to produce light in lasers, as seen in **Figure 15**.

Figure 15 Noble gases are used to produce spectacular laser light shows.

section 2 review

Summary

Properties of Nonmetals

● Nonmetals usually are gases or brittle solids that are not shiny and do not conduct heat or electricity.

Hydrogen

● Hydrogen makes up 90 percent of the atoms in the universe and is highly reactive.

Halogens

● Halogens are in Group 17 and are highly reactive in their elemental form.

The Noble Gases

● Noble gases exist only as isolated atoms because their outer energy levels are full.

Self Check

1. **Describe** two ways in which hydrogen combines with other elements.

2. **Rank** the following nonmetals from lowest number of electrons in the outer level to highest: Cl^-, H^+, He, H.

3. **Explain** how solid nonmetals are different from solid metals.

4. **Describe** how you can tell that a gas is a halogen.

5. **Think Critically** What is the process of a solid changing directly into a vapor? Which element undergoes this process at room temperature?

Applying Math

6. **Interpret data** by identifying the nonmetal with its oxidation number in these compounds: MgO, NaH, $AlBr_3$, and FeS.

 Science Online gpscience.com/self_check_quiz

What Type is it?

Suppose you want an element for a certain use. You might be able to use a metal but not a nonmetal. In this lab, you will test several metals and nonmetals and compare their properties.

Real-World Question

How can you use properties to distinguish metals from nonmetals?

Goals

- **Observe** physical properties.
- **Test** the malleability of the materials.
- **Identify** electrical conductivity in the given materials.

Materials

samples of C, Mg, Al, S, and Sn
dishes for the samples
paper towels

conductivity tester
spatula
small hammer

Safety Precautions

Procedure

1. **Prepare** a table in your Science Journal like the one shown above.

2. **Observe** and record the appearance of each element sample. Include its physical state, color, and whether it is shiny or dull.

3. **Remove** a small sample of one of the elements. Gently tap the sample with a hammer. The sample is malleable if it flattens when tapped and brittle if it shatters. Clean the hammer between testing using a paper towel.

Observing Properties

Element	Appearance	Malleable or Brittle	Electrical Conductivity	Shiny or Dull
Carbon				
Magnesium		Do not write in this book.		
Aluminum				
Sulfur				
Tin				

4. Repeat step 3 for each sample.

5. Test the conductivity of each element by touching the electrodes of the conductivity tester to a sample. If the bulb lights, the element conducts electricity.

Conclude and Apply

1. **Compare and Contrast** Locate each element you used on the periodic table. Compare your results with what you would expect from an element in that location.

2. **Explain** Locate palladium, Pd, on the periodic table. Use the results you obtained during the activity to predict some of the properties of palladium.

Communicating Your Data

Compare your results with those of other students. **For more help, refer to the Science Skill Handbook.**

Mixed Groups

What You'll Learn

- **Distinguish** among metals, nonmetals, and metalloids.
- **Describe** the nature of allotropes.
- **Recognize** the significance of differences in crystal structure in carbon.
- **Understand** the importance of synthetic elements.

Why It's Important

The elements in mixed groups affect your life every day, because they are in everything from the computer you use to the air you breathe.

Review Vocabulary

substance: element or compound that cannot be broken down into simpler components

New Vocabulary

- metalloid
- allotrope
- semiconductor
- transuranium element

The Boron Group

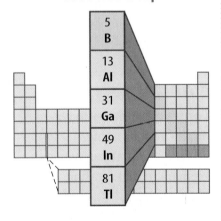

5	B
13	Al
31	Ga
49	In
81	Tl

Figure 16 Aluminum is used frequently in the construction of airplanes because it is light and strong.

Properties of Metalloids

Can an element be a metal and a nonmetal? In a sense, some elements called metalloids are. Metalloids share unusual characteristics. **Metalloids** can form ionic and covalent bonds with other elements and can have metallic and nonmetallic properties. Some metalloids can conduct electricity better than most nonmetals, but not as well as some metals, giving them the name semiconductor. With the exception of aluminum, the metalloids are the elements in the periodic table that are located along the stair-step line. The mixed groups—13, 14, 15, 16, and 17—contain metals, nonmetals, and metalloids.

The Boron Group

Boron, a metalloid, is the first element in Group 13. If you look around your home, you might find two compounds of boron. One of these is borax, which is used in some laundry products to soften water. The other is boric acid, a mild antiseptic. Boron also is used as a grinding material and as boranes, which are compounds used for jet and rocket fuel.

Aluminum, a metal in Group 13, is the most abundant metal in Earth's crust. It is used in soft-drink cans, foil wrap, cooking pans, and as siding. Aluminum is strong and light and is used in the construction of airplanes such as the one in **Figure 16.**

Figure 17 Elements in Group 14 have many uses.

Silicon is used to make the chips that allow this computer to run.

These tin cans are made of steel with a tin coating.

The Carbon Group

Each element in Group 14, the carbon family, has four electrons in its outer energy level, but this is where much of the similarity ends. Carbon is a nonmetal, silicon and germanium are metalloids, and tin and lead are metals. Carbon occurs as an element in coal and as a compound in oil, natural gas, and foods. Carbon in these materials can combine with oxygen to produce carbon dioxide, CO_2. In the presence of sunlight, plants utilize CO_2 to make food. Carbon compounds, many of which are essential to life, can be found in you and all around you. All organic compounds contain carbon, but not all carbon compounds are organic.

Silicon is second only to oxygen in abundance in Earth's crust. Most silicon is found in sand, SiO_2, and almost all rocks and soil. The crystal structure of silicon dioxide is similar to the structure of diamond. Silicon occurs as two allotropes. **Allotropes,** which are different forms of the same element, have different molecular structures. One allotrope of silicon is a hard, gray substance, and the other is a brown powder.

Reading Check *What are allotropes?*

Silicon is the main component in **semiconductors**—elements that conduct an electric current under certain conditions. Many of the electronics that you use every day, like the computer in **Figure 17,** need semiconductors to run. Germanium, the other metalloid in the carbon group, is used along with silicon in making semiconductors. Tin is used to coat other metals to prevent corrosion, like the tin cans in **Figure 17.** Tin also is combined with other metals to produce bronze and pewter. Lead was used widely in paint at one time, but because it is toxic, lead no longer is used.

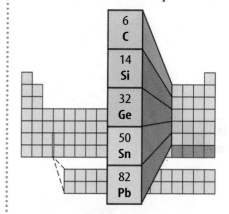

The Carbon Group

| 6 C |
| 14 Si |
| 32 Ge |
| 50 Sn |
| 82 Pb |

Topic:
Buckminsterfullerene
Visit gpscience.com for Web links
to information about this
compound.

Activity Research this compound and describe some of the
qualities that make it unique.

Allotropes of Carbon What do the diamond in a diamond ring and the graphite in your pencil have in common? They are both carbon. Diamond, graphite, and buckminsterfullerene, shown in **Figure 18,** are allotropes of an element.

A diamond is clear and extremely hard. In a diamond, each carbon atom is bonded to four other carbon atoms at the vertices, or corner points, of a tetrahedron. In turn, many tetrahedrons join together to form a giant molecule in which the atoms are held tightly in a strong crystalline structure. This structure accounts for the hardness of diamond.

Graphite is a black powder that consists of hexagonal layers of carbon atoms. In the hexagons, each carbon atom is bonded to three other carbon atoms. The fourth electron of each atom is bonded weakly to the layer next to it. This structure allows the layers to slide easily past one another, making graphite an excellent lubricant. In the mid-1980s, a new allotrope of carbon called buckminsterfullerene was discovered. This soccer-ball-shaped molecule, informally called a buckyball, was named after the architect-engineer R. Buckminster Fuller, who designed structures with similar shapes.

In 1991, scientists were able to use the buckyballs to synthesize extremely thin, graphitelike tubes. These tubes, called nanotubes, are about 1 billionth of a meter in diameter. That means you could stack tens of thousands of nanotubes just to get the thickness of one piece of paper. Nanotubes might be used someday to make computers that are smaller and faster and to make strong building materials.

Figure 18 Three allotropes of carbon are depicted here. **Identify** *the geometric shapes that make up each allotrope.*

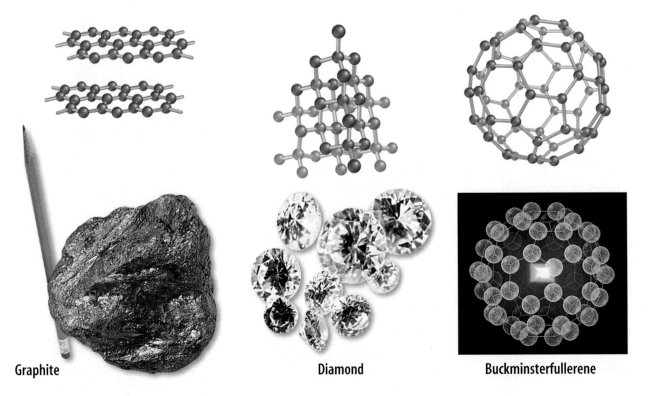

Graphite

Diamond

Buckminsterfullerene

The Nitrogen Group

The nitrogen family makes up Group 15. Each element has five electrons in its outer energy level. These elements tend to share electrons and to form covalent compounds with other elements. Nitrogen often is used to make nitrates (which are compounds that contain the nitrate ion, NO_3^-) and ammonia, NH_3, both of which are used in fertilizers. Nitrogen is the fourth most abundant element in your body. Each breath you take is about 80 percent gaseous nitrogen in the form of diatomic molecules, N_2. Yet you and other animals and plants can't use nitrogen in its diatomic form. The nitrogen must be combined into compounds, such as amino acids.

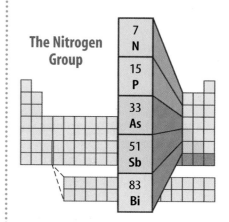

The Nitrogen Group

| 7 N |
| 15 P |
| 33 As |
| 51 Sb |
| 83 Bi |

Applying Math   Use Circle Graphs

CRUST ELEMENTS Oxygen, the predominant element in Earth's crust, makes up approximately 46.6 percent of the crust. If you were to show this information on a circle graph, how many degrees would be used to represent oxygen?

IDENTIFY known values and unknown values

Identify the known values:

oxygen is approximately 46.6%
circle contains 360°

Identify the unknown values:

how many degrees oxygen represents

SOLVE the problem

$$\frac{46.6}{100} = \frac{x}{360°}$$

$$x = \frac{46.6 \times 360°}{100} = 167.76° \text{ or } 168°$$

CHECK the answer

Does your answer seem reasonable? Check your answer by dividing your answer by 360, then multiplying by 100.

Practice Problems

1. The percentages of remaining elements in Earth's crust are: silicon, 27.7; aluminum, 8.1; iron, 5.0; calcium, 3.6; sodium, 2.8; potassium, 2.6; magnesium, 2.1; and other elements, 1.5. Illustrate in a circle graph.

2. The approximate percentages of the elements in the human body are: oxygen, 65; carbon, 18; hydrogen, 10; nitrogen, 3; calcium, 2; and other elements which account for 2. Illustrate these percentages in a circle graph.

For more practice problems, go to page 834, and visit gpscience.com/extra_problems.

Uses of the Nitrogen Group Phosphorus is a nonmetal that has three allotropes. Phosphorous compounds can be used for many things from water softeners to fertilizers, match heads, and even in fine china. Antimony is a metalloid, and bismuth is a metal. Both elements are used with other metals to lower their melting points. Because of this property, the metal in automatic fire-sprinkler heads contains bismuth.

✔ **Reading Check** *Why is bismuth used in fire-sprinkler heads?*

The Oxygen Group

Group 16 on the periodic table is the oxygen group. You can live for only a short time without oxygen, which makes up about 21 percent of air. Oxygen, a nonmetal, exists in the air as diatomic molecules, O_2. During electrical storms, some oxygen molecules, O_2, change into ozone molecules, O_3. Oxygen also has several uses in compound form, including the one shown at left in **Figure 19.**

Nearly all living things on Earth need O_2 for respiration. Living things also depend on a layer of O_3 around Earth for protection from some of the Sun's radiation.

The second element in the oxygen group is sulfur. Sulfur is a nonmetal that exists in several allotropic forms. It exists as different-shaped crystals and as a noncrystalline solid. Sulfur combines with metals to form sulfides of such distinctive colors that they are used as pigments in paints.

The nonmetal selenium and two metalloids—tellurium and polonium—are the other Group 16 elements. Selenium is the most common of these three. This element is one of several that you need in trace amounts in your diet. Many multivitamins contain this nonmetal as an ingredient. But selenium is toxic if too much of it gets into your system. Selenium also is used in photocopiers like the one in **Figure 19.**

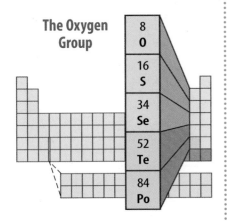

The Oxygen Group

| 8 O |
| 16 S |
| 34 Se |
| 52 Te |
| 84 Po |

Figure 19 Group 16 compounds have a variety of uses.

Solutions of hydrogen peroxide, H_2O_2, are used to clean minor wounds.

Selenium is used in xerography to make photocopies.

Figure 20 The americium used in smoke detectors is a synthetic element that has saved lives.

Smoke detector

Americium-241

Current

+

Battery

Smoke particle

−

Battery

Alarm

Smoke

Cover

Fire

Synthetic Elements

If you made something that always fell apart, you might think you were not successful. However, nuclear scientists are learning to do just that. By smashing existing elements with particles accelerated in a heavy ion accelerator, they have been successful in creating elements not typically found on Earth. Except for technetium 43 and promethium 61, each synthetic element has more than 92 protons.

Bombarding uranium with neutrons can make neptunium, element 93. Half of the synthesized atoms of neptunium disintegrate in about two days. This may not sound useful, but when neptunium atoms disintegrate, they form plutonium. This highly toxic element has been produced in control rods of nuclear reactors and is used in bombs. Plutonium also can be changed to americium, element 95. This element is used in home smoke detectors such as the one in **Figure 20.** In smoke detectors, a small amount of americium emits charged particles. An electric plate in the smoke detector attracts some of these charged particles. When a lot of smoke is in the air, it interferes with the electric current, which immediately sets off the alarm in the smoke detector.

Transuranium Elements Elements having more than 92 protons, the atomic number of uranium, are called **transuranium elements.** These elements do not belong exclusively to the metal, nonmetal, or metalloid group. These are the elements toward the bottom of the periodic table. Some are in the actinide series, and some are on the bottom row of the main periodic table. All of the transuranium elements are synthetic and unstable, and many of them disintegrate quickly.

The Transuranium Elements

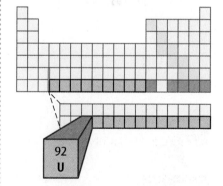

92
U

Figure 21

Some elements, such as gold, silver, tin, carbon, copper, and lead, have been known and used for thousands of years. Most others were discovered much more recently. Even at the time of the American Revolution in 1776, only 24 elements were known. The timeline below shows the dates of discovery of selected elements, ancient and modern.

Au - GOLD
Prized since the Stone Age

Ag - SILVER
Found in tombs dating to 4000 B.C.

A.D. 1774
Cl - CHLORINE
Pale green, toxic gas

A.D. 1700

1817
Cd - CADMIUM
Used to color yellow and red paint

1825
Al - ALUMINUM
Most abundant element in Earth's crust

1868
He - HELIUM
Lighter-than-air gas used to fill balloons

1898
Po - POLONIUM and Ra - RADIUM
Radioactive elements discovered by Marie and Pierre Curie

1898
Ne - NEON
Glows when electricity flows through it

A.D. 1800

1981–1996
Bh- BOHRIUM, Ds - DARMSTADTIUM
Elements isolated by a heavy ion accelerator such as the UNILAC, below

1952
Es - EINSTEINIUM
Radioactive gas named after Albert Einstein

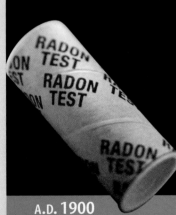

A.D. 1900

A.D. 2000

1900
Rn - RADON
Radioactive gas that may cause cancer

Why make elements? **Figure 21** shows when some of the elements were discovered throughout history. The processes used to discover these elements have varied widely. The most recently discovered elements are synthetic. By studying how the synthesized elements form and disintegrate, you can gain an understanding of the forces holding the nucleus together. When these atoms disintegrate, they are said to be radioactive.

Radioactive elements can be useful. For example, technetium's radioactivity makes it ideal for many medical applications. At this time, many of the synthetic elements last only small fractions of seconds after they are constructed and can be made only in small amounts. However, the value of applications that might be discovered easily could offset their costs.

Seeking Stability Element 114, discovered in 1999, appears to be much more stable than most synthetic elements of its size. It lasted for 30 s before it broke apart. This may not seem like long, but it lasts 100,000 times longer than an atom of element 112. Perhaps this special combination of 114 protons and 175 neutrons allows the nucleus to hold together despite the enormous repulsion between the protons.

In the 1960s, scientists theorized that stable synthetic elements exist. Finding one might help scientists understand how the forces inside the atom work. Perhaps someday you'll read about some of the everyday uses this discovery has brought.

Science Online

Topic: Synthetic Elements
Visit gpscience.com for Web links to information and an online update about synthetic elements.

Activity Find what some of the latest developments are in synthetic elements. Collect information on the one that interests you most and explain what you think is the most intriguing property of this element.

---

section 3 review

Summary

Properties of Metalloids

- Metalloids are elements that can form ionic and covalent bonds with other elements and can have metallic and nonmetallic properties.

Carbon Group

- The elements in Group 14 have four electrons in their outer energy levels.

Nitrogen Group

- The elements in Group 15 tend to share electrons and form covalent bonds.

Synthetic Elements

- Synthetic elements are elements that are not typically found on Earth.

- By synthesizing elements, scientists may understand how the forces inside the atom work.

Self Check

1. **Explain** why Groups 14 and 15 are better representatives of mixed groups than Groups 13 and Group 16.

2. **Describe** how allotropes of silicon differ in appearance.

3. **Explain** how an element is classified as a transuranium element.

4. **Describe** what type of structure a diamond has. How would you build a model of this?

5. **Think Critically** Graphite and a diamond are both made of the element carbon. Why is graphite a lubricant and diamond the hardest gem known?

Applying Math

6. **Calculate** Element 114 lasted 30s before falling apart. It lasted 100,000 times longer than element 112. How long did element 112 last?

LAB Design Your Own

Slippery Carbon

Goals
■ **Make a model** that will demonstrate the molecular structure of graphite.
■ **Compare and contrast** the strength of the different bonds in graphite.
■ **Infer** the relationship between bonding and physical properties.

Possible Materials
thin spaghetti
small gumdrops
thin polystyrene sheets
flat cardboard
scissors

Safety Precautions

Use care when working with scissors and uncooked spaghetti.

◯ *Real-World Question*

Often, a lubricant is needed when two metals touch each other. For example, a sticky lock sometimes works better with the addition of a small amount of graphite. What gives this allotrope of carbon the slippery property of a lubricant? Why do certain arrangements of atoms in a material cause the material to feel slippery?

◯ *Form a Hypothesis*

Based on your understanding of how carbon atoms bond, form a hypothesis about the relationship of graphite's molecular structure to its physical properties.

▶ Test Your Hypothesis

Make A Plan

1. As a group, agree upon a logical hypothesis statement.

2. As a group, sequence and list the steps you need to take to test your hypothesis. Be specific, describing exactly what you will do at each step to make a model of the types of bonding present in graphite.

3. Remember from **Figure 18** that graphite consists of rings of six carbons bonded in a flat hexagon. These rings are bonded to each other. In addition, the flat rings in one layer are weakly attached to other flat layers.

4. List possible materials you plan to use.

5. Read over the experiment to make sure that all steps are in logical order.

6. Will your model be constructed with materials that show weak and strong attractions?

Follow Your Plan

1. Make sure your teacher approves your plan before you start.

2. Have you selected materials to use in your model that demonstrate weak and strong attractions? Carry out the experiment as planned.

3. Once your model has been constructed, list any observations that you make and include a sketch in your Science Journal.

▶ Analyze Your Data

1. **Compare** your model with designs and results of other groups.

2. How does your model illustrate two types of attractions found in the graphite structure?

3. How does the bonding of graphite that you explored in the lab explain graphite's lubricating properties? Write your answer in your Science Journal.

▶ Conclude and Apply

1. **Describe** the results you obtained from your experiment. Did the results support your hypothesis?

2. **Describe** why graphite makes a good lubricant.

3. **Explain** what kinds of bonds you think a diamond has.

Communicating Your Data

Explain to a friend why graphite makes a good lubricant and how the two types of bonds make a difference.

The GAS that glows

A neon sign in the making. Workers carefully twist the light tubes into different shapes.

Neon has made the world a more colorful place

"Nothing in the world gave a glow such as we had seen." With these words, two British chemists recorded their discovery of neon in 1898. Neon is a noble gas that emits a spectacular red-orange glow when an electric current is passed through it. It also makes up a tiny portion of the air we breathe.

But neon's presence remained undetected until a technology called spectroscopy allowed the chemists to view that "blaze of crimson light" in their lab—a light that soon lit up the world in fantastical ways.

Signs of Change

Pink flamingos, cowboys on bucking broncos, deep-sea fish afloat in the air—neon signs make any building or billboard come alive in a kaleidoscope of colors. Barely a decade after neon was discovered, a chemist developed the first neon sign. The chemist took the air out of a glass tube and replaced it with neon gas. When the gas was jolted with electricity, it glowed like a fiery sunset. The chemist sold the light to a barber, who hung it over his storefront. By the 1920s, neon lights were used to advertise everything from cars to diners.

When a touch of mercury is added to neon, it glows a tropical blue. The other colors seen in "neon lights" actually come from other noble gases. Krypton, for instance, glows yellow. Xenon shines like a bluish-white star.

Other Uses for Neon

The vivid light emitted by neon can penetrate the densest fog, making it a natural choice for airplane beacons. Neon also is used to manufacture lasers and television tubes. Neon definitely helps light up our lives!

Identify As a group, brainstorm a new product or business, then design a neon sign to advertise your idea. See if other groups can correctly guess what your sign represents.

Science Online

For more information, visit gpscience.com/time

Reviewing Main Ideas

Section 1 Metals

1. A typical metal is a hard, shiny solid that, due to metallic bonding, is malleable, ductile, and a good conductor.

2. Groups 1 and 2 are the alkali and alkaline earth metals, which have some similar and some contrasting properties.

3. The iron triad, the coinage metals, and the elements in Group 12 are examples of transition elements.

4. The lanthanides and actinides have atomic numbers 58 through 71 and 90 through 103, respectively.

Section 2 Nonmetals

1. Nonmetals can be brittle and dull. They are also poor conductors of electricity.

2. As a typical nonmetal, hydrogen is a gas that forms compounds by sharing electrons with other nonmetals and by forming ionic bonds with metals.

3. All the halogens, Group 17, have seven outer electrons and form covalent and ionic compounds, but each halogen has some properties that are unlike each of the others in the group.

4. The noble gases, Group 18, are elements whose properties and uses are related to their chemical stability.

Section 3 Mixed Groups

1. Groups 13 through 16 include metals, non-metals, and metalloids.

2. Allotropes are forms of the same element having different molecular structures.

3. The properties of three forms of carbon—graphite, diamond, and buckminsterfullerene—depend upon the differences in their crystal structures.

4. All synthetic elements are short-lived. Except for technetium-43 and promethium-61, they have atomic numbers greater than 92 and are referred to as transuranium elements. These elements are found toward the bottom of the periodic table.

FOLDABLES Use the Foldable that you made at the beginning of this chapter to help you review elements and their properties.

Using Vocabulary

allotrope p. 585	radioactive element p. 572
diatomic molecule p. 579	salt p. 580
ductile p. 570	semiconductor p. 585
malleable p. 570	sublimation p. 581
metal p. 570	transition element p. 574
metallic bonding p. 571	transuranium element
metalloid p. 584	p. 589
nonmetal p. 578	

Complete each sentence with the correct vocabulary word(s).

1. The _____ are located to the left of the stair–step line on the periodic table.

2. Different structural forms of the same element are called _____.

3. Positively charged ions are surrounded by freely moving electrons in _____.

4. A(n) _____ is a molecule comprised of two atoms.

5. The _____ are in Groups 3 through 12 on the periodic table.

Checking Concepts

Choose the word or phrase that best answers the question.

6. When magnesium and fluorine react, what type of bond is formed?
   A) metallic   **C)** covalent
   B) ionic   **D)** diatomic

7. What type of bond is found in a piece of pure gold?
   A) metallic   **C)** covalent
   B) ionic   **D)** diatomic

8. Because electrons move freely in metals, which property describes metals?
   A) brittle   **C)** dull
   B) hard   **D)** conductors

9. Which set of elements makes up the most reactive group of all metals?
   A) iron triad
   B) coinage metals
   C) alkali metals
   D) alkaline earth metals

10. Which element is the most reactive of all nonmetals?
    A) fluorine   **C)** hydrogen
    B) uranium   **D)** oxygen

11. Which element is always found in nature combined with other elements?
    A) copper   **C)** magnesium
    B) gold   **D)** silver

12. Which elements are least reactive?
    A) metals   **C)** noble gases
    B) halogens   **D)** actinides

13. What element is formed when neptunium disintegrates?
    A) ytterbium   **C)** americium
    B) promethium   **D)** plutonium

Interpreting Graphics

14. Copy and complete the concept map using the following: *transition elements, hydrogen, metals, inner transition metals, noble gases.*

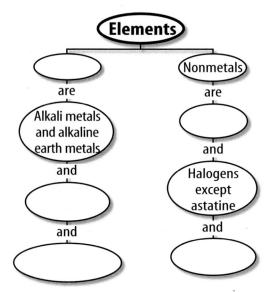

Science Online gpscience.com/vocabulary_puzzlemaker

Thinking Critically

15. Concept Map Copy and complete the concept map using the following: *Na, Fe, Actinides, Hg, Ba, Alkali, and Inner transition.*

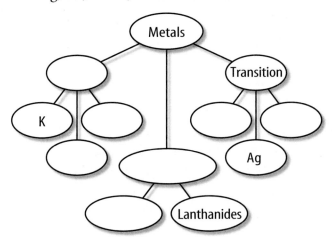

16. Make and Use Tables Use the periodic table to classify each of the following as a lanthanide or actinide: californium, europium, cerium, nobelium, terbium, and uranium.

17. Explain why mercury is rarely used in thermometers that take body temperatures.

18. Explain The density of hydrogen is lower than air and can be used to fill balloons. Why is helium used instead of hydrogen?

19. Explain Copper is a good choice for use in electrical wiring. What type of elements would not work well for this purpose? Why?

20. Explain why various silver compounds are used in photography.

21. Describe Like selenium, chromium is poisonous but is needed in trace amounts in your diet. How would you apply this information in order to use vitamin and mineral pills safely?

22. Compare and Contrast Explain why aluminum is a metal and carbon is not.

23. Explain What is metallic bonding? Explain how this affects conductivity.

24. Describe the geometric shapes of the carbon allotropes.

Applying Math

Use the following table to answer question 25.

Gas Analysis	
Gas	**Volume %**
CO	6.8
H_2	47.3
CH_4	33.9
CO_2	2.2
N_2	6
Other	3.8

25. Interpret Data When coke-oven gas is burned in an industrial process, several gases are produced in the reaction. If 385 grams of coke-oven gas are consumed in this reaction, how many grams of CH_4 (methane) are produced?

26. Use Percentage Chloroform has the chemical formula, $CHCl_3$, and a molecular weight of 119.39 g. What is the percentage of Cl (chlorine) present in this compound?

27. Use Numbers Calculate the molecular weight of gallium bromide ($GaBr_3$).

Part 1 Multiple Choice

Record your answers on the answer sheet provided by your teacher or on a sheet of paper.

1. Which of these elements is the main component of steel, and the most widely used of all metals?
   A. iron
   B. aluminum
   C. cadmium
   D. magnesium

2. What term describes the Group 1 elements lithium, sodium, and potassium?
   A. alkali metals
   B. radioactive elements
   C. lanthanides
   D. transition metals

Use the illustration below to answer questions 3 and 4.

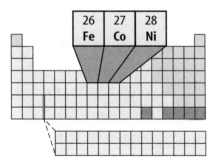

3. What name is given to these three elements which are used in processes that create steel and other metal mixtures?
   A. halogens
   B. the coin metals
   C. actinides
   D. the iron triad

4. To which major group do these elements belong?
   A. nonmetals
   B. noble gases
   C. transition elements
   D. alkali metals

Test-Taking Tip

Eliminate Choices If you don't know the answer to a multiple-choice question eliminate as many incorrect choices as possible.

Use the illustration below to answer questions 5 and 6.

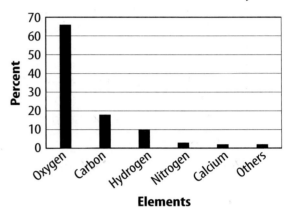

Elements in the Human Body

5. Which of these is a property of the elements that make up 98 percent of the human body?
   A. malleability
   B. poor electrical and heat conductivity
   C. shiny appearance
   D. ductility

6. In which of these phases does the element present in the highest percentage in the human body exist?
   A. gas
   B. solid
   C. liquid
   D. plasma

7. Which element is present in all organic compounds?
   A. silicon
   B. oxygen
   C. nitrogen
   D. carbon

8. Which of these is NOT a property of transuranium elements?
   A. occur naturally
   B. have greater than 92 protons
   C. are synthetic
   D. are unstable

Part 2 | Short Response/Grid In

Record your answers on the answer sheet provided by your teacher or on a sheet of paper.

9. Define the general properties of metals which make them useful and versatile materials.

10. Use the electron configuration of the elements sodium and potassium to explain why these elements do not occur in nature in elemental form.

Use the illustrations below to answer questions 11 and 12.

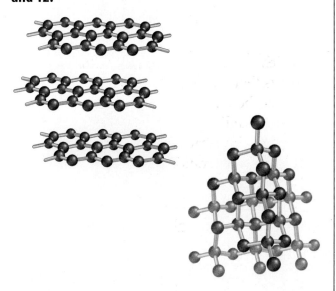

11. Define the term *allotrope*, and identify these allotropes of carbon.

12. Compare the structures of these carbon allotropes and relate the structures to the properties of these materials.

13. Describe some unique properties of hydrogen.

14. Identify and describe the uses of some of the halogens obtained from seawater.

15. Compare the two types of bonds which nonmetals can form.

Part 3 | Open Ended

Record your answers on the answer sheet provided by your teacher or on a sheet of paper.

16. Recent Federal Drug Administration statements advise limiting consumption of tuna and salmon. Which transitional element is the source of the problem? Explain why this element poses a potential risk.

17. Use the properties of metallic bonds to explain why metal hammered into sheets does not break, as well as why metals conduct electricity.

18. Based on its electron configuration and position in the periodic table, explain why fluorine is the most chemically active of all elements.

Use the illustration below to answer question 19.

19. Identify the gas which enables this blimp to remain suspended in the atmosphere. Why would it be dangerous to use hydrogen for this purpose?

20. Explain the importance of organisms that convert nitrogen from its diatomic form into other compounds.

Chemical Bonds

Elements Form Chemical Bonds

Just like these skydivers are linked together to make a stable formation, the atoms in elements can link together with chemical bonds to form a compound. You will read about how chemical bonds form and learn how to write chemical formulas and equations.

Science Journal Describe what makes some bonds more stable than others.

Start-Up Activities

Chemical Bonds and Mixing

You have probably noticed that some liquids like oil and vinegar salad dressings will not stay mixed after the bottle is shaken. However, rubbing alcohol and water will mix. The compounds that make up the liquids are different. This lab will demonstrate the influence the types of chemical bonds have on how the compounds mix.

1. Pour 20 mL of water into a 100-mL graduated cylinder.
2. Pour 20 mL of vegetable oil into the same cylinder. Vigorously swirl the two liquids together, and observe for several minutes.
3. Add two drops of food dye and observe.
4. After several minutes, slowly pour 30 mL of rubbing alcohol into the cylinder.
5. Add two more drops of food dye and observe.
6. **Think Critically** In your Science Journal, write a paragraph describing how the different liquids mixed. Would your final results be different if you added the liquids in a different order? Explain.

Chemical Formulas Every compound has a chemical formula that tells exactly which elements are present in that compound and exactly how many atoms of each element are present in that compound. Make the following Foldable to help identify the chemical formulas from this chapter.

STEP 1 Fold a vertical sheet of notebook paper from side to side.

STEP 2 Cut along every third line of only the top layer to form tabs.

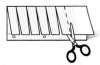

STEP 3 Label each tab.

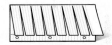

Read and Write Go through the chapter, find ten chemical formulas, and write them on the front of the tabs. As you read the chapter, write what compound each formula represents under the appropriate tab.

Science Online

Preview this chapter's content and activities at gpscience.com

601

Stability in Bonding

Figure 1 The difference between the elemental copper metal and the copper compound formed on the Statue of Liberty is striking.

Elemental copper

Surface coated with a copper compound

Combined Elements

Have you ever noticed the color of the Statue of Liberty? Why is it green? Did the sculptor purposely choose green? Why wasn't white, or tan, or even some other color like purple chosen? Was it painted that way? No, the Statue of Liberty was not painted. The Statue of Liberty is made of the metal copper, which is an element. Pennies, too, are made of copper. Wait a minute, you say. Copper isn't green—it's . . . well, copper colored.

You are right. Uncombined, elemental copper is a bright, shiny copper color. So again the question arises: Why is the Statue of Liberty green?

Compounds Some of the matter around you is in the form of uncombined elements such as copper, sulfur, and oxygen. But, like many other sets of elements, these three elements unite chemically to form a compound when the conditions are right. The green coating on the Statue of Liberty and some old pennies is a result of this chemical change. One compound in this coating, seen in contrast with elemental copper in **Figure 1,** is a new compound called copper sulfate. Copper sulfate isn't shiny and copper colored like elemental copper. Nor is it a pale-yellow solid like sulfur or a colorless, odorless gas like oxygen. It has its own unique properties.

New Properties One interesting observation you will make is that the compound formed when elements combine often has properties that aren't anything like those of the individual elements. Sodium chloride, for example, shown in **Figure 2**, is a compound made from the elements sodium and chlorine. Sodium is a shiny, soft, silvery metal that reacts violently with water. Chlorine is a poisonous greenish-yellow gas. Would you have guessed that these elements combine to make ordinary table salt?

Sodium   +   Chlorine   →   Sodium chloride

Figure 2 Sodium is a soft, silvery metal that combines with chlorine, a greenish-yellow gas represented here as only one atom, to form sodium chloride, which is a white crystalline solid.

Describe *how the properties of table salt are different from those of sodium and chlorine.*

Formulas

The chemical symbols Na and Cl represent the elements sodium and chlorine. When written as NaCl, the symbols make up a formula, or chemical shorthand, for the compound sodium chloride. A **chemical formula** tells what elements a compound contains and the exact number of the atoms of each element in a unit of that compound. The compound that you are probably most familiar with is H_2O, more commonly known as water. This formula contains the symbols H for the element hydrogen and O for the element oxygen. Notice the subscript number 2 written after the H for hydrogen. *Subscript* means "written below." A subscript written after a symbol tells how many atoms of that element are in a unit of the compound. If a symbol has no subscript, the unit contains only one atom of that element. A unit of H_2O contains two hydrogen atoms and one oxygen atom.

Look at the formulas for each compound listed in **Table 1.** What elements combine to form each compound? How many atoms of each element are required to form each of the compounds?

 Reading Check *Describe what a chemical formula tells you.*

Table 1 Some Familiar Compounds

Familiar Name	Chemical Name	Formula
Sand	Silicon dioxide	SiO_2
Milk of magnesia	Magnesium hydroxide	$Mg(OH)_2$
Cane sugar	Sucrose	$C_{12}H_{22}O_{11}$
Lime	Calcium oxide	CaO
Vinegar	Acetic acid	CH_3COOH
Laughing gas	Dinitrogen oxide	N_2O
Grain alcohol	Ethanol	C_2H_5OH
Battery acid	Sulfuric acid	H_2SO_4
Stomach acid	Hydrochloric acid	HCl

20. Writ
    form
    pou
    the

Use the ta
question

21. Elen
    um
    com
    fam
    ratio
    com
    CaC
    two
    fluo
    guid
    pou
    abo
    the

22. Draw
    usec
    diag

23. Use
    usin
    an
    forr

24. Com
    mol

Science

binary comp
chemical bor
chemical fori
covalent bon
hydrate p. 6
ion p. 608

Match each

1. a charge

2. a comp

3. a molec

4. a positi

5. a chemi
charged

6. a bond

7. crystall

8. a partic
atoms

9. shows a

10. tells wh
and the

Choose the
question.

11. Which
with ot
A) meta
B) nob

12. What i
A) cop]
B) cop]
C) cop]
D) cop]

Part 1 Multiple Choice

Record your answers on the answer sheet
provided by your teacher or on a sheet of paper.

1. Which element is NOT part of the compound NH_4NO_3?
   A. nitrogen   C. oxygen
   B. nickel   D. hydrogen

2. When an atom is chemically stable, how many electrons are in its outer energy level?
   A. 0   C. 4
   B. 7   D. 8

Use the figure below to answer questions 3 and 4.

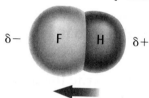

$\delta-$ F   H $\delta+$

3. What type of bond holds the atoms of this molecule together?
   A. covalent   C. triple
   B. ionic   D. double

4. Which statement about this molecule is TRUE?
   A. This is a nonpolar molecule.
   B. The electrons are shared equally in the bonds of this molecule.
   C. This molecule does not have oppositely charged ends.
   D. This is a polar molecule.

Test-Taking Tip

Comprehension Be sure you understand the question before you read the answer choices. Make special note of words like NOT or EXCEPT. Read and consider all the answer choices before you mark your answer sheet.

Question 1 Try to identify each of the elements in the compound before reading the answer choices.

---

5. What do the group 7A elements become when they react with group 1A elements?
   A. negative ions   C. positive ions
   B. neutral   D. polyatomic ions

Use the illustration below to answer questions 6–8.

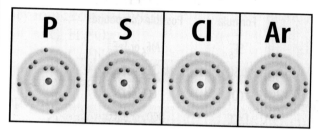

6. Which element is least likely to form an ionic bond with sodium?
   A. phosphorous   C. chlorine
   B. sulfur   D. argon

7. How many electrons are required to complete the outer energy level of a phosphorous atom?
   A. 1   C. 3
   B. 2   D. 4

8. How many electrons are in an argon atom?
   A. 8   C. 18
   B. 10   D. 26

9. What is the oxidation number of sodium in the compound Na_3PO_4?
   A. −1   C. −3
   B. +3   D. +1

10. What is the name of $KC_2H_3O_2$?
    A. potassium carbide
    B. potassium acetate
    C. potassium hydroxide
    D. potassium oxide

11. What is the chemical formula for lead (ll) oxide?
    A. PbO   C. PbO_2
    B. Pb_2O D. Pb_2O_2

Part 2 | Short Response/Grid In

Record your answers on the answer sheet provided by your teacher or on a sheet of paper.

Use the illustration below to answer questions 12 and 13.

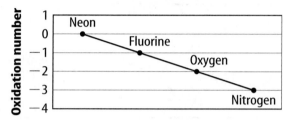

Oxidation Numbers of Some Period 2 Elements

12. Describe the trend in the oxidation numbers of these period 2 elements.

13. Compare the oxidation numbers of nitrogen and fluorine. Why do they differ?

14. Draw electron dot diagrams for carbon and hydrogen. Draw a dot diagram for methane, CH_4, one of many compounds formed by these two elements.

15. Give several examples of ionic compounds. What are two properties often shared by these substances?

16. A compound has the formula $MgSO_4 \cdot 7H_2O$. Identify and define this type of compound. Using the appropriate prefix, write its name.

17. What information is given in a chemical formula?

18. The bonding of atoms and molecules is the result of oppositely charged electrons and protons being held together by electric forces within the atom. Using this information, explain the bonding of NaCl.

Part 3 | Open Ended

Record your answers on a sheet of paper.

Use the illustration below to answer questions 19 and 20.

$$\cdot \ddot{N} \cdot \ + \ \cdot \ddot{N} \cdot \ \rightarrow \ :N::N:$$

19. Describe the bond holding the nitrogen atoms together in this molecule.

20. Nitrogen occurs naturally as a diatomic molecule because N_2 molecules are more stable than nitrogen atoms. H_2, O_2, F_2, Cl_2, Br_2, and I_2 are other diatomic molecules. Draw dot diagrams for three of these molecules.

21. Explain why elements in Group 4A, which have four electrons in the outer energy level, are unlikely to lose all of the electrons in the outer energy level.

22. What factors affect how strongly an atom is attracted to its electrons?

23. Create a chart which compares the properties of polar and nonpolar molecules. Your chart should include several examples of each type of molecule.

24. Scientists have created a compound which combines xenon and fluorine. Why is this compound so unusual and difficult to create? Why is fluorine a good choice for scientists attempting to form a compound with xenon?

25. What is the difference between nitrogen oxide and dinitrogen pentoxide? Why are prefixes used in this situation?

26. KCl is an example of ionic bonding. HCl is an example of covalent bonding. Describe the difference in the bonds in terms of electrons and outer energy levels.

Chemical Reactions

All-American Chemistry

Few things are as American as fireworks on the Fourth of July. The explosions of color and deafening booms get huge reactions from crowds across the country. These sights and sounds are the results of chemical reactions involving various substances with oxygen.

Science Journal Describe several cause-and-effect types of events that might happen in your refrigerator. Later, decide which of the events are chemical reactions.

Start-Up Activities

Rusting—A Chemical Reaction

Like exploding fireworks, rusting is a chemical reaction in which iron metal combines with oxygen. Other metals combine with oxygen, too—some more readily than others. In this lab, you will compare how iron and aluminum react with oxygen.

1. Place a clean iron or steel nail in a dish prepared by your teacher.

2. Place a clean aluminum nail in a second dish. These dishes contain agar gel and an indicator that detects a reaction with oxygen.

3. Observe both nails after one hour. Record any changes around the nails in your Science Journal.

4. Carefully examine both of the dishes the next day.

5. **Think Critically** Record any differences you noticed between the two dishes. Predict if a reaction occurred. How can you tell? What might have caused the differences you observed between the two nails. Explain.

 **Preview this chapter's content and activities at** gpscience.com

 **Study Organizer**

Chemical Reactions Make the following Foldable to help you classify chemical reactions.

STEP 1 Fold a sheet of paper in half lengthwise.

STEP 2 Mark four lines evenly spaced at even intervals down the page.

STEP 3 Cut only the top layer along the four marks to make five tabs.

STEP 4 Label the tabs as shown.

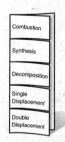

Combustion
Synthesis
Decomposition
Single Displacement
Double Displacement

Classify As you read, record examples of each type of reaction from the book, then review the chapter and list other examples mentioned in the text or from classroom discussions.

Chemical Changes

What You'll Learn

- **Identify** the reactants and products in a chemical reaction.
- **Determine** how a chemical reaction satisfies the law of conservation of mass.
- **Determine** how chemists express chemical changes using equations.

Why It's Important

Chemical reactions cook our food, warm our homes, and provide energy for our bodies.

Review Vocabulary

equation: a statement of the equality or equivalence of mathematical or logical quantities

New Vocabulary

- chemical reaction
- reactant
- product
- chemical equation
- coefficient

Describing Chemical Reactions

Dark mysterious mixtures react, gas bubbles up and expands, and powerful aromas waft through the air. Where are you? Are you in a chemical laboratory carrying out a crucial experiment? No. You are in the kitchen baking a chocolate cake. Nowhere in the house do so many chemical reactions take place as in the kitchen.

Actually, chemical reactions are taking place all around you and even within you. A **chemical reaction** is a change in which one or more substances are converted into new substances. The substances that react are called **reactants.** The new substances produced are called **products.** This relationship can be written as follows:

$$\text{reactants} \xrightarrow{\text{produce}} \text{products}$$

Before burning

After burning

Figure 1 The mass of the candles and oxygen before burning is exactly equal to the mass of the remaining candle and gaseous products.

Conservation of Mass

By the 1770s, chemistry was changing from the art of alchemy to a true science. Instead of being satisfied with a superficial explanation of unknown events, scientists began to study chemical reactions more thoroughly. Through such study, the French chemist Antoine Lavoisier established that the total mass of the products always equals the total mass of the reactants. This principle is demonstrated in **Figure 1.**

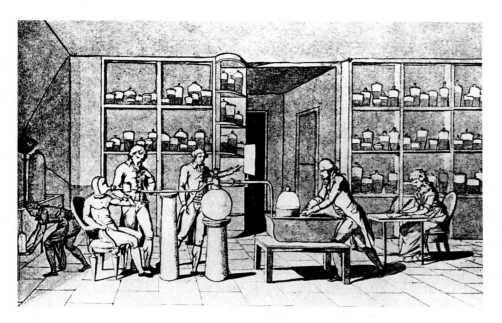

Figure 2 Antoine Lavoisier's wife, Marie-Anne, drew this view of Lavoisier in his laboratory performing studies on oxygen. She depicted herself at the right taking notes.

Lavoisier's Contribution One of the questions that motivated Lavoisier was the mystery of exactly what happened when substances changed form. He began to answer this question by experimenting with mercury. In one experiment, Lavoisier placed a carefully measured mass of solid mercury(II) oxide, which he knew as mercury calx, into a sealed container. When he heated this container, he noted a dramatic change. The red powder had been transformed into a silvery liquid that he recognized as mercury metal, and a gas was produced. When he determined the mass of the liquid mercury and gas, their combined masses were exactly the same as the mass of the red powder he had started with.

$$\text{mercury(II) oxide} = \text{oxygen} \text{ plus } \text{mercury}$$
$$10.0 \text{ g} = 0.7 \text{ g} + 9.3 \text{ g}$$

Lavoisier also established that the gas produced by heating mercury(II) oxide, which we call oxygen, was a component of air. He did this by heating mercury metal with air and saw that a portion of the air combined to give red mercury(II) oxide. He studied the effect of this gas on living animals, including himself. Hundreds of experiments carried out in his laboratory, as shown in **Figure 2,** confirmed that in a chemical reaction, matter is not created or destroyed, but is conserved. This principle became known as the law of conservation of mass. This means that the total starting mass of all reactants equals the total final mass of all products.

Topic: Antoine Lavoisier
Visit gpscience.com for Web links to information about Antoine Lavoisier and his contributions to chemistry.

Activity In your Science Journal, write a brief biography of Antoine Lavoisier that includes some of his non-scientific activities and political interests, as well as his scientific contributions.

 **Reading Check** *What does the law of conservation of mass state?*

The Father of Modern Chemistry When Lavoisier demonstrated the law of conservation of mass, he set the field of chemistry on its modern path. In fact, Lavoisier is known today as the father of modern chemistry for his more accurate explanation of the conservation of mass and for describing a common type of chemical reaction called combustion, which you will learn about later in this chapter. Lavoisier also pioneered early experimentation on the biological phenomena of respiration and metabolism that contributed early milestones in the study of biochemistry, medicine, and even sports science.

Nomenclature Lavoisier's work led him to the conclusion that language and terminology would be critical to communicate novel scientific ideas. In his book *Elements of Chemistry* (1790), Lavoisier wrote, "...we cannot improve a science without improving the language or nomenclature which belongs to it..." With that recognition, Lavoisier began to develop the system of naming substances based on their composition that we still use today. In 1787, Lavoisier and several colleagues published *Méthode de Nomenclature Chimique* as one of the first sets of nomenclature guidelines. Since then, the guidelines have continued to evolve with scientific discovery, and in 1919 the International Union of Pure and Applied Chemistry (IUPAC) was formed. The primary mission of the IUPAC is to coordinate guidelines for naming chemical compounds systematically. Before a new element gets a permanent name, it has a IUPAC name. Element 110, which is now called darmstadium, was previously named ununnilium. **Figure 3** illustrates some of the early key events in nomenclature development.

Reading Check *How did Lavoisier's contributions earn him the name Father of Modern Chemistry?*

Figure 3 This time line of nomenclature development and publications does not end in 1957. In fact, today there are nomenclature organizations for almost every branch of scientific study, and the rules and guidelines for naming substances continue to evolve.

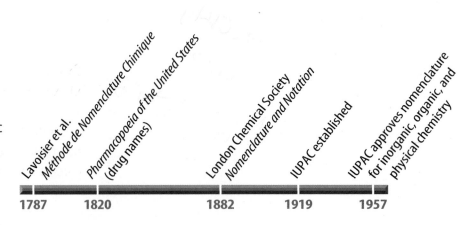

Writing Equations

If you wanted to describe the chemical reaction shown in **Figure 4,** you might write something like this:

Nickel(II) chloride, dissolved in water, plus sodium hydroxide, dissolved in water, produces solid nickel(II) hydroxide plus sodium chloride, dissolved in water.

This series of words is rather cumbersome, but all of the information is important. The same is true of descriptions of most chemical reactions. Many words are needed to state all the important information. As a result, scientists have developed a shorthand method to describe chemical reactions. A **chemical equation** is a way to describe a chemical reaction using chemical formulas and other symbols. Some of the symbols used in chemical equations are listed in **Table 1.**

The chemical equation for the reaction described above in words and shown in **Figure 4** looks like this:

$$NiCl_2(aq) + 2NaOH(aq) \rightarrow Ni(OH)_2(s) + 2NaCl(aq)$$

It is much easier to tell what is happening by writing the information in this form. Later, you will learn how chemical equations make it easier to calculate the quantities of reactants that are needed and the quantities of products that are formed.

Table 1 Symbols Used in Chemical Equations

Symbol	Meaning
$\rightarrow$	produces or forms
+	plus
(s)	solid
(l)	liquid
(g)	gas
(aq)	aqueous, a substance is dissolved in water
heat $\rightarrow$	the reactants are heated
light $\rightarrow$	the reactants are exposed to light
elec. $\rightarrow$	an electric current is applied to the reactants

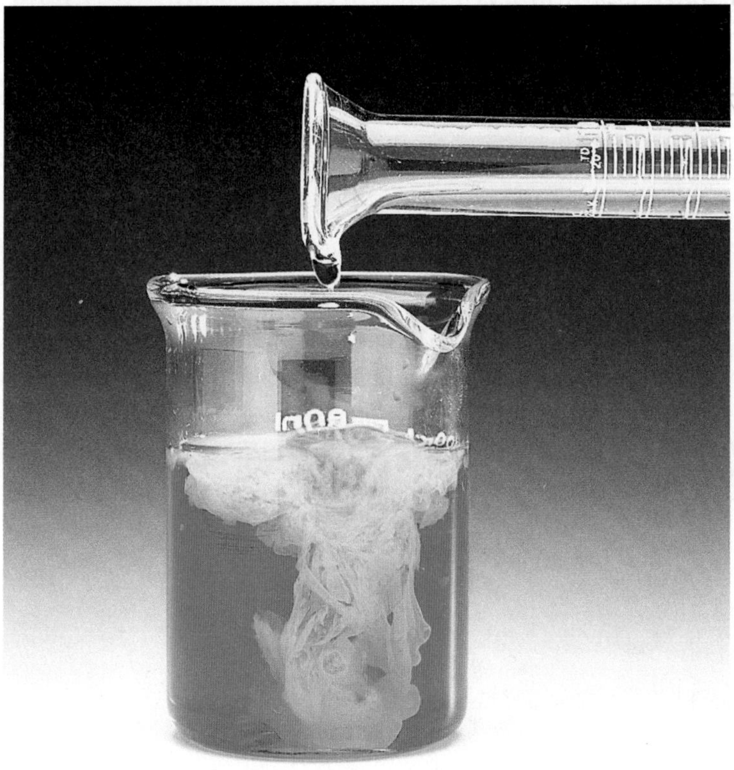

Figure 4 A white precipitate of nickel(II) hydroxide forms when sodium hydroxide is added to a green solution of nickel(II) chloride. Sodium chloride, the other product formed, is in solution.

Procedure

1. Obtain **15 index cards** and mark each as follows: five with *Guard,* five with *Forward,* and five with *Center.*

2. Group the cards to form as many complete basketball teams as possible. Each team needs two guards, two forwards, and one center.

Analysis

1. Write the formula for a team. Write the formation of a team as an equation. Use coefficients in front of each type of player needed for a team.

2. How is this equation like a chemical equation? Why can't you use the remaining cards?

3. How do the remaining cards illustrate the law of conservation of matter in this example?

Try at Home

Unit Managers

What do the numbers to the left of the formulas for reactants and products mean? Remember that according to the law of conservation of mass, matter is neither made nor lost during chemical reactions. Atoms are rearranged but never lost or destroyed. These numbers, called **coefficients,** represent the number of units of each substance taking part in a reaction. Coefficients can be thought of as unit managers.

✔ **Reading Check** *What is the function of coefficients in a chemical equation?*

Imagine that you are responsible for making sandwiches for a picnic. You have been told to make a certain number of three kinds of sandwiches, and that no substitutions can be made. You would have to figure out exactly how much food to buy so that you had enough without any food left over. You might need two loaves of bread, four packages of turkey, four packages of cheese, two heads of lettuce, and ten tomatoes. With these supplies you could make exactly the right number of each kind of sandwich.

In a way, your sandwich-making effort is like a chemical reaction. The reactants are your bread, turkey, cheese, lettuce, and tomatoes. The number of units of each ingredient are like the coefficients of the reactants in an equation. The sandwiches are like the products, and the numbers of each kind of sandwich are like coefficients, also.

Knowing the number of units of reactants enables chemists to add the correct amounts of reactants to a reaction. Also, these units, or coefficients, tell them exactly how much product will form. An example of this is the reaction of one unit of $NiCl_2$ with two units of NaOH to produce one unit of $Ni(OH)_2$ and two units of NaCl. You can see these units in **Figure 5.**

Figure 5 Each coefficient in the equation represents the number of units of each type in this reaction.

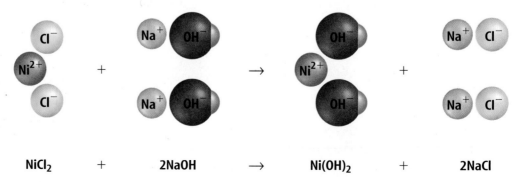

$$NiCl_2 \quad + \quad 2NaOH \quad \rightarrow \quad Ni(OH)_2 \quad + \quad 2NaCl$$

Metals and the Atmosphere

When iron is exposed to air and moisture, it corrodes or rusts, forming hydrated iron(III) oxide. Rust can seriously damage iron structures because it crumbles and exposes more iron to the air. This leads to more breakdown of the iron and eventually can destroy the structure. However, not all reactions of metals with the atmosphere are damaging like rust. Some are helpful.

Aluminum also reacts with oxygen in the air to form aluminum oxide. Unlike rust, aluminum oxide adheres to the aluminum surface, forming an extremely thin layer that protects the aluminum from further attack. You can see this thin layer of aluminum oxide on aluminum outdoor furniture. It makes the once shiny aluminum look dull.

Copper is another metal that corrodes when it is exposed to air, forming a blue-green coating called a patina. You can see this type of corrosion on many public monuments and also on the Statue of Liberty, shown in **Figure 6.**

Figure 6 The blue-green patina that coats the Statue of Liberty contains copper(II) sulfate, among other copper corrosion products.

---

section 1 review

Summary

Describing Chemical Reactions

- A chemical reaction is a process that involves one or more reactants changing into one or more products.

Conservation of Mass

- A basic principle of chemistry is that matter, during a chemical change, can neither be created nor destroyed.

- Antoine Lavoisier is often considered to be "the father of modern chemistry" for his work in defining the law of conservation of mass.

Writing Equations

- Chemical equations describe the change of reactants to products and obey the law of conservation of mass.

Unit Managers

- Coefficients represent how many units of each substance are involved in a chemical reaction.

Self Check

1. **Identify** the reactants and the products in the following chemical equation.

   $Cd(NO_3)_2(aq) + H_2S(g) \longrightarrow CdS(s) + 2HNO_3(aq)$

2. **Identify** the state of matter of each substance in the following reaction.

   $Zn(s) + 2HCl(aq) \longrightarrow H_2(g) + ZnCl_2(aq)$

3. **Explain** why the reaction of oxygen with iron is a problem, but the reaction of oxygen with aluminum is not.

4. **Explain** the importance of the law of conservation of mass.

5. **Think Critically** Why do you think the copper patina was kept when the Statue of Liberty was restored?

Applying Math

6. **Solve One-Step Equations** When making soap, if 890 g of a specific fat react completely with 120 g of sodium hydroxide, the products formed are soap and 92 g of glycerin. Calculate the mass of soap formed to satisfy the law of conservation of mass.

Chemical Equations

What You'll Learn
- **Identify** what is meant by a balanced chemical equation.
- **Determine** how to write balanced chemical equations.

Why It's Important
Chemical equations are the language used to describe chemical change, which allow scientists to develop products for our world.

⊙ **Review Vocabulary**
subscript: in a chemical formula, a number below and to the right of a symbol indicating number of atoms

New Vocabulary
● balanced chemical equation

Balanced Equations

Lavoisier's mercury(II) oxide reaction, shown in **Figure 7,** can be written as:

$$HgO(s) \xrightarrow{\text{heat}} Hg(l) + O_2(g)$$

Notice that the number of mercury atoms is the same on both sides of the equation but that the number of oxygen atoms is not the same. One oxygen atom appears on the reactant side of the equation and two appear on the product side.

Atoms	HgO	→	Hg	+	O$_2$
Hg	1		1		
O	1				2

But according to the law of conservation of mass, one oxygen atom cannot just become two. Nor can you simply add the subscript 2 and write HgO_2 instead of HgO. The formulas HgO_2 and HgO do not represent the same compound. In fact, HgO_2 does not exist. The formulas in a chemical equation must accurately represent the compounds that react.

Fixing this equation requires a process called balancing. Balancing an equation doesn't change what happens in a reaction—it simply changes the way the reaction is represented. The balancing process involves changing coefficients in a reaction to achieve a **balanced chemical equation,** which has the same number of atoms of each element on both sides of the equation.

Figure 7 Mercury metal forms when mercury oxide is heated. Because mercury is poisonous, this reaction is never performed in a classroom laboratory.

Choosing Coefficients Finding out which coefficients to use to balance an equation is often a trial-and-error process. In the equation for Lavoisier's experiment, the number of mercury atoms is balanced, but one oxygen atom is on the left and two are on the right. If you put a coefficient of 2 before the HgO on the left, the oxygen atoms will be balanced, but the mercury atoms become unbalanced. To balance the equation, also put a 2 in front of mercury on the right. The equation is now balanced.

Atoms	$2HgO$	$\rightarrow$	$2Hg$	$+$	O_2
Hg	2		2		
O	2				2

Try Your Balancing Act Magnesium burns with such a brilliant white light that it is often used in emergency flares as shown in **Figure 8.** Burning leaves a white powder called magnesium oxide. To write a balanced chemical equation for this and most other reactions, follow these four steps.

Step 1 Write a chemical equation for the reaction using formulas and symbols. Recall that oxygen is a diatomic molecule.

$$Mg(s) + O_2(g) \rightarrow MgO(s)$$

Step 2 Count the atoms in reactants and products.

Atoms	Mg	$+$	O_2	$\rightarrow$	MgO
Mg	1				1
O			2		1

The magnesium atoms are balanced, but the oxygen atoms are not. Therefore, this equation isn't balanced.

Step 3 Choose coefficients that balance the equation. Remember, never change subscripts of a correct formula to balance an equation. Try putting a coefficient of 2 before MgO.

$$Mg(s) + O_2(g) \rightarrow 2MgO(s)$$

Step 4 Recheck the numbers of each atom on each side of the equation and adjust coefficients again if necessary. Now two Mg atoms are on the right side and only one is on the left side. So a coefficient of 2 is needed for Mg to balance the equation.

$$2Mg(s) + O_2(g) \rightarrow 2MgO(s)$$

 **Reading Check** *How can you balance a chemical equation using coefficients?*

Science online

Topic: Balancing Chemical Equations
Visit gpscience.com for Web links to information about balancing chemical equations.

Activity Using the Web links, locate a website that offers practice problems for balancing chemical equations. Copy several of the unbalanced equations in your Science Journal and try to balance them. Check your work against the answers on the web site when you are done.

Figure 8 Magnesium combines with oxygen, giving an intense white light.

Figure 9 When lithium metal is added to water, it reacts, producing a solution of lithium hydroxide and bubbles of hydrogen gas.

Polish Your Skill When lithium metal is treated with water, hydrogen gas and lithium hydroxide are produced, as shown in **Figure 9**.

Step 1 Write the chemical equation.

$$Li(s) + H_2O \rightarrow LiOH(aq) + H_2(g)$$

Step 2 Check for balance by counting the atoms.

Atoms	Li	+	H_2O	→	LiOH	+	H_2
Li	1				1		
H			2		1		2
O			1		1		

This equation is not balanced. There are three hydrogen atoms on the right and only two on the left. Complete steps 3 and 4 to balance the equation. After each step, count the atoms of each element. When equal numbers of atoms of each element are on both sides, the equation is balanced. The balanced chemical equation looks like this:

$$2Li + 2H_2O \rightarrow 2LiOH + H_2$$

This accurate statement tells chemists how much lithium metal to use to produce a certain amount of hydrogen gas.

section 2 review

Summary

Balanced Equations

- A chemical equation is a way to indicate reactants and products and relative amounts of each.
- A balanced chemical equation tells the exact number of atoms involved in the reaction.
- Balanced chemical equations must satisfy the law of conservation of matter; no atoms of reactant or product can be lost from one side to the other.
- Coefficients are used to achieve balance in a chemical equation.
- Chemical equations cannot be balanced by adjusting the subscript numerals in compound names because doing so would change the compounds.

Self Check

1. **Describe** two reasons for balancing chemical equations.
2. **Balance** this chemical equation: $Fe(s) + O_2(g) \rightarrow FeO(s)$.
3. **Explain** why oxygen gas must always be written as O_2 in a chemical equation.
4. **Infer** What coefficient is assumed if no coefficient is written before a formula in a chemical equation?
5. **Think Critically** Explain why the sum of the coefficients on the reactant side of a balanced equation does not have to equal the sum of the coefficients on the product side of the equation.

Applying Math

6. **Use Numbers** Balance the equation for the reaction $Fe(s) + Cl_2(g) \rightarrow FeCl_3(s)$.

Science Online gpscience.com/self_check_quiz

Classifying Chemical Reactions

Reading Guide

What You'll Learn
- **Identify** the five general types of chemical reactions.
- **Define** the terms *oxidation* and *reduction*.
- **Identify** redox reactions.
- **Predict** which metals will replace other metals in compounds.

Why It's Important
Classifying chemical reactions helps to understand what is happening and predict the outcome of reactions.

Review Vocabulary
states of matter: the physical forms in which all matter naturally exists, most commonly solid, liquid, and gas

New Vocabulary
- combustion reaction
- synthesis reaction
- decomposition reaction
- single-displacement reaction
- double-displacement reaction
- precipitate
- oxidation
- reduction

Types of Reactions

You might have noticed that there are all sorts of chemical reactions. In fact, there are literally millions of chemical reactions that occur every day, and scientists have described many of them and continue to describe more. With all these reactions, it would be impossible to use the information without first having some type of organization. With this in mind, chemists have defined five main categories of chemical reactions: combustion, synthesis, decomposition, single displacement, and double displacement.

Combustion Reactions If you have ever observed something burning, you have observed a combustion reaction. As mentioned previously, Lavoisier was one of the first scientists to accurately describe combustion. He deduced that the process of burning (combustion) involves the combination of a substance with oxygen. Our definition states that a **combustion reaction** occurs when a substance reacts with oxygen to produce energy in the form of heat and light. Combustion reactions also produce one or more products that contain the elements in the reactants. For example, the reaction between carbon and oxygen produces carbon dioxide. Many combustion reactions also will fit into other categories of reactions. For example, the reaction between carbon and oxygen also is a synthesis reaction.

Figure 10 Rust has accumulated on the *Titanic* since it sank in 1912.

Figure 11 Water decomposes into hydrogen and oxygen when an electric current is passed through it. A small amount of sulfuric acid is added to increase conductivity. Notice the proportions of the gases collected. **Describe** *how this is related to the coefficients of the products in the equation.*

Synthesis Reactions
One of the easiest reaction types to recognize is a synthesis reaction. In a **synthesis reaction,** two or more substances combine to form another substance. The generalized formula for this reaction type is as follows: A + B → AB.

The reaction in which hydrogen burns in oxygen to form water is an example of a synthesis reaction.

$$2H_2(g) + O_2(g) \rightarrow 2H_2O(g)$$

This reaction is used to power some types of rockets. Another synthesis reaction is the combination of oxygen with iron in the presence of water to form hydrated iron(II) oxide or rust. This reaction is shown in **Figure 10.**

Decomposition Reactions
A decomposition reaction is just the reverse of a synthesis. Instead of two substances coming together to form a third, a **decomposition reaction** occurs when one substance breaks down, or decomposes, into two or more substances. The general formula for this type of reaction can be expressed as follows: AB → A + B.

Most decomposition reactions require the use of heat, light, or electricity. For example, an electric current passed through water produces hydrogen and oxygen as shown in **Figure 11.**

$$2H_2O(l) \xrightarrow{\text{elec.}} 2H_2(g) + O_2(g)$$

Single Displacement
When one element replaces another element in a compound, it is called a **single-displacement reaction.** Single-displacement reactions are described by the general equation A + BC → AC + B. Here you can see that atom A displaces atom B to produce a new molecule AC. A single displacment reaction is illustrated in **Figure 12,** where a copper wire is put into a solution of silver nitrate. Because copper is a more active metal than silver, it replaces the silver, forming a blue copper(II) nitrate solution. The silver, which is not soluble, forms on the wire.

$$Cu(s) + 2AgNO_3(aq) \rightarrow Cu(NO_3)_2 (aq) + 2Ag(s)$$

✔ Reading Check *Describe a single-displacement reaction.*

Figure 12 Copper in a wire replaces silver in silver nitrate, forming a blue-tinted solution of copper(II) nitrate.

Sometimes single-displacement reactions can cause problems. For example, if iron-containing vegetables such as spinach are cooked in aluminum pans, aluminum can displace iron from the vegetable. This causes a black deposit of iron to form on the sides of the pan. For this reason, it is better to use stainless steel or enamel cookware when cooking spinach.

The Activity Series We can predict which metal will replace another using the diagram shown in **Figure 13,** which lists metals according to how reactive they are. A metal will replace any less active metal. Notice that copper, silver, and gold are the least active metals on the list. That is why these elements often occur as deposits of the relatively pure element. For example, gold is sometimes found as veins in quartz rock, and copper is found in pure lumps known as native copper. Other metals occur as compounds.

Lithium
Potassium
Calcium
Sodium
Aluminum
Zinc
Iron
Tin
Lead
(Hydrogen)
Copper
Silver
Gold

MOST ACTIVE

LEAST ACTIVE

Figure 13 This figure shows the activity series of metals. A metal will replace any other metal that is less active.

Double Displacement In a **double-displacement reaction,** the positive ion of one compound replaces the positive ion of the other to form two new compounds. A double-displacement reaction takes place if a precipitate, water, or a gas forms when two ionic compounds in solution are combined. A **precipitate** is an insoluble compound that comes out of solution during this type of reaction. The generalized formula for this type of reaction is as follows: $AB + CD \rightarrow AD + CB$.

✔ **Reading Check** *What type of reaction produces a precipitate?*

The reaction of barium nitrate with potassium sulfate is an example of this type of reaction. A precipitate—barium sulfate—forms, as shown in **Figure 14.** The chemical equation is as follows:

$$Ba(NO_3)_2(aq) + K_2SO_4(aq) \rightarrow BaSO_4(s) + 2KNO_3(aq)$$

These are a few examples of chemical reactions classified into types. Many more reactions of each type occur around you.

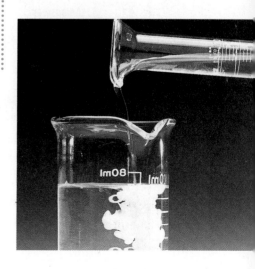

Figure 14 Solid barium sulfate is formed from the reaction of two solutions.
Observe *Has a chemical change occurred in this photo? How can you tell?*

Applying Math — Use Coefficients

BARIUM SULFATE REACTION A sample of barium sulfate is placed on a piece of paper, which is then ignited. Barium sulfate reacts with the carbon from the burned paper producing barium sulfide and carbon monoxide. Write a balanced chemical equation for this reaction.

IDENTIFY known values

We know the substances that are involved in the reaction. From this, we can write a chemical equation showing reactants and products.

$$BaSO_4(s) + C(s) \rightarrow BaS(s) + CO(g)$$

SOLVE the problem

The chemical equation above is not balanced. There are more oxygen atoms on the left side of the equation than there are on the right side. This must be corrected while keeping all other atom counts in balance. Begin to balance the equation by first counting and listing the atoms on the before and after the reaction.

Kind of Atom	Number of Atoms Before Reaction	Number of Atoms After Reaction
Ba	1	1
S	1	1
O	4	1
C	1	1

Next, adjust the coefficients until all atoms are balanced on the left and right sides of the arrow. Try putting a 4 in front of CO. Now you have 4 oxygen atoms on the right, which balances on both sides, but the carbon atoms become unbalanced. To fix this, add a 4 in front of the C in the reactants. The balanced equation looks like this:

$$BaSO_4(s) + 4C(s) \rightarrow BaS(s) + 4CO(g)$$

CHECK the answer

Review the number of atoms on each side of the equation and verify that they are equal.

Practice Problems

1. HCl is slowly added to aqueous Na_2CO_3 forming NaCl, H_2O, and CO_2. Follow the steps above to write a balanced equation for this reaction.

2. Balance this equation: $NaOH(aq) + CaBr_2(aq) \rightarrow Ca(OH)_2(s) + NaBr(aq)$.

For more practice problems, go to page 834 and visit gpscience.com/extra_problems.

Oxidation-Reduction Reactions One characteristic that is common to many chemical reactions is the tendency of the substances to lose or gain electrons. Chemists use the term **oxidation** to describe the loss of electrons and the term **reduction** to describe the gain of electrons. Chemical reactions involving electron transfer of this sort often involve oxygen, which is very reactive, pulling electrons from metallic elements. Corrosion of metal is a visible result, as shown in **Figure 15.**

The cause and effect of oxidation and reduction can be taken one step further by describing the substances after the electron transfer. The substance that gains an electron or electrons obviously becomes more negative, so we say it is reduced. On the other hand, the substance that loses an electron or electrons then becomes more positive, and we say it is oxidized. The electrons that were pulled from one atom were gained by another atom in a chemical reaction called reduction. Reduction is the partner to oxidation; the two always work as a pair, which is commonly referred to as redox.

Figure 15 One of the results of all of these electrons moving from one place to another might show up on the metal body of a tanker.

section 3 review

Summary

Types of Reactions

- Chemical reactions are organized into five basic classes: combustion, synthesis, decomposition, single displacement, and double displacement.

- Lavoisier was one of the first scientists to accurately describe a combustion reaction.

- For single-displacement reactions, we can predict which metal will replace another by comparing the activity characteristic of each.

- Some reactions produce a solid called a precipitate when two ionic substances are combined.

Oxidation-Reduction Reactions

- Oxidation is the loss of electrons and reduction is the corresponding gain of electrons.

- Redox reactions often result in corrosion and rust.

- A substance that gains elections is reduced, and a substance that loses elections is oxidized.

Self Check

1. **Classify** each of the following reactions:
   a. $CaO(s) + H_2O \longrightarrow Ca(OH)_2\ (aq)$
   b. $Fe(s) + CuSO_4(aq) \longrightarrow FeSO_4(aq) + Cu(s)$
   c. $NH_4NO_3(s) \longrightarrow N_2O(g) + 2H_2O(g)$

2. **Describe** what happens in a combustion reaction.

3. **Explain** the difference between synthesis and decomposition reactions.

4. **Determine,** using **Figure 13,** if zinc will displace gold in a chemical reaction and explain why or why not.

5. **Think Critically** Describe one possible economic impact of redox reactions. How might that impact be lessened?

Applying Math

6. **Use Proportions** The following chemical equation is balanced, but the coefficients used are larger than necessary. Rewrite this balanced equation using the smallest coefficients.

   $9Fe(s) + 12H_2O(g) \longrightarrow 3Fe_3O_4(s) + 12H_2(g)$

7. **Use Coefficients** Sulfur trioxide, (SO_3), a pollutant released by coal-burning plants, can react with water in the atmosphere to produce sulfuric acid, H_2SO_4. Write a balanced equation for this reaction.

Chemical Reactions and Energy

Reading Guide

What You'll Learn

■ **Identify** the source of energy changes in chemical reactions.
■ **Compare and contrast** exergonic and endergonic reactions.
■ **Examine** the effects of catalysts and inhibitors on the speed of chemical reactions.

Why It's Important

Chemical reactions provide energy to cook your food, keep you warm, and transform the food you eat into substances you need to live and grow.

⊘ Review Vocabulary

chemical bond: the force that holds two atoms together

New Vocabulary

● exergonic reaction
● exothermic reaction
● endergonic reaction
● endothermic reaction
● catalyst
● inhibitor

Figure 16 When its usefulness is over, a building is sometimes demolished using dynamite. Dynamite charges must be placed carefully so that the building collapses inward, where it cannot harm people or property.

Chemical Reactions—Energy Exchanges

Often a crowd gathers to watch a building being demolished using dynamite. In a few breathtaking seconds, tremendous structures of steel and cement that took a year or more to build are reduced to rubble and a large cloud of dust. A dynamite explosion, as shown in **Figure 16,** is an example of a rapid chemical reaction.

Most chemical reactions proceed more slowly, but all chemical reactions release or absorb energy. This energy can take many forms, such as heat, light, sound, or electricity. The heat produced by a wood fire and the light emitted by a glow stick are two examples of reactions that release energy.

Chemical bonds are the source of this energy. When most chemical reactions take place, some chemical bonds in the reactants are broken, which requires energy. In order for products to be produced, new bonds must form. Bond formation releases energy. Reactions such as dynamite combustion require much less energy to break chemical bonds than the energy released when new bonds are formed. The result is a release of energy and sometimes a loud explosion. Another release of energy is used to power rockets, as shown in **Figure 17.**

Figure 17

Rockets burn fuel to provide the thrust necessary to propel them upward. In 1926, engineer Robert Goddard used gasoline and liquid oxygen to propel the first ever liquid-fueled rocket. Although many people at the time ridiculed Goddard's space travel theories, his rockets eventually served as models for those that have gone to the Moon and beyond. A selection of rockets—including Goddard's—is shown here. The number below each craft indicates the amount of thrust—expressed in newtons (N)—produced during launch.

◀ SPACE SHUTTLE The main engines produce enormous amounts of energy by combining liquid hydrogen and oxygen. Coupled with solid rocket boosters, they produce over 32.5 million newtons (N) of thrust to lift the system's 2 million kg off the ground.

▶ JUPITER C This rocket launched the first United States satellite in 1958. It used a fuel called hydyne plus liquid oxygen.

▼ GODDARD'S MODEL ROCKET Although his first rocket rose only 12.6 m, Goddard successfully launched 35 rockets in his lifetime. The highest reached an altitude of 2.7 km.

▼ LUNAR MODULE Smaller rocket engines, like those used by the Lunar Module to leave the Moon, use hydrazine-peroxide fuels. The number shown below indicates the fixed thrust from one of the module's two engines; the other engine's thrust was adjustable.

400 N   369,350 N   32,500,000 N   15,920 N

Mini LAB

Creating a Colorful Chemical Reaction

Procedure

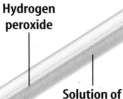

1. Pour 5 mL of **water** into a **test tube.**
2. Sprinkle a few crystals of **copper(II) bromide** into the test tube and observe the color change of the crystals.
3. Slowly add more water and observe what happens.

Analysis

1. What color were the copper(II) bromide crystals after you added them to the test tube of water?
2. What color were they when you added more water?
3. What caused this color change?

More Energy Out

You have probably seen many reactions that release energy. Chemical reactions that release energy are called **exergonic** (ek sur GAH nihk) **reactions.** In these reactions less energy is required to break the original bonds than is released when new bonds form. As a result, some form of energy, such as light or heat, is given off by the reaction. The familiar glow from the reaction inside a glow stick, shown in **Figure 18,** is an example of an exergonic reaction, which produces visible light. In other reactions however, the energy given off can produce heat. This is the case with some heat packs that are used to treat muscle aches and other problems that require heat.

Heat Release When the energy given off in a reaction is primarily in the form of heat, the reaction is called an **exothermic reaction.** The burning of wood and the explosion of dynamite are exothermic reactions. Iron rusting is also exothermic, but, under typical conditions, the reaction proceeds so slowly that it's difficult to detect any temperature change.

✓ Reading Check *Why is a log fire considered to be an exothermic reaction?*

Exothermic reactions provide most of the power used in homes and industries. Fossil fuels that contain carbon, such as coal, petroleum, and natural gas, combine with oxygen to yield carbon dioxide gas and energy. Unfortunately impurities in these fuels, such as sulfur, burn as well, producing pollutants such as sulfur dioxide. Sulfur dioxide combines with water in the atmosphere, producing acid rain.

Figure 18 Glow sticks contain three different chemicals—an ester and a dye in the outer section and hydrogen peroxide in a center glass tube. Bending the stick breaks the tube and mixes the three components. The energy released is in the form of visible light.

Hydrogen peroxide

Solution of dye and ester

More Energy In

Sometimes a chemical reaction requires more energy to break bonds than is released when new ones are formed. These reactions are called **endergonic reactions.** The energy absorbed can be in the form of light, heat, or electricity.

Electricity is often used to supply energy to endergonic reactions. For example, electroplating deposits a coating of metal onto a surface, as shown in **Figure 19.** Also, aluminum metal is obtained from its ore using the following endergonic reaction.

$$2Al_2O_3(l) \xrightarrow{\text{elec.}} 4Al(l) + 3O_2(g)$$

In this case, electrical energy provides the energy needed to keep the reaction going.

Heat Absorption When the energy needed is in the form of heat, the reaction is called an **endothermic reaction.** The term *endothermic* is not just related to chemical reactions. It also can describe physical changes. The process of dissolving a salt in water is a physical change. If you ever had to soak a swollen ankle in an Epsom salt solution, you probably noticed that when you mixed the Epsom salt in water, the solution became cold. The dissolving of Epsom salt absorbs heat. Thus, it is a physical change that is endothermic.

Some reactions are so endothermic that they can cause water to freeze. One such endothermic reaction is that of barium hydroxide ($BaOH)_2$ and ammonium chloride (NH_4Cl) in water, shown in **Figure 20.** Several drops of water were placed on the board, and when the reaction had taken place for several minutes, the temperature of the water in the beaker was cold enough to freeze the water drops and adhere the wood to the beaker. A cold pack, which contains ammonium nitrate crystals and water, is another example of an endothermic reaction.

Figure 19 Electroplating of a metal is an endergonic reaction that requires electricity. A coating of copper was plated onto this coin.

Figure 20 As an endothermic reaction happens, such as the reaction of barium hydroxide and ammonium chloride, energy from the surrounding environment is absorbed, causing a cooling effect. Here, the reaction absorbs so much heat that a drop of water freezes and the beaker holding the reaction sticks to the wood.

Catalysts Metals, such as platinum and palladium, are used as catalysts in the exhaust systems of automobiles. What reactions do you think they catalyze?

Catalysts and Inhibitors Some reactions proceed too slowly to be useful. To speed them up, a catalyst can be added. A **catalyst** is a substance that speeds up a chemical reaction without being permanently changed itself. When you add a catalyst to a reaction, the mass of the product that is formed remains the same, but it will form more rapidly. The catalyst remains unchanged and often is recovered and reused. Catalysts are used to speed many reactions in industry, such as polymerization to make plastics and fibers.

 Reading Check *Why would a catalyst be needed for a chemical reaction?*

At times, it is worthwhile to prevent certain reactions from occurring. Substances called **inhibitors** are used to slow down a chemical reaction. The food preservatives BHT and BHA are inhibitors that prevent spoilage of certain foods, such as cereals and crackers.

One thing to remember when thinking about catalysts and inhibitors is that they do not change the amount of product produced. They only change the rate of production. Catalysts increase the rate and inhibitors decrease the rate. Other factors, including concentration, pressure, and temperature, also affect the rate of reaction and must be considered when catalyzing or inhibiting a reaction.

section 4 review

Summary

Chemical Reactions—Energy Exchanges

- Chemical reactions release or absorb energy as chemical bonds are broken and formed.
- The energy of chemical reactions can be in the form of heat, light, sound, and/or electricity.
- Catalysts are used to increase the chemical reaction rate.

Chemical Energy

- Chemical reactions that release energy are called exergonic. Chemical reactions that absorb energy are called endergonic.
- Exothermic reactions give off heat energy.
- Endothermic reactions absorb energy in the form of heat.
- Exothermic reactions provide most of the power used in homes and industries.

Self Check

1. **Classify** the chemical reaction photosynthesis, which requires energy to proceed, as endergonic or exergonic.

2. **Explain** why a catalyst is not considered a reactant or product in a chemical reaction.

3. **Explain** why crackers containing BHT stay fresh longer than those without it.

4. **Classify** the reaction that makes a firefly glow in terms of energy input or output.

5. **Think Critically** To develop a product that warms people's hands, would you choose an exothermic or endothermic reaction to use? Why?

Applying Math

6. **Calculate** If an endothermic reaction begins at 26°C and loses 2°C per minute, how long will it take to reach 0°C?

7. **Use Graphs** Create a graph of the data in question 6. After 5 min, what is the temperature of the reaction?

 gpscience.com/self_check_quiz

CATALYZED Reaction

A balanced chemical equation tells nothing about the rate of a reaction. One way to affect the rate is to use a catalyst.

▶ Real-World Question

How does the presence of a catalyst affect the rate of a chemical reaction?

Goals

■ **Observe** the effect of a catalyst on the rate of reaction.

■ **Conclude,** based on your observations, whether the catalyst remained unchanged.

Possible Materials

test tubes (3)
test-tube rack
3% hydrogen peroxide,
   H_2O_2 (15 mL)
10-mL graduated cylinder
small plastic teaspoon
sand (¼ tsp)

hot plate
wooden splint
beaker of hot
   water
manganese
   dioxide, MnO_2
   (¼ tsp)

Safety Precautions

WARNING: *Hydrogen peroxide can irritate skin and eyes. Wipe up spills promptly. Point test tubes away from other students.*

▶ Procedure

1. Label three test tubes and set them in a test-tube stand. Pour 5 mL of hydrogen peroxide into each tube.

2. Place about 1/4 teaspoon of sand in tube 2 and the same amount of MnO_2 in tube 3.

3. In the presence of a catalyst, H_2O_2 decomposes rapidly producing oxygen gas, O_2.

Test each tube by: Lighting a wooden splint, blowing out the flame, and inserting the glowing splint into the tube. The splint will relight if oxygen is present.

4. Place all three tubes in a beaker of hot water. Heat on a hot plate until all of the remaining H_2O_2 is driven away and no liquid remains.

▶ Conclude and Apply

1. **Observe** the changes that happened when the solids were added to the tubes.

2. **Infer** which substance, sand or MnO_2, was the catalyst.

3. **Identify** what remained in each tube after the H_2O_2 was driven away.

*C*ommunicating
Your Data

Compare your results with those of your classmates and discuss any differences observed. **For more help refer to the Science Skill Handbook.**

Fossil Fuels and Greenhouse Gases

Goals

- **Observe** how you use fossil fuels in your daily life.
- **Gather data** on the process of burning fossil fuels and how greenhouse gases are released.
- **Research** the chemical reactions that produce greenhouse gases.
- **Identify** the importance of fossil fuels and their effect on the environment.
- **Communicate** your findings to other students.

Data Source

Visit gpscience. com/ internet_lab for more information about fossil fuels, the chemical reactions that produce greenhouse gases, uses of fossil fuels, their effects on the environment, and data from other students.

◎ *Real-World Question*

You've probably heard a lot about global warming and the greenhouse effect. According to one theory, certain gases in the atmosphere might be causing Earth's average global temperature to rise. The gases carbon dioxide, nitrous oxide, and methane, known as greenhouse gases, result from chemical reactions with oxygen when fossil fuels, such as coal, oil, and gas, are burned. What are some everyday activities that you do that might involve energy from fossil fuels? Form a hypothesis about how certain activities add greenhouse gases to our atmosphere.

◎ *Make a Plan*

1. **Observe** the activities of your daily life. How are fossil fuels used each day?
2. **Develop** a way to categorize the different chemical reactions and the greenhouse gases they produce.
3. **Search** reference sources to learn which chemical reactions produce greenhouse gases.
4. **Identify** some activities and functions that do not use fossil fuels.
5. **Infer** if it is possible to never use fossil fuels.

▶ Follow Your Plan

1. Make sure your teacher approves your plan before you start.

2. **Research** the chemical reactions that are commonly understood to produce greenhouse gases.

3. **Compare** the different reactions and their products.

4. **Record** your data in your Science Journal.

▶ Analyze Your Data

1. **Record** in your Science Journal the activities that scientists believe contribute the greatest amount of greenhouse gases to our atmosphere.

2. **Analyze** the types of chemical reactions that produce greenhouse gases. What types of reactions are they?

3. **Compare** your results with other students. Do your results agree with those of environmental scientists? Why might you have identified different contributors to the greenhouse effect?

4. **Make a table** of your data.

▶ Conclude and Apply

1. **Predict** How do you think your data would be affected if you had performed this experiment 100 years ago?

2. **Infer** What processes in nature might also contribute to the release of greenhouse gases? Compare their impact to that made by fossil fuels.

Communicating
Your Data

Find this lab at the link below. Post your data in the table provided. Compare your data to that of other students. Combine your data with that of other students and write an entry in your Science Journal that explains how the production of greenhouse gases could be reduced.

Science Online

gpscience.com/internet_lab

A Clumsy Move Pays Off

Hilaire de Chardonnet

Great scientific discoveries can happen in some very unlikely ways. Most people might not think that an accidental spill left uncleaned would become significant, but that's exactly what led a chemist named Hilaire de Chardonnet (hee LAYR • duh • shar doh NAY) to his discovery. In 1878, Chardonnet accidentally knocked over some nitrate chemicals. He put off cleaning up the mess and ended up inventing artificial silk.

Silk is produced naturally by silkworms. In the mid-1800s, though, silkworms were dying from disease and the silk industry was suffering. Businesses were going under and people were put out of work. Many scientists were working to develop a solution to this problem. Chardonnet had been searching for a silk substitute for years—he just didn't plan to find it by knocking it over!

A Messy Discovery

Chardonnet was in his darkroom developing photographs when the accidental spill took place. He decided to clean up the spill later and finish what he was working on. By the time he returned to wipe up the spill, the chemical solution had turned into a thick, gooey mess. When he pulled the cleaning cloth away, the goop formed long, thin strands of fiber that stuck to the cloth. The chemicals had reacted with the cellulose in the wooden table and liquefied it. The strands of fiber looked just like the raw silk made by silkworms.

Within six years, Chardonnet had developed a way to make the fibers into an artificial silk. Other scientists extended his work, developing a fiber called rayon. Today's rayon is made from sodium hydroxide mixed with wood fibers, which is then stranded and woven into cloth.

Rayon has another real-world application. To help prevent counterfeiting, dollars are printed on paper that contains red and blue rayon fibers. If you can scratch off the red or blue, that means it's ink and your bill is counterfeit. If you can pick out the red or blue fiber with a needle, it's a real bill.

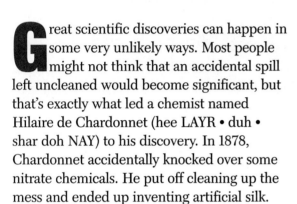

Rayon fiber

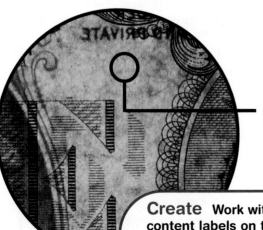

Create Work with a partner to examine the fabric content labels on the inside collars of your clothes. Research the materials, then make a data table that identifies their characteristics.

Science online

For more information, visit gpscience.com/oops

Reviewing Main Ideas

Chemical Changes

1. In a chemical reaction, one or more substances are changed to new substances.

2. The substances that react are called reactants, and the new substances formed are called products. Charcoal, the reactant shown below, is almost pure carbon.

3. The law of conservation of mass states that in chemical reactions, matter is neither created nor destroyed, just rearranged.

4. Chemical equations efficiently describe what happens in chemical reactions.

Section 2 **Chemical Equations**

1. Balanced chemical equations give the exact number of atoms involved in the reaction.

2. A balanced chemical equation has the same number of atoms of each element on both sides of the equation. This satisfies the law of conservation of mass.

3. When balancing chemical equations, change only the coefficients of the formulas, never the subscripts. To change a subscript would change the compound.

Section 3 **Classifying Chemical Reactions**

1. In synthesis reactions, two or more substances combine to form another substance.

2. In single-displacement reactions, one element replaces another in a compound.

3. In double-displacement reactions, ions in two compounds switch places, often forming a gas or insoluble compound.

4. Using the activity series chart, scientists can determine which metal can replace another metal.

Section 4 **Chemical Reactions and Energy**

1. Energy in the form of light, heat, sound or electricity is released from some chemical reactions known as exergonic reactions. This flame releases light and heat energy.

2. Reactions that absorb more energy than they release are called endergonic reactions.

3. Reactions may be sped up by adding catalysts and slowed down by adding inhibitors.

4. When energy is released in the form of heat, the reaction is exothermic.

FOLDABLES Use the Foldable that you made at the beginning of this chapter to help you review chemical reactions.

Using Vocabulary

balanced chemical equation p. 638	endothermic reaction p. 649
catalyst p. 650	exergonic reaction p. 648
chemical equation p. 635	exothermic reaction p. 648
chemical reaction p. 632	inhibitor p. 650
coefficient p. 636	oxidation p. 645
combustion reaction p. 641	precipitate p. 643
	product p. 632
decomposition reaction p. 642	reactant p. 632
	reduction p. 645
double-displacement reaction p. 643	single-displacement reaction p. 642
endergonic reaction p. 649	synthesis reaction p. 642

For each set of vocabulary words below, explain the relationship that exists.

1. coefficient—balanced chemical equation

2. synthesis reaction—decomposition reaction

3. reactant—product

4. catalyst—inhibitor

5. exothermic reaction—endothermic reaction

6. chemical reaction—product

7. endergonic reaction—exergonic reaction

8. single-displacement reaction—double-displacement reaction

9. chemical reaction—synthesis reaction

10. oxidation—reduction

Checking Concepts

Choose the word or phrase that best answers the question.

11. Oxygen gas is always written as O_2 in chemical equations. What term is used to describe the "2" in this formula?
    A) product
    B) coefficient
    C) catalyst
    D) subscript

12. What law is based on the experiments of Lavoisier?
    A) coefficients
    B) gravity
    C) chemical reaction
    D) conservation of mass

13. What must an element be in order to replace another element in a compound?
    A) more reactive
    B) less reactive
    C) more inhibiting
    D) less inhibiting

14. How do you indicate that a substance in an equation is a solid?
    A) (l)
    B) (g)
    C) (s)
    D) (aq)

15. What term is used to describe the "4" in the expression $4\ Ca(NO_3)_2$?
    A) coefficient
    B) formula
    C) subscript
    D) symbol

16. What type of compound is the food additive BHA?
    A) catalyst
    B) oxidized
    C) inhibitor
    D) reduced

17. How do you show that a substance is dissolved in water when writing an equation?
    A) (aq)
    B) (s)
    C) (g)
    D) (l)

18. What word would you use to describe HgO in the reaction that Lavoisier used to show conservation of mass?
    A) catalyst
    B) inhibitor
    C) product
    D) reactant

19. When hydrogen burns, what is oxygen's role?
    A) catalyst
    B) inhibitor
    C) product
    D) reactant

20. What kind of chemical reaction involves one substance losing an electron and another substance gaining an electron?
    A) combustion
    B) decomposition
    C) redox
    D) synthesis

Science Online gpscience.com/vocabulary_puzzlemaker

Interpreting Graphics

21. Copy and complete the concept map using the following terms: *oxidized, redox reactions, lost, reduced, oxidation, gained,* and *reduction.*

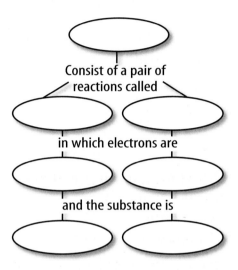

Consist of a pair of reactions called

in which electrons are

and the substance is

22. Sometimes a bond formed in a chemical reaction is weak and the product breaks apart as it forms. This is shown by a double arrow in chemical equations. Copy and complete the concept map, using the words *product(s)* and *reactant(s).* In the blank in the center, fill in the formulas for the substances appearing in the reversible reaction.

$$H_2(g) + I_2(g) \rightleftharpoons 2HI(g)$$

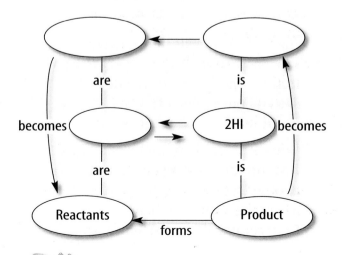

are   is

becomes   2HI   becomes

are   is

Reactants   Product

forms

Thinking Critically

23. Write a balanced chemical equation for the reaction of propane $C_3H_8(g)$ burning in oxygen to form carbon dioxide and water vapor.

24. Interpret the balanced chemical equation from question 23 to explain the law of conservation of mass.

25. Hypothesize Zn is placed in a solution of $Cu(NO_3)_2$ and Cu is placed in a $Zn(NO_3)_2$ solution. In which of these will a reaction occur?

26. Predict what kind of energy process happens when lye, $NaOH(s)$, is put in water and the water gets hot.

27. Recognize Cause and Effect Sucrose, or table sugar, is a disaccharide. This means that sucrose is composed of two simple sugars chemically bonded together. Sucrose can be separated into its components by heating it in an aqueous sulfuric acid solution. Research what products are formed by breaking up sucrose. What role does the acid play?

28. Classify Make an outline with the general heading *Chemical Reactions.* Include the five types of reactions, with a description and example of each.

Applying Math

29. Interpret Data When 46 g of sodium were exposed to dry air, 62 g of sodium oxide formed. How many grams of oxygen from the air were used?

30. Calculate Mass Chromium is produced by reacting its oxide with aluminum. If 76 g of Cr_2O_3 and 27 g of Al completely react to form 51 g of Al_2O_3, how many grams of Cr are formed?

Part 1 | Multiple Choice

Record your answers on the answer sheet provided by your teacher or on a sheet of paper.

Use the photograph below to answer questions 1 and 2.

1. The photograph above shows a chemical reaction in which water decomposes into hydrogen gas and oxygen gas when an electric current is passed through it. Which of the following is the correct chemical equation for this reaction?
   A. $H_2O(l) \rightarrow H_2(g) + O(g)$
   B. $H_2O(l) \rightarrow 2H(g) + O(g)$
   C. $2H_2O(l) \rightarrow 2H_2(g) + 2O(g)$
   D. $2H_2O(l) \rightarrow 2H_2(g) + O_2(g)$

2. Which of the following is the correct classification for the chemical reaction shown in the photograph?
   A. synthesis
   B. decomposition
   C. single displacement
   D. double displacement

Test-Taking Tip

Missing Information Questions often will ask about missing information. Notice what is missing as well as what is given.

3. Which of the following types of reaction is the opposite of a synthesis reaction?
   A. displacement   **C.** combustion
   B. reversible   **D.** decomposition

4. Which substance is the precipitate in the following reaction?

   $Ba(NO_3)_2(aq) + K_2SO_4(aq) \rightarrow$
   $\quad BaSO_4(s) + 2KNO_3(aq)$

   A. $Ba(NO_3)_2$ **C.** $BaSO_4$
   B. K_2SO_4 **D.** KNO_3

5. Which of the following reactions is endothermic?
   A. iron rusting
   B. burning wood
   C. exploding dynamite
   D. mixing Epsom salt in water

Use the figure below to answer questions 6 and 7.

6. Which of the metals in the activity series shown above would you expect to be mostly found in nature as a deposit of a relatively pure element?
   A. copper   **C.** silver
   B. lithium   **D.** iron

7. Which of the following metals would most likely replace lead in a solution?
   A. potassium   **C.** silver
   B. copper   **D.** gold

Part 2 | Short Response/Grid In

Record your answers on the answer sheet provided by your teacher or on a sheet of paper.

8. What is a synthesis reaction?

9. What is the source of the heat, light, sound, and electricity that can be produced during a chemical reaction?

Use the photograph below to answer questions 10 and 11.

10. The photograph above shows the reaction of aqueous nickel(II) chloride, $NiCl_2$, and aqueous sodium hydoxide, $NaOH$, to form solid nickel(II) hydroxide, $Ni(OH)_2$, and aqueous sodium chloride, $NaCl$. Write a balanced chemical equation for this reaction.

11. State the conservation of mass as it applies to the chemical reaction in the photograph above.

12. What do the symbols *s, aq, g,* and *l* mean when they are placed in parentheses next to the formulas for substances in chemical equations?

13. Food preservatives are a type of inhibitor. Explain why this is useful in foods.

14. What are the substances that react and the substances that are produced in a chemical reaction called?

Part 3 | Open Ended

Record your answers on a sheet of paper.

Use the photograph below to answer questions 15 and 16.

15. The photograph above shows a chemical reaction between, Mg, and oxygen gas, O_2. This reaction is exergonic and exothermic. Explain what these terms mean and how you can tell that a chemical reaction is exergonic or exothermic.

16. The reaction of magnesium and oxygen gas forms magnesium oxide, MgO. Write chemical equation for this reaction and explain the process you use to balance the equation.

17. Name and describe three notations that may be used above the arrow in a chemical equation.

18. Explain what is wrong with the following balanced equation:

$$4Al(s) + 6O(g) \rightarrow 2Al_2O_3(s)$$

What is the correct form of the equation?

19. What is a double-displacement reaction? Describe the double-displacement reaction shown in the following chemical equation in which lead nitrate, $Pb(NO_3)_2$, and potassium iodide, KI, react to form lead iodide, PbI_2, and potassium nitrate, KNO_3.

$$Pb(NO_3)_2 + 2KI \rightarrow PbI_2 + 2KNO_3$$

How Are Algae & Photography Connected?

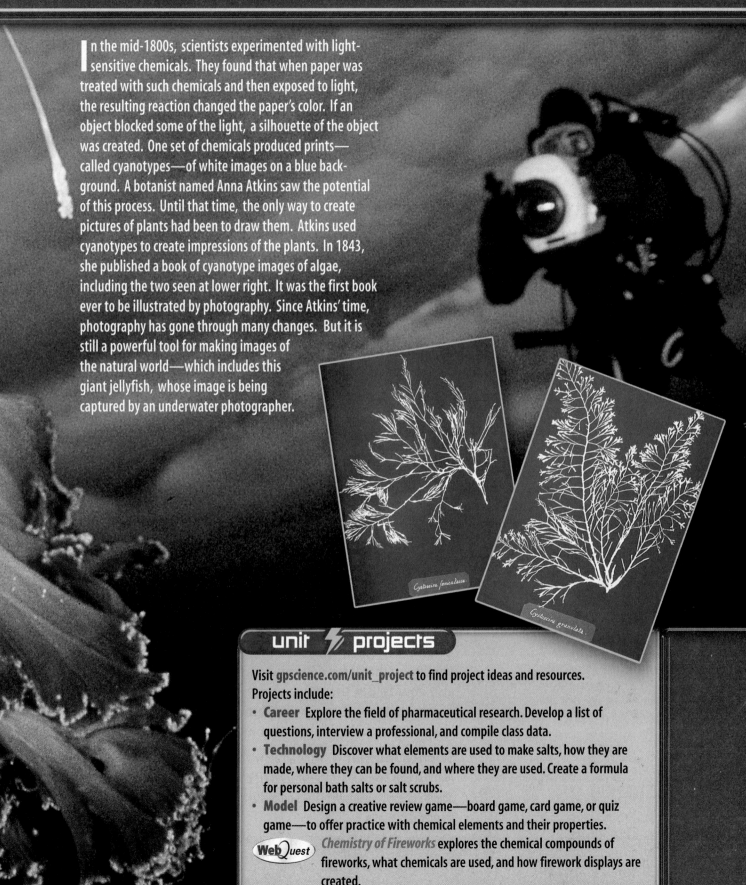

In the mid-1800s, scientists experimented with light-sensitive chemicals. They found that when paper was treated with such chemicals and then exposed to light, the resulting reaction changed the paper's color. If an object blocked some of the light, a silhouette of the object was created. One set of chemicals produced prints—called cyanotypes—of white images on a blue background. A botanist named Anna Atkins saw the potential of this process. Until that time, the only way to create pictures of plants had been to draw them. Atkins used cyanotypes to create impressions of the plants. In 1843, she published a book of cyanotype images of algae, including the two seen at lower right. It was the first book ever to be illustrated by photography. Since Atkins' time, photography has gone through many changes. But it is still a powerful tool for making images of the natural world—which includes this giant jellyfish, whose image is being captured by an underwater photographer.

unit ⚡ projects

Visit **gpscience.com/unit_project** to find project ideas and resources.
Projects include:

- **Career** Explore the field of pharmaceutical research. Develop a list of questions, interview a professional, and compile class data.
- **Technology** Discover what elements are used to make salts, how they are made, where they can be found, and where they are used. Create a formula for personal bath salts or salt scrubs.
- **Model** Design a creative review game—board game, card game, or quiz game—to offer practice with chemical elements and their properties.

WebQuest *Chemistry of Fireworks* explores the chemical compounds of fireworks, what chemicals are used, and how firework displays are created.

Solutions

Mixed-Up Chemistry

You can recognize that seawater is a liquid, and you know that lemonade and suntan lotion are also liquids. But they have something else in common, too—they are all solutions. In this chapter, you will learn about solutions.

Science Journal Write an answer to this question: *Are all liquids necessarily solutions, and are all solutions liquids?* Check your answer later and revise it if you've learned differently.

Start-Up Activities

Solution Identification by Solvent Subtraction

What do you like to drink when you're thirsty? Do you prefer water from the faucet, bottled water, or a sports drink that contains substances added to replace those lost during sweating? What do these thirst quenchers contain? Try the following lab to find out.

1. Obtain three solution samples from your teacher and place equal amounts of each in separate, marked, 100-mL beakers.

2. Carefully, boil each solution on a hot plate. As soon as the liquid is gone, remove each beaker to a heat-proof surface using a thermal mitt. Let cool.

3. Examine the inside of your cooled beakers. What do you see? Guess the identity of each solution.

4. **Think Critically** Describe in your Science Journal what remained in each of the three containers and explain how solutions may look alike but contain different substances.

Solvent-Solute Comparison Make the following Foldable to compare and contrast the characteristics of solvents and solutes.

STEP 1 Fold one sheet of paper lengthwise.

STEP 2 Fold into thirds.

STEP 3 Unfold and draw overlapping ovals. Cut the top sheet along the folds.

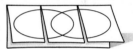

STEP 4 Label the ovals as shown.

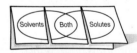

Construct a Venn Diagram As you read the chapter, list characteristics that are unique to solvents under the left tab, those unique to solutes under the right tab, and characteristics common to both under the middle tab.

Preview this chapter's content and activities at gpscience.com

How Solutions Form

Reading Guide

What You'll Learn
- **Determine** how things dissolve.
- **Examine** the factors that affect the rates at which solids and gases dissolve in liquids.

Why It's Important
Many chemical reactions take place in solution—the food you eat is digested, or chemically changed, by the solution that is in your stomach.

❓ Review Vocabulary
alloy: a mixture of elements that has metallic properties

New Vocabulary
- solution
- solute
- solvent
- polar

What is a solution?

Hummingbirds are fascinating creatures. They can hover for long periods while they sip nectar from flowers through their long beaks. To attract hummingbirds, many people use feeder bottles containing a red liquid, as shown in **Figure 1.** The liquid is a solution of sugar and red food coloring in water.

Suppose you are making some hummingbird food. When you add sugar to water and stir, the sugar crystals disappear. When you add a few drops of red food coloring and stir, the color spreads evenly throughout the sugar water. Why does this happen?

Hummingbird food is one of many solutions. A **solution** is a mixture that has the same composition, color, density, and even taste throughout. The reason you no longer see the sugar crystals and the reason the red dye spreads out evenly is that they have formed a completely homogeneous mixture. The sugar crystals broke up into sugar molecules, the red dye into its molecules, and both mixed evenly among the water molecules.

Figure 1 Liquid solutions, like this hummingbird food, which has sugar and food coloring, may contain gases, other liquids, or solids.

Liquid phase

Solutes and Solvents

To describe a solution, you may say that one substance is dissolved in another. The substance being dissolved is the **solute,** and the substance doing the dissolving is the **solvent.** When a solid dissolves in a liquid, the solid is the solute and the liquid is the solvent. Thus, in salt water, salt is the solute and water is the solvent. In carbonated soft drinks, carbon dioxide gas is one of the solutes and water is the solvent. When a liquid dissolves in another liquid, the substance present in the larger amount is usually called the solvent.

> ✔ **Reading Check** *How do you know which substance is the solute in a solution?*

Nonliquid Solutions Solutions can also be gaseous or even solid. Examples of all three solution phases are shown in **Figure 1** and **Figure 2.** Did you know that the air you breathe is a solution? In fact, all mixtures of gases are solutions. Air is a solution of 78 percent nitrogen, 21 percent oxygen, and small amounts of other gases such as argon, carbon dioxide and hydrogen. The sterling silver and brass used in musical instruments is an example of a solid solution. The sterling silver contains 92.5 percent silver and 7.5 percent copper. The brass is a solution of copper and zinc metals. Solid solutions are known as alloys. They are made by melting the metal solute and solvent together. Most coins, as shown in **Figure 3,** are alloys.

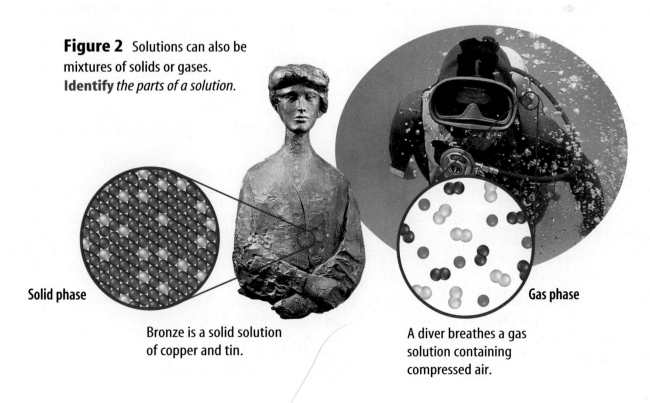

Figure 2 Solutions can also be mixtures of solids or gases. **Identify** *the parts of a solution.*

Solid phase

Bronze is a solid solution of copper and tin.

Gas phase

A diver breathes a gas solution containing compressed air.

Figure 3

Susan B. Anthony dollar

Sacagawea dollar

Have you ever accidentally put a non-United States coin into a vending machine? Of course, the vending machine didn't accept it. If a vending machine is that selective, how can it be fooled by two coins that look and feel very different? This is exactly the case with the silver Susan B. Anthony dollar and the new golden Sacagawea dollar. Vending machines can't tell them apart.

◀ Vending machines recognize coins by size, weight, and electrical conductivity. The size and weight of the Susan B. Anthony coin were easy to copy. Copying the coin's electrical conductivity was more difficult.

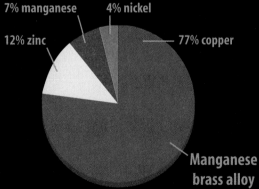

7% manganese

4% nickel

12% zinc

77% copper

Manganese brass alloy

▲ Over 30,000 samples of coin coatings were tested to find an alloy and thickness that would copy the conductivity of the Susan B. Anthony dollar. The final composition of the alloy is shown in the graph above. The key ingredient? Manganese.

Manganese brass alloy
Copper core
Manganese brass alloy

▲ The dollar's copper core is half the coin's thickness. It is sandwiched between two layers of manganese brass alloy.

How Substances Dissolve

Fruit drinks and sports drinks are examples of solutions made by dissolving solids in liquids. Like hummingbird food, both contain sugar as well as other substances that add color and flavor. How do solids such as sugar dissolve in water?

The dissolving of a solid in a liquid occurs at the surface of the solid. To understand how water solutions form, keep in mind two things you have learned about water. Like the particles of any substance, water molecules are constantly moving. Also, water molecules are **polar,** which means they have a positive area and a negative area. Molecules of sugar are also polar.

How It Happens **Figure 4** shows molecules of sugar dissolving in water. First, water molecules cluster around sugar molecules with their negative ends attracted to the positive ends of the sugar. Then, the water molecules pull the sugar molecules into solution. Finally, the water molecules and the sugar molecules mix evenly, forming a solution.

✔ **Reading Check** *How do water molecules help sugar molecules dissolve?*

The process described in **Figure 4** repeats as layer after layer of sugar molecules moves away from the crystal, until all the molecules are evenly spread out. The same three steps occur for any solid solute dissolving in a liquid solvent.

Dissolving Liquids and Gases A similar but more complex process takes place when a gas dissolves in a liquid. Particles of liquids and gases move much more freely than do particles of solids. When gases dissolve in gases or when liquids dissolve in liquids, this movement spreads solutes evenly throughout the solvent, resulting in a homogenous solution.

Dissolving Solids in Solids How can you mix solids to make alloys? Although solid particles do move a little, this movement is not enough to spread them evenly throughout the mixture. The solid metals are first melted and then mixed together. In this liquid state, the metal atoms can spread out evenly and will remain mixed when cooled.

Figure 4 Dissolving sugar in water can be thought of as a three-step process.

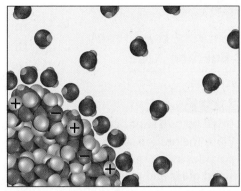

Step 1 Moving water molecules cluster around the sugar molecules as their negative ends are attracted to the positive ends of the sugar molecules.

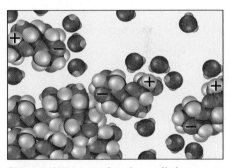

Step 2 Water molecules pull the sugar molecules into solution.

Step 3 Water molecules and sugar molecules spread out to form a homogeneous mixture.

Observing the Effect of Surface Area

Procedure

1. Grind up two **sugar cubes.**
2. Place the ground sugar particles into a **medium-sized glass** and place two **unground sugar cubes** into a similar glass.
3. Add an equal amount of **distilled water** at room temperature to each glass.

Analysis

1. Compare the times required to dissolve each.
2. What do you conclude about the dissolving rate and surface area?

Try at Home

Figure 5 Crystal size affects solubility. Large crystals dissolve in water slowly because the amount of surface area is limited. Increasing the amount of surface area by creating smaller particles increases the rate of dissolving.

Rate of Dissolving

If two substances will form a solution, they will do so at a particular rate. Sometimes the rate at which a solute dissolves into a solvent is fast and other times slow. There are several things you can do to speed up the rate of dissolving—stirring, reducing crystal size, and increasing temperature are three of the most effective techniques.

Stirring How can you speed up the dissolving process? Think about how you make a drink from a powdered mix. After you add the mix to water, you stir it. Stirring a solution speeds up dissolving because it brings more fresh solvent into contact with more solute. The fresh solvent attracts the particles of solute, causing the solid solute to dissolve faster.

Crystal Size Another way to speed the dissolving of a solid in a liquid is to grind large crystals into smaller ones. Suppose you want to use a 5-g crystal of rock candy to sweeten your water. If you put the whole crystal into a glass of water, it might take several minutes to dissolve, even with stirring. However, if you first grind the crystal of rock candy into a powder, it will dissolve in the same amount of water in a few seconds.

Why does breaking up a solid cause it to dissolve faster? Breaking the solid into smaller pieces greatly increases its surface area, as you can see in **Figure 5.** Because dissolving takes place at the surface of the solid, increasing the surface area allows more solvent to come into contact with more solid solute. Therefore, the speed of the dissolving process increases.

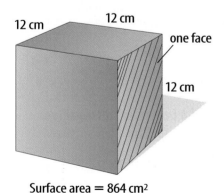

12 cm
12 cm
12 cm
one face

Surface area = 864 cm²

A face of a cube is the outer surface that has four edges.

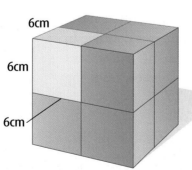

6cm
6cm
6cm

Surface area = 1,728 cm²

Pull apart the cube into smaller cubes of equal size. You now have eight cubes and forty-eight faces.

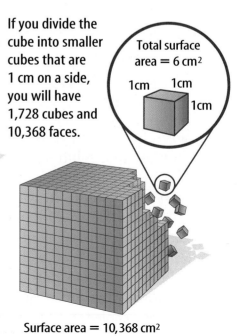

If you divide the cube into smaller cubes that are 1 cm on a side, you will have 1,728 cubes and 10,368 faces.

Total surface area = 6 cm²
1cm 1cm
1cm

Surface area = 10,368 cm²

Applying Math   Calculate

CALCULATING SURFACE AREA The length, height, and width of a cube are each 1 cm. If the cube is cut in half to form two rectangles, what is the total surface area of the new pieces?

IDENTIFY the known values and the unknown value

Identify the known values:

- The cube has dimensions of $l = h = w = 1$ cm.
- The rectangular solid has a width $w = 0.5$ cm.
- The rectangular solid has a length and height $l = h = 1$ cm.

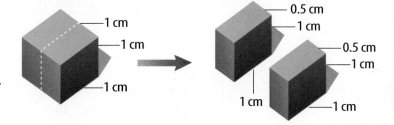

- The cube and the rectangular solids have six faces: front and back ($h \times w$); left and right ($h \times l$); and top and bottom ($w \times l$). The total surface area of the cube or the rectangular solid is the sum of these areas, or $2(h \times w) + 2(h \times l) + 2(w \times l)$.

Identify the unknown value:

Find the total surface area of the two rectangular solids.

SOLVE the problem

The surface area of the cube is:
$2(1 \text{ cm} \times 1 \text{ cm}) + 2(1 \text{ cm} \times 1 \text{ cm}) + 2(1 \text{ cm} \times 1 \text{ cm}) = 6 \text{ cm}^2$

The surface area of the rectangular solid is:
$2(1 \text{ cm} \times 0.5 \text{ cm}) + 2(1 \text{ cm} \times 1 \text{ cm}) + 2(0.5 \text{ cm} \times 1 \text{ cm}) = 4 \text{ cm}^2$

Because there are two rectangular solids, their total surface area is:
$4 \text{ cm}^2 + 4 \text{ cm}^2 = 8 \text{ cm}^2$

To find out how much new surface area has been created, compare the two results:
$8 \text{ cm}^2 - 6 \text{ cm}^2 = 2 \text{ cm}^2$

CHECK the answer

Does your answer seem reasonable? Consider what you've done by splitting the cube in two. You have created two new faces, each with a surface area of 1 cm^2; therefore, the answer 2 cm^2 is correct.

Practice Problems

1. A cube of salt with a length, height, and width of 5 cm is attached along a face to another cube of salt with the same dimensions. What is the combined surface area of the new rectangular solid?

2. How much surface area has been lost?

For more practice problems go to page 834, and visit gpscience.com/extra_problems.

Temperature In addition to stirring and decreasing particle size, a third way to increase the rate at which most solids dissolve is to increase the temperature of the solvent. Think about making hot chocolate from a mix. You can make the sugar in the chocolate mix dissolve faster by putting it in hot water instead of cold water. Increasing the temperature of a solvent speeds up the movement of its particles. This increase causes more solvent particles to bump into the solute. As a result, solute particles break loose and dissolve faster.

Controlling the Process Think about how the three factors you just learned affect the rate of dissolving. Can these factors combine to further increase the rate or perhaps control the rate of dissolving? Each technique, stirring, crushing, and heating, is known to speed up the rate of dissolving by itself. However, when two or more techniques are combined, the rate of dissolving is even faster. Consider a sugar cube placed in cold water. You know that the sugar cube will eventually dissolve. You can predict that heating the water will increase the rate by some amount. You can also predict that heat and stirring will increase the rate further. Finally, you can predict that crushing the cube combined with heating and stirring will result in the fastest rate of dissolving. Knowing how much each technique affects the rate will allow you to control the rate of dissolving more precisely.

section 1 review

Summary

What is a solution?

- A solution is a uniform mixture.
- Solutions have the same composition, color, density, and taste throughout.

Solutes and Solvents

- In a solution, the solute is the substance that is being dissolved; the solvent is the substance that is doing the dissolving.

How Substances Dissolve

- The process of dissolving happens at the surface and is aided by polarity and molecular movement.

Rate of Dissolving

- Stirring, surface area, and temperature affect the rate of dissolving.

Self Check

1. **List** possible ways that phases of matter could combine to form a solution.
2. **Describe** how temperature affects the rate of dissolving.
3. **Describe** how the metal atoms in an alloy are mixed.
4. **Think Critically** Amalgrams, which are sometimes used in tooth fillings, are alloys of mercury with other metals. Is an amalgam a solution? Explain.

Applying Math

5. **Find Surface Area** Calculate the surface area of a rectangular solid with dimensions $l = 2$ cm, $w = 1$ cm, and $h = 0.5$ cm.
6. **Calculate Percent Increase** If the length of the rectangle in question 5 is increased by 10%, by how much will the surface area increase?

 Science Online gpscience.com/self_check_quiz

Solubility and Concentration

Reading Guide

What You'll Learn

- **Define** the concept of solubility.
- **Identify** how to express the concentration of solutions.
- **List** and define three types of solutions.
- **Describe** the effects of pressure and temperature on the solubility of gases.

Why It's Important

Solutions such as medicine and lemonade work and taste a particular way because of the specific solution concentrations

⊙ Review Vocabulary

concentration: describes how much solute is present in a solution compared to the amount of solvent

New Vocabulary

- solubility
- saturated solution
- unsaturated solution
- supersaturated solution

How much can dissolve?

You can stir several teaspoons of sugar into lemonade, and the sugar will dissolve. However, if you continue adding sugar, eventually the point is reached when no more sugar dissolves and the excess granules sink to the bottom of the glass. This indicates how soluble sugar is in water. **Solubility** (sol yuh BIH luh tee) is the maximum amount of a solute that can be dissolved in a given amount of solvent at a given temperature.

✔ **Reading Check** *What is solubility?*

Comparing Solubilities The amount of a substance that can dissolve in a solvent depends on the nature of these substances. **Figure 6** shows two beakers with the same volume of water and two different solutes. In one beaker, 1 g of solute A dissolves completely, but additional solute does not dissolve and falls to the bottom of the beaker. On the other hand, 1 g of solute B dissolves completely, and two more grams also dissolve before solute begins to fall to the bottom. If the temperature of the water is the same in both beakers, you can conclude that substance B is more soluble than substance A. **Table 1** shows how the solubility of several substances varies at 20°C. For solutes that are gases, the pressure also must be given.

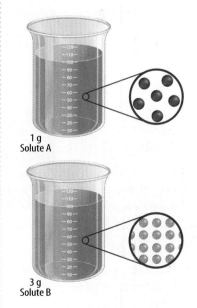

Figure 6 Substance B is more soluble in water than substance A at the same temperature.

1 g
Solute A

3 g
Solute B

Table 1 Solubility of Substances in Water at 20°C

Substance	Solubility in g/100 g of Water
Solid Substances	
Salt (sodium chloride)	35.9
Baking soda (sodium bicarbonate)	9.6
Washing soda (sodium carbonate)	21.4
Lye (sodium hydroxide)	109.0
Sugar (sucrose)	203.9
Gaseous Substances*	
Hydrogen	0.00017
Oxygen	0.005
Carbon dioxide	0.16

*at normal atmospheric pressure

Concentration

Suppose you add one teaspoon of lemon juice to a glass of water to make lemonade. Your friend adds four teaspoons of lemon juice to another glass of water the same size. You could say that your glass of lemonade is dilute and your friend's lemonade is concentrated, because your friend's drink now has more lemon flavor than yours. A concentrated solution is one in which a large amount of solute is dissolved in the solvent. A dilute solution is one that has a small amount of solute in the solvent.

Precise Concentrations How much real fruit juice is there in one of those boxed fruit drinks? You can read the label to find out. *Concentrated* and *dilute* are not precise terms. However, concentrations of solutions can be described precisely. One way is to state the percentage by volume of the solute. The percentage by volume of the juice in the drink shown in **Figure 7** is 10 percent. Adding 10 mL of juice to 90 mL of water makes 100 mL of this drink. Commonly, fruit-flavored drinks can contain from ten percent to 100 percent fruit juice. Generally, if two or more liquids are being mixed, the concentration is given in percentage by volume. To be certain of the concentration of your beverage, chose a product that is 100% juice.

Figure 7 The concentrations of fruit juices often are given in percent by volume like these. Concentrations commonly range from 10 percent to 100 percent juice.
Identify *the product that has the highest concentration.*

Types of Solutions

How much solute can dissolve in a given amount of solvent? That depends on a number of factors, including the solubility of the solute. Here you will examine the types of solutions based on the amount of a solute dissolved.

Saturated Solutions If you add 35 g of copper(II) sulfate, $CuSO_4$, to 100 g of water at 20°C, only 32 g will dissolve. You have a saturated solution because no more copper(II) sulfate can dissolve. A **saturated solution** is a solution that contains all the solute it can hold at a given temperature. However, if you heat the mixture to a higher temperature, more copper(II) sulfate can dissolve. Generally, as the temperature of a liquid solvent increases, the amount of solid solute that can dissolve in it also increases. **Table 2** shows the amounts of a few solutes that can dissolve in 100 g of water at different temperatures, forming saturated solutions. Some of these data also are shown on the accompanying graph.

Solubility Curves Each line on the graph from **Table 2** is called a solubility curve for a particular substance. You can use a solubility curve to figure out how much solute will dissolve at any temperature given on the graph. For example, about 78 g of KBr (potassium bromide) will form a saturated solution in 100 g of water at 47°C. How much NaCl (sodium chloride) will form a saturated solution with 100 g of water at the same temperature?

Unsaturated Solutions An **unsaturated solution** is any solution that can dissolve more solute at a given temperature. Each time a saturated solution is heated to a higher temperature, it becomes unsaturated. The term *unsaturated* isn't precise. If you look at **Table 2,** you'll see that at 20°C, 35.9 g of NaCl (sodium chloride) forms a saturated solution in 100 g of water. However, an unsaturated solution of NaCl could be any amount less than 35.9 g in 100 g of water at 20°C.

Table 2 Solubility of Compounds in g/100 g of Water			
Compound	**0°C**	**20°C**	**100°C**
Copper(II) sulfate	23.1	32.0	114
Potassium bromide	53.6	65.3	104
Potassium chloride	28.0	34.0	56.3
Potassium nitrate	13.9	31.6	245
Sodium chlorate	79.6	95.9	204
Sodium chloride	35.7	35.9	39.2
Sucrose (sugar)	179.2	203.9	487.2

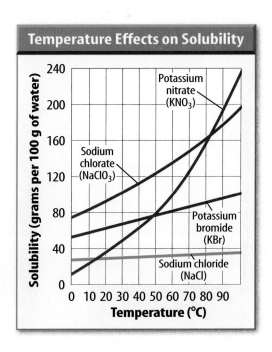

Temperature Effects on Solubility

![Reading Check] **What happens to a saturated solution if it is heated?**

Supersaturated Solutions If you make a saturated solution of potassium nitrate at 100°C and then let it cool to 20°C, part of the solute comes out of solution. This is because, at the lower temperature, the solvent cannot hold as much solute. Most other saturated solutions behave in a similar way when cooled. However, if you cool a saturated solution of sodium acetate from 100°C to 20°C without disturbing it, no solute comes out. At this point, the solution is supersaturated. A **supersaturated solution** is one that contains more solute than a saturated solution at the same temperature. Supersaturated solutions are unstable. For example, if a seed crystal of sodium acetate is dropped into the supersaturated solution, excess sodium acetate crystallizes out, as shown in **Figure 8.**

Solution Energy As the supersaturated solution of sodium acetate crystallizes, the solution becomes hot. Energy is given off as new bonds form between the ions and the water molecules. Some portable heat packs use crystallization from supersaturated solutions to produce heat. After crystallization, the heat pack can be reused by heating it to again dissolve all the solute.

Another result of solution energy is to reduce the temperature of the solution. Some substances, such as ammonium nitrate, must draw energy from the surroundings to dissolve. This is what happens when a cold pack is activated to treat minor injuries or to reduce swelling. When the inner bags of ammonium nitrate and water are broken, the ammonium nitrate draws energy from the water, which causes the temperature of the water to drop and the pack cools.

Figure 8 A supersaturated solution is unstable.
Explain *why this is so.*

A seed crystal of sodium acetate is added to a supersaturated solution of sodium acetate.

Excess solute immediately crystallizes from solution.

The crystallization reaction continues to draw solute from the solution.

Solubility of Gases

When you shake an opened bottle of soda, it bubbles up and may squirt out. Shaking or pouring a solution of a gas in a liquid causes gas to come out of solution. Agitating the solution exposes more gas molecules to the surface, where they escape from the liquid.

Pressure Effects What might you do if you want to dissolve more gas in a liquid? One thing you can do is increase the pressure of that gas over the liquid. Soft drinks are bottled under increased pressure. This increases the amount of carbon dioxide that dissolves in the liquid. When the pressure is released, the carbon dioxide bubbles out.

Temperature Effects Another way to increase the amount of gas that dissolves in a liquid is to cool the liquid. This is just the opposite of what you do to increase the speed at which most solids dissolve in a liquid. Imagine what happens to the carbon dioxide when a bottle of soft drink is opened. Even more carbon dioxide will bubble out of a soft drink as it gets warmer.

Figure 9 Solutions of gases behave differently from those of solids or liquids. This soda is bottled under pressure to keep carbon dioxide in solution. When the bottle is opened, pressure is released and carbon dioxide bubbles out of solution.

section 2 review

Summary

How much can dissolve?

- Solubility tells how much solute can dissolve in a solvent at a particular temperature.

Concentration

- A concentrated solution has a large amount of dissolved solute. A dilute solution has a small amount of dissolved solute.

Types of Solutions

- Saturated, unsaturated, and supersaturated solutions are defined by how much solute is dissolved.

- Solubility curves help predict how much solute can dissolve at a particular temperature.

- Some supersaturated solutions absorb or give off energy.

Gases in Solution

- Pressure and temperature affect gases in solution. High pressure and low temperature allow more gas to dissolve.

Self Check

1. **Explain** Do all solutes dissolve to the same extent in the same solvent? How do you know?

2. **Interpret** from **Table 2** the mass of sugar that would have to be dissolved in 100 g of water to form a saturated solution at 20°C.

3. **Determine** which is more soluble in water: 17 g of solute X dissolved in 100 mL of water at 23°C or 26 g of solute Z dissolved in 100 mL of water at 23°C.

4. **Identify** the type of solution you have if, at 35°C, solute continues to dissolve as you add more.

5. **Think Critically** Explain how keeping a carbonated beverage capped helps keep it from going "flat."

Applying Math

6. **Calculate Cost** By volume, orange drink is ten percent each of orange juice and corn syrup. A 1.5-L can of the drink costs $0.95. A 1.5-L can of orange juice is $1.49, and 1.5 L of corn syrup is $1.69. Per serving, does it cost less to make your own orange drink or buy it ?

section 3

Particles in Solution

Reading Guide

What You'll Learn

- **Examine** how some solutes break apart in water solutions to form positively and negatively charged particles.
- **Determine** how some solutions conduct electricity.
- **Describe** how antifreeze works.

Why It's Important

Many of the products we use every day rely on the effects of solutes in solution.The name and the formula convey information about the compound.

Review Vocabulary

conductivity: property of metals and alloys that allows them to be good conductors of heat and electricity

New Vocabulary

- ion
- electrolyte
- nonelectrolyte
- ionization
- dissociation

Particles with a Charge

Did you know that there are charged particles in your body that conduct electricity? In fact, you could not live without them. Some help nerve cells transmit messages. Each time you blink your eyes or wave your hand nerves control how muscles will respond. These charged particles, called **ions,** are in the fluids that are in and around all the cells in your body. The compounds that produce solutions of ions that conduct electricity in water are known as **electrolytes.** Some substances, like sodium chloride, are strong electrolytes and conduct a strong current. Strong electrolytes exist completely in the form of ions in solution. Other substances, like acetic acid in vinegar, remain mainly in the form of molecules when they dissolve in water. They produce few ions and conduct current only weakly. They are called weak electrolytes. Substances that form no ions in water and cannot conduct electricity are called **nonelectrolytes.** Among these are organic molecules like ethyl alcohol and sucrose.

Ionization Ionic solutions form in two ways. Electrolytes, such as hydrogen chloride, are molecules made up of neutral atoms. To form ions, the molecules must be broken apart in such a way that the atoms take on a charge. This process of forming ions is called **ionization.** The process is shown in **Figure 10,** using hydrogen chloride as a model.

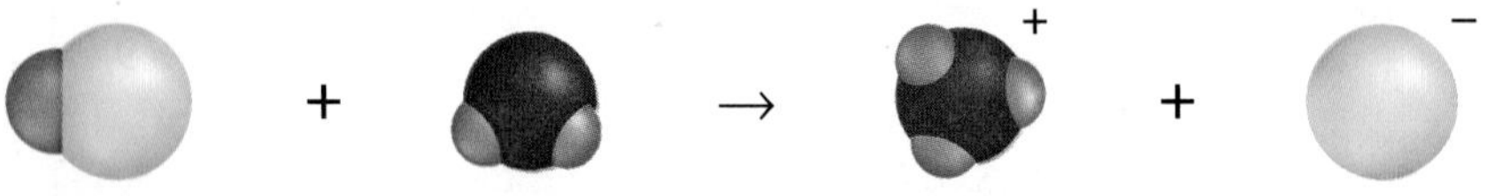

$HCl + H_2O \rightarrow H_3O^+ + Cl^-$

Figure 10 Both hydrogen chloride and water are polar molecules. Water surrounds the hydrogen chloride molecules and pulls them apart, forming positive hydrogen ions and negative chloride ions. Hydrogen ions are often shown as H_3O^+ to emphasize the role water plays in ionization.

Dissociation The second way that ionic solutions form is by the separation of ionic compounds. The ions already exist in the ionic compound and are attracted into the solution by the surrounding polar water molecules. **Dissociation** is the process in which an ionic solid, such as sodium chloride, separates into its positive and negative ions. A model of a sodium chloride crystal is shown in **Figure 11.** In the crystal, each positive sodium ion is attracted to six negative chloride ions. Each of the negative chloride ions is attracted to six sodium ions, a pattern that exists throughout the crystal.

When placed in water, the crystal begins to break apart under the influence of water molecules. Remember that water is polar, which means that the positive areas of the water molecules are attracted to the negative chloride ions. Likewise the negative oxygen part of the water molecules is attracted to the sodium ions.

In **Figure 12,** water molecules are approaching the sodium and chloride ions in the crystal. The water molecules surround the sodium and chloride ions, having pulled them away from the crystal and into solution. The sodium and chloride ions have dissociated from one another. The solution now consists of sodium and chloride ions mixed with water. The ions move freely through the solution and are capable of conducting an electric current.

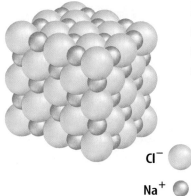

Figure 11 This is a model of a sodium chloride crystal. Each chloride ion is surrounded by six sodium ions and vice versa.

Cl⁻ ○
Na⁺ ○

✔ **Reading Check** *What are the differences and similarities between dissociation and ionization?*

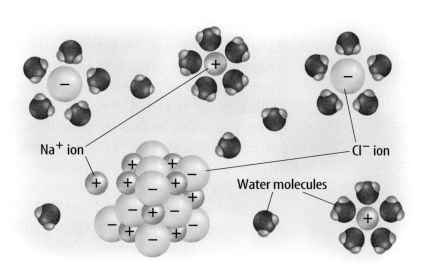

Na⁺ ion

Cl⁻ ion

Water molecules

Figure 12 Sodium chloride dissociates as water molecules attract and pull the sodium and chloride ions from the crystal. Water molecules then surround and separate the Na⁺ and Cl⁻ ions. **Predict** *Will sodium chloride in solution conduct electricity?*

Effects of Solute Particles

All solute particles—polar and nonpolar, electrolyte and nonelectrolyte—affect the physical properties of the solvent, such as its freezing point and its boiling point. These effects can be useful. For example, adding antifreeze to water in a car radiator lowers the freezing point of the radiator fluid. Sugar and salt also would do the same thing, however, both would damage the cooling system. The effect that a solute has on the freezing point or boiling point of a solvent depends on the number of solute particles in solution, not on the chemical nature of the particles.

Lowering Freezing Point Adding a solute such as antifreeze to a solvent lowers the freezing point of the solvent. How much the freezing point goes down depends upon how many solute particles you add. How does this work?

As a substance freezes, its particles arrange themselves in an orderly pattern. The added solute particles interfere with the formation of this pattern, making it harder for the solvent to freeze as shown in **Figure 13.** To overcome this interference, a lower temperature is needed to freeze the solvent.

Animal Antifreeze Certain animals that live in extremely cold climates have their own kind of antifreeze. Caribou, for example, contain substances in the lower section of their legs that prevent freezing in subzero temperatures. The caribou can stand for long periods of time in snow and ice with no harm to their legs. Fish that live in polar waters also have a natural chemical antifreeze called glycoprotein in their bodies. Glycoprotein prevents ice crystals from forming in the moist tissues. Many insects also have a similar antifreeze chemical to protect them from freezing temperatures.

Raising Boiling Point Surprisingly, antifreeze also raises the boiling point of the water. How can it do this? The amount the boiling point is raised depends upon the number of solute molecules present. Solute particles interfere with the evaporation of solvent particles. Thus, more energy is needed for the solvent particles to escape from the liquid surface, and so the boiling point of the solution will be higher than the boiling point of solvent alone.

Figure 13 Solute molecules interfere with the freezing process by blocking molecules of solvent as they try to join the growing crystal lattice. For example, antifreeze molecules added to water block the formation of ice crystals.

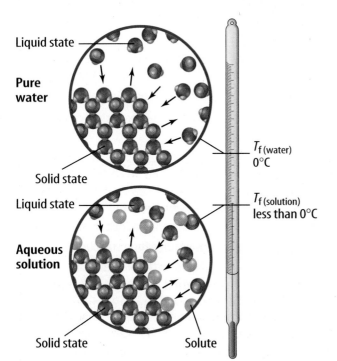

Liquid state

Pure water

Solid state

$T_{f\,(water)}$
$0°C$

Liquid state

$T_{f\,(solution)}$
less than $0°C$

Aqueous solution

Solid state   Solute

Car Radiators The beaker in **Figure 14** represents a car radiator when it contains water molecules only—no antifreeze. Some of those molecules on the surface will vaporize, and the number of molecules that do vaporize depends upon the temperature of the solvent. As temperature increases, water molecules move faster, and more molecules vaporize. Finally, when the pressure of the water vapor equals atmospheric pressure, the water boils. Have you ever seen a vehicle at the side of the road with vapors rising from the radiator?

The result of adding antifreeze is shown in **Figure 14.** Particles of solute are distributed evenly throughout the solution, including the surface area. Now fewer water molecules can reach the surface and evaporate, making the vapor pressure of the solution lower than that of the solvent. This means that it will take a higher temperature to make the car's radiator boil over.

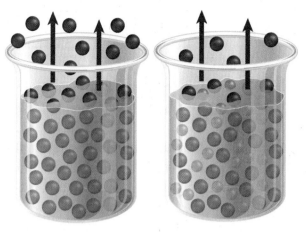

Solvent particles vaporize freely from the surface.

Solute particles block part of the surface, making it more difficult for solvent to vaporize.

Figure 14 Solute particles raise the boiling point of a solution. **Describe** *how antifreeze works in a car.*

section 3 review

Summary

Particles with a Charge

- Ions, charged particles, are formed from neutral compounds in such a way that the atoms take on a charge.
- Electrolytes can conduct electricity.
- Dissociation is the process of breaking an ionic compound into its positive and negative ions.
- Ions are found in the fluids that are in and around the cells in your body.

Effects of Solute Particles

- Solutes affect the physical properties of a solution by the number of solute particles, not by the chemical nature of the solute.
- Solute particles change the freezing and boiling points of solutions.
- Fish that live in polar waters have a natural chemical antifreeze called glycoprotein.
- When a substance freezes, its particles arrange themselves in an orderly pattern.

Self Check

1. **Determine** what has taken place, ionization or dissociation, if calcium phosphate ($Ca_3(PO_4)_2$) breaks into Ca^{2+} and PO_4^{3-}.
2. **Identify** what kinds of solute particles are present in water solutions of electrolytes and nonelectrolytes.
3. **Describe** how an ionic substance dissociates in water.
4. **Describe** how the concentration of a solution influences its boiling point.
5. **Think Critically** In cold weather, people often put salt on ice that forms on sidewalks and driveways. The salt helps melt the ice, forming a saltwater solution. Explain why this solution may not refreeze.

Applying Math

6. **Graph Data** Use the following data points (0,12), (10,8),(20,4), and (30,0), to graph the effect of a solute on the freezing point of a solvent. Label the *x*-axis *Grams solute* and the *y*-axis *Freezing point*.
7. **Calculate Slope** Find the slope of the line you graphed in question 6.

Boiling Points of Solutions

Adding small amounts of salt to water that is being boiled and adding antifreeze to a radiator have a common result—increasing the boiling point.

▶ Real-World Question

How much can the boiling point of a solution be changed?

Goals
■ **Determine** how adding salt affects the boiling point of water.

Possible Materials
distilled water (400 mL)   ring stand
Celsius thermometer   hot plate
table salt, NaCl (72 g)   250-mL beaker

Safety Precautions

▶ Procedure

1. Copy the data table as shown in the next column. Bring 100 mL of distilled water to a gentle boil in a 250-mL beaker. Record the temperature. Do not touch the hot plate surface.

2. Dissolve 12 g of NaCl in 100 mL of distilled water. Bring this solution to a gentle boil and record its boiling point.

3. Repeat step 2, using 24 g of NaCl, then 36 g.

4. Make a graph of your results. Put boiling point on the *x*-axis and grams of NaCl on the *y*-axis.

Effects of Solute on Boiling Point

Grams of NaCl Solute	Boiling Point (°C)
0	
12	Do not write in this book.
24	
36	

▶ Conclude and Apply

1. **Explain** the difference between the boiling points of pure water and a water solution.

2. **Predict** what would have been the effect of doubling the amount of water instead of the amount of NaCl in step 3.

3. **Predict** what would happen if you continued to add more salt. Would your graph continue in the same pattern or eventually level off? Explain your prediction.

𝒞ommunicating Your Data

Compare your results with those of other groups and discuss any differences in the results obtained. **For more help, refer to the** Science Skill Handbook.

Dissolving Without Water

When Water Won't Work

Water often is referred to as the universal solvent because it can dissolve so many things. However, there are some things, such as oil, that it can't dissolve. Why?

As you learned in the first section, water has positive and negative areas that allow it to attract polar solutes. However, **nonpolar** materials have no separated positive and negative areas. Because of this, they are not attracted to polar materials, which means they are not attracted to water molecules. Nonpolar materials do not dissolve in water except to a small extent, if at all.

Nonpolar Solutes An example of a nonpolar substance that does not dissolve in water can be seen on many dinner tables. The vinegar-and-oil salad dressing shown in **Figure 15** has two distinct layers—the bottom layer is vinegar, which is a solution of acetic acid in water, and the top layer is salad oil.

Most salad oils contain large molecules made of carbon and hydrogen atoms, which are called hydrocarbons. In hydrocarbons, carbon and hydrogen atoms share electrons in a nearly equal manner. This equal distribution of electrons means that the molecule has no separate positive and negative areas. Therefore, the nonpolar oil molecule is not attracted to the polar water molecules in the vinegar solution. That's why you must shake this kind of dressing to mix it just before you pour it on your salad.

Figure 15 Oil and vinegar do not form a solution.
Infer *How do the particles in this substance disperse so it is tasteful to eat?*

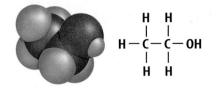

Figure 16 Ethanol, C_2H_5OH, has a polar —OH group at one end but the $-C_2H_5$ section is nonpolar. **Determine** *the molecular weight for ethanol.*

Versatile Alcohol Some substances form solutions with polar as well as nonpolar solutes because their molecules have a polar and a nonpolar end. Ethanol, shown in **Figure 16,** is such a molecule. The polar end dissolves polar substances, and the nonpolar end dissolves nonpolar substances. For example, ethanol dissolves iodine, which is nonpolar, as well as water, which is polar.

✔ **Reading Check** *How can alcohol dissolve both polar and nonpolar substances?*

Useful Nonpolar Molecules

Some materials around your house may be useful as nonpolar solvents. For example, mineral oil may be used as a solvent to remove candle wax from glass or metal candleholders. Both the mineral oil and the candle wax are nonpolar materials. Mineral oil can also aid in removing bubble gum from some surfaces for the same reason. Oil-based paints contain pigments that are dissolved in oils. In order to thin or remove such paints, a nonpolar solvent must be used. The gasoline you use in your car and lawnmower is a solution of hydrocarbons, which are nonpolar substances.

Dry cleaners use nonpolar solvents when removing oily stains. The word *dry* refers to the fact that no water is used in the process. Molecules of a nonpolar solute can slip easily among molecules of a nonpolar solvent. That is why dry cleaning can remove stains of grease and oil that you cannot clean easily yourself. A general statement that describes which substance dissolves which is the phrase "like dissolves like."

Many nonpolar solvents are connected with specific jobs. People who paint pictures using oil-based paints probably use the solvent turpentine. It comes from the sap of a pine tree. **Figure 17** shows how well turpentine dissolves non-polar paint.

Figure 17 With no polarity to interfere, paint molecules slide smoothly among molecules of turpentine.

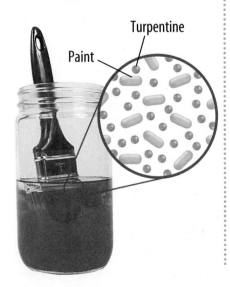

Turpentine

Paint

Drawbacks of Nonpolar Solvents Although nonpolar solvents have many uses, they have some drawbacks, too. First, many nonpolar solvents are flammable. Also, some are toxic, which means they are dangerous if they come into contact with the skin or if their vapors are inhaled. For these reasons, you must always be careful when handling these materials and never use them in a closed area. Good ventilation is critical, because nonpolar solvents tend to evaporate more readily than water, and even small amounts of a nonpolar liquid can produce high concentrations of harmful vapor in the air.

Figure 18 The long hydrocarbon tail of sodium stearate is nonpolar. The head is ionic.

How Soap Works The oils on human skin and hair keep them from drying out, but the oils can also attract and hold dirt. The oily dirt is a nonpolar mixture, so washing with water alone won't clean away the dirt. This is where soap comes in. Soaps, you might say, have a split personality. They are substances that have polar and nonpolar properties. Soaps are salts of fatty acids, which are long hydrocarbon molecules with a carboxylic acid group –COOH at one end. When a soap is made, the hydrogen atom of the acid group is removed, leaving a negative charge behind, and a positive ion of sodium or potassium is attached. This is shown in **Figure 18.**

Thus, soap has an ionic end that will dissolve in water and a long hydrocarbon portion that will dissolve in oily dirt. In this way, the dirt is removed from your skin, hair, or a fabric, suspended in the wash water, and washed away, as shown in **Figure 19.**

Reading Check *Why doesn't water alone clean oily dirt?*

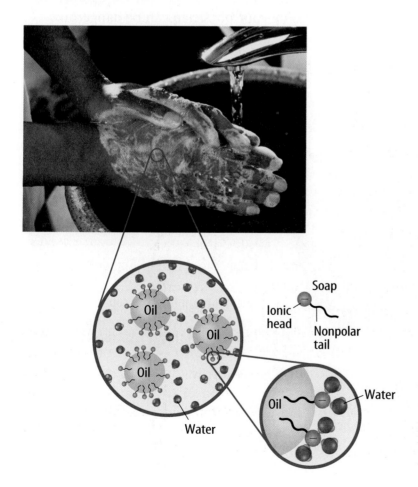

Figure 19 Soap cleans because its nonpolar hydrocarbon part dissolves in oily dirt and its ionic part interacts strongly with water. The oil and water mix and the dirt is washed away.

Mini LAB

Observing Clinging Molecules

Procedure
1. Lay **two clean pennies** side by side and heads up on a **paper towel.**
2. Slowly place drops of **water** from a **dropper** onto the head of one penny. Count each drop and continue until the accumulated water spills off the edge of the penny.
3. With adult supervision, repeat step 2 using **rubbing alcohol,** which is approximately 30 percent isopropyl alcohol, and the other penny.

Analysis
1. Which penny held the most drops before liquid spilled over the edge of the penny?
2. Isopropyl alcohol has the formula C_3H_7OH. How polar do you think it is?
3. How do the results of the experiment support the concept of polarity and molecules sticking to each other?

Figure 20 The structural formula of vitamin A shows a long hydrocarbon chain that makes it nonpolar. Foods such as liver, lettuce, cheese, eggs, carrots, sweet potatoes, and milk are good sources of this fat-soluble vitamin. Vitamins D, E, and K are also fat-soluble vitamins.

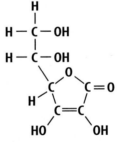

Polarity and Vitamins

Having the right kinds and amounts of vitamins is important for your health. Some of the vitamins you need, such as vitamin A, shown in **Figure 20,** are nonpolar and can dissolve in fat, which is another nonpolar substance. Because fat and fat-soluble vitamins do not wash away with the water that is present in the cells throughout your body, the vitamins can accumulate in your tissues. Some fat-soluble vitamins are toxic in high concentrations, so taking large doses or taking doses that are not recommended by your physician can be dangerous.

Figure 21 Although vitamin C has carbon-to-carbon bonds, it also has polar groups and is water soluble. Foods such as those shown here are good sources of vitamin C, which helps heal wounds and helps the body absorb iron.

Table 3 Sources of Vitamin C		
Food	**Amount**	**mg**
Orange juice, fresh	1 cup	124
Green peppers, raw	1/2 cup	96
Broccoli, raw	1/2 cup	70
Cantaloupe	1/4 melon	70
Strawberries	1/2 cup	42

Other vitamins, such as vitamins B and C, are polar compounds. When you look at the structure of vitamin C, shown in **Figure 21,** you will see that it has several carbon-to-carbon bonds. This might make you think that it is nonpolar. But, if you look again, you will see that it also has several oxygen-to-hydrogen bonds that resemble those found in water. This makes vitamin C polar.

Polar vitamins dissolve readily in the water that is in your body. These vitamins do not accumulate in tissue because any excess vitamin is washed away with the water in the body. For this reason, you must replace water-soluble vitamins by eating enough of the foods that contain them or by taking vitamin supplements. **Table 3** lists some good sources of vitamin C. In general, the best way to stay healthy is to eat a variety of healthy foods. Such a diet will supply the vitamins you need with no risk of overdoses.

 **Reading Check** *Why do you need to replace some of the vitamins used in your body?*

INTEGRATE Health

Vitamins in Excess
Drinking too much carrot juice, which is rich in beta carotene, a substance related to vitamin A, can cause the palms of your hands and soles of your feet to turn orange. Though this condition is not serious and the color fades in time, taking too much of some vitamins can be dangerous. Research what might result from taking too much vitamin B-6, vitamin D, and niacin.

section 4 review

Summary

When Water Won't Work
- Water cannot dissolve all substances.
- Nonpolar molecules are not attracted to polar molecules.

Useful Nonpolar Molecules
- Nonpolar solvents have many household and industrial uses.
- Nonpolar solvents may have drawbacks including flammability and toxicity.
- Some molecules have both polar and nonpolar properties. Solvents made from these kinds of substances can dissolve things that water alone cannot.

Polarity and Vitamins
- Nonpolar vitamins may dissolve in fat and accumulate to sometimes dangerous levels in your body.
- Polar vitamins dissolve readily in water and can be flushed from the body before absorption.

Self Check

1. **Explain** how a solute can dissolve in polar and nonpolar solvents.
2. **Explain** the phrase "like dissolves like" and give an example of two polar "like" substances.
3. **Describe** how soap cleans greasy dirt from your hands.
4. **Infer** Some small engines require a mixture of oil and gasoline. Gasoline evaporates easily. What conclusion can be drawn about the polarity of the engine oil?
5. **Think Critically** What might happen to your skin if you washed with soap too often?

Applying Math

6. **Calculate Mass** If 60 mg of vitamin C in a multivitamin provides only 75 percent of the recommended daily dosage for children, how much is recommended?
7. **Interpret Data** To get the recommended 80 mg of vitamin C, refer to **Table 3** to determine approximately how much fresh orange juice you must drink.

Saturated S🫙lutions

🜂 Real-World Question

Two major factors to consider when you are dissolving a solute in water are temperature and the ratio of solute to solvent. What happens to a solution as the temperature changes? To be able to draw conclusions about the effect of temperature, you must keep other variables constant. For example, you must be sure to stir each solution in a similar manner. How does solubility change as temperature is increased?

🜂 Procedure

1. Place 20 mL of distilled water in a test tube.

2. Add 30 g of sugar.

3. Stir. Does this dissolve?

4. If it dissolves completely, add another 5 g of sugar to the test tube. Does it dissolve?

5. Continue adding 5-g amounts of sugar until no more sugar dissolves.

6. Now place the beaker of water on the hot plate and hang the thermometer from the ring stand so that the bulb is immersed about halfway into the beaker, making sure it does not touch the sides or bottom. Record the starting temperature.

Goals

- **Observe** the effects of temperature on the amount of solute that dissolves.

Materials

distilled water at room temperature
large test tubes
Celsius thermometer
table sugar
copper wire stirrer, bent into a spiral as shown on the next page
test-tube holder
graduated cylinder (25-mL)
beaker (250-mL) with 150 mL of water
electric hot plate
test-tube rack
ring stand

Safety Precautions

WARNING: *Do NOT touch the test tubes or hot plate surface when hot plate is turned on or cooling down. When heating a solution in a test tube, keep it pointed away from yourself and others. Do NOT remove goggles until clean up including washing hands is completed.*

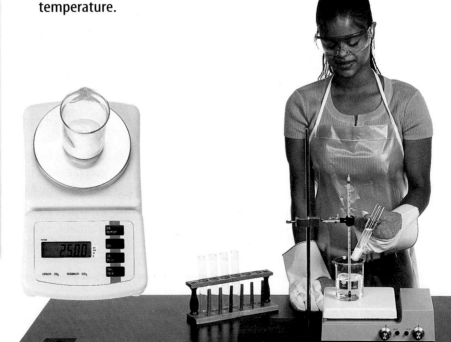

7. Using a test-tube holder, place the test tube into the water.

8. Gradually increase the temperature of the hot plate, while stirring the solution in the tube, until all the sugar dissolves.

9. Note the temperature at which this happens.

10. Add another 5 g of sugar and continue. Note the temperature at which this additional sugar dissolves.

11. Continue in this manner until you have at least four data points. Note the total amount of sugar that has dissolved. Record your data on the data table.

◉ *Analyze Your Data*

1. **Graph** your results using a line graph. Place grams of solute per 100 g of water (multiply the number of grams by five because you used only 20 mL of water) on the *y*-axis and place temperature on the *x*-axis.

2. **Interpret Data** Using your graph, estimate the solubility of sugar at 100°C and at 0°C, the boiling and freezing point of water, respectively.

Dissolving Sugar in Water	
Temperature	Total Grams of Sugar Dissolved
Do not write in this book.	

◉ *Conclude and Apply*

1. How did the saturation change as the temperature was increased?

2. **Compare** your results with those given in **Table 2.**

*C*ommunicating **Your Data**

Compare your results with those of other groups and discuss any differences noted. Why might these differences have occurred? **For more help, refer to the Science Skill Handbook.**

SCIENCE Stats

Weird Solutions

Did you know...

... The "brightest" solutions can glow like a streetlight. Glowing rods called light sticks are an example. Each rod contains two liquids in separate glass or plastic containers. When you flex the rod, the containers break, and the solutions mix and react to produce luminescence, or glowing light. A similar process called bioluminescence allows some living organisms, like fireflies, to glow.

... One of the hardest solutions is steel, a solid solution of iron, carbon, and other elements. When you add some chromium and nickel to the mix, you get stainless steel, which is a tough, rust-resistant solution. In 1998, the United States produced nearly 100 million metric tons of raw steel—enough to make more than 1,800 Empire State Buildings.

... The saltiest and largest body of solution in the western hemisphere is the Great Salt Lake in Utah. If all the salt in the lake dried out and hardened, the result would be a rock with a mass of about 4 1/2 trillion kg—as heavy as 300 million large trucks.

Applying Math

1. Great Salt Lake's salinity is 5 percent when the water is highest and 30 percent when the water is lowest. What is the lake's salinity when the water level is halfway between its highest and lowest levels?
2. Glow sticks shine for about 10 h. If you kept a glow stick glowing continuously in your window for seven days, how many sticks would you need?

Reviewing Main Ideas

Section 1 How Solutions Form

1. A solution is a mixture that has the same composition, color, density, and taste throughout.

2. The substance being dissolved is called a solute, and the substance that does the dissolving is called a solvent.

3. The rate of dissolving can be increased by stirring, increasing surface area, or increasing temperature.

4. Under similar conditions, small particles of solute dissolve faster than large particles.

Section 2 Solubility and Concentration

1. Some compounds are more soluble than others, and this can be measured.

2. *Concentrated* and *dilute* are not precise terms used to describe concentration of solutions.

3. Concentrations can be expressed as percent by volume.

4. An unsaturated solution can dissolve more solute, and a saturated solution, like this tea, cannot. A supersaturated solution is made by raising the temperature of a saturated solution and adding more solute. If it is cooled carefully, the supersaturated solution will retain the dissolved solute.

Section 3 Particles in Solution

1. Substances that dissolve in water to produce solutions that conduct electricity are called electrolytes.

2. When water pulls apart the molecules of a polar substance, forming ions, the process is called ionization.

3. When ionic solids dissolve in water, the process is called dissociation, because the ions are already present in the solid.

Section 4 Dissolving Without Water

1. Water cannot dissolve all solutes.

2. Nonpolar solvents are needed to dissolve nonpolar solutes.

3. Some vitamins are nonpolar and dissolve in the fat contained in some body cells.

4. Nonpolar solvents can be dangerous as well as helpful. Many products, including the substances shown here, are packaged with cautions of flammability and toxicity.

FOLDABLES Use the Foldable that you made at the beginning of this section to help you review the characteristics of solvents and solutes.

Using Vocabulary

dissociation p. 677
electrolyte p. 676
ion p. 676
ionization p. 676
nonelectrolyte p. 676
nonpolar p. 681
polar p. 667
saturated solution p. 673

solubility p. 671
solution p. 664
solute p. 665
solvent p. 665
supersaturated solution
     p. 674
unsaturated solution p. 673

Fill in the blanks with correct vocabulary or words.

1. In lemonade, sugar is the _____ and water is the _____.

2. During _____, particles in an ionic solid are separated and drawn into solution.

3. If more of substance B dissolves in water than substance A, then substance B has a higher _____ than substance A.

4. Adding a seed crystal may cause solute to crystallize from a(n) _____.

5. More solute can be added to a(n) _____.

6. Nonpolar solutes in a solution are called _____.

Checking Concepts

Choose the word or phrase that best answers the question.

7. Which of the following is NOT a solution?
   A) glass of flat soda
   B) air in a scuba tank
   C) bronze alloy
   D) mud in water tank

8. What term is NOT appropriate to use when describing solutions?
   A) heterogeneous   C) liquid
   B) gaseous   D) solid

9. When iodine is dissolved in alcohol, what term is used to describe the alcohol?
   A) alloy   C) solution
   B) solvent   D) solute

10. What word is used to describe a mixture that is 85 percent copper and 15 percent tin?
    A) alloy   C) saturated
    B) solvent   D) solute

11. Solvents such as paint thinner and gasoline evaporate more readily than water because they are what type of compounds?
    A) ionic
    B) nonpolar
    C) dilute
    D) polar

12. What can a polar solvent dissolve?
    A) any solute   C) a nonpolar solute
    B) a polar solute   D) no solute

13. If a water solution conducts electricity, what must the solute be?
    A) gas   C) liquid
    B) electrolyte D) nonelectrolyte

14. In forming a water solution, what process does an ionic compound undergo?
    A) dissociation   C) ionization
    B) electrolysis   D) no change

15. What can you increase to make a gas more soluble in a liquid?
    A) particle size   C) stirring
    B) pressure   D) temperature

16. If a solute crystallizes out of a solution when a seed crystal is added, what kind of solution is it?
    A) unsaturated   C) supersaturated
    B) saturated   D) dilute

Interpreting Graphics

17. Copy and complete the following concept map on solutions.

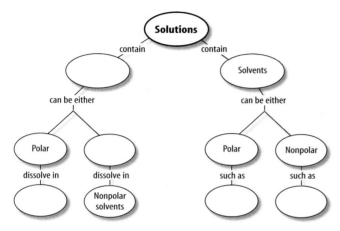

Use the table below to answer question 18.

Limits of Solubility

Compound	Type of Solution	Solubility in 100 g Water at 20°C
$CuSO_4$		32.0 g
KCl	Do not write in this book.	34.0 g
KNO_3		31.6 g
$NaClO_3$		95.9 g

18. Identify Using the data in **Table 2,** fill in the following table. Use the terms *saturated,* *unsaturated,* and *supersaturated* to describe the type of solution.

Thinking Critically

19. Explain why potatoes might cook more quickly in salted water than in unsalted water.

20. Explain what happens when an ionic compound such as copper(II) sulfate, $CuSO_4$, dissolves in water.

21. Explain why the term *dilute* is not precise.

22. Explain why the statement, "Water is the solvent in a solution," is not always true.

Applying Math

23. Measure in SI 153 g of potassium nitrate have been dissolved in enough water to make 1 L of this solution. You use a graduated cylinder to measure 80 mL of solution. What mass of potassium nitrate is in the 80-mL sample?

Use the graph below to answer question 24.

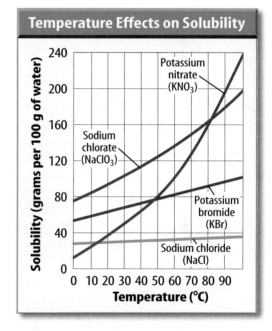

24. Interpret Data Determine the temperature at which a solution of 80 g potassium nitrate (KNO_3) in 100 mL of water is saturated.

25. Use Numbers How would you make a 25 percent solution by volume of apple juice?

26. Make and Use Graphs Using **Table 2,** make a graph of solubility versus temperature for $CuSO_4$ (copper(II) sulfate) and KCl. How would you make a saturated solution of each substance at 80°C?

Record your answers on the answer sheet provided by your teacher or on a sheet of paper.

Use the graph below to answer questions 1–3.

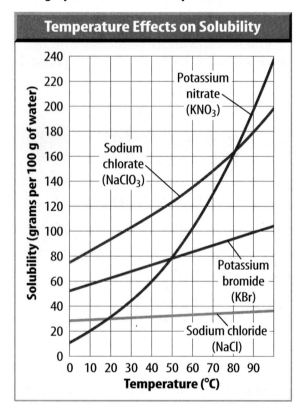

Temperature Effects on Solubility

1. How much potassium nitrate will you have to add to 100 g of water at 40°C to make a saturated solution?
A. 60 g   **C.** 100 g
B. 60 g   **D.** 240 g

2. If 25 g of sodium chlorate are dissolved in 100 g of water at 70°C, how would you describe the solution?
A. concentrated   **C.** saturated
B. supersaturated   **D.** dilute

Test-Taking Tip

Answer Bubbles Double check that you are filling in the correct answer bubble for the question number you are working on.

3. Which of the following will make a saturated solution if added to 100 g of water?
A. 20 g of NaCl if the water is 50°C
B. 100 g of KBr if the water is 90°C
C. 80 g of $NaClO_3$ if the water is 30°C
D. 60 g of KNO_3 if the water is 100°C

4. Which of the following statements about solubility is true as the temperature increases?
A. The solubility of both gases and solids increases.
B. The solubility of both gases and solids decreases.
C. The solubility of gases increases, while the solubility of solids decreases.
D. The solubility of gases decreases, while the solubility of solids increases.

Use the illustration below to answer questions 5 and 6.

5. Which of the following will NOT make the crystal of rock candy dissolve faster in water?
A. stirring   **C.** cooling
B. heating   **D.** shaking

6. Which of the following statements is true about how grinding the crystal would affect its dissolving rate?
A. Grinding would increase the surface area and slow down dissolving.
B. Grinding would increase the surface area and speed up dissolving.
C. Grinding would decrease the surface are and slow down dissolving.
D. Grinding would decrease the surface area and speed up dissolving.

Part 2 | **Short Response/Grid In**

Record your answers on the answer sheet provided by your teacher or on a sheet of paper.

7. A girl mixes a saturated solution of sugar in water in science lab on a Friday. On Monday, the open container has particles of sugar on the bottom. Explain possible reasons to explain why this happened.

Use the illustration below to answer questions 8 and 9.

8. The drawing above shows carbon dioxide gas dissolved in water. What are two ways you could make more gas dissolve in the water?

9. Why does shaking or pouring a carbonated drink cause gas to come out of solution?

10. The solubility of potassium chloride in water is 34 g per 100 g of water at 20°C. A warm solution containing 100 g of potassium chloride in 200 g of water is cooled to 20°C. How many grams of potassium chloride will come out of solution?

11. When water is heated slowly, small bubbles form in the liquid. These bubbles do not contain water vapor. What is in the bubbles and why do they form?

12. Why will turpentine remove oil-based paint from a paint brush while water will not?

13. Why is salt mixed with the ice in a hand-crank ice cream maker?

Part 3 | **Open Ended**

Record your answers on a sheet of paper.

14. Why can carp, catfish, and other fish with low oxygen needs live in warmer waters than can trout, which need large amounts of oxygen?

Use the illustration below to answer questions 15 and 16.

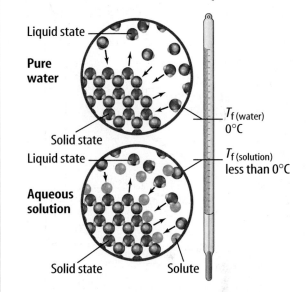

15. The drawing shows what happens to the freezing point of water when antifreeze is dissolved in the water to form a solution. Explain how this happens.

16. How can antifreeze also raise the boiling point of water?

17. When you take clothing to the dry cleaner, it is important to identify any stains that are on the clothing. Why does the dry cleaner need this information?

18. You are given a clear water solution containing potassium nitrate. How could you determine whether the solution is unsaturated, saturated, or supersaturated?

19. A solution conducts electricity. What do you know about the solution?

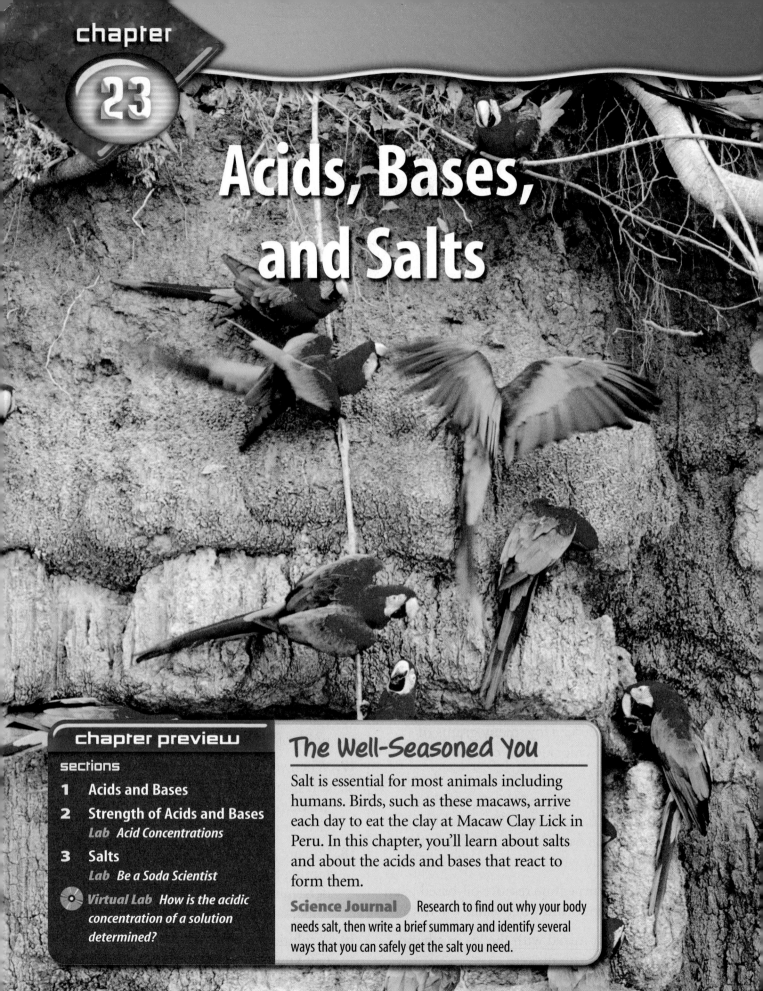

Acids, Bases, and Salts

The Well-Seasoned You

Salt is essential for most animals including humans. Birds, such as these macaws, arrive each day to eat the clay at Macaw Clay Lick in Peru. In this chapter, you'll learn about salts and about the acids and bases that react to form them.

Science Journal Research to find out why your body needs salt, then write a brief summary and identify several ways that you can safely get the salt you need.

Start-Up Activities

The Effects of Acid Rain

Many limestone caves and rock formations are shaped by water containing carbon dioxide. Higher levels of carbon dioxide in acid rain can damage marble structures. Observe this reaction using soda water to represent acid rain and chalk, which like limestone and marble, is calcium carbonate.

1. Measure approximately 5 g of classroom chalk.
2. Crush it slightly and place it in a 100-mL beaker.
3. Add 50 mL of plain, bottled, carbonated water to the beaker.
4. After several minutes, stir the mixture.
5. When the mixture stops reacting, filter it using a paper filter in a glass funnel.
6. Dry the residue overnight and determine its mass.
7. **Think Critically** Record your observations in your Science Journal. How did the mass change? Write your conclusions about the effect of acid rain on marble buildings and monuments.

Acids, Bases, and Salts The very essence of life, DNA, is an acid. You also may be familiar with ascorbic acid, or vitamin C. Make the following Foldable to compare and contrast the characteristics of acids, bases, and salts.

STEP 1 **Fold** one sheet of paper lengthwise.

STEP 2 **Fold** into thirds.

STEP 3 **Unfold** and **draw** overlapping ovals. Cut the top sheet along the folds.

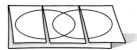

STEP 4 **Label** the ovals *Acids, Salts,* and *Bases*.

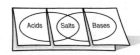

Construct a Venn Diagram As you read the chapter, list the characteristics of acids, bases, and salts under the appropriate tabs.

Preview this chapter's content and activities at gpscience.com

Acids and Bases

Reading Guide

What You'll Learn

- **Compare and contrast** acids and bases and identify the characteristics they have.
- **Examine** some formulas and uses of common acids and bases.
- **Determine** how the process of ionization and dissociation apply to acids and bases.

Why It's Important

Acids and bases are found almost everywhere—from fruit juice and gastric juice to soaps.

Review Vocabulary

electrolyte: compound that breaks apart in water, forming charged particles (ions) that can conduct electricity

New Vocabulary

- acid
- hydronium ion
- indicator
- base

Acids

What comes to mind when you hear the word *acid?* Do you think of a substance that can burn your skin or even burn a hole through a piece of metal? Do you think about sour foods like those shown in **Figure 1?** Although some acids can burn and are dangerous to handle, most acids in foods are safe to eat. What acids have in common, however, is that they contain at least one hydrogen atom that can be removed when the acid is dissolved in water.

Properties of Acids When an acid dissolves in water, some of the hydrogen is released as hydrogen ions, H^+. An **acid** is a substance that produces hydrogen ions in a water solution. It is the ability to produce these ions that gives acids their characteristic properties. When an acid dissolves in water, H^+ ions interact with water molecules to form H_3O^+ ions, which are called **hydronium ions** (hi DROH nee um • I ahnz).

Acids have several common properties. For one thing, all acids taste sour. The familiar, sour taste of many foods is due to acids. However, taste never should be used to test for the presence of acids. Some acids can damage tissue by producing painful burns. Acids are corrosive. Some acids react strongly with certain metals, seeming to eat away the metals as metallic compounds and hydrogen gas form. Acids also react with indicators to produce predictable changes in color. An **indicator** is an organic compound that changes color in acid and base. For example, the indicator litmus paper turns red in acid.

Figure 1 The acids in these common foods give each their distinctive sour taste.

Common Acids Many foods contain acids. In addition to citric acid in citrus fruits, lactic acid is found in yogurt and buttermilk, and any pickled food contains vinegar, also known as acetic acid. Your stomach uses hydrochloric acid to help digest your food. At least four acids (sulfuric, phosphoric, nitric, and hydrochloric) play vital roles in industrial applications.

✔ Reading Check *Which four acids are important for industry?*

Table 1 lists the names and formulas of a few acids, their uses, and some properties. Three acids are used to make fertilizers—most of the nitric acid and sulfuric acid and approximately 90 percent of phosphoric acid produced are used for this purpose. Many acids can burn, but sulfuric acid can burn by removing water from your skin as easily as it takes water from sugar, as shown in **Figure 2.**

Figure 2 When sulfuric acid is added to sugar the mixture foams, removing hydrogen and oxygen atoms as water and leaving air-filled carbon.

Table 1 Common Acids and Their Uses

Name, Formula	Use	Other Information
Acetic acid, CH_3COOH	Food preservation and preparation	When in solution with water, it is known as vinegar.
Acetylsalicylic acid, $HOOC-C_6H_4-OOCCH_3$	Pain relief, fever relief, to reduce inflammation	Known as aspirin
Ascorbic acid, $H_2C_6H_6O_6$	Antioxidant, vitamin	Called vitamin C
Carbonic acid, H_2CO_3	Carbonated drinks	Involved in cave, stalactite, and stalagmite formation and acid rain
Hydrochloric acid, HCl	Digestion as gastric juice in stomach, to clean steel in a process called pickling	Commonly called muriatic acid
Nitric acid, HNO_3	To make fertilizers	Colorless, yet yellows when exposed to light
Phosphoric acid, H_3PO_4	To make detergents, fertilizers and soft drinks	Slightly sour but pleasant taste, detergents containing phosphates cause water pollution
Sulfuric acid, H_2SO_4	Car batteries, to manufacture fertilizers and other chemicals	Dehydrating agent, causes burns by removing water from cells

Observing Acid Relief

WARNING: *Do not eat antacid tablets.*

Procedure

1. Add 150 mL of **water** to a **250-mL beaker.**
2. Add three drops **1M HCl** and 12 drops of **universal indicator.**
3. Observe the color of the solution.
4. Add an **antacid tablet** and observe for 15 minutes.

Analysis

1. Describe any changes that took place in the solution.
2. Explain why these changes occurred.

Bases

You might not be as familiar with bases as you are with acids. Although you can eat some foods that contain acids, you don't consume many bases. Some foods, such as egg whites, are slightly basic. Other examples of basic materials are baking powder and amines found in some foods. Medicines, such as milk of magnesia and antacids, are basic, too. Still, you come in contact with many bases every day. For example, each time you wash your hands using soap, you are using a base. One characteristic of bases is that they feel slippery, like soapy water. Bases are important in many types of cleaning materials, as shown in **Figure 3.** Bases are important in industry, also. For example, sodium hydroxide is used in the paper industry to separate fibers of cellulose from wood pulp. The freed cellulose fibers are made into paper.

Bases can be defined in two ways. Any substance that forms hydroxide ions, OH^-, in a water solution is a **base.** In addition, a base is any substance that accepts H^+ from acids. The definitions are related, because the OH^- ions produced by some bases do accept H^+ ions.

Properties of Bases One way to think about bases is as the complements, or opposites, of acids. Although acids and bases share some common features, the bases have their own characteristic properties. In the pure, undissolved state, many bases are crystalline solids. In solution, bases feel slippery and have a bitter taste. Like strong acids, strong bases are corrosive, and contact with skin can result in severe burns. Therefore, taste and touch never should be used to test for the presence of a base. Finally, like acids, bases react with indicators to produce changes in color. The indicator litmus turns blue in bases.

Figure 3 Bases are commonly found in many cleaning products used around the home. **Identify** *the property of bases evident in soaps.*

Figure 4 Two applications of bases are shown here.

Aluminum hydroxide is a base used in water-treatment plants. Its sticky surface collects impurities, making them easier to filter from the water.

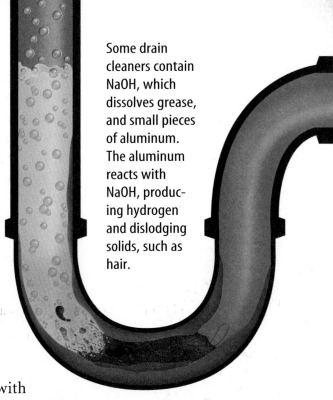

Some drain cleaners contain NaOH, which dissolves grease, and small pieces of aluminum. The aluminum reacts with NaOH, producing hydrogen and dislodging solids, such as hair.

Common Bases You probably are familiar with many common bases because they are found in cleaning products used in the home. These and some other bases are shown in **Table 2,** which also includes their uses and some information about them. **Figure 4** shows two uses of bases that you might not be familiar with.

Table 2 Common Bases and Their Uses		
Name, Formula	**Use**	**Other Information**
Aluminum hydroxide, $Al(OH)_3$	Color-fast fabrics, antacid, water purification as shown in **Figure 4**	Sticky gel that collects suspended clay and dirt particles on its surface
Calcium hydroxide, $Ca(OH)_2$	Leather-making, mortar and plaster, lessen acidity of soil	Called caustic lime
Magnesium hydroxide, $Mg(OH)_2$	Laxative, antacid	Called milk of magnesia when in water
Sodium hydroxide, NaOH	To make soap, oven cleaner drain cleaner, textiles, and paper	Called lye and caustic soda; generates heat (exothermic) when combined with water, reacts with metals to form hydrogen
Ammonia, NH_3	Cleaners, fertilizer, to make rayon and nylon	Irritating odor that is damaging to nasal passages and lungs

Acidic Stings *Some ants add sting to their bite by injecting a solution of formic acid. In fact, formic acid was named for ants, which make up the genus Formica. Still, ants are considered tasty treats by many animals. For example, one woodpecker called a flicker has saliva that is basic enough to take the sting out of ants.*

Solutions of Acids and Bases

Many of the products that rely on the chemistry of acids and bases are solutions, such as the cleaning products and food products mentioned previously. Because of its polarity, water is the main solvent in these products.

Dissociation of Acids You have learned that substances such as HCl, HNO_3, and H_2SO_4 are acids because of their ability to produce hydrogen ions (H^+) in water. *When an acid dissolves in water, the negative areas of nearby water molecules attract the positive hydrogen in the acid.* The acid dissociates—or separates—into ions and the hydrogen atom combines with a water molecule to form hydronium ions (H_3O^+). Therefore, an acid can more accurately be described as a compound that produces hydronium ions when dissolved in water. This process is shown in **Figure 5.**

Dissociation of Bases Compounds that can form hydroxide ions (OH^-) in water are classified as bases. If you look at **Table 2,** you will find that most of the substances listed contain –OH in their formulas. *When bases that contain –OH dissolve in water, the negative areas of nearby water molecules attract the positive ion in the base.* The positive areas of nearby water molecules attract the –OH of the base. *The base dissociates into a positive ion and a negative ion—a hydroxide ion (OH⁻).* This process also is shown in **Figure 5.** Unlike acid dissociation, *water molecules do not combine with the ions formed from the base.*

$$NaOH(s) \xrightarrow{H_2O} Na^+(aq) + OH^-(aq)$$

Figure 5 Acids and bases are classified by the ions they produce when they dissolve in water. Acids produce hydronium ions in water. Bases produce hydroxide ions in water.

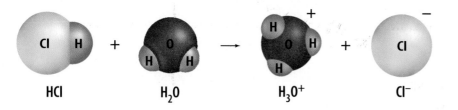

$$HCl + H_2O \rightarrow H_3O^+ + Cl^-$$

HCl H_2O H_3O^+ Cl^-

When hydrogen chloride dissolves in water, a hydronium ion and a chloride ion are produced.

$$NaOH + H_2O \rightarrow Na^+ + OH^- + H_2O$$

NaOH H_2O Na^+ OH^- H_2O

When sodium hydroxide dissolves in water, a sodium ion and a hydroxide ion are produced.

Figure 6 Ammonia reacts with water to produce some hydroxide ions, therefore, it is a base.

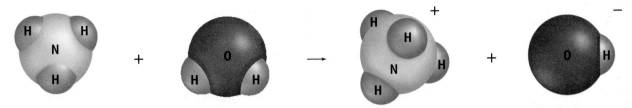

Ammonia Ammonia is a base that does not contain −OH. In a water solution, dissociation takes place when the ammonia molecule attracts a hydrogen ion from a water molecule, forming an ammonium ion (NH_4^+). This leaves a hydroxide ion (OH^-), as shown in **Figure 6.**

✔ Reading Check *How does ammonia react in a water solution?*

Ammonia is a common household cleaner. However, products containing ammonia never should be used with other cleaners that contain chlorine (sodium hypochlorite), such as some bathroom bowl cleaners and bleach. A reaction between sodium hypochlorite and ammonia produces the toxic gases hydrazine and chloramine. Breathing these gases can severely damage lung tissues and cause death.

Solutions of both acids and bases produce some ions that are capable of carrying electric current to some extent. Thus, they are said to be electrolytes.

Science Online

Topic: Cleaner Chemistry
Visit gpscience.com for Web links to information about the dangers of mixing ammonia cleaners with chlorine or hydrochloric acid cleaners.

Activity Visit the cleaning products and laundry sections of the grocery store. Read the labels on several products. Make a list of products that include warnings on the labels and those that do not. Share your findings with the class.

section 1 review

Summary

Acids

- Acids, when dissolved in water, release H^+, which forms hydronium ions (H_3O^+).
- Acids are sour tasting, corrosive, and reactive with indicators.

Bases

- Bases, when dissolved in water, form OH^-.
- Bases exist as crystals in the solid state, are slippery, have a bitter taste, are corrosive, and are reactive with indicators.

Solutions of Acids and Bases

- The polar nature of water allows acids and bases to dissolve in water.
- Dissociation is the separation of substances, such as acids and bases, into ions in water.

Self Check

1. **Identify** three important acids and three important bases and describe their uses.
2. **Describe** an indicator.
3. **Predict** what metallic compound forms when sulfuric acid reacts with magnesium metal.
4. **Infer** If an acid donates H^+ and a base produces OH^-, what compound is likely to be produced when acids react with bases?
5. **Think Critically** Vinegar contains acetic acid, CH_3COOH. Is acetic acid organic or inorganic? How do you know?

Applying Math

6. **Calculate** the molecular weight of acetylsalicylic acid, $C_9H_8O_4$.

Strength of Acids and Bases

What You'll Learn

- **Determine** what is responsible for the strength of an acid or a base.
- **Compare and contrast** strength and concentration.
- **Examine** the relationship between pH and acid or base strength.
- **Examine** electrical conductivity.

Why It's Important

Understanding the strength of acids and bases helps you use them safely.

ⓘ Review Vocabulary

acid strength: the ability of an acid to dissociate completely

New Vocabulary

- strong acid
- weak acid
- strong base
- weak base
- pH
- buffer

Figure 7 Nearly all molecules of HCl, a strong acid, dissociate into ions in water. The bulb burns brightly. Only a few molecules of acetic acid, a weak acid, dissociate. The bulb is dimmer.

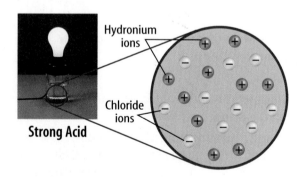

Hydronium ions

Chloride ions

Strong Acid

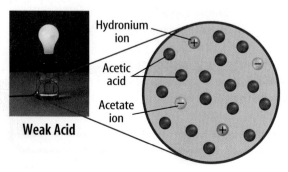

Hydronium ion

Acetic acid

Acetate ion

Weak Acid

Strong and Weak Acids and Bases

Some acids must be handled with great care. For example, sulfuric acid found in car batteries can burn your finger, yet you drink acids such as citric acid in orange juice and carbonic acid in soft drinks. Obviously, some acids are stronger than others. One measure of acid strength is the ability to dissociate in solution.

The strength of an acid or base depends on how many acid or base particles dissociate into ions in water. When a **strong acid** dissolves in water, nearly all the acid molecules dissociate into ions. HCl, HNO_3, and H_2SO_4 are examples of strong acids. When a **weak acid** dissolves in water, only a small fraction of the molecules dissolve in water. Acetic acid and carbonic acid are examples of weak acids.

Ions in solution can conduct an electric current. The more ions a solution contains, the more current it can conduct. The ability of a solution to conduct a current can be demonstrated using a lightbulb connected to a battery with leads placed in the solution, as shown in **Figure 7.** The strong acid solution conducts more current and the lightbulb burns brightly. The weak acid solution does not conduct as much current as a strong acid solution and the bulb burns less brightly.

Strong and Weak Acids Equations describing dissociation can be written in two ways. In strong acids, such as HCl, nearly all the acid dissociates. This is shown by writing the equation using a single arrow pointing toward the ions that are formed.

$$HCl(g) + H_2O(l) \rightarrow H_3O^+(aq) + Cl^-(aq)$$

Almost 100 percent of the particles in solution are H_3O^+ and Cl^- ions, and only a negligible number of HCl molecules are present.

Equations describing the dissociation of weak acids, such as acetic acid, are written using double arrows pointing in opposite directions. This means that only some of the CH_3COOH dissociates and the reaction does not go to completion.

$$CH_3COOH(l) + H_2O(l) \rightleftharpoons H_3O^+(aq) + CH_3COO^-(aq)$$

In an acetic acid solution, most of the particles are CH_3COOH molecules, and only a few CH_3COO^- and H^+ ions are in solution.

Strong and Weak Bases Remember that many bases are ionic compounds that dissociate to produce ions when they dissolve. A **strong base** dissociates completely in solution. The following equation shows the dissociation of sodium hydroxide, a strong base.

$$NaOH(s) \rightarrow Na^+(aq) + OH^-(aq)$$

The dissociation of ammonia, which is a weak base, is shown using double arrows to indicate that not all the ammonia ionizes. A **weak base** is one that does not dissociate completely.

$$NH_3(aq) + H_2O(l) \rightleftharpoons NH_4^+(aq) + OH^-(aq)$$

Because ammonia produces only a few ions and most of the ammonia remains in the form of NH_3, ammonia is a weak base.

Strength and Concentration Sometimes, when talking about acids and bases, the terms *strength* and *concentration* can be confused. The terms *strong* and *weak* are used to classify acids and bases. The terms refer to the ease with which an acid or base dissociates in solution. *Strong* acids and bases dissociate completely; *weak* acids and bases dissociate only partially. In contrast, the terms *dilute* and *concentrated* are used to indicate the concentration of a solution, which is the amount of acid or base dissolved in the solution. It is possible to have dilute solutions of strong acids and bases and concentrated solutions of weak acids and bases, as shown in **Figure 8.**

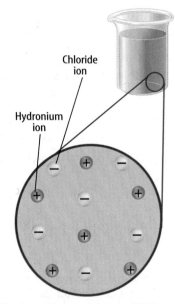

Figure 8 You can have a dilute solution of a strong acid and a concentrated solution of a weak acid.

This is a dilute solution of HCl.

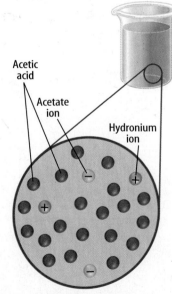

This is a concentrated solution of acetic acid.

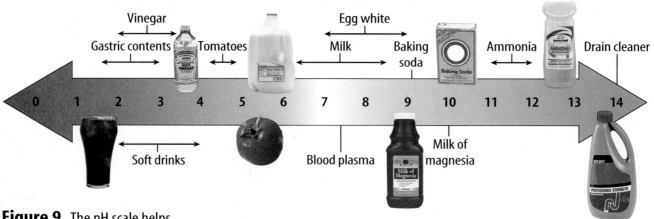

Vinegar

Gastric contents Tomatoes

Egg white

Milk   Baking
          soda

Ammonia   Drain cleaner

0   1   2   3   4   5   6   7   8   9   10   11   12   13   14

Soft drinks

Blood plasma

Milk of
magnesia

Figure 9 The pH scale helps classify solutions as acidic or basic.

pH of a Solution

If you have a swimming pool or keep tropical fish, you know that the pH of the water must be controlled. Also, many products such as shampoos claim to control pH so it suits your type of hair. The **pH** of a solution is a measure of the concentration of H^+ ions in it. The greater the H^+ concentration is, the lower the pH is and the more acidic the solution is. The pH measures how acidic or basic a solution is. To indicate pH, a scale ranging from 0 to 14 has been devised, as shown in **Figure 9.**

As the scale shows, solutions with a pH lower than 7 are described as acidic, and the lower the value is, the more acidic the solution is. Solutions with a pH greater than 7 are basic, and the higher the pH is, the more basic the solution is. A solution with a pH of exactly 7 indicates that the concentrations of H^+ ions and OH^- ions are equal. These solutions are considered neutral. Pure water at 25°C has a pH of 7.

One way to determine pH is by using a universal indicator paper. This paper undergoes a color change in the presence of H_3O^+ ions and OH^- ions in solution. The final color of the pH paper is matched with colors in a chart to find the pH, as shown in **Figure 10.** Is this an accurate way to determine pH?

Figure 10 The pH of a sample can be measured in several ways. Indicator paper gives an approximate value quickly, however, a pH meter is quick and more precise.

An instrument called a pH meter is another tool to determine the pH of a solution. This meter is operated by immersing the electrodes in the solution to be tested and reading the dial. Small, battery-operated pH meters with digital readouts are precise and convenient for use outside the laboratory when testing the pH of soils and streams, as shown in **Figure 10.**

Blood pH Your blood circulates throughout your body carrying oxygen, removing carbon dioxide, and absorbing nutrients from food that you have eaten. In order to carry out its many functions properly, the pH of blood must remain between 7.0 and 7.8. The main reason for this is that enzymes, the protein molecules that act as catalysts for many reactions in the body, cannot work outside this pH range. Yet you can eat foods that are acidic without changing the pH of your blood. How can this be? The answer is that your blood contains compounds called buffers that enable small amounts of acids or bases to be absorbed without harmful effects.

Buffers are solutions containing ions that react with additional acids or bases to minimize their effects on pH. One buffer system in blood involves bicarbonate ions, HCO_3^-. Because of these buffer systems, small amounts of even concentrated acid will not change pH much, as shown in **Figure 11.** Buffers help keep your blood close to a nearly constant pH of 7.4.

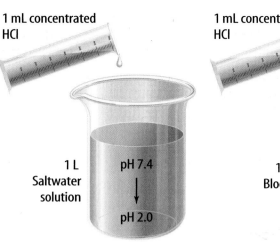
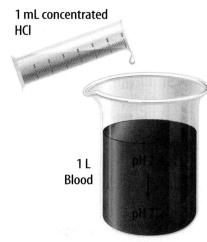

1 mL concentrated HCl

1 L Saltwater solution — pH 7.4 → pH 2.0

1 mL concentrated HCl

1 L Blood — pH 7.4, pH 7.2

Figure 11 This experiment shows how well blood plasma acts as a buffer. Adding 1 mL of concentrated HCl to 1 L of salt water changes the pH from 7.4 to 2.0. Adding the same amount of concentrated HCl to 1 L of blood plasma changes the pH from 7.4 to 7.2.

 **Reading Check** *What are buffers and how are they important for health?*

section 2 review

Summary

Strong and Weak Acids and Bases

- When strong acids dissolve in water, nearly all the acid molecules dissociate into ions. When weak acids dissolve in water, few molecules dissociate.

- When strong bases dissolve in water, nearly all base particles dissociate. When weak bases dissove, only a few particles dissociate.

- Ions in solution can conduct electricity.

- Strength refers to the ability of an acid or base to dissociate in water; concentration refers to how much acid or base is in solution.

pH of a Solution

- pH describes a substance as acidic or basic.

- Buffers are substances that minimize the effects of an acid or base on pH.

Self Check

1. **Describe** what determines the strength of an acid. A base?

2. **Explain** how to make a dilute solution of a strong acid.

3. **Explain** how electricity can be conducted by solutions.

4. **Describe** pH values of 9.1, 1.2, and 5.7 as basic, acidic, or very acidic.

5. **Think Critically** The proper pH range for a swimming pool is between 7.2 and 7.8. Most pools use two substances, Na_2CO_3 and HCl, to maintain this range. How would you adjust the pH if you found it was 8.2? 6.9?

Applying Math

6. **Use Equations** To determine the difference in pH strength, calculate 10^n, where n = difference between pHs. How much more acidic is a solution of pH 2.4 than a solution of pH 4.4?

Acid Concentrations

◉ Real-World Question

The science of acids and bases is not practiced only in a high-tech laboratory by degreed scientists. You can investigate the acidic concentrations of things in your own home using a simple home-made indicator solution. How can you tell if a substance is a strong or weak acid?

Goals

■ **Determine** the relative concentrations of common acid substances.

Materials

home-made cabbage indicator (indicates both acids and bases)
coffee filter
wax paper
grease pencil or masking tape
teaspoons (3)
alum
cream of tartar
fruit preservative

Safety Precautions

◉ Procedure

1. Use the grease pencil or masking tape and a pencil to label three areas on the wax paper *alum, cream of tartar,* and *fruit preservative.* These areas should be about 8 cm apart.

2. Place approximately 1/2 teaspoon of each of the three powders on the wax paper where labeled. Use a separate teaspoon for each substance.

3. Cut three strips from the coffee filter, about 1 cm wide by 8 cm long.

4. Dip the end of one of the strips into the cabbage indicator solution, then lay the wet end on top of the alum.

5. Wet a second strip and lay it in on top of the cream of tartar.

6. Wet the third strip and lay on top of the fruit preservative.

7. Wait 5 minutes, then check the indicator strips and record your observations.

◉ Conclude and Apply

1. Determine if all three substances were acids. Did the indicator strips turn a similar color?

2. Explain why each substance produced a different color.

3. Propose a possible rank of the concentrations.

4. Predict what you would have observed if you used sodium hydroxide instead of alum.

𝒞ommunicating Your Data

Compare your results with other groups in the class. Discuss any differences in the results you obtained.

Salts

Reading Guide

***What* You'll Learn**
- **Identify** a neutralization reaction.
- **Determine** what a salt is and how salts form.
- **Compare and contrast** soaps and detergents.
- **Examine** how esters are made and what they are used for.

***Why* It's Important**
You need salt to live and soaps and detergents to keep yourself and your clothing clean.

Review Vocabulary
ester: organic compounds made from acids and alcohols

New Vocabulary
- neutralization
- salt
- titration
- soap

Neutralization

Advertisements for antacids claim that these products neutralize the excess stomach acid that causes indigestion. Normally, gastric juice contains a dilute solution of hydrochloric acid. Too much acid can produce discomfort. Antacids contain bases or other compounds containing sodium, potassium, calcium, magnesium, or aluminum that react with acids to lower acid concentration. **Figure 12** shows what happens when you take an antacid tablet containing sodium bicarbonate—$NaHCO_3$. The equation is:

$$HCl(aq) + NaHCO_3(s) \rightarrow NaCl(aq) + CO_2(g) + H_2O(l)$$

In this case, the acid (HCl) is neutralized by the base ($NaHCO_3$).
Neutralization is a chemical reaction between an acid and a base that takes place in a water solution. For example, when HCl is neutralized by NaOH, hydronium ions from the acid combine with hydroxide ions from the base to produce neutral water.

$$H_3O^+(aq) + OH^- \rightarrow 2H_2O(l)$$

Salt Formation The equation above accounts for only half of the ions in the solution. The remaining ions react to form a salt. A **salt** is a compound formed when the negative ions from an acid combine with the positive ions from a base. In the reaction between HCl and NaOH the salt formed in water solution is sodium chloride.

$$Na^+(aq) + Cl^-(aq) \rightarrow NaCl(aq)$$

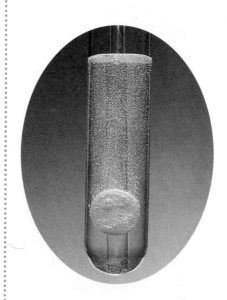

Figure 12 An antacid tablet reacts in your stomach much as it does in this dilute HCl. Usually, people chew antacid tablets before swallowing them. **Explain** *how this would affect the rate of the reaction.*

Figure 13 Like many animals, elephants get salt from natural deposits. Salt helps to maintain body processes.

Acid-Base Reactions The following general equation represents acid-base reactions in water.

> ### Acid-Base Reactions
>
> acid + base → salt + water

Another neutralization reaction occurs between HCl, an acid, and $Ca(OH)_2$, a base producing water and the salt $CaCl_2$.

$$2HCl(aq) + Ca(OH)_2(aq) \rightarrow CaCl_2(aq) + 2H_2O(l)$$

Salts

Salt is essential for many animals large and small. Some animals find it at natural deposits, as shown in **Figure 13.** Even insects, such as butterflies, need salt and often are found clustered on moist ground. You need salt too, especially because you lose salt in perspiration. How humans obtain one salt—sodium chloride—is shown in **Figure 14.**

There are many other salts, however, a few of which are shown in **Table 3.** Most salts are composed of a positive metal ion and an ion with a negative charge, such as Cl^- or CO_3^{2-}. Ammonium salts contain the ammonium ion, NH_4^+, rather than a metal.

Table 3 Some Common Salts and Their Uses		
Name, Formula	**Common Name**	**Uses**
Sodium chloride, NaCl	Salt	Food, manufacture of chemicals
Sodium hydrogen carbonate, $NaHCO_3$	Sodium bicarbonate Baking soda	Food, antacids
Calcium carbonate, $CaCO_3$	Calcite, chalk	Manufacture of paint and rubber tires
Potassium nitrate, KNO_3	Saltpeter	Fertilizers
Potassium carbonate, K_2CO_3	Potash	Manufacture of soap and glass
Sodium phosphate, Na_3PO_4	TSP	Detergents
Ammonium chloride, NH_4Cl	Sal ammoniac	Dry-cell batteries

Figure 14

The salt you use every day comes from both the land and the sea. Some salt can be mined from the ground in much the same way as coal, or salt can be obtained by the process of evaporation in crystallizing ponds.

◀ **EVAPORATION PROCESS** Workers fill evaporation ponds, like these near San Francisco Bay, California, with salt water, or brine. They move the brine from pond to pond as it becomes saltier through evaporation. (Red-tinted ponds have a higher salt content.) The saltiest water is then pumped from evaporation ponds into crystallizing ponds, where the remaining water is drained off. In the five years it takes to produce a crop of salt, brine may move through as many as 23 different ponds.

▲ **MINING SALT** Underground salt deposits are found where there was once a sea. Salt mines can be located deep underground or near Earth's surface in salt domes. Salt domes, such as the one above on Avery Island, Louisiana, form when pressure from Earth pushes buried salt deposits close to the surface, where they are easily mined.

Unit cell of sodium chloride (NaCl)

◀ **TABLE SALT** Raw sodium chloride is washed in chemicals and water to remove impurities before it appears on your dining-room table as salt. Iodine is added to table salt to ensure against iodine deficiency in the diet.

▼ **SALT MOUNDS** When the crystallizing ponds are drained, the result is huge piles of salt, like these on the Caribbean island of Bonaire.

Titration

Sometimes you need to know the concentration of an acidic or basic solution; for example, to determine the purity of a commercial product. This can be done using a process called **titration** (ti TRAY shun), in which a solution of known concentration is used to determine the concentration of another solution. **Figure 15** shows a titration experiment.

Titration involves a solution of known concentration, called the standard solution. This is added slowly and carefully to a solution of unknown concentration to which an acid/base indicator has been added. If the solution of unknown concentration is a base, a standard acid solution is used. If the unknown is an acid, a standard base solution is used.

The Endpoint Has a Color The titration shown in **Figure 15** shows how you could find the concentration of an acid solution. First, you would add a few drops of an indicator, such as phenolphthalein (fee nul THAY leen), to a carefully measured amount of the solution of unknown concentration. Phenolphthalein is colorless in an acid but turns bright pink in the presence of a base.

Then, you would slowly and carefully add a base solution of known concentration to this acid-and-indicator mixture. Toward the end of the titration you must add a base drop by drop until one last drop of the base turns the solution pink and the color persists. The point at which the color persists is known as the end point, the point at which the acid is completely neutralized by the base. When you know what volume of base was used, you use that value and the known concentration of the base to calculate the concentration of the acid solution.

Science Online

Topic: Acid/Base Indicators
Visit gpscience.com for Web links to information about acids, bases, and indicators.

Activity Obtain several strips of pH indicator paper and test various liquids around your house to determine if they are acidic, basic, or neutral. You might try the liquids in your refrigerator, rain from a puddle, or swimming pool water.

Figure 15 In this titration, a base of known concentration is being added to an acid of unknown concentration. The swirl of pink color shows that the end point is near.
Explain *How do you know when the endpoint has been reached?*

Figure 16 Natural indicators include red cabbage, radishes, and roses.

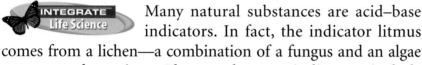

Many natural substances are acid–base indicators. In fact, the indicator litmus comes from a lichen—a combination of a fungus and an algae or a cyanobacterium. Flowers that are indicators include hydrangeas, which produce blue blossoms when the pH of the soil is acidic and pink blossoms when the soil is basic. This is just the opposite of litmus.

Other natural indicators possess a range of color. For example, the color of red cabbage varies from deep red at pH 1 to lavender at pH 7 and yellowish green at pH 10. Grape juice is also an indicator, as you can find out by doing the Try at Home MiniLAB.

Applying Science

How can you handle an upsetting situation?

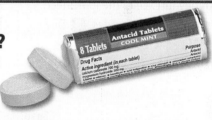

Most of us have, at some time, experienced an upset stomach. Often, the cause is the excess acid within our stomachs. For digestive purposes, our stomachs contain dilute hydrochloric acid with a pH between 1.6 and 3.0. A doctor might recommend an antacid treatment for an upset stomach. What type of compound is "anti acid"?

Identifying the Problem

You have learned that neutralization reactions change acids and bases into salts. Antacids typically contain small amounts of $Ca(OH)_2$, $Al(OH)_3$, or $NaHCO_3$, which are bases. Whereas having an excess of acid lowers the pH

of your stomach contents, these compounds raise the pH of your stomach contents. How does this change of pH make you feel better?

Solving the Problem
1. What compounds are produced from a reaction of HCl and $Mg(OH)_2$?
2. Why is it important to have some acid in your stomach?
3. How could you compare how well antacid products neutralize acid? Describe the procedure you would use.

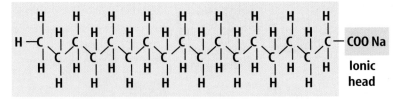

Figure 17 Soaps that contain sodium like this one made from stearic acid are solids, those that contain potassium are liquids. **Write** *the chemical formula for this soap.*

Nonpolar hydrocarbon tail

Soaps and Detergents

The next time you are in a supermarket, go to the aisle with soaps and detergents. You'll see all kinds of products—solid soaps, liquid soaps, and detergents for washing clothes and dishes. What are all these products? Do they differ from one another? Yes, they do differ slightly in how they are made and in the ingredients included for color and aroma. Still, all these products are classified into two types—soaps and detergents.

Soaps The reason soaps clean so well is explained by polar and nonpolar molecules. **Soaps** are organic salts. They have a nonpolar organic chain of carbon atoms on one end and either a sodium or potassium salt of a carboxylic acid (kar bahk SIHL ihk), $-COOH$, group at the other end. Look at **Figure 17.** The nonpolar, hydrocarbon end interacts with oils and dirt so that they can be removed readily, and the ionic end, COONa or COOK, helps them dissolve in water.

To make an effective soap, the acid must contain 12 to 18 carbon atoms. If it contains fewer than 12 atoms, it will not be able to mix well with and clean oily dirt. If it has too many carbon atoms, its sodium or potassium salt will not be soluble in water. **Figure 18** shows how soap interacts with dirt particles to clean your hands.

Figure 18 This is how soaps clean. **A** The long hydrocarbon tail of a soap molecule mixes well with oily dirt while the ionic head attracts water molecules. **B** Dirt now linked with the soap rinses away as water flows over it.

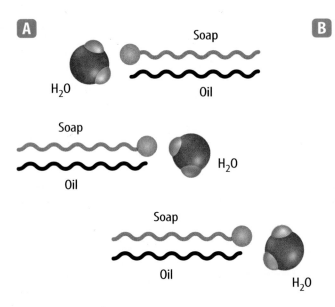

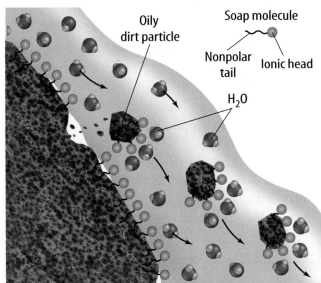

Commercial Soaps A simple soap like the one shown in **Figure 17** can be made by reacting a long-chain fatty acid with sodium or potassium hydroxide. The fatty acids used to make commercial soaps come from natural sources, such as canola, palm, and coconut oils. One problem with all soaps, however, is that the sodium and potassium ions can be replaced by ions of calcium, magnesium, and iron found in some water known as hard water. When this happens, the salts formed are insoluble. They precipitate out of solution in the form of soap scum. Detergents were developed to avoid this problem.

☑ Reading Check *How are simple soaps made?*

Detergents Detergents are synthetic products that are made from petroleum molecules, instead of from natural fatty acids like their soap counterparts. Similar to soaps, detergents have long hydrocarbon chains, but instead of a carboxylic acid group (–COOH) at the end, they may contain instead a sulfonic acid group. These acids form more soluble salts with the ions in hard water and thereby lessen the problem of soap scum. Detergents can also be used in cold water. Most detergents contain additional ingredients called builders and surfactants to enhance sudsing and further improve cleaning in hard water.

Despite solving the problem of cleaning in hard water, detergents are not the complete solution to our needs. Some detergents contained phosphates, the use of which has been restricted or banned in many states, and these are no longer produced because they cause water pollution. Certain sulfonic acid detergents also present problems in the form of excess foaming in water treatment plants and streams, as shown in **Figure 19.** These detergents do not break down easily by bacteria and remain in the environment for long periods of time.

Ecology Before the environmental impact of phosphates was understood, phosphates were added to detergents. Eventually water/detergent mixtures would be washed into streams where the phosphates acted like strong fertilizers, causing algae and water plants to grow uncontrollably. Research this problem and, in your Science Journal, write a pamphlet or speech that an environmental activist might have used to convince law makers of the need for change.

Figure 19 Foam from non-biodegradable detergents can build up in waterways.

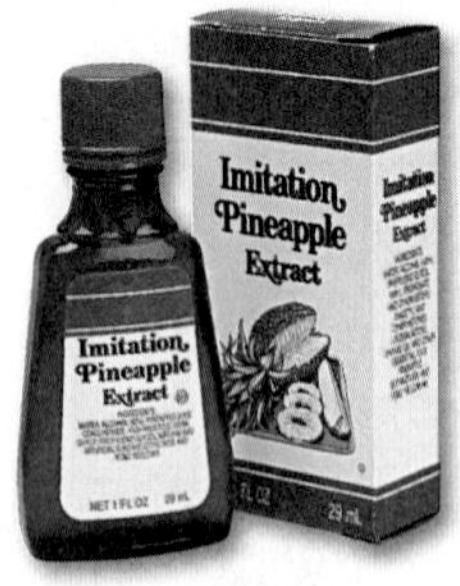

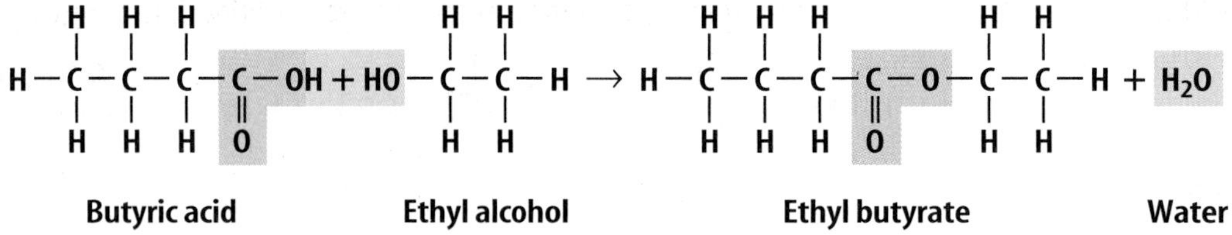

Figure 20 This structural equation shows the formation of the ester ethyl butyrate, an ester that tastes and smells like pineapple. **Predict** *which alcohol you would use to prepare butyl butyrate.*

Versatile Esters

In a way esters can be thought of as the organic counterparts of salts. Like salts, esters are made from acids, and water is formed in the reaction used to prepare them. The difference is that salts are made from bases and esters come from alcohols that are not bases but have a hydroxyl group.

Esters have many different applications. Esters of the alcohol glycerine are used commercially to make soaps. Other esters are used widely in flavors and perfumes, and still others can be transformed into fibers to make clothing.

Reading Check ***What are three types of products that are made from esters?***

Esters for Flavor Many fruit-flavored soft drinks and desserts taste like the real fruit. If you look at the label though, you might be surprised to find that no fruit was used—only artificial flavor. Most likely this artificial flavor contains some esters.

The reaction to prepare esters involves removing a molecule of water from an acid and an alcohol. Often concentrated sulfuric acid is added to aid this reaction. **Figure 20** shows the reaction of butyric (byew TIHR ihk) acid and ethyl alcohol to produce water and the ester, ethyl butyrate, which is a component in pineapple flavor.

Although natural and artificial flavors often contain a blend of many esters, the odor of some individual esters immediately makes you think of particular fruits, as shown in **Figure 21.** For example, octyl acetate smells much like oranges, and both pentyl and butyl acetates smell like bananas.

Making realistic synthetic flavors is an art, in which chemists vary the composition to achieve the desired taste. Strawberry flavor, for example, may contain several esters.

Figure 21 These esters have strong fruity aromas.

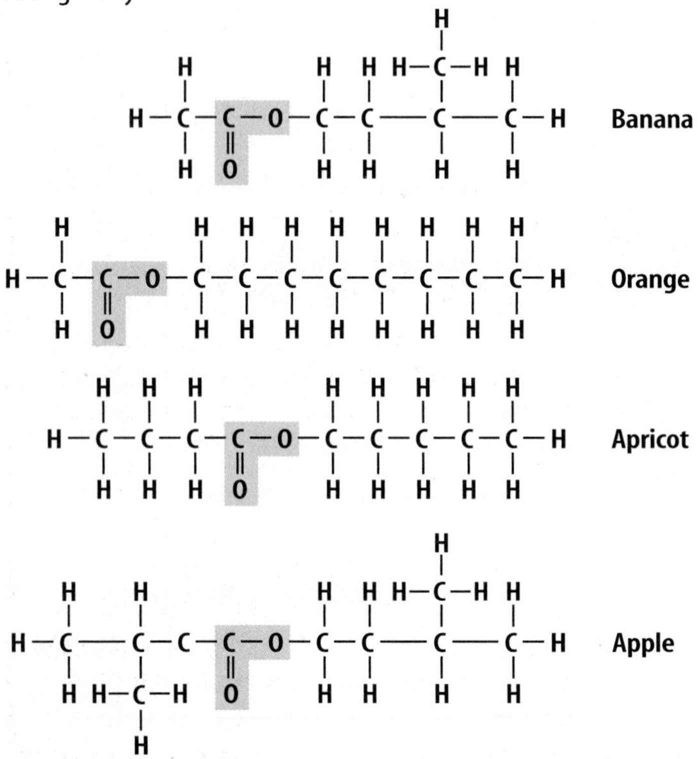

Figure 22 Polyesters and nylons are polymers most often used for clothing fibers.

$$HO-\overset{O}{\underset{}{C}}-\text{⬡}-\overset{O}{\underset{}{C}}-OH \;+\; HO-\overset{H}{\underset{H}{C}}-\overset{H}{\underset{H}{C}}-OH \;\longrightarrow\; \sim\!\!\!\left[\overset{O}{\underset{}{C}}-\text{⬡}-\overset{O}{\underset{}{C}}-O-\overset{H}{\underset{H}{C}}-\overset{H}{\underset{H}{C}}-O\right]\!\!\!\sim \;+\; 2H_2O$$

Organic acid   Alcohol   Polymer (1 unit)   Water

Polyesters Synthetic fibers known as polyesters are polymers; that is, they are chains containing many or *poly* esters. They are made from an organic acid that has two −COOH groups and an alcohol that has two −OH groups, as shown in **Figure 22.** The two compounds form long nonpolar chains that are closely packed together. This adds strength to the polymer fiber. Many varieties of polyesters can be made, depending on what alcohols and acids are used. They can be woven or knitted into fabrics that are durable, water repellent, colorfast, and do not wrinkle easily. Because of their low moisture content however, they tend to build up a static electric charge that causes them to cling. Polyesters often are combined with natural fibers, as shown in **Figure 22.**

Blends of polyester and cotton fibers make comfortable activewear.

section 3 review

Summary

Salts

- Salts are solid compounds formed from the negative ions of an acid and the positive ions from a base.
- Salt is a dietary essential.

Neutralization and Titration

- Acids and bases in solution can combine to bring pH closer to neutral. Products such as stomach antacids use this principle.
- Titration is a method used to determine the concentration of an acidic or basic solution.

Soaps, Detergents, and Esters

- Soaps and detergents are polar, which allows one end to attract dirt and grease molecules and the other end to attract water to wash the dirt away.
- Esters are organic compounds that are made from acids and alcohols.

Self Check

1. **Describe** a neutralization reaction. What are the products of such reactions?

2. **Identify** the purpose of an indicator in a titration experiment.

3. **Explain** how the composition of detergents differs from that of soaps.

4. **Identify** the molecule that is produced in the reaction between an alcohol and an acid to form an ester.

5. **Think Critically** Give the names and formulas of the salts formed in these neutralizations: sulfuric acid and calcium hydroxide, nitric acid and potassium hydroxide, and carbonic acid and aluminum hydroxide.

Applying Math

6. **Calculate Ratios** In the following reaction:
   $$2HCl(aq) + Ca(OH)_2(aq) \rightarrow CaCl_2(aq) + 2H_2O(l)$$
   acid reacts with base in what ratio? How many molecules of HCl are needed to produce four molecules of H_2O?

Design Your Own

Be a Soda Scientist

Goals

■ **Observe** evidence of a neutralization reaction using an indicator.

■ **Compare** the acidity levels in soft drinks.

■ **Design** an experiment that uses the independent variable of acid content of soft drinks and the dependent variable of amount of base added to determine the relative acidity of the drinks.

Possible Materials

different colorless soft drinks (3)
test tubes (3)
25-mL graduated cylinder
droppers (2)
1% phenolphthalein
dilute NaOH solution (0.1*M*)

Safety Precautions

WARNING: *Sodium hydroxide is caustic. Wear eye protection and avoid any skin contact with the solution. Flush thoroughly under a stream of water if any of the NaOH touches your skin. Keep your hands away from your face.*

◗ Real-World Question

The next time you drink a can of soda, take a look at the ingredients label. Carbonated soft drinks contain carbonic acid and sometimes phosphoric acid. You have learned that bases can neutralize acids. Using a proper indicator and a base solution, how could you compare the acidity levels in soft drinks?

◗ Form a Hypothesis

Based on your knowledge of acids and bases, develop a hypothesis about how neutralization reactions can be used to rank the acidity of soft drinks.

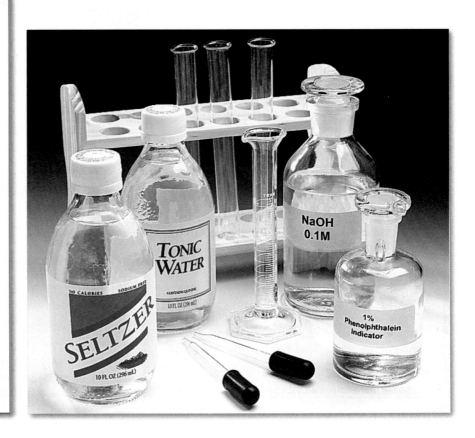

▶ Test Your Hypothesis

Make a Plan

1. As a group, agree upon and write the hypothesis statement.

2. In a logical manner, list the specific steps that you will use to test your hypothesis.

3. **List** all of the materials that you will need to test your hypothesis.

4. **Design** a data table in your Science Journal that will allow you to record the amount of NaOH that was required to neutralize each soda sample.

5. **Decide** the amount of soda to be tested in each trial as a control. Decide also how many times to repeat each trial.

6. **Predict** whether you can test only colorless solutions with this procedure and explain why.

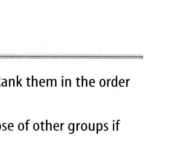

Follow Your Plan

1. Make sure your teacher approves your plan before you start.

2. **Observe** the color change that the indicator phenolphthalein undergoes in a solution that changes from an acidic pH to a basic pH.

3. While doing the experiment, write your observations and complete the data table in your Science Journal.

▶ Analyze Your Data

1. **Classify** the sodas you tested based on their acidities. Rank them in the order of most acidic to least acidic.

2. **Predict** if your acidity values can be compared with those of other groups if they used different amounts of soda.

▶ Conclude and Apply

1. **Evaluate** the results. Do they support your hypothesis? Explain why or why not.

2. **Predict** At warmer temperatures less gas dissolves in a liquid. How would this affect the results of an experiment comparing two sodas stored at different temperatures?

Communicating
Your Data

Compare your soda rankings with those of other class groups. **Discuss** possible reasons for any differences observed.

Acid Rain

Protecting Earth from the damaging effects of chemically loaded precipitation

A cid rain is rain, snow, or sleet that is more acidic than unpolluted precipitation. It's caused by the burning of fossil fuels, such as coal, oil, and natural gas. In the United States, most gasoline and electricity come from fossil fuels. People burn fossil fuels each time they drive a car, heat a building, or turn on a light.

Normally, raindrops pick up particles and natural chemicals in the air. When rain falls, it mixes with the carbon dioxide in the atmosphere, giving clean rain a slightly acidic pH of 5.6. Then, natural chemicals found in the air and soil balance out the acidity, giving most lakes and streams a pH between 6.0 and 8.0. But when pollutants are introduced, these natural bases are not strong enough to neutralize these solutions. Wind can carry this acidic moisture for hundreds of miles before it falls to Earth as acid rain.

Eating Away at History

Like all acids, acid rain can corrode, or eat away at, substances. Many historical monuments, such as the Mayan temples in Mexico and the Parthenon in Greece, have been slowly but steadily damaged by acid rain. This kind of damage can be fixed, though it costs billions of dollars to ensure that ancient monuments and buildings are not destroyed.

Some Solutions

In some countries, high acid levels in lakes and streams have been lowered by adding lime to the water. Lime, a natural base, balances out the damaging chemicals. In the United States, all new cars must have catalytic converters, which help reduce the amount of exhaust pollution that vehicles give off.

You also can make a difference. Turning off the lights when you are not using them means a power plant does not have to produce as much electricity. By carpooling, using public transportation, and walking, there is less pollution from cars. The results of all these individual actions can make a huge difference in preserving our environment.

Ride a bike! It saves fuel, is nonpolluting, and helps preserve the environment.

List Go to a local park or forest. List any effects of acid rain that you see. Make a list of the things you do that use energy or cause pollution. Think about what your family can do to reduce pollution and save energy. Share your list with an adult.

Science Online

For more information, visit gpscience.com/time

Reviewing Main Ideas

Section 1 Acids and Bases

1. An acid is a substance that produces hydrogen ions, H^+, in solution. A base produces hydroxide ions, OH^-, in solution.

2. Some foods can be classified as acidic or basic. Properties of acids and bases are due, in part, to the presence of the H^+ and OH^- ions.

3. Common acids include hydrochloric acid, sulfuric acid, nitric acid, and phosphoric acid. Common bases include sodium hydroxide, calcium hydroxide, and ammonia.

4. Acidic solutions form when certain polar compounds ionize as they dissolve in water. Except for ammonia, basic solutions form when certain ionic compounds dissociate upon dissolving in water.

Section 2 Strength of Acids and Bases

1. The strength of an acid or base is determined by how completely it forms ions when it is in solution.

2. Strength and concentration are not the same thing. Concentration involves the relative amounts of solvent and solute in a solution, whereas strength is related to the extent to which a substance dissociates.

3. pH measures the concentration of hydronium ions in water solution using a scale ranging from 0 to 14.

4. For acidic solutions of equal concentration, the stronger the acid is, the lower its pH is. For basic solutions of equal concentration, the stronger the base is, the higher its pH is.

Section 3 Salts

1. In a neutralization reaction, the H_3O^+ ions from an acid react with the OH^- ions from a base to produce water molecules. The products of a neutralization reaction are a salt and water.

2. Salts form when negative ions from an acid combine with positive ions from a base.

3. Soaps and detergents are organic salts. Unlike soaps, detergents do not react with compounds in hard water to form soap scum as shown here.

4. Esters are organic compounds formed by the reaction of an organic acid and an alcohol.

FOLDABLES Use the Foldable that you made at the beginning of this chapter to help you review acids, bases, and salts.

Using Vocabulary

acid p.696	salt p.707
base p.698	soap p.712
buffer p.705	strong acid p.702
hydronium ion p.696	strong base p.703
indicator p.696	titration p.710
neutralization p.707	weak acid p.702
pH p.704	weak base p.703

Explain the differences between each set of vocabulary words given below. Then explain how the words are related.

1. acid—base

2. acid—salt

3. salt—soap

4. base—soap

5. neutralization—salt

6. strong acid—pH

7. hydronium ion—acid

8. indicator—titration

9. pH—buffer

10. weak base—strong base

Checking Concepts

Choose the word or phrase that best answers the question.

11. What best describes solutions of equal concentrations of HCl and CH_3COOH?
    A) do not have the same pH
    B) will react the same with metals
    C) will make the same salts
    D) have the same amount of ionization

12. What is hydrochloric acid also known as?
    A) battery acid   C) stomach acid
    B) citric acid   D) vinegar

13. Which of the following acids ionizes only partially in water?
    A) HCl   C) HNO_3
    B) H_2SO_4 D) CH_3COOH

14. Which of the following is another name for sodium hydroxide (NaOH)?
    A) ammonia   C) lye
    B) caustic lime   D) milk of magnesia

15. Carrots have a pH of 5.0, so how would you describe them?
    A) acidic   C) neutral
    B) basic   D) an indicator

16. What is the pH of pure water at 25°C?
    A) 0   C) 7
    B) 5.2   D) 14

17. A change of what property permits certain materials to act as indicators?
    A) acidity   C) concentration
    B) color   D) taste

18. Which of the following might you use to titrate an oxalic acid solution?
    A) HBr   C) NaOH
    B) $Ca(NO_3)_2$ D) NH_4Cl

Interpreting Graphics

Use the table below to answer question 19.

19. Which of the substances listed in the table would be most effective for neutralizing battery acid?

pH Readings	
Substance	pH
Battery acid	1.5
Lemon juice	2.5
Apple	3
Milk	6.7
Seawater	8.5
Ammonia	12

Science Online   gpscience.com/vocabulary_puzzlemaker

Use the illustration below to answer question 20.

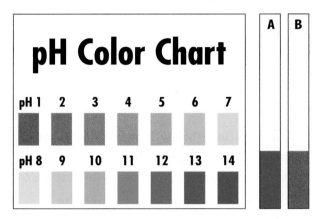

20. Compare the pH test strips for Sample A and Sample B and determine which sample is the acid.

Thinking Critically

21. **Describe** what happens to hydrogen chloride, HCl, when dissolved in water to form hydrochloric acid.

22. **Explain** how the hydroxide ion in NaOH differs from the $-OH$ group in an alcohol.

23. **Explain** why ammonia is considered a base, even though it contains no hydroxide ions. Is it a strong or weak base?

24. **Explain** why a concentrated acid is not necessarily a strong acid.

25. **Explain** why the substances CH_4 and SiO_2 do not conduct electricity.

26. **Compare and Contrast** How would the pH of a dilute solution of HCl compare with the pH of a concentrated solution of the same acid?

27. **Recognize Cause and Effect** Ramón often saw his mother cleaning white deposits from inside her teakettle using vinegar. When she added vinegar, bubbles formed. When she finished, all the white deposits were gone. What do you think these white deposits might be? Do you think dish detergent would have worked as well?

28. **Draw Conclusions** You have equal amounts of three colorless liquids: A, B, and C. You add several drops of phenolphthalein to each liquid. A and B remain colorless, but C turns pink. Next, you add some C to A and the pink color disappears. Then, you add the rest of C to B and the mixture remains pink. What can you infer about each of these liquids? Which original liquid could have had a pH of 7?

Applying Math

29. **Calculate pH** If an acid is added to a solution of pH 10 and the solution changes 4 pH units, what is the new pH?

30. **Use Proportions** To make an indicator solution, a student mixes 3 mL of a concentrated solution to 100 mL water. How much concentrate is needed to make 3 liters of the indicator?

Use the graph below to answer question 31.

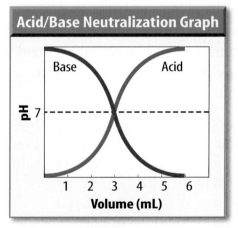

31. **Interpret Graphs** The graph illustrates an acid-base neutralization reaction. Which line (red or blue) represents a base being neutralized by an acid?

32. **Interpret Graphs** Using the graph above, how much acid must be added to the base to neutralize it?

Part 1 Multiple Choice

Record your answers on the answer sheet provided by your teacher or on a sheet of paper.

Use the table below to answer questions 1 and 2.

Acid Solution Data		
Solution	pH	Dissociation
W	7	none
X	2	complete
Y	6	partial
Z	4	??

1. Which word best describes the dissociation of acid solution Z?
   A. complete
   B. partial
   C. none
   D. exactly 50%

2. Which solution contains a strong acid?
   A. solution W
   B. solution X
   C. solution Y
   D. solution Z

3. Hard water often contains various amounts of metallic substances. Which of the following ions does NOT contribute to hard water?
   A. calcium
   B. iron
   C. magnesium
   D. sodium

4. An unknown substance in solution is slippery to the touch, dissolves easily in water, and turns litmus paper blue. The substance is most likely a(n)
   A. acid.
   B. base.
   C. salt.
   D. ester.

Test-Taking Tip

Answer All Questions Never skip a question. If you are unsure of an answer, mark your best guess on another sheet of paper and mark the question in your test booklet to remind you to come back to it at the end of the test.

5. Which chemical formula below describes a hydronium ion?
   A. H_3O^+
   B. OH^-
   C. COOH
   D. H_2O

The titration curve (below) indicates the changes that happened to the solution as drops of a strong base are added. Use the graph below to answer questions 6–9.

Acids, Bases, and Salts

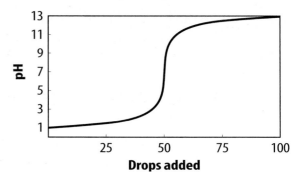

6. Before any drops were added, what was the pH of the solution?
   A. 1
   B. 3
   C. 9
   D. 13

7. At what drop count are the acid and base exactly neutralized?
   A. 0
   B. 25
   C. 50
   D. 75

8. At the instant of neutralization, what is in the beaker besides water?
   A. acid only
   B. base only
   C. salt only
   D. equal amounts of each

9. If the chemical equation for this reaction is $2HCl + Ca(OH)_2 \rightarrow CaCl_2 + 2H_2O$, how many water molecules are formed as x molecules of $CaCl_2$ form?
   A. 2
   B. twice as many, $2x$
   C. half as many, $x/2$
   D. an equal number, x

Part 2 | Short Response/Grid In

Record your answers on the answer sheet provided by your teacher or on a sheet of paper.

Use the model below of hydrogen sulfide to answer questions 10–13.

10. What is the chemical formula for this substance?

11. How many hydronium ions can it form in water?

12. Write the chemical equation for this dissociation reaction.

13. Hydrogen sulfide is a weak acid. Describe how much this substance will dissociate in water.

14. A conductivity apparatus is inserted into a beaker of water, but the light bulb does not glow. Describe what will happen if NaCl crystals are added with stirring.

15. When the pH of a solution drops from 3.0 to 1.0, the hydronium ion concentration increases by a factor of one hundred fold from 0.0010. What is the concentration at pH = 1.0?

16. Compare and contrast the terms *strength* and *concentration* as they apply to acids and bases in solution.

17. An environmental scientist tested rain puddles with pH test strips and a pH meter. The test strips indicated pH = 2.0 and the meter indicated pH = 2.9. By percent, calculate by how much these results differ.

18. Name the salt that is produced by each of the following acid-base pairs. $HCl + NaOH$; $HNO_3 + KOH$; $H_2SO_4 + Ca(OH)_2$

Part 3 | Open Ended

Record your answers on a sheet of paper.

One of the solutions has a pH of 10, the other pH of 12. In one beaker the bulb glows more brightly than does the other. Use the figures below to answer questions 19 and 20.

19. Are the two solutions acids or bases? Explain how you know.

20. Explain why the bulbs glow with different intensities. Use the words *strong* and *weak* in your answer.

21. Describe how dishwashing liquid cleans dirty plates.

22. Na_2SO_4 is a soluble salt. Write the chemical equation describing its dissociation in water.

23. Identify the acid and base that are neutralized to form the salt Na_2SO_4 in a titration reaction.

24. Assume you have a HCl solution of unknown concentration. If 25.0 mL of this solution requires 50.0 mL of a known concentration of a NaOH solution to neutralize, how much more concentrated is the HCl solution than the NaOH solution?

25. Explain why a weak acid in solution has a higher pH than a strong acid of the same concentration.

Organic Compounds

What's in the willows?

The bark of willow trees was used to treat pain and fever long before its active ingredient was known. Now we know that willow bark contains a compound related to aspirin. Today, aspirin and thousands of other useful substances are synthesized from compounds found in petroleum.

Science Journal Can you think of any other medicines that come from natural sources, such as plants?

Start-Up Activities

Carbon, the Organic Element

The element carbon exists in three very different forms: dull, black charcoal; slippery, gray graphite; and bright, sparkling diamond. However, this is nothing compared with the millions of different compounds that carbon can form. In this lab, you will seek out the carbon hidden in two common substances.

WARNING: *Always use extreme caution around an open flame. Point test tubes away from yourself and others.*

1. Place a small piece of bread in a test tube.
2. Using a test-tube holder, hold the tube over the flame of a laboratory burner until you observe changes in the bread.
3. Using a clean test tube and a small amount of paper instead of bread, repeat step 2.
4. **Think Critically** Based on what you observed and what remained in the test tubes, infer what these residues might be.

Organic Compounds Make the following Foldable to help you understand the vocabulary terms in this chapter.

STEP 1 Fold a vertical sheet of notebook paper from side to side.

STEP 2 Cut along every third line of only the top layer to form tabs.

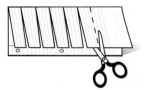

Build Vocabulary As you read the chapter, list the vocabulary words about organic compounds on the tabs. As you learn the definitions, write them under the tab for each vocabulary word.

 Preview this chapter's content and activities at gpscience.com

Simple Organic Compounds

Reading Guide

What You'll Learn

- **Identify** the difference between organic and inorganic carbon compounds.
- **Examine** the structures of some organic compounds.
- **Differentiate** between saturated and unsaturated hydrocarbons.
- **Identify** isomers of organic compounds.

Why It's Important

Carbon compounds surround you—they're in your food, your body, and most materials you use every day.

Review Vocabulary

compound: substance formed from two or more elements

New Vocabulary

- organic compound
- hydrocarbon
- saturated hydrocarbon
- isomer
- unsaturated hydrocarbon

Organic Compounds

What do you have in common with your athletic shoes, sunglasses, and backpack? All the items shown in **Figure 1** contain compounds of the element carbon—and so do you. Most compounds containing the element carbon are **organic compounds.**

At one time, scientists thought that only living organisms could make organic compounds, which is how they got their name. By 1830, scientists could make organic compounds in laboratories, but they continued to call them organic.

Of the millions of carbon compounds known today, more than 90 percent of them are considered organic. The others, including carbon dioxide and the carbonates, are considered inorganic.

Bonding You may wonder why carbon can form so many organic compounds. The main reason is that a carbon atom has four electrons in its outer energy level. This means that each carbon atom can form four covalent bonds with atoms of carbon or with other elements. As you have learned, a covalent bond is formed when two atoms share a pair of electrons. This large number of bonds allows carbon to form many types of compounds ranging from small compounds used as fuel, to complex compounds found in medicines and dyes, and the polymers used in plastics and textile fibers.

Figure 1 Most items used every day contain carbon.

Arrangement Another reason carbon can form so many compounds is that carbon can link together with other carbon atoms in many different arrangements—chains, branched chains, and even rings. It also can form double and triple bonds as well as single bonds. In addition, carbon can bond with atoms of many other elements, such as hydrogen and oxygen. **Figure 2** shows some possible arrangements for carbon compounds.

Hydrocarbons

Carbon forms an enormous number of compounds with hydrogen alone. A compound made up of only carbon and hydrogen atoms is called a **hydrocarbon.** Does the furnace, stove, or water heater in your home burn natural gas? A main component of the natural gas used for these purposes is the hydrocarbon methane. The chemical formula of methane is CH_4.

Methane can be represented in two other ways, as shown in **Figure 3.** The structural formula uses lines to show that four hydrogen atoms are bonded to one carbon atom in a methane molecule. Each line between atoms represents a single covalent bond. The second way, the space-filling model, shows a more realistic picture of the relative size and arrangement of the atoms in the molecule. Most often, however, chemists use chemical and structural formulas to write about reactions.

☑ Reading Check *Name three ways that chemists represent organic compounds.*

Another hydrocarbon used as fuel is propane. Some stoves, most outdoor grills, and the heaters in hot-air balloons burn this hydrocarbon, which is found in bottled gas. Propane's structural formula and space-filling model also are shown in **Figure 3.**

Methane and other hydrocarbons produce more than 90 percent of the energy humans use. Carbon compounds also are important in medicines, foods, and clothing. To understand how carbon can play so many roles, you must understand how it forms bonds.

Figure 2 Carbon atoms bond to form straight, branched, and cyclic chains.

Heptane is found in gasoline.

Isoprene exists in natural rubber.

Vanillin is found in vanilla flavoring.

Figure 3 Natural gas is mostly methane, CH_4, but bottled gas is mostly propane, C_3H_8.
Compare and contrast *the two gases.*

Methane
CH_4

Propane
C_3H_8

Table 1 Some Hydrocarbons

Name	Chemical Formula	Structural Formula
Methane	CH_4	H—C—H (with H above and H below)
Ethane	C_2H_6	H—C—C—H (with H above and below each C)
Propane	C_3H_8	H—C—C—C—H (with H above and below each C)
Butane	C_4H_{10}	H—C—C—C—C—H (with H above and below each C)

Single Bonds

In some hydrocarbons, the carbon atoms are joined by single covalent bonds. Hydrocarbons containing only single-bonded carbon atoms are called **saturated hydrocarbons.** Saturated means that a compound holds as many hydrogen atoms as possible—it is saturated with hydrogen atoms.

Reading Check *What are saturated hydrocarbons?*

Table 1 lists four saturated hydrocarbons. Notice how each carbon atom appears to be a link in a chain connected by single covalent bonds. **Figure 4** shows a graph of the boiling points of some hydrocarbons. Notice the relationship between boiling points and the addition of carbon atoms.

Structural Isomers Perhaps you have seen or know about butane, which is a gas that sometimes is burned in camping stoves and lighters. The chemical formula of butane is C_4H_{10}. Another hydrocarbon called isobutane has exactly the same chemical formula. How can this be? The answer lies in the arrangement of the four carbon atoms. Look at **Figure 5.** In a molecule of butane, the carbon atoms form a continuous chain. The carbon chain of isobutane is branched. The arrangement of carbon atoms in each compound changes the shape of the molecule, and very often affects its physical properties, as you will soon see. Isobutane and butane are isomers.

Figure 4 Boiling points of hydrocarbons increase as the number of carbon atoms in the chain increases.
Predict *the approximate boiling point of hexane.*

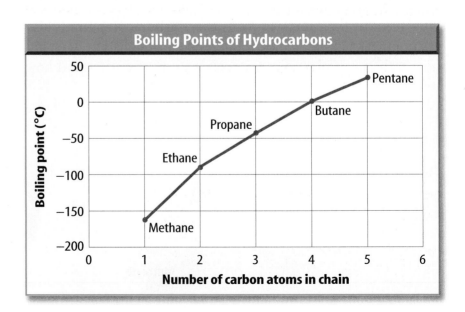

Part 2 | Short Response/Grid In

Record your answers on the answer sheet provided by your teacher or on a sheet of paper.

10. Describe the type of bonds carbon can form.

Use the illustrations below to answer question 11.

11. These molecules are isomers. Given the information that these are hydrocarbons, write their chemical formulas.

12. Describe the general relationship between melting point, boiling point, and the amount of branching in an isomer.

13. Describe some properties and uses of alcohols.

14. What is the process used to separate petroleum compounds called? On what physical property is this process based?

15. Identify some fractions into which crude petroleum is separated.

Test-Taking Tip

Formulas Think about structural formulas before answering the question.

Question 11 Remember how many bonds each carbon atom can form.

Part 3 | Open Ended

Record your answers on a sheet of paper.

16. Describe useful properties of polymers. List several objects made of polymer material that would likely have been made of wood or metal in the past.

17. Identify the polymer material used to make CD cases and foam drinking cups. How can it be used to create two types of containers that have such different properties?

Use the illustration below to answer questions 18 and 19.

18. How is this paper chain a good representation of a protein? Describe the importance of proteins in the human body.

19. Draw a section of this polymer with three or four links using the formulas of the two amino acids given in the chapter and label each link.

20. When you read or hear about cholesterol in the news, it is usually associated with negative effects on the heart and blood vessels. Why does the body make a substance that can potentially damage the circulatory system?

21. Plastic polymers can be prepared cheaply to replace more expensive natural substances. However, disposal presents problems because they do not decompose readily in landfills. Describe two ways of solving this problem.

New Materials Through Chemistry

Chemistry on the Slopes

If you enjoy boarding down a snowy slope, you'll appreciate that most of your equipment is made from materials that have been engineered specifically to meet the challenges of a demanding sport.

Science Journal How do manufacturers find materials to meet their needs?

Start-Up Activities

Chemistry and Properties of Materials

When an engineer designs a vehicle, bridge, or building, the materials used for construction must be selected to match the function. Can the manufacturing process affect a material's performance?

WARNING: *Use proper protection when handling hot objects or working near an open flame. Tie back hair; roll up sleeves.*

1. Using tongs, hold a 5-cm piece of steel wire in a lab burner flame until the wire glows red-hot for 30 seconds.

2. Quickly drop the hot wire into a beaker of cold water.

3. Repeat step 1 with another 5-cm piece of steel wire, but place this hot wire on a heat-proof surface to cool instead of in water.

4. After both pieces of wire are cool, compare the flexibility of the wires.

5. **Think Critically** Write what you observe about the flexibility of the two wires. Suggest reasons.

Materials Classification Make the following Foldable to help you organize materials into groups based on their common features.

 STEP 1 Draw a mark at the midpoint of a sheet of paper along the side edge. Then **fold** the top and bottom edges in to touch the midpoint.

 STEP 2 **Fold** in half from side to side.

STEP 3 **Turn** the paper horizontally. **Open and cut** along the inside fold lines to form four tabs.

 STEP 4 **Label** the tabs *Alloys, Ceramics, Polymers,* and *Composites.*

Classify As you read Chapter 25, list three or more examples of common materials for each group.

 **Preview this chapter's content and activities at** gpscience.com

Materials with a Past

Reading Guide

What You'll Learn
- **Identify** how different alloys are used.
- **Explain** how the properties of alloys determine their use.

Why It's Important
Alloys make modern cities, space travel, and many other things possible.

⊘ Review Vocabulary
physical properties: characteristics that can be observed or measured without changing the composition of the material

New Vocabulary
- alloy
- malleability
- luster
- conductivity
- ductility

Alloys

For ages, people have searched for better materials to use to make their lives more comfortable and their tasks easier. Ancient cultures used stone tools until methods for processing metals became known. Today, advances in metal processing are still occurring as scientists continue to improve the art of blending metals, or making alloys, to make better metal products. An **alloy** is a mixture of elements that has metallic properties. For example, pewter is a mixture of the elements tin, copper, and antimony. If you were to see a pewter mug or a pewter figurine, you probably would not hesitate to say that the objects are metallic. Alloys can produce materials with improved properties such as greater hardness, strength, lightness, or durability.

Alloys Through Time In about 3500 B.C., historians believe that ancient Sumerians in the Tigris-Euphrates Valley (now Iraq) accidentally discovered bronze. They believe that Sumerians used rocks rich in copper and tin ore to make fire rings to keep their campfires from spreading. The hot campfire melted the copper and tin ores within the rocks, creating bronze. This first known mixture of metals became so popular and widely used that a 2,000-year span of history is known as the Bronze Age. The ancients did not have a chemical language for their discovery, but the bronze tools and objects, such as those shown in **Figure 1,** helped change the history of civilization.

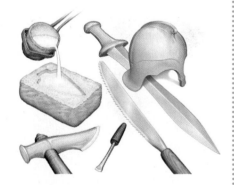

Figure 1 Artifacts from the Bronze Age prove that alloys of metal were used as early as 3500 B.C.

Figure 2 This roll of coated copper wire has the metallic properties of ductility and conductivity. The French horn has the properties of malleability and luster.
Explain *the difference between malleability and ductility.*

Materials Change Typical objects from the Bronze Age and the following Iron Age include spearheads, tools, and even body armor. Bronze and iron are still used today, but it is doubtful that ancient people would recognize them. The methods of processing these alloys have undergone many changes. Other alloys also have been developed through the ages, giving people a large selection of materials to choose from today.

Properties of Metals and Alloys

Alloys retain the metallic properties of metals, some of which are shown in **Figure 2,** but what are the properties of metals? Metals have **luster,** which means they reflect light or have a shiny appearance. The shiny appearance of aluminum foil and a new copper coin demonstrates the property of luster. **Ductility** (duk TIH luh tee) means the metal can be pulled into wires. The copper electrical wire in your home demonstrates the ductility of metals and alloys. **Malleability** (mal yuh BIH luh tee) is the property that allows metals and alloys to be hammered or rolled into thin sheets. Aluminum foil that is used in food preparation and food storage demonstrates the malleability of aluminum. The French horn above demonstrates the luster and malleability of brass. **Conductivity** (kahn duk TIH vuh tee) means that heat or electrical charges can move easily through the material. Metals and alloys have high conductivity because some of their electrons are not tightly held by their atoms. Metals and alloys usually are good conductors of heat and electricity because of these loosely bound electrons. Copper is used to carry electricity because it is conductive and ductile.

 **Reading Check** *What are five other examples of items that you know of that have metallic properties?*

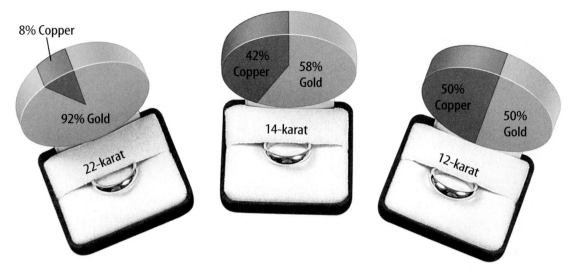

Figure 3 These gold rings appear to look alike, but they vary in the amount of copper that has been added to the gold. The composition of the alloy that was used to make the ring will determine its properties.

Panning for Gold During the California gold rush, prospectors obtained gold by panning. Panning is a process in which dirt and gravel are washed away with water, leaving gold in the bottom of the pan. In your Science Journal describe the properties of gold that were important in making this technique useful.

Choosing an Alloy What properties of an alloy are most important? The answer depends upon how the alloy will be used and which characteristics are the most desirable. Look at the characteristics of familiar objects made from alloys such as the gold jewelry shown in Figure 3. The rings appear to be made of pure gold, but they are made from alloys.

Gold is a bright, expensive metal that is soft and bends easily. Copper, on the other hand, is an inexpensive metal that is harder than gold. When gold and copper are melted, mixed, and allowed to cool, an alloy forms. The properties will vary depending upon the amount of each metal that is added. A ring made with a higher percentage of gold will bend easily due to gold's softness. This ring will be more valuable because it contains a higher percentage of gold. A ring with a higher percentage of copper will not bend as easily because copper is harder than gold. This ring will be less valuable because it contains more copper, a less-expensive metal.

Which properties are needed? The alloy chosen for jewelry and the alloy chosen for a drill bit probably will not be the same. However, the characteristics of the final product must be considered in both situations before the product is constructed.

How hard does the alloy have to be to prevent the object from breaking when it is used? Will the object be exposed to chemicals that will react with the alloy and cause the alloy to fail? These questions relate to the properties of the alloy and its intended use. This represents only two of the many possible questions that must be answered while a product is being designed.

Uses of Alloys

Alloys are used in a variety of products, as shown in **Table 1.** If you see an object that looks metallic, it is most likely an alloy. Alloys that are exceptionally strong are used to manufacture industrial machinery, construction beams, and railroad cars and rails. Automobile and aircraft bodies that require strong materials are constructed of alloys that are corrosion resistant and lightweight but able to carry heavy loads. Other types of alloys are used in products such as food cans, carving knives, and roller skates. **Figure 4** shows some additional examples of alloys.

Table 1 Common Alloys		
Name	**Composition**	**Use**
Bronze	copper, tin	jewelry, marine hardware
Brass	copper, zinc	hardware, musical instruments
Sterling silver	silver, copper	tableware
Pewter	tin, copper, antimony	tableware
Solder	lead, tin	plumbing
Wrought iron	iron, carbon	porch railings, fences, sculpture

If you have a tooth filling, your dentist might have used a silver and mercury alloy to fill it, preventing further tooth decay. Other alloys that are resistant to tissue rejection can be used inside the human body. Special pins and screws made from alloys are used by surgeons to connect broken bones, as shown in **Figure 4.** Alloys also are used as metal plates to repair damage to the skull. These plates protect the brain from injury and are safe to use inside the body.

Reading Check *What are several uses for alloys?*

Figure 4 The fork and saw blade, on the right, are both steel alloys, but they differ in chemical composition. Surgical steel, shown below, can be used to join bones.
Identify *the properties of steel that are most important.*

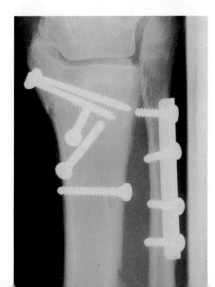

Steel—An Important Alloy There are various classes of steel. They are classified by the amount of carbon and other elements present, as well as by the manufacturing process that is used to refine the iron ore. The classes of steel have different properties and therefore different uses. Steel is a strong alloy and is used often if a great deal of strength is required. Office buildings have steel beams to support the weight of the structure. Bridges, overpasses, and streets also are reinforced with steel. Ship hulls, bedsprings, and automobile gears and axles are made from steel. Another class of steel, called stainless steel, is used in surgical instruments, cooking utensils, and large vessels where food products are prepared.

✔ **Reading Check** *Why is steel an important alloy?*

New Alloys

Steel is not the only common type of alloy. Aluminum is familiar because it is used to make soda cans and cooking foil. Did you know that engineers also are using new aluminum and titanium alloys to build large commercial aircraft? The aircraft shown in **Figure 5** shows how extensively alloys are used in new aircraft construction. The new alloys are strong, lightweight and last longer than alloys used in the past. Also, a lighter plane is less expensive to fly.

Space-Age Alloys Titanium alloy panels, developed for the space shuttle heat shield, might be used on future reusable launch vehicles that are designed to carry payloads to the International Space Station. Titanium and metallic alloys with similar heat-resistive and strength properties may prove to be key materials for other space applications as well.

Science Online

Topic: Materials for Space Vehicles

Visit gpscience.com for Web links to recent news on research for new materials that can be used for space travel.

Activity Write a newspaper article or present a newscast story about your findings.

Figure 5 This commercial aircraft uses new alloys in its construction. Notice that the aircraft skin is mostly alloy construction.
Infer *why alloys are important to manufacturers.*

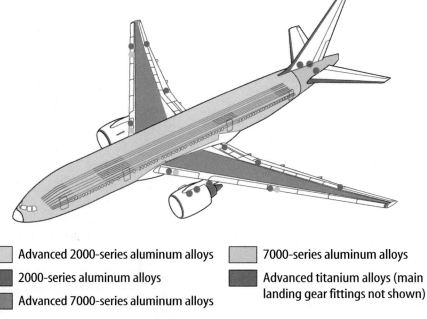

Advanced 2000-series aluminum alloys

2000-series aluminum alloys

Advanced 7000-series aluminum alloys

7000-series aluminum alloys

Advanced titanium alloys (main landing gear fittings not shown)

New Titanium Alloy Heat Shield

The original heat shield on the space shuttle uses ceramic tiles that are prone to cracking as a result of the high temperature and stress they experience during reentry into Earth's atmosphere. Each broken ceramic tile must be removed carefully and a new one glued into place before the shuttle can be used on another mission. This maintenance is expensive, and damaged tiles are physically difficult to replace.

The new titanium alloy tiles, shown in **Figure 6,** are much larger and easier to attach to the heat shield than the ceramic tiles are. A lower maintenance cost for the heat shield is expected by using the new alloy. Scientists and engineers also predict that the new alloy will protect the space shuttle as well as the old ceramic tiles did.

Figure 6 New alloys are being tested for use on the space shuttle. An experimental titanium alloy heat-resistant tile similar to this example may cover space shuttles in future flights.

section 1 review

Summary

Alloys

- Alloy metals defined several historical time periods, including the Bronze Age and the Iron Age.
- An alloy has characteristics that are different from, and often improved upon, the individual elements of the alloy.

Properties of Metals and Alloys

- Alloy materials are defined by their physical properties.

Uses of Alloys

- Alloy materials are used in industry, food service, medicine, and aeronautics.

New Alloys

- New alloys that are strong and lightweight can be used for high-tech applications including aircraft and spacecraft.

Self Check

1. **Identify** two medical uses of alloys.
2. **List** the properties of metals and alloys.
3. **Describe** how steels are classified.
4. **Explain** the effects of adding small amounts of another substance to a material.
5. **Think Critically** If you were designing a skyscraper in an earthquake zone, what properties would the structural materials need?

Applying Math

6. **Calculate** Use the information from **Figure 3** to calculate the actual amount of gold in a 65-g, 14-karat gold necklace.
7. **Find Mass** If a 7.6-g sample of copper can be hammered into a 2-cm $\times$ 2-cm sheet, calculate the number of grams necessary to hammer a 17-cm $\times$ 17-cm sheet under the same manufacturing conditions.

Versatile Materials

Reading Guide

What You'll Learn
- **Examine** the versatile properties of ceramics.
- **Identify** how ceramic materials are used.
- **Explain** what a semiconductor is.

Why It's Important
Ceramics and semiconductors are classes of materials that make computers, electronic games, and many medical devices practical.

Review Vocabulary
compound: a chemical combination of two or more different elements into a substance that has properties different than the component elements

New Vocabulary
- ceramics
- doping
- semiconductor
- integrated circuit

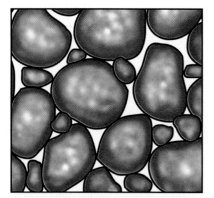

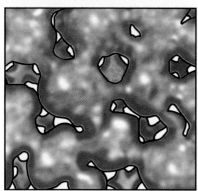

Figure 7 Ceramics are molded and then heated to high temperatures to force the particles to merge. The object shrinks and the structure becomes more dense.

Ceramics

Do you think of floor tiles, pottery, or souvenir nicknacks when you see the word *ceramic*? By definition, **ceramics** are materials that are made from dried clay or claylike mixtures. Ceramics have been around for centuries—in fact, pieces of clay pottery from 10,000 B.C. have been found. The first walled town, Jericho, was built about 8,000 B.C. The wall surrounding Jericho, as well as the homes inside the walls were constructed of bricks made from mud and straw that were baked in the Sun. Around 1,500 B.C., the first glass vessels were made and kilns were used to fire and glaze pottery. By 50 B.C. the Romans developed concrete and used it as a building material. Some of the structures built by the Romans still stand today. About the same time that Romans were developing concrete, the Syrians were developing glass-blowing techniques to make glass vessels. Pottery, bricks, glass, and concrete are examples of ceramics.

How are ceramics made? Traditional ceramics are made from easily obtainable raw materials—clay, silica (sand), and feldspar (crystalline rocks). These raw materials were used by ancient civilizations to make ceramic materials and still are used today. However, some of the more recent ceramics are made from compounds of metallic elements and carbon, nitrogen, or sulfur.

 **Reading Check** *What raw materials are used to make traditional ceramic objects?*

Figure 8 Ceramic materials are used for a wide range of products. Glass, pottery, bricks, and tile are all made from ceramics. **Explain** *how ceramics and alloys are similar.*

Firing Ceramics After the raw materials are processed, ceramics usually are made by molding the ceramic into the desired shape, then heating it to temperatures between 1,000°C and 1,700°C. The heating process, called firing, causes the spaces between the particles to shrink, as shown in **Figure 7.** The entire object shrinks as the spaces become smaller. This extremely dense internal structure gives ceramics their strength. This is demonstrated by the use of ceramics on the space shuttle heat shield. They are able to withstand the high temperatures and stress of reentry into Earth's atmosphere. However, these same ceramics also are fragile and will break if they are dropped or if the temperature changes too quickly.

Traditional Ceramics Ceramics are known also for their chemical resistance to oxygen, water, acids, bases, salts, and strong solvents. These qualities make ceramics useful for applications where they may encounter these substances. For instance, ceramics are used for tableware because your foods contain acids, water, and salts. Ceramic tableware is not damaged by contact with foods containing these substances.

Traditional ceramics also are used as insulators because they do not conduct heat or electricity. You may have seen electric wires attached to poles or posts with ceramic insulators. These insulators keep the current flowing through the wire instead of into the ground.

The properties of ceramics can be customized, which makes them useful for a wide variety of applications as shown in **Figure 8.** Changing the composition of the raw materials or the manufacturing process changes the properties of the ceramic. Manufacturing ceramics is similar to manufacturing alloys because scientists determine which properties are required and then attempt to create the ceramic material.

Mini LAB

Modeling a Composite Material

Procedure

WARNING: *Wear goggles and an apron while doing this lab. Wash your hands before leaving the lab.*

1. Mix four tablespoons of **sand,** four tablespoons of **aquarium gravel or small pebbles,** and six tablespoons of **white glue** in a paper cup.
2. Add enough water to thoroughly mix the ingredients.
3. Stir the mixture until it is smooth.
4. Allow the mixture to sit for several days and observe.
5. Dispose of the cup as instructed by your teacher.

Analysis

1. Describe what happened to your mixture after several days. Is it a ceramic?
2. How is your mixture similar to concrete?
3. What are some of the properties of your product?

Try at Home

Figure 9 This ceramic hip socket is used to replace damaged sockets in the human body.
Identify *other parts of the body that can be helped by ceramics.*

Modern Ceramics Ceramics can be customized to have non-traditional properties, too. Ceramics traditionally are used as insulators, but there are exceptions. For instance, chromium dioxide conducts electricity as well as most metals, and some copper-based ceramics have superconductive properties. One application of nontraditional ceramics uses a transparent, electrically conductive ceramic in aircraft windshields to keep them free of ice and snow.

Ceramics have medical uses. **Figure 9** shows a ceramic replacement hip socket for use in the human body. Ceramics can be used in the body because they are strong and resistant to body fluids, which can damage other materials. In the medical field, surgeons use ceramics for the repair and replacement of joints such as hips, knees, shoulders, elbows, fingers, and wrists. Dentists use ceramics for tooth replacements, repair, and braces.

Applying Science

Can you choose the right material?

Scientists continue to learn about atoms and how they bond. With this new knowledge, chemists today are able to create substances with a wide range of properties. This is especially evident in the production of specialized ceramics.

Ceramic Properties

Material	Wear Resistant	Conducts Electricity	React with Chemicals	Melting Point
ceramic A	highly resistant	no	no	3,000°C
ceramic B	wears easily	no	yes	100°C
ceramic C	moderately resistant	yes	no	1,500°C
ceramic D	resistant	yes	no	500°C

Identifying the Problem

As an engineer working on the design of a new car, you need to select the right ceramic materials to build parts of the car's engine and its onboard computer. The table above shows the materials you have to choose from. Using the properties given in the table, decide which materials should be used for the engine parts and the onboard computer. Be prepared to explain your answer.

Solving the Problem

1. Which of the above materials would you use when you build the engine? Explain the factors that you considered to make your decision.
2. Which of the above materials would you select when building the onboard computer? Explain your selection.
3. If you had to choose a material for building the car's bumper, what factors would you consider? Do you think that a ceramic material would be the best choice? Explain your answer.

Semiconductors

Another class of versatile materials is semiconductors. Semiconductors are the materials that make computers and other electronic devices possible.

The Periodic Table What are semiconductors? To answer this, think about the periodic table. The elements on the left side and in the center of the table are metals. Metals are good conductors of electricity. Nonmetals, located on the right side of the table, are poor conductors of electricity and are electrical insulators. The small number of elements found along the staircase-shaped border shown in **Figure 10** between the metals and nonmetals are metalloids. Some metalloids, such as silicon (Si) and germanium (Ge), are semiconductors. **Semiconductors** are poorer conductors of electricity than metals but better conductors than nonmetals, and their electrical conductivity can be controlled. This property makes semiconductor devices useful.

Controlling Conductivity Adding other elements to some metalloids can change their electrical conductivities. For example, the conductivity of silicon can be increased by replacing silicon atoms with atoms of other elements, such as arsenic (As) or gallium (Ga), as shown in **Figure 11.** If the added atoms, called impurities, have fewer electrons than silicon atoms, the silicon crystals will contain holes, or areas with fewer electrons. Electrons now can move from hole to hole across the crystal, increasing conductivity.

Adding even a single atom of one of these elements to a million silicon atoms significantly changes the conductivity. By controlling the type and number of atoms added, the conductivity of silicon can vary over a wide range.

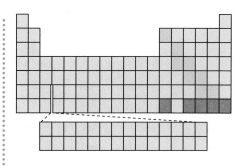

Figure 10 This outline of the periodic table clearly shows the metalloids, which appear in green.

Figure 11 Pure silicon is a poor conductor of electricity. Adding another element with fewer outer electrons, as an impurity, creates an area of fewer electrons called a hole.

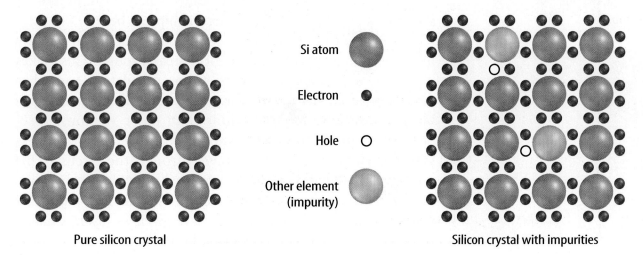

Si atom

Electron

Hole

Other element (impurity)

Pure silicon crystal

Silicon crystal with impurities

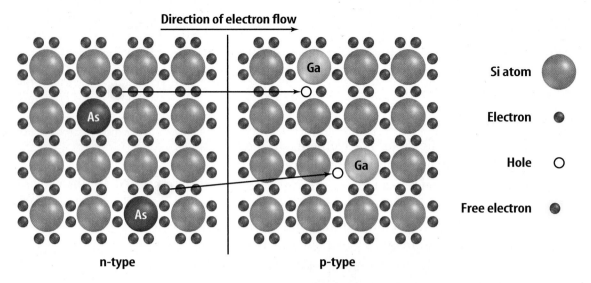

Direction of electron flow

Si atom

Electron

Hole

Free electron

n-type   p-type

Figure 12 The electrons flow from the arsenic-doped silicon to the germanium-doped silicon, filling the available holes. The electron flow is controlled by the sequence of n-type or p-type semiconductors.
Summarize *the difference between n-type and p-type semiconductors.*

Doping The process of adding impurities or other elements to a semiconductor to modify the conductivity is called **doping.** Depending on the element added, the overall number of electrons in the semiconductor is increased or decreased. If the impurity causes the overall number of electrons to increase, the semiconductor is called an *n-type* semiconductor. If doping reduces the overall number of electrons, the semiconductor is called a *p-type* semiconductor.

Integrated Circuits By placing n-type and p-type semiconductors together, semiconductor devices such as transistors and diodes can be made. These devices are used to control the flow of electrons in electrical circuits, as shown in **Figure 12.** During the 1960s, methods were developed for making these components extremely small. At the same time, the integrated circuit was developed.

An **integrated circuit** contains many semiconducting devices. Integrated circuits as small as 1 cm on a side can contain millions of semiconducting devices. Because of their small size, integrated circuits are sometimes called microchips. **Figure 13** shows how small an integrated circuit chip can be.

Being able to pack so many circuit components onto a tiny integrated circuit was a technological breakthrough. This makes it possible for today's televisions, radios, calculators, and other devices to be smaller in size, cheaper to manufacture, and capable of more advanced functions than older versions. Also, because the circuit components are so close together, it takes less time for electric current to travel through the circuit. This enables electronic signals to be processed more rapidly by computers, cell phones, and other electronic appliances. **Figure 14** illustrates how integrated circuits have given us faster, smaller, and more capable computers since the 1940s.

Figure 13 A single integrated circuit fits easily on a fingertip. This small size allows computers and other electronic devices to be compact.

Figure 14

The earliest, room-size computers relied on vacuum tubes to store data. Today's computers use microchips, tiny flakes of silicon engraved with millions of circuit components. A selection of computers is shown here, beginning with the Electronic Numerical Integrator and Computer (ENIAC), developed by the Army in 1946.

A A technician programs the ENIAC, the first electronic computer. Some of the 18,000 vacuum tubes that ran the ENIAC are shown at right.

B A young woman operates a 1960s-era computer. The inset photo shows an integrated circuit from such a computer.

C Teenagers surf the Internet on a modern personal computer. The microchips that store computer programs are now smaller than a fingernail.

Figure 15 Desktop computers use semiconductors to perform their tasks.

Infer *what you think computers will look like in twenty years.*

Semiconductors and Computers

Semiconductors make today's computers possible. A desktop computer is an example of a device that uses semiconductors. A computer has three main jobs. First, it must be able to receive and store the information that is needed to solve a problem. Next, it must be able to follow instructions to perform tasks in a logical way. Finally, a computer must communicate information to the outside world. All three jobs can be done with a combination of hardware and software components.

Computer hardware refers to the major permanent components of a computer, such as the keyboard, monitor, mouse, and central processing unit (CPU). These components are shown in **Figure 15.** Software refers to the instructions that tell the computer what to do. When a computer system is functioning properly, the hardware and software work together to perform tasks.

section 2 review

Summary

Ceramics

- Ceramics are dense composites that are strong and heat resistant.

- Resistance to oxygen, water, acids, bases, and solvents makes ceramics useful for a wide range of applications.

- The properties of ceramics can be customized by changing raw materials and manufacturing processes.

- Ceramics are used as insulators because they do not conduct heat or electicity.

Semiconductors

- Semiconductors conduct electricity moderately well.

- The degree of semiconductor conductivity can be controlled, making these materials versatile.

- Doping is the process of adding impurities to a semiconductor to change its conductivity.

- Integrated circuits are microchips that contain many semiconducting devices.

Self Check

1. **Describe** how ceramic materials are made.

2. **List** five uses of ceramic materials.

3. **Describe** electrical conductivity of ceramics.

4. **Explain** what semiconductors are and where they are used.

5. **Think Critically** Computers and software have changed the way businesses operate. If you operated a distribution center for a manufacturer, how would you use computers to assist you?

Applying Math

6. **Calculate** Ceramic A forms when heated to 1,400°C and has a density of 5.3 g/cm³. Ceramic B forms at a temperature 675°C cooler and is four times as dense. What temperature is required to form Ceramic B and what is its density?

7. **Solve a Problem** A developmental ceramic is designed to be 35% silica and 65% sulfur. If a researcher needs 75 g of this material for a test, how many grams of each component will she need?

 Science online gpscience.com/self_check_quiz

Polymers and Composites

What You'll Learn

- **Identify** what a polymer is and the variety of polymers around us.
- **Explain** what a composite material is and why composites are used.

Why It's Important

Synthetic polymers and composite materials can replace natural materials such as engine oil, wood, and paper to conserve natural resources.

Review Vocabulary

polymerization: a chemical reaction in which two or more molecules combine to form a larger molecule with repeating structural units (polymer)

New Vocabulary

- polymer
- synthetic
- monomer
- composite

Polymers

Polymers are a class of natural or manufactured substances that are composed of molecules arranged in large chains with small, simple, repeating units called monomers. A **monomer** is one specific molecule that is repeated in the polymer chain. Each link in the chain is a monomer. Polypropylene, for example, might have 50,000 to 200,000 monomers in its chain. Several examples of manufactured polymers are shown in **Table 2.**

Not all polymers are manufactured. Some occur naturally. Proteins, cellulose, and nucleic acids are polymers found in living things. In this section, the focus will be on manufactured or synthetic polymers. **Synthetic** means that the polymer does not occur naturally, but it was manufactured in a laboratory or chemical plant. Many synthetic polymers are designed to out perform their natural counterparts.

Table 2 Common Polymers		
Polymer	**Monomer**	**Uses**
Polyethylene	$\left[CH_2 - CH_2 \right]$	bottles, garment bags
Polyvinyl chloride (PVC)	$\left[CH_2 - CHCl \right]$	pipe, bottles, compact discs, computer housing
Polypropylene	$\left[CH_2 - CH \atop \quad\quad CH_3 \right]$	rope, luggage, carpet, film
Polystyrene	$\left[CH - CH_2 \right]$	toys, packaging, egg cartons, flotation devices

Synthetic Polymers Many disposable items such as plates, diapers, trash bags, and utensils are made from synthetic polymers. These products are used once, then thrown away. Most synthetic polymers do not decompose in landfills. In your Science Journal, infer the problems that this might cause and suggest solutions to these problems.

History of Synthetic Polymers
Humankind has used natural polymers for centuries. The ancient Egyptians soaked their burial wrappings in natural resins to help preserve their dead. Animal horns and turtle shells, which contain natural resins, were used to make combs and buttons for many years. In the 1800s, scientists began developing processes to improve natural polymers and to create new ones in the laboratory.

In 1839, Charles Goodyear, an American inventor, found that heating sulfur and natural rubber together improved the qualities of natural rubber. By treating the rubber with sulfur, the natural rubber was no longer brittle when it became cold or soft when it became hot. In the late 1860s, John Hyatt developed celluloid as a replacement for ivory in billiard balls. Celluloid was used in other products such as umbrella handles and toys. These early polymers had many drawbacks, but they were the beginning of the development of a huge class of materials now referred to as polymers. Today, so many types of synthetic polymers exist that they tend to be divided into groups such as plastics, synthetic fibers, adhesives, surface coatings, and synthetic rubbers. **Figure 16** shows a time line of when some of these materials were created.

Hydrocarbons
Today, synthetic polymers usually are made from fossil fuels such as oil, coal, or natural gas. Fossil fuels are composed primarily of carbon and hydrogen and are referred to as hydrocarbons. Because synthetic polymers are made from hydrocarbons, carbon and hydrogen are the primary components of most synthetic polymers.

Figure 16 Polymers were developed in the late 1800s, but they did not become widely used until after World War II in 1945.

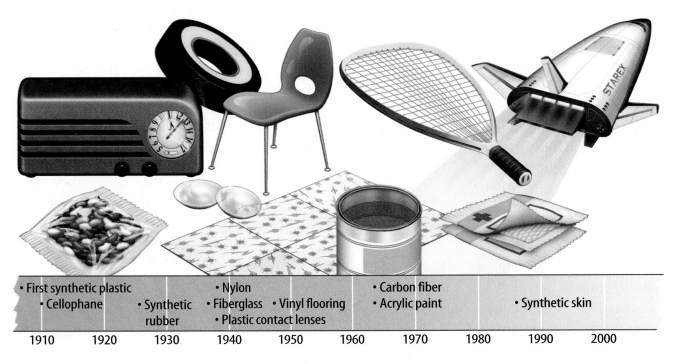

• First synthetic plastic			• Nylon			• Carbon fiber			
• Cellophane		• Synthetic	• Fiberglass	• Vinyl flooring		• Acrylic paint		• Synthetic skin	
		rubber	• Plastic contact lenses						
1910	1920	1930	1940	1950	1960	1970	1980	1990	2000

Changing Properties Polymers are a class of materials with a wide range of uses. The reason that polymers can be used for so many applications is directly related to the ease with which their properties can be modified. Polymers are long chains of monomers. If the composition or arrangement of monomers is changed, then the properties of the material will change.

Figure 17 shows that the monomer ethylene can be modified to produce a polymer with different properties and uses. Ethylene has only two carbon atoms and six bonding sites. The number of carbon atoms in the polymer can be high, and each bonding site represents a possibility of a change in properties. Polyethylene can be high density or low density depending upon how the molecules are attached to the monomer. One of the substances on the monomer can be replaced by another substance or a group of substances and the properties will change, too. The possibilities for creating new materials are almost limitless.

The Plastics Group

Plastics are widely used for many products because they have desirable properties. Plastics are usually lightweight, strong, impact resistant, waterproof, moldable, chemical resistant, and inexpensive. Examples of plastics are easy to find. They are used to make toys, computer housing, telephones, containers, plates, and so on. The properties of plastics vary widely within this group. Some plastics are clear, some melt at high temperature, and some are flexible. Transparency, melting temperature, and flexibility are properties of plastics that relate to the composition of the polymer.

Science Online

Topic: Changing Properties
Visit gpscience.com for Web links to information about how changing the monomer changes the properties of the polymer.

Activity Write a paragraph describing a scientific experiment in which you test the theory that changing the monomer changes the polymer.

Figure 17 The arrangement of the branches along the chain can affect the properties of the polymer. **Compare and contrast** *the different types of plastics.*

High-density polyethylene, HDPE, is firmer, stronger, and less translucent than LDPE. This chain has little side-branching, which allows the chain to pack closer together, thus giving it a higher density and different properties.

Low-density polyethylene, LDPE, is flexible, tough, and chemical resistant. The chain has a great deal of side-branching, which causes low density.

Polyvinyl chloride (PVC) is used in building materials. The substitution of chlorine for a hydrogen in the polyethylene chain makes the polymer harder and more heat resistant.

Synthetic Fibers Nylon, polyester, acrylic, and polypropylene are examples of polymers that can be manufactured as fibers. Most synthetic fibers are composed of carbon chains because they are produced from petroleum or natural gas. Synthetic fibers can be mass-produced to almost any set of desired properties. Nylon is often used in wind and water-resistant clothing such as lightweight jackets. Polyester and polyester blended with natural fibers such as cotton often are used in clothing. Polyester fiber also is used to fill pillows and quilts. Polyurethane is the foam used in mattresses and pillows.

Synthetic fibers called aramids are a family of nylons with special properties. **Figure 18** shows some uses for these materials. Aramids are used to make fireproof clothing. Firefighters, military pilots, and race car drivers are examples of professionals that make use of this special fabric. Another aramid fiber is used to make bulletproof vests, race car survival cells, puncture-resistant gloves, and motorcycle clothing. Although they are lightweight, these aramids are five times stronger than steel.

Adhesives Synthetic polymers are used to make adhesives that can be modified to provide the best properties for a particular application. Contact cements are used in the manufacture of automobile parts, furniture, leather goods, and decorative laminates. They adhere instantly and the bond gets stronger after it dries. Structural adhesives are used in construction projects. One structural adhesive, silicone, is used to seal windows and doors to prevent heat loss in homes and other buildings. Ultraviolet-cured adhesives are used by orthodontists to adhere brace brackets to teeth. These adhesives bond after exposure to ultraviolet light. Other types of adhesives are hot-melts and transparent, pressure-sensitive tape.

✔ **Reading Check** *What are five uses of adhesives?*

Figure 18 Some synthetic polymers make hazardous conditions safer. Firefighters' jackets are made with an aramid fiber that is fireproof. Race car survival cells are manufactured from another aramid, which provides protection for the driver during a crash.

Surface Coatings and Elastic Polymers Many surface coatings use synthetic polymers. Polyurethane is a popular polymer that is used to protect and enhance wood surfaces. Many paints use synthetic polymers in their composition, too.

Synthetic rubber is a synthetic elastic polymer. It is used to manufacture tires, gaskets, belts, and hoses. The soles of some shoes also are made from this rubber.

INTEGRATE Life Science

Taking a Cue from Nature Spinning long fibers into threads and fabrics is not an original idea. Spiders spun fibers for their webs long before humans copied the idea and began spinning fibers themselves. Nylon fiber is another idea borrowed from nature. The silkworm produces a highly desirable fiber that is woven into fabric for items such as blouses and stockings. Can you imagine how long it would take a silkworm to produce enough silk for one blouse? Nylon was produced in the laboratory as a possible substitute for silk. Why do you think natural silk fabric is more expensive than nylon fabric?

Composites

The properties of a synthetic polymer can be altered by using more than one material. A **composite** is a mixture of two or more materials—one embedded or layered in the other. Composite materials of plastic are used to construct boat and car bodies, as shown in **Figure 19.** These bodies are made of a glass-fiber composite that is a mixture of small threads or fibers of glass embedded in a plastic. The structure of the fiberglass reinforces the plastic, making a strong, lightweight composite. If a substance is lightweight but brittle, such as some plastics, embedding flexible fibers into it can alter the brittleness property. After the substance has the flexible fibers embedded, the product is less brittle and can withstand greater forces before it breaks. Glass fibers are used often to reinforce plastics because glass is inexpensive, but other materials can be used as well.

Figure 19 Composite materials are used to make some cars and boats. The glass fiber embedded in the plastic reinforces the plastic structure. This composite material is strong, lightweight, and corrosion resistant.
Define *the term* glass-fiber composite.

Science nline

Topic: Polymers
Visit gpscience.com for Web links to recent news on newly created polymers and new uses for existing polymers.

Activity Research the use of polymer adhesives for use on spacecraft and other high-performance applications. Present your finding to the class.

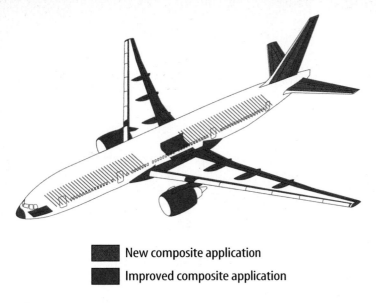

Figure 20 Commercial aircraft use composite materials in some locations. Composites provide corrosion resistance and are easy to repair.

New composite application

Improved composite application

Composites in Flight Composite materials are used in the construction of satellites. Lighter-weight satellites are less expensive to launch into orbit, yet the structure still is able to withstand the stress of the launch. Carbon fibers are used to strengthen the plastic body, creating a material that is four times more firm and 40 percent stronger than aluminum. Satellites made of graphite composites are about 13 percent lighter than satellites made of aluminum. The composite material is stronger and lighter in weight than aluminum, therefore it is less expensive to launch and can endure the stress of the launch better.

Commercial aircraft use composite materials in their construction, as shown in **Figure 20.** Aircraft made of composites also benefit from the strong yet lightweight properties of composite materials. The weight of this aircraft was reduced by more than 2,600 kg by using advanced alloys and composite materials. The lower weight results in cost savings by reducing the amount of fuel required to operate the aircraft.

✔ Reading Check *Why are composites used in aircraft?*

section 3 review

Summary

Polymers

- The composition and chemistry of the polymers allows almost limitless modifications.
- Synthetic polymers are those that have been developed in the lab or manufacturing facility; they do not occur naturally.
- Some of the common classes of synthetic polymers include plastics, fibers, adhesives, and surface coatings.

Composites

- Composites are a class of materials made by embedding one or more components into another.
- Composite materials often are selected for products because their properties offer savings and performance benefits.

Self Check

1. **Explain** what a polymer is and give three examples of items that are made from polymers.
2. **Identify** the raw materials that are used to make most synthetic polymers.
3. **Explain** what a composite material is and give three examples of items that are made from composites.
4. **Classify** synthetic polymers into groups based upon their uses.
5. **Think Critically** How are synthetic polymers creating waste-disposal problems? Discuss possible solutions to this problem.

Applying Math

6. **Find Mass** A telecommunications company launches 10,000-kg satellites. A new satellite made from composites promises to reduce that mass by 25%. What is the mass of the new satellite?

Science Online gpscience.com/self_check_quiz

WHAT CAN YOU DO WITH THIS STUFF?

This substance is fun to play with. But how do you describe its properties?

◉ Real-World Question

What are the properties of this new material and what can it be used for?

Goals

- **Predict** the properties of this material.
- **Determine** possible uses for the material.

Materials

white glue	100-mL beaker or cup
borax laundry soap	graduated cylinder
warm water	craft stick for mixing
250-mL beaker or cup	

Safety Precautions

WARNING: *Never eat lab materials.*

◉ Procedure

1. Prepare a data table to record your observations of the following: stretched slowly, stretched quickly, rolled into a ball and left alone, pressed onto newspaper ink, dropped on a hard surface.

2. Put about 100 mL of warm water in the larger beaker and add borax laundry soap until soap no longer dissolves.

3. Put 5 mL of water and 10 mL of white glue into the smaller beaker and mix completely.

4. Add 5 mL of the borax solution to the glue solution and continue mixing for a couple of minutes.

5. When the substance firms up, remove it from the container and continue to mix it by pressing with your fingers until it is like soft clay.

6. **Examine** the properties of this material and record them in your data table.

◉ Conclude and Apply

1. **Identify** the properties of this material.

2. **Evaluate** Get together with other students and brainstorm. What could this material be used for, and which of its properties would make it useful for that purpose?

3. **Apply** You're in charge of marketing this product. Prepare an advertisement with text and graphics on a sheet of notebook paper. Which magazine would you place this ad in and why?

𝒞 ommunicating Your Data

Compare your conclusions with those of other students in your class. **For more help, refer to the** Science Skill Handbook.

Can polymer composites be stronger than STEEL?

Goals

- **Model** appropriate equipment to test wood, steel, and fiberglass composite rods.
- **Measure** the force required to flex the test rods.
- **Calculate** the relative flexibility of each rod.
- **Estimate** the performance of each material.

Possible Materials

meterstick

spring scale (0–12-kg range and 0–2-kg range)

wood, steel, and fiberglass composite rods (6.35 mm in diameter by 50 cm long)

supports to hold the test rods

graph paper

Safety Precautions

WARNING: *Wear safety goggles at all times during this lab.*

▶ Real-World Question

Why are composite materials used instead of wood or metal in high-performance applications? Scientists and engineers test many materials before selecting the best one for a specific use. Composites are used in aircraft parts, sports equipment, and space vehicles because of their strength and low weight. What other factors might be important? How do you measure performance and choose the best material for an application?

▶ Procedure

1. Hook the spring scale to the center of the fiberglass rod. Have a team member pull down on the spring scale until the top of the test rod moves down 1 cm from the zero point. Record the scale reading on the data table.

2. Pull down on the spring scale until the rod flexes 2 cm, then 3 cm. Record both of the spring scale readings on the data chart.

3. Repeat steps 1 and 2 on the steel and wood rods. Record the data in your table. Refer to the example table shown here.

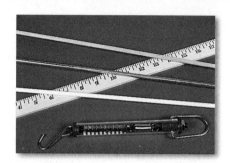

▶ Analyze Your Data

1. **Graph** For each of the rods, graph the force measured on the *y*-axis and the distances on the *x*-axis.

2. **Calculate** the slope of each line in kilograms per centimeter. The slope is a relative measure of the flexibility of the samples.

3. **Determine** the specific performance number, which is used to compare different materials, by dividing the slope of each line by the density of the corresponding material. The densities are: composite = 1.2 g/cm^3, steel = 7.9 g/cm^3, and wood = 0.5 g/cm^3.

Data Table

Material	Distance Flexed (cm)	Force Required (kg)
composite	1	
composite	2	
composite	3	Do not write in this book.
steel	1	
steel	2	
steel	3	
wood	1	
wood	2	
wood	3	

▶ Conclude and Apply

1. **Identify** which rod had the highest specific performance number. What is meant by the statement that a polymer composite is twice as strong as steel?

2. **Analyze** which variables could affect the flexibility measurement.

3. **Model** Using the data that you have gathered, create a model exhibit showing possible construction uses for each of these materials. Indicate the reason the specific material was chosen.

Communicating **Your Data**

Give an oral presentation on choosing the best material for a specific application to another class of students using your model.

TIME

SCIENCE AND
Society

SCIENCE
ISSUES
THAT AFFECT
YOU!

Wonder Fiber

In 1964, Stephanie Kwolek was a chemist working at a research laboratory. Her assignment? Create a new type of tough, lightweight fiber. Kwolek's routine at the lab was about the same each day. She combined different substances in test tubes. She stirred them. She heated them. Then she would have any new substance spun into fibers and tested.

A Shocking Discovery

At one point, Kwolek was working with two polymers. She wanted to use heat to combine them, but they would not melt. So she decided to use a solvent to dissolve them. But when she poured the solvent onto one of the polymers, she got something unlike anything she had ever seen. Not only did it look different, it behaved differently when she stirred it. It separated into two distinct layers.

Kwolek thought this strange liquid might be something special. She asked one of her coworkers to spin it into fibers using a machine called a spinneret. The other chemist refused at first,

saying the liquid wouldn't form fibers. And besides, it would probably gum up the equipment. But Kwolek had a hunch about this liquid. So she persisted until the other chemist agreed to try to spin the liquid into fibers.

Stephanie Kwolek

A New Type of Fiber

What they found was shocking. The fibers that formed in the spinneret were very lightweight, but also extremely stiff and strong. Kwolek had accidentally discovered a new type of synthetic fiber—a fiber made from a new substance called a liquid-crystal solution.

This new fiber was five times stronger than steel, and over the decades since its discovery, it has been put to many uses, such as in bulletproof vests, boat hulls, fiber-optic cables, cut-resistant gloves, airplane parts, skis, tennis rackets, and parts of spacecraft. The discovery was a huge accomplishment for Kwolek and has benefited many people in the form of bulletproof vests used by police officers and the tough clothing used by firefighters.

This police dog can thank Stephanie Kwolek for its bulletproof vest!

Research Visit your school's media center or the link to the right to find out more about the superfiber Kwolek discovered. Compare what you uncover with what others in the class find.

Science Online

For more information, visit gpscience.com/time

Reviewing Main Ideas

Section 1 Materials with a Past

1. People have been making and using alloys for thousands of years. Some common alloys include bronze, brass, and various alloys of iron.

2. An alloy is a mixture of a metal with one or more other elements. Metals and alloys, like these shown here, have the properties of luster, ductility, malleability, and conductivity.

Section 2 Versatile Materials

1. Ceramics are used in a wide range of products, such as aircraft windshields. This is due to the ability of scientists to customize the properties of ceramics.

2. Semiconductors are made from silicon doped with other elements.

3. Ceramic materials are made by molding the object, and then heating the object to high temperatures. This process increases the density of the material.

4. Integrated circuits contain n-type and p-type semiconducting devices.

Section 3 Polymers and Composites

1. Polymers are a class of natural or human-made substances that are composed of molecules that are in large chains with simple repeating units called monomers.

2. Synthetic polymers can be produced in many forms, ranging from thin films to thick slabs or blocks. Synthetic fibers are produced in thin strands that can be woven into fabrics.

3. A composite is a mixture of two materials, one embedded in the other. Reinforced concrete and fiberglass are examples of composites. The skateboard in the figure to the right is constructed of a fiberglass composite. The composite material is strong and flexible.

FOLDABLES Use the Foldable that you made at the beginning of this chapter to help you review classifications of materials.

Using Vocabulary

alloy p. 758
ceramics p. 764
composite p. 775
conductivity p. 759
doping p. 768
ductility p. 759
integrated circuit p. 768

luster p. 759
malleability p. 759
monomer p. 771
polymer p. 771
semiconductor p. 767
synthetic p. 771

Fill in the blank with the correct word or words.

1. _____ is the property of metals and alloys that describes their ability to be hammered or rolled into thin sheets.

2. _____ are used to make heat shield tiles for the space shuttle, but may be replaced by an alloy that is less fragile.

3. Fiberglass is a(n) _____ material that is used to make boats and skateboards.

4. Fiberglass is a(n) _____.

5. Chrome and other shiny, reflective surfaces illustrate the property of _____.

6. _____ are used to make plastics for products such as food containers, toys and electronic cases.

Checking Concepts

Choose the word or phrase that best answers the question.

7. Which metal replaces bronze as a widely used metal?
   A) copper
   B) tin
   C) zinc
   D) iron

8. Why are metals and alloys good conductors of heat and electricity?
   A) They have loosely bound electrons within the atom.
   B) They have luster and malleability.
   C) They are composed of mixtures.
   D) They have a shiny appearance.

9. An alloy of steel will contain iron and what element?
   A) mercury
   B) tin
   C) zinc
   D) carbon

10. What raw materials are many synthetic polymers made from?
    A) hydrocarbons
    B) iron ore
    C) fiberglass
    D) ceramics

Use the photo below to answer question 11.

11. What type of fibers, shown above, are often used to reinforce polymers in automobile bodies?
    A) ceramic
    B) metal alloy
    C) hydrocarbon
    D) glass

12. Which element below is found in both brass and bronze?
    A) mercury
    B) copper
    C) tin
    D) zinc

13. Which of the following is a natural fiber?
    A) nylon
    B) polyester
    C) silk
    D) acrylic

14. Which group of materials below is not classified as synthetic?
    A) ceramics
    B) alloys
    C) composites
    D) metal ores

15. Customizing properties is NOT likely in which of the following?
    A) alloys
    B) synthetic polymers
    C) ceramics
    D) pure metals

16. Which of the following elements is used to dope silicon crystals?
    A) carbon
    B) zinc
    C) copper
    D) gallium

Science Online gpscience.com/vocabulary_puzzlemaker

Interpreting Graphics

17. Copy and complete the following concept map using the terms *composites, hydrocarbons, polymers, adhesives, plastics, synthetic fibers,* and *surface coatings.*

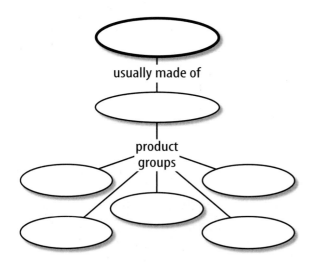

usually made of

product groups

18. Look at the polymer below. Draw the monomer upon which the polymer is based.

$$
\begin{bmatrix}
& H & H & H & H & H & H \\
& | & | & | & | & | & | \\
-C&-&C&-&C&-&C&-&C&-&C- \\
& | & | & | & | & | & | \\
& H & CN & H & CN & H & CN
\end{bmatrix}
$$

Thinking Critically

19. **Infer** A lower-karat gold has less gold in it than a higher-karat gold. Why might you prefer a ring that is 10-karat gold over a ring that is 20-karat gold?

20. **Explain** the advantages and disadvantages of using composites in the world of sports.

21. **Explain** A synthetic fiber might be preferred over a natural fiber for use outdoors because it will not rot. How could this negatively affect the environment?

22. **Compare and contrast** alloys and ceramics.

23. **Recognize Cause and Effect** A student performing the two-page lab did not see any difference in the flexing of the rods. What are some possible causes of this result?

24. **Measure in SI** A bronze trophy has a mass of 952 g. If the bronze is 85 percent copper, how many grams of tin are contained in the trophy?

Applying Math

Use the graph below to answer questions 25 and 26.

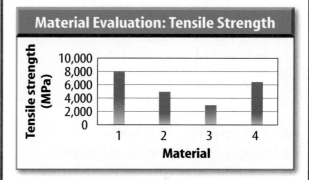

25. **Interpret Graphs** Tensile strength is a measure of the amount of "pulling" stress an object can withstand before it breaks or becomes damaged. The graph above shows a comparison of tensile strength for four materials that an engineer is considering for a new product. Which material should be considered if the product must be tear-resistant?

26. **Compare Materials** Refer to the chart above and calculate, in percent, how much more stress material 4 can withstand than material 3.

27. **Find Mass** An experimental alloy is made up of 28 percent gold and equal parts of two other elements, X and Y. How many grams of the other elements are in a 75-g sample of the alloy?

Part 1 Multiple Choice

Record your answers on the answer sheet provided by your teacher or on a sheet of paper.

Use the photo below to answer questions 1–3.

1. Which property of metals and alloys makes the French horn in the photograph above appear shiny?
   A. conductivity   C. luster
   B. ductility   D. malleability

2. The French horn is made of brass. What element was combined with copper to make the brass?
   A. antimony   C. tin
   B. silver   D. zinc

3. What property allowed the metal from which the instrument is made to be shaped into a French horn?
   A. conductivity   C. luster
   B. ductility   D. malleability

4. Approximately when was the alloy bronze first discovered?
   A. 3500 B.C.   C. 1400 A.D.
   B. 350 B.C.   D. 1800 A.D.

Test-Taking Tip

Relax Stay calm during the test. If you feel yourself getting nervous, close your eyes and take five slow, deep breaths.

5. A 14-karat gold ring is 58% gold and 42% copper by mass. If the ring has a mass of 5.3 g, what is the mass of the gold used to make the ring?
   A. 2.2 g   C. 5.3 g
   B. 3.1 g   D. 5.8 g

6. Which of the following terms refers to substances and materials that are created in a laboratory or chemical plant?
   A. component   C. integrated
   B. composite   D. synthetic

Use the photo below to answer questions 7 and 8.

7. The photograph above shows a roll of copper electrical wire with a polymer coating. Which of the following properties are needed for the wire?
   A. ductility and malleability
   B. ductility and conductivity
   C. malleability and luster
   D. malleability and conductivity

8. What property should the polymer coating on the wire have?
   A. low melting point
   B. high conductivity
   C. high malleability
   D. high resistivity

Part 2 | Short Response/Grid In

Record your answers on the answer sheet provided by your teacher or on a sheet of paper.

Use the illustration below to answer questions 9 and 10.

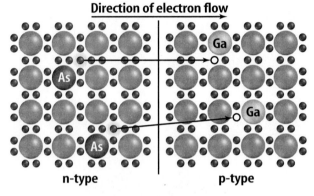

Direction of electron flow

Ga
As
Ga
As

n-type   p-type

● Si atom   • Electron   ○ Hole   • Free electron

9. Which of the two semiconductors shown in the illustration above is an n-type and which is a p-type? How can you tell?

10. Explain how adding an impurity increases the conductivity of the semiconductor shown in the illustration.

11. What properties of ceramics makes them suitable for use as heat shields on the space shuttles? What properties of ceramics are drawbacks to their use as shields?

12. The production of fibers from nylon is an idea that was borrowed from nature. Name two animals that produce fibers.

13. Where on the periodic table are semiconductors located?

14. Name three properties of traditional ceramics that make them useful as food serving bowls.

15. Why do manufacturers frequently use alloys when making different products, rather than just using metals?

Part 3 | Open Ended

Record your answers on a sheet of paper.

16. Describe the process of firing traditional ceramics. Explain how this process affects the properties of the ceramics.

17. Explain what polymers are. There are so many types of polymers that they are divided into different groups. What are some of these groups?

18. Explain what is meant by a *hole* in a semiconductor. Describe what happens to holes as current flows through the semiconductor.

19. Describe some properties of modern ceramics that traditional ceramics do not have. What are some uses of modern ceramics?

Use the illustration below to answer questions 20 and 21.

20. The pet food and water bowl shown in the figure above is made of high-density polyethylene, HDPE. Describe the structure of HDPE and explain how its properties make it a good material for use as a pet bowl.

21. Explain why the pet bowl could not be made from low-density polyethylene, LDPE. What are some products that could be made from LDPE?

22. Name and describe the glass-fiber composite that is often used to make boat and car bodies. What are some properties of this composite?

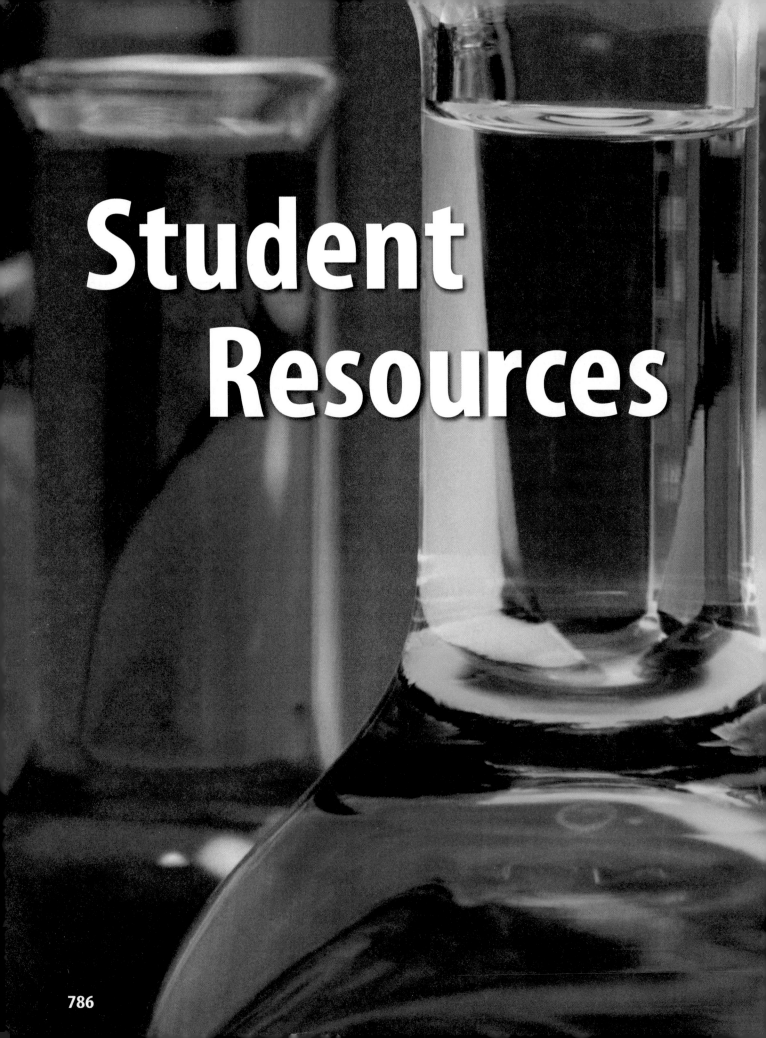

Student Resources

CONTENTS

Test the Hypothesis

Now that you have formed your hypothesis, you need to test it. Using an investigation, you will make observations and collect data, or information. This data might either support or not support your hypothesis. Scientists collect and organize data as numbers and descriptions.

Follow a Procedure In order to know what materials to use, as well as how and in what order to use them, you must follow a procedure. **Figure 8** shows a procedure you might follow to test your hypothesis.

Procedure

1. Use regular gasoline for two weeks.
2. Record the number of kilometers between fill-ups and the amount of gasoline used.
3. Switch to premium gasoline for two weeks.
4. Record the number of kilometers between fill-ups and the amount of gasoline used.

Figure 8 A procedure tells you what to do step by step.

Identify and Manipulate Variables and Controls In any experiment, it is important to keep everything the same except for the item you are testing. The one factor you change is called the independent variable. The change that results is the dependent variable. Make sure you have only one independent variable, to assure yourself of the cause of the changes you observe in the dependent variable. For example, in your gasoline experiment the type of fuel is the independent variable. The dependent variable is the efficiency.

Many experiments also have a control—an individual instance or experimental subject for which the independent variable is not changed. You can then compare the test results to the control results. To design a control you can have two cars of the same type. The control car uses regular gasoline for four weeks. After you are done with the test, you can compare the experimental results to the control results.

Collect Data

Whether you are carrying out an investigation or a short observational experiment, you will collect data, as shown in **Figure 9.** Scientists collect data as numbers and descriptions and organize it in specific ways.

Observe Scientists observe items and events, then record what they see. When they use only words to describe an observation, it is called qualitative data. Scientists' observations also can describe how much there is of something. These observations use numbers, as well as words, in the description and are called quantitative data. For example, if a sample of the element gold is described as being "shiny and very dense" the data are qualitative. Quantitative data on this sample of gold might include "a mass of 30 g and a density of 19.3 g/cm^3."

Figure 9 Collecting data is one way to gather information directly.

Figure 10 Record data neatly and clearly so it is easy to understand.

When you make observations you should examine the entire object or situation first, and then look carefully for details. It is important to record observations accurately and completely. Always record your notes immediately as you make them, so you do not miss details or make a mistake when recording results from memory. Never put unidentified observations on scraps of paper. Instead they should be recorded in a notebook, like the one in **Figure 10.** Write your data neatly so you can easily read it later. At each point in the experiment, record your observations and label them. That way, you will not have to determine what the figures mean when you look at your notes later. Set up any tables that you will need to use ahead of time, so you can record any observations right away. Remember to avoid bias when collecting data by not including personal thoughts when you record observations. Record only what you observe.

Estimate Scientific work also involves estimating. To estimate is to make a judgment about the size or the number of something without measuring or counting. This is important when the number or size of an object or population is too large or too difficult to accurately count or measure.

Sample Scientists may use a sample or a portion of the total number as a type of estimation. To sample is to take a small, representative portion of the objects or organisms of a population for research. By making careful observations or manipulating variables within that portion of the group, information is discovered and conclusions are drawn that might apply to the whole population. A poorly chosen sample can be unrepresentative of the whole. If you were trying to determine the rainfall in an area, it would not be best to take a rainfall sample from under a tree.

Measure You use measurements everyday. Scientists also take measurements when collecting data. When taking measurements, it is important to know how to use measuring tools properly. Accuracy also is important.

Length To measure length, the distance between two points, scientists use meters. Smaller measurements might be measured in centimeters or millimeters.

Length is measured using a metric ruler or meter stick. When using a metric ruler, line up the 0-cm mark with the end of the object being measured and read the number of the unit where the object ends. Look at the metric ruler shown in **Figure 11.** The centimeter lines are the long, numbered lines, and the shorter lines are millimeter lines. In this instance, the length would be 4.50 cm.

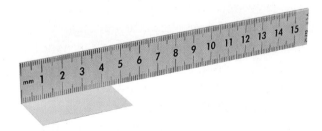

Figure 11 This metric ruler has centimeter and millimeter divisions.

Mass The SI unit for mass is the kilogram (kg). Scientists can measure mass using units formed by adding metric prefixes to the unit gram (g), such as milligram (mg). To measure mass, you might use a triple-beam balance similar to the one shown in **Figure 12.** The balance has a pan on one side and a set of beams on the other side. Each beam has a rider that slides on the beam.

When using a triple-beam balance, place an object on the pan. Slide the largest rider along its beam until the pointer drops below zero. Then move it back one notch. Repeat the process for each rider proceeding from the larger to smaller until the pointer swings an equal distance above and below the zero point. Sum the masses on each beam to find the mass of the object. Move all riders back to zero when finished.

Instead of putting materials directly on the balance, scientists often take a tare of a container. A tare is the mass of a container into which objects or substances are placed for measuring their masses. To mass objects or substances, find the mass of a clean container. Remove the container from the pan, and place the object or substances in the container. Find the mass of the container with the materials in it. Subtract the mass of the empty container from the mass of the filled container to find the mass of the materials you are using.

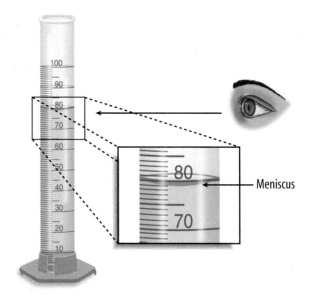

Figure 13 Graduated cylinders measure liquid volume.

Liquid Volume To measure liquids, the unit used is the liter. When a smaller unit is needed, scientists might use a milliliter. Because a milliliter takes up the volume of a cube measuring 1 cm on each side it also can be called a cubic centimeter ($cm^3 = cm \times cm \times cm$).

You can use beakers and graduated cylinders to measure liquid volume. A graduated cylinder, shown in **Figure 13,** is marked from bottom to top in milliliters. In lab, you might use a 10-mL graduated cylinder or a 100-mL graduated cylinder. When measuring liquids, notice that the liquid has a curved surface. Look at the surface at eye level, and measure the bottom of the curve. This is called the meniscus. The graduated cylinder in **Figure 13** contains 79.0 mL, or 79.0 cm^3, of a liquid.

Temperature Scientists often measure temperature using the Celsius scale. Pure water has a freezing point of 0°C and boiling point of 100°C. The unit of measurement is degrees Celsius. Two other scales often used are the Fahrenheit and Kelvin scales.

Figure 12 A triple-beam balance is used to determine the mass of an object.

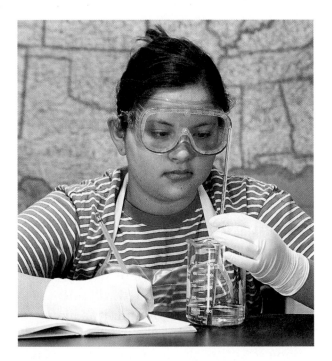

Figure 14 A thermometer measures the temperature of an object.

Scientists use a thermometer to measure temperature. Most thermometers in a laboratory are glass tubes with a bulb at the bottom end containing a liquid such as colored alcohol. The liquid rises or falls with a change in temperature. To read a glass thermometer like the thermometer in **Figure 14,** rotate it slowly until a red line appears. Read the temperature where the red line ends.

Form Operational Definitions An operational definition defines an object by how it functions, works, or behaves. For example, when you are playing hide and seek and a tree is home base, you have created an operational definition for a tree.

Objects can have more than one operational definition. For example, a ruler can be defined as a tool that measures the length of an object (how it is used). It can also be a tool with a series of marks used as a standard when measuring (how it works).

Analyze the Data

To determine the meaning of your observations and investigation results, you will need to look for patterns in the data. Then you must think critically to determine what the data mean. Scientists use several approaches when they analyze the data they have collected and recorded. Each approach is useful for identifying specific patterns.

Interpret Data The word *interpret* means "to explain the meaning of something." When analyzing data from an experiment, try to find out what the data show. Identify the control group and the test group to see whether or not changes in the independent variable have had an effect. Look for differences in the dependent variable between the control and test groups.

Classify Sorting objects or events into groups based on common features is called classifying. When classifying, first observe the objects or events to be classified. Then select one feature that is shared by some members in the group, but not by all. Place those members that share that feature in a subgroup. You can classify members into smaller and smaller subgroups based on characteristics. Remember that when you classify, you are grouping objects or events for a purpose. Keep your purpose in mind as you select the features to form groups and subgroups.

Compare and Contrast Observations can be analyzed by noting the similarities and differences between two more objects or events that you observe. When you look at objects or events to see how they are similar, you are comparing them. Contrasting is looking for differences in objects or events.

Recognize Cause and Effect A cause is a reason for an action or condition. The effect is that action or condition. When two events happen together, it is not necessarily true that one event caused the other. Scientists must design a controlled investigation to recognize the exact cause and effect.

Draw Conclusions

When scientists have analyzed the data they collected, they proceed to draw conclusions about the data. These conclusions are sometimes stated in words similar to the hypothesis that you formed earlier. They may confirm a hypothesis, or lead you to a new hypothesis.

Infer Scientists often make inferences based on their observations. An inference is an attempt to explain observations or to indicate a cause. An inference is not a fact, but a logical conclusion that needs further investigation. For example, you may infer that a fire has caused smoke. Until you investigate, however, you do not know for sure.

Apply When you draw a conclusion, you must apply those conclusions to determine whether the data supports the hypothesis. If your data do not support your hypothesis, it does not mean that the hypothesis is wrong. It means only that the result of the investigation did not support the hypothesis. Maybe the experiment needs to be redesigned, or some of the initial observations on which the hypothesis was based were incomplete or biased. Perhaps more observation or research is needed to refine your hypothesis. A successful investigation does not always come out the way you originally predicted.

Avoid Bias Sometimes a scientific investigation involves making judgments. When you make a judgment, you form an opinion. It is important to be honest and not to allow any expectations of results to bias your judgments. This is important throughout the entire investigation, from researching to collecting data to drawing conclusions.

Communicate

The communication of ideas is an important part of the work of scientists. A discovery that is not reported will not advance the scientific community's understanding or knowledge. Communication among scientists also is important as a way of improving their investigations.

Scientists communicate in many ways, from writing articles in journals and magazines that explain their investigations and experiments, to announcing important discoveries on television and radio. Scientists also share ideas with colleagues on the Internet or present them as lectures, like the student is doing in **Figure 15.**

Figure 15 A student communicates to his peers about his investigation.

SAFETY SYMBOLS

SAFETY SYMBOLS	HAZARD	EXAMPLES	PRECAUTION	REMEDY
DISPOSAL	Special disposal procedures need to be followed.	certain chemicals, living organisms	Do not dispose of these materials in the sink or trash can.	Dispose of wastes as directed by your teacher.
BIOLOGICAL	Organisms or other biological materials that might be harmful to humans	bacteria, fungi, blood, unpreserved tissues, plant materials	Avoid skin contact with these materials. Wear mask or gloves.	Notify your teacher if you suspect contact with material. Wash hands thoroughly.
EXTREME TEMPERATURE	Objects that can burn skin by being too cold or too hot	boiling liquids, hot plates, dry ice, liquid nitrogen	Use proper protection when handling.	Go to your teacher for first aid.
SHARP OBJECT	Use of tools or glassware that can easily puncture or slice skin	razor blades, pins, scalpels, pointed tools, dissecting probes, broken glass	Practice common-sense behavior and follow guidelines for use of the tool.	Go to your teacher for first aid.
FUME	Possible danger to respiratory tract from fumes	ammonia, acetone, nail polish remover, heated sulfur, moth balls	Make sure there is good ventilation. Never smell fumes directly. Wear a mask.	Leave foul area and notify your teacher immediately.
ELECTRICAL	Possible danger from electrical shock or burn	improper grounding, liquid spills, short circuits, exposed wires	Double-check setup with teacher. Check condition of wires and apparatus.	Do not attempt to fix electrical problems. Notify your teacher immediately.
IRRITANT	Substances that can irritate the skin or mucous membranes of the respiratory tract	pollen, moth balls, steel wool, fiberglass, potassium permanganate	Wear dust mask and gloves. Practice extra care when handling these materials.	Go to your teacher for first aid.
CHEMICAL	Chemicals can react with and destroy tissue and other materials	bleaches such as hydrogen peroxide; acids such as sulfuric acid, hydrochloric acid; bases such as ammonia, sodium hydroxide	Wear goggles, gloves, and an apron.	Immediately flush the affected area with water and notify your teacher.
TOXIC	Substance may be poisonous if touched, inhaled, or swallowed.	mercury, many metal compounds, iodine, poinsettia plant parts	Follow your teacher's instructions.	Always wash hands thoroughly after use. Go to your teacher for first aid.
FLAMMABLE	Flammable chemicals may be ignited by open flame, spark, or exposed heat.	alcohol, kerosene, potassium permanganate	Avoid open flames and heat when using flammable chemicals.	Notify your teacher immediately. Use fire safety equipment if applicable.
OPEN FLAME	Open flame in use, may cause fire.	hair, clothing, paper, synthetic materials	Tie back hair and loose clothing. Follow teacher's instruction on lighting and extinguishing flames.	Notify your teacher immediately. Use fire safety equipment if applicable.

 Eye Safety Proper eye protection should be worn at all times by anyone performing or observing science activities.

 Clothing Protection This symbol appears when substances could stain or burn clothing.

 Animal Safety This symbol appears when safety of animals and students must be ensured.

 Handwashing After the lab, wash hands with soap and water before removing goggles.

Safety in the Science Laboratory

The science laboratory is a safe place to work if you follow standard safety procedures. Being responsible for your own safety helps to make the entire laboratory a safer place for everyone. When performing any lab, read and apply the caution statements and safety symbol listed at the beginning of the lab.

General Safety Rules

1. Obtain your teacher's permission to begin all investigations and use laboratory equipment.

2. Study the procedure. Ask your teacher any questions. Be sure you understand safety symbols shown on the page.

3. Notify your teacher about allergies or other health conditions which can affect your participation in a lab.

4. Learn and follow use and safety procedures for your equipment. If unsure, ask your teacher.

5. Never eat, drink, chew gum, apply cosmetics, or do any personal grooming in the lab. Never use lab glassware as food or drink containers. Keep your hands away from your face and mouth.

6. Know the location and proper use of the safety shower, eye wash, fire blanket, and fire alarm.

Prevent Accidents

1. Use the safety equipment provided to you. Goggles and a safety apron should be worn during investigations.

2. Do NOT use hair spray, mousse, or other flammable hair products. Tie back long hair and tie down loose clothing.

3. Do NOT wear sandals or other open-toed shoes in the lab.

4. Remove jewelry on hands and wrists. Loose jewelry, such as chains and long necklaces, should be removed to prevent them from getting caught in equipment.

5. Do not taste any substances or draw any material into a tube with your mouth.

6. Proper behavior is expected in the lab. Practical jokes and fooling around can lead to accidents and injury.

7. Keep your work area uncluttered.

Laboratory Work

1. Collect and carry all equipment and materials to your work area before beginning a lab.

2. Remain in your own work area unless given permission by your teacher to leave it.

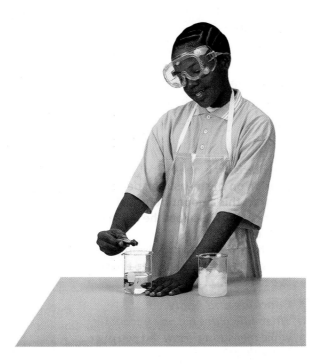

3. Dispose of chemicals and other materials as directed by your teacher. Place broken glass and solid substances in the proper containers. Never discard materials in the sink.

4. Clean your work area.

5. Wash your hands with soap and water thoroughly BEFORE removing your goggles.

Emergencies

1. Report any fire, electrical shock, glassware breakage, spill, or injury, no matter how small, to your teacher immediately. Follow his or her instructions.

2. If your clothing should catch fire, STOP, DROP, and ROLL. If possible, smother it with the fire blanket or get under a safety shower. NEVER RUN.

3. If a fire should occur, turn off all gas and leave the room according to established procedures.

4. In most instances, your teacher will clean up spills. Do NOT attempt to clean up spills unless you are given permission and instructions to do so.

5. If chemicals come into contact with your eyes or skin, notify your teacher immediately. Use the eyewash or flush your skin or eyes with large quantities of water.

6. The fire extinguisher and first-aid kit should only be used by your teacher unless it is an extreme emergency and you have been given permission.

7. If someone is injured or becomes ill, only a professional medical provider or someone certified in first aid should perform first-aid procedures.

3. Always slant test tubes away from yourself and others when heating them, adding substances to them, or rinsing them.

4. If instructed to smell a substance in a container, hold the container a short distance away and fan vapors towards your nose.

5. Do NOT substitute other chemicals/substances for those in the materials list unless instructed to do so by your teacher.

6. Do NOT take any materials or chemicals outside of the laboratory.

7. Stay out of storage areas unless instructed to be there and supervised by your teacher.

Laboratory Cleanup

1. Turn off all burners, water, and gas, and disconnect all electrical devices.

2. Clean all pieces of equipment and return all materials to their proper places.

⑨ Energy Graphs

▶ Real-World Question

How much energy does the United States use compared to the rest of the world?

Possible Materials 📝
- calculator
- colored pencils
- white paper
- metric ruler
- compass and pencil

▶ Procedure

1. Study the energy consumption chart.
2. Use the data to make a bar graph of the oil consumption of the countries in the chart.
3. Construct a circle graph showing the total energy consumption of the countries in the chart.

▶ Conclude and Apply

1. Calculate the percentage of the world's oil that the USA uses.
2. Calculate the percentage of the world's total energy the USA uses.
3. The population of the United States is 290,300,000, and the world population is 6,300,000,000. Calculate what percentage of the world's population is made up of the U.S. population.

2002 Energy Consumption (Equivalent of Millions of Metric Tons of Oil)						
	Oil	Natural Gas	Coal	Hydroelectric	Nuclear	Total Energy Use
USA	894.3	600.7	553.8	58.2	185.8	2,293.0
China	245.7	27.0	663.4	55.8	5.9	997.8
Russia	122.9	349.6	98.5	37.2	32.0	640.2
Japan	242.6	69.7	105.3	20.5	71.3	509.4
Germany	127.2	74.3	84.6	5.9	37.5	329.4
Rest of the world	1,889.9	1,160.7	892.3	414.5	278.3	4,635.2

⑩ Measuring Refraction

▶ Real-World Question

Do some liquids refract light more than others?

Possible Materials 📝 🔧 🥽 🧪 📐
- glasses (3)
- straws or pencils (3)
- water
- white vinegar
- vegetable oil
- metric ruler or protractor

▶ Procedure

1. Pour 300 mL of water into a glass, 300 mL of vinegar into a second glass, and 300 mL of vegetable oil into a third glass.
2. Place a straw into each glass so that each straw is resting at the same angle.
3. Set the glasses side by side, view them from eye level, and observe the angle of refraction caused by each liquid.
4. Use a metric ruler or protractor to measure the refraction caused by the water, vinegar, and oil.

▶ Conclude and Apply

1. List the amount of refraction created by each liquid.
2. Define *refraction*.

Adult supervision required for all labs.

11 It Sounds Different

▷ Real-World Question
How do sounds change when heard through different mediums?

Possible Materials
- wood block
- balloon
- water
- ticking watch

▷ Procedure
1. Hold a wood block next to your ear and have a partner hold a ticking watch next to the block. Note the sound of the watch through the wood.
2. Blow up a balloon and hold the balloon next to your ear. Have a friend hold the ticking watch against the other side of the balloon and note the sound of the watch through the air in the balloon.

3. Fill the balloon with water and securely tie its neck. Hold the water balloon next to your ear and have a partner hold the ticking watch against the other side of the balloon. Note the sound of the watch through the water in the balloon.

▷ Conclude and Apply
1. Compare the sound of the watch when it traveled through the three different mediums.
2. Infer why the watch sounded different in the different mediums.

12 Electromagnetic Waves

▷ Real-World Question
How do polarized sunglasses stop electromagnetic waves and prevent glare?

Possible Materials
- 2 sets of polarized sunglasses (or one set, broken in half)

▷ Procedure
1. On a sunny day, observe the bright glare from a shiny object without sunglasses.
2. Now close one eye and use the other to look through the polarized lens. Hold the lens in front of your eye, then turn it a quarter turn. Record your observations of how the electromagnetic rays of the sunlight changed in each case.

3. Hold one lens in front of another. Look through both lenses at the shiny object and slowly rotate one lens. Record your observations of what happened to the light.

▷ Conclude and Apply
1. Describe what happened to the sunlight glare when you looked through the double lenses of the glasses and rotated them.
2. Why do you think the brightness is different when holding the glasses horizontally and vertically?
3. Infer how polarized sunglasses block electromagnetic energy from the Sun.

Extra Try at Home Labs

13 Light Show

Real-World Question

What does light look like when it passes through different materials?

Possible Materials

- aluminum foil
- clear plastic wrap
- wax paper
- flashlight

Procedure

1. Have a partner hold a 30-cm × 30-cm square of plastic wrap about 30 cm from a white wall.
2. Darken the room and shine a flashlight through the plastic wrap. Observe the amount of light that passes through the wrap and shines on the wall. Be sure to keep the flashlight location constant.

3. Hold a 30-cm × 30-cm square of aluminum foil in front of the wall, darken the room, and shine the light on the foil. Observe what happens to the light.
4. Hold a 30-cm × 30-cm square of wax paper in front of the wall, darken the room, and shine the light on the paper. Observe the amount of light that strikes the wall.
5. Repeat step 4 after folding the wax paper once, then twice, and then several times.

Conclude and Apply

1. Describe your observations of the light when you shined it on the different materials.
2. Identify translucent materials in your home that are used to partially block light.

14 Mirror, Mirror on the Car

Real-World Question

Why are car side-view mirrors convex mirrors?

Possible Materials

- plane mirror
- tennis balls, cans, or other objects (15)
- meterstick

Procedure

1. Measure a distance of 10 m directly behind your family car. Be certain you are not walking into traffic. (If you do not have access to a car, set up chairs and mirrors to simulate where they are positioned in a car.)
2. Line up 15 objects behind the car. The objects should be perpendicular to the car and about 0.5 m apart. Place the first object in line with the back bumper and line up the other objects so that they extend beyond the rear side view of the driver.

3. Sit in the driver's seat, with a parent or guardian present, and look in the side view mirror. Count the number of objects you can see.
4. Sit in the same position and have a partner place a plane mirror over the side-view mirror. Count the number of objects you can see.

Conclude and Apply

1. Compare the number of objects you saw in the convex, side-view mirror with the number of objects you saw in the plane mirror.
2. Infer why convex mirrors are used for side view mirrors on cars.

Adult supervision required for all labs.

15 Lemon Clean

▶ **Real-World Question**

What chemical changes happen to coins?

Possible Materials 🌀 🔪 🔥
- lemon
- paring knife
- tarnished penny
- tarnished nickel
- tarnished dime
- tarnished quarter
- metric ruler

▶ **Procedure**

1. Cut a slit in a lemon 1 cm wide and 1 cm deep. Insert a tarnished penny halfway into the slit.

2. On the same side of the lemon, repeat step 1 for the nickel and the dime.

3. Cut a 1.5 cm wide and 1.5 cm deep slit on the same side of the lemon and insert the quarter halfway into the slit.

4. Leave the coins in the lemon for two days before removing them. Observe the chemical change that happened to the sides of the coins that were in the lemon.

▶ **Conclude and Apply**

1. Describe the change that happened to the coins.

2. Infer why this change happened.

16 Overflowing Ice

▶ **Real-World Question**

What happens to water when it freezes?

Possible Materials 🌀 🔥
- plastic drink bottle
- plate
- water
- freezer

▶ **Procedure**

1. Fill a clean, plastic drink bottle with water. The water should come to the top brim of the bottle.

2. Place a plate in a freezer. Be certain the plate is level and not tilted to one side.

3. Carefully place the bottle on the plate without spilling any of the water. If water spills, refill the bottle.

4. Leave the bottle in the freezer overnight and observe the ice that forms the next day.

▶ **Conclude and Apply**

1. Describe what the ice looks like.

2. Infer why the ice formed this way.

3. Infer how the results of your experiment would be different if you had used rubbing alcohol instead of water. *Hint: Look up the freezing point of rubbing alcohol.*

17 How big is an atom?

Real-World Question
If an atom's nucleus were as big as the head of a pin, how far away would the nearest electron be?

Possible Materials
- pins
- measuring tapes
- outdoor playing field
- masking or duct tape

Procedure
1. The diameter of an atom's nucleus is about 1×10^{-15} m. The orbit of an electron is about 1×10^{-10} m. Calculate how big the orbit would be if the nucleus were the size of the head of a pin (about 0.0001 m).

2. Put your pin through a piece of tape so you can find it later. Measure out the distance to the first electron, and mark the spot with a second pin and tape.

Conclude and Apply
1. Earth orbits the Sun at about 150 million km. This is 214 times the Sun's radius. How many km away would Earth orbit the Sun, if it were on the same scale as an atom's first electron?
2. How many times the nucleus' radius is the orbit of an electron?

18 Get a Half-life

Real-World Question
How would you determine the half-life of a radioactive substance?

Possible Materials
- pennies (200)
- shoe box

Procedure
1. To model the half-life of 200 atoms of a radioactive substance, place 200 pennies in a shoe box, with the "heads" side up.
2. Close the shoe box and shake it for 3 s.
3. Open the shoe box, shift the pennies around until they are all flat, and remove all pennies that are now "tails" side up. Record the number of pennies that you removed from the box and the number of pennies that are left in the box.

4. Repeat steps 2 and 3 until all pennies are removed from the box or you have done this process ten times. Record each shake-and-remove step as increments of 3 s—3 s, 6 s, 9 s, etc. This is the time interval.
5. Graph the data as number of pennies left versus time.

Conclude and Apply
1. According to the graph, how much decay (shaking) time was required for half of your atoms (pennies) to decay (go "tails" up)?
2. If you increased the number of atoms (pennies), would your results change?
3. How would you define the term *half-life*? How would you measure half-life?

Adult supervision required for all labs.

19 Mining for Metals

▶ **Real-World Question**

How do miners get metal from ore?

Possible Materials 🥽 🧤 🧪 🥄 📋

- potato chips
- rolling pin or heavy book
- plastic bags (2)
- water
- hotplate, kettle, or stove
- tea leaf strainer, flour sifter, or other mesh device

▶ **Procedure**

1. Pretend that the potato chips represent ore taken from the ground, and the fat represents a metal compound.

2. Research the mining process and develop a procedure to process the "ore" to refine the "metal."

▶ **Conclude and Apply**

1. Were you satisfied with your procedure and results? What could you do better next time?

2. How does your procedure compare to the real mining process?

20 Disappearing Peanuts

▶ **Real-World Question**

How can you observe chemical bonds breaking?

Possible Materials 🥽 🧤 🚫 🧫 🧪 ☠️ 🚫 📋

- polystyrene packing peanuts or polystyrene cups
- acetone fingernail polish remover
- glass jar or shallow dish
- measuring cup

▶ **Procedure**

1. Work in a well-ventilated area.

2. Pour 30 mL of acetone into a glass jar or shallow dish.

3. Drop a polystyrene packing peanut into the acetone and observe how the polystyrene and acetone react.

4. Drop several peanuts into the acetone and observe what happens to them.

5. Drop a handful of peanuts into the acetone so that they stack up above the liquid observe the reaction that occurs.

▶ **Conclude and Apply**

1. Describe what happened to the polystyrene peanuts.

2. Infer why this happened to the peanuts.

Adult supervision required for all labs.

21 Balanced Reactions

▶ Real-World Question

What would a balanced chemical reaction look like, in terms of atoms and molecules?

Possible Materials 📁 📧

- round fruit (grapes, oranges, apples), marshmallows, foam balls, or any other suitable objects to represent atoms
- sharp toothpicks or straightened paper clips to represent bonds

▶ Procedure

1. Look through the chapter to find two examples of chemical reactions. Balance the equations, if necessary.
2. Make a key for your modeling set. For example, grape = carbon, marshmallow = oxygen, apple = magnesium.

3. Model each balanced chemical reaction that you have written down by bonding the "atoms" together with toothpicks or straightened paper clips to make the reactants. Sketch what you have modeled.
4. Using only the atoms from the reactants, break bonds and make new bonds to form the products. Sketch what you have modeled.

▶ Conclude and Apply

1. How do you know that the law of conservation of mass is followed in the reactions you modeled?
2. What would you do to fix a reaction that did not follow the conservation of mass?
3. What is the evidence that a reaction has occurred?

22 Sticky Solution

▶ Real-World Question

How can heat change a solution?

Possible Materials 🍲 🔥 🥄 🧤 📧

- cornstarch
- pot
- kitchen stove or hotplate
- glass
- measuring cup
- tablespoon
- wooden spoon
- oven mitt

▶ Procedure

1. Pour 300 mL of water into a clean glass.
2. Add a tablespoon of cornstarch to the water and stir the water until a solution is formed. Observe what the solution looks like.

3. Pour your solution into a pot and boil the solution over a hotplate or stovetop burner.
4. Once the solution is boiling, stir it with a wooden spoon.
5. Boil the solution for 2 min and observe how the solution changes.

▶ Conclude and Apply

1. Describe the water and cornstarch solution before it boiled.
2. Describe how heat changed the water and cornstarch solution.

Adult supervision required for all labs.

23 Kitchen Indicator

▶ Real-World Question

How many pHs can you measure around your home?

Possible Materials
- purple cabbage
- water
- pot
- hotplate or stove
- knife
- several clear glasses
- spoons
- baking soda, juice, soda, vinegar, and milk

▶ Procedure

1. Chop up the purple cabbage and put it in the pot. Boil it until the water turns purple.
2. Discard the cabbage, but keep the water.
3. Add a spoonful of the cabbage water to a spoonful of each household substance you plan to test. What color is the mixture? (The color change indicates if the substance is acidic or basic. If the color is red or pink, the substance is an acid. If the color is blue, green, or yellow, it is a base.)
4. Record your colors on a chart.
5. Ask an adult to select other substances for you to test. By adding more of the acidic and basic substances, you can get several interesting colors from this indicator.

▶ Conclude and Apply

How can you prove that vinegar is between drinking soda and baking soda on the pH scale by using the cabbage indicator?

24 Organic Bonding

▶ Real-World Question

How can you and your family represent organic bonding?

Possible Materials
- family members or friends
- large construction paper rings
- pins or tape

▶ Procedure

1. You are a carbon atom. Each of your arms and legs is a place for a bond. Link the paper rings to your arms or legs to represent the correct number of hydrogen atoms.
2. Get together with friends to form ethane, propane, butane, and isobutane.
3. With five friends, make a benzene ring. To make a double bond, touch a neighbor's foot with yours while holding his or her hand. Make a single bond with your other neighbor by holding hands.
4. After forming each molecule, try to move from one side of the room to the other. Which molecules twist and bend easily, and which don't? Make a table of your observations.

▶ Conclude and Apply

1. Use your observations to explain why the boiling point of hydrocarbons increases with the number of carbon atoms.
2. Use your observations to explain why benzene is so stable.
3. How many people would you need to form a protein molecule? Are there enough students in your school?

Isobutane

Extra Try at Home Labs

25 Quick Dry

▶ Real-World Question
Do synthetic fibers dry more quickly than natural fibers?

Possible Materials 🥽 🧤
- measuring cup
- drinking glasses (4)
- water
- 3-cm × 3-cm squares of:
  - cotton cloth
  - wool cloth
  - polyester cloth
  - nylon cloth

▶ Procedure
1. Pour 400 mL of water into each of the four glasses.
2. Submerge a square of fabric in each beaker and soak the squares for 3 min.
3. Remove the fabric squares, lay them flat on several layers of paper towels, and place them in direct sunlight.
4. Check the dampness of each cloth square every 3 min for 15 min.

▶ Conclude and Apply
1. Describe the results of your activity.
2. Infer why athletes wear polyester or nylon clothing.

Adult supervision required for all labs.

Computer Skills

People who study science rely on computers, like the one in **Figure 16,** to record and store data and to analyze results from investigations. Whether you work in a laboratory or just need to write a lab report with tables, good computer skills are a necessity.

Using the computer comes with responsibility. Issues of ownership, security, and privacy can arise. Remember, if you did not author the information you are using, you must provide a source for your information. Also, anything on a computer can be accessed by others. Do not put anything on the computer that you would not want everyone to know. To add more security to your work, use a password.

Use a Word Processing Program

A computer program that allows you to type your information, change it as many times as you need to, and then print it out is called a word processing program. Word processing programs also can be used to make tables.

Figure 16 A computer will make reports neater and more professional looking.

Learn the Skill To start your word processing program, a blank document, sometimes called "Document 1," appears on the screen. To begin, start typing. To create a new document, click the *New* button on the standard tool bar. These tips will help you format the document.

- The program will automatically move to the next line; press *Enter* if you wish to start a new paragraph.
- Symbols, called non-printing characters, can be hidden by clicking the *Show/Hide* button on your toolbar.
- To insert text, move the cursor to the point where you want the insertion to go, click on the mouse once, and type the text.
- To move several lines of text, select the text and click the *Cut* button on your toolbar. Then position your cursor in the location that you want to move the cut text and click *Paste.* If you move to the wrong place, click *Undo.*
- The spell check feature does not catch words that are misspelled to look like other words, like "cold" instead of "gold." Always reread your document to catch all spelling mistakes.
- To learn about other word processing methods, read the user's manual or click on the *Help* button.
- You can integrate databases, graphics, and spreadsheets into documents by copying from another program and pasting it into your document, or by using desktop publishing (DTP). DTP software allows you to put text and graphics together to finish your document with a professional look. This software varies in how it is used and its capabilities.

Use a Database

A collection of facts stored in a computer and sorted into different fields is called a database. A database can be reorganized in any way that suits your needs.

Learn the Skill A computer program that allows you to create your own database is a database management system (DBMS). It allows you to add, delete, or change information. Take time to get to know the features of your database software.

- Determine what facts you would like to include and research to collect your information.
- Determine how you want to organize the information.
- Follow the instructions for your particular DBMS to set up fields. Then enter each item of data in the appropriate field.
- Follow the instructions to sort the information in order of importance.
- Evaluate the information in your database, and add, delete, or change as necessary.

Use the Internet

The Internet is a global network of computers where information is stored and shared. To use the Internet, like the students in **Figure 17,** you need a modem to connect your computer to a phone line and an Internet Service Provider account.

Learn the Skill To access internet sites and information, use a "Web browser," which lets you view and explore pages on the World Wide Web. Each page is its own site, and each site has its own address, called a URL. Once you have found a Web browser, follow these steps for a search (this also is how you search a database).

Figure 17 The Internet allows you to search a global network for a variety of information.

- Be as specific as possible. If you know you want to research "gold," don't type in "elements." Keep narrowing your search until you find what you want.
- Web sites that end in *.com* are commercial Web sites; *.org, .edu,* and *.gov* are non-profit, educational, or government Web sites.
- Electronic encyclopedias, almanacs, indexes, and catalogs will help locate and select relevant information.
- Develop a "home page" with relative ease. When developing a Web site, NEVER post pictures or disclose personal information such as location, names, or phone numbers. Your school or community usually can host your Web site. A basic understanding of HTML (hypertext mark-up language), the language of Web sites, is necessary. Software that creates HTML code is called authoring software, and can be downloaded free from many Web sites. This software allows text and pictures to be arranged as the software is writing the HTML code.

Use a Spreadsheet

A spreadsheet, shown in **Figure 18,** can perform mathematical functions with any data arranged in columns and rows. By entering a simple equation into a cell, the program can perform operations in specific cells, rows, or columns.

Learn the Skill Each column (vertical) is assigned a letter, and each row (horizontal) is assigned a number. Each point where a row and column intersect is called a cell, and is labeled according to where it is located—Column A, Row 1 (A1).

- Decide how to organize the data, and enter it in the correct row or column.
- Spreadsheets can use standard formulas or formulas can be customized to calculate cells.
- To make a change, click on a cell to make it activate, and enter the edited data or formula.
- Spreadsheets also can display your results in graphs. Choose the style of graph that best represents the data.

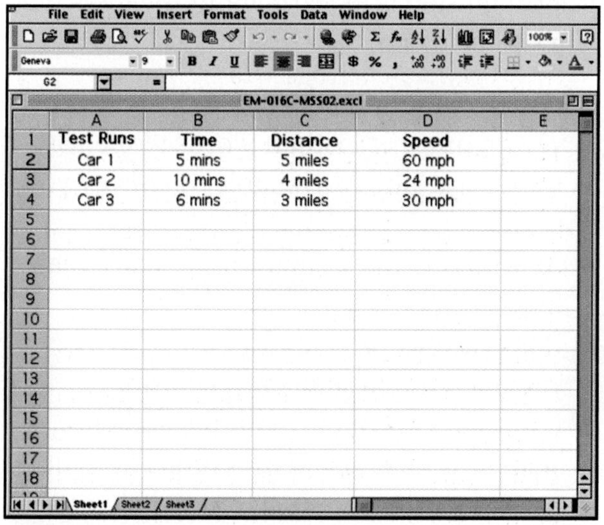

	A	B	C	D	E
1	Test Runs	Time	Distance	Speed	
2	Car 1	5 mins	5 miles	60 mph	
3	Car 2	10 mins	4 miles	24 mph	
4	Car 3	6 mins	3 miles	30 mph	

Figure 18 A spreadsheet allows you to perform mathematical operations on your data.

Use Graphics Software

Adding pictures, called graphics, to your documents is one way to make your documents more meaningful and exciting. This software adds, edits, and even constructs graphics. There is a variety of graphics software programs. The tools used for drawing can be a mouse, keyboard, or other specialized devices. Some graphics programs are simple. Others are complicated, called computer-aided design (CAD) software.

Learn the Skill It is important to have an understanding of the graphics software being used before starting. The better the software is understood, the better the results. The graphics can be placed in a word-processing document.

- Clip art can be found on a variety of internet sites, and on CDs. These images can be copied and pasted into your document.
- When beginning, try editing existing drawings, then work up to creating drawings.
- The images are made of tiny rectangles of color called pixels. Each pixel can be altered.
- Digital photography is another way to add images. The photographs in the memory of a digital camera can be downloaded into a computer, then edited and added to the document.
- Graphics software also can allow animation. The software allows drawings to have the appearance of movement by connecting basic drawings automatically. This is called in-betweening, or tweening.
- Remember to save often.

Presentation Skills

Develop Multimedia Presentations

Most presentations are more dynamic if they include diagrams, photographs, videos, or sound recordings, like the one shown in **Figure 19.** A multimedia presentation involves using stereos, overhead projectors, televisions, computers, and more.

Learn the Skill Decide the main points of your presentation, and what types of media would best illustrate those points.

- Make sure you know how to use the equipment you are working with.
- Practice the presentation using the equipment several times.
- Enlist the help of a classmate to push play or turn lights out for you. Be sure to practice your presentation with him or her.
- If possible, set up all of the equipment ahead of time, and make sure everything is working properly.

Figure 19 These students are engaging the audience using a variety of tools.

Computer Presentations

There are many different interactive computer programs that you can use to enhance your presentation. Most computers have a compact disc (CD) drive that can play both CDs and digital video discs (DVDs). Also, there is hardware to connect a regular CD, DVD, or VCR. These tools will enhance your presentation.

Another method of using the computer to aid in your presentation is to develop a slide show using a computer program. This can allow movement of visuals at the presenter's pace, and can allow for visuals to build on one another.

Learn the Skill In order to create multimedia presentations on a computer, you need to have certain tools. These may include traditional graphic tools and drawing programs, animation programs, and authoring systems that tie everything together. Your computer will tell you which tools it supports. The most important step is to learn about the tools that you will be using.

- Often, color and strong images will convey a point better than words alone. Use the best methods available to convey your point.
- As with other presentations, practice many times.
- Practice your presentation with the tools you and any assistants will be using.
- Maintain eye contact with the audience. The purpose of using the computer is not to prompt the presenter, but to help the audience understand the points of the presentation.

Math Review

Use Fractions

A fraction compares a part to a whole. In the fraction $\frac{2}{3}$, the 2 represents the part and is the numerator. The 3 represents the whole and is the denominator.

Reduce Fractions To reduce a fraction, you must find the largest factor that is common to both the numerator and the denominator, the greatest common factor (GCF). Divide both numbers by the GCF. The fraction has then been reduced, or it is in its simplest form.

Example Twelve of the 20 chemicals in the science lab are in powder form. What fraction of the chemicals used in the lab are in powder form?

Step 1 Write the fraction.

$$\frac{\text{part}}{\text{whole}} = \frac{12}{20}$$

Step 2 To find the GCF of the numerator and denominator, list all of the factors of each number.

Factors of 12: 1, 2, 3, 4, 6, 12 (the numbers that divide evenly into 12)

Factors of 20: 1, 2, 4, 5, 10, 20 (the numbers that divide evenly into 20)

Step 3 List the common factors.

1, 2, 4.

Step 4 Choose the greatest factor in the list. The GCF of 12 and 20 is 4.

Step 5 Divide the numerator and denominator by the GCF.

$$\frac{12 \div 4}{20 \div 4} = \frac{3}{5}$$

In the lab, $\frac{3}{5}$ of the chemicals are in powder form.

Practice Problem At an amusement park, 66 of 90 rides have a height restriction. What fraction of the rides, in its simplest form, has a height restriction?

Add and Subtract Fractions To add or subtract fractions with the same denominator, add or subtract the numerators and write the sum or difference over the denominator. After finding the sum or difference, find the simplest form for your fraction.

Example 1 In the forest outside your house, $\frac{1}{8}$ of the animals are rabbits, $\frac{3}{8}$ are squirrels, and the remainder are birds and insects. How many are mammals?

Step 1 Add the numerators.

$$\frac{1}{8} + \frac{3}{8} = \frac{(1+3)}{8} = \frac{4}{8}$$

Step 2 Find the GCF.

$$\frac{4}{8} \quad (\text{GCF, 4})$$

Step 3 Divide the numerator and denominator by the GCF.

$$\frac{4}{4} = 1, \ \frac{8}{4} = 2$$

$\frac{1}{2}$ of the animals are mammals.

Example 2 If $\frac{7}{16}$ of the Earth is covered by freshwater, and $\frac{1}{16}$ of that is in glaciers, how much freshwater is not frozen?

Step 1 Subtract the numerators.

$$\frac{7}{16} - \frac{1}{16} = \frac{(7-1)}{16} = \frac{6}{16}$$

Step 2 Find the GCF.

$$\frac{6}{16} \quad (\text{GCF, 2})$$

Step 3 Divide the numerator and denominator by the GCF.

$$\frac{6}{2} = 3, \ \frac{16}{2} = 8$$

$\frac{3}{8}$ of the freshwater is not frozen.

Practice Problem A bicycle rider is going 15 km/h for $\frac{4}{9}$ of his ride, 10 km/h for $\frac{2}{9}$ of his ride, and 8 km/h for the remainder of the ride. How much of his ride is he going over 8 km/h?

Unlike Denominators To add or subtract fractions with unlike denominators, first find the least common denominator (LCD). This is the smallest number that is a common multiple of both denominators. Rename each fraction with the LCD, and then add or subtract. Find the simplest form if necessary.

Example 1 A chemist makes a paste that is $\frac{1}{2}$ table salt (NaCl), $\frac{1}{3}$ sugar ($C_6H_{12}O_6$), and the rest water (H_2O). How much of the paste is a solid?

Step 1 Find the LCD of the fractions.

$$\frac{1}{2} + \frac{1}{3} \text{ (LCD, 6)}$$

Step 2 Rename each numerator and each denominator with the LCD.

$$1 \times 3 = 3, \ 2 \times 3 = 6$$
$$1 \times 2 = 2, \ 3 \times 2 = 6$$

Step 3 Add the numerators.

$$\frac{3}{6} + \frac{2}{6} = \frac{(3+2)}{6} = \frac{5}{6}$$

$\frac{5}{6}$ of the paste is a solid.

Example 2 The average precipitation in Grand Junction, CO, is $\frac{7}{10}$ inch in November, and $\frac{3}{5}$ inch in December. What is the total average precipitation?

Step 1 Find the LCD of the fractions.

$$\frac{7}{10} + \frac{3}{5} \text{ (LCD, 10)}$$

Step 2 Rename each numerator and each denominator with the LCD.

$$7 \times 1 = 7, \ 10 \times 1 = 10$$
$$3 \times 2 = 6, \ 5 \times 2 = 10$$

Step 3 Add the numerators.

$$\frac{7}{10} + \frac{6}{10} = \frac{(7+6)}{10} = \frac{13}{10}$$

$\frac{13}{10}$ inches total precipitation, or $1\frac{3}{10}$ inches.

Practice Problem On an electric bill, about $\frac{1}{8}$ of the energy is from solar energy and about $\frac{1}{10}$ is from wind power. How much of the total bill is from solar energy and wind power combined?

Example 3 In your body, $\frac{7}{10}$ of your muscle contractions are involuntary (cardiac and smooth muscle tissue). Smooth muscle makes $\frac{3}{15}$ of your muscle contractions. How many of your muscle contractions are made by cardiac muscle?

Step 1 Find the LCD of the fractions.

$$\frac{7}{10} - \frac{3}{15} \text{ (LCD, 30)}$$

Step 2 Rename each numerator and each denominator with the LCD.

$$7 \times 3 = 21, \ 10 \times 3 = 30$$
$$3 \times 2 = 6, \ 15 \times 2 = 30$$

Step 3 Subtract the numerators.

$$\frac{21}{30} - \frac{6}{30} = \frac{(21-6)}{30} = \frac{15}{30}$$

Step 4 Find the GCF.

$$\frac{15}{30} \text{ (GCF, 15)}$$

$$\frac{1}{2}$$

$\frac{1}{2}$ of all muscle contractions are cardiac muscle.

Example 4 Tony wants to make cookies that call for $\frac{3}{4}$ of a cup of flour, but he only has $\frac{1}{3}$ of a cup. How much more flour does he need?

Step 1 Find the LCD of the fractions.

$$\frac{3}{4} - \frac{1}{3} \text{ (LCD, 12)}$$

Step 2 Rename each numerator and each denominator with the LCD.

$$3 \times 3 = 9, \ 4 \times 3 = 12$$
$$1 \times 4 = 4, \ 3 \times 4 = 12$$

Step 3 Subtract the numerators.

$$\frac{9}{12} - \frac{4}{12} = \frac{(9-4)}{12} = \frac{5}{12}$$

$\frac{5}{12}$ of a cup of flour.

Practice Problem Using the information provided to you in Example 3 above, determine how many muscle contractions are voluntary (skeletal muscle).

Multiply Fractions To multiply with fractions, multiply the numerators and multiply the denominators. Find the simplest form if necessary.

Example Multiply $\frac{3}{5}$ by $\frac{1}{3}$.

Step 1 Multiply the numerators and denominators.

$$\frac{3}{5} \times \frac{1}{3} = \frac{(3 \times 1)}{(5 \times 3)} = \frac{3}{15}$$

Step 2 Find the GCF.

$$\frac{3}{15} \quad (\text{GCF, 3})$$

Step 3 Divide the numerator and denominator by the GCF.

$$\frac{3}{3} = 1, \quad \frac{15}{3} = 5$$

$$\frac{1}{5}$$

$\frac{3}{5}$ multiplied by $\frac{1}{3}$ is $\frac{1}{5}$.

Practice Problem Multiply $\frac{3}{14}$ by $\frac{5}{16}$.

Find a Reciprocal Two numbers whose product is 1 are called multiplicative inverses, or reciprocals.

Example Find the reciprocal of $\frac{3}{8}$.

Step 1 Inverse the fraction by putting the denominator on top and the numerator on the bottom.

$$\frac{8}{3}$$

The reciprocal of $\frac{3}{8}$ is $\frac{8}{3}$.

Practice Problem Find the reciprocal of $\frac{4}{9}$.

Divide Fractions To divide one fraction by another fraction, multiply the dividend by the reciprocal of the divisor. Find the simplest form if necessary.

Example 1 Divide $\frac{1}{9}$ by $\frac{1}{3}$.

Step 1 Find the reciprocal of the divisor.

The reciprocal of $\frac{1}{3}$ is $\frac{3}{1}$.

Step 2 Multiply the dividend by the reciprocal of the divisor.

$$\frac{\frac{1}{9}}{\frac{1}{3}} = \frac{1}{9} \times \frac{3}{1} = \frac{(1 \times 3)}{(9 \times 1)} = \frac{3}{9}$$

Step 3 Find the GCF.

$$\frac{3}{9} \quad (\text{GCF, 3})$$

Step 4 Divide the numerator and denominator by the GCF.

$$\frac{3}{3} = 1, \quad \frac{9}{3} = 3$$

$$\frac{1}{3}$$

$\frac{1}{9}$ divided by $\frac{1}{3}$ is $\frac{1}{3}$.

Example 2 Divide $\frac{3}{5}$ by $\frac{1}{4}$.

Step 1 Find the reciprocal of the divisor.

The reciprocal of $\frac{1}{4}$ is $\frac{4}{1}$.

Step 2 Multiply the dividend by the reciprocal of the divisor.

$$\frac{\frac{3}{5}}{\frac{1}{4}} = \frac{3}{5} \times \frac{4}{1} = \frac{(3 \times 4)}{(5 \times 1)} = \frac{12}{5}$$

$\frac{3}{5}$ divided by $\frac{1}{4}$ is $\frac{12}{5}$ or $2\frac{2}{5}$.

Practice Problem Divide $\frac{3}{11}$ by $\frac{7}{10}$.

Use Ratios

When you compare two numbers by division, you are using a ratio. Ratios can be written 3 to 5, 3:5, or $\frac{3}{5}$. Ratios, like fractions, also can be written in simplest form.

Ratios can represent probabilities, also called odds. This is a ratio that compares the number of ways a certain outcome occurs to the number of outcomes. For example, if you flip a coin 100 times, what are the odds that it will come up heads? There are two possible outcomes, heads or tails, so the odds of coming up heads are 50:100. Another way to say this is that 50 out of 100 times the coin will come up heads. In its simplest form, the ratio is 1:2.

Example 1 A chemical solution contains 40 g of salt and 64 g of baking soda. What is the ratio of salt to baking soda as a fraction in simplest form?

Step 1 Write the ratio as a fraction.
$$\frac{salt}{baking\ soda} = \frac{40}{64}$$

Step 2 Express the fraction in simplest form.
The GCF of 40 and 64 is 8.
$$\frac{40}{64} = \frac{40 \div 8}{64 \div 8} = \frac{5}{8}$$

The ratio of salt to baking soda in the sample is 5:8.

Example 2 Sean rolls a 6-sided die 6 times. What are the odds that the side with a 3 will show?

Step 1 Write the ratio as a fraction.
$$\frac{number\ of\ sides\ with\ a\ 3}{number\ of\ sides} = \frac{1}{6}$$

Step 2 Multiply by the number of attempts.
$$\frac{1}{6} \times 6\ attempts = \frac{6}{6}\ attempts = 1\ attempt$$

1 attempt out of 6 will show a 3.

Practice Problem Two metal rods measure 100 cm and 144 cm in length. What is the ratio of their lengths in simplest form?

Use Decimals

A fraction with a denominator that is a power of ten can be written as a decimal. For example, 0.27 means $\frac{27}{100}$. The decimal point separates the ones place from the tenths place.

Any fraction can be written as a decimal using division. For example, the fraction $\frac{5}{8}$ can be written as a decimal by dividing 5 by 8. Written as a decimal, it is 0.625.

Add or Subtract Decimals When adding and subtracting decimals, line up the decimal points before carrying out the operation.

Example 1 Find the sum of 47.68 and 7.80.

Step 1 Line up the decimal places when you write the numbers.
$$\begin{array}{r} 47.68 \\ +\ 7.80 \\ \hline \end{array}$$

Step 2 Add the decimals.
$$\begin{array}{r} 47.68 \\ +\ 7.80 \\ \hline 55.48 \end{array}$$

The sum of 47.68 and 7.80 is 55.48.

Example 2 Find the difference of 42.17 and 15.85.

Step 1 Line up the decimal places when you write the number.
$$\begin{array}{r} 42.17 \\ -15.85 \\ \hline \end{array}$$

Step 2 Subtract the decimals.
$$\begin{array}{r} 42.17 \\ -15.85 \\ \hline 26.32 \end{array}$$

The difference of 42.17 and 15.85 is 26.32.

Practice Problem Find the sum of 1.245 and 3.842.

Multiply Decimals To multiply decimals, multiply the numbers like any other number, ignoring the decimal point. Count the decimal places in each factor. The product will have the same number of decimal places as the sum of the decimal places in the factors.

Example Multiply 2.4 by 5.9.

Step 1 Multiply the factors like two whole numbers.
$24 \times 59 = 1416$

Step 2 Find the sum of the number of decimal places in the factors. Each factor has one decimal place, for a sum of two decimal places.

Step 3 The product will have two decimal places.
14.16

The product of 2.4 and 5.9 is 14.16.

Practice Problem Multiply 4.6 by 2.2.

Divide Decimals When dividing decimals, change the divisor to a whole number. To do this, multiply both the divisor and the dividend by the same power of ten. Then place the decimal point in the quotient directly above the decimal point in the dividend. Then divide as you do with whole numbers.

Example Divide 8.84 by 3.4.

Step 1 Multiply both factors by 10.
$3.4 \times 10 = 34$, $8.84 \times 10 = 88.4$

Step 2 Divide 88.4 by 34.

$$\begin{array}{r} 2.6 \\ 34\overline{)88.4} \\ -68 \\ \hline 204 \\ -204 \\ \hline 0 \end{array}$$

8.84 divided by 3.4 is 2.6.

Practice Problem Divide 75.6 by 3.6.

Use Proportions

An equation that shows that two ratios are equivalent is a proportion. The ratios $\frac{2}{4}$ and $\frac{5}{10}$ are equivalent, so they can be written as $\frac{2}{4} = \frac{5}{10}$. This equation is a proportion.

When two ratios form a proportion, the cross products are equal. To find the cross products in the proportion $\frac{2}{4} = \frac{5}{10}$, multiply the 2 and the 10, and the 4 and the 5. Therefore $2 \times 10 = 4 \times 5$, or $20 = 20$.

Because you know that both proportions are equal, you can use cross products to find a missing term in a proportion. This is known as solving the proportion.

Example The heights of a tree and a pole are proportional to the lengths of their shadows. The tree casts a shadow of 24 m when a 6-m pole casts a shadow of 4 m. What is the height of the tree?

Step 1 Write a proportion.
$$\frac{\text{height of tree}}{\text{height of pole}} = \frac{\text{length of tree's shadow}}{\text{length of pole's shadow}}$$

Step 2 Substitute the known values into the proportion. Let h represent the unknown value, the height of the tree.
$$\frac{h}{6} = \frac{24}{4}$$

Step 3 Find the cross products.
$$h \times 4 = 6 \times 24$$

Step 4 Simplify the equation.
$$4h = 144$$

Step 5 Divide each side by 4.
$$\frac{4h}{4} = \frac{144}{4}$$
$$h = 36$$

The height of the tree is 36 m.

Practice Problem The ratios of the weights of two objects on the Moon and on Earth are in proportion. A rock weighing 3 N on the Moon weighs 18 N on Earth. How much would a rock that weighs 5 N on the Moon weigh on Earth?

Use Percentages

The word *percent* means "out of one hundred." It is a ratio that compares a number to 100. Suppose you read that 77 percent of the Earth's surface is covered by water. That is the same as reading that the fraction of the Earth's surface covered by water is $\frac{77}{100}$. To express a fraction as a percent, first find the equivalent decimal for the fraction. Then, multiply the decimal by 100 and add the percent symbol.

Example Express $\frac{13}{20}$ as a percent.

Step 1 Find the equivalent decimal for the fraction.

$$\begin{array}{r} 0.65 \\ 20\overline{)13.00} \\ \underline{12\ 0} \\ 1\ 00 \\ \underline{1\ 00} \\ 0 \end{array}$$

Step 2 Rewrite the fraction $\frac{13}{20}$ as 0.65.

Step 3 Multiply 0.65 by 100 and add the % sign.
$$0.65 \times 100 = 65 = 65\%$$

So, $\frac{13}{20} = 65\%$.

This also can be solved as a proportion.

Example Express $\frac{13}{20}$ as a percent.

Step 1 Write a proportion.
$$\frac{13}{20} = \frac{x}{100}$$

Step 2 Find the cross products.
$$1300 = 20x$$

Step 3 Divide each side by 20.
$$\frac{1300}{20} = \frac{20x}{20}$$
$$65\% = x$$

Practice Problem In one year, 73 of 365 days were rainy in one city. What percent of the days in that city were rainy?

Solve One-Step Equations

A statement that two things are equal is an equation. For example, $A = B$ is an equation that states that A is equal to B.

An equation is solved when a variable is replaced with a value that makes both sides of the equation equal. To make both sides equal the inverse operation is used. Addition and subtraction are inverses, and multiplication and division are inverses.

Example 1 Solve the equation $x - 10 = 35$.

Step 1 Find the solution by adding 10 to each side of the equation.
$$x - 10 = 35$$
$$x - 10 + 10 = 35 + 10$$
$$x = 45$$

Step 2 Check the solution.
$$x - 10 = 35$$
$$45 - 10 = 35$$
$$35 = 35$$

Both sides of the equation are equal, so $x = 45$.

Example 2 In the formula $a = bc$, find the value of c if $a = 20$ and $b = 2$.

Step 1 Rearrange the formula so the unknown value is by itself on one side of the equation by dividing both sides by b.
$$a = bc$$
$$\frac{a}{b} = \frac{bc}{b}$$
$$\frac{a}{b} = c$$

Step 2 Replace the variables a and b with the values that are given.
$$\frac{a}{b} = c$$
$$\frac{20}{2} = c$$
$$10 = c$$

Step 3 Check the solution.
$$a = bc$$
$$20 = 2 \times 10$$
$$20 = 20$$

Both sides of the equation are equal, so $c = 10$ is the solution when $a = 20$ and $b = 2$.

Practice Problem In the formula $h = gd$, find the value of d if $g = 12.3$ and $h = 17.4$.

Use Statistics

The branch of mathematics that deals with collecting, analyzing, and presenting data is statistics. In statistics, there are three common ways to summarize data with a single number—the mean, the median, and the mode.

The **mean** of a set of data is the arithmetic average. It is found by adding the numbers in the data set and dividing by the number of items in the set.

The **median** is the middle number in a set of data when the data are arranged in numerical order. If there were an even number of data points, the median would be the mean of the two middle numbers.

The **mode** of a set of data is the number or item that appears most often.

Another number that often is used to describe a set of data is the range. The **range** is the difference between the largest number and the smallest number in a set of data.

A **frequency table** shows how many times each piece of data occurs, usually in a survey. **Table 2** below shows the results of a student survey on favorite color.

Table 2 Student Color Choice		
Color	**Tally**	**Frequency**
red	‖‖	4
blue	₩	5
black	‖	2
green	‖‖	3
purple	₩ ‖	7
yellow	₩ ‖	6

Based on the frequency table data, which color is the favorite?

Example The speeds (in m/s) for a race car during five different time trials are 39, 37, 44, 36, and 44.

To find the mean:

Step 1 Find the sum of the numbers.

$$39 + 37 + 44 + 36 + 44 = 200$$

Step 2 Divide the sum by the number of items, which is 5.

$$200 \div 5 = 40$$

The mean is 40 m/s.

To find the median:

Step 1 Arrange the measures from least to greatest.

36, 37, 39, 44, 44

Step 2 Determine the middle measure.

36, 37, <u>39</u>, 44, 44

The median is 39 m/s.

To find the mode:

Step 1 Group the numbers that are the same together.

44, 44, 36, 37, 39

Step 2 Determine the number that occurs most in the set.

<u>44, 44</u>, 36, 37, 39

The mode is 44 m/s.

To find the range:

Step 1 Arrange the measures from largest to smallest.

44, 44, 39, 37, 36

Step 2 Determine the largest and smallest measures in the set.

<u>44</u>, 44, 39, 37, <u>36</u>

Step 3 Find the difference between the largest and smallest measures.

$$44 - 36 = 8$$

The range is 8 m/s.

Practice Problem Find the mean, median, mode, and range for the data set 8, 4, 12, 8, 11, 14, 16.

Use Geometry

The branch of mathematics that deals with the measurement, properties, and relationships of points, lines, angles, surfaces, and solids is called geometry.

Perimeter The **perimeter** (P) is the distance around a geometric figure. To find the perimeter of a rectangle, add the length and width and multiply that sum by two, or $2(l + w)$. To find perimeters of irregular figures, add the length of the sides.

Example 1 Find the perimeter of a rectangle that is 3 m long and 5 m wide.

Step 1 You know that the perimeter is 2 times the sum of the width and length.
$$P = 2(3 \text{ m} + 5 \text{ m})$$

Step 2 Find the sum of the width and length.
$$P = 2(8 \text{ m})$$

Step 3 Multiply by 2.
$$P = 16 \text{ m}$$

The perimeter is 16 m.

Example 2 Find the perimeter of a shape with sides measuring 2 cm, 5 cm, 6 cm, 3 cm.

Step 1 You know that the perimeter is the sum of all the sides.
$$P = 2 + 5 + 6 + 3$$

Step 2 Find the sum of the sides.
$$P = 2 + 5 + 6 + 3$$
$$P = 16$$

The perimeter is 16 cm.

Practice Problem Find the perimeter of a rectangle with a length of 18 m and a width of 7 m.

Practice Problem Find the perimeter of a triangle measuring 1.6 cm by 2.4 cm by 2.4 cm.

Area of a Rectangle The **area** (A) is the number of square units needed to cover a surface. To find the area of a rectangle, multiply the length times the width, or $l \times w$. When finding area, the units also are multiplied. Area is given in square units.

Example Find the area of a rectangle with a length of 1 cm and a width of 10 cm.

Step 1 You know that the area is the length multiplied by the width.
$$A = (1 \text{ cm} \times 10 \text{ cm})$$

Step 2 Multiply the length by the width. Also multiply the units.
$$A = 10 \text{ cm}^2$$

The area is 10 cm².

Practice Problem Find the area of a square whose sides measure 4 m.

Area of a Triangle To find the area of a triangle, use the formula:

$$A = \frac{1}{2}(\text{base} \times \text{height})$$

The base of a triangle can be any of its sides. The height is the perpendicular distance from a base to the opposite endpoint, or vertex.

Example Find the area of a triangle with a base of 18 m and a height of 7 m.

Step 1 You know that the area is $\frac{1}{2}$ the base times the height.
$$A = \frac{1}{2}(18 \text{ m} \times 7 \text{ m})$$

Step 2 Multiply $\frac{1}{2}$ by the product of 18×7. Multiply the units.
$$A = \frac{1}{2}(126 \text{ m}^2)$$
$$A = 63 \text{ m}^2$$

The area is 63 m².

Practice Problem Find the area of a triangle with a base of 27 cm and a height of 17 cm.

Circumference of a Circle The **diameter** (*d*) of a circle is the distance across the circle through its center, and the **radius** (*r*) is the distance from the center to any point on the circle. The radius is half of the diameter. The distance around the circle is called the **circumference** (C). The formula for finding the circumference is:

$$C = 2\pi r \ \ or \ \ C = \pi d$$

The circumference divided by the diameter is always equal to 3.1415926... This nonterminating and nonrepeating number is represented by the Greek letter π (pi). An approximation often used for π is 3.14.

Example 1 Find the circumference of a circle with a radius of 3 m.

Step 1 You know the formula for the circumference is 2 times the radius times π.
$$C = 2\pi(3)$$

Step 2 Multiply 2 times the radius.
$$C = 6\pi$$

Step 3 Multiply by π.
$$C = 19 \text{ m}$$

The circumference is 19 m.

Example 2 Find the circumference of a circle with a diameter of 24.0 cm.

Step 1 You know the formula for the circumference is the diameter times π.
$$C = \pi(24.0)$$

Step 2 Multiply the diameter by π.
$$C = 75.4 \text{ cm}$$

The circumference is 75.4 cm.

Practice Problem Find the circumference of a circle with a radius of 19 cm.

Area of a Circle The formula for the area of a circle is:
$$A = \pi r^2$$

Example 1 Find the area of a circle with a radius of 4.0 cm.

Step 1 $A = \pi(4.0)^2$

Step 2 Find the square of the radius.
$$A = 16\pi$$

Step 3 Multiply the square of the radius by π.
$$A = 50 \text{ cm}^2$$

The area of the circle is 50 cm².

Example 2 Find the area of a circle with a radius of 225 m.

Step 1 $A = \pi(225)^2$

Step 2 Find the square of the radius.
$$A = 50625\pi$$

Step 3 Multiply the square of the radius by π.
$$A = 158962.5$$

The area of the circle is 158,962 m².

Example 3 Find the area of a circle whose diameter is 20.0 mm.

Step 1 You know the formula for the area of a circle is the square of the radius times π, and that the radius is half of the diameter.
$$A = \pi\left(\frac{20.0}{2}\right)^2$$

Step 2 Find the radius.
$$A = \pi(10.0)^2$$

Step 3 Find the square of the radius.
$$A = 100\pi$$

Step 4 Multiply the square of the radius by π.
$$A = 314 \text{ mm}^2$$

The area is 314 mm².

Practice Problem Find the area of a circle with a radius of 16 m.

Volume The measure of space occupied by a solid is the **volume** (V). To find the volume of a rectangular solid multiply the length times width times height, or $V = l \times w \times h$. It is measured in cubic units, such as cubic centimeters (cm^3).

Example Find the volume of a rectangular solid with a length of 2.0 m, a width of 4.0 m, and a height of 3.0 m.

Step 1 You know the formula for volume is the length times the width times the height.
$$V = 2.0 \text{ m} \times 4.0 \text{ m} \times 3.0 \text{ m}$$

Step 2 Multiply the length times the width times the height.
$$V = 24 \text{ m}^3$$

The volume is 24 m^3.

Practice Problem Find the volume of a rectangular solid that is 8 m long, 4 m wide, and 4 m high.

To find the volume of other solids, multiply the area of the base times the height.

Example 1 Find the volume of a solid that has a triangular base with a length of 8.0 m and a height of 7.0 m. The height of the entire solid is 15.0 m.

Step 1 You know that the base is a triangle, and the area of a triangle is $\frac{1}{2}$ the base times the height, and the volume is the area of the base times the height.
$$V = \left[\frac{1}{2}(b \times h)\right] \times 15$$

Step 2 Find the area of the base.
$$V = \left[\frac{1}{2}(8 \times 7)\right] \times 15$$
$$V = \left(\frac{1}{2} \times 56\right) \times 15$$

Step 3 Multiply the area of the base by the height of the solid.
$$V = 28 \times 15$$
$$V = 420 \text{ m}^3$$

The volume is 420 m^3.

Example 2 Find the volume of a cylinder that has a base with a radius of 12.0 cm, and a height of 21.0 cm.

Step 1 You know that the base is a circle, and the area of a circle is the square of the radius times π, and the volume is the area of the base times the height.
$$V = (\pi r^2) \times 21$$
$$V = (\pi 12^2) \times 21$$

Step 2 Find the area of the base.
$$V = 144\pi \times 21$$
$$V = 452 \times 21$$

Step 3 Multiply the area of the base by the height of the solid.
$$V = 9490 \text{ cm}^3$$

The volume is 9490 cm^3.

Example 3 Find the volume of a cylinder that has a diameter of 15 mm and a height of 4.8 mm.

Step 1 You know that the base is a circle with an area equal to the square of the radius times π. The radius is one-half the diameter. The volume is the area of the base times the height.
$$V = (\pi r^2) \times 4.8$$
$$V = \left[\pi\left(\frac{1}{2} \times 15\right)^2\right] \times 4.8$$
$$V = (\pi 7.5^2) \times 4.8$$

Step 2 Find the area of the base.
$$V = 56.25\pi \times 4.8$$
$$V = 176.63 \times 4.8$$

Step 3 Multiply the area of the base by the height of the solid.
$$V = 847.8$$

The volume is 847.8 mm^3.

Practice Problem Find the volume of a cylinder with a diameter of 7 cm in the base and a height of 16 cm.

Science Applications

Measure in SI

The metric system of measurement was developed in 1795. A modern form of the metric system, called the International System (SI), was adopted in 1960 and provides the standard measurements that all scientists around the world can understand.

The SI system is convenient because unit sizes vary by powers of 10. Prefixes are used to name units. Look at **Table 3** for some common SI prefixes and their meanings.

Table 3 Common SI Prefixes			
Prefix	**Symbol**	**Meaning**	
kilo-	k	1,000	thousand
hecto-	h	100	hundred
deka-	da	10	ten
deci-	d	0.1	tenth
centi-	c	0.01	hundredth
milli-	m	0.001	thousandth

Example How many grams equal one kilogram?

Step 1 Find the prefix *kilo* in **Table 3.**

Step 2 Using **Table 3,** determine the meaning of *kilo*. According to the table, it means 1,000. When the prefix *kilo* is added to a unit, it means that there are 1,000 of the units in a "*kilo*unit."

Step 3 Apply the prefix to the units in the question. The units in the question are grams. There are 1,000 grams in a kilogram.

Practice Problem Is a milligram larger or smaller than a gram? How many of the smaller units equal one larger unit? What fraction of the larger unit does one smaller unit represent?

Dimensional Analysis

Convert SI Units In science, quantities such as length, mass, and time sometimes are measured using different units. A process called dimensional analysis can be used to change one unit of measure to another. This process involves multiplying your starting quantity and units by one or more conversion factors. A conversion factor is a ratio equal to one and can be made from any two equal quantities with different units. If 1,000 mL equal 1 L then two ratios can be made.

$$\frac{1{,}000 \text{ mL}}{1 \text{ L}} = \frac{1 \text{ L}}{1{,}000 \text{ mL}} = 1$$

One can covert between units in the SI system by using the equivalents in **Table 3** to make conversion factors.

Example 1 How many cm are in 4 m?

Step 1 Write conversion factors for the units given. From **Table 3,** you know that 100 cm = 1 m. The conversion factors are

$$\frac{100 \text{ cm}}{1 \text{ m}} \quad and \quad \frac{1 \text{ m}}{100 \text{ cm}}$$

Step 2 Decide which conversion factor to use. Select the factor that has the units you are converting from (m) in the denominator and the units you are converting to (cm) in the numerator.

$$\frac{100 \text{ cm}}{1 \text{ m}}$$

Step 3 Multiply the starting quantity and units by the conversion factor. Cancel the starting units with the units in the denominator. There are 400 cm in 4 m.

$$4 \text{ m} \times \frac{100 \text{ cm}}{1 \text{ m}} = 400 \text{ cm}$$

Practice Problem How many milligrams are in one kilogram? (Hint: You will need to use two conversion factors from **Table 3.**)

Table 4 Unit System Equivalents	
Type of Measurement	**Equivalent**
Length	1 in = 2.54 cm
	1 yd = 0.91 m
	1 mi = 1.61 km
Mass and Weight*	1 oz = 28.35 g
	1 lb = 0.45 kg
	1 ton (short) = 0.91 tonnes (metric tons)
	1 lb = 4.45 N
Volume	$1 \text{ in}^3 = 16.39 \text{ cm}^3$
	1 qt = 0.95 L
	1 gal = 3.78 L
Area	$1 \text{ in}^2 = 6.45 \text{ cm}^2$
	$1 \text{ yd}^2 = 0.83 \text{ m}^2$
	$1 \text{ mi}^2 = 2.59 \text{ km}^2$
	1 acre = 0.40 hectares
Temperature	$°C = \dfrac{(°F - 32)}{1.8}$
	$K = °C + 273$

*Weight is measured in standard Earth gravity.

Convert Between Unit Systems Table 4 gives a list of equivalents that can be used to convert between English and SI units.

Example If a meterstick has a length of 100 cm, how long is the meterstick in inches?

Step 1 Write the conversion factors for the units given. From **Table 4,** 1 in = 2.54 cm.

$$\frac{1 \text{ in}}{2.54 \text{ cm}} \quad and \quad \frac{2.54 \text{ cm}}{1 \text{ in}}$$

Step 2 Determine which conversion factor to use. You are converting from cm to in. Use the conversion factor with cm on the bottom.

$$\frac{1 \text{ in}}{2.54 \text{ cm}}$$

Step 3 Multiply the starting quantity and units by the conversion factor. Cancel the starting units with the units in the denominator. Round your answer based on the number of significant figures in the conversion factor.

$$100 \text{ cm} \times \frac{1 \text{ in}}{2.54 \text{ cm}} = 39.37 \text{ in}$$

The meterstick is 39.4 in long.

Practice Problem A book has a mass of 5 lbs. What is the mass of the book in kg?

Practice Problem Use the equivalent for in and cm (1 in = 2.54 cm) to show how $1 \text{ in}^3 = 16.39 \text{ cm}^3$.

Precision and Significant Digits

When you make a measurement, the value you record depends on the precision of the measuring instrument. This precision is represented by the number of significant digits recorded in the measurement. When counting the number of significant digits, all digits are counted except zeros at the end of a number with no decimal point such as 2,050, and zeros at the beginning of a decimal such as 0.03020. When adding or subtracting numbers with different precision, round the answer to the smallest number of decimal places of any number in the sum or difference. When multiplying or dividing, the answer is rounded to the smallest number of significant digits of any number being multiplied or divided.

Example The lengths 5.28 and 5.2 are measured in meters. Find the sum of these lengths and record your answer using the correct number of significant digits.

Step 1 Find the sum.

5.28 m	2 digits after the decimal
+ 5.2 m	1 digit after the decimal
10.48 m	

Step 2 Round to one digit after the decimal because the least number of digits after the decimal of the numbers being added is 1.

The sum is 10.5 m.

Practice Problem How many significant digits are in the measurement 7,071,301 m? How many significant digits are in the measurement 0.003010 g?

Practice Problem Multiply 5.28 and 5.2 using the rule for multiplying and dividing. Record the answer using the correct number of significant digits.

Scientific Notation

Many times numbers used in science are very small or very large. Because these numbers are difficult to work with scientists use scientific notation. To write numbers in scientific notation, move the decimal point until only one non-zero digit remains on the left. Then count the number of places you moved the decimal point and use that number as a power of ten. For example, the average distance from the Sun to Mars is 227,800,000,000 m. In scientific notation, this distance is 2.278×10^{11} m. Because you moved the decimal point to the left, the number is a positive power of ten.

The mass of an electron is about 0.000 000 000 000 000 000 000 000 000 000 911 kg. Expressed in scientific notation, this mass is 9.11×10^{-31} kg. Because the decimal point was moved to the right, the number is a negative power of ten.

Example Earth is 149,600,000 km from the Sun. Express this in scientific notation.

Step 1 Move the decimal point until one non-zero digit remains on the left.
1.496 000 00

Step 2 Count the number of decimal places you have moved. In this case, eight.

Step 3 Show that number as a power of ten, 10^8.

The Earth is 1.496×10^8 km from the Sun.

Practice Problem How many significant digits are in 149,600,000 km? How many significant digits are in 1.496×10^8 km?

Practice Problem Parts used in a high performance car must be measured to 7×10^{-6} m. Express this number as a decimal.

Practice Problem A CD is spinning at 539 revolutions per minute. Express this number in scientific notation.

Make and Use Graphs

Data in tables can be displayed in a graph—a visual representation of data. Common graph types include line graphs, bar graphs, and circle graphs.

Line Graph A line graph shows a relationship between two variables that change continuously. The independent variable is changed and is plotted on the *x*-axis. The dependent variable is observed, and is plotted on the *y*-axis.

Example Draw a line graph of the data below from a cyclist in a long-distance race.

Table 5 Bicycle Race Data	
Time (h)	Distance (km)
0	0
1	8
2	16
3	24
4	32
5	40

Step 1 Determine the *x*-axis and *y*-axis variables. Time varies independently of distance and is plotted on the *x*-axis. Distance is dependent on time and is plotted on the *y*-axis.

Step 2 Determine the scale of each axis. The *x*-axis data ranges from 0 to 5. The *y*-axis data ranges from 0 to 40.

Step 3 Using graph paper, draw and label the axes. Include units in the labels.

Step 4 Draw a point at the intersection of the time value on the *x*-axis and corresponding distance value on the *y*-axis. Connect the points and label the graph with a title, as shown in **Figure 20.**

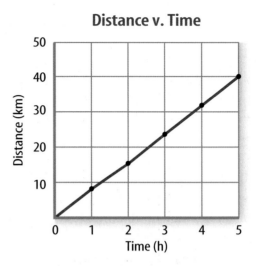

Distance v. Time

Figure 20 This line graph shows the relationship between distance and time during a bicycle ride.

Practice Problem A puppy's shoulder height is measured during the first year of her life. The following measurements were collected: (3 mo, 52 cm), (6 mo, 72 cm), (9 mo, 83 cm), (12 mo, 86 cm). Graph this data.

Find a Slope The slope of a straight line is the ratio of the vertical change, rise, to the horizontal change, run.

$$\text{Slope} = \frac{\text{vertical change (rise)}}{\text{horizontal change (run)}} = \frac{\text{change in } y}{\text{change in } x}$$

Example Find the slope of the graph in **Figure 20.**

Step 1 You know that the slope is the change in *y* divided by the change in *x*.
$$\text{Slope} = \frac{\text{change in } y}{\text{change in } x}$$

Step 2 Determine the data points you will be using. For a straight line, choose the two sets of points that are the farthest apart.
$$\text{Slope} = \frac{(40-0)\text{ km}}{(5-0)\text{ hr}}$$

Step 3 Find the change in *y* and *x*.
$$\text{Slope} = \frac{40 \text{ km}}{5 \text{ h}}$$

Step 4 Divide the change in *y* by the change in *x*.
$$\text{Slope} = \frac{8 \text{ km}}{\text{h}}$$

The slope of the graph is 8 km/h.

Bar Graph To compare data that does not change continuously you might choose a bar graph. A bar graph uses bars to show the relationships between variables. The *x*-axis variable is divided into parts. The parts can be numbers such as years, or a category such as a type of animal. The *y*-axis is a number and increases continuously along the axis.

Example A recycling center collects 4.0 kg of aluminum on Monday, 1.0 kg on Wednesday, and 2.0 kg on Friday. Create a bar graph of this data.

Step 1 Select the *x*-axis and *y*-axis variables. The measured numbers (the masses of aluminum) should be placed on the *y*-axis. The variable divided into parts (collection days) is placed on the *x*-axis.

Step 2 Create a graph grid like you would for a line graph. Include labels and units.

Step 3 For each measured number, draw a vertical bar above the *x*-axis value up to the *y*-axis value. For the first data point, draw a vertical bar above Monday up to 4.0 kg.

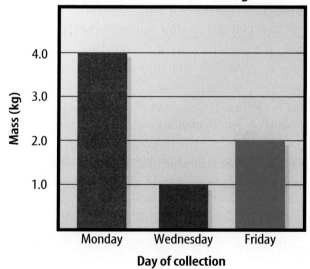

Aluminum Collected During Week

Practice Problem Draw a bar graph of the gases in air: 78% nitrogen, 21% oxygen, 1% other gases.

Circle Graph To display data as parts of a whole, you might use a circle graph. A circle graph is a circle divided into sections that represent the relative size of each piece of data. The entire circle represents 100%, half represents 50%, and so on.

Example Air is made up of 78% nitrogen, 21% oxygen, and 1% other gases. Display the composition of air in a circle graph.

Step 1 Multiply each percent by 360° and divide by 100 to find the angle of each section in the circle.

$$78\% \times \frac{360°}{100} = 280.8°$$

$$21\% \times \frac{360°}{100} = 75.6°$$

$$1\% \times \frac{360°}{100} = 3.6°$$

Step 2 Use a compass to draw a circle and to mark the center of the circle. Draw a straight line from the center to the edge of the circle.

Step 3 Use a protractor and the angles you calculated to divide the circle into parts. Place the center of the protractor over the center of the circle and line the base of the protractor over the straight line.

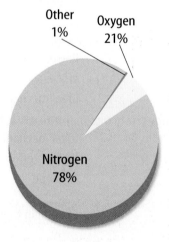

Practice Problem Draw a circle graph to represent the amount of aluminum collected during the week shown in the bar graph to the left.

Formulas

Chapter 1
The Nature of Science

Density = mass/volume

Kelvin = °Celsius + 273

% Error = [(Accepted value − Experimental value)/Accepted value] × 100

Chapter 2
Motion

Speed = distance/time

Acceleration = change in velocity/time

Change in velocity = final velocity − initial velocity

Chapter 3
Forces

Acceleration = net force/mass

Force = mass × acceleration

Gravitational force = mass × (acceleration due to gravity)

Weight = mass × 9.8 m/s^2

Momentum (p) = mass × velocity

Force = $(mv_f − mv_i)$/time

change in displacement = initial velocity (change in time) + $\frac{1}{2}$ acceleration(change in time)^2

average velocity = change in displacement/change in time = (final velocity + initial velocity)/2

average acceleration = change in velocity/change in time

Chapter 4
Energy

Kinetic energy = $\frac{1}{2}$ mass × (velocity)^2

Gravitational potential energy (GPE) = mass × 9.8 m/s^2 × height

Mechanical energy = gravitational potential energy + kinetic energy

Chapter 5
Work and Machines

Work = force × distance

Power = work/time

Efficiency = (work$_{out}$/work$_{in}$) × 100%

Ideal mechanical advantage (IMA) = length of effort arm/length of resistance arm = L_e/L_r

Ideal mechanical advantage (IMA) = radius of wheel/radius of axle = r_w/r_a

IMA = effort distance/resistance distance = length of slope/height of slope = l/h

Chapter 6
Thermal Energy

Change in thermal energy = mass × change in temperature × specific heat

or

$Q = m × (T_{final} − T_{initial}) × C_p$

Q = mass × heat of fusion = mH_f

Q = mass × heat of vaporization = mH_v

Chapter 7
Electricity

Electric current $=$ voltage difference/resistance   or $I = V/R$

Electric power $=$ current $\times$ voltage difference   or $P = I \times V$

Electric energy $=$ power $\times$ time   or $E = P \times t$

Series Circuits

$I_t = I_1 = I_2 = I_3 = \ldots$

$V_t = V_1 + V_2 + V_3 + \ldots$

$R_t = R_1 + R_2 + R_3 + \ldots$

Parallel Circuits

$I_t = I_1 + I_2 + I_3 + \ldots$

$V_t = V_1 = V_2 = V_3 = \ldots$

$1/R_t = 1/R_1 + 1/R_2 + 1/R_3 + \ldots$

Chapter 10
Waves

Wave velocity $=$ wavelength $\times$ frequency   or $v_w = \lambda \times f$

Chapter 13
Light

Index of refraction $=$

     speed of light in a vacuum/speed of light in a substance   or $n = c/v$

Chapter 16
Solids, Liquids, and Gases

Pressure $=$ force/area   or $P = F/A$

Boyle's law $P_1 \times V_1 = P_2 \times V_2$

Charles's law $V_1/T_1 = V_2/T_2$

Chapter 22
Solutions

Surface area of a rectangular solid $= 2(h \times w) + 2(h \times l) + 2(w \times l)$

EXTRA Math Problems

For help and hints with these problems, visit gpscience.com/extra_problems.

Chapter 1 The Nature of Science

1. How many centimeters are in four meters?

2. How many deciliters are in 500 mL?

3. How many liters are in 2540 cm^3?

4. A young child has a mass of 40 kg. What is the mass of the child in grams?

5. Iron has a density of 7.9 g/cm^3. What is the mass in kg of an iron statue that has a volume of 5.4 L?

6. A 2-L bottle of soda has a volume of 2000 cm^3. What is the volume of the bottle in cubic meters?

7. A big summer movie has a running time of 96 minutes. What is the movie's running time in seconds?

8. The temperature in space is approximately 3 K. What is this temperature in degrees Celsius?

9. The x-axis of a certain graph is distance traveled in meters and the y-axis is time in seconds. Two points are plotted on this graph with coordinates (2, 43) and (5, 68). What is the elapsed time between the two points?

10. A circle graph has labeled segments of: 57%, 21%, 13%, and 6%. What percentage does the unlabeled segment have?

Chapter 2 Motion

11. John rides his bike 2.3 km to school. After school, he rides an additional 1.4 km to the mall in the opposite direction. What is his total distance traveled?

12. A squirrel runs 4.8 m across a lawn, stops, then runs 2.3 m back in the opposite direction. What is the squirrel's displacement from its starting point?

13. An ant travels 75 cm in 5 s. What was the ant's speed?

14. It took you 6.5 h to drive 550 km. What was your speed?

15. A bus leaves at 9 A.M. with a group of tourists. They travel 350 km before they stop for lunch. Then they travel an additional 250 km until the end of their trip at 3 P.M. What was the average speed of the bus?

16. Halfway through a cross-country meet, a runner's speed is 4 m/s. In the last stretch, she increases her speed to 7 m/s. What is her change in speed?

17. It takes a car one minute to go from rest to 30 m/s. What is the acceleration of this car?

18. You are running at a speed of 10 km/h and hit a patch of mud. Two seconds later your speed is 8 km/h. What is your acceleration in units of m/s^2?

19. A weight lifter is trying to lift a 1500-N weight but can apply a force of only 1200 N on the weight. One of his friends helps him lift it all the way. What force was applied to the weight by the weight lifter's friend?

20. During a tug-of-war, Team A is applying a force of 5000 N while Team B is applying a force of 8000 N. What is the net force applied to the rope?

21. You are in a car traveling an average speed of 60 km/h. The total trip is 240 km. How long does the trip take?

22. You are riding in a train that is traveling at a speed of 120 km/h. How long will it take to travel 950 km?

23. A car goes from rest to a speed of 90 km/h in 10 s. What is the car's acceleration in m/s^2?

24. A cart rolling at a speed of 10 m/s comes to a stop in 2 s. What is the cart's acceleration?

Chapter 3 Forces

25. A 85-kg mass has an acceleration of 5.5 m/s^2. What is the net force applied?

26. A 3200-N force is applied to a 160-kg mass. What is the acceleration of the mass?

27. If you are pushing on a box with a force of 20 N and there is a force of 7 N on the box due to sliding friction, what is the net force on the box?

28. A 2-kg object is dropped from a height of 1000 m. What is the force of air resistance on the object when it reaches terminal velocity?

29. How much force is needed to lift a 25-kg mass?

30. A person is on an elevator that moves downward with an acceleration of 1.8 m/s^2. If the person weighs 686 N, what is the net force on the person?

31. The acceleration due to gravity on the moon is about 1.6 m/s^2. If you weigh 539 N on Earth, how much would you weigh on the moon?

32. If a 5000-kg mass is moving at a speed of 43 m/s, what is its momentum?

33. How fast must a 50-kg mass travel to have a momentum of 1500 kg m/s?

34. What is the net force on a 4000-kg car that doubles its speed from 15 m/s to 30 m/s over 10 seconds?

35. A book with a mass of 1 kg is sliding on a table. If the frictional force on the book is 5 N, calculate the book's acceleration. Is it speeding up or slowing down?

36. What is the weight of a person with a mass of 80 kg?

37. A car with a mass of 1,200 kg has a speed of 30 m/s. What is the car's momentum?

Chapter 4 Energy

38. What is the kinetic energy of a 5-kg object moving at 7 m/s?

39. An object has kinetic energy of 600 J and a speed of 10 m/s. What is its mass?

40. If you throw a 0.4-kg ball at a speed of 20 m/s, what is the ball's kinetic energy?

41. A rollercoaster car moving around a high turn has 100,000 J of GPE and 23,000 J of KE. What is its mechanical energy?

42. If you have a mass of 80 kg and you are standing on a platform 3 m above the ground, what is your gravitational potential energy?

43. A 2-kg book is moved from a shelf that is 2 m off the ground to a shelf that is 1.5 m off the ground. What is its change in GPE?

44. A car is traveling at 30 m/s with a kinetic energy of 900 kJ. What is its mass?

45. At top of a hill, a rollercoaster has 67,500 J of kinetic energy and 290,000 J of potential energy. Gradually the roller coaster comes to a stop due to friction. If the roller coaster has 30,000 J of potential when it stops, how much heat energy is generated by friction from the top of the hill until it stops?

46. A system has a total mechanical energy of 350 J and kinetic energy of 220 J. What is its potential energy?

47. An object held in the air has a GPE of 470 J. The object then is dropped. Halfway down, what is the object's kinetic energy?

48. A car with a mass of 900 kg is traveling at a speed of 25 m/s. What is the kinetic energy of the car in joules?

49. What is the gravitational potential energy of a diver with a mass of 60 kg who is 10 m above the water?

50. If your weight is 500 N, and you are standing on a floor that is 20 m above the ground, what is your gravitational potential energy?

Chapter 5 Work and Machines

51. When moving a couch, you exert a force of 400 N and push it 4 m. How much work have you done?

52. How much work is needed to lift a 50-kg weight to a shelf 3 m above the floor?

53. By applying a force of 50 N, a pulley system can lift a box with a mass of 20 kg. What is the mechanical advantage of the pulley system?

54. How much energy do you save per hour if you replace a 60-watt lightbulb with a 55-watt lightbulb?

55. Suppose you supply energy to a machine at a rate of 700 W and that the machine converts 560 J into heat every second. At what rate does the machine do work?

56. You exert a force of 200 N on a machine over a distance of 0.3 m. If the machine moves an object a distance of 0.5 m, how much force does the machine exert on the object? Assume friction can be ignored.

57. What is the efficiency of a machine if you do work on the machine at a rate of 1200 W and the machine does work at a rate of 300 W?

58. What is the IMA of a seesaw with a 1.6-m effort arm and a 1.2-m resistance arm?

59. What is the IMA of a wheel with a radius of 0.35 m and an axle radius of 0.04 m?

60. An inclined plane has an IMA of 1.5 and a height of 2.0 m. How long is this inclined plane?

61. What power is used by a machine to perform 800 J of work in 25 s?

62. A person pushes a box up a ramp that is 3 m long, and 1 m high. If the box has a mass of 20 kg, and the person pushes with a force of 80 N, what is the efficiency of the ramp?

63. A first class lever has a mechanical advantage of 5. How large would a force need to be to lift a rock with a mass of 100 kg?

Chapter 6 Thermal Energy

64. Water has a specific heat of 4184 J/(kg K). How much energy is needed to increase the temperature of a kilogram of water 5°C?

65. The temperature of a block of iron, which has a specific heat of 450 J/(kg K), increases by 3 K when 2700 J of energy are added to it. What is the mass of this block of iron?

66. How much energy is needed to heat 1 kg of sand, which has a specific heat of 664 J/(kg K), from 30°C to 50°C?

67. 1 kg of water (specific heat = 4184 J/(kg K)) is heated from freezing (0°C) to boiling (100°C). What is the change in thermal energy?

68. A concrete statue (specific heat = 600 J/(kg K)) sits in sunlight and warms up to 40°C. Overnight, it cools to 15°C and loses 90,000 J of thermal energy. What is its mass?

69. A glass of water has temperature of 70°C. What is its temperature in K?

70. A substance with a mass of 10 kg loses 106.5 kJ of heat when its temperature drops 15°C. What is this substance's specific heat?

71. Air is cooled from room temperature (25°C) to 100 K. What is the temperature change in K?

72. To remove 800 J of heat from a refrigerator, the compressor in the refrigerator does 500 J of work. How much heat is released into the surrounding room?

73. A calorimeter contains 1 kg of water [specific heat = 4184 J/(kg K)]. An object with a mass of 4.23 kg is added to the water. If the water temperature increases by 3 K and the temperature of the object decreases by 1 K, what is the specific heat of the object?

74. How much heat is needed to raise the temperature of 100 g of water by 50 K, if the specific heat of water is 4,184 J/kg K?

75. A sample of an unknown metal has a mass of 0.5 kg. Adding 1,985 J of heat to the metal raises its temperature by 10 K. What is the specific heat of the metal?

Chapter 7 Electricity

76. A circuit has a resistance of 4 Ω. What voltage difference will cause a current of 1.4 A to flow in the circuit?

77. How many amperes of current will flow in a circuit if the voltage difference is 9 V and the resistance in the circuit is 3 Ω?

78. If a voltage difference of 3 V causes a 1.5 A current to flow in a circuit, what is the resistance in the circuit?

79. The current in an appliance is 3 A and the voltage difference is 120 V. How much power is being supplied to the appliance?

80. What is the current into a microwave oven that requires 700 W of power if the voltage difference is 120 V?

81. What is the voltage difference in a circuit that uses 2420 W of power if 11 A of current flows into the circuit?

82. How much energy is used when a 110 kW appliance is used for 3 hours?

83. A television has a power rating of 210 W. If the television uses 1.68 kWh of energy, for how long has the television been on?

84. How much does it cost to light six 100-W lightbulbs for six hours if the price of electrical energy is $0.09/kWh?

85. An electric clothes dryer uses 4 kW of electric power. How long did it take to dry a load of clothes if electric power costs $0.09/kWh, and the cost of using the dryer was $0.27?

86. What is the resistance of a lightbulb that draws 0.5 amp of current when plugged into a 120-V outlet?

87. How much current flows through a 100-W lightbulb that is plugged into a 120-V outlet?

88. Eight amps of current flow through a hair dryer connected to a 120-V outlet. How much electrical power does the hair dryer use?

89. Compare the electrical energy that is used by a 100-W lightbulb that burns for 10 h, and a 1,200-W hair dryer that is used for 15 min.

Chapter 8 Magnetism and Its Uses

90. How many turns are in the secondary coil of a step-down transformer that reduces a voltage from 900 V to 300 V and has 15 turns in the primary coil?

91. A step-down transformer reduces voltage from 2400 V to 120 V. What is the ratio of the number of turns in the primary coil to the number of turns in the secondary coil of the transformer?

92. The current produced by an AC generator switches direction twice for each revolution of the coil. How many times does a 110-Hz alternating current switch direction each second?

93. What is the output voltage from a step-down transformer with 200 turns in the primary coil and 100 turns in the secondary coil if the input voltage was 750 V?

94. What is the output voltage from a step-up transformer with 25 turns in the primary coil and 75 turns in the secondary coil if the input voltage was 120 V?

95. How many turns are in the primary coil of a step-down transformer that reduces a voltage from 400 V to 100 V and has 80 turns in the secondary coil?

96. How many turns are in the secondary coil of a step-up transformer that increases voltage from 30 V to 150 V and has seven turns in the primary coil?

97. The coil of a 60-Hz generator makes 60 revolutions each second. How many revolutions does the coil make in five minutes?

98. If a generator coil makes 6000 revolutions in two minutes, how many revolutions does it make each second?

Chapter 9 Energy Sources

99. A gallon of gasoline contains about 2800 g of gasoline. If burning one gram of gasoline releases about 48 kJ of energy, how much energy is released when a gallon of gasoline is burned? (1 kJ = 1000 J)

100. An automobile engine converts the energy released by burning gasoline into mechanical energy with an efficiency of about 25%. If burning 1 kg of gasoline releases about 48,000 kJ of energy, how much mechanical energy is produced by the engine when 1 kg of gasoline is burned?

101. You heat a cup of water in a 750-W microwave oven for 40 s, and warm the water by 20°C. If it takes about 20 kJ of energy to raise the temperature of a cup of water by 20°C, what is the efficiency of the microwave oven?

102. On average, solar energy strikes Earth's surface with an intensity of about 200 W/m². If solar cells are 10% efficient, how large an area would have to be covered by solar cells to generate enough electrical power to light a 100-W lightbulb?

103. What is the overall efficiency of a hydroelectric plant if the process of falling water turning a turbine is 80% efficient, the turbine spinning an electric generator is 95% efficient, and the transmission through power lines is 90% efficient?

104. When a certain $^{235}_{92}U$ nucleus is struck by a neutron, it forms the two nuclei $^{91}_{36}Kr$ and $^{142}_{56}Ba$. How many neutrons are emitted when this occurs?

105. A nuclear reactor contains 100,000 kg of enriched uranium. About 4% of the enriched uranium is the isotope uranium-235. What is the mass of uranium-235 in the reactor core?

106. Suppose the number of uranium-235 nuclei that are split doubles at each stage of a chain reaction. If the chain reaction starts with one nucleus split in the first stage, how many nuclei will have been split after six stages?

107. From 1970 to 1995 the carbon dioxide concentration in Earth's atmosphere increased from about 325 parts per million to about 360 parts per million. What was the percentage change in the concentration of carbon dioxide?

108. About 85% of the energy used in the U.S. comes from fossil fuels. How many times greater is the amount of energy used from fossil fuel than the amount used from all other energy sources?

Chapter 10 Waves

109. What is the wavelength of a wave with a frequency of 0.4 kHz traveling at 16 m/s?

110. Two waves are traveling in the same medium with a speed of 340 m/s. What is the difference in frequency of the waves if the one has a wavelength of 5 m and the other has a wavelength of 0.2 m?

111. Transverse wave A has an amplitude of 7 cm. This wave constructively interferes with wave B. While the two waves overlap, the amplitude of the resulting wave is 10 cm. What is the amplitude of wave B?

112. What is the wavelength of a wave with a frequency of 5 Hz traveling at 15 m/s?

113. What is the velocity of a wave that has a wavelength of 6 m and a frequency of 3 Hz?

114. A ray of light hits a mirror at an angle of 35° to the normal. What is the angle of the reflected ray to the normal?

115. A wave has a wavelength of 250 cm and a frequency of 4 Hz. What is its speed?

116. A wave has a frequency of 5.6 MHz. What is the frequency of this wave in Hz?

117. A light ray strikes a mirror and is reflected. The angle between the incident and reflected rays is 86°. What is the angle of the incident ray to the normal?

118. What is the frequency of a wave with a wavelength of 7 m traveling at 21 m/s?

Chapter 11 Sound

119. What is the wavelength of a 440-Hz sound wave traveling with a speed of 347 m/s?

120. A sound wave with a frequency of 440 Hz travels in steel with a speed of 5200 m/s. What is the wavelength of the sound wave?

121. A wave traveling in water has a wavelength of a 750 m and a frequency of 2 Hz. How fast is this wave moving?

122. At 0°C sound travels through air with a speed of about 331 m/s and through aluminum with a speed of 4877 m/s. How many times longer is the wavelength of a sound wave in aluminum compared to the wavelength of a sound wave in air if both waves have the same frequency?

123. The speed of sound in air at 0°C is 331 m/s, and at 20°C is 344 m/s. What is the percentage change in the speed of sound at 20°C compared to 0°C?

124. In a lab experiment, measurements of the speed of sound in air were 329.7 m/s, 333.6 m/s, 330.8 m/s, 331.7 m/s, and 332.2 m/s. What is the average value of these measurements?

125. What is the frequency of the first overtone of a 440-Hz wave?

126. The wreck of the *Titanic* is at a depth of about 3800 m. A sonar unit on a ship above the *Titanic* emits a sound wave that travels at a speed of 1500 m/s. How long does it take a sound wave reflected from the *Titanic* to return to the ocean surface?

127. A sonar unit on a ship emits a sound wave. The echo from the ocean floor is detected two seconds later. If the speed of sound in water is 1500 m/s, how deep is the ocean beneath the ship?

128. One flute plays a note with a frequency of 443 Hz, and another flute plays a note with a frequency of 440 Hz. What is the frequency of the beats that the flute players hear?

129. A sound wave has a wavelength of 50 m and a frequency of 22 cycles per second. What is the speed of the sound wave?

130. A tsunami travels across the ocean at a speed of 500 km/h. If the distance between the wave crests is 200 km, what is the frequency of the wave?

Chapter 12 Electromagnetic Waves

131. Express the number 20,000 in scientific notation.

132. An electromagnetic wave has a wavelength of 0.054 m. What is the wavelength in scientific notation?

133. Earth is about 4,500,000,000 years old. Express this number in scientific notation.

134. The speed of electromagnetic waves in air is 300,000 km/s. What is the frequency of electromagnetic waves that have a wavelength of 5×10^{-3} km?

135. The speed of radio waves in water Is about 2.26×10^5 km/s. What is the frequency of radio waves that have a wavelength of 3.0 km?

136. Radio waves with a frequency of 125,000 Hz have a wavelength of 1.84 km when traveling in ice. What is the speed of the radio waves in ice?

137. Some infrared waves have a frequency of 10,000,000,000,000 Hz. Express this frequency in scientific notation.

138. An infrared wave has a frequency of 1×10^{13} Hz and a wavelength of 3×10^{-5} m. Express this wavelength as a decimal number.

139. An AM radio station broadcasts at a frequency of 620 kHz. Express this frequency in Hz using scientific notation.

140. An FM radio station broadcasts at a frequency of 101 MHz. Express this frequency in Hz using scientific notation.

Chapter 13 Light

141. A ray of light hits a plane mirror at 35° from the normal. What angle does the reflected ray make with the normal?

142. A light ray strikes a plane mirror. The angle between the light ray and the surface of the mirror is 25°. What angle does the reflected ray make with the normal?

143. About 8% of men and 0.5% of women have some form of color blindness. The number of men who experience color blindness is how many times larger than the number of women who experience color blindness?

144. The index of refraction of a material is the speed of light in a vacuum divided by the speed of light in the material. If the index of refraction of the mineral rock salt is 1.52, and the speed of light in a vacuum is 300,000 km/s, what is the speed of light in rock salt?

145. A laser is used to measure the distance from Earth to the Moon. The laser beam is reflected from a mirror on the Moon's surface. If the time needed for the laser to reach the Moon and be reflected back is 2.56 s, and the laser beam travels at 300,000 km/s, what is the distance to the Moon?

146. A light ray is reflected from a plane mirror. If the angle between the incident ray and the reflected ray is 104°, what is the angle of incidence?

147. In the human eye, there are about 7,000,000 cone cells distributed over an area of 5 cm². If cone cells are evenly distributed over this region, how many cone cells are distributed over an area of 2 cm²? Express your answer in scientific notation.

148. What will happen to a ray of light leaving water and entering air if it hits the boundary at an angle of 49° to the normal? (The critical angle for water and air is 49°.)

149. A ray of light hits a plane mirror at 60° from the normal. What is the angle between the reflected ray and the surface of the mirror?

150. When a light beam is reflected from a glass surface, only 4% of the energy carried by the beam is reflected. If a light beam is reflected from one glass surface and then another, what is the ratio of the energy carried by the beam after the second reflection, compared to the energy carried by the beam before the first reflection?

Chapter 14 Mirrors and Lenses

151. A light ray strikes a plane mirror. The angle between the incident light ray and the normal to the mirror is 55°. What is the angle between the reflected ray and the normal?

152. The magnification of a mirror or lens equals the image size divided by the object size. If a plant cell with a diameter of 0.0035 mm is magnified so that the diameter of the image is 0.028 cm, what is the magnification?

153. A convex lens in a magnifying glass has a focal length of 5 cm. How far should the lens be from an object if the image formed is virtual, enlarged, and upright?

154. A concave mirror forms a real image that is 3/4 the size of the object. How far is the object from the mirror?

155. Magnification equals the image size divided by the object size. Magnification also equals the distance of the image from the lens divided by the distance of the object from the lens. A penny has a diameter of 2.0 cm. A convex lens forms an image with a diameter of 5.2 cm and is 6.0 cm from the lens. What is the distance between the penny and the lens?

156. Light enters the human eye through the pupil. In the dark, the pupil is dilated and has a diameter of about 1 cm. The Keck telescope has a mirror with a diameter of 10 m. If both the pupil and the Keck mirror are circles, what is the ratio of the area of the Keck telescope mirror to the area of a dilated human pupil?

157. A small insect is viewed in a compound microscope. The objective lens of the microscope forms a real image 20 times larger than the insect. The eyepiece lens then magnifies this real image by 10 times. What is the magnification of the microscope?

158. A light source is placed a distance of 1.2 m from a concave mirror on the optical axis. The reflected light rays are parallel and form a light beam. What is the focal length of the mirror?

159. In some types of reflecting telescopes the eyepiece is located behind the concave mirror. A small curved mirror in front of the concave mirror reflects light through a hole in the concave mirror to the eyepiece. Suppose a circular concave mirror with a diameter of 50 cm has a hole with a diameter of 10 cm. What is the ratio of the reflecting area of the mirror with the 10-cm hole to the reflecting area of the same mirror without the hole?

160. Astronomers have proposed building the OWL telescope (**o**ver**w**helmingly **l**arge telescope) with a mirror 100 m in diameter. The diameter of the *Hubble Space Telescope* mirror is 2.4 m. What percentage of the surface area of the OWL mirror would be covered by the surface area of the *Hubble* mirror?

Chapter 15 Classification of Matter

161. Two solutions, one with a mass of 450 g and the other with a mass of 350 g, are mixed. A chemical reaction occurs and 125 g of solid crystals are produced that settle on the bottom of the container. What is the mass of the remaining solution?

162. Carbon reacts with oxygen to form carbon dioxide according to the following equation: $C + O_2 \rightarrow CO_2$. When 120 g of carbon reacts with oxygen, 440 g of carbon dioxide are formed. How much oxygen reacted with the carbon?

163. Salt water is distilled by boiling it and condensing the vapor. After distillation, 1,164 g of water have been collected and 12 g of salt are left behind in the original container. What was the original mass of the salt water?

164. Calcium carbonate, $CaCO_3$, decomposes according to the reaction: $CaCO_3 \rightarrow CaO + CO_2$. When 250 g of $CaCO_3$ decompose completely, the mass of CaO is 56% of the mass of the products of this reaction. What is the mass of CO_2 produced?

165. Water breaks down into hydrogen gas and oxygen gas according to the reaction: $2H_2O \rightarrow 2H_2 + O_2$. In this reaction the mass of oxygen produced is eight times greater than the mass of hydrogen produced. If 36 g of water form hydrogen and oxygen gas, what is the mass of hydrogen gas produced?

166. The size of particles in a solution is about 1 nm (1 nm = 0.000000001 m). Write 0.000000001 m in scientific notation.

167. A chemical reaction produces two new substances, one with a mass of 34 g and the other with a mass of 39 g. What was the total mass of the reactants?

168. The human body is about 65% oxygen. If a person has a mass of 75.0 kg, what is the mass of oxygen in their body?

169. A 112-g serving of ice cream contains 19 g of fat. What percentage of the serving is fat?

170. The mass of the products produced by a chemical reaction is measured. The reaction is repeated five times, with the same mass of reactants used each time. The measured product masses are 50.17 g, 50.12 g, 50.17 g, 50.10 g, and 50. 14 g. What is the average of these measurements?

Chapter 16 Solids, Liquids, and Gases

171. A book is sitting on a desk. The area of contact between the book and the desk is 0.06 m². If the book's weight is 30 N, what is the pressure the book exerts on the desk?

172. A skater has a weight of 500 N. The skate blades are in contact with the ice over an area of 0.001 m². What is the pressure exerted on the ice by the skater?

173. The weight of the water displaced by a person floating in the water is 686 N. What is the person's mass?

174. The pressure on a balloon that has a volume of 7 L is 100 kPa. If the temperature stays the same and the pressure on the balloon is increased to 250 kPa, what is the new volume of the balloon?

175. Two cylinders contain pistons that are connected by fluid in a hydraulic system. A force of 1,300 N is exerted on one piston with an area of 0.05 m². What is the force exerted on the other piston which has an area of 0.08 m²?

176. A gas-filled weather balloon floating in the atmosphere has an initial volume of 850 L. The weather balloon rises to a region the pressure is 56 kPa, and its volume expands to 1700 L. If the temperature remains the same, what was the initial pressure on the weather balloon?

177. The air in a tire pump has a volume of 1.50 L at a temperature of 5°C. If the temperature is increased to 30°C and the pressure remains constant, what is the new volume?

178. A block of wood with a mass of 1.2 kg is floating in a container of water. If the density of water is 1.0 g/cm³, what is the volume of water displaced by the floating wood?

179. In a hydraulic system, a force of 7,500 N is exerted on a piston with an area of 0.05 m². If the force exerted on a second piston in the hydraulic system is 1,500 N, what is the area of this second piston?

180. A gold bar weighs 17.0 N. If the density of gold is 19.3 g/cm³, what is the volume of the gold bar?

181. A book is sitting on a desk. If the surface area of the book's cover is 0.05 m², and atmospheric pressure is 100.0 kPa, what is the downward force of the atmosphere on the book?

182. A piston applies a pressure of 5,000 N/m². If the piston has a surface area of 0.1 m², how much force can the piston apply?

Chapter 17 Properties of Atoms and the Periodic Table

183. Boron has a mass number of 11 and an atomic number of 5. How many neutrons are in a boron atom?

184. A magnesium atom has 12 protons and 12 neutrons. What is its mass number?

185. Iodine-127 has a mass number of 127 and 74 neutrons. What percentage of the particles in an iodine-127 nucleus are protons?

186. How many neutrons are in an atom of phosphorus-31?

187. What is the ratio of neutrons to protons in the isotope radium-234?

188. About 80% of all magnesium atoms are magnesium-24, about 10% are magnesium-25, and about 10% are magnesium-26. What is the average atomic mass of magnesium?

189. The half-life of the radioactive isotope rubidium-87 is 48,800,000,000 years. Express this half-life in scientific notation.

190. The radioactive isotope nickel-63 has a half-life of 100 years. How much of a 10.0-g sample of nickel-63 is left after 300 years?

191. A sample of the radioactive isotope cobalt-62 is prepared. The sample has a mass of 1.00 g. After three minutes, the mass of cobalt-62 remaining is 0.25 g. What is the half-life of cobalt-62?

192. A neutral phosphorus atom has 15 electrons. How many electrons are in the third energy level?

Chapter 18 Radioactivity and Nuclear Reactions

193. How many protons are in the nucleus $^{81}_{36}Kr$?

194. How many neutrons are in the nucleus $^{56}_{26}Fe$?

195. What is the ratio of neutrons to protons in the nucleus $^{241}_{95}Am$?

196. How many alpha particles are emitted when the nucleus $^{222}_{86}Rn$ decays to $^{218}_{84}Po$?

197. How many beta particles are emitted when the nucleus $^{40}_{19}K$ decays to the nucleus $^{40}_{20}Ca$?

198. An alpha particle is the same as the helium nucleus $^{4}_{2}He$. What nucleus is produced when the nucleus $^{226}_{88}Ra$ decays by emitting an alpha particle?

199. How long will it take a sample of $^{194}_{84}Po$ to decay to 1/8 of its original amount if $^{194}_{84}Po$ has a half-life of 0.7 s?

200. The half-life of $^{131}_{53}I$ is 8.04 days. How much time would be needed to reduce 1 g of $^{131}_{53}I$ to 0.25 g?

201. A sample of radioactive carbon-14 sample has decayed to 12.5% of its original amount. If the half-life of carbon-14 is 5730 years, how old is this sample?

202. A sample of $^{38}_{17}Cl$ is observed to decay to 25% of the original amount in 74.4 minutes. What is the half-life of $^{38}_{17}Cl$?

Chapter 19 Elements and Their Properties

203. In seawater the concentration of fluoride ions, F^-, is 1.3×10^{-3} g/L. How many liters of seawater would contain 1.0 g of F^-?

204. There are three isotopes of hydrogen. The isotope deuterium, with one proton and one neutron in the nucleus, makes up 0.015% of all hydrogen atoms. Of every million hydrogen atoms, how many are deuterium?

205. A vitamin and mineral supplement pill contains 1.0×10^{-5} g of selenium. According to the label on the bottle, this amount is 18% of the recommended daily value. What is the recommended daily value of selenium in g?

206. The density of silver is 10.5 g/cm^3 and the density of copper is 8.9 g/cm^3. What is the difference in mass between a piece of silver with a volume of 5 cm^3 and a piece of copper with a volume of 5 cm^3?

207. A person has a mass of 68.3 kg. If 65% of the mass of a human body is oxygen, what is the mass of oxygen in this person's body?

208. The melting point of aluminum is 660.0°C. What is the melting point of aluminum on the Fahrenheit temperature scale?

209. A certain gold ore produces about 5 g of gold for every 1,000 kg of ore that is mined. If one ounce = 28.3 g, how many kg of ore must be mined to produce an ounce of gold?

210. A metal bolt with a mass of 26.6 g is placed in a 50-mL graduated cylinder containing water. The water level in the cylinder rises from 27.0 mL to 30.5 mL. What is the density of the bolt in g/cm^3?

211. On a circle graph showing the percentage of elements in the human body, the wedge representing nitrogen takes up 10.8°. What is the percentage of nitrogen in the human body?

212. The synthetic element hassium-261 has a half-life of 9.3 s. The synthetic element fermium-255 has a half-life of 20.1 h. How many times longer is the half-life of fermium-255 than the half-life of hassium-261?

Chapter 20 Chemical Bonds

213. What is the formula of the compound formed when ammonium ions, NH_4^+, and phosphate ions, PO_4^{3-}, combine?

214. Show that the sum of positive and negative charges in a unit of calcium chloride ($CaCl_2$) equals zero.

215. What is the formula for iron(III) oxide?

216. How many hydrogen atoms are in three molecules of ammonium phosphate, $(NH_4)_3PO_4$?

217. The overall charge on the polyatomic phosphate ion, PO_4^{3-}, is 3−. What is the oxidation number of phosphorus in the phosphate ion?

218. The overall charge on the polyatomic dichromate ion, $Cr_2O_7^{2-}$, is 2−. What is the oxidation number of chromium in this polyatomic ion?

219. What is the formula for lead(IV) oxide?

220. What is the formula for potassium chlorate?

221. What is the formula for carbon tetrachloride?

222. What percentage of the mass of a sulfuric acid molecule, H_2SO_4, is sulfur?

Chapter 21 Chemical Reactions

223. Lithium reacts with oxygen to form lithium oxide according to the equation: $4Li + O_2 \rightarrow 2Li_2O$. If 27.8 g of Li react completely with 32.0 g of O_2, how many grams of Li_2O are formed?

224. What coefficients balance the following equation: $\_Zn(OH)_2 + \_H_3PO_4 \rightarrow \_Zn_3(PO_4)_2 + \_H_2O$?

225. Aluminum hydroxide, $Al(OH)_3$, decomposes to form aluminum oxide, Al_2O_3, and water according to the reaction: $2Al(OH)_3 \rightarrow Al_2O_3 + 3H_2O$. If 156.0 g of $Al(OH)_3$ decompose to from 102.0 g of Al_2O_3, how many grams of H_2O are formed?

226. In the following balanced chemical reaction one of the products is represented by the symbol X: $BaCO_3 + C + H_2O \rightarrow Ba(OH)_2 + H_2O + 2X$. What is the formula for the compound represented by X?

227. When propane, C_3H_8, is burned, carbon dioxide and water vapor are produced according to the following reaction: $C_3H_8 + 5O_2 \rightarrow 3CO_2 + 4H_2O$. How much propane is burned if 160.0 g of O_2 are used and 132.0 g of CO_2 and 72.0 g of H_2O are produced?

228. Increasing the temperature usually causes the rate of a chemical reaction to increase. If the rate of a chemical reaction doubles when the temperature increases by 10°C, by what factor does the rate of reaction increase if the temperature increases by 30°C?

229. When acetylene gas, C_2H_2, is burned, carbon dioxide and water are produced. Find the coefficients that balance the chemical equation for the combustion of acetylene: $\_C_2H_2 + \_O_2 \rightarrow \_CO_2 + \_H_2O$.

230. What coefficients balances the following equation: $\_CS_2 + \_O_2 \rightarrow \_CO_2 + \_SO_2$?

231. When methane, CH_4, is burned, 50.1 kJ of energy per gram are released. When propane, C_3H_8, is burned, 45.8 kJ of energy are released. If a mixture of 1 g of methane and 1 g of propane is burned, how much energy is released per gram of mixture?

232. A chemical reaction produces 0.050 g of a product in 0.18 s. In the presence of a catalyst, the reaction produces 0.050 g of the same product in 0.007 s. How much faster is the rate of reaction in the presence of the enzyme?

Extra Math Problems

Chapter 22 Solutions

233. A cup of orange juice contains 126 mg of vitamin C and 1/2 cup of strawberries contain 42 mg of vitamin C. How many cups of strawberries contain as much vitamin C as one cup of orange juice?

234. A Sacagawea dollar coin is made of manganese brass alloy that is 1/25 nickel. Express this number as a percentage.

235. What is the total surface area of a 2-cm cube?

236. A cube has 2-cm sides. If it is split in half, what is the total surface area of the two pieces?

237. What is the increase in surface area when a cube with 2-cm sides is divided into eight equal parts?

238. How much surface area is lost if two 4-cm cubes are attached at one face?

239. At 20°C, the solubility in water of potassium bromide, KBr, is 65.3 g/100 mL. What is the maximum amount of potassium bromide that will dissolve in 237 mL of water?

240. At 20°C, the solubility of sodium chloride, NaCl, in water is 35.9 g/100 mL. If the maximum amount of sodium chloride is dissolved in 500 mL of water at 20°C, the mass of the dissolved sodium chloride is what percentage of the mass of the solution?

241. At 60°C, the solubility of sucrose (sugar) in water is 287.3 g/100 mL. At this temperature, what is the minimum amount of water needed to dissolve 50.0 g of sucrose?

242. A fruit drink contains 90% water and 10% fruit juice. How much fruit juice does 500 mL of fruit drink contain?

Chapter 23 Acids, Bases, and Salts

243. The difference between the pH of an acidic solution and the pH of pure water is 3. What is the pH of the solution?

244. The pH of rain that fell over a region had measured values of 4.6, 5.1, 4.8, 4.5, 4.5, 4.9, 4.7, and 4.8. What was the mean value of the measured pH?

245. If 5.5% of 473.0 mL of vinegar is acetic acid, how many milliliters of acetic acid are there?

246. The difference between the pH of a basic solution and the pH of pure water is 2. What is the pH of the solution?

247. On the pH scale, a decrease of one unit means that the concentration of H^+ ions increases 10 times. If the pH of a solution changes from 6.5 to 4.5, how has the concentration of H^+ ions changed?

248. Write the balanced chemical equation for the neutralization of H_2SO_4, sulfuric acid, by KOH, potassium hydroxide.

249. Write the balanced chemical equation for the neutralization of HBr, hydrobromic acid, by $Al(OH)_3$, aluminum hydroxide.

250. A molecule of acetylsalicylic acid, or aspirin, has the chemical formula $COOHC_6H_4COOCH_3$. What is the mass of a molecule of acetylsalicylic acid in amu?

251. Write the equation for the reaction when HNO_3, nitric acid, ionizes in water.

252. When Na_2O, sodium oxide, reacts with water, the base NaOH, sodium hydroxide, is formed. Write the balanced equation for this reaction.

Chapter 24 Organic Compounds

253. The hydrocarbon octane, C_8H_{18}, has a boiling point of 259°F. What is its boiling point on the Celsius temperature scale?

254. A barrel of oil is 42.0 gallons. About 45% of a barrel of oil is turned into gasoline during the fractional distillation process. In 2001, about 19.6 million barrels of crude oil were refined each day. How many gallons of gasoline were produced each day?

255. In 2001, about 56% of the crude oil used by the United States was imported. If the United States used 20.6 million barrels of crude oil a day, how many million barrels of crude oil were imported in 2001?

256. Four molecules of a hydrocarbon contain carbon atoms and 56 hydrogen atoms. What is the formula for a molecule of this hydrocarbon?

257. For saturated hydrocarbons, the number of hydrogen atoms in a molecule can be calculated by the formula $N_H = 2N_C + 2$, where N_H is the number of hydrogen atoms and N_C is the number of carbon atoms in the molecule. If a molecule of the saturated hydrocarbon decane has 22 hydrogen atoms, how many carbon atoms does a decane molecule contain?

258. Fats supply 9 Calories per gram, carbohydrates and proteins each supply 4 Calories per gram. If 100 g of potato chips contain 7 g of protein, 53 g of carbohydrates, and 35 g of fats, how many Calories are in 100 g of potato chips?

259. The basal metabolism rate (BMR) is the amount of energy required to maintain basic body functions. The BMR is approximately 1.0 Calories/hr per kilogram of body mass. For a person with a mass of 65 kg, how many Calories are needed each day to maintain basic body functions?

260. A food Calorie is an energy unit equal to 4,184 joules. If a person uses 2,070 Calories in one day, what is the power being used? Express your answer in watts.

261. In each 100 g of cheddar cheese there are 33 g of fat. Calculate how many grams of fat are in 250 g of cheddar cheese.

262. A car gets 25 miles per gallon of gas. If the car is driven 12,000 miles in one year and gasoline costs $1.55 per gallon, what was the cost of the gasoline used in one year?

Chapter 25 New Materials Through Chemistry

263. A stainless steel spoon contains 30.0 g of iron, 6.8 g of chromium, and 3.2 g of nickel. What percentage of the stainless steel is chromium?

264. A 14-karat gold earring has a mass of 10 g. What is the mass of gold in the earring?

265. In 1997, about 6,400,000,000 kg of polyvinyl chloride were used in the United States. About 6% of the PVC used was for packaging. Express in scientific notation how many kilograms of PVC were used for packaging in 1997.

266. The molecules in a sample of polypropylene have an average length of 60,000 monomers. The monomer of polypropylene has the formula CH_2CHCH_3. Express in scientific notation the mass, in amu, of a polypropylene molecule made of 60,000 monomers.

267. A certain process for manufacturing integrated circuits packs 47,600,000 transistors into an area of 340 mm^2. If this process is used to produce an integrated circuit with an area of 1 cm^2, express in scientific notation the number of transistors in this integrated circuit.

268. The melting points of five different samples of a new aluminum alloy have measured values of 631.5°C, 632.3°C, 636.1°C, 637.4°C, and 630.2°C. What is the mean of these measurements?

269. Rounded to the nearest degree, eight measured values of the melting point of a stainless steel alloy are 1,421°C, 1,420°C, 1,421°C, 1,423°C, 1,423°C, 1,421°C, 1,424°C, and 1,419°C. What is the mode of these measurements?

270. The measured values of the copper content of seven bronze buttons found at an archaeological site are 83%, 90%, 91%, 72%, 79%, 87%, and 89%. What is the median of these measurements?

271. The number of transistors and other components per mm^2 on an integrated circuit has doubled, on average, every two years. If integrated circuits contained 100,000 transistors in 1982, estimate how many transistors an integrated circuit of the same size contained in 1998.

272. A car contains 200 kg of plastic parts instead of steel parts. The density of steel is twice the density of plastic. If the volume of the plastic parts equals the volume of the same parts made of steel, how much less is the mass (kg) of the car by using plastic parts instead of steel?

Physical Science Reference Tables

Standard Units

Symbol	Name	Quantity
m	meter	length
kg	kilogram	mass
Pa	pascal	pressure
K	kelvin	temperature
mol	mole	amount of a substance
J	joule	energy, work, quantity of heat
s	second	time
C	coulomb	electric charge
V	volt	electric potential
A	ampere	electric current
Ω	ohm	resistance

Physical Constants and Conversion Factors

Acceleration due to gravity	g	9.8 m/s/s or m/s^2
Avogadro's Number	N$_A$	6.02×10^{23} particles per mole
Electron charge	e	1.6×10^{-19} C
Electron rest mass	m$_e$	9.11×10^{-31} kg
Gravitation constant	G	6.67×10^{-11} N $\times$ m^2/kg^2
Mass-energy relationship		1 u (amu) = 9.3×10^2 MeV
Speed of light in a vacuum	c	3.00×10^8 m/s
Speed of sound at STP		331 m/s
Standard Pressure		1 atmosphere
		101.3 kPa
		760 Torr or mmHg
		14.7 lb/in.^{2}

Wavelengths of Light in a Vacuum

Violet	$4.0 - 4.2 \times 10^{-7}$ m
Blue	$4.2 - 4.9 \times 10^{-7}$ m
Green	$4.9 - 5.7 \times 10^{-7}$ m
Yellow	$5.7 - 5.9 \times 10^{-7}$ m
Orange	$5.9 - 6.5 \times 10^{-7}$ m
Red	$6.5 - 7.0 \times 10^{-7}$ m

The Index of Refraction for Common Substances
($\lambda = 5.9 \times 10^{-7}$ m)

Air	1.00
Alcohol	1.36
Canada Balsam	1.53
Corn Oil	1.47
Diamond	2.42
Glass, Crown	1.52
Glass, Flint	1.61
Glycerol	1.47
Lucite	1.50
Quartz, Fused	1.46
Water	1.33

Heat Constants

	Specific Heat (average) (kJ/kg $\times$ °C) (J/g $\times$ °C)	Melting Point (°C)	Boiling Point (°C)	Heat of Fusion (kJ/kg) (J/g)	Heat of Vaporization (kJ/kg) (J/g)
Alcohol (ethyl)	2.43 (liq.)	−117	79	109	855
Aluminum	0.90 (sol.)	660	2467	396	10500
Ammonia	4.71 (liq.)	−78	−33	332	1370
Copper	0.39 (sol.)	1083	2567	205	4790
Iron	0.45 (sol.)	1535	2750	267	6290
Lead	0.13 (sol.)	328	1740	25	866
Mercury	0.14 (liq.)	−39	357	11	295
Platinum	0.13 (sol.)	1772	3827	101	229
Silver	0.24 (sol.)	962	2212	105	2370
Tungsten	0.13 (sol.)	3410	5660	192	4350
Water (solid)	2.05 (sol.)	0	–	334	–
Water (liquid)	4.18 (liq.)	–	100	–	–
Water (vapor)	2.01 (gas)	–	–	–	2260
Zinc	0.39 (sol.)	420	907	113	1770

Standard Units

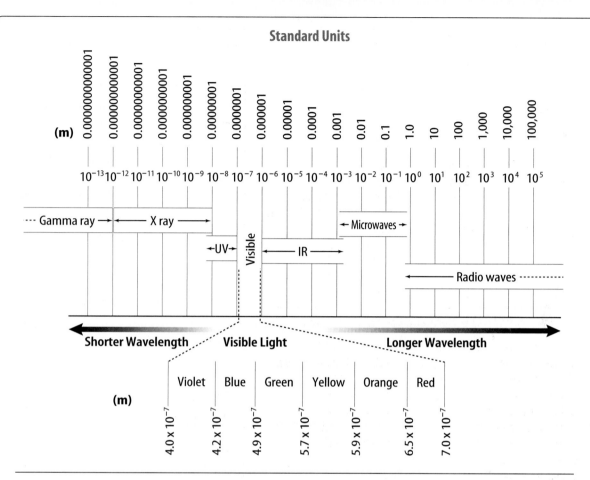

Uranium Disintegration Series

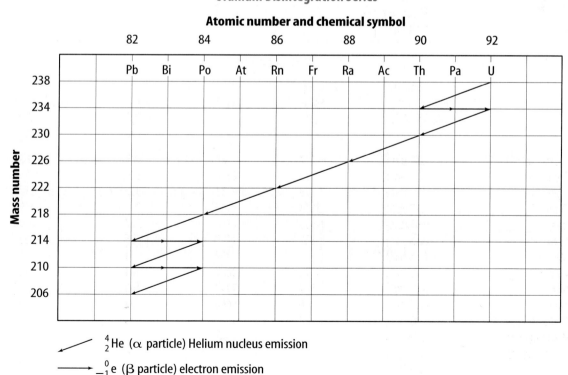

PERIODIC TABLE OF THE ELEMENTS

Columns of elements are called groups. Elements in the same group have similar chemical properties.

Gas

Liquid

Solid

Synthetic

Element —— Hydrogen
Atomic number —— 1
Symbol —— H
Atomic mass —— 1.008
State of matter

The first three symbols tell you the state of matter of the element at room temperature. The fourth symbol identifies elements that are not present in significant amounts on Earth. Useful amounts are made synthetically.

	1	2	3	4	5	6	7	8	9
1	Hydrogen 1 **H** 1.008								
2	Lithium 3 **Li** 6.941	Beryllium 4 **Be** 9.012							
3	Sodium 11 **Na** 22.990	Magnesium 12 **Mg** 24.305							
4	Potassium 19 **K** 39.098	Calcium 20 **Ca** 40.078	Scandium 21 **Sc** 44.956	Titanium 22 **Ti** 47.867	Vanadium 23 **V** 50.942	Chromium 24 **Cr** 51.996	Manganese 25 **Mn** 54.938	Iron 26 **Fe** 55.845	Cobalt 27 **Co** 58.933
5	Rubidium 37 **Rb** 85.468	Strontium 38 **Sr** 87.62	Yttrium 39 **Y** 88.906	Zirconium 40 **Zr** 91.224	Niobium 41 **Nb** 92.906	Molybdenum 42 **Mo** 95.94	Technetium 43 **Tc** (98)	Ruthenium 44 **Ru** 101.07	Rhodium 45 **Rh** 102.906
6	Cesium 55 **Cs** 132.905	Barium 56 **Ba** 137.327	Lanthanum 57 **La** 138.906	Hafnium 72 **Hf** 178.49	Tantalum 73 **Ta** 180.948	Tungsten 74 **W** 183.84	Rhenium 75 **Re** 186.207	Osmium 76 **Os** 190.23	Iridium 77 **Ir** 192.217
7	Francium 87 **Fr** (223)	Radium 88 **Ra** (226)	Actinium 89 **Ac** (227)	Rutherfordium 104 **Rf** (261)	Dubnium 105 **Db** (262)	Seaborgium 106 **Sg** (266)	Bohrium 107 **Bh** (264)	Hassium 108 **Hs** (277)	Meitnerium 109 **Mt** (268)

The number in parentheses is the mass number of the longest-lived isotope for that element.

Rows of elements are called periods. Atomic number increases across a period.

The arrow shows where these elements would fit into the periodic table. They are moved to the bottom of the table to save space.

Lanthanide series	Cerium 58 **Ce** 140.116	Praseodymium 59 **Pr** 140.908	Neodymium 60 **Nd** 144.24	Promethium 61 **Pm** (145)	Samarium 62 **Sm** 150.36
Actinide series	Thorium 90 **Th** 232.038	Protactinium 91 **Pa** 231.036	Uranium 92 **U** 238.029	Neptunium 93 **Np** (237)	Plutonium 94 **Pu** (244)

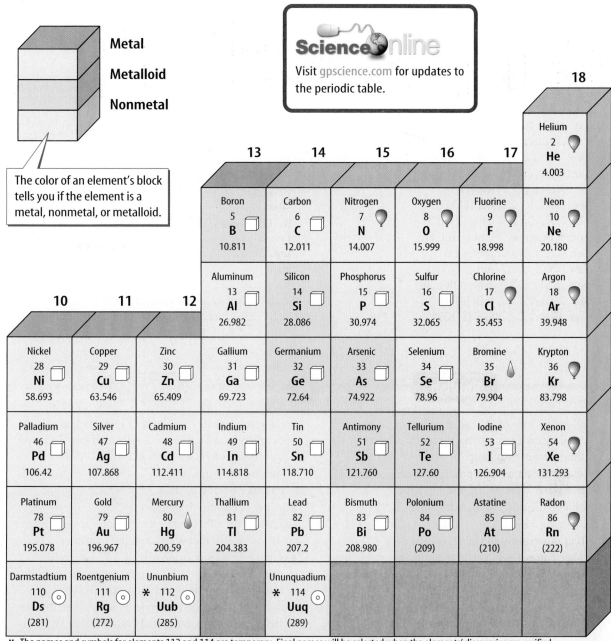

Metal

Metalloid

Nonmetal

The color of an element's block tells you if the element is a metal, nonmetal, or metalloid.

Science Online

Visit gpscience.com for updates to the periodic table.

* The names and symbols for elements 112 and 114 are temporary. Final names will be selected when the elements' discoveries are verified.

18
Helium
2
He
4.003

13	14	15	16	17
Boron 5 B 10.811	Carbon 6 C 12.011	Nitrogen 7 N 14.007	Oxygen 8 O 15.999	Fluorine 9 F 18.998	Neon 10 Ne 20.180
Aluminum 13 Al 26.982	Silicon 14 Si 28.086	Phosphorus 15 P 30.974	Sulfur 16 S 32.065	Chlorine 17 Cl 35.453	Argon 18 Ar 39.948

10	11	12	13	14	15	16	17
Nickel 28 Ni 58.693	Copper 29 Cu 63.546	Zinc 30 Zn 65.409	Gallium 31 Ga 69.723	Germanium 32 Ge 72.64	Arsenic 33 As 74.922	Selenium 34 Se 78.96	Bromine 35 Br 79.904	Krypton 36 Kr 83.798
Palladium 46 Pd 106.42	Silver 47 Ag 107.868	Cadmium 48 Cd 112.411	Indium 49 In 114.818	Tin 50 Sn 118.710	Antimony 51 Sb 121.760	Tellurium 52 Te 127.60	Iodine 53 I 126.904	Xenon 54 Xe 131.293
Platinum 78 Pt 195.078	Gold 79 Au 196.967	Mercury 80 Hg 200.59	Thallium 81 Tl 204.383	Lead 82 Pb 207.2	Bismuth 83 Bi 208.980	Polonium 84 Po (209)	Astatine 85 At (210)	Radon 86 Rn (222)
Darmstadtium 110 Ds (281)	Roentgenium 111 Rg (272)	Ununbium * 112 Uub (285)		Ununquadium * 114 Uuq (289)			

Europium 63 Eu 151.964	Gadolinium 64 Gd 157.25	Terbium 65 Tb 158.925	Dysprosium 66 Dy 162.500	Holmium 67 Ho 164.930	Erbium 68 Er 167.259	Thulium 69 Tm 168.934	Ytterbium 70 Yb 173.04	Lutetium 71 Lu 174.967
Americium 95 Am (243)	Curium 96 Cm (247)	Berkelium 97 Bk (247)	Californium 98 Cf (251)	Einsteinium 99 Es (252)	Fermium 100 Fm (257)	Mendelevium 101 Md (258)	Nobelium 102 No (259)	Lawrencium 103 Lr (262)

Glossary/Glosario

Cómo usar el glosario en español:
1. Busca el término en inglés que desees encontrar.
2. El término en español, junto con la definición, se encuentran en la columna de la derecha.

Pronunciation Key

Use the following key to help you sound out words in the glossary.

a..............back (BAK)		**ew**............foo**d** (FEWD)	
ay.............**d**ay (DAY)		**yoo**............**p**ure (PYOOR)	
ah.............**f**ather (FAH thur)		**yew**............**f**ew (FYEW)	
ow............**fl**ower (FLOW ur)		**uh**.............com**m**a (CAH muh)	
ar.............**c**ar (CAR)		**u (+ con)**......**r**ub (RUB)	
e..............**l**ess (LES)		**sh**.............**sh**elf (SHELF)	
ee.............**l**eaf (LEEF)		**ch**.............na**t**ure (NAY chur)	
ih.............**tr**ip (TRIHP)		**g**..............**g**ift (GIHFT)	
i (i + con + e)..i**d**ea (i DEE uh)		**j**...............**g**em (JEM)	
oh............**g**o (GOH)		**ing**............**s**ing (SING)	
aw............**s**oft (SAWFT)		**zh**.............vi**s**ion (VIH zhun)	
or.............**orb**it (OR buht)		**k**..............**c**ake (KAYK)	
oy.............**c**oin (COYN)		**s**..............**s**eed, **c**ent (SEED, SENT)	
oo............**f**oot (FOOT)		**z**..............**z**one, rai**s**e (ZOHN, RAYZ)	

English **A** **Español**

acceleration: rate of change of velocity; can be calculated by dividing the change in the velocity by the time it takes the change to occur. (p. 47)

acid: any substance that produces hydrogen ions, H^+, in a water solution. (p. 696)

acoustics: the study of sound. (p. 339)

air resistance: force that opposes the motion of objects that move through the air. (p. 73)

alcohol: compound, such as ethanol, that is formed when $-OH$ groups replace one or more hydrogen atoms in a hydrocarbon. (p. 733)

allotropes: different forms of the same element having different molecular structures. (p. 585)

alloy: a mixture of elements that has metallic properties. (p. 758)

alpha particle: particle consisting of two protons and two neutrons that is emitted from a decaying atomic nucleus. (p. 541)

alternating current (AC): electric current that reverses its direction of flow in a regular pattern. (p. 242)

amplitude: a measure of the energy carried by a wave. (p. 300)

aromatic compound: an organic compound that contains the benzene ring structure and may have a pleasant or unpleasant odor and flavor. (p. 731)

atom: the smallest particle of an element that still retains the properties of the element. (p. 507)

aceleración: tasa de cambio de la velocidad; se calcula dividiendo el cambio en la velocidad por el tiempo que toma para que ocurra el cambio. (p. 47)

ácido: sustancia que produce iones de hidrógeno, H^+, en una solución de agua. (p. 696)

acústica: el estudio del sonido. (p. 339)

resistencia del aire: fuerza que se opone al movimiento de los objetos que se mueven por el aire. (p. 73)

alcohol: compuesto, como el etanol, que se forma cuando grupos $-OH$ reemplazan a uno o más átomos de hidrógeno en un hidrocarburo. (p. 733)

alótropos: formas diferentes del mismo elemento que tienen diferentes estructuras moleculares. (p. 585)

aleación: una mezcla de elementos que tiene propiedades metálicas. (p. 758)

partícula alfa: partícula compuesta por dos protones y dos neutrones y que es emitida por un núcleo atómico en descomposición. (p. 541)

corriente alterna (CA): corriente eléctrica que invierte su dirección de flujo en un patrón regular. (p. 242)

amplitud: medida de la energía transportada por una onda. (p. 300)

compuesto aromático: compuesto orgánico que contiene la estructura del anillo bencénico y que puede tener un olor y un sabor agradables o desagradables. (p. 731)

átomo: la partícula más pequeña de un elemento que mantiene las propiedades del elemento. (p. 507)

Glossary/Glosario

atomic number: number of protons in an atom's nucleus. (p. 513)

average atomic mass: weighted-average mass of the mixture of an element's isotopes. (p. 515)

average speed: total distance an object travels divided by the total time it takes to travel that distance. (p. 42)

número atómico: número de protones en el núcleo de un átomo. (p. 513)

masa atómica promedio: masa de peso promedio resultado de la mezcla de los isótopos de un elemento. (p. 515)

velocidad promedio: distancia que recorre un objeto dividida por el tiempo que dura en recorrer dicha distancia. (p. 42)

B

balanced chemical equation: chemical equation with the same number of atoms of each element on both sides of the equation. (p. 638)

balanced forces: forces on a object that combine to give a zero net force and do not change the motion of the object. (p. 53)

base: any substance that forms hydroxide ions, OH$^-$, in a water solution. (p. 698)

beta particle: electron that is emitted from a decaying atomic nucleus. (p. 543)

bias: occurs when a scientist's expectations change how the results of an experiment are viewed. (p. 10)

binary compound: compound that is composed of two elements. (p. 615)

biomass: renewable organic matter from plants and animals, such as wood and animal manure, that can be burned to provide heat. (p. 276)

boiling point: the temperature at which the pressure of the vapor in the liquid is equal to the external pressure acting on the surface of the liquid. (p. 479)

bubble chamber: radiation detector, consisting of a container of superheated liquid under high pressure, that is used to detect the paths of charged particles. (p. 547)

buffer: solution containing ions that react with added acids or bases and minimize their effects on pH. (p. 705)

buoyancy: ability of a fluid—a liquid or a gas—to exert an upward force on an object immersed in the fluid. (p. 485)

ecuación química Fnceada: ecuación química con el mismo número de átomos de cada elemento en los dos lados de la ecuación. (p. 638)

fuerzas equilibradas: fuerzas en un objeto que se combinan para dar una fuerza neta de cero y no cambiar el movimiento del objeto. (p. 53)

base: sustancia que forma iones de hidróxido, OH$^-$, en una solución de agua. (p. 698)

partícula beta: electrón emitido por un núcleo atómico en descomposición. (p. 543)

predisposición: ocurre cuando las expectativas de un científico cambian la forma en que son vistos los resultados de un experimento. (p. 10)

compuesto binario: compuesto conformado por dos elementos. (p. 615)

biomasa: materia orgánica renovable proveniente de plantas y animales, tales como madera y estiércol animal, que puede ser incinerada para producir calor. (p. 276)

punto de ebullición: temperatura a la cual la presión del vapor de un líquido es igual a la presión externa que actúa sobre la superficie del líquido. (p. 479)

cámara de burbujas: detector de radiación que consiste de un contenedor de un líquido sobrecalentado a alta presión, usado para detectar la trayectoria de las partículas cargadas. (p. 547)

buffer: solución que contiene iones que reaccionan con los ácidos o bases agregados y que minimiza los efectos de éstos en el pH. (p. 705)

fuerza flotante: capacidad de un fluido, líquido o gas, para ejercer una fuerza ascendente sobre un objeto inmerso en un fluido. (p. 485)

C

carbohydrates: group of biological compounds, such as sugars and starches, with twice as many hydrogen atoms as oxygen atoms. (p. 745)

carbohidratos: grupo de compuestos biológicos tales como azúcares y almidones que contienen el doble de átomos de hidrógeno que de oxígeno. (p. 745)

carrier wave: specific frequency that a radio station is assigned and uses to broadcast signals. (p. 367)

catalyst: substance that speeds up a chemical reaction without being permanently changed itself. (p. 650)

cathode-ray tube: sealed vacuum tube that produces one or more beams of electrons that produce an image when they strike the coating on the inside of a TV screen. (p. 370)

centripetal acceleration: acceleration of an object toward the center of a curved or circular path. (p. 81)

centripetal force: a net force that is directed toward the center of a curved or circular path. (p. 81)

ceramics: versatile materials made from dried clay or clay-like mixtures with customizable properties; produced by a process in which an object is molded and then heated to high temperatures, increasing its density. (p. 764)

chain reaction: ongoing series of fission reactions. (p. 552)

charging by contact: process of transferring charge between objects by touching or rubbing. (p. 195)

charging by induction: process of rearranging electrons on a neutral object by bringing a charged object close to it. (p. 196)

chemical bond: force that holds atoms together in a compound. (p. 606)

chemical change: change of one substance into a new substance. (p. 462)

chemical equation: shorthand method to describe chemical reactions using chemical formulas and other symbols. (p. 635)

chemical formula: chemical shorthand that uses symbols to tell what elements are in a compound and their ratios. (p. 603)

chemical potential energy: energy stored in chemical bonds. (p. 103)

chemical property: any characteristic of a substance, such as flammability, that indicates whether it can undergo a certain chemical change. (p. 461)

chemical reaction: process in which one or more substances are changed into new substances. (p. 632)

circuit: closed conducting loop through which an electric current can flow. (p. 201)

cloud chamber: radiation detector that uses water or ethanol vapor to detect the paths of charged particles. (p. 546)

cochlea: spiral-shaped, fluid-filled structure in the inner ear that converts sounds waves to nerve impulses. (p. 326)

onda transportadora: frecuencia específica que se le asigna a una estación de radio y que la usa para emitir señales. (p. 367)

catalizador: sustancia que acelera una reacción química sin cambiar el mismo permanentemente. (p. 650)

tubo de rayos catódicos: tubo vacío sellado que produce uno o más haces de electrones para producir una imagen al chocar con el revestimiento del interior de una pantalla de televisor. (p. 370)

aceleración centrípeta: aceleración de un objeto dirigida hacia el centro de un trayecto curvo o circular. (p. 81)

fuerza centrípeta: fuerza neta dirigida hacia el centro de un trayecto curvo o circular. (p. 81)

cerámicas: materiales versátiles hechos con arcilla seca o mezclas parecidas a la arcilla con propiedades adaptables, producidos mediante un proceso en el cual un objeto es moldeado y luego sujeto a altas temperaturas, aumentando su densidad. (p. 764)

reacción en cadena: serie continua de reacciones de fisión. (p. 552)

carga por contacto: proceso de transferir carga entre objetos por contacto o frotaciòn. (p. 195)

carga por inducción: proceso de redistribución de los electrones en un objeto neutro acercándoles un objeto con carga. (p. 196)

enlace químico: fuerza que mantiene a los átomos juntos dentro de un compuesto. (p. 606)

cambio químico: transformación de una sustancia en una nueva sustancia. (p. 462)

ecuación química: método simplificado para describir reacciones químicas usando fórmulas químicas y otros símbolos. (p. 635)

fórmula química: nomenclatura química que usa símbolos para expresar qué elementos están en un compuesto y en qué proporción. (p. 603)

energía química potencial: energía almacenada en los enlaces químicos. (p. 103)

propiedad química: cualquier característica de una sustancia, como por ejemplo la combustibilidad, que indique si puede someterse a determinado cambio químico. (p. 461)

reacción química: proceso en el cual una o más sustancias son cambiadas por nuevas sustancias. (p. 632)

circuito: circuito conductor cerrado a través del cual puede fluir una corriente eléctrica. (p. 201)

cámara de vapor: detector de radiaciones que usa vapor de agua o de etanol para detectar la trayectoria de las partículas cargadas. (p. 546)

cóclea: estructura en el oído interno, con forma de espiral y llena de un fluido, la cual convierte las ondas sonoras en impulsos nerviosos. (p. 326)

coefficient: number in a chemical equation that represents the number of units of each substance taking part in a chemical reaction. (p. 636)

coherent light: light of a single wavelength that travels in a single direction with its crests and troughs aligned. (p. 398)

colloid (KAHL oyd): heterogeneous mixture whose particles never settle. (p. 454)

combustion reaction: a type of chemical reaction that occurs when a substance reacts with oxygen to produce energy in the form of heat and light. (p. 641)

composite: mixture of two materials, one of which is embedded in the other. (p. 775)

compound: substance formed from two or more elements in which the exact combination and proportion of elements is always the same. (p. 452)

compound machine: machine that is a combination of two or more simple machines. (p. 146)

compressional wave: a wave for which the matter in the medium moves back and forth along the direction that the wave travels. (p. 292)

concave lens: a lens that is thicker at the edges than in the middle; causes light rays to diverge and forms reduced, upright, virtual images; and is usually used in combination with other lenses. (p. 426)

concave mirror: a reflective surface that curves inward and can magnify objects or create beams of light. (p. 418)

conduction: transfer of thermal energy by collisions between particles in matter at a higher temperature and particles in matter at a lower temperature. (p. 164)

conductivity (kahn duk TIHV ut ee): property of metals and alloys that allows heat or electrical charges to pass through the material easily. (p. 759)

conductor: material, such as copper wire, in which electrons can move easily. (p. 194)

constant: in an experiment, a variable that does not change when other variables change. (p. 9)

control: standard used for comparison of test results in an experiment. (p. 9)

convection: transfer of thermal energy in a fluid by the movement of warmer and cooler fluid from one place to another. (p. 165)

convex lens: a lens that is thicker in the middle than at the edges and can form real or virtual images. (p. 424)

coeficiente: número en una ecuación química que representa el número de unidades de cada una de las sustancias que participan en una reacción química. (p. 636)

luz coherente: luz de una sola longitud de onda que viaja en una sola dirección con sus crestas y sus depresiones alineadas. (p. 398)

coloide: mezcla heterogénea cuyas partículas nunca se sedimentan. (p. 454)

reacción de combustión: un tipo de reacción química que ocurre cuando una sustancia reacciona con oxígeno para producir energía en forma de calor y luz. (p. 641)

compuesto: mezcla de dos materiales, uno de los cuales está embebido en el otro. (p. 775)

compuesto: sustancia formada por dos o más elementos en la que la combinación y proporción exacta de los elementos es siempre la misma. (p. 452)

máquina compuesta: máquina compuesta por dos o más máquinas simples. (p. 146)

onda de compresión: onda para la cual la materia en el medio se mueve hacia adelante y hacia atrás en la dirección en que viaja la onda. (p. 292)

lente cóncavo: lente que es más delgado en los bordes que en el centro; hace que los rayos de luz se desvíen y forma imágenes reducidas, verticales y virtuales, y generalmente se utiliza en combinación con otros lentes. (p. 426)

espejo cóncavo: superficie reflexiva que se curva hacia el interior y que puede amplificar los objetos o crear rayos de luz. (p. 418)

conducción: transferencia de energía térmica por colisiones entre partículas de materia a una temperatura alta y partículas de materia a una temperatura más baja. (p. 164)

conductividad: propiedad de los metales y aleaciones que permite fácilmente el paso de calor o cargas eléctricas a través del material. (p. 759)

conductor: material, como el alambre de cobre, a través del cual los electrones se pueden mover con facilidad. (p. 194)

constante: en un experimento, una variable que no cambia cuando cambian otras variables. (p. 9)

control: estándar usado para la comparación de resultados de pruebas en un experimento. (p. 9)

convección: transferencia de energía térmica en un fluido por el movimiento de fluidos con mayores y menores temperaturas de un lugar a otro. (p. 165)

lente convexo: lente que es más delgado en el centro que en los bordes y que puede formar imágenes reales o virtuales. (p. 424)

convex mirror: a reflective surface that curves outward and forms a reduced, upright, virtual image. (p. 421)

cornea: transparent covering on the eyeball through which light enters the eye. (p. 427)

covalent bond: attraction formed between atoms when they share electrons. (p. 611)

crest: the highest points on a transverse wave. (p. 296)

critical mass: amount of fissionable material required so that each fission reaction produces approximately one more fission reaction. (p. 552)

espejo convexo: una superficie reflexiva que se curva hacia el exterior y forma una imagen reducida, vertical y virtual. (p. 421)

córnea: cubierta transparente del globo ocular a través de la cual entra la luz al ojo. (p. 427)

enlace covalente: atracción formada entre átomos que comparten electrones. (p. 611)

cresta: los puntos más altos en una onda transversal. (p. 296)

masa crítica: cantidad de material fisionable requerido de manera que cada reacción de fisión produzca aproximadamente una reacción de fisión adicional. (p. 552)

D

decibel: unit for sound intensity; abbreviated dB. (p. 329)

decomposition reaction: chemical reaction in which one substance breaks down into two or more substances. (p. 642)

density: mass per unit volume of a material. (p. 19)

deoxyribonucleic (dee AHK sih ri boh noo klay ihk) acid: a type of essential biological compound found in the nuclei of cells that codes and stores genetic information and controls the production of RNA. (p. 744)

dependent variable: factor that changes as a result of changes in the other variables. (p. 9)

depolymerization: process using heat or chemicals to break a polymer chain into its monomers. (p. 741)

diatomic molecule: a molecule that consists of two atoms of the same element. (p. 579)

diffraction: the bending of waves around an obstacle; can also occur when waves pass through a narrow opening. (p. 306)

diffusion: spreading of particles throughout a given volume until they are uniformly distributed. (p. 479)

direct current (DC): electric current that flows in only one direction. (p. 242)

displacement: distance and direction of an object's change in position from the starting point. (p. 39)

dissociation: process in which an ionic compound separates into its positive and negative ions. (p. 677)

distance: how far an object moves. (p. 39)

distillation: process than can separate two substances in a mixture by evaporating a liquid and recondensing its vapor. (p. 461)

decibel: unidad que mide la intensidad del sonido; se abrevia dB. (p. 329)

reacción de descomposición: reacción química en la cual una sustancia se descompone en dos o más sustancias. (p. 642)

densidad: masa por unidad de volumen de un material. (p. 19)

ácido desoxirribonucleico: compuesto biológico esencial encontrado en el núcleo de células que codifican y almacenan información genética y que controla la producción de ARN. (p. 744)

variable dependiente: factor que varía como resultado de los cambios en las otras variables. (p. 9)

despolimerización: proceso en el que se utilizan calor o químicos para descomponer una cadena de polímeros en sus monómeros. (p. 741)

molécula diatómica: molécula formada por dos átomos del mismo elemento. (p. 579)

difracción: curvatura de las ondas alrededor de un obstáculo, la cual también puede ocurrir cuando éstas pasan a través de una abertura angosta. (p. 306)

difusión: propagación de partículas en la totalidad de un volumen determinado hasta que se distribuyen de manera uniforme. (p. 479)

corriente directa (CD): corriente eléctrica que fluye en una sola dirección. (p. 242)

desplazamiento: distancia y dirección del cambio de posición de un objeto desde el punto inicial. (p. 39)

disociación: proceso en el cual un compuesto iónico se separa en sus iones positivos y negativos. (p. 677)

distancia: qué tan lejos se mueve un objeto. (p. 39)

destilación: proceso que puede separar dos sustancias de una mezcla por medio de la evaporación de un líquido y la recondensación de su vapor. (p. 461)

Glossary/Glosario

doping: process of adding impurities to a semiconductor to increase its conductivity. (p. 768)

Doppler effect: change in pitch or frequency that occurs when a source of a sound is moving relative to a listener. (p. 331)

double-displacement reaction: chemical reaction that produces a precipitate, water, or a gas when two ionic compounds in solution are combined. (p. 643)

ductile: ability of metals to be drawn into wires. (p. 570)

ductility (duk TIHL uh tee): ability of metals or alloys to be pulled into wires. (p. 759)

dopaje: proceso que consiste en añadir impurezas a un semiconductor para aumentar su conductividad. (p. 768)

efecto Doppler: cambio en la altura o frecuencia que ocurre cuando una fuente de sonido se mueve en relación con un oyente. (p. 331)

reacción de doble desplazamiento: reacción química que produce un precipitado, agua o gas cuando se combinan dos compuestos iónicos en una solución. (p. 643)

ductibilidad: capacidad de los metales para convertirse en alambres. (p. 570)

ductilidad: capacidad de los metales o aleaciones para ser convertidos en alambres. (p. 759)

E

eardrum: tough membrane in the outer ear that is about 0.1 mm thick and transmits sound vibrations into the middle ear. (p. 325)

echolocation: process in which objects are located by emitting sounds and interpreting sound waves that are reflected. (p. 339)

efficiency: ratio of the output work done by the machine to the input work done on the machine, expressed as a percentage. (p. 136)

elastic potential energy: energy stored when an object is compressed or stretched. (p. 103)

electrical power: rate at which electrical energy is converted to another form of energy; expressed in watts (W). (p. 210)

electric current: the net movement of electric charges in a single direction, measured in amperes (A). (p. 201)

electric motor: device that converts electrical energy to mechanical energy by using the magnetic forces between an electromagnet and a permanent magnet to make a shaft rotate. (p. 235)

electrolyte: compound that breaks apart in water, forming charged particles (ions) that can conduct electricity. (p. 676)

electromagnet: temporary magnet made by wrapping a wire coil, carrying a current, around an iron core. (p. 232)

electromagnetic induction: process in which electric current is produced in a wire loop by a changing magnetic field. (p. 238)

electromagnetic waves: waves created by vibrating electric charges, can travel through a vacuum or through matter, and have a wide variety of frequencies and wavelengths. (p. 354)

tímpano: membrana fuerte del oído externo que tiene más o menos 0.1 mm de grosor y transmite las vibraciones del sonido al oído medio. (p. 325)

ecolocalización: proceso en el cual los objetos son localizados emitiendo sonidos e interpretando ondas de sonido que se reflejan. (p. 339)

eficiencia: relación del trabajo efectuado por una máquina y el trabajo hecho en ésta, expresada en porcentaje. (p. 136)

energía elástica potencial: energía almacenada cuando un objeto es comprimido o estirado. (p. 103)

potencia eléctrica: proporción a la cual la energía eléctrica se convierte en otra forma de energía; se expresa en vatios (V). (p. 210)

corriente eléctrica: movimiento neto de cargas eléctricas en una sola dirección, medido en amperios (A). (p. 201)

motor eléctrico: dispositivo que convierte la energía eléctrica en energía mecánica usando las fuerzas magnéticas entre un electroimán y un imán permanente para que el eje gire. (p. 235)

electrolito: compuesto que se descompone en agua formando partículas cargadas (iones) que pueden conducir electricidad. (p. 676)

electroimán: imán temporal que se hace envolviendo una bobina de cable que conduce una corriente, alrededor de un núcleo de hierro. (p. 232)

inducción electromagnética: proceso en el cual una corriente eléctrica es producida en un circuito cerrado de cable mediante un campo magnético cambiante. (p. 238)

ondas electromagnéticas: ondas creadas por la vibración de cargas eléctricas, que pueden viajar a través del vacío o de la materia y que tienen una amplia variedad de frecuencias y de longitudes de onda. (p. 354)

electron cloud: area around the nucleus of an atom where the atom's electrons are most likely to be found. (p. 511)

electron dot diagram: uses the symbol for an element and dots representing the number of electrons in the element's outer energy level. (p. 522)

electrons: particles surrounding the center of an atom that have a charge of $1-$. (p. 507)

element: substance with atoms that are all alike. (p. 450)

endergonic reaction: chemical reaction that requires energy input (heat, light, or electricity) in order to proceed. (p. 649)

endothermic reaction: chemical reaction that requires heat energy in order to proceed. (p. 649)

exergonic reaction: chemical reaction that releases some form of energy, such as light or heat. (p. 648)

exothermic reaction: chemical reaction in which energy is primarily given off in the form of heat. (p. 648)

experiment: organized procedure for testing a hypothesis; tests the effect of one thing on another under controlled conditions. (p. 8)

nube de electrones: área alrededor del núcleo de un átomo en donde hay más probabilidad de encontrar los electrones de los átomos. (p. 511)

diagrama de punto de electrones: usa el símbolo de un elemento y puntos que representan el número de electrones en el nivel de energía externo del elemento. (p. 522)

electrones: partículas que rodean el centro de un átomo que tienen la carga de $1-$. (p. 507)

elemento: sustancia en la cual todos los átomos son iguales. (p. 450)

reacción endergónica: reacción química que requiere entrada de energía (calor, luz o electricidad) para poder proceder. (p. 649)

reacción endotérmica: reacción química que requiere energía de calor para proceder. (p. 649)

reacción exergónica: reacción química que libera una forma de energía, tal como, luz o calor. (p. 648)

reacción exotérmica: reacción química en la cual la energía es inicialmente emitida en forma de calor. (p. 648)

experimento: procedimiento organizado para probar una hipótesis; prueba el efecto de una cosa sobre otra bajo condiciones controladas. (p. 8)

F

first law of thermodynamics: states that the increase in thermal energy of a system equals the work done on the system plus the heat added to the system. (p. 175)

fluorescent light: light that results when ultraviolet radiation produced inside a fluorescent bulb causes the phosphor coating inside the bulb to glow. (p. 395)

focal length: distance from the center of a lens or mirror to the focal point. (p. 418)

focal point: the point on the optical axis of a concave mirror or convex lens where light rays, that are initially parallel to the optical axis, pass through after they strike the mirror or lens. (p. 418)

force: a push or pull exerted on an object. (p. 52)

fossil fuels: oil, natural gas, and coal; formed from the decayed remains of ancient plants and animals. (p. 257)

frequency: the number of wavelengths that pass a fixed point each second; is expressed in hertz (Hz). (p. 297)

friction: force that opposes the sliding motion between two touching surfaces. (p. 70)

primera ley de la termodinámica: establece que el aumento en la energía térmica de un sistema es igual al trabajo realizado sobre el sistema más el calor agregado a éste. (p. 175)

luz fluorescente: luz que resulta cuando una radiación ultravioleta producida dentro de una bombilla fluorescente hace que brille el revestimiento de fósforo dentro de la bombilla. (p. 395)

longitud focal: distancia desde el centro de un lente o espejo al punto focal. (p. 418)

punto focal: el punto en el eje óptico de un espejo cóncavo o lente convexo en el cual los rayos de luz, que inicialmente son paralelos al eje óptico, cruzan luego de chocar con el espejo o lente. (p. 418)

fuerza: impulso o tracción sobre un objeto. (p. 52)

combustibles fósiles: petróleo, gas natural y carbón formados por los restos descompuestos de plantas y animales ancestrales. (p. 257)

frecuencia: el número de longitudes de onda que pasan por un punto fijo en un segundo; se expresa en hertz (Hz). (p. 297)

fricción: fuerza que se opone al movimiento deslizante entre dos superficies en contacto. (p. 70)

G

galvanometer: a device that uses an electromagnet to measure electric current. (p. 234)

gamma ray: electromagnetic wave with no mass and no charge that travels at the speed of light and is usually emitted with alpha or beta particles from a decaying atomic nucleus; has a wavelength less than about ten trillionths. (pp. 365, 543)

Geiger counter: radiation detector that produces a click or a flash of light when a charged particle is detected. (p. 548)

generator: device that uses electromagnetic induction to convert mechanical energy to electrical energy. (p. 238)

geothermal energy: thermal energy in hot magma; can be converted by a power plant into electrical energy. (p. 275)

Global Positioning System (GPS): a system of satellites and ground monitoring stations that enable a receiver to determine its location at or above Earth's surface. (p. 373)

graph: visual display of information or data that can provide a quick way to communicate a lot of information and allow scientists to observe patterns. (p. 22)

gravitational potential energy: energy stored by objects due to their position above Earth's surface; depends on the distance above Earth's surface and the object's mass. (p. 104)

gravity: attractive force between two objects that depends on the masses of the objects and the distance between them. (p. 75)

group: vertical column in the periodic table. (p. 520)

galvanómetro: dispositivo que usa un electroimán para medir la corriente eléctrica. (p. 234)

rayo gama: onda electromagnética sin masa ni carga que viaja a la velocidad de la luz y que usualmente es emitida con partículas alfa o beta a partir de un núcleo atómico en descomposición; tiene una electromagnética con longitudes de onda menores a diez trillonésimas de metro. (pp. 365, 543)

contador Geiger: detector de radiación que produce un sonido seco o un destello de luz al detectar una partícula cargada. (p. 548)

generador: dispositivo que usa inducción electromagnética para convertir energía mecánica en energía eléctrica. (p. 238)

energía geotérmica: energía térmica en el magma caliente, la cual se puede convertir mediante una planta industrial en energía eléctrica. (p. 275)

Sistema de Posicionamiento Global (GPS): sistema de satélites y estaciones de monitoreo en tierra que permiten que un receptor determine su ubicación en o sobre la superficie terrestre. (p. 373)

gráfica: presentación visual de información que puede suministrar una forma rápida de comunicar gran cantidad de información y que permite que los científicos puedan observar los patrones. (p. 22)

energía gravitacional potencial: energía almacenada por objetos debido a su posición sobre la superficie terrestre, la cual depende de la distancia sobre la superficie terrestre y de la masa del objeto. (p. 104)

gravedad: fuerza de atracción entre dos objetos que depende de las masas de los objetos y de la distancia entre ellos. (p. 75)

grupo: columna vertical en la tabla periódica. (p. 520)

H

half-life: amount of time it takes for half the nuclei in a sample of a radioactive isotope to decay. (p. 544)

heat: thermal energy that flows from a warmer material to a cooler material. (p. 160)

heat engine: device that converts thermal energy into work. (p. 176)

heat of fusion: amount of energy required to change a substance from the solid phase to the liquid phase. (p. 478)

vida media: tiempo requerido para que se descomponga la mitad de los núcleos de una muestra de isótopo radiactivo. (p. 544)

calor: energía térmica que fluye de un material caliente a uno frío. (p. 160)

motor de calor: dispositivo que convierte la energía térmica en trabajo. (p. 176)

calor de fusión: cantidad de energía necesaria para cambiar una sustancia del estado sólido al líquido. (p. 478)

heat of vaporization: the amount of energy required for the liquid at its boiling point to become a gas. (p. 479)

heterogeneous (het uh ruh JEE nee us) mixture: mixture, such as mixed nuts or a dry soup mix, in which different materials are unevenly distributed and are easily identified. (p. 453)

holography: technique that produces a complete three-dimensional photographic image of an object. (p. 401)

homogeneous (hoh moh JEE nee us) mixture: solid, liquid, or gas that contains two or more substances blended evenly throughout. (p. 454)

hydrate: compound that has water chemically attached to its ions and written into its chemical formula. (p. 620)

hydrocarbon: saturated or unsaturated compound containing only carbon and hydrogen atoms. (p. 727)

hydroelectricity: electricity produced from the energy of falling water. (p. 273)

hydronium ions (hi DROH nee um • I ahnz): H_3O^+ ions, which form when an acid dissolves in water and H^+ ions interact with water. (p. 696)

hypothesis: educated guess using what you know and what you observe. (p. 8)

calor de vaporización: cantidad de energía necesaria para que un líquido en su punto de ebullición se convierta en gas. (p. 479)

mezcla heterogénea: mezcla, tal como una mezcla de nueces o una mezcla seca para hacer sopa, en la cual diferentes materiales están distribuidos en forma desigual y se pueden identificar fácilmente. (p. 453)

holografía: técnica que produce una imagen fotográfica tridimensional completa de un objeto. (p. 401)

mezcla homogénea: sólido, liquido, o gas que contiene dos o más sustancias mezcladas de manera uniforme en toda la mezcla. (p. 454)

hidrato: compuesto que contiene agua químicamente conectada a sus iones y representada en su fórmula química. (p. 620)

hidrocarburos: compuestos saturados o no saturados que contienen únicamente átomos de carbono e hidrógeno. (p. 727)

hidroelectricidad: electricidad producida a partir de la energía generada por una caída de agua. (p. 273)

iones de hidronio: iones H_3O^+ que se forman cuando un ácido se disuelve en agua y los iones H^+ interactúan con el agua. (p. 696)

hipótesis: suposición fundamentada que se basa en lo que se sabe y lo que se observa. (p. 8)

I

incandescent light: light produced by heating a piece of metal, usually tungsten, until it glows. (p. 394)

inclined plane: simple machine that consists of a sloping surface, such as a ramp, that reduces the amount of force needed to lift something by increasing the distance over which the force is applied. (p. 144)

incoherent light: light that contains more than one wavelength, and travels in many directions with its crests and troughs unaligned. (p. 398)

independent variable: factor that, as it changes, affects the measure of another variable. (p. 9)

index of refraction: property of a material indicating how much light slows down when traveling in the material. (p. 386)

indicator: organic compound that changes color in acids and bases. (p. 696)

inertia: resistance of an object to a change in its motion. (p. 54)

infrared waves: electromagnetic waves that have a wavelength between about 1 mm and 750 billionths of a meter. (p. 362)

luz incandescente: luz que se produce al calentar una pieza de metal, generalmente tungsteno, hasta que brille. (p. 394)

plano inclinado: máquina simple que consiste de una superficie inclinada, tal como una rampa, que reduce la fuerza necesaria para levantar un objeto aumentando la distancia sobre la cual se aplica dicha fuerza. (p. 144)

luz incoherente: luz que contiene más de una longitud de onda y que viaja en varias direcciones con sus crestas y depresiones no alineadas. (p. 398)

variable independiente: factor que, a medida que cambia, afecta la medida de otra variable. (p. 9)

índice de refracción: propiedad de un material para indicar la cantidad de luz que se frena al pasar a través del material. (p. 386)

indicador: compuesto orgánico que cambia de color en presencia de ácidos y bases. (p. 696)

inercia: resistencia de un objeto a cambiar su movimiento. (p. 54)

ondas infrarrojas: ondas electromagnéticas que tienen una longitud de onda entre aproximadamente 1 mm y 750 billonésimas de metro. (p. 362)

inhibitor: substance that slows down a chemical reaction or prevents it from occurring by combining with a reactant. (p. 650)

instantaneous speed: speed of an object at a given point in time; is constant for an object moving with constant speed, and changes with time for an object that is slowing down or speeding up. (p. 42)

insulator: material in which electrons are not able to move easily. (p. 195)

insulator: material in which heat flows slowly. (p. 169)

integrated circuit: tiny chip of semiconductor material that can contain millions of transistors, diodes, and other components. (p. 768)

intensity: amount of energy that flows through a certain area in a specific amount of time. (p. 328)

interference: occurs when two or more waves overlap and combine to form a new wave. (p. 308)

internal combustion engine: heat engine that burns fuel inside the engine in chambers or cylinders. (p. 176)

ion: charged particle that has either more or fewer electrons than protons. (pp. 608, 676)

ionic bond: attraction formed between oppositely charged ions in an ionic compound. (p. 610)

ionization: process in which electrolytes dissolve in water and separate into charged particles. (p. 676)

isomers: compounds with identical chemical formulas but different molecular structures and shapes. (p. 729)

isotopes: atoms of the same element that have different numbers of neutrons. (p. 514)

inhibidor: sustancia que reduce una reacción química o previene que ocurra por una combinación con un reactivo. (p. 650)

velocidad instantánea: velocidad de un objeto en un punto dado en el tiempo; es constante para un objeto que se mueve a una velocidad constante y cambia con el tiempo en un objeto que está reduciendo o aumentando su velocidad. (p. 42)

aislador: material a través del cual los electrones no se pueden mover con facilidad. (p. 195)

aislador: material en el cual el calor fluye lentamente. (p. 169)

circuito integrado: pedazo minúsculo de material semiconductor que puede contener millones de transistores, diodos y otros componentes. (p. 768)

intensidad: cantidad de energía que fluye a través de cierta área en un tiempo específico. (p. 328)

interferencia: ocurre cuando dos o más ondas se sobreponen y combinan para formar una nueva onda. (p. 308)

motor de combustión interna: motor de calor que quema combustible en su interior en cámaras o cilindros. (p. 176)

ion: partícula cargada que tiene ya sea más o menos electrones que protones. (pp. 608, 676)

enlace iónico: atracción formada entre iones con cargas opuestas en un compuesto iónico. (p. 610)

ionización: proceso en el cual los electrolitos se disuelven en agua y se separan en partículas cargadas. (p. 676)

isómeros: compuestos con fórmulas químicas idénticas pero diferentes estructuras y formas moleculares. (p. 729)

isótopos: átomos del mismo elemento que tienen diferente número de neutrones. (p. 514)

J

joule: SI unit of energy. (p. 102)

julio: unidad SI de energía. (p. 102)

K

kinetic energy: energy a moving object has because of its motion; depends on the mass and speed of the object. (p. 102)

kinetic theory: explanation of the behavior of molecules in matter; states that all matter is made of constantly moving particles that collide without losing energy. (p. 476)

energía cinética: energía que tiene un cuerpo debido a su movimiento, la cual depende de la masa y velocidad del objeto. (p. 102)

teoría cinética: explicación del comportamiento de las moléculas en la materia, la cual establece que todas las sustancias están compuestas de partículas en constante movimiento que colindan sin perder energía. (p. 476)

law of conservation of charge: states that charge can be transferred from one object to another but cannot be created or destroyed. (p. 193)

law of conservation of energy: states that energy can never be created or destroyed. (p. 111)

law of conservation of mass: states that the mass of all substances present before a chemical change equals the mass of all the substances remaining after the change. (p. 465)

lever: simple machine consisting of a bar free to pivot about a fixed point called the fulcrum. (p. 138)

lipids: group of biological compounds that contains the same elements as carbohydrates but in different arrangements and combinations, and includes saturated and unsaturated fats and oils. (p. 746)

loudness: human perception of sound intensity. (p. 329)

luster: property of metals and alloys that describes having a shiny appearance or reflecting light. (p. 759)

ley de la conservación de carga: establece que la carga puede ser transferida entre un objeto y otro pero no puede ser creada o destruida. (p. 193)

ley de la conservación de energía: establece que la energía nunca puede ser creada ni destruida. (p. 111)

ley de conservación de la masa: establece que la masa de todas las sustancias presente antes de un cambio químico es igual a la masa de todas las sustancias resultantes después del cambio. (p. 465)

palanca: máquina simple que consiste de una barra que puede girar sobre un punto fijo llamado pivote. (p. 138)

lípidos: grupo de compuestos biológicos que contienen los mismos elementos que los carbohidratos pero en diferentes disposiciones y combinaciones, y que incluye grasas y aceites saturados o no saturados. (p. 746)

volumen de sonido: percepción humana de la intensidad del sonido. (p. 329)

lustre: propiedad de los metales y aleaciones que describe que tienen una apariencia brillante o que reflejan la luz. (p. 759)

machine: device that makes doing work easier by increasing the force applied to an object, changing the direction of an applied force, or increasing the distance over which a force can be applied. (p. 132)

magnetic domain: group of atoms in a magnetic material with the magnetic poles of the atoms pointing in the same direction. (p. 229)

magnetic field: surrounds a magnet and exerts a force on other magnets and objects made of magnetic materials. (p. 225)

magnetic pole: region on a magnet where the magnetic force exerted by a magnet is strongest; like poles repel and opposite poles attract. (p. 225)

magnetism: the properties and interactions of magnets. (p. 224)

malleability (mal yuh BIHL yt ee): ability of metals and alloys to be rolled or hammered into thin sheets. (pp. 570, 759)

mass: amount of matter in an object. (p. 19)

mass number: sum of the number of protons and neutrons in an atom's nucleus. (p. 513)

máquina: artefacto que facilita la ejecución del trabajo aumentando la fuerza que se aplica a un objeto, cambiando la dirección de una fuerza aplicada o aumentando la distancia sobre la cual se puede aplicar una fuerza. (p. 132)

dominio magnético: grupo de átomos en un material magnético en el cual los polos magnéticos de los átomos apuntan en la misma dirección. (p. 229)

campo magnético: rodea a un imán y ejerce una fuerza sobre otros imanes y objetos hechos de materiales magnéticos. (p. 225)

polo magnético: zona en un imán en donde la fuerza magnética ejercida por un imán es la más fuerte; los polos iguales se repelen y los polos opuestos se atraen. (p. 225)

magnetismo: propiedades e interacciones de los imanes. (p. 224)

maleabilidad: capacidad de los metales y aleaciones de ser rolados o martillados para formar láminas delgadas. (pp. 570, 759)

masa: cantidad de materia en un objeto. (p. 19)

número de masa: suma del número de protones y neutrones en el núcleo de un átomo. (p. 513)

mechanical advantage (MA): ratio of the output force exerted by a machine to the input force applied to the machine. (p. 136)

mechanical energy: sum of the potential energy and kinetic energy in a system. (p. 108)

medium: matter in which a wave travels. (p. 291)

melting point: temperature at which a solid begins to liquefy. (p. 478)

metal: element that typically is a hard, shiny solid, is malleable, and is a good conductor of heat and electricity. (p. 570)

metallic bonding: occurs because electrons move freely among a metal's positively charged ions and explains properties such as ductility and the ability to conduct electricity. (p. 571)

metalloid: element that shares some properties with metals and some with nonmetals. (p. 584)

microscope: uses convex lenses to magnify small, close objects. (p. 435)

microwaves: radio waves with wavelengths of between about 1 m and 1 mm. (p. 361)

mirage: image of a distant object produced by the refraction of light through air layers of different densities. (p. 388)

model: can be used to represent an idea, object, or event that is too big, too small, too complex, or too dangerous to observe or test directly. (p. 11)

molecule: a neutral particle that forms as a result of electron sharing. (p. 611)

momentum: property of a moving object that equals its mass times its velocity. (p. 86)

monomer: small molecule that forms a link in a polymer chain and can be made to combine with itself repeatedly. (pp. 739, 771)

music: sounds that are deliberately used in a regular pattern. (p. 333)

ventaja mecánica (MA): relación de la fuerza ejercida por una máquina y la fuerza aplicada a dicha máquina. (p. 136)

energía mecánica: suma de la energía potencial y energía cinética en un sistema. (p. 108)

medio: materia a través de la cual viaja una onda. (p. 291)

punto de fusión: temperatura a la cual un sólido comienza a licuarse. (p. 478)

metal: elemento típicamente duro, sólido brillante, maleable y buen conductor del calor y la electricidad. (p. 570)

enlace metálico: ocurre debido a que los electrones se mueven libremente entre los iones de un metal cargados positivamente y explica propiedades tales como la ductibilidad y la capacidad para conducir electricidad. (p. 571)

metaloide: elemento que tiene algunas propiedades de los metales y algunas de los no metales. (p. 584)

microscopio: instrumento que usa lentes convexos para amplificar objetos pequeños cercanos. (p. 435)

microondas: ondas de radio con longitudes de onda entre aproximadamente 1 mm y 1 m. (p. 361)

espejismo: imagen de un objeto distante producida por la refracción de la luz a través de capas de aire de diferentes densidades. (p. 388)

modelo: puede ser usado para representar una idea, objeto o evento que es demasiado grande, demasiado pequeño, demasiado complejo o demasiado peligroso para ser observado o probado directamente. (p. 11)

molécula: partícula neutra que se forma al compartir electrones. (p. 611)

inercia: propiedad de un objeto en movimiento que es igual a su masa por su velocidad. (p. 86)

monómero: pequeña molécula que forma una conexión en una cadena de polímeros y que se puede combinar consigo misma repetidamente. (pp. 739, 771)

música: sonidos que se usan deliberadamente en un patrón regular. (p. 333)

N

net force: sum of the forces that are acting on an object. (p. 53)

neutralization: chemical reaction that occurs when the H_3O^+ ions from an acid react with the OH^- ions from a base to produce water molecules. (p. 707)

neutron: neutral particle, composed of quarks, inside the nucleus of an atom. (p. 507)

fuerza neta: suma de fuerzas que actúan sobre un objeto. (p. 53)

neutralización: reacción química que ocurre cuando los iones H_3O^+ de un ácido reaccionan con los iones OH^- de una base para producir moléculas de agua. (p. 707)

neutrón: partícula neutra, compuesta por quarks, dentro del núcleo de un átomo. (p. 507)

Newton's second law of motion: states that the acceleration of an object is in the same direction as the net force on the object, and that the acceleration equals the net force divided by the mass. (p. 69)

Newton's third law of motion: states that when one object exerts a force on a second object, the second object exerts a force on the first object that is equal in strength and in the opposite direction. (p. 83)

nonelectrolyte: substance that does not ionize in water and cannot conduct electricity. (p. 676)

nonmetal: element that usually is a gas or brittle solid at room temperature, is not malleable or ductile, is a poor conductor of heat and electricity, and typically is not shiny. (p. 578)

nonpolar: not having separated positive and negative areas; nonpolar materials do not attract water molecules and do not dissolve easily in water. (p. 681)

nonpolar molecule: molecule that shares electrons equally and does not have oppositely charged ends. (p. 614)

nonrenewable resources: natural resource, such as fossil fuels, that cannot be replaced by natural processes as quickly as it is used. (p. 263)

nuclear fission: process of splitting an atomic nucleus into two or more nuclei with smaller masses. (p. 551)

nuclear fusion: reaction in which two or more atomic nuclei form a nucleus with a larger mass. (p. 553)

nuclear reactor: uses energy from a controlled nuclear chain reaction to generate electricity. (p. 264)

nuclear waste: radioactive by-product that results when radioactive materials are used. (p. 268)

nucleic acids: essential organic polymers that control the activities and reproduction of cells. (p. 744)

nucleotides: complex, organic molecules that make up RNA and DNA; contain an organic base, a phosphoric acid unit, and a sugar. (p. 744)

nucleus: positively charged center of an atom that contains protons and neutrons and is surrounded by a cloud of electrons. (p. 507)

segunda ley de movimiento de Newton: establece que la aceleración de un objeto es en la misma dirección que la fuerza neta del objeto y que la aceleración es igual a la fuerza neta dividida por su masa. (p. 69)

tercera ley de movimiento de Newton: establece que cuando un objeto ejerce una fuerza sobre un segundo objeto, el segundo objeto ejerce una fuerza igual de fuerte sobre el primer objeto y en dirección opuesta. (p. 83)

no electrolito: sustancia que no se ioniza en el agua y no puede conducir electricidad. (p. 676)

no metal: elemento que por lo general es un gas o un sólido frágil a temperatura ambiente, no es maleable o dúctil, es mal conductor del calor y la electricidad, y por lo general no es brillante. (p. 578)

no polar: sustancia que no tiene áreas positivas y negativas separadas; los materiales no polares no atraen las moléculas de agua y no se disuelven fácilmente en ésta. (p. 681)

molécula no polar: molécula que comparte equitativamente los electrones y que no tiene extremos con cargas opuestas. (p. 614)

recursos no renovables: recursos naturales, tales como combustibles fósiles, que no pueden ser reemplazados por procesos naturales tan pronto como son usados. (p. 263)

fisión nuclear: proceso de división de un núcleo atómico en dos o más núcleos con masas más pequeñas. (p. 551)

fusión nuclear: reacción en la cual dos o más núcleos atómicos forman un núcleo con mayor masa. (p. 553)

reactor nuclear: usa energía de una reacción nuclear controlada en cadena para generar electricidad. (p. 264)

desperdicio nuclear: subproducto radioactivo que resulta del uso de materiales radiactivos. (p. 268)

ácidos nucleicos: polímeros orgánicos esenciales que controlan las actividades y la reproducción de las células. (p. 744)

nucleótidos: moléculas orgánicas complejas que componen el ARN y el ADN y que contienen una base orgánica, una unidad de ácido fosfórico y un azúcar. (p. 744)

núcleo: centro de un átomo con carga positiva que contiene protones y neutrones y está rodeado por una nube de electrones. (p. 507)

O

Ohm's law: states that the current in a circuit equals the voltage difference divided by the resistance. (p. 205)

ley de Ohm: establece que la corriente en un circuito es igual a la diferencia de voltaje dividida por la resistencia. (p. 205)

opaque: material that absorbs or reflects all light and does not transmit any light. (p. 384)

optical axis: imaginary straight line that is perpendicular to the center of a concave mirror or convex lens. (p. 418)

organic compounds: large number of compounds containing the element carbon. (p. 726)

overtone: vibration whose frequency is a multiple of the fundamental frequency. (p. 334)

oxidation: the loss of electrons from the atoms of a substance. (p. 645)

oxidation number: positive or negative number that indicates how many electrons an atom has gained, lost, or shared to become stable. (p. 615)

opaco: material que absorbe o refleja toda la luz pero no la transmite. (p. 384)

eje óptico: línea recta imaginaria perpendicular al centro de un espejo cóncavo o lente convexo. (p. 418)

compuestos orgánicos: un gran número de compuestos que contienen el elemento carbono. (p. 726)

armónico: vibración cuya frecuencia es un múltiplo de la frecuencia fundamental. (p. 334)

oxidación: la pérdida de electrones de los átomos de una sustancia. (p. 645)

número de oxidación: número positivo o negativo que indica cuántos electrones ha ganado, perdido o compartido un átomo para poder ser estable. (p. 615)

P

parallel circuit: circuit in which electric current has more than one path to follow. (p. 208)

pascal: SI unit of pressure. (p. 490)

period: horizontal row in the periodic table. (p. 523); the amount of time it takes one wavelength to pass a fixed point; is expressed in seconds. (p. 297)

periodic table: organized list of all known elements that are arranged by increasing atomic number and by changes in chemical and physical properties. (p. 516)

petroleum: liquid fossil fuel formed from decayed remains of ancient organisms; can be refined into fuels and used to make plastics. (p. 259)

pH: a measure of the concentration of hydronium ions in a solution using a scale ranging from 0 to 14, with 0 being the most acidic and 14 being the most basic. (p. 704)

photon: particle that electromagnetic waves sometimes behave like; has energy that increases as the frequency of the electromagnetic wave increases. (p. 358)

photovoltaic cell: device that converts solar energy into electricity; also called a solar cell. (p. 271)

physical change: any change in size, shape, or state of matter in which the identity of the substance remains the same. (p. 460)

physical property: any characteristic of a material, such as size or shape, that you can observe or attempt to observe without changing the identity of the material. (p. 458)

circuito paralelo: circuito en el cual la corriente eléctrica tiene más de una trayectoria para seguir. (p. 208)

pascal: unidad SI de presión. (p. 490)

período: fila horizontal en la tabla periódica. (p. 523); la cantidad de tiempo que requiere una longitud de onda para pasar un punto fijo; se expresa en segundos. (p. 297)

tabla periódica: lista organizada de todos los elementos conocidos y que han sido ordenados de manera ascendente por número atómico y por cambios en sus propiedades químicas y físicas. (p. 516)

petróleo: combustible fósil líquido que se forma a partir de residuos en descomposición de organismos ancestrales y que puede ser refinado para producir combustibles y usado para hacer plásticos. (p. 259)

pH: medida de la concentración de iones de hidronio en una solución, usando una escala de 0 a 14, en la cual 0 es la más ácida y 14 la más básica. (p. 704)

fotón: partícula como la cual algunas veces se comportan las ondas electromagnéticas; tiene energía que aumenta a medida que la frecuencia de la onda electromagnética aumenta. (p. 358)

células fotovoltaicas: dispositivo que convierte la energía solar en electricidad; también llamada celda solar. (p. 271)

cambio físico: cualquier cambio en tamaño, forma o estado de una sustancia en la cual la identidad de la sustancia sigue siendo la misma. (p. 460)

propiedad física: cualquier característica de un material, tal como tamaño o forma, que se puede haber observar o tratado de observar sin cambiar la identidad del material. (p. 458)

pigment: colored material that is used to change the color of other substances. (p. 392)

pitch: how high or low a sound seems; related to the frequency of the sound waves. (p. 330)

plane mirror: flat, smooth mirror that reflects light to form upright, virtual images. (p. 473)

plasma: matter consisting of positively and negatively charged particles. (p. 480)

polar: having separated positive and negative areas; polar materials attract water molecules and dissolve easily in water. (p. 667)

polarized light: light whose waves vibrate in only one direction. (p. 400)

polar molecule: molecule with a slightly positive end and a slightly negative end as a result of electrons being shared unequally. (p. 614)

polyatomic ion: positively or negatively charged, covalently bonded group of atoms. (p. 619)

polyethylene: polymer formed from a chain containing many ethylene units; often used in plastic bags and plastic bottles. (p. 739)

polymer: class of natural or synthetic substances made up of many smaller, simpler molecules, called monomers, arranged in large chains. (pp. 739, 771)

potential energy: stored energy an object has due to its position. (p. 103)

power: amount of work done, or the amount of energy transferred, divided by the time required to do the work or transfer the energy; measured in watts (W). (p. 129)

precipitate: insoluble compound that comes out of solution during a double-displacement reaction. (p. 643)

pressure: amount of force exerted per unit area; SI unit is the pascal (Pa). (p. 486)

product: in a chemical reaction, the new substance that is formed. (p. 632)

proteins: large, complex, biological polymers formed from amino acid units; make up many body tissues such as muscles, tendons, hair, and fingernails. (p. 742)

proton: particle, composed of quarks, inside the nucleus of an atom that has a charge of 1+. (p. 507)

pulley: simple machine that consists of a grooved wheel with a rope, chain, or cable running along the groove; can be either fixed or movable. (p. 141)

pigmento: material de color que se usa para cambiar el color de otras sustancias. (p. 392)

altura: qué tan alto o bajo parece un sonido; tiene relación con la frecuencia de las ondas sonoras. (p. 330)

espejo plano: espejo plano y liso que refleja la luz para formar imágenes verticales y virtuales. (p. 473)

plasma: materia consistente de partículas con cargas positivas y negativas. (p. 480)

polar: sustancia que tiene áreas positivas y negativas separadas; los materiales polares atraen las moléculas de agua y se disuelven fácilmente en ésta. (p. 667)

luz polarizada: luz cuyas ondas vibran en una sola dirección. (p. 400)

molécula polar: molécula con un extremo ligeramente positivo y otro ligeramente negativo como resultado de un compartir desigual de los electrones. (p. 614)

ion poliatómico: grupo de átomos enlazados covalentemente, con carga positiva o negativa. (p. 619)

polietileno: polímero formado por una cadena que contiene varias unidades de etileno; es comúnmente usado en la fabricación de bolsas y envases plásticos. (p. 739)

polímero: clase de sustancias naturales o sintéticas compuestas por muchas moléculas más simples y pequeñas, llamadas monómeros, ordenadas en largas cadenas. (pp. 739, 771)

energía potencial: energía almacenada que un objeto tiene debido a su posición. (p. 103)

potencia: cantidad de trabajo realizado o cantidad de energía transferida, dividida por el tiempo requerido para realizar el trabajo o transferir la energía; medida en vatios (V). (p. 129)

precipitado: compuesto insoluble que resulta de una solución durante una reacción de doble desplazamiento. (p. 643)

presión: cantidad de fuerza ejercida por unidad de área; la unidad SI es el pascal (Pa). (p. 486)

producto: es la nueva sustancia que se forma en una reacción química. (p. 632)

proteínas: polímeros biológicos extensos y complejos formados por unidades de aminoácidos; conforman muchos tejidos del cuerpo como los músculos, los tendones, el pelo y las uñas. (p. 742)

protón: partícula, compuesta por quarks, dentro del núcleo de un átomo que tiene una carga de 1+. (p. 507)

polea: máquina simple que consiste de una rueda acanalada con una cuerda, cadena o cable que se desliza por el canal y que puede ser fija o móvil. (p. 141)

Q

quarks: particles of matter that make up protons and neutrons. (p. 507)

quarks: partículas de materia que constituyen los protones y neutrones. (p. 507)

R

radiant energy: energy carried by an electromagnetic wave. (p. 357)

radiation: transfer of thermal energy by electromagnetic waves. (p. 167)

radioactive element: element, such as radium, whose nucleus breaks down and emits particles and energy. (p. 572)

radioactivity: process that occurs when a nucleus decays and emits alpha, beta, or gamma radiation. (p. 538)

radio waves: electromagnetic waves with wavelengths longer than about 1 mm, used for communications. (p. 361)

rarefaction: the least dense regions of a compressional wave. (p. 296)

reactant: in a chemical reaction, the substance that reacts. (p. 632)

real image: an image formed by light rays that converge to pass through the place where the image is located. (p. 419)

reduction: the gain of electrons by the atoms of a substance. (p. 645)

reflecting telescope: uses a concave mirror, plane mirror, and convex lens to collect and focus light from distant objects. (p. 433)

refracting telescope: uses two convex lenses to gather and focus light from distant objects. (p. 433)

refraction: the bending of a wave as it changes speed in moving from one medium to another. (p. 304)

renewable resource: energy source that is replaced almost as quickly as it is used. (p. 271)

resistance: tendency for a material to oppose electron flow and change electrical energy into thermal energy and light; measured in ohms (Ω). (p. 203)

resonance: the process by which an object is made to vibrate by absorbing energy at its natural frequencies. (p. 311)

resonator: hollow, air-filled chamber that amplifies sound when the air inside it vibrates. (p. 335)

energía radiante: energía transportada por una onda electromagnética. (p. 357)

radiación: transferencia de energía térmica mediante ondas electromagnéticas. (p. 167)

elemento radiactivo: elemento, como el radio, cuyo núcleo se divide y emite partículas y energía. (p. 572)

radiactividad: proceso que ocurre cuando un núcleo se descompone y emite radiación alfa, beta o gama. (p. 538)

ondas de radio: ondas electromagnéticas con longitudes de onda más largas de aproximadamente 1 mm y que se usan en las comunicaciones. (p. 361)

rarefacción: las regiones menos densas de una onda de compresión. (p. 296)

reactante: es la sustancia que reacciona en una reacción química. (p. 632)

imagen real: imagen formada por rayos de luz que convergen para pasar a través del sitio donde está localizada la imagen. (p. 419)

reducción: la obtención de electrones por los átomos de una sustancia. (p. 645)

telescopio reflexivo: usa un espejo cóncavo, un espejo plano y lentes convexos para recolectar y enfocar la luz proveniente de objetos distantes. (p. 433)

telescopio refractivo: usa dos lentes convexos para reunir y enfocar la luz proveniente de objetos distantes. (p. 433)

refracción: curvatura de una onda al cambiar su velocidad al pasar de un medio a otro. (p. 304)

recursos renovables: fuente de energía que es reemplazada casi tan pronto como es usada. (p. 271)

resistencia: tendencia de un material de oponerse al fluido de los electrones y convertir la energía eléctrica en energía térmica y luz; se mide en ohmios (Ω). (p. 203)

resonancia: el proceso por el cual un objeto vibra al absorber energía en sus frecuencias naturales. (p. 311)

resonador: cámara hueca, llena de aire, que amplifica el sonido cuando vibra el aire en su interior. (p. 335)

Glossary/Glosario

retina: inner lining of the eye that has cells which convert light images into electrical signals for interpretation by the brain. (p. 427)

retina: capa interna del ojo que posee células que convierten imágenes iluminadas en señales eléctricas para que el cerebro las interprete. (p. 427)

S

---

salt: compound formed when negative ions from an acid combine with positive ions from a base. (pp. 580, 707)

saturated hydrocarbon: compound, such as propane or methane, that contains only single bonds between carbon atoms. (p. 728)

saturated solution: any solution that contains all the solute it can hold at a given temperature. (p. 673)

scientific law: statement about what happens in nature that seems to be true all the time; does not explain why or how something happens. (p. 12)

scientific method: organized set of investigation procedures that can include stating a problem, forming a hypothesis, researching and gathering information, testing a hypothesis, analyzing data, and drawing conclusions. (p. 7)

screw: simple machine that consists of an inclined plane wrapped in a spiral around a cylindrical post. (p. 145)

second law of thermodynamics: states that is impossible for heat to flow from a cool object to a warmer object unless work is done. (p. 175)

semiconductor: materials having conductivity properties between that of metals (good conductors) and nonmetals (insulators) and having controllable conductivity parameters. (pp. 585, 767)

series circuit: circuit in which electric current has only one path to follow. (p. 207)

SI: International System of Units—the improved, universally accepted version of the metric system that is based on multiples of ten and includes the meter (m), liter (L), and kilogram (kg). (p. 15)

simple machine: machine that does work with only one movement—lever, pulley, wheel and axle, inclined plane, screw, and wedge. (p. 138)

single-displacement reaction: chemical reaction in which one element replaces another element in a compound. (p. 643)

sliding friction: frictional force that opposes the motion of two surfaces sliding past each other. (p. 72)

sal: compuesto iónico que se forma cuando un halógeno adquiere un electrón de un metal. (pp. 580, 707)

hidrocarburo saturado: compuesto, como el propano y el metano, que contiene únicamente enlaces simples entre los átomos de carbono. (p. 728)

solución saturada: cualquier solución que contiene todo el soluto que puede retener a una temperatura determinada. (p. 673)

ley científica: enunciado acerca de lo que ocurre en la naturaleza, lo cual parece ser cierto en todo momento, sin explicar cómo o por qué algo ocurre. (p. 12)

método científico: conjunto organizado de procedimientos de investigación que puede incluir el planteamiento de un problema, formulación de una hipótesis, investigación y recopilación de información, comprobación de la hipótesis, análisis de datos y elaboración de conclusiones. (p. 7)

tornillo: máquina simple que consiste de un plano inclinado envuelto en espiral alrededor de un poste cilíndrico. (p. 145)

segunda ley de la termodinámica: establece que es imposible que el calor fluya de un objeto frío a uno caliente, a menos que se realice un trabajo. (p. 175)

semiconductor: materiales que tienen propiedades de conductividad entre aquellas de los metales (buenos conductores) y los no metales (aisladores) y que tienen parámetros de conductividad controlables. (pp. 585, 767)

circuito en serie: circuito en el cual la corriente eléctrica tiene una sola trayectoria para seguir. (p. 207)

SI: Sistema Internacional de Unidades: la versión mejorada y aprobada universalmente del sistema métrico que se basa en múltiplos de diez e incluye el metro (m), el litro (L) y el kilogramo (Kg). (p. 15)

máquina simple: máquina que realiza el trabajo con un solo movimiento: palanca, polea, rueda y eje, plano inclinado, tornillo y cuña. (p. 138)

reacción de un solo desplazamiento: reacción química en la cual un elemento reemplaza a otro elemento en un compuesto. (p. 643)

fricción deslizante: fuerza de fricción que se opone al movimiento de dos superficies que se deslizan entre sí. (p. 72)

Glossary/Glosario

soaps: organic salts with nonpolar, hydrocarbon ends that interact with oils and dirt and polar ends that helps them dissolve in water. (p. 712)

solar collector: device used in an active solar heating system that absorbs radiant energy from the Sun. (p. 174)

solenoid: a wire wrapped into a cylindrical coil. (p. 232)

solubility: maximum amount of a solute that can be dissolved in a given amount of solvent at a given temperature. (p. 671)

solute: in a solution, the substance being dissolved. (p. 665)

solution: homogeneous mixture that remains constantly and uniformly mixed and has particles that are so small they cannot be seen with a microscope. (pp. 454, 664)

solvent: in a solution, the substance in which the solute is dissolved. (p. 665)

sonar: system that uses the reflection of sound waves to detect objects underwater. (p. 341)

sound quality: difference between sounds having the same pitch and loudness. (p. 334)

specific heat: amount of thermal energy needed to raise the temperature of 1 kg of a material 1°C. (p. 161)

speed: distance an object travels per unit of time. (p. 39)

standard: exact, agreed-upon quantity used for comparison. (p. 14)

standing wave: a wave pattern that forms when waves of equal wavelength and amplitude, but traveling in opposite directions, continuously interfere with each other; has points called nodes that do not move. (p. 310)

static electricity: the accumulation of excess electric charge on an object. (p. 192)

static friction: frictional force that prevents two surfaces from sliding past each other. (p. 71)

strong acid: any acid that dissociates almost completely in solution. (p. 702)

strong base: any base that dissociates completely in solution. (p. 703)

strong force: attractive force that acts between protons and neutrons in an atomic nucleus. (p. 537)

sublimation: the process of a solid changing directly to a vapor without forming a liquid. (p. 581)

jabones: sales orgánicas con extremos de hidrocarburos no polares que interactúan con aceites y suciedad, y con extremos polares que ayudan a disolverlos en agua. (p. 712)

recolector solar: dispositivo utilizado en un sistema activo de calefacción solar, el cual absorbe la energía radiante del sol. (p. 174)

solenoide: cable envuelto en forma de bobina cilíndrica. (p. 232)

solubilidad: máxima cantidad de soluto que puede ser disuelto en una cantidad dada de solvente a una temperatura determinada. (p. 671)

soluto: en una solución, la sustancia que está disuelta. (p. 665)

solución: mezcla homogénea que permanece constante y uniformemente mezclada y que tiene partículas tan pequeñas que no pueden ser vistas en un microscopio. (pp. 454, 664)

solvente: en una solución, la sustancia en la cual se disuelve el soluto. (p. 665)

sonar: sistema que usa la reflexión de las ondas sonoras para detectar objetos bajo el agua. (p. 341)

calidad del sonido: diferencia entre sonidos que tienen la misma altura e intensidad sonora. (p. 334)

calor específico: cantidad de energía térmica necesaria para aumentar un grado centígrado la temperatura de un kilogramo de material. (p. 161)

velocidad: distancia que recorre un objeto por unidad de tiempo. (p. 39)

estándar: cantidad exacta y acordada, usada para hacer comparaciones. (p. 14)

onda estacionaria: patrón de una onda que se forma cuando ondas con la misma longitud de onda y amplitud, pero que viajan en direcciones opuestas, interfieren continuamente entre sí; tiene puntos llamados nodos que no se mueven. (p. 310)

electricidad estática: la acumulación del exceso de carga eléctrica en un objeto. (p. 192)

fricción estática: fuerza que evita que dos superficies en contacto se deslicen una sobre otra. (p. 71)

ácido fuerte: cualquier ácido que se disocie casi por completo en una solución. (p. 702)

base fuerte: cualquier base que se disocie completamente en una solución. (p. 703)

fuerza de atracción: fuerza de atracción que actúa entre protones y neutrones en un núcleo atómico. (p. 537)

sublimación: proceso mediante el cual un sólido se convierte directamente en vapor sin pasar por estado líquido. (p. 581)

substance: element or compound that cannot be broken down into simpler components and maintain the properties of the original substance. (p. 450)

substituted hydrocarbon: hydrocarbon with one or more of its hydrogen atoms replaced by atoms or groups of other elements. (p. 732)

supersaturated solution: any solution that contains more solute than a saturated solution at the same temperature. (p. 674)

suspension: heterogeneous mixture containing a liquid in which visible particles settle. (p. 456)

synthesis reaction: chemical reaction in which two or more substances combine to form a different substance. (p. 642)

synthetic: describes polymers, such as plastics, adhesives, and surface coatings, that are made from hydrocarbons. (p. 771)

sustancia: elemento o compuesto que no se puede descomponer en componentes más simples y que mantiene las propiedades de la sustancia original. (p. 450)

hidrocarburo sustituido: un hidrocarburo en el cual uno o más de sus átomos de hidrógeno son reemplazados por átomos o grupos de otros elementos. (p. 732)

solución sobresaturada: cualquier solución que contenga más solutos que una solución saturada a la misma temperatura. (p. 674)

suspensión: mezcla heterogénea que contiene un líquido en el cual las partículas visibles se sedimentan. (p. 456)

reacción síntesis: reacción química en la cual se combinan dos o más sustancias para formar una sustancia diferente. (p. 642)

sintético: describe a los polímeros, tales como plásticos, adhesivos y recubrimientos de superficies, hechos de hidrocarburos. (p. 771)

technology: application of science to help people. (p. 13)

temperature: measure of the average kinetic energy of all the particles in an object. (p. 159)

theory: explanation of things or events that is based on knowledge gained from many observations and investigations. (p. 12)

thermal energy: sum of the kinetic and potential energy of the particles in an object; is transferred by conduction, convection, and radiation. (p. 159)

thermal expansion: increase in the size of a substance when the temperature is increased. (p. 481)

thermodynamics: study of the relationship between thermal energy, heat, and work. (p. 174)

titration (ti TRAY shun): process in which a solution of known concentration is used to determine the concentration of another solution. (p. 710)

total internal reflection: occurs when light strikes a boundary between two materials and is completely reflected. (p. 402)

tracer: radioactive isotope, such as iodine-131, that can be detected by the radiation it emits after it is absorbed by a living organism. (p. 554)

transceiver: device that transmits one radio signal and receives another radio signal at the same time, allowing a cordless phone user to talk and listen at the same time. (p. 371)

tecnología: aplicación de la ciencia en beneficio de la población. (p. 13)

temperatura: medida de la energía cinética promedio de todas las partículas en un objeto. (p. 159)

teoría: explicación de las cosas o eventos que se basa en el conocimiento obtenido a partir de numerosas observaciones e investigaciones. (p. 12)

energía térmica: suma de la energía cinética y potencial de las partículas en un objeto, la cual se transfiere por conducción, convección y radiación. (p. 159)

expansión térmica: aumento del tamaño de una sustancia al aumentar la temperatura. (p. 481)

termodinámica: estudio de la relación entre la energía térmica, el calor y el trabajo. (p. 174)

titulación: proceso mediante el cual una solución con una concentración conocida es usada parea determinar la concentración de otra solución. (p. 710)

reflexión interna total: ocurre cuando la luz choca con el límite entre dos materiales y se refleja completamente. (p. 402)

indicador radiactivo: isótopo radioactivo, tal como el yodo-131, que puede ser detectado por la radiación que emite después de ser absorbido por un organismo vivo. (p. 554)

radio transmisor-receptor: dispositivo que transmite y recibe una señal de radio al mismo tiempo, permitiendo que un usuario de un teléfono inalámbrico pueda hablar y escuchar al mismo tiempo. (p. 371)

transformer: device that uses electromagnetic induction to increase or decrease the voltage of an alternating current. (p. 243)

transition elements: elements in Groups 3 through 12 of the periodic table; occur in nature as uncombined elements and include the iron triad and coinage metals. (p. 574)

translucent: material that transmits some light but not enough to see objects clearly through it. (p. 384)

transmutation: process of changing one element to another through radioactive decay. (p. 542)

transparent: material that transmits almost all the light striking it so that objects can be clearly seen through it. (p. 384)

transuranium elements: elements having more than 92 protons, all of which are synthetic and unstable. (p. 589)

transverse wave: wave for which the matter in the medium moves back and forth at right angles to the direction the wave travels; has crests and troughs. (p. 292)

trough: the lowest points on a transverse wave. (p. 296)

turbine: large wheel that rotates when pushed by steam, wind, or water and provides mechanical energy to a generator. (p. 240)

Tyndall effect: scattering of a light beam as it passes through a colloid. (p. 455)

transformador: dispositivo que usa inducción electromagnética para aumentar o disminuir el voltaje de una corriente alterna. (p. 243)

elementos de transición: los elementos de los grupos 3 al 12 de la tabla periódica que se encuentran en la naturaleza como elementos sin combinar e incluyen la tríada de hierro y los metales con los que se fabrican las monedas. (p. 574)

translúcido: material que transmite alguna luz pero no la suficiente para ver claramente los objetos a través del mismo. (p. 384)

transmutación: proceso de cambio de un elemento a otro mediante la descomposición radioactiva. (p. 542)

transparente: material que transmite casi toda la luz, chocándola de tal manera que los objetos pueden ser claramente vistos a través del mismo. (p. 384)

elementos transuránicos: elementos que tienen más de 92 protones, todos los cuales son sintéticos e inestables. (p. 589)

onda transversal: onda para la cual la materia en el medio se mueve hacia adelante y hacia atrás en ángulos rectos respecto a la dirección en que viaja la onda; ésta tiene cresta y depresiones. (p. 292)

depresión: los puntos más bajos en una onda transversal. (p. 296)

turbina: rueda grande que gira al ser impulsada por vapor, viento o agua y que suministra energía mecánica a un generador. (p. 240)

efecto Tyndall: difusión de un rayo de luz al pasar a través de un coloide. (p. 455)

ultrasonic: sound waves with frequencies above 20,000 Hz. (p. 330)

ultraviolet waves: electromagnetic waves with wavelengths between about 400 billionths and 10 billionths of a meter. (p. 363)

unsaturated hydrocarbon: compound, such as ethene or ethyne, that contains at least one double or triple bond between carbon atoms. (p. 730)

unsaturated solution: any solution that can dissolve more solute at a given temperature. (p. 673)

ultrasónico: ondas de sonido con frecuencia superiores a 20,000 Hz. (p. 330)

ondas ultravioleta: ondas electromagnéticas con longitudes de onda entre aproximadamente 10 y 400 billonésimas de metro. (p. 363)

hidrocarburo no saturado: compuesto, como el etileno, que contiene al menos un enlace doble o triple entre los átomos de carbono. (p. 730)

solución no saturada: cualquier solución que puede disolver más solutos a una temperatura determinada. (p. 673)

variable: factor that can cause a change in the results of an experiment. (p. 9)

variable: factor que puede causar un cambio en los resultados de un experimento. (p. 9)

Glossary/Glosario

velocity: the speed and direction of a moving object. (p. 44)

virtual image: an image formed by diverging light rays that is perceived by the brain, even though no actual light rays pass through the place where the image seems to be located. (p. 418)

viscosity: a fluid's resistance to flow. (p. 489)

visible light: electromagnetic waves with wavelengths of 750 to 400 billionths of a meter that can be detected by human eyes. (p. 363)

voltage difference: related to the force that causes electric charges to flow; measured in volts (V). (p. 200)

volume: amount of space occupied by an object. (p. 18)

velocidad direccional: la rapidez y dirección de un objeto en movimiento. (p. 44)

imagen virtual: la imagen que se forma al divergir los rayos de luz, la cual es percibida por el cerebro, aún cuando ningún rayo de luz real pase por el sitio donde la imagen parezca estar localizada. (p. 418)

viscosidad: resistencia de un fluido al flujo. (p. 489)

luz visible: ondas electromagnéticas con longitudes de onda entre 400 y 750 billonésimas de metro y que pueden ser detectadas por el ojo humano. (p. 363)

diferencia de voltaje: se refiere a la fuerza que causa que las cargas eléctricas fluyan; se mide en voltios (V). (p. 200)

volumen: espacio ocupado por un objeto. (p. 18)

W

wave: a repeating disturbance or movement that transfers energy through matter or space. (p. 290)

wavelength: distance between one point on a wave and the nearest point just like it. (p. 297)

weak acid: any acid that only partly dissociates in solution. (p. 702)

weak base: any base that does not dissociate completely in solution. (p. 703)

wedge: simple machine that is an inclined plane with one or two sloping sides. (p. 145)

weight: gravitational force exerted on an object. (p. 77)

wheel and axle: simple machine that consists of a shaft or axle attached to the center of a larger wheel, so that the shaft and the wheel rotate together. (p. 143)

work: transfer of energy that occurs when a force makes an object move; measured in joules. (p. 126)

onda: alteración o movimiento repetitivo que transfiere energía a través de la materia o el espacio. (p. 290)

longitud de onda: distancia entre un punto en una onda y el punto semejante más cercano. (p. 297)

ácido débil: cualquier ácido que solamente se disocie parcialmente en una solución. (p. 702)

base débil: cualquier base que no se disocie completamente en una solución. (p. 703)

cuña: máquina simple que consiste de un plano inclinado con uno o dos lados en declive. (p. 145)

peso: fuerza gravitacional ejercida sobre un objeto. (p. 77)

rueda y eje: máquina simple que consiste de una barra o eje sujeto al centro de una rueda de mayor tamaño de manera que el eje y la rueda giran juntos. (p. 143)

trabajo: transferencia de energía que se produce cuando una fuerza hace mover un objeto y que se mide en julios. (p. 126)

X

X rays: electromagnetic waves with wavelengths between about 10 billionths of a meter and 10 trillionths of a meter, that are often used for medical imaging. (p. 365)

rayos X: ondas electromagnéticas con longitudes de onda entre 10 billonésimas de metro y 10 trillonésimas de metro, las cuales se utilizan con frecuencia para producir imágenes de uso médico. (p. 365)

Glossary/Glosario

A

Absolute zero, 477

Acceleration, 47–51; calculating, 48, 50, 69 *act;* centripetal, **81,** *81;* equation for, 48; and force, 57 *lab,* 68, *68,* 69; graphing, *49;* gravitational, *77,* 77–78; negative, 47, *47, 49,* 50, *50;* positive, 47, *47, 49,* 50, *50;* of roller coasters, 51, *51;* and speed, 47, *47,* 48; and velocity, 48, 50

Acetic acid, *696,* 697, *697,* 702, 703, *703,* 733, *733*

Acetylene (ethyne), 730, *730,* 733, 734

Acetyl salicylic acid (aspirin), 697, 731, *731*

Acid(s), **696,** *696;* common, 697, *697;* concentration of, 703, *703,* 706, 710 *lab;* indicator of, 696, 710 *act,* 711, *711;* neutralization of, 707, *707,* 716–717 *lab;* organic, *700,* 735 *lab;* properties of, 696; solutions of, 700, *700;* strength of, 703, 716–717 *lab;* strong, *702,* **702**–703, *703;* and titration, 710; weak, *702,* **702**–703, *703*

Acid precipitation, 648, 695 *lab,* 718, *718*

Acid rain, 648; effect of, 695 *lab,* 718, *718*

Acoustics, **339,** 339

Acrylic, 774

Actinides, 576

Actinium, 576

Action, and reaction, 83, *83*

Active solar heating, 174, *174*

Activities, Applying Math, 16, 24, 40, 69, 86, 102, 104, 128, 131, 162, 211, 212, 299, 342, 357, 463, 487, 493, 548, 587, 617, 644, 669; Applying Science, 228, 269, 426, 514, 618, 711, 744, 766; Integrate, 9, 11, 17, 30, 39, 45, 48, 76, 79, 84, 104, 111, 115, 136, 162, 176, 179, 202, 205, 208, 227, 228, 240, 267, 275, 295, 299, 309, 324, 325, 331, 343, 363, 364, 370, 391, 392, 403, 408, 417, 427, 429, 456, 459, 462, 463, 480, 489, 514, 520, 524, 540, 549, 554, 573, 576, 608, 609, 617, 637, 650, 678, 685, 700, 713, 733, 743, 760, 772; Science Online, 7, 12, 18, 43, 53, 69, 76, 102, 113, 160, 174, 196, 212, 237, 242, 268, 275, 295, 307, 329, 334, 371, 372, 396, 401, 428, 434, 464, 479, 507, 519, 521, 523, 540, 553, 591, 605, 610, 639, 674, 684, 701, 710, 733, 740, 744, 762, 770, 773, 775; Standardized Test Practice, 34–35, 64–65, 96–97, 122–123, 154–155, 186–187, 220–221, 252–253, 284–285, 318–319, 350–351, 380–381, 412–413, 444–445, 472–473, 502–503, 564–565, 598–599, 628–629, 658–659, 692–693, 722–723, 754–755, 784–785

Adhesives, 774

Air conditioners, 178

Aircraft, aluminum in, 584, *584;* composite materials in, 776, *776;* jumbo jet, 92, *92*

Aircraft carriers, 48

Airplane simulator, 11, *11*

Air pollution, ozone depletion, 364, *364*

Air resistance, *73,* 73–74, *74,* 89 *lab*

Alcohol, 682, *682,* **733,** *733,* 735 lab

Alkali metals, 572, *572*

Alkaline earth metals, 573, *573*

Allotropes, **585,** 586, *586,* 592–593 *lab*

Alloys, 665, *666,* **758**–763; aluminum, 762, *762;* ceramics, 762–763, *763,* 764, 765, 766, 766 *act;* properties of, 759, 759 *lab,* 760, *760, 761;* steel, 761, *762,* 778–779 *lab;* titanium, 762–763, *763;* uses of, 761, *761*

Alpha particles, **541**–542, *542*

Alternating current (AC), **242,** 243, 244, *244*

Alternator, 239

Aluminum, *451,* 570, 584, *584, 590,* 649, 762, *762;* chemical reaction with oxygen, 631 *lab,* 637; modeling an atom of, 509 *lab;* symbol for, 506

Aluminum hydroxide, 699, *699*

Aluminum oxide, 616, 637

Americium, *451,* 589, *589*

Amine group, 743

Amino acids, 742

Ammonia, 579, 587, 699, 701, *701*

Ammonium chloride, 708

Amorphous solids, 482

Ampere, 15

Amplitude, *300,* **300**–301, *301,* 327

AM radio, 367, *368,* 368, *369*

Angelou, Maya, 60

Angle(s), of incidence, 304, *304;* of reflection, 304, *304*

Animal(s), in cold climates, 678; control of heat flow by, 168, *168;* speed of, 41

Antacid, 698, 698 *lab,* 707, *707,* 711 *act*

Antarctica, ozone hole in, 364

Antifreeze, 678, 679

Antimony, 588

Anvil, 325, *325*

Apollo 11, 85

Appearance, 458, *458*

Appliances, cost of using, 213; and electric current, 242, *242;* energy used by, 211

Applying Math, The Acceleration of a Sled, 69; Barium Sulfate Reaction, 644; Calculate a Change in Thermal Energy, 162;

Index

Index

Index

Index

Index

Index

Index

Index

Washing soda, 672

Waste(s), nuclear, 268 *act,* **268**–269, *269*

Water, boiling point of, *20,* 21, *21;* chemical stability of, 606, *606;* as compound, 452; compounds with, 620, *620,* 620 *lab;* decomposition of, 642, *642;* dissolving without, 681–685; distilling, 449 *lab;* energy from, 273, *273,* 274, *274;* freezing point of, *20,* 21, *21;* in generation of electricity, 240, 273, *273,* 274, *274;* heated by solar energy, 255 *lab,* 259 *lab;* molecules in, 482, *482;* refraction of light in, 305, *305;* specific heat of, 161, 162; states of, 476, *476;* symbol for, 506

Water heaters, 259 *lab*

Water waves, 292, *292,* 293, *293,* 294, 306, *306,* 354, *354, 359*

Watt, James, 130

Watt (W), 130

Wave(s), 288–313, **290;** amplitude of, *300,* **300**–301, *301,* 327; behavior of, 303–311; carrier, **367,** *367, 368, 369;* compressional, **292,** *292,* 296, *296,* 300, *300,* 322, *322,* 327, 330; crest of, 296, *296;* in different mediums, 291, 302 *lab;* diffraction of, *306,* 306–307, *307,* 307 *act;* electromagnetic. *See* Electromagnetic waves; and

energy, 289 *lab,* 290, 290–291, *291;* frequency of, 297–298, *298,* 358, 360, *360,* 374–375 *lab;* infrared, *360, 362, 362;* infrasonic, 330; and interference, *308,* 308–309, *309;* mechanical, 291–295, *292, 293, 294;* microwaves, *360,* **361,** *361,* 361 *lab,* 371; ocean, *294,* 299; and particles, 359, *359;* period of, 297; properties of, 312–313 *lab;* radio, 307, *360,* 360–362, **361,** *367,* 367–373, *368, 369,* 371 *act;* reflection of, *303,* 303–304, *304;* refraction of, *304,* 304–305, *305;* and resonance, 311, 311 *lab;* seismic, 295, *295,* 295 *act;* sound, 292–293, 303, 311, *322,* 322–323, 324, 325–326, 354, *354;* speed of, 298–299, 299 *act,* 358; standing, **310,** *310;* subsonic, 330; transverse, **292,** *292,* 301, *301;* trough of, 296, *296;* tsunami, 299; ultrasonic, **330,** 341–343, *342, 343;* ultraviolet, 353 *lab,* 363, **363**–364; water, 292, *292,* 293, *293,* 294, 306, *306,* 354, *354, 359*

Wavelength, **297,** *297,* 297 *lab, 298,* 307, *307,* 358, *358*

Weak acids, *702,* **702**–703, *703*

Weak bases, 703

Weather balloon, 492, *492*

Weathering, chemical, 464, *464;* physical, 464, *464*

Weather satellites, 160 *act*

Wedge, 145, *145*

Weight, 77–78, *78*

Weightlessness, 78–79, *79*

Well, *736*

Wet cell batteries, *202,* 202–203

Wheel and axle, *143,* 143–144, *144*

White Cliffs of Dover, 464, *464*

Wide-angle lenses, 436, *436*

Wind energy, 274, *274,* 274, *274*

Windmills, 240, *240,* 274, *274*

Wintergreen, 731, *731*

Wood, specific heat of, 161

Woodwind instruments, 334, 335

Work, *126,* **126**–131, *127, 129,* 148–149 *lab;* calculating, 128–129, 128 *act,* 129 *lab;* changing direction of force, 133, *133,* 134, *134,* 136; converting heat to, 176, *176;* and energy, 127, *127;* increasing distance, 133, *133;* increasing force, 132, *132,* 134, *134;* increasing thermal energy by, 174, *175;* and machines, 125 *lab, 132,* 132–135, *133, 134, 135*

Xenon, 582

X ray, *360,* 363, **365,** *365,* 560, *560,* 573

Xylophone, 336, *336*

Zero, absolute, 477

Zinc, 468, 575

Magnification Key: Magnifications listed are the magnifications at which images were originally photographed.
LM–Light Microscope
SEM–Scanning Electron Microscope
TEM–Transmission Electron Microscope

Acknowledgments: Glencoe would like to acknowledge the artists and agencies who participated in illustrating this program: Absolute Science Illustration; Andrew Evansen; Argosy; Articulate Graphics; Craig Attebery represented by Frank & Jeff Lavaty; CHK America; John Edwards and Associates; Gagliano Graphics; Pedro Julio Gonzalez represented by Melissa Turk & The Artist Network; Robert Hynes represented by Mendola Ltd.; Morgan Cain & Associates; JTH Illustration; Laurie O'Keefe; Matthew Pippin represented by Beranbaum Artist's Representative; Precision Graphics; Publisher's Art; Rolin Graphics, Inc.; Wendy Smith represented by Melissa Turk & The Artist Network; Kevin Torline represented by Berendsen and Associates, Inc.; WILDlife ART; Phil Wilson represented by Cliff Knecht Artist Representative; Zoo Botanica.

Photo Credits

Cover Roger Ressmeyer/CORBIS; **i ii** Roger Ressmeyer/CORBIS; **viii** SuperStock; **ix** Peter Ardito/Index Stock; **x** Russell D. Curtis/Photo Researchers; **xi** CMCD/PhotoDisc; **xiii** (tr)Richard Megna/Fundamental Photographs, (bl)Alfred Pasieka/Science Photo Library/Photo Researchers; **xiv** George B. Diebold/The Stock Market/CORBIS; **xv** Ginger Chih/Peter Arnold, Inc.; **xvi** NASA; **xvii** AISI/Visuals Unlimited; **xxi** Dominic Oldershaw; **1** James H. Karales/Peter Arnold, Inc.; **2–3** John Terence Turner/FPG/Getty Images; **3** (tl)Artville, (br)Charles L. Perrin; **4–5** Roger Ressmeyer/CORBIS; **6** Shuttle Mission Imagery/Liaison Agency/Getty Images; **7** Will McIntyre/Photo Researchers; **9** James L. Amos/CORBIS; **10** David Young-Wolff/PhotoEdit, Inc.; **11** (l)Roger Ressmeyer/CORBIS, (r)Douglas Mesney/The Stock Market/CORBIS; **12** J. Marshall/The Image Works; **13** (t)Jonathan Nourok/PhotoEdit, Inc., (b)Tony Freeman/PhotoEdit, Inc.; **14** Mark Burnett; **15** courtesy Bureau International Des Poids et Mesures; **16** Matt Meadows; **17** (t)Amanita Pictures, (bl)Runk/Schoenberger from Grant Heilman, (br)CORBIS; **24** Mark Burnett; **28** Bob Daemmrich; **29** Icon Images; **30** Rosalie Winard; **31** (l)TSADO/NCDC/NOAA/Tom Stack & Assoc., (r)David Ball/The Stock Market/CORBIS; **32** Amanita Pictures; **35** courtesy Bureau International Des Poids et Mesures; **36–37** Lester Lefkowitz/CORBIS; **38** Icon Images; **42** Paul Silverman/Fundamental Photographs; **44** (t)SuperStock, (b)Robert Holmes/CORBIS; **49** (t)RDF/Visuals Unlimited, (c)Ron Kimball, (b)Richard Megna/Fundamental Photographs; **50** (l)The Image Finders, (r)Richard Hutchings; **51** Dan Feicht/Cedar Point Amusement Park; **52** Globus Brothers Studios, NYC; **53** Tim Courlas/Horizons Companies; **54** Neal Haynes/Rex USA, Ltd.; **55** (t)Paul Kennedy/Liaison Agency/Getty Images, (b)courtesy Insurance Institute for Highway Safety; **56** Donald Johnston/Stone/Getty Images; **57** Mark Burnett; **58 59** Icon Images; **60** Sylvain Grandadam/Stone/Getty Images; **61** (tl bl)Tony Freeman/PhotoEdit, Inc., (br)Peter Newton/Stone/Getty Images; **66–67** Tim Wright/CORBIS; **68** David Young-Wolff/PhotoEdit, Inc.; **70** (t)Pictor, (c)M.W. Davidson/Photo Researchers, (b)PhotoDisc; **71** Bob Daemmrich; **72** (t)Bob Daemmrich, (b)Peter Fownes/Stock South/PictureQuest; **73** (t)James Sugar/Black Star/PictureQuest, (b)Michael Newman/PhotoEdit, Inc.; **74** Keith Kent/Peter Arnold, Inc.; **76** StockTrek/CORBIS; **77** Peticolas-Megna/Fundamental Photographs; **78** NASA; **80** (tl)KS Studios, (tr)David Young-Wolff/PhotoEdit, Inc., (bl)Richard Megna/Fundamental Photographs; **81** Pictor; **83** Steven Sutton/Duomo; **84** Philip Bailey/The Stock Market/CORBIS; **85** (l)NASA, (tr br)North American Rockwell, (cr)Teledyne Ryan Aeronautical; **87** (l)Jeff Smith/Fotosmith, (r)Richard Megna/Fundamental Photographs; **88** Amanita Pictures; **90** Matt Meadows; **91** Tim Courlas/Horizons Companies; **92** (tr)Tony Craddock/Science Photo Library/Photo Researchers, Inc., (bl)Bettmann/CORBIS; **93** (l)Keith Kent/Peter Arnold, Inc., (r)Richard Megna/Fundamental Photographs; **98–99** Jim Cummins/CORBIS; **100** Runk/Schoenberger from Grant Heilman; **101** (l)Tony Walker/PhotoEdit, Inc., (c)Mark Burnett, (r)D. Boone/CORBIS; **105** KS Studios; **109** Walter H. Hodge/Peter Arnold, Inc.; **110** RFD/Visuals Unlimited; **114** Rudi Von Briel; **117** Matt Meadows; **118** (tl)Brompton Studios, (cr)Hank Morgan/Photo Researchers, (bl)TIME; **119** (tl)SuperStock, (cr)Telegraph Colour Library/FPG/Getty Images, (bl)Jana R. Jirak/Visuals Unlimited; **124–125** Jakob Helbig/Getty Images; **125** Timothy Fuller; **126** Tim Courlas/Horizons Companies; **127** (tr)Tim Courlas/Horizons Companies, (bl)Michael Newman/PhotoEdit, Inc.; **129** Michelle Bridwell/PhotoEdit, Inc.; **131** Jules Frazier/PhotoDisc; **132** Michael Newman/PhotoEdit, Inc.; **133** (t)Mark Burnett, (b)Tony Freeman/PhotoEdit, Inc.; **134** Joseph P. Sinnot/Fundamental Photographs; **135** Richard Megna/Fundamental Photographs; **139** (tl)Tom Pantages, (tr)Mark Burnett, (br)Tony Freeman/PhotoEdit, Inc.; **140** (tr)Richard T. Nowitz, (cl)David Madison/Stone/Getty Images, (br)John Henley/The Stock Market/CORBIS; **141** A.J. Copley/Visuals Unlimited; **143** Mark Burnett; **144** Tom Pantages; **145** (tl)file photo, (tr)Mark Burnett, (b)Amanita Pictures; **146** SuperStock; **147** Mark Burnett; **149** Hickson-Bender; **150** (tr)courtesy D. Carr & H. Craighead, Cornell University, (bl)Joe Lertola; **151** (tl)Ed Lallo/Liaison Agency/Getty Images, (r)C. Squared Studios/PhotoDisc, (bl)file photo; **155** Michael Newman/PhotoEdit, Inc.; **156–157** Charles E. Rotkin/CORBIS; **160** Aaron Haupt; **165** Peter Ardito/Index Stock; **166** (t)Earth Imaging/Stone/Getty Images, (bl)K&K Arman/Bruce Coleman, Inc./PictureQuest, (br)Charles & Josette Lenars/CORBIS; **168** (l)Doug Cheeseman/Peter Arnold, Inc., (c)Tim Davis/The Stock Market/CORBIS, (r)Ed Reschke/Peter Arnold, Inc.; **169** David Madison; **171** Leonard Lee Rue III/Bruce Coleman, Inc./PictureQuest; **174** Stu Rosner/Stock Boston; **180** Tim Courlas/Horizons Companies; **182** (tl bl)NASA, (cr)Roy Johnson/Tom Stack & Assoc.; **183** (l)Dean Conger/CORBIS, (r)Michael Newman/PhotoEdit, Inc.; **186** Doug Martin; **188–189** National Geographic Society; **189** National Geographic Society; **190–191** John Lawrence/Getty Images; **191** Michael Newman/PhotoEdit, Inc.; **193** Geoff Butler; **195** KS Studios; **196** Tim Courlas/Horizons Companies; **197** T. Wiewandt/DRK Photo; **198** Dale Sloat/PhotoTake NYC/PictureQuest; **203** Ray Ellis/Photo Researchers; **204** Thomas Veneklasen; **206 207 208** Geoff Butler; **210** (tl)Geoff Butler, (tr)Tony Freeman/PhotoEdit, Inc., (cl bl)Aaron Haupt; **214 215** Geoff Butler; **216** Bernard Gotfryd/Woodfin Camp & Associates; **217** (t)file photo, (bl)Lester Lefkowitz/The Stock Market/CORBIS, (br)Harold Stucker/Black Star; **222–223** Kennan Ward/CORBIS;

Freeman/PhotoEdit, Inc., (br)Icon Images; **473** (tl)Doug Martin, (tr)Chris Rogers/The Stock Market/CORBIS, (bl br)Doug Martin; **474–475** The Stock Market/CORBIS; **475** Matt Meadows; **476** Icon Images; **479** Doug Martin; **480** JISAS/Lockheed/Science Photo Library/Photo Researchers; **481** (t)CORBIS, (b)Mark E. Gibson/Visuals Unlimited; **483** Amanita Pictures; **484** Matt Meadows; **488** (t)Bob Daemmrich, (b)Tim Courlas/Horizons Companies; **490** file photo; **492** Mark Burnett/Photo Researchers; **495** Matt Meadows; **496** John Evans; **498** (t)Geoff Tomkinson/Science Photo Library/Photo Researchers, (b)Michael Dwyer/Stock Boston; **499** (tl)Breck P. Kent/Animals Animals, (bl)Icon Images, (br)Aaron Haupt; **503** (l)Mark Burnett/Photo Researchers, (r)Mark E. Gibson/ Visuals Unlimited; **504–505** Walter Bibikow/Index Stock Imagery, NY; **508** (tl)Science Photo Library/Photo Researchers, (tr)Hank Morgan/Science Source/Photo Researchers, (b)Photo Researchers; **509** Sheila Terry/Science Photo Library/Photo Researchers; **512** Matt Meadows; **516** Science Museum/Science & Society Picture Library; **524** CORBIS; **525** The Image Bank/Getty Images; **526** Tom Pantages; **527** PhotoEdit, Inc.; **528** (t)Roger Ressmeyer/ CORBIS, (bl)Maria Stenzel/National Geographic Image Collection; **534–535** Paul Souders/Getty Images; **535** Aaron Haupt; **536** (t)Mark Burnett, (b)David Frazier/The Image Works; **540** American Institute of Physics/Emilio Segrè Visual Archives; **542** Amanita Pictures; **546** (l)courtesy Supersaturated Environments, (r)Science Photo Library/ Photo Researchers; **547** (t)CERN, P. Loiez/Science Photo Library/Photo Researchers, (others)Richard Megna/ Fundamental Photographs; **549** Hank Morgan/Photo Researchers; **554** Oliver Meckes/Nicole Ottawa/Photo Researchers; **555** UCLA School of Medicine; **556** J.L. Carson/Custom Medical Stock Photo; **557** Tim Courlas/ Horizons Companies; **558** (l)Fermilab/Visuals Unlimited, (r)Matt Meadows; **559** Matt Meadows; **560** (t)C. Powell, P. Fowler & D. Perkins/Science Photo Library/Photo Researchers, Inc., (cr bl)CORBIS; **561** (l)Martha Cooper/ Peter Arnold, Inc., (r)Oliver Meckes/Nicole Ottawa/Photo Researchers; **566–567** Jon Feingersh/The Stock Market/ CORBIS; **567** (inset)Dorling Kindersley; **568–569** Georg Gerster/Photo Researchers; **570** (t)Henry Groskinsky/ TimePix, (b)Nubar Alexanian/CORBIS; **572** (l)Stephen Frisch/Stock Boston, (r)Jerry Mason/Science Photo Library/Photo Researchers; **573** Firefly Productions/The Stock Market/CORBIS; **574** Mark Burnett/Photo Researchers; **575** (tl)Matt Meadows, (tr)Icon Images, (br)Jeff J. Daly/ Fundamental Photographs; **577** Melvyn P. Lawes/Papilio/ CORBIS; **579** (l to r)Russ Lappa/Science Source/Photo Researchers, Gregory G. Dimijian/Photo Researchers, Stephen Frisch/Stock Boston, Andrew J. Martinez/Photo Researchers, Doug Martin, Charles Gupton/Stock Boston; **580** (t)Spencer Grant/PhotoEdit, Inc./PictureQuest, (b)Dick Frank/The Stock Market/CORBIS; **581** (t)Norris Blake/Visuals Unlimited, (b)Steven Senne/AP/Wide World Photos; **582** Gene J. Puskar/AP/Wide World Photos; **584** George Hall/CORBIS; **585** (l)Icon Images, (r)Charles D. Winters/ Photo Researchers; **586** (l)Aaron Haupt, (c)Rick Gayle Studio/The Stock Market/CORBIS, (r)Alfred Pasieka/Science Photo Library/Photo Researchers; **588** (l)RMIP/Richard Haynes, (r)PhotoDisc; **590** (tl)Gianni Dagli Orti/CORBIS, (tc)Jonathan Blair/CORBIS, (tr)Patricia Lanza, (cl)CORBIS, (ccl)Tony Freeman/CORBIS, (c)PhotoDisc, (ccr)Paul A. Souders/CORBIS, (cr)Bettmann/CORBIS, (bl)Steve Cole/

CORBIS, (bc)Artville, (br)Achim Zschau; **592** (t)Charles D. Winters/Photo Researchers, (b)Richard Hutchings; **593** Richard Hutchings; **594** Lynn Johnson/Aurora; **595** (cw from left)Charles D. Winters/Photo Researchers, Stephen Frisch/Stock Boston, Charles E. Zirkle, Doug Martin, Karl Hartmann/Sachs/PhotoTake NYC, Russ Lappa/Science Source/Photo Researchers; **599** George Hall/CORBIS; **600–601** Tom Sanders/CORBIS; **601** Amanita Pictures; **602** (t)John Evans, (b)CORBIS; **603** (l)Runk/Schoenberger from Grant Heilman, (c)Richard Megna/Fundamental Photographs, (r)Syndicated Features Limited/The Image Works; **607** Mark Burnett; **608** John Paul Kay/Peter Arnold, Inc.; **612** John Evans; **613** Patricia Lanza; **615** Bettmann/ CORBIS; **620** Amanita Pictures; **622** (l)Amanita Pictures, (r)Richard Hutchings; **623** Richard Hutchings; **624** Daniel Belknap; **625** Syndicated Features Limited/The Image Works; **630–631** Transglobe/Index Stock; **633** Mansell Collection/ TimePix; **635** Richard Megna/Fundamental Photographs; **637** CORBIS; **638** Richard Megna/Fundamental Photographs; **639** Doug Martin/Photo Researchers; **640** Richard Megna/Fundamental Photographs; **641** Emory Kristof/National Geographic Image Collection; **642** (t)Charles D. Winters/Photo Researchers, (b)Stephen Frisch/Stock Boston/PictureQuest; **643** Richard Megna/ Fundamental Photographs; **645** Royalty-Free/CORBIS; **646** Tim Matsui/Liaison Agency/Getty Images; **647** (l cl)Bettmann/CORBIS, (cr)Roger Ressmeyer/NASA/ CORBIS, (r)NASA/CORBIS; **648** Matt Meadows; **649** (t)Jeff J. Daly/Fundamental Photographs, (c b)Matt Meadows; **651** Timothy Fuller; **652** UNEP-Topham/The Image Works; **653** Michael Newman/PhotoEdit, Inc.; **654** (tr)Bibliotheque du Museum D'Histoire Naturelle, (bl)Brad Maushart/ Graphistock; **655** (tr)Icon Images, (l)Charles D. Winters/ Photo Researchers, (br)Joseph P. Sinnot/Fundamental Photographs; **658** Charles D. Winters/Photo Researchers; **659** (l)Richard Megna/Fundamental Photographs, (r)Firefly Productions/The Stock Market/CORBIS; **660–661** Norbert Wu; **661** (inset)Spencer Collection, New York Public Library, Astor, Lenox & Tilden Foundation; **662–663** Stephen Frink/ Index Stock Imagery; **664** Annie Griffiths Belt/CORBIS; **665** (l)Christie's Images, (r)Flip Schulke/Black Star; **666** (t)Bettmann/CORBIS, (c)Brian Gordon Green, (b)Michael Newman/PhotoEdit, Inc./PictureQuest; **672** Mark Thayer Photography, Inc.; **674** Richard Megna/Fundamental Photographs; **680** Tim Courlas/Horizons Companies; **681 682** Icon Images; **683** Steve Mason/PhotoDisc; **684** KS Studios; **686** (l)Matt Meadows, (r)Tim Courlas/Horizons Companies; **687** Matt Meadows; **688** (tr)Tony Freeman/ PhotoEdit, Inc., (cl)AISI/Visuals Unlimited, (br)Standard-Examiner/Kort Duce/AP/Wide World Photos; **689 690 692** Icon Images; **694–695** Frans Lanting/Minden Pictures; **696** KS Studios; **697** Matt Meadows; **698** Geoff Butler; **699** Rick Poley/Index Stock; **702** Matt Meadows; **704** (cw from tl)Elaine Shay, Brent Turner/BLT Productions, Matt Meadows, Elaine Shay, StudiOhio, Icon Images, CORBIS, (c)Dominic Oldershaw, (bl)Mark Burnett/Stock Boston/PictureQuest, (br)Matt Meadows; **706** Tim Courlas/ Horizons Companies; **707** Charles D. Winters/Photo Researchers; **708** The Purcell Team/CORBIS; **709** (tl b)Kevin Schafer/CORBIS, (tr)Wolkmar Wentzel, (c)CORBIS; **710** Richard Megna/Fundamental Photographs; **711** (t)Matt Meadows, (b)Amanita Pictures; **713** Chinch Gryniewicz/ Ecoscene/CORBIS; **714** Amanita Pictures; **715** PhotoDisc; **716** (t)Bill Lyons/Liaison Agency/Getty Images, (b)Richard

PERIODIC TABLE OF THE ELEMENTS

Columns of elements are called groups. Elements in the same group have similar chemical properties.

Gas
Liquid
Solid
Synthetic

Element — Hydrogen
Atomic number — 1
Symbol — **H**
Atomic mass — 1.008

State of matter

The first three symbols tell you the state of matter of the element at room temperature. The fourth symbol identifies elements that are not present in significant amounts on Earth. Useful amounts are made synthetically.

1

2

	3	**4**	**5**	**6**	**7**	**8**	**9**

1
Hydrogen
1
H
1.008

2
Lithium
3
Li
6.941

Beryllium
4
Be
9.012

3
Sodium
11
Na
22.990

Magnesium
12
Mg
24.305

4
Potassium
19
K
39.098

Calcium
20
Ca
40.078

Scandium
21
Sc
44.956

Titanium
22
Ti
47.867

Vanadium
23
V
50.942

Chromium
24
Cr
51.996

Manganese
25
Mn
54.938

Iron
26
Fe
55.845

Cobalt
27
Co
58.933

5
Rubidium
37
Rb
85.468

Strontium
38
Sr
87.62

Yttrium
39
Y
88.906

Zirconium
40
Zr
91.224

Niobium
41
Nb
92.906

Molybdenum
42
Mo
95.94

Technetium
43
Tc
(98)

Ruthenium
44
Ru
101.07

Rhodium
45
Rh
102.906

6
Cesium
55
Cs
132.905

Barium
56
Ba
137.327

Lanthanum
57
La
138.906

Hafnium
72
Hf
178.49

Tantalum
73
Ta
180.948

Tungsten
74
W
183.84

Rhenium
75
Re
186.207

Osmium
76
Os
190.23

Iridium
77
Ir
192.217

7
Francium
87
Fr
(223)

Radium
88
Ra
(226)

Actinium
89
Ac
(227)

Rutherfordium
104
Rf
(261)

Dubnium
105
Db
(262)

Seaborgium
106
Sg
(266)

Bohrium
107
Bh
(264)

Hassium
108
Hs
(277)

Meitnerium
109
Mt
(268)

The number in parentheses is the mass number of the longest-lived isotope for that element.

Rows of elements are called periods. Atomic number increases across a period.

The arrow shows where these elements would fit into the periodic table. They are moved to the bottom of the table to save space.

Lanthanide series

Cerium
58
Ce
140.116

Praseodymium
59
Pr
140.908

Neodymium
60
Nd
144.24

Promethium
61
Pm
(145)

Samarium
62
Sm
150.36

Actinide series

Thorium
90
Th
232.038

Protactinium
91
Pa
231.036

Uranium
92
U
238.029

Neptunium
93
Np
(237)

Plutonium
94
Pu
(244)